LANDSLIDES

INVESTIGATION AND MITIGATION

SPECIAL REPORT 247

A. Keith Turner
Robert L. Schuster

Editors

TRANSPORTATION RESEARCH BOARD
National Research Council

National Academy Press
Washington, D. C. 1996

Transportation Research Board Special Report 247

Subscriber Category
IIIA soils, geology, and foundations

NOTICE: The project that is the subject of this report was approved by the Governing Board of the National Research Council, whose members are drawn from the councils of the National Academy of Sciences, the National Academy of Engineering, and the Institute of Medicine. The members of the committee responsible for the report were chosen for their special competence and with regard for appropriate balance.

This report has been reviewed by a group other than the authors according to procedures approved by a Report Review Committee consisting of members of the National Academy of Sciences, the National Academy of Engineering, and the Institute of Medicine.

The Transportation Research Board does not endorse products or manufacturers; trade and manufacturers' names may appear in this Special Report because they are considered essential to its object.

This report was sponsored in part by the National Science Foundation.

Transportation Research Board publications are available by ordering directly from TRB. They may also be obtained on a regular basis through organizational or individual affiliation with TRB; affiliates and library subscribers are eligible for substantial discounts. For further information, write to the Transportation Research Board, National Research Council, 2101 Constitution Avenue, N.W., Washington, D.C. 20418.

Library of Congress Cataloging-in-Publication Data

Landslides : investigation and mitigation / A. Keith Turner, Robert L. Schuster, editors.
    p.   cm. — (Special report / Transportation Research Board, National Research Council ; 247)
    Includes bibliographical references and index.
    ISBN 0-309-06151-2
    ISBN 0-309-06208-X (pbk.)
    1. Landslide hazard analysis. 2. Landslides. I. Turner, A. Keith, 1941– . II. Schuster, Robert L. III. Series: Special report (National Research Council (U.S.) Transportation Research Board) ; 247.
QE599.2.L36 1996
551.3'07—dc20   95-40780
                          CIP

COVER Damage to railroad line in Olympia, Washington, caused by slope failure after 1965 Seattle-Tacoma earthquake (photograph courtesy of G.W. Thorsen, Washington Department of Natural Resources). *Top inset:* October 1986 reactivation of Cucaracha landslide in Gaillard Cut, Panama Canal, which originally extended across canal (dredges had removed much of toe by time of photograph); white material in upper center is lime added to surface as remedial measure (photograph courtesy of Panama Canal Commission). *Bottom inset:* Recurring landslide on Colorado State Highway 133 at McClure Pass west of Aspen, spring 1993 (photograph courtesy of T.E. Taylor, Colorado State Patrol).

FRONTISPIECE Derailment of *California Zephyr* passenger train near Granby, Colorado, in April 1985 caused by minor landslide (see Figure 2-6, p. 18) (photograph courtesy of R. L. Schuster).

TITLE PAGE *Inset:* Railway and highway at Howe Sound, British Columbia, blocked by October 1990 rock fall, which originated approximately 300 meters above highway (photograph courtesy of D. Howard, *Vancouver Sun*). *Background:* detail from drawing by Alexandre Collin (1846) of landslides in clay slopes along French canals (see Figure 1-5, page 9).

PREFACE

Transportation Research Board reports that synthesize information related to landslides and rock fall span more than four decades. Previous reports have been widely used in the geotechnical engineering community as comprehensive, practical sources of information on landslides and their control. Among the most widely distributed of TRB publications, these reports have enjoyed wide international appeal and have been translated into several languages.

Recognizing the lack of a single source in the English language that covers the entire spectrum of issues related to landslides, the Transportation Research Board (then the Highway Research Board) created the Committee on Landslide Investigations in 1951. The efforts of that committee resulted in Special Report 29, *Landslides and Engineering Practice*, which was published in 1958.

In 1972 a task force was organized with members drawn from several of the committees within the Soils and Geology Group to undertake the revision of Special Report 29. This task force concluded that because of the large amount of new technical information available since the 1958 publication, the best course of action was to prepare a completely new report. Six years was required to prepare and publish TRB Special Report 176, *Landslides: Analysis and Control*.

Special Report 176 was reprinted a number of times, and by 1989 the Board was once again faced with the decision of whether to continue to reprint the existing text or to undertake a further revision.

A polling of the members of the TRB committees within the Soil Mechanics and the Geology and Earth Materials sections clearly indicated that the report should be revised to address the latest advances in the methods for investigation and mitigation of landslides.

Accordingly, a study committee was established in 1990 with the charge to review the 1978 report, identify needed changes and additions, and prepare a new report for publication. Although aware of the heavy workload that such a revision would entail, a number of members of the task force responsible for developing the 1978 report agreed to serve on the new committee under the chairmanship of A. Keith Turner of the Colorado School of Mines.

At the first meeting of the study committee in January 1990, members were assigned the preparation of specific chapters. Responsibilities included developing chapter outlines, identifying expertise outside the committee membership to provide specific material or assist in writing chapters, writing material, coordinating the efforts of multiple authors, and reviewing chapter drafts before submittal for approval by the entire committee membership. In addition to holding meetings in Washington, D.C., during the TRB Annual Meeting, the committee held three other formal meetings. These discussions greatly improved the quality of the individual chapters. Although the material contained in each chapter is solely attributed to the author(s) of the chapter, committee members also were responsible for reviewing the report as a whole.

This latest volume in the series of TRB Special Reports on landslides contains 25 chapters written by 30 authors. Prepared by experts from the United States, Canada, and the Netherlands, this report has a broader international scope and considerably more extensive coverage than its predecessors. The authors of several of the chapters are involved in international landslide coordination programs, and this new report has been designed to reinforce those international efforts.

Measurements of the International System of Units (SI) are used in this report. A table of conversion factors for SI and inch/pound (U.S. customary) units of measurements is provided in Appendix B.

ACKNOWLEDGMENTS

Coordination and technical editing of the efforts of multiple authorship are not easy tasks. The Transportation Research Board and the future users of this report worldwide are indebted to the members of the study committee for their efforts. Sincere appreciation is expressed to all those who contributed information in the form of data and photographs, ideas, and advice, without which this volume could not have been completed. To list all contributors would be impossible; to list the most important would be unfair to the others.

This book was produced by busy individuals who volunteered their time and talents to its creation. The committee members were aided by several individuals who agreed to write, review, and supervise the development of particular chapters or portions of chapters. Brief biographies for all members of the study committee and the chapter authors are provided at the end of the report. The editors wish to thank this dedicated group for their encouragement, unfailing goodwill, and patience during the lengthy editorial process.

The funding assistance provided by the National Science Foundation in support of this project is gratefully acknowledged.

Finally, special thanks go to TRB staff members G. P. Jayaprakash, Engineer of Soils, Geology, and Foundations; Robert E. Spicher, Director, Technical Activities; Nancy A. Ackerman, Director, Reports and Editorial Services; and Naomi Kassabian, Associate Editor.

A. Keith Turner and Robert L. Schuster
Editors

CONTENTS

PRINCIPLES, DEFINITIONS, AND ASSESSMENT

Chapter 1

A. KEITH TURNER AND
G. P. JAYAPRAKASH

INTRODUCTION

This Special Report is organized into five major sections with 25 chapters. Although considerable efforts were expended to eliminate repetition of material in different chapters, some reiteration was necessary to provide continuity of thought and to allow adequate explanation of specific topics. Such repetition was judged more acceptable than excessive referral within the text to other sections and chapters.

In accordance with the evolution of this series of reports, the new title, *Landslides: Investigation and Mitigation,* was selected to reflect the increased knowledge of landslide processes, the procedures for landslide investigation that are now available, and the much more complex regulatory and economic climate under which landslide investigations and corrective actions must be undertaken. In fact, the titles of these reports since the first in 1958 mirror changes in societal values at least as much as evolution in scientific knowledge and engineering technologies. The 1958 report reflected engineering practice in resolving landslide instabilities along transportation facilities; the report published in 1978 reflected the evolving strategies for analysis and control of landslides.

In the years since the last volume was published in 1978, there have been many advances in the way landslide investigation and mitigation are conducted. Chief among these advances are the advent of the personal computer, the availability of new geotextile products, and new understandings of the behavior of earth materials. Personal computers have allowed numerical stability analysis methods to become commonplace; the use of geotextiles presents options for better and more economical mitigation procedures; and the improved methods for field investigations coupled with new understanding of landslide processes supply better data and concepts to the landslide analysis process.

However, landslide investigation and mitigation have been even more greatly affected by the imposition of environmental regulations and economic considerations. Throughout the world there has evolved a much greater appreciation of the impact of human activities on the natural environment. Consequently, the investigation of slope instabilities has been increasingly integrated with broader land use planning and land development activities. New transportation facilities, and the renovation or improvement of existing facilities, are frequently required to incorporate design elements that reflect natural landscape conditions and minimize visual impacts. In many hilly or mountainous terrains, such requirements translate into sophisticated landslide investigation and mitigation actions.

1. INTENDED AUDIENCE

Although slope stability problems related to transportation facilities are stressed, most of the discussions and examples in this report apply equally well to all cases of slope instability. As noted by Eckel in his introduction to the first TRB report on landslides:

The factors of geology, topography, and climate that interact to cause landslides are the same regardless of the use to which man puts a given piece of land. The methods for examination of landslides are equally applicable to problems in all kinds of natural or human environment. And the known methods for prevention or correction of landslides are, within economic limits, independent of the use to which the land is put. It is hoped, therefore, that despite the narrow range of much of its exemplary material, this volume will be found useful to any engineer whose practice leads him to deal with landslides. (Eckel 1958, 2-3)

Those statements are still true. The contents of this volume include several aspects that were not addressed in the earlier editions, and the text has been written and organized to appeal to a diverse audience, including

- Transportation engineers responsible for land-slide investigations throughout the world,
- Students in geoscience and geotechnical fields with an interest in landslides, and
- Researchers needing a definitive source for land-slide investigation and mitigation procedures.

Each of these groups has different needs, and this report attempts to address them while maintaining a balance and some brevity in the presentation.

For example, the report contains comprehensive, practical discussions of field investigations, laboratory testing, and stability analysis procedures and technologies. These topics are important to both practicing engineers and students of landslides. It was assumed that many engineers would require a reasonably complete single source of much of the information concerning both investigation and mitigation activities. This volume addresses that need.

In addition, it was expected that many students and researchers would desire comprehensive references to the literature and discussions of case studies, state-of-the-art techniques, and research directions. Accordingly, considerable effort was expended in identifying suitable literature citations and in providing some discussion of recent developments. References to specialized and hard-to-obtain sources were avoided as much as possible; most of the cited references will be readily available through university and special libraries.

2. DEFINITIONS AND RESTRICTIONS

In this report the term *landslide* is used to denote "the movement of a mass of rock, debris or earth down a slope" (Cruden 1991). As it is now used in North America, the term has a much more extensive meaning than its component parts suggest because the phenomena described as landslides are not limited either to the land or to sliding (Cruden 1991). In accordance with the practice in previous reports, ground subsidence and collapse are excluded, and snow avalanches and ice falls are not discussed.

In the period since 1978, the Commission on Landslides and Other Mass Movements of the International Association of Engineering Geology (IAEG) has continued its work on terminology. The declaration by the United Nations of the International Decade for Natural Disaster Reduction (1990–2000) prompted the Commission's Suggested Nomenclature for Landslides (IAEG 1990) and the creation of the International Geotechnical Societies' UNESCO Working Party on the World Landslide Inventory (WP/WLI). The Working Party has prepared the *Multilingual Landslide Glossary* to encourage use of standard terminology in describing landslides (WP/WLI and Canadian Geotechnical Society 1993). The terminology used in this report and defined at some length in Chapter 3 is consistent with the suggested methods and the glossary of the UNESCO Working Party (WP/WLI 1990, 1991, 1993; WP/WLI and Canadian Geotechnical Society 1993).

3. HISTORICAL INFORMATION CONCERNING LANDSLIDES

3.1 Importance

Throughout the world, valleys in mountainous regions have experienced accelerated economic development in response to general population growth and associated demands for increased mining, forestry, and agricultural activities. In some areas, such as parts of North America and Europe, the growth of skiing and other recreational activities has spurred development in mountain regions. This economic growth has demanded expansion of transportation and communication facilities. The short history of extensive human development in many of these areas makes the evaluation of potential landslide hazards and appropriate countermeasures very difficult. A large body of documented evidence concerning landsliding events in long-inhabited mountain regions, notably the Alps of Europe, does exist. In a study conducted by the Geological Survey of Canada, 137 landslide case histories in the Alps were collected and used to formulate the appropriate roles of various active or passive mitigation measures, monitoring, and risk acceptance to guide development in the mountains of western Canada (Eisbacher and Clague 1984). Such studies have not been widely emulated, but it appears that major landslide disasters in mountain regions can be avoided if historical experience is evaluated and used wisely.

In many regions large landslides are infrequent events. In comparison with the length of human lifetimes, their occurrence is so low as to lull many into a false sense of security concerning landslide hazards, especially in areas of lower topographic relief. An appreciation of historical experiences with landslides is a frequently neglected but important component of landslide investigation and mitigation studies.

Historical descriptions of landslides often provide insight into other aspects of the development of scientific and engineering knowledge. Few useful descriptions of landslides predate the Industrial Revolution. There was neither an economic incentive nor a scientific basis to support such studies until the late 1700s and early 1800s. The development and construction of canals, and subsequently railways, placed new importance on slope instability.

3.2 Early Historical Studies

It is beyond the scope of this report to present a detailed historical review of landslide investigations. Several reviews of historical landslides have been published (Voight 1978; Eisbacher and Clague 1984).

However, four examples of early studies from the 1800s are briefly presented to provide the reader with some concept of the insights that these historical documents may provide. All happen to be European examples; they were chosen because they illustrate the evolution of concepts concerning landslide processes. Three examples refer to large and spectacular natural landslides that were subjects of great popular interest and

debate. The fourth example concerns what appears to have been the earliest application of soil mechanics methods to slope stability analysis.

3.2.1 Rossberg Landslide of 1806

On September 2, 1806, a very large, extremely rapid rock fall–rock slide, or *sturzstrom*, occurred in central Switzerland. As described by Eisbacher and Clague (1984), a large section of the Rossberg Massif, estimated to involve 10×10^6 to 20×10^6 m³ of rock, rapidly moved down and away from the mountain and buried much of the small town of Goldau, destroying about 300 houses and killing 457 people (Figure 1-1). Part of the material filled about one-seventh the volume of the Lauerzer See, producing a wave 20 m high that surged over some lakeside villages. Zay (1807)

FIGURE 1-1
Sketch map of the Rossberg landslide of 1806 near Goldau, Switzerland (Eisbacher and Clague 1984).

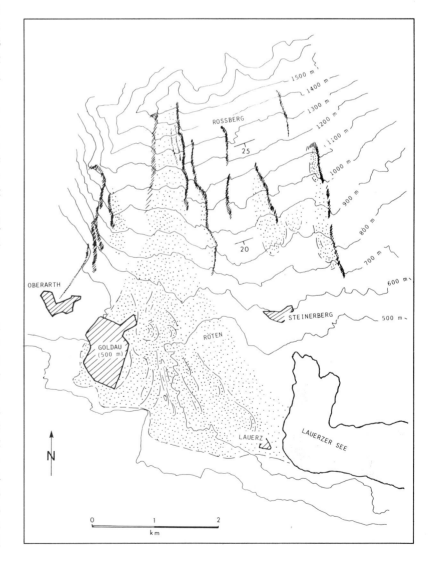

wrote an extremely important, early technical monograph. The landslide was subsequently studied by many others [e.g., Heim (1932)].

This catastrophe attracted wide attention throughout Europe, and the site of the disaster was visited by many notable persons, including artists such as the landscape painter Turner and writers such as Lord Byron. Its cause was debated by many scientists. Evidence pointed to groundwater conditions as the major cause. The winter of 1805–1806 was exceptionally snowy in central Switzerland, and the heavy snowpack was retained by a cold spring. The delayed but rapid snowmelt was augmented by heavy rains in July and August (Eisbacher and Clague 1984). It was thus logical to suggest that exceptional saturation of the rocks was the primary cause. Conybeare et al. (1840) explicitly referred to the Rossberg landslide as "far too well known to require any detail. . . . It occurred in the summer of 1806 after a season of excessive wetness, and is universally ascribed to the undermining agency of land-springs."

3.2.2 Bindon Landslide of 1839

On Christmas Day 1839, a very different type of landslide occurred along the south coast of England. Although there was no loss of life and only minor property damage, the date of the landslide's occurrence provoked wide public atten-

tion and heated debate concerning possible causes and religious significance. As a consequence of this interest, the "Bindon landslip" is among the most documented of all landslides to have occurred in Britain.

The landslide was subjected to extensive scientific investigation by several of the most eminent geologists of the period. Reports based on eyewitness accounts and geological observations at the site were quickly published by Conybeare et al. (1840), Roberts (1840), and many others. These reports included numerous engraved illustrations (Figure 1-2) that gained wide distribution. Conybeare et al. (1840) opened their account as follows:

> The following memoir has been undertaken in order to lay before the reader a distinct account of the most remarkable example ever recorded to have occurred within this island of that class of disturbances affecting the configuration of portions of the earth's surface which results from the undermining agency of water. (Conybeare et al. 1840, 1)

The landslide attracted enormous crowds of curious visitors during the following years, and the local farmers levied a charge of sixpence on visitors wishing to pass through their lands to view it. It has been reported that it was accorded the sin-

FIGURE 1-2
Contempory view of the Bindon landslide of 1839 in England. View is to east along zone of subsidence that marks landward extent of landslide. Area to right of this valley moved laterally toward the sea (Roberts 1840).

gular honor of having a popular musical score, "The Landslip Quadrille," written to celebrate it.

The Bindon landslide also materially affected the evolving science of geology. The early reports (Conybeare et al. 1840; Roberts 1840) were the first to explain the significance of climate and groundwater conditions in promoting slope instability. The clear intent of most investigators was to demonstrate that water could cause such slope instabilities and that such landslides were not related to volcanism or earthquakes. Roberts (1840) stated that "the summer, autumn, and winter of 1839 will long be remembered as the wettest that has almost been known" and quoted the engineer of the Southampton Railway as saying that "so large a quantity of rain has not fallen within the memory of any living person."

Roberts (1840) suggested that these same landslide processes had potential for disrupting roads and that there was a real danger in not understanding them. He provided an example to support his claims:

> It is surprising, often almost incredible how soon and how completely all recollection of natural phenomena, storms, slips, etc., unless attended by unusual features is erased. When the deep cutting . . . was about to commence in 1825, an elderly gentleman, Mr. John Warren, told his brother commissioners of turnpike, that the whole of that highly elevated valley had subsided forty years before; and prognosticated that a road would not long remain without accident. . . [M]any disbelieved the statement. The road was accordingly made, and soon slipped down from twenty feet at one end, to eight feet at the other, towards the sea. (Roberts 1840)

Conybeare et al. (1840) also provided analyses of the mechanisms of the failure, including calculations of the weights of the failed masses and the effects of hydrostatic pressures in promoting instability. These reports and concepts had a major impact on those responsible for constructing earthworks for the rapidly expanding railway system.

3.2.3 Elm Landslide of 1881

The catastrophe at Elm, Switzerland, in 1881 became famous because the events leading up to and accompanying the failure were carefully documented in German by Buss and Heim (1881) and

Heim (1882, 1932). An excellent modern review of these historical reports was provided in English by Hsu (1978), and Heim's 1932 report was translated into English by Skermer (1989). The original reports (Buss and Heim 1881; Heim 1882) included interviews with eyewitnesses as well as geological observations. One of the eyewitnesses used a stopwatch to time the initial failure (Eisbacher and Clague 1984).

The failure was a very large and extremely rapid rock fall–rock slide, or *sturzstrom*, similar to but somewhat smaller than the Rossberg landslide of 1806. In this case the failure of the slope was precipitated partly by natural causes and partly by the extraction of slate from the Plattenberg quarry located at the foot of the cliff. This quarry was developed by local farmers with no mining experience (Hsu 1978). The quarrying undermined a large mass of rock on the mountainside above the quarry. The sudden failure of the mountain slope caused a mass of rock, estimated to have been 10×10^6 m³, to fall onto the Plattenberg quarry platform. From there the rock mass was expelled horizontally at velocities estimated to have exceeded 80 m/sec (Heim 1932) across the valley and toward the town of Elm, claiming the lives of 115 people (Figures 1-3 and 1-4).

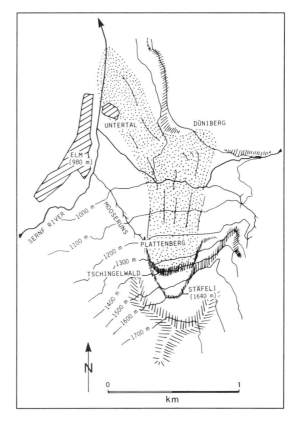

FIGURE 1-3
Sketch map of Elm landslide of 1881 in Switzerland (Eisbacher and Clague 1984).

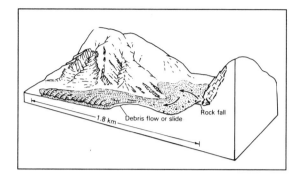

The initial reports (Buss and Heim 1881; Heim 1882) emphasized the unexpected ways in which these sturzstroms move. For example, Heim (1882) reported that several people lost their lives when they ran uphill toward the hamlet of Düniberg (see Figure 1-3) and were overwhelmed by material that surged up the opposite valley slope to a height of about 100 m (Hsu 1978). Heim also reported the observations of several survivors in Elm, in particular their impressions of the flowing nature of the rock mass and the suddenness with which it stopped moving. Heim's detailed observations led to conclusions concerning the hazards resulting from these sturzstroms, especially the large horizontal distances over which they move (Eisbacher and Clague 1984). In his later work, Heim (1932) included calculations concerning the kinematic behavior of the Elm sturzstrom (Hsu 1978). Hsu stated that Heim's interpretations of the mechanisms of sturzstrom movement did not get the recognition they deserved, perhaps because Heim's work was not translated into English until more than 50 years later (Skermer 1989).

3.2.4 Studies of Slope Stability Along French Canals

In 1846, Alexandre Collin, a French engineer with extensive experience in the construction of canals, published his treatise on the stability of clay slopes (Collin 1846). Unfortunately, perhaps because it was not translated into English until more than a century later (Schriever 1956), Collin's report did not become widely known to civil engineers.

The instability of clay slopes was a relatively new problem to engineers in the mid-1800s. Early canals in both England and France did not involve deep cuts and high fills; such heavy earthworks were not associated with canals until the 1820s. Slope failures resulted, and subsequently railway

engineers also encountered widespread slope failures in clay materials forming both cuts and fills. They clearly recognized the deep rotational type of movement and adopted gravel-filled trenches passing through the slip surfaces as their chief remedial measure.

Collin's report presented valuable field data, including surveys of the slip surface for about 15 failures (Figure 1-5). He concluded that the cause of failure was inadequate shear strength. Because he was working with materials that today would be classified either as stiff-fissured clay (in the cut slopes) or as poorly compacted clay (in the fills), he noted that in many cases failure occurred some years after initial construction. He attributed this failure to a process causing progressive softening of the clay and suggested water saturation as the most common cause. To reduce the probability of failures, he recommended drainage and establishment of grass cover on slopes, methods that today are recognized as appropriate for cut slopes in stiff-fissured clays.

Collin conducted the first documented shear tests on clays, which demonstrated the importance of water content and what are now referred to as the rheological properties of clays. He advocated the inductive approach: working from observation to theory. In this he was at odds with many of his contemporaries, who favored the deductive approach: the derivation of theoretical conclusions from oversimplified assumptions without reference to field observations. Using the inductive approach, Collin outlined an approximate method for analyzing the stability of clay slopes based on the shape of the slip surface and the strength of the clay.

Skempton (1946) presented a review of the historical significance of Collin's work and provided a list of references to Collin's report that he had found. The list is very short; there are only four references by English-speaking scientists and engineers to Collin's work in the century following the publication of Collin's report. It is unfortunate that Collin's observations and recommendations concerning the stability of clay slopes did not receive much wider exposure.

However, the investigators of the disastrous Panama Canal landslides, which are discussed further in Chapter 2, apparently were aware of Collin's report because they referred to it (Reid 1924). Just before World War II, this same reference came to

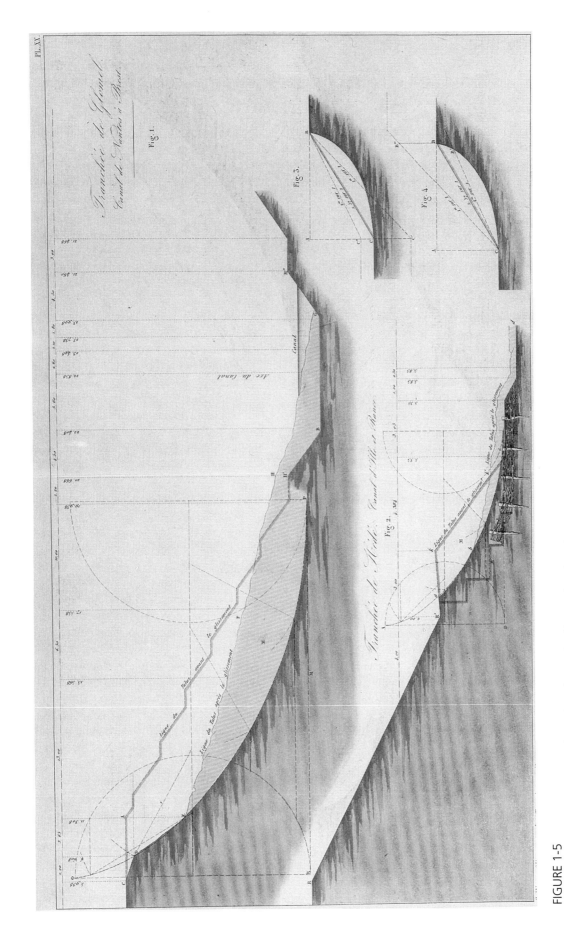

FIGURE 1-5
Two landslides in clay slopes along French canals. Upper drawing shows
Gomel cut failure of May 1838 on canal from Nantes to Brest. Lower
drawing shows 1838 failure of Hédé cut on Ille-et-Rance canal (Collin
1846, Plate XX).
COURTESY OF DAVID W. CORSON, DIRECTOR, OLIN ● KROCH ● URIS LIBRARIES, CORNELL UNIVERSITY

the attention of Robert F. Leggett, Director of the Division of Building Research of the Canadian National Research Council in Ottawa. With some difficulty, Leggett obtained a copy of the Collin report. Although his war duties intervened, Leggett initially assisted with the report's translation into English and ultimately encouraged and supported others in completing such a translation, which was finally published after about a decade of part-time efforts (Schriever 1956). The translation also contains a memoir concerning Collin written by Skempton (1956), which is an updated version of his earlier review (Skempton 1946).

4. OVERVIEW OF REPORT

The 25 chapters forming this report are organized into Parts 1 through 5. This arrangement was adopted to group chapters according to related landslide investigation and mitigation topics. It is hoped that this grouping will assist readers in identifying those chapters most likely to address their immediate needs.

Part 1, Principles, Definitions, and Assessment, contains six chapters. In addition to this introductory chapter, topics covered are the socioeconomic significance of landslides, landslide types and processes, landslide triggering mechanisms, principles of landslide hazard reduction, and the application of hazard and risk assessment and decision making under uncertainty to landslide management.

Many of these topics are either new to this report or greatly expanded compared with the discussions contained in previous reports. The socioeconomic significance of landslides is emphasized because landslide losses continue to grow as human development expands into unstable hillside areas under the pressures of increasing populations. A significant proportion of world landslide losses involves transportation facilities: highways, railways, canals, and pipelines. The nation most severely affected by landslides is Japan, which suffers estimated total (direct plus indirect) landslide losses of $4 billion annually. In the United States, Italy, and India, total annual economic losses due to landslides have been estimated to range from $1 billion to $2 billion. Many other countries have lesser, but major, annual landslide losses.

Chapter 3 includes further development of the landslide classification principles introduced in previous reports, introduces current international stan-

dards proposed for the terminology and description of landslides, and links these standards to the landslide classification. In a similar fashion, the chapters on landslide triggering mechanisms, principles of landslide hazard reduction, and landslide hazard and risk assessment methods represent considerable expansions of earlier presentations.

Part 2, Investigation, includes five chapters that collectively review, in some detail, the entire landslide investigation process. This section begins with the organization of the investigation process and the importance of providing adequate field investigation. Subsequent chapters focus on various aspects of an ideal landslide investigation:

- Initial reconnaissance methods, including aerial photography, remote sensing, and geographic information systems;
- Surface observation and geologic mapping, including the use of surveying methods for identifying and monitoring landslide movements;
- Subsurface investigation, including geophysical explorations, field tests, sample collection methods, and groundwater monitoring; and
- Specialized field instrumentation to monitor landslide movements.

Four chapters form Part 3, Strength and Stability Analysis. The principles of soil and rock mechanics are presented in separate chapters, and stability analysis methods for both soil and rock slopes are presented in two other chapters. A considerable effort has been made to explain and contrast the most appropriate methods for both soil and rock materials.

Part 4 comprises three chapters on landslide mitigation issues. Chapter 16 introduces this section with a review of important considerations and constraints that affect the slope design process. Design methods for the stabilization of soil slopes are provided in Chapter 17, and rock slope stabilization and protection measures are discussed in Chapter 18. In this report an attempt has been made to treat soil and rock slopes on a more equal basis and to compare the best mitigation procedures for each class of slopes.

Part 5, Special Cases and Materials, represents a major addition to the coverage in previous reports: the issues and concerns of landslide investigations in specific environmental or geotechnical conditions. Such special aspects include tropical and residual soils, colluvium and talus, shales and other degradable materials, hydraulic tailings, loess, soft sensitive clays, and permafrost.

REFERENCES

ABBREVIATIONS

HRB   Highway Research Board (now Transportation Research Board)
IAEG   International Association of Engineering Geology
UNESCO   United Nations Educational, Scientific, and Cultural Organization
WP/WLI   Working Party on the World Landslide Inventory (International Geotechnical Societies and UNESCO)

Buss, E., and A. Heim. 1881. *Der Bergsturz von Elm, den 11 September 1881.* Zurich, 163 pp.

Collin, A. 1846. *Recherches Expérimentales sur les Glissements Spontanés des Terrains Argileux, accompagnées de Considerations sur Quelques Principes de la Méchanique Terrestre.* Carilian-Goeury and Dalmont, Paris. Translated by W.R. Schriever under the title *Landslides in Clays by Alexandre Collin, 1846,* University of Toronto Press, Canada, 1956, 161 pp., 21 plates.

Conybeare, W.D., (Mrs.) Buckland, and W. Dawson. 1840. *Ten Plates Comprising a Plan, Sections and Views, Representing the Changes Produced on the Coast of East Devon, between Axmouth and Lyme Regis, by the Subsidence of the Land and the Elevation of the Bottom of the Sea, on 26th December, 1839, and 3rd February, 1840.* John Murray, Albemarle Street, London, 14 pp., 10 plates.

Cruden, D.M. 1991. A Simple Definition of a Landslide. *Bulletin of the International Association of Engineering Geology,* No. 43, pp. 27–29.

Eckel, E.B., ed. 1958. *Special Report 29: Landslides and Engineering Practice.* HRB, National Research Council, Washington, D.C., 232 pp.

Eisbacher, G.H., and J.J. Clague. 1984. *Destructive Mass Movements in Mountains: Hazard and Management.* Paper 84-16. Geological Survey of Canada, Ottawa, Ontario, 230 pp.

Heim, A. 1882. Der Bergsturz von Elm. *Zeitschrift Deutsche Geologische Gesellschaft,* Vol. 34, pp. 74–115.

Heim, A. 1932. Bergsturz und Menschenleben. *Beiblatt zur Vierteljahrsschrift der Naturforschenden Gesellschaft in Zurich,* Vol. 77, pp. 1–217. Translated by N. Skermer under the title *Landslides and Human Lives,* BiTech Publishers, Vancouver, British Columbia, Canada, 1989, 195 pp.

Hsu, K.J. 1978. Albert Heim: Observations on Landslides. In *Rockslides and Avalanches, 1: Natural Phenomena* (B. Voight, ed.), Developments in Geotechnical Engineering, Vol. 14A, Elsevier, Amsterdam, Netherlands, pp. 72–93.

IAEG Commission on Landslides. 1990. Suggested Nomenclature for Landslides. *Bulletin of the International Association of Engineering Geology,* No. 41, pp. 13–16.

Reid, H.F. 1924. The Movement of the Slides. In *Report of the Committee of the National Academy of Sciences on Panama Canal Slides,* National Academy of Sciences, Washington, D.C., pp. 79–84.

Roberts, G. 1840. *An Account of and a Guide to the Mighty Land-slip of Dowlands and Bindon, near Lyme Regis, December 25th, 1839,* 5th ed. Daniel Dunster, Lyme Regis, 38 pp.

Schriever, W.R. 1956. *Landslides in Clays by Alexandre Collin, 1846.* University of Toronto Press, Canada, 161 pp., 21 plates. (See Collin, A.)

Schuster, R.L., and R.J. Krizek, eds. 1978. *Special Report 176: Landslides: Analysis and Control.* TRB, National Research Council, Washington, D.C., 234 pp.

Skempton, A.W. 1946. Alexandre Collin, 1808–1890, Pioneer Soil Mechanics. *Transactions of the Newcomen Society,* London, Vol. 25, pp. 91–103.

Skempton, A.W. 1956. Alexandre Collin (1808–1890) and His Pioneer Work in Soil Mechanics. In *Landslides in Clays by Alexandre Collin, 1846* (W.R. Schriever, translator), University of Toronto Press, Toronto, Canada, pp. xi–xxxiv.

Skermer, N. 1989. *Landslides and Human Lives.* BiTech Publishers, Vancouver, British Columbia, Canada, 195 pp. (See Heim, A.)

Voight, B., ed. 1978. *Rockslides and Avalanches, 1: Natural Phenomena.* Developments in Geotechnical Engineering, Vol. 14A, Elsevier, Amsterdam, Netherlands, 833 pp.

WP/WLI. 1990. A Suggested Method for Reporting a Landslide. *Bulletin of the International Association of Engineering Geology,* No. 41, pp. 5–12.

WP/WLI. 1991. A Suggested Method for a Landslide Summary. *Bulletin of the International Association of Engineering Geology,* No. 43, pp. 101–110.

WP/WLI. 1993. A Suggested Method for Describing the Activity of a Landslide. *Bulletin of the International Association of Engineering Geology,* No. 47, pp. 53–57.

WP/WLI and Canadian Geotechnical Society. 1993. *Multilingual Landslide Glossary.* BiTech Publishers, Richmond, British Columbia, Canada.

Zay, K. 1807. *Goldau und seine Gegend, wie sie war und was sie geworden.* Zurich, 390 pp.

Chapter 2

ROBERT L. SCHUSTER

SOCIOECONOMIC SIGNIFICANCE OF LANDSLIDES

1. INTRODUCTION

Landslides have been recorded for several centuries in Asia and Europe. The oldest landslides on record occurred in Honan Province in central China in 1767 B.C., when earthquake-triggered landslides dammed the Yi and Lo rivers (Xue-Cai and An-ning 1986).

The following note by Marinatos may well refer to a catastrophic landslide resulting in serious social and economic losses:

> In the year 373/2 B.C., during a disastrous winter night, a strange thing happened in central Greece. Helice, a great and prosperous town on the north coast of the Peloponnesus, was engulfed by the waves after being leveled by a great earthquake. Not a single soul survived. (Marinatos 1960)

Research by Marinatos indicated that Helice probably was engulfed as the ground slipped toward the sea a distance of about 1 km. Seed (1968) concluded that this was a major landslide, resulting from soil liquefaction caused by the earthquake.

Slope failures have caused untold numbers of casualties and huge economic losses. In many countries, economic losses due to landslides are great and apparently are growing as development expands into unstable hillside areas under the pressures of expanding populations. In addition to killing people and animals (both livestock and wildlife), landslides destroy or damage residential and industrial developments as well as agricultural and forest lands and negatively affect water quality in rivers and streams.

Landslides are responsible for considerably greater socioeconomic losses than is generally recognized; they represent a significant element of many major multiple-hazard disasters. Much landslide damage is not documented because it is considered to be a result of the triggering process (i.e., part of a multiple hazard) and thus is included by the news media in reports of earthquakes, floods, volcanic eruptions, or typhoons, even though the cost of damage from landslides may exceed all other costs from the overall multiple-hazard disaster. For example, it was not generally recognized by the media that most of the losses due to the 1964 Alaska earthquake resulted from ground failure rather than from shaking of structures.

Government agencies and those who formulate policy need to develop a better understanding of the socioeconomic significance of landslides. That knowledge will allow officials at all levels of government to make rational decisions on allocation of funds needed for landslide research; for avoidance, prevention, control, and warning; and for postfailure repair and reconstruction.

2. FUTURE LANDSLIDE ACTIVITY

In spite of improvements in recognition, prediction, mitigative measures, and warning systems, worldwide landslide activity is increasing; this trend is expected to continue in the 21st century. The factors causing this expected augmented activity are

1. Increased urbanization and development in landslide-prone areas,
2. Continued deforestation of landslide-prone areas, and
3. Increased regional precipitation caused by changing climate patterns.

2.1 Increased Urbanization and Development

Population pressures are increasing in most of the world today and have resulted in rapid urbanization and development. For example, in the United States the land areas of the 142 cities with populations greater than 100,000 increased by 19 percent in the 15-year period from 1970 to 1985. Legget (1973) estimated that by the year 2000, 360 000 km^2 in the 48 conterminous United States will have been paved or built upon. This is an area about the size of the state of Montana. As a result of these population pressures, human activities have disturbed large volumes of geologic materials in housing development and in construction of industrial structures, transportation routes and facilities, mines and quarries, dams and reservoirs, and communications systems. Because of the huge extent of these activities, they increasingly have expanded into landslide-prone areas; thus, these developments have been a major factor in the recent increase in damaging slope failures.

In other countries, particularly developing countries, this pattern is being repeated, but with even more serious consequences. As development occurs, more and more of it is on hillside slopes that are susceptible to landsliding. All predictions are that worldwide slope distress due to urbanization and development will accelerate in the 21st century.

Population pressures are also causing increased landslide losses in other ways. An obvious example is the necessary construction of transportation facilities required by expanding populations. In landslide-prone areas, these facilities are often at risk.

2.2 Continued Deforestation

In many of the developing nations of the world, forests are being destroyed at ever-increasing rates.

Removal of forest cover increases flooding, erosion, and landslide activity. Deforestation, which is expected to continue unimpeded into the 21st century, is causing serious landslide problems in many of these countries, Nepal being the best-documented example. According to the World Resources Institute (*Facts on File Yearbook* 1990), approximately 15 to 20 million ha of tropical forest is currently being destroyed annually, an area the size of the state of Washington.

2.3 Increased Regional Precipitation

For a period of about 3 years in the early 1980s, El Niño caused regional weather changes in western North America that resulted in much heavier-than-normal precipitation in mountainous areas. One of the results was a tremendous increase in landslide activity in California, Colorado, Nevada, Oregon, Utah, and Washington. Climatologists do not know what to expect from future El Niños except that these climatic perturbations will also change climate patterns, certainly increasing precipitation in some areas of the world and thus causing landslide activity.

Scientists do not know what to expect from the much-publicized greenhouse effect either. Will it cause an overall increase in temperature and decrease in precipitation (as occurred in central North America in the late 1980s) or will it disrupt climate patterns, resulting in drought in some areas and increased precipitation in others (as occurred in western North America at the same time)? If areas that are prone to landsliding are subjected to greater-than-normal precipitation, they are apt to experience increased landslide activity.

3. ECONOMIC LOSSES CAUSED BY LANDSLIDES

In this discussion of the expense of landslides at national and local levels, the costs are given in U.S. dollars for the time at which they were originally determined, except where noted. In addition, the original values adjusted to 1990 U.S. dollars are presented in parentheses; the adjustments were made on the basis of yearly cost-of-living indexes for the United States (Council of Economic Advisers 1991).

3.1 Categories of Damage Costs

There are significant advantages to the ability of government officials, land use planners, and others to distinguish between direct and indirect landslide costs and to determine whether these costs affect public or private entities.

3.1.1 Direct Versus Indirect Costs

Landslide costs include both direct and indirect losses that affect public and private properties. Direct costs are the repair, replacement, or maintenance resulting from damage to property or installations within the boundaries of the responsible landslides or from landslide-caused flooding. An outstanding example of direct costs resulting from a single major landslide is the $200 million ($260 million) loss due to the 1983 Thistle, Utah, landslide (University of Utah 1984). This 21-million-m³ debris slide (Figure 2-1) severed major transportation arteries, and the lake it impounded inundated the town of Thistle and railroad switching yards.

All other costs of landslides are indirect. Examples of indirect costs are

1. Loss of industrial, agricultural, and forest productivity and tourist revenues as a result of damage to land or facilities or interruption of transportation systems;
2. Reduced real estate values in areas threatened by landslides;
3. Loss of tax revenues on properties devalued as the result of landslides;
4. Measures to prevent or mitigate additional landslide damage;
5. Adverse effects on water quality in streams and irrigation facilities outside the landslide;
6. Loss of human or animal productivity because of injury, death, or psychological trauma; and
7. Secondary physical effects, such as landslide-caused flooding, for which losses are both direct and indirect.

Indirect costs may exceed direct costs; unfortunately, however, most indirect costs are difficult to evaluate and thus are often ignored or, when estimated, are too conservative.

3.1.2 Public Versus Private Costs

Of possibly greater importance than whether costs are directly or indirectly attributable to a landslide is attribution of the costs on the basis of who is actually faced with the losses. On this basis, landslide losses can be separated into costs to public

FIGURE 2-1
Aerial view in September 1983 of April 1983 Thistle debris slide, Utah, showing Thistle Lake, which was impounded by the landslide; realignment of Denver and Rio Grande Western Railroad in lower center; and large cut for rerouting US-6/50 in extreme lower left.

and private entities (Fleming and Taylor 1980). The possibility of a major landslide that could destroy port facilities and create a wave that might inundate downtown Kodiak, Alaska, is an example of a landslide threat during the 1970s and 1980s that is alleged to have caused indirect costs relating to planning for expansion of the port area (Schuster and Fleming 1988).

Public costs are those that must be met by government agencies; all others are private costs. The largest direct public costs commonly have been for rebuilding or repairing government-owned highways and railroads and appurtenant structures such as sidewalks and storm drains. Other examples of direct public costs resulting from landslides are those for repair or replacement of public buildings, dams and reservoirs, canals, harbor and port facilities, and communications and electrical power systems. Indirect public costs include losses of tax revenues, reduction of potential for productivity of government forests, impact on quality of sport and commercial fisheries, and so forth. An interesting example of indirect public costs due to the impact on fisheries of mass movement and erosion was presented by a study of Tomiki Creek, Mendocino County, California, in the early 1980s. This study found that production of steelhead trout and salmon in Tomiki Creek was reduced 80 percent by landslide, gully, and streambank erosion, resulting in a continuing loss in fisheries potential of $844,000 ($1 million) annually (Soil Conservation Service 1986). In the case of major landslide events, public costs are sustained by all levels of government from federal to local and often by more than one agency within a particular level.

Private costs consist mainly of damage to real estate and structures, either private homes or industrial facilities. In the United States, most railroads are privately owned. Severe landslide problems can result in financial ruin for affected private property owners because of the general unavailability of landslide insurance or other means to distribute damage costs.

3.2 Difficulties in Determining Losses

Although it often is possible to determine the costs of individual landslides, reliable estimates of the total costs of landslides of large geographic entities, such as nations, provinces and states, or even counties, are generally very difficult to obtain. In the public sector, accounting for landslide costs is often lost within general maintenance operations; this seems to be particularly common for transportation agencies. To separate out landslide costs is in itself a costly and complicated operation. In the private sector, the costs incurred by natural hazards are often downplayed as much as possible in order to minimize negative publicity for the company involved.

Landslide cost data commonly are more readily available for industrialized nations than for developing countries. For this reason, most of the economic data presented here are for industrialized countries, such as the United States, Japan, and those in Europe. However, because the severity of the worldwide landslide problem is becoming more widely recognized, the collection of economic data for landslide damages is spreading to all affected nations.

3.3 Losses in the United States

Landslides occur in every one of the United States and are widespread in the island territories of American Samoa, Guam, Puerto Rico, and the U.S. Virgin Islands (Committee on Ground Failure Hazards 1985). They constitute a significant hazard in more than half the states, including Alaska and Hawaii. In the conterminous United States, the areas most seriously affected are the Pacific Coast, the Rocky Mountains, and the Appalachian Mountains (Figure 2-2).

Most of the loss estimates presented here for the United States can be related directly to other industrialized nations with similar terrains and mixes of urban and rural habitats. However, the costs are somewhat higher than might be expected in developing countries, where property and labor values commonly are lower than they are in the United States and other industrialized nations.

Although no cost-reporting mechanism is in use nationally, the U.S. Geological Survey has developed a method for estimating the cost of landslide damage (Fleming and Taylor 1980). Application of this method to smaller geographic areas has suggested that incomplete and inaccurate records have resulted in reported costs that are much lower than those actually incurred. It also appears that losses are on the increase in most regions in spite of an improved understanding of landslide processes and a rapidly developing technical capability for

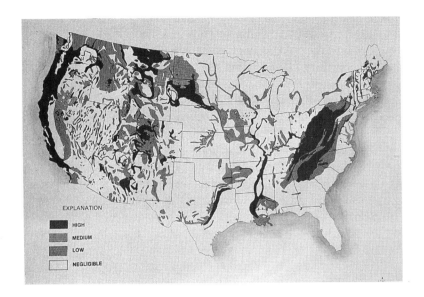

FIGURE 2-2
Map showing
relative potential
for landsliding
in conterminous
United States
(modified from
Radbruch-Hall
et al. 1982).

landslide prediction and mitigation (Committee on Ground Failure Hazards 1985).

In perhaps the first national estimation of U.S. landslide costs, Smith (1958) reported that "the average annual cost of landslides in the U.S. runs to hundreds of millions of dollars," which was probably a realistic figure for that time. However, in the 37 years since Smith assembled his cost data, inflation, residential and commercial development that continues to expand into landslide-susceptible areas, and the use of larger cuts and fills in construction have increased the annual costs of landslides.

On the basis of their analysis of landslide loss data for southern California, Krohn and Slosson (1976) established a figure of $20 per year in 1971 dollars ($46 per year) for damage to each private home in that area. Extrapolating this figure to the estimated 20 million people who at that time resided in areas of the United States with moderate to high landslide susceptibility, they estimated the annual national costs of landslides to private dwellings at about $400 million in 1971 dollars ($1.3 billion). This figure did not include indirect costs or costs to public property, forest or agricultural lands, mines, transportation facilities, and so on. Also in 1976, Jones (at the National Workshop on Natural Hazards, Institute for Behavioral Sciences, University of Colorado, Boulder) estimated that direct costs of landslides for buildings and their sites were about $500 million ($1.2 billion) annually. These estimates were substantiated by Wiggins et al. (1978), who arrived at a total of $370 million in 1970 dollars

($1.2 billion) for annual losses to buildings in the United States due to landslides.

Using the above information, previously unpublished data, inflationary trends, and rough estimates of indirect costs, Schuster (1978) estimated that the total direct and indirect costs of slope failures in the United States exceeded $1 billion per year. Schuster and Fleming (1986) believed that by 1985 this annual figure had increased to nearly $1.5 billion ($1.8 billion), most of the increase being due to inflation. In 1985 the National Research Council (Committee on Ground Failure Hazards 1985) estimated that annual landslide costs in the United States were about $1 billion to $2 billion ($1.2 billion to $2.4 billion), a figure of about $5 to $10 per capita per year ($6 to $12 per capita per year) averaged over the entire nation. Slosson (1987) estimated that total landslide losses in the state of California alone were as high as $2 billion for the decade from 1977 to 1987.

Brabb (1984) used unpublished data based on interviews of personnel from state highway departments and geological surveys to come up with a much lower figure, about $250 million per year (about $315 million per year). However, Brabb's study did not include indirect costs or the costs of infrequent catastrophic events, such as those for landslides from the 1964 Alaska earthquake, because of the difficulty in establishing initial costs and recurrence intervals (Brabb 1989). In addition, interviews of this type often underestimate true costs because the personnel providing the information do not know the total landslide costs within their areas of jurisdiction. Such costs can be determined with reasonable accuracy only by means of rigorous study programs that include delineation of costs to private corporations and property owners.

Total annual costs of landslides for transportation systems in the United States are difficult to determine because of the difficulty in analyzing the following:

1. Smaller slides that are routinely corrected by maintenance forces;
2. Slides on non-federal-aid public highways and roads;
3. Slides on privately owned transportation routes, such as railroads; and
4. Indirect costs related to landslide damage, such as traffic disruptions and delays, inconvenience

to travelers and shippers, and analysis and prevention of landslides.

In spite of these handicaps, attempts were made during the 1970s to estimate annual landslide losses to the U.S. highway system. Chassie and Goughnour (1976a, 1976b) reported on a survey by the Federal Highway Administration in which it was indicated that approximately $50 million per year ($115 million per year) was spent at that time to repair landslides on the federally financed portion of the national highway system. This system includes federal and state highways but not most county and city roads and streets, private roads and streets, or roads built by other federal agencies, such as the U.S. Forest Service, the Bureau of Land Management, or the National Park Service. If indirect costs, costs to non-federal-aid highways, and the other factors noted above were added, Chassie and Goughnour (1976b) estimated that $100 million ($230 million) was a conservative value of the annual landslide damage to highways and roads in the United States in the 1970s.

The 1976 Federal Highway Administration survey of landslide costs for U.S. highways was duplicated by Walkinshaw (1992), who obtained repair and maintenance costs for landslide damage to 1.3 million km of state highways for the 5-year period from 1986 to 1990. Walkinshaw found that the total average annual cost of contract landslide repairs on state highways for this period was $65.4 million (Figure 2-3), and annual average landslide maintenance costs (repairs by highway department maintenance forces) were reported as $41.4 million, for a total average annual direct cost of nearly $106 million, a figure nearly equal to the 1990 equivalent of $115 million that Chassie and Gougnour (1976b) found for annual repair and maintenance costs in the 1970s. Thus, direct landslide costs to highways have remained nearly constant (when noted in 1990 dollars) in spite of the near-completion of the Interstate highway construction program and drier-than-normal weather in the western United States during the 1986–1990 survey period.

It should be remembered that the cost figures presented in both the Chassie and Goughnour (1976b) and the Walkinshaw (1992) surveys do not represent total landslide costs, either direct or indirect, for the U.S. highway system. One deficiency of these surveys is that the state and federal

highways for which the surveys were conducted represent only about 20 percent of the 6 239 000 km in the entire U.S. highway and road system. However, this 20 percent probably is subject to the major part of landslide costs because it has been constructed to higher standards than the rest of the system (i.e., larger cuts and fills were used).

Another deficiency of these surveys is that many state transportation departments do not maintain satisfactory inventories of their highway landslide maintenance costs. Several states that have kept good maintenance records (particularly Maine, West Virginia, Kentucky, Missouri, Texas, Colorado, and California) have found that the maintenance costs for landslides have exceeded their contract repair costs (Walkinshaw 1992). California distinguished itself by reporting the highest annual cost for landslide maintenance of all the states—more than $15 million per year, even during 5 years of well-below-normal precipitation.

Such landslide cost surveys have not attempted to determine indirect costs of landslides. A cost item that often is large but is extremely difficult to determine accurately is the indirect cost of loss of business in communities whose commerce is hindered by the closure of transportation routes because of landslides. An example of the magnitude of such indirect costs in relationship to direct actual repair costs was provided by the 1983 landslide closure of US-50 by landslides both west (in California) and east (in Nevada) of south Lake Tahoe. The total cost of repairs to the heavily traveled highway was $3.6 million ($4.7 million) (Walkinshaw 1992), but the estimated economic

FIGURE 2-3
Average annual costs (in thousands of dollars) of contracted highway repairs on U.S. state highway systems for 1986–1990 (Walkinshaw 1992).

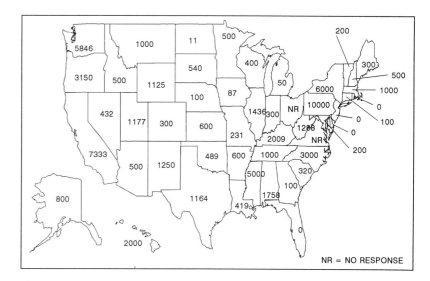

FIGURE 2-4
(top right)
August 1989 rock
fall from gneiss cliff
onto Washington
State Highway 20,
North Cascades
National Park,
blocking highway
for about 2 weeks
(note tunnel portal
behind debris);
example of highway
landslide for which
indirect costs due
to forced traffic
diversion are
difficult to
determine.

FIGURE 2-5
(top far right)
Damage to railroad
line in Olympia,
Washington, caused
by slope failure
after 1965
Seattle-Tacoma
earthquake.
G. W. THORSEN, DIVISION
OF GEOLOGY AND EARTH
RESOURCES, WASHINGTON
DEPARTMENT OF NATURAL
RESOURCES

FIGURE 2-6
(bottom)
Landslide resulting
in derailment of
California Zephyr
passenger train near
Granby, Colorado,
in April 1985. Small
landslide started as
earth slide through
base of railway
embankment in
middle background
and rapidly moved
about 70 m as
debris flow, partially
damming Fraser
River in foreground.

loss to the area from 2 1/2 months of access disruption and the resulting loss of tourist revenues was $70 million ($92 million) (*San Francisco Chronicle* 1983), nearly 20 times as much as the direct expenditures for repair. A lesser, but more common, example is the August 1989 rock fall (Figure 2-4) that blocked Washington State Highway 20 in North Cascades National Park for 2 weeks. During this period traffic from the northern Puget Sound area to north-central Washington had to be directed farther south to US-2 and Interstate 90 at a cost in both mileage and time.

There is no firm information on landslide losses by U.S. railroads because nearly all U.S. railroads are private corporations that do not commonly release such data. However, it is estimated that direct losses to railroads from landslides in the Rocky Mountain states for the period 1982–1985, during which precipitation was much greater than normal, exceeded $100 million ($120 million). During periods of normal precipitation, landslide losses to U.S. railroads are much lower than they were during this unusual period. An economic impact analysis by the University of Utah (1984) noted that the largest single loss caused by the 1983 Thistle, Utah, landslide (Figure 2-1) was the $81 million ($107 million) in revenue lost by the Denver and Rio Grande Western Railroad (D&RGW) because of temporary closure of their main line by the slide. Figures 2-5 and 2-6 show examples of the effects of landslides on railway operations.

3.4 Losses in Other Nations

Japan probably has the dubious honor of being the nation with the world's greatest total land-

slide costs. In 1982, N. Ohhira (personal communication, Director-General, Japanese National Research Center for Disaster Prevention, Tsukuba City) noted that annual losses in Japan totaled about $1.5 billion ($2 billion), a figure comparable with that for the United States. However, on the basis of data provided by the Japanese Ministry of Construction, Oyagi (1989) estimated that the costs of landslide control works constructed in Japan in 1987 and 1988 were in excess of $4 billion per year ($4.4 billion per year). Similarly, Moriyama and Horiuchi (1993) and Nishimoto (1993) reported a total cost for Japanese landslide control works in fiscal year 1992 of approximately $4.7 billion. Of this figure, $3.4 billion went to the Sabo erosion control works (mainly check dams to prevent debris flow damage), $50 million for "landslide prevention works," and $850 million for "slope failure remedies" (all three of which are considered here as landslide control works). An additional $54 million was spent to control snow avalanches.

The Alpine nations of western Europe, particularly Italy, Austria, Switzerland, and France, have been subject to significant landslide activity from the beginning of recorded history. For the Alpine countries, Eisbacher and Clague (1984) described 137 landslide case histories that represent "the most interesting, costly and tragic mass movements witnessed in 2000 years of Alpine settlement." The unpublished results of a 1976 United Nations Educational, Scientific, and Cultural Organization (UNESCO) survey indicated that annual landslide losses in Italy were about $1.14 billion ($2.6 billion) (M. Arnould, personal communication, 1982, Ecole Nationale Supérieure des Mines, Paris). No similar information has been encountered for landslide costs of other Alpine nations, but it is estimated that they would be somewhat lower than those for Italy.

On the basis of estimated annual landslide damages of $100 million ($135 million) to 10 000 km of highways and roads in the hilly and mountainous topography of northern India, Mathur (1982) arrived at an annual cost for landslide damages of nearly $1 billion ($1.35 billion) for the total 89 000 km of roads in this landslide-prone area. In addition to the commonly used reconstruction and maintenance costs, Mathur's estimates included indirect costs, such as loss of tourist trade, loss of person-hours and vehicle-hours resulting from road blockages, and failure of communications, that may not have been included in the estimates for the United States, Japan, and Italy. Besides Mathur's cost data, Chopra (1977) noted that catastrophic damage to roads in north Bengal and Sikkim occurred in 1968 and 1973; restoration was estimated to cost $14 million ($53 million) and $8 million ($24 million), respectively.

Thus, landslide costs (direct plus indirect) in the United States, Japan, Italy, and India seem to be roughly comparable, somewhere between $1 billion per year and $5 billion per year for each country. Although there have been few other published estimates of landslide costs, the data that are available indicate that landslide costs for other countries are considerably lower. Li (1989) reported that annual losses for China are about $500 million. Ayala and Ferrer (1989) arrived at a figure of $220 million for yearly landslide costs in Spain. S. G. Evans (personal communication, 1989, Geological Survey of Canada, Ottawa) estimated that annual landslide costs for Canada are about $50 million. In

Hong Kong, a small, densely populated area with serious landslide problems, the government spends $25 million per year for landslide studies and remedial works (Brand 1989). Hawley (1984) estimated that annual landslide costs for New Zealand are approximately $12 million ($15 million). In 1982, Swedish costs were about $10 million per year to $20 million per year ($13 million per year to $27 million per year) (Cato 1982), and those for Norway are estimated at $6 million per year (Gregersen and Sandersen 1989).

For the industrialized nations of central and eastern Europe, little information is available on national landslide costs. However, it is well known that several republics of the former Soviet Union have serious landslide problems in their far-ranging hill and mountain areas. Because of the huge area involved, total landslide costs in the republics of the former Soviet Union are estimated to be on the same order as those previously given for China. As an example of these costs, Khegai and Popov (1989) estimated that landslide activity (mainly debris flows) in the vicinity of Alma-Ata, Kazakhstan, has caused total damage of about $500 million in the past few decades. The central European mountains of Czechoslovakia, Poland, Hungary, Romania, and Bulgaria have also proved to be susceptible to landsliding (Kotarba 1989), but no cost figures have been published for these countries.

Very few national landslide cost estimates are available for developing countries because little research has been done on this subject. However, landslide disasters are common in many of these countries, particularly in mountainous areas. Especially hard hit have been the Himalayan and Andean nations and the island nations around the Pacific Rim of Fire, particularly Papua New Guinea (Figure 2-7), Indonesia, the Philippines, and Taiwan. For example, Charma (1974) reported that landslides in Nepal have killed hundreds of people, displaced more than 1,000 families, and cost hundreds of thousands of dollars in damage, much of which has been to roads and highways. For developing countries in these areas, landslide losses probably represent a larger part of the gross national product than for the industrialized nations discussed earlier.

3.5 Losses in Smaller Geographic Areas

In the United States there was little documentation of major landslide damages until the early

FIGURE 2-7
Bairaman River rock slide–debris avalanche, 1985, triggered by magnitude 7.1 earthquake on island of New Britain, Papua New Guinea. This 180-million-m³ landslide caused little direct financial loss because it occurred in unpopulated mountainous area. However, it formed 210-m-high blockage of Bairaman River. When this natural dam failed 14 months after quake, huge flood destroyed village of Bairaman, 40 km downstream. Residents of village had been evacuated before dam breached; consequently, no lives were lost.
P. LOWENSTEIN, GEOLOGICAL SURVEY OF PAPUA NEW GUINEA

FIGURE 2-8
Wreckage of Government Hill School, Anchorage, Alaska, which was destroyed by 700 000-m³ landslide triggered by March 1964 earthquake (Hansen 1965). There were no casualties because earthquake occurred on Sunday when school was not in session. Graben 4 m deep in foreground is at head of slide.
W.R. HANSEN, U.S. GEOLOGICAL SURVEY

part of the 20th century. Although economic data are sketchy, landslides caused by the 1906 San Francisco earthquake had a significant socioeconomic effect on northern California, an effect that received little publicity because of the enormity of direct earthquake damages in the city of San Francisco itself. The 1906 earthquake triggered ground failures (primarily slope failures) over a 600-km zone extending along the northern California coast from Eureka on the north to southern Monterey County and as far as 100 km inland (Youd and Hoose 1978). Hillside landslides triggered by the earthquake were too numerous for documentation of each occurrence. Many were in unpopulated areas; however, where slope failures impinged on human works, the results were generally disastrous. For example, 10 men were killed and two lumber mills destroyed by landslides in the Santa Cruz Mountains, and 5 km of the Ocean Shore Railroad was destroyed along the coastal bluffs south of San Francisco. Roadways, bridges, pipelines, and buildings suffered considerable damage from lateral spreads. Pipeline breaks were particularly critical in San Francisco, cutting off the water supply to a city that was soon devastated by fire.

Some 500 reservoir-induced landslides along the shore of Franklin D. Roosevelt Lake, the reservoir impounded by Grand Coulee Dam on the Columbia River in Washington State, caused at least $20 million (about $150 million) in damages between 1934 and 1952 (Jones et al. 1961). Landslide activity along the shores of Lake Roosevelt has continued since then, particularly in 1969 and 1974 when the reservoir was drawn down 40 m for construction of the Grand Coulee Dam third power plant (Schuster 1979).

The most economically devastating landslides in the United States in recent decades were those triggered by the 1964 Alaska earthquake, the 1980 landslides in southern California, the 1982 landslides in the San Francisco Bay area, and the 1983–1984 landslides in Utah and surrounding states. Youd (1978) estimated that ground failure caused 60 percent of the $300 million ($1.26 billion) total damage from the 1964 Alaska earthquake; nearly all of the ground failure consisted of landslides, including lateral spreads. Five major landslides caused about $50 million ($210 million) in damage to nonmilitary facilities in Anchorage, Alaska's largest city (Figure 2-8). The total damage to Alaskan highways, railways, and bridges caused by lateral spreads also amounted to about $50 million. Flow failures in the coastal communities of Valdez (Figure 2-9), Seward, and Whittier carried away port facilities that originally cost about $15 million ($63 million).

Within the United States greater effort at detailing the costs of slope movements has been expended in California than in any other state.

Landslide-causing storms have plagued southern California for the past three decades. Exceptional landslide activity occurred in 1951–1952, 1956, 1957–1958, 1961–1962, 1968–1969, 1977–1978, 1979–1980, and 1982. The Portuguese Bend landslide (Palos Verdes Hills, California) was estimated to have cost in excess of $10 million ($45 million) in damages to roads, homes, and other structures between 1956 and 1959 (Merriam 1960). It was necessary to raze 127 residential dwellings and a privately owned recreational club that were located on the slide. Subsequent litigation resulted in an award of approximately $9.5 million ($41 million) by the County of Los Angeles to property owners in the affected area on the grounds that road construction by the county was responsible for initiating the failure (Vonder Linden 1989).

Since the time of the Portuguese Bend landslide, there have been many costly landslides in southern California. The 1978 Bluebird Canyon landslide (Figure 2-10) caused an estimated direct loss of $15 million ($30 million) to private property in Laguna Beach, south of Los Angeles (Tan 1980). The estimated total losses in the six southern counties of California in 1980 due to all types of landslides caused by heavy winter rainfall approximated $500 million ($800 million) (Slosson and Krohn 1982). As another example of southern California slope failure costs, a study by the U.S. Geological Survey during the winter rainy seasons of 1978–1979 and 1979–1980 documented 120 landslides in San Diego County, which caused damages of about $19 million ($34 million) (Shearer et al. 1983).

The most recent major losses in southern California were caused by the Big Rock Mesa landslide along the Malibu coast west of Los Angeles. This large creeping mass movement, which began in the late summer of 1983 and threatened "to dump 120 acres overlooking the Pacific Coast Highway into the ocean" (*Los Angeles Times* 1984), by March 1984 resulted in condemnation of 13 houses and threatened more than 300 others. The individual homes ranged in value from $400,000 to more than $1 million ($500,000 to more than $1.25 million). During 1984 many lawsuits related to this landslide were filed by property owners against Los Angeles County and a number of consultants. According to a deputy county counsel, the total of legal claims against Los Angeles County as a result of the Big Rock Mesa landslide by July 1984

FIGURE 2-9

(*a*) Destruction of port facilities at Valdez, Alaska, by submarine landslide triggered by March 1964 earthquake (Coulter and Migliaccio 1966). Dashed lines indicate dock area destroyed by slide. DAVID LANEVILLE, U.S. GEOLOGICAL SURVEY (*b*) Valdez dock area after reconstruction, September 1965. U.S. BUREAU OF LAND MANAGEMENT

FIGURE 2-10
Damage caused by
1978 Bluebird
Canyon landslide,
Laguna Beach,
California, which
destroyed 25
homes, parts of
three streets, and
area's public utilities.
WOODY HIGDON,
COURTESY OF LEIGHTON
AND ASSOCIATES, INC.

was more than $500 million ($630 million) (Association of Engineering Geologists 1984).

The San Francisco Bay area of northern California also has been hit hard by landslides. In a classic study of landslide costs in that area, Taylor and Brabb (1972) documented losses amounting to $25 million ($90 million) in the nine Bay-area counties for the rainy season of 1968–1969, a large expense for the relatively small area involved. Of this total, about $10 million ($36 million) consisted of loss or damage to public property, mainly for relocation or repair of roads and utilities; about $9 million ($32 million) was for loss or damage to private property, primarily because of reduced market value; and about $6 million ($21 million) consisted of miscellaneous costs that could not be classified in either the public or private sector. The intense storms of January 1982 in the San Francisco Bay area triggered thousands of debris flows and a few large landslides. About 30 people were killed and hundreds left homeless in this catastrophe. About 6,500 homes and 1,000 businesses were damaged or destroyed. Most of the fatal or damaging landslides were debris flows. Creasey (1988) documented total direct costs of the landslides as being in excess of $66 million ($90 million). In the wake of these damages, 930 lawsuits and claims in excess of $298 million ($404 million) were filed against city and county agencies in the San Francisco Bay region as of May 1982 (Smith 1982), an amount considerably greater than the total property losses.

In 1980 a massive rock slide–debris avalanche (Figure 2-11) with a volume of 2.8 km³ descended

at high velocity from the north slope of Mount St. Helens, Washington, as a result of the eruption of the volcanic peak (Voight et al. 1983). The debris avalanche traveled about 22 km westward, burying about 60 km² of the valley of the North Fork Toutle River under poorly sorted earth-and-timber debris. It destroyed nine highway bridges, many kilometers of highways and roads, and numerous private and public buildings (Schuster 1983). The debris avalanche also formed several new lakes by damming the North Fork Toutle River and its tributaries. These lakes and their natural dams posed downstream hazards because of the possibility of failure of the natural dams, which could have resulted in catastrophic downstream flooding. The largest landslide-dammed lake is 260-million-m³ Spirit Lake, which was prevented from overtopping its natural dam by a bedrock drainage tunnel 2.9 km long that was completed in 1985 at a cost of $29 million ($35 million) (Sager and Budai 1989).

Mud flows continued downstream for 95 km beyond the toe of the debris avalanche, modifying a total of more than 120 km of river channel, including the Toutle River and sections of the Cowlitz and Columbia rivers (Schuster 1983). The mud flows destroyed or badly damaged about 200 homes on the floodplain of the Toutle River. About half of the 27 km of Washington State

FIGURE 2-11
Mount St. Helens
debris avalanche in
valley of North Fork
Toutle River, May
1980: view east
from distal margin
of avalanche toward
devastated cone
of volcano.
R. M. KRIMMEL, U.S.
GEOLOGICAL SURVEY

Highway 504 along the Toutle River was buried under as much as 2 m of sediment (Figure 2-12). Mud flows also buried many kilometers of private logging roads and county roads and destroyed 27 km of logging railway (Figure 2-13). The mud flows and giant logjams they were carrying destroyed or badly damaged 27 highway and railroad bridges (Figure 2-14).

Abnormally high precipitation in 1982–1984 caused thousands of landslides in the western mountain areas of the United States. Anderson et al. (1984) estimated that total direct costs of landslides in the state of Utah in spring 1983 exceeded $250 million ($330 million). Estimates of direct costs of the 1984 Utah landslides were as high as $50 million ($63 million) (B.N. Kaliser, personal communication, 1984, Utah Geological and Mineral Survey, Salt Lake City). The April 1983 Thistle debris slide (Figure 2-1), Utah's single most destructive slope failure, and the lake it

formed by damming the Spanish Fork River severed three major transportation arteries: US-6/50, US-89, and the main transcontinental line of the Denver and Rio Grande Western Railroad (D&RGW) (Kaliser 1983). The D&RGW spent about $40 million ($53 million) to reestablish its line outside the devastated area, mostly to construct a twin-bore tunnel about 900 m long that bypassed the landslide and lake (Malone 1983). Before the lake was drained, it inundated the town of Thistle, resulting in destruction of 10 homes, 15 businesses, and D&RGW switching yards.

An economic impact analysis prepared by the University of Utah (1984) evaluated direct and indirect costs of the Thistle landslide. The direct costs totaled $200 million ($260 million). In addition, numerous indirect costs were reported; most of these involved temporary or permanent closure of highway and railroad facilities to the detriment of local coal, uranium, and petroleum industries; several types of businesses; and tourism. A branch line of the D&RGW that joined the main line at Thistle was closed by the lake and has not been reopened. Numerous private enterprises, six communities, and two counties were directly affected by this railroad closure, and other businesses and communities were indirectly affected because of reduced production, unemployment, and reduced income. Of the jobs lost directly as a result of the Thistle landslide, 196 were abolished permanently. More than 2,500 jobs were temporarily lost in the mining industry alone. During 1983, unemployment in the two affected counties increased more than 300 percent. Perhaps the largest single loss due to the Thistle landslide was $81 million ($107 million) in revenue lost by the D&RGW in 1983 as a result of the slide. The indirect effects of the Thistle landslide disaster have produced temporary

FIGURE 2-14
Remains of steel-girder bridge across Toutle River, Washington State, May 1980. Bridge was destroyed by mud flow associated with eruption of Mount St. Helens.

FIGURE 2-12
(top left)
Mud flow from Mount St. Helens covering Washington State Highway 504 to depth of 2 m in valley of North Fork Toutle River. Mud flow material was easily removed from highway surface by means of front-end loaders.

FIGURE 2-13
(bottom left)
Railway cars and logs overturned by 1980 Toutle River mud flow, Washington State.

and permanent losses that may perhaps exceed the direct costs. Although there were no casualties in the Thistle landslide, it ranks as the most economically expensive individual landslide (in terms of both direct and total costs) in North America.

There have been few estimates of financial losses due to landslides in the eastern United States. However, studies of landslide costs in Pittsburgh, Pennsylvania, and Cincinnati, Ohio, indicate that costs in the Appalachian Mountains, and particularly in urban areas, are significant. Expenditures for landslide damages in Allegheny County (Pittsburgh), Pennsylvania, for the period 1970–1976 were estimated at about $4 million per year ($12 million per year) for an annual per capita outlay of about $2.50 ($7.00) (Fleming and Taylor 1980). In Hamilton County (Cincinnati), Ohio, landslide damage costs for the 6-year period from 1973 to 1978 averaged $5.1 million per year ($12.4 million per year), an annual per capita outlay of $5.80 ($14). Not included in this total was at least $22 million ($53 million) that was spent to stabilize a single landslide in Cincinnati (Fleming 1981).

With the exception of these estimates of landslide costs for Pittsburgh and Cincinnati, little attempt has been made to determine landslide losses for the eastern United States. The data for Pittsburgh and Cincinnati suggest that significant damages occur there each year as opposed to the western United States, where landslide activity is closely associated with years of above-average precipitation or with single high-intensity storms (Schuster and Fleming 1986). The general pattern of damages in areas susceptible to landsliding in the eastern United States is one of consistently large annual costs punctuated by a few years of extreme damages caused by severe hurricanes on the East Coast.

The many major slope failures that occurred during construction of the 12-km long Gaillard Cut in the Continental Divide segment of the Panama Canal (Lutton et al. 1979) constituted one of the world's most extreme cases of damage to a transportation system (Figures 2-15 and 2-16). Slope failures not only severely disrupted construction, delaying completion of the canal by nearly 2 years, but also caused closing of the canal on seven different occasions after it was opened to traffic in 1914. In 1915 the two largest landslides, the East and West Culebra slides, with volumes of 13 and 10 million m³, respectively, occurred simul-taneously, completely blocking the canal (Berman 1991). As noted by MacDonald (1942), "The confidence of the American people and its Congress was shaken by the delay in achieving continuous service." Although detailed costs of damages resulting from Panama Canal landslides from the construction period to present are not available, the following data published by the Panama Canal Company indicate the economic severity of the effects of the slope failures (MacDonald 1942):

1. During construction, excavation was disrupted for days and weeks at a time because landslides blocked haulage railroad tracks;
2. Steam shovels, locomotives, drilling equipment, railway cars, and other equipment were destroyed during construction (Figure 2-16);
3. Construction costs were millions of dollars higher than they would have been if the landslides had not occurred;
4. Between the beginning of construction and 1940, 57 million m³ of landslide material was removed from the canal; and
5. Many millions of dollars in shipping tolls were lost by delay in opening the canal and by periods of enforced closure due to landslides.

Although landslides have not closed the canal since 1920, they still threaten navigation and pose a continuing and expensive maintenance problem for the Panama Canal Commission, which is now a binational agency representing the Republic of Panama and the United States. A 4.6-million-m³ reactivation of the Cucaracha landslide (Figure 2-17) nearly closed the canal in 1986 (Berman 1991).

The original width of the channel in the Gaillard Cut was 91 m; by 1970 it had been widened to 152 m. By the mid-1980s the increase in large-beam ships proved to be an obstacle to navigation through the cut. Thus, in May 1991 the Panama Canal Commission implemented a widening program; the width of the channel is being increased to 192 m in the straight portions of the cut and to 213 to 223 m on the curves (Schuster and Alfaro 1992). Approximately 27×10 m³ of material will be excavated, which could result in increased slope failure hazards. To reduce the risk, slopes are being geotechnically designed and drainage systems installed to alleviate the rainfall-induced pore pressures that cause the slopes to fail.

FIGURE 2-15
Cucaracha landslide,
February 2, 1913,
during construction
of Gaillard Cut,
Panama Canal
(MacDonald 1942).
Construction
railroad tracks are
on bottom of future
canal. Activity of
this 2.2-million-m³
earth slide–earth
flow continued
during and after
filling of canal.
COURTESY OF PANAMA
CANAL COMMISSION

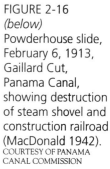

FIGURE 2-16
(below)
Powderhouse slide,
February 6, 1913,
Gaillard Cut,
Panama Canal,
showing destruction
of steam shovel and
construction railroad
(MacDonald 1942).
COURTESY OF PANAMA
CANAL COMMISSION

As much as 650 mm of torrential rain fell on parts of the island of Oahu, Hawaii, in 24 hr over New Year's Eve 1987–1988. Resultant flooding and debris flows caused an estimated $34 million in damages (Dracup et al. 1991). The Niu, Kuliouou, and Hahallone valleys were the most severely affected; most of the damage in these valleys was due to debris flows. In addition, the storm resulted in an estimated 10 to 15 soil slides in each of the valleys. Some of the slides caused heavy damage in residential areas. Total cost data are not available for these slides except for a state-sponsored residential subdivision in Kuliouou Valley, where a single slow-moving landslide severely damaged 17 family residences, which were replaced at a total cost of $5.7 million (*Honolulu Star-Bulletin* 1989).

Physical measures to protect structures or developments from actual or potential landsliding can be very expensive. Recently the control of landslides that threaten important dams and reservoirs in Canada, Peru, and New Zealand has resulted in large expenditures. In Canada mitigation has prevented reactivation of the prehistoric 1.5-km³ Downie rock slide along the Columbia River north of Revelstoke, British Columbia (Schuster 1979). Beginning in 1977, drainage measures, augmented by instrumentation, were installed to improve the stability of this massive landslide, the toe of which

was to be partially inundated by the reservoir of the under-construction Revelstoke Dam. Costs of preventive measures totaled $21 million Canadian (A.S. Imrie, personal communication, 1984, B.C. Hydro, Vancouver, British Columbia, Canada). Although preventive measures are not landslide damages per se, the funds were spent to reduce the threat of much larger damages.

FIGURE 2-17
October 1986
reactivation of
Cucaracha landslide
in Gaillard Cut,
Panama Canal. Slide
extended nearly
across canal, but
much of toe had
been removed by
dredges by time
this photograph
was taken.
COURTESY OF PANAMA
CANAL COMMISSION

FIGURE 2-18
(below) Creeping
rock slide (*arrows*)
endangering
Tablachaca Dam and
Reservoir, Mantaro
River, Peru, February
1982, before costly
control measures,
consisting primarily of
reservoir-level earth
buttress, surface and
subsurface drainage,
and rock anchors,
were used to reduce
threat of catastrophic
slope failure to Peru's
largest hydropower
dam.

An even larger cost was entailed in controlling movement of a 3-million-m³ creeping mass of rock and colluvium endangering Tablachaca Dam on the Mantaro River in Peru (Figure 2-18). This dam is Peru's largest producer of electric power. Deere and Perez (1985) noted that approximately $40 million ($50 million) was spent by the Peruvian government in landslide stabilization measures consisting of (*a*) a 460 000-m³ toe buttress founded on densified river sediments; (*b*) 405 prestressed rock anchors; (*c*) 1300 m of drainage tunnels, 190 radial drains, 21 horizontal drains, and 3300 m of surface ditches; (*d*) 68 500 m³ of

rock excavation; (*e*) numerous inclinometers, piezometers, extensometers, and other instrumentation; and (*f*) improvement of the river-channel flow pattern (Morales Arnao et al. 1984).

The most expensive landslide stabilization program yet undertaken has been the $220 million works that were recently constructed along the shoreline of proposed Lake Dunstan in southern New Zealand (Bell 1992). This hydroelectric storage reservoir is to be impounded behind the completed Clyde Dam in the schist terrain of Central Otago where large ancient landslide complexes, which border about 25 percent of the reservoir shoreline, pose a potential threat to the long-term operation of the power scheme (Bell 1992; Gillon and Hancox 1992).

On March 5, 1987, two earthquakes shook the eastern slopes of the Andes Mountains in eastern Ecuador, triggering landslides and flooding that resulted in destruction or local severing of nearly 70 km of the Trans-Ecuadorian oil pipeline and the main highway from Quito to the eastern rain forests and oil fields (Figure 2-19) (Nieto and Schuster 1988; Schuster 1991). Economic losses were estimated at $1 billion ($1.15 billion), most of which were due to temporary loss of transport capacity for petroleum and petroleum products. The effects of the widespread denudation on the agricultural and hydroelectric development of the area were impossible to evaluate, but undoubtedly were very large. An estimated 1,000 to 2,000 deaths occurred as a direct result of the landslides and related floods.

The Val Pola rock avalanche, one of the most catastrophic landslides in European history, occurred in the central Italian Alps in July 1987. Following a prolonged period of heavy summer precipitation in the Alps, 35 million m³ of rock avalanched into the Valtellina south of Bormio (Figure 2-20) (Cambiaghi and Schuster 1989; Govi 1989). Although no cost data for this event have been reviewed, the Val Pola landslide killed 27 people, destroyed four villages that had been evacuated in anticipation of the event, completely buried the main north-south highway in this part of the Alps, and dammed the Adda River. Because of the danger of overtopping and failure of the natural dam, some 25,000 people were evacuated from the valley downstream. The lake was lowered by pumps and siphons, and two 3.5-km-long permanent diversion tunnels were constructed through

FIGURE 2-19
Aerial view of destruction of Trans-Ecuadorian oil pipeline and adjacent Trans-Ecuadorian highway by earthquake-triggered debris flow issuing from minor tributary of Coca River, eastern Ecuador, March 1987.

FIGURE 2-20 *(below)* Val Pola rock avalanche, northern Italy, July 1987, which destroyed major segment of Highway 38 between Bormio and Sondrio and dammed Adda River to form Lake Val Pola. Note emergency spillway under construction across landslide dam *(bottom left).*

the left bedrock abutment of the natural dam (Cambiaghi and Schuster 1989). The highway was reconstructed above the toe of the landslide. Because of continuing danger at the site after the 1987 failure, the upper part of the Val Pola landslide was heavily instrumented with microseismic networks, inclinometers, extensometers, piezometers, and meteorological equipment at a total cost of $18 million (an indirect cost of the landslide) (Experimental Institute for Models and Structures, written communication, 1988, Bergamo, Italy).

4. LANDSLIDE CASUALTIES

Human casualties due to landslides have been recorded since people began to congregate and build in areas subject to slope failure. The world's most devastating landslide disasters in terms of numbers of casualties have been triggered by earthquakes; the two worst cases occurred in central China. In 1786 the earthquake of Kangding-Louding in Sichuan Province caused a huge landslide that dammed the Dadu River for 10 days (Li 1989). When the landslide dam was overtopped and failed, the resulting flood extended 1400 km downstream and drowned about 100,000 people. In an even greater tragedy, 200,000 people were killed in the 1920 Gansu Province earthquake from a combination of landslides, collapsed cave homes, fallen buildings, and exposure to the harsh winter climate (Close and McCormick 1922). The description by Close and McCormick suggests that as many as 100,000 were killed by the landslides. In an event similar to that in which the Dadu River was dammed, a 1933 earthquake

near Deixi in northwestern Sichuan Province caused landslides that killed 6,800 people directly and drowned at least 2,500 more when the resulting landslide dam failed (Li et al. 1986).

In a similar tragedy in southern Italy, the 1786 Calabria earthquake triggered landslides that killed approximately 50,000 people (Cotecchia et al. 1969). Landslides caused by the quake formed about 250 lakes in the area; secondary deaths occurred for several years after the quake from malaria spread by mosquitoes that bred in the new lakes.

In this century the problem of deaths and injuries due to landslides has been exacerbated by the burgeoning population in landslide-prone areas. Varnes (1981) estimated that during the period 1971–1974 nearly 600 people per year were killed worldwide by slope failures. About 90 percent of these deaths occurred within the Circum-Pacific region (i.e., in or on the margins of the Pacific Basin). Probably the best-known recent catastrophic landslides of the Circum-Pacific region are the debris avalanches of 1962 and 1970 on the slopes of Mt. Huascaran in the Cordillera Blanca of Peru. In January 1962 a large debris avalanche that started on the north peak of Mt. Huascaran obliterated mountain villages, killing some 4,000 to 5,000 people (Cluff 1971). Eight years later an even greater number were killed when a magnitude (**M**) 7.75 earthquake off the coast of Peru triggered another disastrous debris avalanche on the slopes of Mt. Huascaran. This landslide descended into the same valley at average speeds of about 320 km/hr but devastated a much larger area than in 1962, burying the towns of Yungay and Ranrahirca and killing more than 18,000 people (Cluff 1971; Plafker et al. 1971).

Another multiple-hazard catastrophe hit South America in 1985 when volcanic mud flows triggered by a minor eruption of Nevado del Ruiz volcano in Colombia destroyed the city of Armero (pre-eruption population: 29,000) (Voight 1990). More than 20,000 were entombed and 5,000 more were injured. The catastrophic loss of life was partially due to failure in emergency response. The disaster occurred in spite of the fact that Colombian and international scientists, alerted by nearly a year of precursory activity by the volcano, had warned that Ruiz might erupt and had prepared a hazard zoning map that accurately predicted the tragic effect of the eruption weeks before it occurred.

Among industrialized nations, Japan has probably suffered the largest continuing loss of life and property from landslides. Although some landslides in Japan are caused by earthquakes, most are a direct result of heavy rains during the typhoon season. When urban areas are in the path of rapid landslides, extensive damage occurs. For example, in July 1938, Kobe, one of Japan's largest cities, was swept by debris flows generated by torrential rainfall, resulting in 505 deaths and destruction of more than 100,000 homes (Nakano et al. 1974; Fukuoka 1982; Ministry of Construction 1983). In 1983 a heavy rainstorm in Nagasaki and northern Kyushu caused 5,000 slides and debris flows that killed 333 people (National Research Center for Disaster Prevention 1983). Table 2-1 summarizes socioeconomic losses due to catastrophic landslides in Japan from 1938 to 1981; note that Japan has been affected almost annually by catastrophic slope failures resulting in large losses of life and property.

Eisbacher and Clague (1984) presented a fascinating account of fatalities, injuries, and property damage due to landslides in the European Alps for the past 2,000 years in the form of 137 case histories derived from historical and technical records. Details on 17 of the most catastrophic landslides since the 13th century are presented in Table 2-2.

The most disastrous landslide in Europe occurred in 1963 at Vaiont Reservoir in northeastern Italy (Table 2-2, Figure 2-21). This 250-million-m³ reservoir-induced rock slide traveled at high velocity into the reservoir, sending a wave 260 m up the opposite slope and at least 100 m over the crest of the thin-arch Vaiont Dam into the valley below, where it destroyed five villages and took about 2,000 lives (Kiersch 1964). On the basis of Italian reports, Hendron and Patton (1985) estimated the following economic losses from the slide:

1. Loss of the dam and reservoir: $100 million ($425 million),
2. Other property damage: tens of millions of dollars (in 1990 dollars, approaching $100 million), and
3. Civil suits for personal injury and loss of life: $16 million ($68 million).

Thus, the 1990 equivalent economic loss would be about $600 million without taking into account the value of the lives lost.

The republics of the former Soviet Union have also experienced large loss of life due to landslides; most of the deaths occurred in isolated

Table 2-1
Socioeconomic Losses in Major Landslide Disasters, Japan, 1938–1981 [Ministry of Construction (Japan) 1983]

Date	Prefecture	Landslide Area	No. of Residents Dead or Missing	No. of Homes Destroyed or Badly Damaged
July 1938	Hyogo	Mount Rokko (Kobe and vicinity)	505	130,192
July 1945	Hiroshima	Kure and vicinity	1,154	1,954
Sept. 1947	Gumma	Mount Akagi	271	1,538
July 1951	Kyoto	Kameoka	114	15,141
June 1953	Kumamoto	Mount Aso	102	—[a]
July 1953	Wakayama	Arita River	460	4,772
Aug. 1953	Kyoto	Minamiyamashiro	336	5,122
Sept. 1958	Shizuoka	Kanogawa River	1,094	19,754
Aug. 1959	Yamanashi	Kamanashi River	43	277
June 1961	Nagano	Ina Valley Region	130	3,018
Sept. 1966	Yamanashi	Lake Saiko	32	81
July 1967	Hyogo	Mount Rokko	92	746
July 1967	Hiroshima	Kure and vicinity	88	289
July 1972	Kumamoto	Amakusa Island	115	750
Aug. 1972	Niigata	Kurokawa Village	31	1,102
July 1974	Kagawa	Shodo-shima Island	29	1,139
Aug. 1975	Aomori	Mount Iwaki	22	28
Aug. 1975	Kochi	Niyodo River	68	536
Sept. 1976	Kagawa	Shodo-shima Island	119	2,001
May 1978	Niigata	Myoko-Kogen	13	25
Oct. 1978	Hokkaido	Mount Usu	3	144
Aug. 1979	Gifu	Horadani	3	16
Aug. 1981	Nagano	Ubara	10	56

NOTE: Each of these catastrophic mass movements was caused by heavy rainfall, commonly related to typhoons. None was caused by earthquakes or volcanic activity.
[a] No data.

catastrophes. The greatest of these in this century occurred in 1949, when the **M** 7.5 Khait earthquake in Soviet Tadzhikistan triggered a series of debris avalanches and flows that buried 33 villages. Estimates of the number of deaths from these landslides ranged from 12,000 (Jaroff 1977) to 20,000 (Wesson and Wesson 1975). Some large cities of the former USSR, such as Alma-Ata, Dushanbe, Frunze, and Yerevan, are located in valleys that are subject to dangerous debris flows (Gerasimov and Zvonkova 1974). For example, in 1921 a large debris flow passed through Alma-Ata, the capital of the Kazakh Republic, killing 500 people and inflicting considerable damage on the city (Yesenov and Degovets 1982).

Landslide deaths in the United States have been estimated at 25 to 50 per year (Committee on Ground Failure Hazards 1985). About five people per year are killed by landslides in Canada (S. G. Evans, personal communication, 1987, Geological Survey of Canada, Ottawa). Most of

FIGURE 2-21
Vaiont rock slide, northeastern Italy, 1963: 250-million-m³ rock slide caused Vaiont Reservoir to catastrophically overtop its dam, resulting in extensive property damage and loss of life.

Table 2-2
Major Landslide Disasters in European Alps Since 13th Century (Eisbacher and Clague 1984)

YEAR	LOCATION	TYPE OF SLOPE FAILURE	NO. OF DEATHS
1219	Plaine d'Oisans (Romanche River valley), France	Failure of landslide dam, resulting in downstream flooding	"Thousands"
1248	Mount Granier, France	Rock avalanche	1,500–5,000
1348	Dobratsch Massif, Austria	Earthquake-triggered rock falls and rock avalanches	Heavy losses
1419	Ganderberg-Passeier Wildsee (Passer valley), Italy	Failure of rock-slide dam, resulting in downstream flooding	~400
1486	Zarera (Val Lagune), Switzerland	Rock avalanche	300
1499	Kienholz (Brienzer See), Switzerland	Debris flow	~400
1515	Biasca (Val Blenio), Switzerland	Failure of rock-avalanche dam, resulting in downstream flooding	~600
1569	Hofgastein (Gastein Valley), Austria	Debris flow	147
1569	Schwaz (Inn Valley), Austria	Debris flow	140
1584	Corbeyrier-Yvorne (Tour d'Ai), Switzerland	Debris flow	328
1618	Piuro (Val Bregaglia), Italy	Rock-debris avalanche	~1,200
1669	Salzburg, Austria	Rock topple–rock fall	250
1806	Goldau (Rossberg Massif), Switzerland	Rock avalanche	457
1814	Antelao Massif (Boite Valley), Italy	Rock avalanche	300
1881	Elm (Sernf Valley), Switzerland	Rock avalanche	115
1892	St. Gervais (Arve Valley), France	Ice-debris flow	177
1963	Vaiont Reservoir (Piave Valley), Italy	Rock slide caused flooding along shore of reservoir and downstream	~1,900

these are killed by relatively small events, most commonly by rock falls. Although there have been some very large and catastrophic landslides in North America, most have occurred in mountainous, relatively unpopulated areas; thus, these failures commonly have not resulted in major losses of life. However, there have been several notable exceptions. In 1903 a great rock slide killed about 70 people in the coal mining town of Frank, Alberta, Canada (McConnell and Brock 1904). A more recent Canadian catastrophe was the 1971 flow failure in sensitive clay that demolished part of the town of Saint-Jean-Vianney, Quebec, destroying 40 homes and killing 31 people (Tavenas et al. 1971). By far the most disastrous landslide (in terms of lives lost) to occur within the territory of the United States occurred on the island of Puerto Rico in October 1985 when heavy rainfall from Tropical Storm Isabel caused a major rock slide (Figure 2-22) that obliterated much of the Mameyes district of the city of Ponce. The slide killed at least 129 people and destroyed about 120 houses (Jibson 1992). The death toll at Mameyes was the greatest from a single slide in North American history.

Interestingly, the world's two largest landslides in modern history have resulted in relatively few casualties. The 1911 Usoy landslide in Soviet Tadzhikistan (then Russia), with an estimated volume of 2.5 km³, was a truly catastrophic event. However, in spite of the great volume and apparently high velocity of this earthquake-triggered landslide, casualties were low because the area was sparsely populated. Most of the deaths occurred in the village of Usoy, whose 54 inhabitants were buried (Bolt et al. 1975, 178–179). This landslide also formed the world's highest historic landslide dam, a 570-m-high blockage of the Murgob River that still impounds 60-km-long Lake Sarez. The natural dam is being considered as the site of a hydroelectric power project. The 2.8-km³ rock slide–debris avalanche (Figure 2-10) that accompanied the 1980 eruption of Mount St. Helens in Washington State is the world's largest historic landslide. However, even though this huge mass moved down valley at high velocity, it killed only 5 to 10 people (Schuster 1983). The low casualty rate was a direct result of the evacuation of residents and visitors in anticipation of a possible eruption of the volcano.

The economic value of loss of life due to landslides has not commonly been included in calculating the costs of landslides because it is difficult to place a specific value on a human life. However, in cost-benefit studies, federal agencies in the United States recently have assigned human-life values with a median of roughly $2 million each (Scanlan 1990). If these values are realistic, the economic losses due to the 25 to 50 annual landslide deaths in the United States are on the order of $50 million per year to $100 million per year.

5. POSITIVE EFFECTS OF LANDSLIDES

Landslides constitute a major element in mass wasting of the continents. Thus, in terms of geologic time, they help to provide stable land that is suitable for agriculture and habitation. In the shorter term, it is difficult to conceive of benefits that might accrue as a result of natural landslide activity. However, in a few cases landslides were purposely triggered to obtain socioeconomic benefits. An example is a 100-m-high sediment-retention dam on the Malaya Alma-Atinka River upstream from Alma-Ata in the Kazakh Republic of the former Soviet Union (Yesenov and Degovets 1982). This dam, which was constructed in 1966–1967, protects Alma-Ata from the debris flows mentioned earlier. It was formed from landslides that were triggered by setting off large explosive charges in the valley walls. The resulting landslide dam was then shaped into a traditional check dam by means of earth-moving equipment. Engineers and scientists from the former Soviet Union plan to use the experience gained from this artificially made landslide dam to construct a much larger hydropower dam on the Naryn River in the Republic of Kyrgyzstan (Adushkin 1993). The three artificially made landslides that will form the 270-m-high landslide dam, which will be the main component of the Kambarata hydropower project, will be triggered by 250 tons of chemical explosives. Power will be generated by means of flow through a penstock that will bypass the landslide dam through its bedrock right abutment.

6. CONCLUSIONS

In many countries, socioeconomic losses due to landslides are great and apparently are growing as human development expands into unstable hillslope areas under the pressures of increasing populations. A significant proportion of world landslide losses affect transportation facilities: highways, railways, canals, and pipelines. The nation most severely affected by landslides is Japan, which suffers estimated total (direct plus indirect) landslide losses of $4 billion annually. In the United States,

FIGURE 2-22
October 1985, Mameyes, Puerto Rico, rock slide: triggered by rainfall, 300 000-m³ landslide killed at least 129 people.
R. W. JIBSON, U.S. GEOLOGICAL SURVEY

Italy, and India, total annual economic losses due to landslides have been estimated to range from $1 billion to $2 billion. Many other countries have lesser, but major, annual landslide losses.

On the basis of worldwide data for 1971–1974, nearly 600 people a year are killed by landslides. However, tens of thousands have been killed in the 20th century in each of a few truly disastrous landslides, a fact that raises the average annual number of worldwide landslide deaths for this century to well above the 600 recorded for 1971–1974. Landslide deaths in the United States are estimated at 25 to 50 a year.

REFERENCES

Adushkin, V.V., L.M. Pernik, and Y.N. Zykov. 1993. Modeling of Explosion-Triggered Rock Slides for Construction of the Kambarata I Hydroelectric Power Dam, Republic of Kyrgyzstan. *Landslide News* (Japan Landslide Society), No. 7, pp. 7–9.

Anderson, L.R., J.R. Keaton, T. Saarinen, and W.G. Wells II. 1984. *The Utah Landslides, Debris Flows, and Floods of May–June 1983.* National Research Council, Washington, D.C., 96 pp.

Association of Engineering Geologists. 1984. Southern California—Section News. *Newsletter,* Vol. 27, No. 3, p. 34.

Ayala, F.J., and M. Ferrer. 1989. Extent and Economic Significance of Landslides in Spain. In *Landslides: Extent and Economic Significance* (E.E. Brabb and B.L. Harrod, eds.), Proc., 28th International Geological Congress: Symposium on Landslides, Washington, D.C., July 17, 1989, A.A. Balkema, Rotterdam, Netherlands, pp. 169–178.

Bell, D.H. 1992. Preface. In *Landslides: Proc., Sixth International Symposium on Landslides* (D.H. Bell, ed.)., Christchurch, New Zealand, Feb. 10–14, A.A. Balkema, Rotterdam, Netherlands, Vol. 1, pp. vii–viii.

Berman, G. 1991. Landslides on the Panama Canal. *Landslide News* (Japan Landslide Society), No. 5, pp. 10–14.

Bolt, B.A., W.L. Horn, G.A. Macdonald, and R.F. Scott. 1975. *Geological Hazards.* Springer-Verlag, New York, 328 pp.

Brabb, E.E. 1984. *Minimum Landslide Damage in the United States.* U.S. Geological Survey Open-File Report 84-486, 4 pp.

Brabb, E.E. 1989. Landslides: Extent and Economic Significance in the United States. In *Landslides: Extent and Economic Significance* (E.E. Brabb and B.L. Harrod, eds.), Proc., 28th International Geological Congress: Symposium on Landslides, Washington, D.C., July 17, A.A. Balkema, Rotterdam, Netherlands, pp. 325–350.

Brand, E.W. 1989. Occurrence and Significance of Landslides in Southeast Asia. In *Landslides: Extent and Economic Significance* (E.E. Brabb and B.L. Harrod, eds.), Proc., 28th International Geological Congress: Symposium on Landslides, Washington, D.C., July 17, 1989, A.A. Balkema, Rotterdam, Netherlands, pp. 303–324.

Cambiaghi, A., and R.L. Schuster. 1989. Landslide Damming and Environmental Protection—A Case Study from Northern Italy. In *Proc., Second International Symposium on Environmental Geotechnology*, Shanghai, May 15–18, 1989, Envo Publishing Company, Inc., Bethlehem, Pa., Vol. 1, pp. 469–480.

Cato, I. 1982. The 1977 Landslide at Tuve and the Complex Origin of Clays in South-western Sweden. In *Landslides and Mudflows: Reports of the Alma-Ata International Seminar, October 1981* (A.I. Sheko, ed.), Centre of International Projects, GKNT, Moscow, pp. 279–289.

Charma, C.K. 1974. *Landslides and Soil Erosion in Nepal.* Sangeeta Publishers, Kathmandu, Nepal, 93 pp.

Chassie, R.G., and R.D. Goughnour. 1976a. National Highway Landslide Experience. *Highway Focus*, Vol. 8, No. 1, pp. 1–9.

Chassie, R.G., and R.D. Goughnour. 1976b. States Intensifying Efforts to Reduce Highway Landslides. *Civil Engineering*, Vol. 46, No. 4, pp. 65–66.

Chopra, B.R. 1977. Landslides and Other Mass Movements Along Roads in Sikkim and North Bengal. *Bulletin of the International Association of Engineering Geology*, No. 16, pp. 162–166.

Close, U., and E. McCormick. 1922. Where the Mountains Walked. *National Geographic Magazine*, Vol. 41, No. 5, pp. 445–464.

Cluff, L.S. 1971. Peru Earthquake of May 31, 1970; Engineering Geology Observations. *Seismological Society of America Bulletin*, Vol. 61, No. 3, pp. 511–521.

Committee on Ground Failure Hazards. 1985. *Reducing Losses from Landslides in the United States.* Commission on Engineering and Technical Systems, National Research Council, Washington, D.C., 41 pp.

Cotecchia, V., G. Travaglini, and G. Melidoro. 1969. I Movimenti Franosi e gli Sconvolgimenti della Rete Idrografica Prodotti in Calabria dal Terremoto del 1783. *Geologia Applicata e Idrogeologia* (Italy), Vol. 4, pp. 1–24.

Coulter, H.W., and R.R. Migliaccio. 1966. *Effects of the Earthquake of March 17, 1964, at Valdez, Alaska.* U.S. Geological Survey Professional Paper 542-C, pp. C1–C36.

Council of Economic Advisers. 1991. *Economic Report of the President, 1991.* Washington, D.C., 411 pp.

Creasey, C.L. 1988. Landslide Damage: A Costly Outcome of the Storm. In *Landslides, Floods, and Marine Effects of the Storm of January 3–5, 1982, in the San Francisco Bay Region, California* (S.D. Ellen and G.F. Wieczorek, eds.), U.S. Geological Survey Professional Paper 1434, pp. 195–203.

Deere, D.U., and J.Y. Perez. 1985. Remedial Measures for Large Slide Movements. In *Proceedings of P.R.C.-U.S.-Japan Trilateral Symposium/Workshop on Engineering for Multiple Natural Hazard Mitigation*, Beijing, People's Republic of China, Jan. 7–12, pp. L-7-1 to L-7-15.

Dracup, J.A., E.D.H. Cheng, J.M. Nigg, and T.A. Schroeder. 1991. *The New Year's Eve Flood on Oahu, Hawaii—December 31, 1987–January 1, 1988.* Natural Disaster Studies, Vol. 1. Committee on Natural Disasters, National Research Council, Washington, D.C., 72 pp.

Eisbacher, G.H., and J.J. Clague. 1984. *Destructive Mass Movements in High Mountains: Hazard and Management.* Geological Survey Canada Paper 84-16, Ottawa, 230 pp.

Facts on File Yearbook. 1990. Tropical Forest Loss Reported. Facts on File Publications, New York, pp. 528–529.

Fleming, R.W. 1981. Stop 1: Landslide Damage in Greater Cincinnati. In *Engineering Geology of the Cincinnati Area* (R.W. Fleming, A.M. Johnson, and J.E. Hough, eds.), Annual Meeting, Geological Society of America, Cincinnati, Field Trip No. 18, pp. 543–570.

Fleming, R.W., and F.A. Taylor. 1980. *Estimating the Costs of Landslide Damage in the United States*. U.S. Geological Survey Circular 832, 21 pp.

Fukuoka, M. 1982. Some Case Studies on Landslides in Japan. In *Landslides and Mudflows—Reports of the Alma-Ata International Seminar, October 1981* (A.I. Sheko, ed.), Centre of International Projects, GKNT, Moscow, pp. 333–352.

Gerasimov, I.P., and T.B. Zvonkova. 1974. Natural Hazards in the Territory of the USSR: Study, Control, and Warning. In *Natural Hazards— Local, National, Global* (G.F. White, ed.), Oxford University Press, New York, pp. 243–251.

Gillon, M.D., and G.T. Hancox. 1992. Cromwell Gorge Landslides—A General Overview. In *Landslides: Proc., Sixth International Symposium on Landslides* (D.H. Bell, ed.), Christchurch, New Zealand, Feb. 10–14, A.A. Balkema, Rotterdam, Netherlands, Vol. 1, pp. 83–102.

Govi, M. 1989. The 1987 Landslide on Mount Zandila in the Valtellina, Northern Italy. *Landslide News* (Japan Landslide Society), No. 3, pp. 1–3.

Gregersen, O., and F. Sandersen. 1989. Landslides: Extent and Economic Significance in Norway. In *Landslides: Extent and Economic Significance* (E.E. Brabb and B.L. Harrod, eds.), Proc., 28th International Geological Congress: Symposium on Landslides, Washington, D.C., July 17, A.A. Balkema, Rotterdam, Netherlands, pp. 133–139.

Hansen, W.R. 1965. *Effects of the Earthquake of March 27, 1964 at Anchorage, Alaska*. U.S. Geological Survey Professional Paper 542-A, pp. A1–A68.

Hawley, J.G. 1984. Slope Instability in New Zealand. In *Natural Hazards in New Zealand* (I. Spedon and M.J. Crozier, eds.), National Commission for UNESCO, Wellington, New Zealand, pp. 88–13.

Hendron, A.J., Jr., and F.D. Patton. 1985. *The Vaiont Slide: A Geotechnical Analysis Based on New Geologic Observations of the Failure Surface*. Waterways Experiment Station Technical Report, U.S. Army Corps of Engineers, Vicksburg, Miss., Vol. 1, 104 pp.

Honolulu Star-Bulletin. 1989. Kuliouou Residents Say a Fond, Sad Aloha. Aug. 8, p. A-3.

Jaroff, L. 1977. Forecasting the Earth's Convulsions. In *Nature/Science Annual*, 1977 Edition. Time-Life Books, New York, pp. 21–33.

Jibson, R.W. 1992. The Mameyes, Puerto Rico, Landslide Disaster of October 7, 1985. In *Landslides/Landslide Mitigation* (J.A. Johnson and J.E. Slosson, eds.), Geological Society of America, Reviews in Engineering Geology, Vol. 9, Chap. 5, pp. 37–54.

Jones, F.O., D.R. Embody, and W.L. Peterson. 1961. *Landslides along the Columbia River Valley, Northeastern Washington*. U.S. Geological Survey Professional Paper 367, 98 pp.

Kaliser, B.N. 1983. Geologic Hazards of 1983. *Survey Notes, Utah Geological and Mineral Survey*, Salt Lake City, Vol. 17, No. 2, pp. 3–8, 14.

Khegai, A.Y., and N.V. Popov. 1989. The Extent and Economic Significance of the Debris Flow and Landslide Problem in Kazakhstan, in the Soviet Union. In *Landslides: Extent and Economic Significance* (E.E. Brabb and B.L. Harrod, eds.), Proc., 28th International Geological Conference: Symposium on Landslides, Washington, D.C., July 17, A.A. Balkema, Rotterdam, Netherlands, pp. 221–225.

Kiersch, G.A. 1964. Vaiont Reservoir Disaster. *Civil Engineering*, Vol. 34, No. 3, pp. 32–39.

Kotarba, A. 1989. Landslides: Extent and Economic Significance in Central Europe. In *Landslides: Extent and Economic Significance* (E.E. Brabb and B.L. Harrod, eds.,), Proc., 28th International Geological Congress: Symposium on Landslides, Washington, D.C., July 17, A.A. Balkema, Rotterdam, Netherlands, pp. 191–202.

Krohn, J.P., and J.E. Slosson. 1976. Landslide Potential in the United States. *California Geology*, Vol. 29, No. 10, pp. 224–231.

Legget, R.F. 1973. *Cities and Geology*. McGraw-Hill, New York, 624 pp.

Li Tianchi. 1989. Landslides: Extent and Economic Significance in China. In *Landslides: Extent and Economic Significance* (E.E. Brabb and B.L. Harrod, eds.), Proc., 28th International Geological Congress: Symposium on Landslides, Washington, D.C., July 17, A.A. Balkema, Rotterdam, Netherlands, pp. 271–287.

Li Tianchi, R.L. Schuster, and Wu Jishan. 1986. Landslide Dams in South-central China. In *Landslide Dams—Processes, Risk, and Mitigation* (R.L. Schuster, ed.), American Society of Civil Engineers Geotechnical Special Publication No. 3, pp. 146–162.

Los Angeles Times. 1984. 13 Malibu Families Flee Landslide—Mesa Homes Inch Toward Cliff Overlooking Coast Highway, Los Angeles, California. March 3, pp. 1 and 6.

Lutton, R.J., D.C. Banks, and W.E. Strohm, Jr. 1979. Slides in Gaillard Cut, Panama Canal Zone. In *Rockslides and Avalanches, 2: Engineering Sites* (B. Voight, ed.), Elsevier, New York, pp. 151-224.

MacDonald, D.F. 1942. *Panama Canal Slides—The Third Locks Project*. Department of Operation and Maintenance, Special Engineering Division, The Panama Canal Company, Balboa Heights, Panama Canal Zone, 73 pp.

Malone, F. 1983. D and RGW Digs Out—Fast. *Railway Age*, Sept., pp. 42–43.

Marinatos, S.N. 1960. Helice: A Submerged Town of Classical Greece. *Archaeology*, Vol. 13, No. 3, pp. 186–193.

Mathur, H.N. 1982. Influence of Human Activities on Landslides, Mudflows and Slope Movements in India and Efforts at Reducing Their Negative Impact. In *Landslides and Mudflows—Reports of the Alma-Ata International Seminar, October 1981* (A.I. Sheko, ed.), Centre of International Projects, GKNT, Moscow, pp. 20–44.

McConnell, R.G., and R.W. Brock. 1904. The Great Landslide at Frank, Alberta. In *Annual Report of the Canada Department of the Interior for the Year 1902-03*, Session Paper 25, Canada Department of the Interior, Ottawa, pp.1–17.

Merriam, R. 1960. Portuguese Bend Landslide, Palos Verdes, California. *Journal of Geology*, Vol. 68, No. 2, pp. 140–153.

Ministry of Construction (Japan). 1983. *Reference Manual on Erosion Control Works* (in Japanese). Erosion Control Department, Japan, 386 pp.

Morales Arnao, B., V.K. Garga, R.S. Wright, and J.Y. Perez. 1984. The Tablachaca Slide No.5, Peru, and Its Stabilization. In *Proc., Fourth International Symposium on Landslides*, Toronto, Vol. 1, pp. 597–604.

Moriyama, Y., and S. Horiuchi. 1993. The Budget for Fiscal 1993 (in Japanese). *Sabo and Flood Control*, Japan Sabo Association, Vol. 25, No. 6, pp. 60–64.

Nakano, T., H. Kadomura, T. Mizutani, M. Okuda, and T. Sekiguchi. 1974. Natural Hazards—Report from Japan. In *Natural Hazards—Local, National, Global* (G.F. White, ed.), Oxford University Press, New York, Chap. 28, pp. 231–243.

National Research Center for Disaster Prevention (Japan). 1983. *Outline of Recent Activities of the National Research Center for Disaster Prevention*. Science and Technology Agency, Japan, 131 pp.

Nieto, A.S., and R.L. Schuster. 1988. Mass Wasting and Flooding Induced by the 5 March 1987 Ecuador Earthquakes. *Landslide News* (Japan Landslide Society), No. 2, pp. 5–8.

Nishimoto, H. 1993. The Supplementary Budget for Fiscal 1992. *Sabo and Flood Control*, Japan Sabo Association, Vol. 25, No. 6, pp. 68–69.

Oyagi, N. 1989. Geological and Economic Extent of Landslides in Japan and Korea. In *Landslides: Extent and Economic Significance* (E.E. Brabb and B.L. Harrod, eds.), Proc., 28th International Geological Congress: Symposium on Landslides, Washington, D.C., July 17, A.A. Balkema, Rotterdam, Netherlands, pp. 289–302.

Plafker, G., G.E. Ericksen, and J. Fernandez Concha. 1971. Geological Aspects of the May 31, 1970, Peru Earthquake. *Seismological Society of America Bulletin*, Vol. 61, No. 3, pp. 543–578.

Radbruch-Hall, D.H., R.B. Colton, W.E. Davies, I. Lucchita, B.A. Skip, and D.J. Varnes. 1982. *Landslide Overview Map of the Conterminous United States*. U.S. Geological Survey Professional Paper 1183, 25 pp.

Sager, J.W., and C.M. Budai. 1989. Geology and Construction of the Spirit Lake Outlet Tunnel, Mount St. Helens, Washington. In *Engineering Geology in Washington*, Bulletin 78, Washington Division of Geology and Earth Resources, Olympia, Vol. 2, pp. 1229–1234.

San Francisco Chronicle. 1983. Highway 50 Reopens and Tahoe Rejoices. June 24, p. 2.

Scanlan, C. 1990. Placing a Value on Life a Vital Part of Federal Cost-Benefit Trade-Offs. *Denver Post*, July 14.

Schuster, R.L. 1978. Introduction. In *Special Report 176: Landslides—Analysis and Control* (R.L. Schuster and R.J. Krizek, eds.), TRB, National Research Council, Washington, D.C., Chap. 1, pp. 1–10.

Schuster, R.L. 1979. Reservoir-Induced Landslides. *Bulletin of the International Association of Engineering Geology*, No. 20, pp. 8–15.

Schuster, R.L. 1983. Engineering Aspects of the 1980 Mount St. Helens Eruptions. *Bulletin of the Association of Engineering Geologists*, Vol. 20, No. 2, pp. 125–143.

Schuster, R.L. 1991. Introduction. In *The March 5, 1987, Ecuador Earthquakes—Mass Wasting and Socioeconomic Effects*, Natural Disaster Studies, Vol. 5, Committee on Natural Disasters, National Research Council, Washington, D.C., pp. 11–22.

Schuster, R.L., and L.D. Alfaro. 1992. Landslide Hazards in the Gaillard Cut, Panama Canal Zone. *Abstracts with Programs*, 1992 Annual Meeting of the Geological Society of America, Vol. 24, No. 7, p. A166.

Schuster, R.L., and R.W. Fleming. 1986. Economic Losses and Fatalities due to Landslides. *Bulletin of the Association of Engineering Geologists*, Vol. 23, No. 1, pp. 11–28.

Schuster, R.L., and R.W. Fleming. 1988. Socioeconomic Significance of Landslides and Mudflows. In *Landslides and Mudflows* (E.A. Kozlovskii, ed.), UNESCO/UNEP, Moscow, pp. 131–141.

Seed, H.B. 1968. Landslides during Earthquakes due to Soil Liquefaction. *Journal of the Soil Mechanics and Foundations Division*, ASCE, Vol. 94, No. SM5, pp. 1055–1122.

Shearer, C.F., F.A. Taylor, and R.W. Fleming. 1983. *Distribution and Costs of Landslides in San Diego County, California, during the Rainfall Years of 1978–79 and 1979–80*. U.S. Geological Survey Open-File Report 83-582, 15 pp.

Slosson, J.E. 1987. Landslides and Other Ground Failure Hazards: Where Have We Failed? *Abstracts with Programs*, 83rd Annual Meeting, Cordilleran Section, Geological Society of America, Vol. 19, No. 6, p. 452.

Slosson, J.E., and J.P. Krohn. 1982. Southern California Landslides of 1978 and 1980. In *Proc., Floods and Debris Flows in Southern California and Arizona, 1978 and 1980*, Sept. 17–18, 1980, National Research Council; Environmental Quality Laboratory, California Institute of Technology, Pasadena, pp. 291–319.

Smith, R. 1958. Economic and Legal Aspects. In *Special Report 29: Landslides and Engineering Practice* (E.B. Eckel, ed.), HRB (now TRB), National Research Council, Washington, D.C., National Research Council Publication 544, pp. 6–19.

Smith, T.C. 1982. Lawsuits and Claims against Cities and Counties Mount after January 1982 Storm. *California Geology*, Vol. 35, No. 7, pp. 163–164.

Soil Conservation Service. 1986. *Tomiki Creek Unit, Redwood Empire Target Area, Mendocino County, California*. U.S. Department of Agriculture, Davis, Calif., 264 pp.

Tan, S.S. 1980. Slope Failures in Orange County due to 1978 Rains. *California Geology*, Vol. 33, No. 9, pp. 202–205.

Tavenas, F., J.Y. Chagnon, and P. LaRochelle. 1971. The Saint-Jean-Vianey Landslide—Observations and Eyewitness Accounts. *Canadian Geotechnical Journal*, Vol. 8, No. 3, pp. 202–205.

Taylor, F.A., and E.E. Brabb. 1972. *Maps Showing Distribution of and Cost by Counties of Structurally Damaging Landslides in the San Francisco Bay Region, California, Winter 1968–69*. U.S. Geological Survey Miscellaneous Field Studies Map MF-327.

University of Utah. 1984. *Flooding and Landslides in Utah—An Economic Impact Analysis*. University of Utah Bureau of Economic and Business Research, Utah Department of Community and Economic Development, and Utah Office of Planning and Budget, Salt Lake City, 123 pp.

Varnes, D.J. 1981. Slope-Stability Problems of the Circum-Pacific Region as Related to Mineral and Energy Resources. In *Energy Resources of the Pacific Region* (M.T. Halbouty, ed.), American Association of Petroleum Geologists Studies in Geology, No. 12, pp. 489–505.

Voight, B. 1990. The 1985 Nevado del Ruiz Volcano Catastrophe: Anatomy and Retrospection. *Journal of Volcanology and Geothermal Research*, Vol. 44, pp. 349–386.

Voight, B., R.J. Janda, H. Glicken, and P.M. Douglass. 1983. Nature and Mechanisms of the Mount St. Helens Rock-Slide Avalanche of 18 May 1980. *Geotechnique*, Vol. 33, No. 3, pp. 243–273.

Vonder Linden, K. 1989. The Portuguese Bend Landslide. *Engineering Geology*, Vol. 27, pp. 301–373.

Walkinshaw, J. 1992. Landslide Correction Costs on U.S. State Highway Systems. In *Transportation Research Record 1343*, TRB, National Research Council, Washington, D.C., pp. 36–41.

Wesson, C.V.K., and R.L. Wesson. 1975. Odyssey to Tadzhik—An American Family Joins a Soviet Seismological Expedition. *U.S. Geological Survey Earthquake Information Bulletin*, Vol. 7, No. 1, pp. 8–16.

Wiggins, J.H., J.E. Slosson, and J.R. Krohn. 1978. *National Hazards—Earthquake, Landslide, Expansive Soil Loss Models*. Technical Report, J.H. Wiggins Co., Redondo Beach, Calif., 162 pp.

Xue-Cai, F., and G. An-ning. 1986. Principal Characteristics of Earthquake Landslides in China. *Geologia Applicata e Idreogeologia* (Italy), Vol. 21, No. 2, pp. 27–45.

Yesenov, Y.E., and A.S. Degovets. 1982. Protection of the City of Alma-Ata from Mud-Rock Flows. In *Landslides and Mudflows—Reports of the Alma-Ata International Seminar, October 1981* (A.I. Sheko, ed.), Centre of International Projects, GKNT, Moscow, pp. 454–465.

Youd, T.L. 1978. Major Cause of Earthquake Damage Is Ground Failure. *Civil Engineering*, ASCE, Vol. 48, No. 4, pp. 47–51.

Youd, T.L., and S.N. Hoose. 1978. *Historic Ground Failures in Northern California Triggered by Earthquakes*. U.S. Geological Survey Professional Paper 993, 177 pp.

Chapter 3

DAVID M. CRUDEN AND
DAVID J. VARNES

LANDSLIDE TYPES AND PROCESSES

1. INTRODUCTION

The range of landslide processes is reviewed in this chapter, and a vocabulary is provided for describing the features of landslides relevant to their classification for avoidance, control, or remediation. The classification of landslides in the previous landslide report (Varnes 1978) has been widely adopted, so departures from it have been minimized and the emphasis is on the progress made since 1978. Although this chapter is complete in itself, particular attention is drawn to changes and additions to the vocabulary used by Varnes in the previous report and the reasons for the changes.

The term *landslide* denotes "the movement of a mass of rock, debris or earth down a slope" (Cruden 1991, 27). The phenomena described as landslides are not limited either to the land or to sliding; the word as it is now used in North America has a much more extensive meaning than its component parts suggest (Cruden 1991). The coverage in this chapter will, however, be identical to that of the previous report (Varnes 1978). Ground subsidence and collapse are excluded, and snow avalanches and ice falls are not discussed.

This chapter also follows Varnes's expressed intention of (1978, 12) "developing and attempting to make more precise a useful vocabulary of terms by which ... [landslides] ... may be described." The terms Varnes recommended in 1978 are largely retained unchanged and a few useful new terms have been added. Eliot (1963, 194) noted:

... Words strain,
Crack and sometimes break, under the burden,
Under the tension, slip, slide, perish,
Decay with imprecision, will not stay in place,
Will not stay still. ...

Such displaced terms are identified in this chapter. Following Varnes (1978), the use of terms relating to the geologic, geomorphic, geographic, or climatic characteristics of a landslide has been discouraged, and the section in the previous report in which these terms are discussed has been deleted.

The viewpoint of the chapter is that of the investigator responding to a report of a landslide on a transportation route. What can be usefully observed and how should these observations be succinctly and unambiguously described?

The technical literature describing landslides has grown considerably since 1978. An important source of landslide information is the proceedings of the International Symposium on Landslides. The third symposium met in New Delhi, India (Swaminathan 1980), and the symposium has since met quadriennially in Toronto, Canada (Canadian Geotechnical Society 1984); in Lausanne, Switzerland (Bonnard 1988); and in Christchurch, New Zealand (Bell 1992); it is scheduled to meet in Trondheim, Norway, in 1996.

Among the other important English language texts and collections of descriptions of landslides have been those by Zaruba and Mencl (1982), Brunsden and Prior (1984), Crozier (1986), and

Costa and Wieczorek (1987). Eisbacher and Clague (1984) and Skermer's translation of Heim's *Bergsturz und Menschenleben* (1932) have made descriptions of the classic landslides of the European Alps more accessible to North Americans.

Important reviews of landsliding around the world were edited by Brabb and Harrod (1989) and Kozlovskii (1988). Kyunttsel (1988) reviewed experience with classification in the USSR and noted "considerable divergences of views between various researchers concerning the mechanisms underlying certain types of landslides. This applies particularly to lateral spreads."

A historical perspective has been added to the discussion of spreading to show that this type of landslide was recognized in North America over 100 years ago and is represented here by some extremely large movements. Both the size and the gentle slopes of these movements command particular attention.

Crozier commented:

> The two generalized classifications most likely to be encountered in the English speaking world are by J.N. Hutchinson (1968; Skempton and Hutchinson, 1969) and D.J. Varnes (1958; 1978). . . . Both authors use type of movement to establish the principal groups. . . . The major distinction between the two classifications is the difference accorded to the status of flow movements . . . slope movements which are initiated by shear failure on distinct, boundary shear surfaces but which subsequently achieve most of their translational movement by flowage . . . this dilemma depends on whether the principal interest rests with analyzing the conditions of failure or with treating the results of movement. Hutchinson's classification appears to be related more closely to this first purpose. . . . Both Hutchinson's and Varnes' classifications have tended to converge over recent years, particularly in terminology. . . . Whereas Varnes' scheme is perhaps easier to apply and requires less expertise to use, Hutchinson's classification has particular appeal to the engineer contemplating stability analysis. (Crozier 1986, Ch. 2)

The synthesis of these two classifications has continued. Hutchinson (1988) included topples, and this chapter has benefited from his comments. In Section 4 of this chapter particularly, which deals with landslide activity, many of Hutchinson's suggestions from the Working Party on the World Landslide Inventory (WP/WLI) have been adopted (WP/WLI 1993a,b).

Under Hutchinson's chairmanship, the International Association of Engineering Geology (IAEG) Commission on Landslides and Other Mass Movements continued its work on terminology. The declaration by the United Nations of the International Decade for Natural Disaster Reduction (1990–2000) prompted the IAEG Commission's Suggested Nomenclature for Landslides (1990) and the creation of the WP/WLI by the International Geotechnical Societies and the United Nations Educational, Scientific, and Cultural Organization (UNESCO). The Working Party, formed from the IAEG Commission, the Technical Committee on Landslides of the International Society for Soil Mechanics and Foundation Engineering, and nominees of the International Society for Rock Mechanics, published *Directory of the World Landslide Inventory* (Brown et al. 1992) listing many workers interested in the description of landslides worldwide. The Working Party has also prepared the *Multilingual Landslide Glossary*, which will encourage the use of standard terminology in describing landslides (WP/WLI 1993b). The terminology suggested in this chapter is consistent with the suggested methods and the glossary of the UNESCO Working Party (WP/WLI 1990, 1991, 1993a,b).

2. FORMING NAMES

The criteria used in the classification of landslides presented here follow Varnes (1978) in emphasizing type of movement and type of material. Any landslide can be classified and described by two nouns: the first describes the material and the second describes the type of movement, as shown in Table 3-1 (e.g., *rock fall, debris flow*).

The names for the types of materials are unchanged from Varnes's classification (1978): *rock, debris,* and *earth*. The definitions for these terms are given in Section 7. Movements have again been divided into five types: *falls, topples, slides, spreads,* and *flows*, defined and described in Section 8. The sixth type proposed by Varnes (1978, Figure 2.2), *complex landslides*, has been dropped from the formal classification, although the term *complex* has been retained as a description of the style of activity of a landslide. Complexity can also be indicated

Table 3-1
Abbreviated Classification of Slope Movements

Type of Movement	Type of Material		
		Engineering Soils	
	Bedrock	Predominantly Coarse	Predominantly Fine
Fall	Rock fall	Debris fall	Earth fall
Topple	Rock topple	Debris topple	Earth topple
Slide	Rock slide	Debris slide	Earth slide
Spread	Rock spread	Debris spread	Earth spread
Flow	Rock flow	Debris flow	Earth flow

by combining the five types of landslide in the ways suggested below. The large classification chart accompanying the previous report (Varnes 1978, Figure 2.1) has been divided into separate figures distributed throughout this chapter.

Table 3-2
Glossary for Forming Names of Landslides

ACTIVITY

State	Distribution	Style
Active	Advancing	Complex
Reactivated	Retrogressive	Composite
Suspended	Widening	Multiple
Inactive	Enlarging	Successive
Dormant	Confined	Single
Abandoned	Diminishing	
Stabilized	Moving	
Relict		

DESCRIPTION OF FIRST MOVEMENT

Rate	Water Content	Material	Type
Extremely rapid	Dry	Rock	Fall
Very rapid	Moist	Soil	Topple
Rapid	Wet	Earth	Slide
Moderate	Very wet	Debris	Spread
Slow			Flow
Very slow			
Extremely slow			

DESCRIPTION OF SECOND MOVEMENT

Rate	Water Content	Material	Type
Extremely rapid	Dry	Rock	Fall
Very rapid	Moist	Soil	Topple
Rapid	Wet	Earth	Slide
Moderate	Very wet	Debris	Spread
Slow			Flow
Very slow			
Extremely slow			

NOTE: Subsequent movements may be described by repeating the above descriptors as many times as necessary.

The name of a landslide can become more elaborate as more information about the movement becomes available. To build up the complete identification of the movement, descriptors are added in front of the two-noun classification using a preferred sequence of terms. The suggested sequence provides a progressive narrowing of the focus of the descriptors, first by time and then by spatial location, beginning with a view of the whole landslide, continuing with parts of the movement, and finally defining the materials involved. The recommended sequence, as shown in Table 3-2, describes activity (including state, distribution, and style) followed by descriptions of all movements (including rate, water content, material, and type).

This sequence is followed throughout the chapter and all terms given in Table 3-2 are highlighted in bold type and discussed. Second or subsequent movements in complex or composite landslides can be described by repeating, as many times as necessary, the descriptors used in Table 3-2. Descriptors that are the same as those for the first movement may then be dropped from the name.

For instance, the very large and rapid slope movement that occurred near the town of Frank, Alberta, Canada, in 1903 (McConnell and Brock 1904) was a *complex, extremely rapid, dry rock fall–debris flow* (Figure 3-1). From the full name of this landslide at Frank, one would know that both the debris flow and the rock fall were extremely rapid and dry because no other descriptors are used for the debris flow.

As discussed in Section 4.3, the addition of the descriptor *complex* to the name indicates the sequence of movement in the landslide and distinguishes this landslide from a *composite rock fall–debris flow*, in which rock fall and debris flow movements were occurring, sometimes simultaneously, on different parts of the displaced mass. The

FIGURE 3-1
Slide at Frank, Alberta, Canada *(oblique aerial photograph from south)*. About 4:10 a.m. on April 29, 1903, about 85 million tonnes of rock moved down east face of Turtle Mountain, across entrance of Frank mine of Canadian American Coal Company, Crowsnest River, southern end of town of Frank, main road from east, and Canadian Pacific mainline through Crowsnest Pass. Displaced mass continued up opposite side of valley before coming to rest 120 m above valley floor. Event lasted about 100 seconds.
PHOTOGRAPH NAPL T31L-213; REPRODUCED FROM COLLECTION OF NATIONAL AIR PHOTO LIBRARY WITH PERMISSION OF NATURAL RESOURCES CANADA

full name of the landslide need only be given once; subsequent references should then be to the initial material and type of movement, for example, "the rock fall" or "the Frank rock fall" for the landslide at Frank, Alberta.

Several noun combinations may be required to identify the multiple types of material and movement involved in a complex or composite landslide. To provide clarity in the description, a dash known as an "en dash" is used to link these stages, as in *rock fall–debris flow* in the example above. (An en dash is half the length of a regular dash and longer than a hyphen; it is used to remove ambiguity by indicating linkages between terms composed of two nouns.)

The full name of a landslide may be cumbersome and there is a natural tendency, particularly among geologists, to establish type examples with which other landslides may be compared. Shreve (1968), for instance, referred to the landslide in Frank, Alberta, as belonging to the *Blackhawk type*. It seems clear that type examples should be historic landslides that have been investigated in detail shortly after their occurrence and are of continuing interest to landslide specialists. In addition, for a type example to be useful, other landslides with the same descriptors should occur in similar material. The Blackhawk landslide (Figure 3-2) was a prehistoric landslide, and thus was not subject to investigation at its occurrence (Shreve 1968). It is therefore not a suitable type example; nevertheless, it may have been a Frank-type landslide.

Although Varnes (1978, 25) discussed "terms relating to geologic, geomorphic, geographic, or

tions may be unlikely. The inclusion of complex and composite landslides would increase the number of type examples to over a billion.

3. LANDSLIDE FEATURES AND GEOMETRY

Before landslide types are discussed, it is useful to establish a nomenclature for the observable landslide features and to discuss the methods of expressing the dimensions and geometry of landslides.

3.1 Landslide Features

Varnes (1978, Figure 2.1t) provided an idealized diagram showing the features for a *complex earth slide–earth flow*, which has been reproduced here as Figure 3-3. More recently, the IAEG Commission on Landslides (1990) produced a new idealized landslide diagram (Figure 3-4) in which the various features are identified by numbers, which are defined in different languages by referring to the accompanying tables. Table 3-3 provides the definitions in English.

The names of the features are unchanged from Varnes's classification (1978). However, Table 3-3 contains explicit definitions for the *surface of rupture* (Figure 3-4, *10*), the *depletion* (*16*), the *depleted mass* (*17*), and the *accumulation* (*18*) and expanded definitions for the *surface of separation* (*12*) and the *flank* (*19*). The sequence of the first nine landslide features has been rearranged to proceed from the crown above the head of the displaced material to the toe at the foot of the

climatic setting," he recommended against the practice of using type examples because the terms "are not informative to a reader who lacks knowledge of the locality" (1978, 26). Moreover, type examples are impractical because of the sheer number required to provide a fairly complete landslide classification. About 100,000 type examples would be required for all the combinations of descriptors and materials with all the types of movement defined in Table 3-2 and in the following sections, although admittedly some combina-

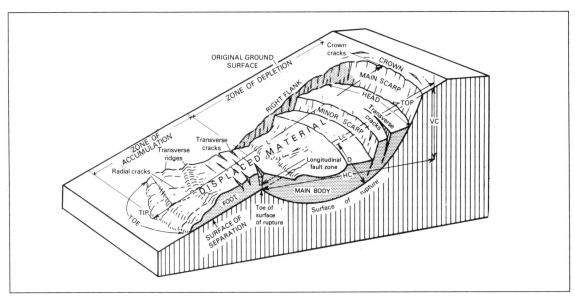

FIGURE 3-3
Block diagram of idealized complex earth slide–earth flow (Varnes 1978, Figure 2.1t).

displaced material. This sequence may make these features easier to remember.

It may also be helpful to point out that in the *zone of depletion* (*14*) the elevation of the ground surface decreases as a result of landsliding, whereas in the *zone of accumulation* (*15*) the elevation of the ground surface increases. If topographic maps or digital terrain models of the landslide exist for both before and after movements, the zones of depletion and accumulation can be found from the differences between the maps or models. The volume decrease over the zone of depletion is, of course, the depletion, and the volume increase over the zone of accumulation is the accumulation. The accumulation can be expected to be larger than the depletion because the ground generally dilates during landsliding.

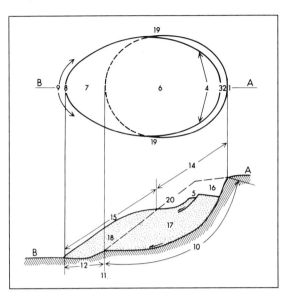

FIGURE 3-4
Landslide features: *upper portion,* plan of typical landslide in which dashed line indicates trace of rupture surface on original ground surface; *lower portion,* section in which hatching indicates undisturbed ground and stippling shows extent of displaced material. Numbers refer to features defined in Table 3-3 (IAEG Commission on Landslides 1990).

Table 3-3
Definitions of Landslide Features

Number	Name	Definition
1	Crown	Practically undisplaced material adjacent to highest parts of main scarp
2	Main scarp	Steep surface on undisturbed ground at upper edge of landslide caused by movement of displaced material (*13*, stippled area) away from undisturbed ground; it is visible part of surface of rupture (*10*)
3	Top	Highest point of contact between displaced material (*13*) and main scarp (*2*)
4	Head	Upper parts of landslide along contact between displaced material and main scarp (*2*)
5	Minor scarp	Steep surface on displaced material of landslide produced by differential movements within displaced material
6	Main body	Part of displaced material of landslide that overlies surface of rupture between main scarp (*2*) and toe of surface of rupture (*11*)
7	Foot	Portion of landslide that has moved beyond toe of surface of rupture (*11*) and overlies original ground surface (*20*)
8	Tip	Point on toe (*9*) farthest from top (*3*) of landslide
9	Toe	Lower, usually curved margin of displaced material of a landslide, most distant from main scarp (*2*)
10	Surface of rupture	Surface that forms (or that has formed) lower boundary of displaced material (*13*) below original ground surface (*20*); mechanical idealization of surface of rupture is called *slip surface* in Chapter 13
11	Toe of surface of rupture	Intersection (usually buried) between lower part of surface of rupture (*10*) of a landslide and original ground surface (*20*)
12	Surface of separation	Part of original ground surface (*20*) now overlain by foot (*7*) of landslide
13	Displaced material	Material displaced from its original position on slope by movement in landslide; forms both depleted mass (*17*) and accumulation (*18*); it is stippled in Figure 3-4
14	Zone of depletion	Area of landslide within which displaced material (*13*) lies below original ground surface (*20*)
15	Zone of accumulation	Area of landslide within which displaced material lies above original ground surface (*20*)
16	Depletion	Volume bounded by main scarp (*2*), depleted mass (*17*), and original ground surface (*20*)
17	Depleted mass	Volume of displaced material that overlies surface of rupture (*10*) but underlies original ground surface (*20*)
18	Accumulation	Volume of displaced material (*13*) that lies above original ground surface (*20*)
19	Flank	Undisplaced material adjacent to sides of surface of rupture; compass directions are preferable in describing flanks, but if left and right are used, they refer to flanks as viewed from crown
20	Original ground surface	Surface of slope that existed before landslide took place

3.2 Landslide Dimensions

The IAEG Commission on Landslides (1990) utilized the nomenclature described in Section 3.1 (including Figure 3-4 and Table 3-3) to provide definitions of some dimensions of a typical landslide. The IAEG Commission diagram is reproduced here as Figure 3-5. Once again, each dimension is identified on the diagram by a number, and these numbers are linked to tables giving definitions in several languages. Table 3-4 gives the definitions in English.

The quantities L_d, W_d, D_d and L_r, W_r, D_r are introduced because, with an assumption about the shape of the landslide, their products lead to estimates of the volume of the landslide that are use-

ful in remedial work. For instance, for many rotational landslides, the surface of rupture can be approximated by half an ellipsoid with semiaxes D_r, $W_r/2$, $L_r/2$. As shown in Figure 3-6(a), the volume of an ellipsoid is (Beyer 1987, 162)

$$VOL_{eps} = \frac{4}{3}\pi a \cdot b \cdot c$$

where a, b, and c are semimajor axes. Thus, the volume of a "spoon shape" corresponding to one-half an ellipsoid is

$$VOL_{ls} = \frac{1}{2}\frac{4}{3}\pi a \cdot b \cdot c = \frac{4}{6}\pi a \cdot b \cdot c$$

But as shown in Figure 3-6(b), for a landslide $a=D_r$, $b=W_r/2$, and $c=L_r/2$. Therefore, the volume of ground displaced by a landslide is approximately

$$VOL_{ls} = \frac{4}{6}\pi a \cdot b \cdot c = \frac{4}{6}\pi D_r \cdot W_r/2 \cdot L_r/2$$
$$= \frac{1}{6}\pi D_r \cdot W_r \cdot L_r$$

This is the volume of material before the landslide moves. Movement usually increases the volume of the material being displaced because the displaced material dilates. After the landslide, the volume of displaced material can be estimated by $\frac{1}{6}\pi D_d W_d L_d$ (WP/WLI 1990, Equation 1).

A term borrowed from the construction industry, the *swell factor*, may be used to describe the increase in volume after displacement as a percentage of the volume before displacement. Church (1981, Appendix 1) suggested that a swell factor of 67 percent "is an average figure obtained from existing data for solid rock" that has been mechan-

FIGURE 3-5
Landslide dimensions: *upper portion,* plan of typical landslide in which dashed line is trace of rupture surface on original ground surface; *lower portion,* section in which hatching indicates undisturbed ground, stippling shows extent of displaced material, and broken line is original ground surface. Numbers refer to dimensions defined in Table 3-4 (IAEG Commission on Landslides 1990).

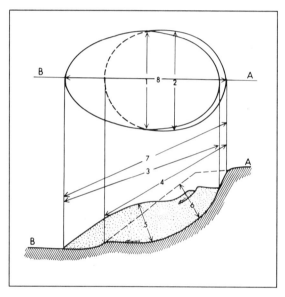

Table 3-4
Definitions of Landslide Dimensions

NUMBER	NAME	DEFINITION
1	Width of displaced mass, W_d	Maximum breadth of displaced mass perpendicular to length, L_d
2	Width of surface of rupture, W_r	Maximum width between flanks of landslide perpendicular to length, L_r
3	Length of displaced mass, L_d	Minimum distance from tip to top
4	Length of surface of rupture, L_r	Minimum distance from toe of surface of rupture to crown
5	Depth of displaced mass, D_d	Maximum depth of displaced mass measured perpendicular to plane containing W_d and L_d
6	Depth of surface of rupture, D_r	Maximum depth of surface of rupture below original ground surface measured perpendicular to plane containing W_r and L_r
7	Total length, L	Minimum distance from tip of landslide to crown
8	Length of center line, L_{cl}	Distance from crown to tip of landslide through points on original ground surface equidistant from lateral margins of surface of rupture and displaced material

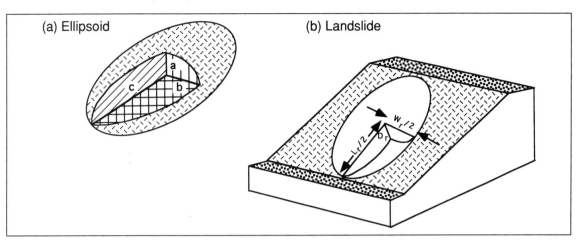

FIGURE 3-6
Estimation of
landslide volume
assuming a
half-ellipsoid
shape.

ically excavated. His estimates may approximate the upper bound for the swell due to landsliding. Nicoletti and Sorriso-Valvo (1991) chose an average dilation of 33 percent, so $4D_rW_rL_r \approx 3D_dW_dL_d$. More precise information is as yet unavailable.

The ground-surface dimensions of the displaced material, L_d, W_d, and of the surface of rupture, W_r, and the total length, L, of the landslide can be measured with an electronic distance-measuring instrument; a rangefinder may be sufficiently precise for a one-person reconnaissance. Measurement of the distance L_r may present problems because the toe of the surface of rupture is often not exposed. Its position can sometimes be estimated from graphical extrapolations of the main scarp supported by measurements of displacements within the displaced mass (Cruden 1986). Although D_d and D_r can also be estimated by these techniques, site investigations provide more precise methods of locating surfaces of rupture under displaced material (Hutchinson 1983).

The total length of the landslide, dimension L (5, Figure 3-5), is identical with length L, "the maximum length of the slide upslope," shown in Figure 3-3 (Varnes 1978); both are the straight-line distances from crown to tip. Readers are cautioned that several writers define the length of a landslide in terms of its horizontal extent and frequently use the letter L to define this horizontal distance in tabulations of observations and in calculations. This use of L is a source of potential confusion and inaccuracy, and readers should make certain that they can identify the dimension being specified by L in every case.

It should also be emphasized that it is unlikely that the displaced material at the tip has traveled

from the crown despite the frequency of this assumption, originally due to Heim (1932). Material displaced from close to the landslide crown usually comes to rest close to the head of the landslide. Nicoletti and Sorriso-Valvo (1991) proposed that an estimate of the "overall runout" of a landslide be determined by measuring the length of a line constructed along the original ground surface equidistant from the lateral margins of the displaced material. However, such measurements may not have immediate physical significance and are also more difficult and imprecise than measurements of L. The length of the landslide measured through these central points is called the *length of center line*, L_{cl}. Note that L_{cl} will increase with the number of points surveyed on the center line, and the ratio L_{cl}/L will increase with the curvature of the center line in plan and section.

The difference in elevation between the crown and the tip of the landslide may be used to determine H, the height of the landslide. Combining estimates of H and L allows computation of the travel angle α, as shown by Figure 3-7. If the tip is visible from the crown, the travel angle can be mea-

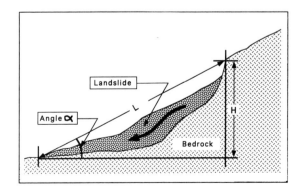

FIGURE 3-7
Definition of
travel angle (α)
of a landslide.

sured directly with a hand clinometer. The *H*-value may be conveniently estimated with an altimeter when tip and crown are accessible but not visible from each other. Hutchinson (1988) compiled data from several different types of debris flows to illustrate how debris-flow mobility appears to be related to the travel angle—and to the volume and lithology of the displaced material (Figure 3-8).

The measurements discussed above are adequate during reconnaissance for defining the basic dimensions of single-stage landslides whose displacement vectors parallel a common plane. Such landslides can be conveniently recorded on a landslide report such as that shown in Figure 3-9. Estimates of landslide volume determined by these methods are imprecise when topography diverts the displacing material from rectilinear paths. More elaborate surveys and analyses are then necessary (Nicoletti and Sorriso-Valvo 1991).

4. LANDSLIDE ACTIVITY

The broad aspects of landslide activity should be investigated and described during initial reconnaissance of landslide movements and before more detailed examination of displaced materials is undertaken. The terms relating to landslide age

and state of activity defined by Varnes (1978) and some of his terms defining sequence or repetition of movement have been regrouped under three headings:

1. **State of Activity,** which describes what is known about the timing of movements;
2. **Distribution of Activity,** which describes broadly where the landslide is moving; and
3. **Style of Activity,** which indicates the manner in which different movements contribute to the landslide.

The terms used to define these three characteristics of landslide activity are given in the top section of Table 3-2 and are highlighted in bold type the first time they are used in the following sections.

The reader is cautioned that the following discussions relate to the terminology proposed by the UNESCO Working Party (WP/WLI 1990, 1991, 1993a,b) and given in Table 3-2. Other reports and authors may use classifications that apply different meanings to apparently identical terms. For example, in Chapter 9 of this report, a Unified Landslide Classification System is introduced that is based on landslide classification concepts presented by McCalpin (1984) and Wieczorek (1984). This sys-

FIGURE 3-8
Mobility of sturzstroms, chalk debris flows, and landslides in mine wastes related to travel angle (α) and debris volume (modified from Hutchinson 1988, Figure 12).

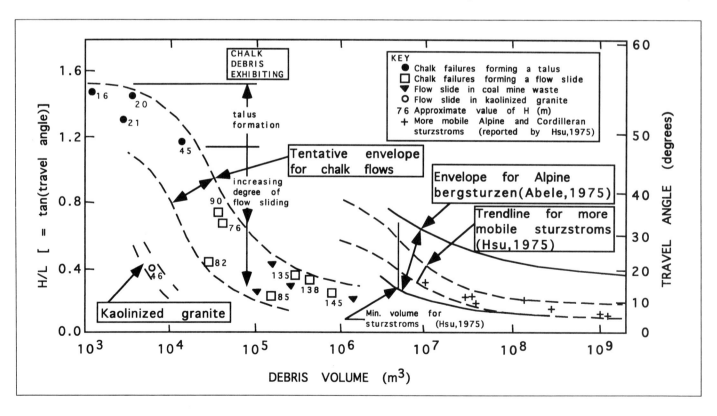

FIGURE 3-9
Proposed standard
landslide report
form.

LANDSLIDE REPORT

Inventory Number: _____

Date of Report: _____
                      day   month   year

Date of Landslide Occurrence: _____
                                     day   month   year

Landslide Locality: _____

Reporter's Name: _____

Affiliation: _____

Address: _____

Phone: _____

Position:		Degrees	Minutes	Seconds
	Latitude			
	Longitude			
	Elevation:	crown		m a.s.l.
	Surface of rupture	toe		m a.s.l.
		tip		m a.s.l.

Geometry:

		Surface of rupture	Displaced Mass	
	Length	$L_r =$	$L_d =$	$L =$
	Width	$W_r =$	$W_d =$	
	Depth	$D_r =$	$D_d =$	

Volume: $V = \pi L_d D_d W_d / 6$ or $V =$ Swell factor =

$V =$ $m^3 \times 10^n$ $n =$

Damage: Value _____

Injuries   Deaths

tem is compared with a stability classification proposed by Crozier (1984). For further information on such alternative systems, the reader should refer to Tables 9-1, 9-2, and 9-6 and the associated text in Chapter 9.

4.1 State of Activity

Figure 3-10 illustrates the several states of activity by using an idealized toppling failure as an example. **Active** landslides are those that are currently moving; they include first-time movements and reactivations. A landslide that is again active after being inactive may be called **reactivated.** Slides that are reactivated generally move on preexisting shear surfaces whose strength parameters approach residual (Skempton 1970) or ultimate (Krahn and Morgenstern 1979) values. They can be distinguished from first-time slides on whose surfaces of rupture initial resistance to shear will generally approximate peak values (Skempton and Hutchinson 1969). Landslides that have moved within the last annual cycle of seasons but that are not moving at present were described by Varnes (1978) as **suspended.**

Inactive landslides are those that last moved more than one annual cycle of seasons ago. This state can be subdivided. If the causes of movement remain apparent, the landslide is **dormant.** However, if the river that has been eroding the toe of the moving slope changes course, the landslide is **abandoned** (Hutchinson 1973; Hutchinson and Gostelow 1976). If the toe of the slope has been protected against erosion by bank armoring or if other artificial remedial measures have stopped the movement, the landslide can be described as **stabilized.**

Landslides often remain visible in the landscape for thousands of years after they have moved and then stabilized. Such landslides were called *ancient* or *fossil* by Zaruba and Mencl (1982, 52), perhaps because they represent the skeletons of once-active movements. When these landslides have been covered by other deposits, they are referred to as *buried* landslides. Landslides that have clearly developed under different geomorphic or climatic conditions, perhaps thousands of years ago, can be called **relict.** Road construction in southern England reactivated relict debris flows that had occurred under periglacial conditions (Skempton and Weeks 1976).

Within regions, standard criteria might be developed to assist in distinguishing suspended landslides from dormant and relict landslides. These criteria would describe the recolonization by vegetation of surfaces exposed by slope movements and the dissection of the new topography by drainage. The rate of these changes depends on both the local climate and the local vegetation, so these criteria must be used with extreme caution. Nevertheless, it is generally true that when the main scarp of a landslide supports new vegetation, the landslide is usually dormant, and when drainage extends across a landslide without obvious discontinuities, the landslide is commonly relict. However, these generalizations must be confirmed by detailed study of typical slope movements under local conditions; Chapter 9 provides a systematic approach for such determinations. Figure 9-7 shows some idealized stages in the evolution of topographic features on suspended, dormant, and relict landslides.

The various states of activity are also defined by an idealized graph of displacement versus time (Figure 3-11). For an actual landslide, such a graph

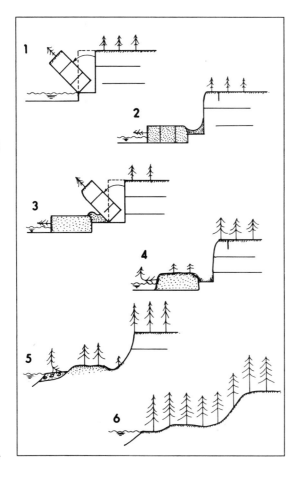

FIGURE 3-10
Sections through topples in different states of activity:
(*1*) *active*—erosion at toe of slope causes block to topple;
(*2*) *suspended*—local cracking in crown of topple;
(*3*) *reactivated*—another block topples;
(*4*) *dormant*—displaced mass begins to regain its tree cover and scarps are modified by weathering;
(*5*) *stabilized*—fluvial deposition stabilizes toe of slope, which begins to regain its tree cover; and
(*6*) *relict*—uniform tree cover over slope.

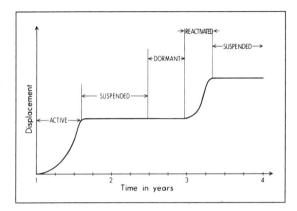

can be created by plotting differences in the position of a target on the displacing material with time. Such graphs are particularly well suited to portraying the behavior of slow-moving landslides because they presuppose that the target is not displaced significantly over the time period during which measurement takes place. The velocity of the target can be estimated by the average rate of displacement of the target over the time period between measurements.

There is some redundancy in using the descriptions of activity state with those for rate of movement (see Section 5). Clearly, if the landslide has a measurable rate of movement, it is either active or reactivated. The state of activity might then be used to refer to conditions before the current movements of the landslide. If, for instance, remedial measures had been undertaken on a landslide that is now moving with moderate velocity, the landslide might be described as a *previously stabilized, moving, moderate-velocity landslide*. Landslides with no discernible history of previous movement would be described as active.

4.2 Distribution of Activity

Varnes (1978) defined a number of terms that can be used to describe the activity distribution in a landslide. Figure 3-12 shows idealized sections through landslides exhibiting various distributions of activity.

If the surface of rupture is extending in the direction of movement, the landslide is *advancing*, whereas if the surface of rupture is extending in the direction opposite the movement of the displaced material, the landslide is said to be *retrogressive*. If the surface of rupture is extending at one or both lateral margins, the landslide is *widening*. Movement may be limited to the displacing material or

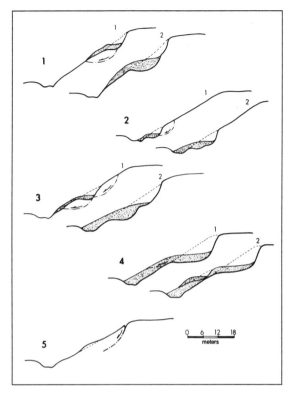

the surface of rupture may be *enlarging,* continually adding to the volume of displacing material. If the surface of rupture of the landslide is enlarging in two or more directions, Varnes (1978, 23) suggested the term *progressive* for the landslide, noting that this term had also been used for both advancing and retrogressive landslides. The term is also currently used to describe the process by which the surface of rupture extends in some landslides (*progressive failure*). The possibility of confusion seems sufficient now to abandon the term *progressive* in favor of describing the landslide as enlarging. Hutchinson (1988, 9) has drawn attention to *confined* movements that have a scarp but no visible surface of rupture in the foot of the displaced mass. He suggested that displacements in the head of the displaced mass are taken up by compression and slight bulging in the foot of the mass.

To complete the possibilities, terms are needed for landslides in which the volume of material being displaced grows less with time and for those landslides in which no trend is obvious. The term *diminishing* for an active landslide in which the volume of material being displaced is decreasing with time seems free of undesired implications. A landslide in which displaced materials continue to move but whose surface of rupture shows no visible changes can be simply described as *moving*. Several types of landslide may exhibit diminishing behavior. Movement may stop in parts of both rotational slides and topples after substantial displacements because the movements themselves reduce the gravitational forces on the displaced masses. Similarly, movements of rock masses may rapidly dilate cracks in the masses, cause decreases in fluid pore pressures within these cracks, and hence decrease rates of movement. However, it may be premature to conclude that the displacing material is stabilizing because the volume being displaced is decreasing with time. Hutchinson (1973) pointed out that the activity of rotational slides caused by erosion at the toe of slopes in cohesive soils is often cyclic.

4.3 Style of Activity

The style of landslide activity, or the way in which different movements contribute to the landslide, can be defined by terms originally established by Varnes (1978, 23). Figure 3-13 shows idealized sec-

tions through landslides exhibiting various styles of activity. Varnes defined **complex** landslides as those with at least two types of movement. However, it is now suggested that the term *complex* be limited to cases in which the various movements occur in sequence. For instance, the topple described by Giraud et al. (1990) and shown as Figure 3-13(*1*), in which some of the displaced mass subsequently slid, is termed a *complex rock topple–rock slide*. Not all the toppled mass slid, but no significant part of the displaced mass slid without first toppling. Some of the displaced mass may be still toppling while other parts are sliding.

The term **composite**, formerly a synonym for *complex*, is now proposed to describe landslides in which different types of movement occur in different areas of the displaced mass, sometimes simultaneously. However, the different areas of the displaced mass show different sequences of movement. For example, the structures shown in Figure 3-13(*2*), first described by Harrison and Falcon (1934, 1936), were called *slide toe topples* by Goodman and Bray (1976), but according to the classification proposed in this chapter, they are *composite rock slide–rock topples*. The term *composite* was introduced by Prior et al. (1968, 65, 76) to describe mudflows in which "slipping and sliding . . . occur in intimate association with flowing" and "the material . . . behaves as a liquid and flows rapidly between confining marginal shears." In the proposed naming convention, such movements are *composite earth slides-earth flows* and the convention of treating the topographically higher of the two movements as the first movement and the lower of the two movements as the second movement was adopted.

A **multiple** landslide shows repeated movements of the same type, often following enlargement of the surface of rupture. The newly displaced masses are in contact with previously displaced masses and often share a surface of rupture with them. In a *retrogressive, multiple rotational slide*, such as that shown in Figure 3-14, two or more blocks have each moved on curved surfaces of rupture tangential to common, generally deep surfaces of rupture (Eisbacher and Clague 1984).

A **successive** movement is identical in type to an earlier movement but in contrast to a multiple movement does not share displaced material or a surface of rupture with it [Figure 3-13(*3*)]. According to Skempton and Hutchinson (1969, 297), "successive rotational slips consist of an

FIGURE 3-13
Sections through landslides showing different styles of activity. (*1*) *Complex:* gneiss (*A*) and migmatites (*I*) are forming topples caused by valley incision; alluvial materials fill valley bottom; after weathering further weakens toppled material, some of displaced mass moves by sliding (modified from Giraud et al. 1990). (*2*) *Composite:* limestones have slid on underlying shales, causing toppling failures below toe of slide rupture surface (modified from Harrison and Falcon 1934). (*3*) *Successive:* later landslide (*AB*) is same type as landslide *CD* but does not share displaced material or rupture surface. (*4*) *Single.*

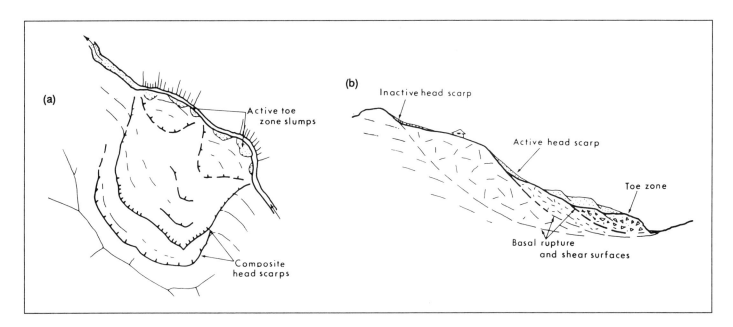

assembly of individual shallow rotational slips." Hutchinson (1967, 116) commented that "irregular successive slips which form a mosaic rather than a stepped pattern in plan are also found."

Single landslides consist of a single movement of displaced material, often as an unbroken block [Figure 3-13(4)]. For instance, Hutchinson (1988) described single topples in which a single block moved and contrasted these with multiple topples (Figure 3-15). Single landslides differ from the other styles of movement, which require disruption of the displaced mass or independent movements of portions of the mass.

5. RATE OF MOVEMENT

The previous rate-of-movement scale provided by Varnes (1978, Figure 2.1u) is shown here as Figure 3-16. This scale is unchanged from Varnes's original scale (1958) except for the addition of the equivalent SI units, which range from meters per second to millimeters per year. Varnes (1958) did not discuss the divisions of the scale, then given in units ranging from feet per second to feet per 5 years; the scale probably represented a codification of informal practice in the United States at the time. Nemčok et al. (1972) suggested a fourfold division of a similar range of velocities.

Figure 3-17 presents a modified scale of landslide velocity classes. The divisions of the scale have been adjusted to increase in multiples of 100 by a slight increase in its upper limit and a decrease

in its lower limit. These two limits now span 10 orders of magnitude. Interpretation of the scale was aided by Morgenstern's (1985) analogy to the Mercalli scale of earthquake intensity. He pointed out that the effects of a landslide can be sorted into six classes corresponding approximately to the six fastest movement ranges of Varnes's scale.

FIGURE 3-14 (*above*) (*a*) Map view and (*b*) cross section of typical retrogressive, multiple rotational slide (Eisbacher and Clague 1984, Figure 10).

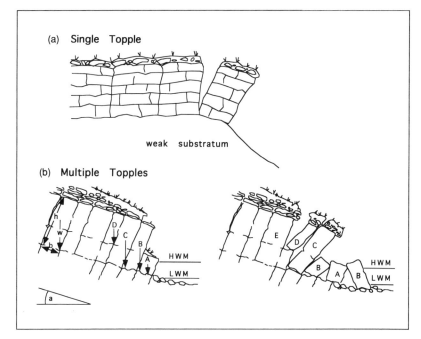

FIGURE 3-15
Comparison of (*a*) single topple (Hutchinson 1988) with (*b*) multiple topples [modified from Varnes 1978, Figure 2.1d1 (de Freitas and Watters 1973)].

FIGURE 3-16
Varnes landslide
movement scale
(Varnes 1978,
Figure 2.1u).

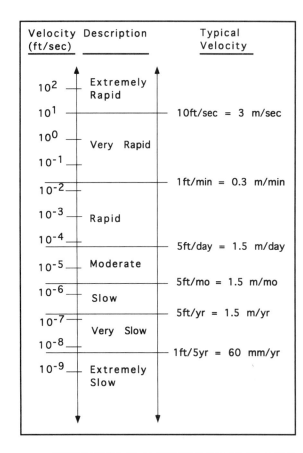

Velocity (ft/sec)	Description	Typical Velocity
10^2	Extremely Rapid	
10^1		10ft/sec = 3 m/sec
10^0	Very Rapid	
10^{-1}		
10^{-2}		1ft/min = 0.3 m/min
10^{-3}	Rapid	
10^{-4}		5ft/day = 1.5 m/day
10^{-5}	Moderate	
10^{-6}	Slow	5ft/mo = 1.5 m/mo
10^{-7}		5ft/yr = 1.5 m/yr
10^{-8}	Very Slow	
10^{-9}	Extremely Slow	1ft/5yr = 60 mm/yr

Velocity Class	Description	Velocity (mm/sec)	Typical Velocity
7	Extremely Rapid		
6	Very Rapid	5×10^3	5 m/sec
5	Rapid	5×10^1	3 m/min
4	Moderate	5×10^{-1}	1.8 m/hr
3	Slow	5×10^{-3}	13 m/month
2	Very Slow	5×10^{-5}	1.6 m/year
1	Extremely Slow	5×10^{-7}	16 mm/year

FIGURE 3-17
Proposed landslide
velocity scale.

An added seventh class brings these effect classes into correspondence with the divisions of the velocity scale.

The Mercalli scale is based on descriptions of local effects of an earthquake; degrees of damage can be evaluated by investigating a house or a section of a street. Yet the intensity value can be correlated with the total energy release of the event because both local damage and the area affected are related to the magnitude of the earthquake. The situation is different for landslides. Small, **rapid** debris avalanches are known to have caused total destruction and loss of lives. In contrast, a large slope movement of **moderate** velocity can have much less serious effects because it can be avoided or the structures affected can be evacuated or rebuilt. It is suggested that a measure of landslide risk should include both the area affected and the velocity; the product of these two parameters is approximately proportional to the power release of the landslide.

Varnes (1984) drew attention to the United Nations Disaster Relief Organization terminology in which the specific risk, R_s, or the expected degree of loss due to landsliding or any other natural phenomenon, can be estimated as the product of the hazard (H) and the vulnerability (V). The hazard is the probability of occurrence of the phenomenon within a given area; the vulnerability is the degree of loss in the given area of elements at risk: population, properties, and economic activities. The vulnerability ranges from 0 to 1. In this terminology the vulnerability of the landslide might well increase with velocity because it can be expected that **extremely rapid** landslides would cause greater loss of life and property than **slow** landslides.

A parameter that is difficult to quantify is the internal distortion of the displaced mass. Structures on a moving mass generally are damaged in proportion to the internal distortion of their foundations. For example, the Lugnez slope in Switzerland (Huder 1976) is a 25-km² area moving steadily downward at a 15-degree angle at a velocity as high as 0.37 m/year. The movements have been observed by surveying since 1887. Yet in the six villages on the slope with 300-year-old stone houses and churches with bell towers, none of these structures have suffered damage when displaced because the block is moving without distortion. Damage will also depend on the type of landslide, and each type may require separate consideration.

Landslide velocity is a parameter whose destructive significance requires independent definition. Table 3-5 defines the probable destructive significance of the seven velocity classes on the new landslide velocity scale (Figure 3-17). Several case histories in which the effects of landslides on humans and their activities have been well described and for which the landslide velocities are also known are given in Table 3-6, which suggests a correlation between vulnerability and landslide velocity. An important limit appears to lie between **very rapid** and **extremely rapid** movement, which approximates the speed of a person running (5 m/sec). Another important boundary is between the **slow** and **very slow** classes (1.6 m/year), below which some structures on the landslide are undamaged. Terzaghi (1950, 84) identified as *creep* those slope movements that were "proceeding at an imperceptible rate. . . . Typical creep is a continuous movement which proceeds at an average rate of less than a foot per decade.

Table 3-5
Definition of Probable Destructive Significance of Landslides of Different Velocity Classes

Landslide Velocity Class	Probable Destructive Significance
7	Catastrophe of major violence; buildings destroyed by impact of displaced material; many deaths; escape unlikely
6	Some lives lost; velocity too great to permit all persons to escape
5	Escape evacuation possible; structures, possessions, and equipment destroyed
4	Some temporary and insensitive structures can be temporarily maintained
3	Remedial construction can be undertaken during movement; insensitive structures can be maintained with frequent maintenance work if total movement is not large during a particular acceleration phase
2	Some permanent structures undamaged by movement
1	Imperceptible without instruments; construction possible with precautions

Table 3-6
Examples of Landslide Velocity and Damage

Landslide Velocity Class	Landslide Name or Location	Reference	Estimated Landslide Velocity	Damage
7	Elm	Heim (1932)	70 m/sec	115 deaths
7	Goldau	Heim (1932)	70 m/sec	457 deaths
7	Jupille	Bishop (1973)	31 m/sec	11 deaths, houses destroyed
7	Frank	McConnell and Brock (1904)	28 m/sec	70 deaths
7	Vaiont	Mueller (1964)	25 m/sec	1,900 deaths by indirect damage
7	Ikuta	Engineering News Record (1971)	18 m/sec	15 deaths, equipment destroyed
7	St. Jean Vianney	Tavenas et al. (1971)	7 m/sec	14 deaths, structures destroyed
6	Aberfan	Bishop (1973)	4.5 m/sec	144 deaths, some buildings damaged
5	Panama Canal	Cross (1924)	1 m/min	Equipment trapped, people escaped
4	Handlova	Zaruba and Mencl (1969)	6 m/day	150 houses destroyed, complete evacuation
3	Schuders	Huder (1976)	10 m/year	Road maintained with difficulty
3	Wind Mountain	Palmer (1977)	10 m/year	Road and railway require frequent maintenance, buildings adjusted periodically
2	Lugnez	Huder (1976)	0.37 m/year	Six villages on slope undisturbed
2	Little Smoky	Thomson and Hayley (1975)	0.25 m/year	Bridge protected by slip joint
2	Klosters	Haefeli (1965)	0.02 m/year	Tunnel maintained, bridge protected by slip joint
2	Ft. Peck Spillway	Wilson (1970)	0.02 m/year	Movements unacceptable, slope flattened

Higher rates of creep movement are uncommon." Terzaghi's rate is about 10⁻⁶ mm/sec. The limit of perceptible movements on the scale given in Figure 3-17 and Table 3-5 is conservatively lower than Terzaghi's. Still lower rates of movement can be detected with appropriate instrumentation (Kostak and Cruden 1990).

Varnes (1978, 17) pointed out that "creep has come to mean different things to different persons, and it seems best to avoid the term or to use it in a well-defined manner. As used here, creep is considered to have a meaning similar to that used in the mechanics of materials; that is, creep is simply deformation that continues under constant stress." The term *creep* should be replaced by the appropriate descriptors from Figure 3-17, either **very slow** or **extremely slow,** to describe the rate of movement of landslides.

Estimates of landslide velocities can be made by repeated surveys of the positions of displaced objects (Thomson and Hayley 1975; Huder 1976), by reconstruction of the trajectories of portions of the displaced mass (Heim 1932; McConnell and Brock 1904; Ter-Stepanian 1980), by eyewitness observations (Tavenas et al. 1971), by instrumentation (Wilson 1970; Wilson and Mikkelsen 1978), and by other means. The Colorado Department of Transportation experimented with the use of time-lapse photography to document the movement of a relatively slow-moving but very large landslide. The estimates reported were usually the peak velocities of substantial portions of the displaced masses; these estimates are suitable for damage assessments. Rates of movement will differ within the displaced mass of the landslide with position, time, and the period over which the velocity is estimated. Such

differences argue against very precise reports of landslide velocity in reconnaissance surveys.

6. WATER CONTENT

Varnes (1978) suggested the following modifications to terms first proposed by Radbruch-Hall (1978) to describe the water content of landslide materials by simple observations of the displaced material:

1. **Dry:** no moisture visible;
2. **Moist:** contains some water but no free water; the material may behave as a plastic solid but does not flow;
3. **Wet:** contains enough water to behave in part as a liquid, has water flowing from it, or supports significant bodies of standing water; and
4. **Very wet:** contains enough water to flow as a liquid under low gradients.

These terms may also provide guidance in estimating the water content of the displaced materials while they were moving. However, soil or rock masses may drain quickly during and after displacement, so this guidance may be qualitative rather than quantitative. Individual rock or soil masses may have water contents that differ considerably from the average water content of the displacing material. For example, Hutchinson (1988) noted that debris slides (which Hutchinson termed *mudslides*) generally were composed of materials that exhibited a fabric or texture consisting of lumps of various sizes in a softened clay matrix. Samples taken from different portions of this fabric had considerably different water contents, with lumps having much lower water contents than that of the matrix (Figure 3-18).

7. MATERIAL

According to Shroder (1971) and Varnes (1978), the material contained in a landslide may be described as either **rock,** a hard or firm mass that was intact and in its natural place before the initiation of movement, or **soil,** an aggregate of solid particles, generally of minerals and rocks, that either was transported or was formed by the weathering of rock in place. Gases or liquids filling the pores of the soil form part of the soil.

Soil is divided into earth and debris (see Table 3-1). **Earth** describes material in which 80 percent

FIGURE 3-18 Mudslide fabric and associated variability in water content (modified from Hutchinson 1988, Figure 9).

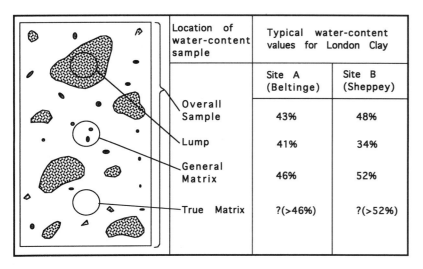

Location of water-content sample	Typical water-content values for London Clay	
	Site A (Beltinge)	Site B (Sheppey)
Overall Sample	43%	48%
Lump	41%	34%
General Matrix	46%	52%
True Matrix	?(>46%)	?(>52%)

or more of the particles are smaller than 2 mm, the upper limit of sand-size particles recognized by most geologists (Bates and Jackson 1987). **Debris** contains a significant proportion of coarse material; 20 to 80 percent of the particles are larger than 2 mm, and the remainder are less than 2 mm. This division of soils is crude, but it allows the material to be named by a swift and even remote visual inspection.

The terms used should describe the displaced material in the landslide before it was displaced. The term *rock fall*, for instance, implies that the displacing mass was a rock mass at the initiation of the landslide. The displaced mass may be debris after the landslide. If the landslide is complex and the type of movement changes as it progresses, the material should be described at the beginning of each successive movement. For instance, a rock fall that was followed by the flow of the displaced material can be described as a *rock fall–debris flow*.

8. TYPE OF MOVEMENT

The kinematics of a landslide—how movement is distributed through the displaced mass—is one of the principal criteria for classifying landslides. However, of equally great importance is its use as a major criterion for defining the appropriate response to a landslide. For instance, occasional falls from a rock cut adjacent to a highway may be contained by a rock fence or similar barrier; in contrast, toppling from the face of the excavation may indicate adversely oriented discontinuities in the rock mass that require anchoring or bolting for stabilization.

In this section the five kinematically distinct types of landslide movement are described in the sequence **fall, topple, slide, spread,** and **flow** (Figure 3-19). Each type of landslide has a number of common modes that are frequently encountered in practice and that are described briefly, often with examples of some complex landslides whose first or initial movements were of that particular type. These descriptions show how landslides of that type may evolve.

8.1 Fall

A **fall** starts with the detachment of soil or rock from a steep slope along a surface on which little or no shear displacement takes place. The material then descends mainly through the air by falling, bouncing, or rolling. Movement is very rapid to

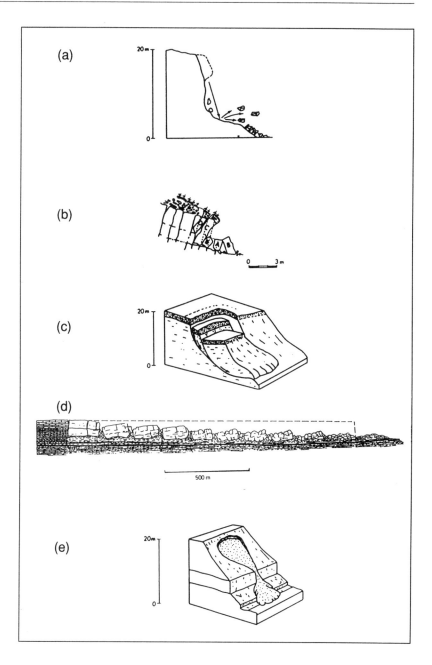

extremely rapid. Except when the displaced mass has been undercut, falling will be preceded by small sliding or toppling movements that separate the displacing material from the undisturbed mass. Undercutting typically occurs in cohesive soils or rocks at the toe of a cliff undergoing wave attack or in eroding riverbanks.

8.1.1 Modes of Falling

Observations show that the forward motion of masses of soil or rock is often sufficient for *free fall* if the slopes below the masses exceed 76 degrees

FIGURE 3-19
Types of landslides: (a) *fall*, (b) *topple*, (c) *slide*, (d) *spread*, (e) *flow*. Broken lines indicate original ground surfaces; arrows show portions of trajectories of individual particles of displaced mass [modified from Varnes 1978, Figure 2.1 (Zaruba and Mencl 1969)].

(0.25:1). The falling mass usually strikes a slope inclined at less than this angle (Ritchie 1963), which causes *bouncing*. Rebound from the impact will depend on material properties, particularly restitution coefficients, and the angle between the slope and the trajectory of the falling mass (Hungr and Evans 1988). The falling mass may also break up on impact.

On long slopes with angles at or below 45 degrees (1:1), particles will have movement paths dominated by *rolling*. There is a gradual transition to rolling from bouncing as bounces shorten and incidence angles decrease. Local steepening of the slope may again project rolling particles into the air, restarting the sequence of free fall, bouncing, and rolling (Hungr and Evans 1988).

8.1.2 Complex Falls

Sturzstroms are extremely rapid flows of dry debris created by large falls and slides (Hsu 1975). These flows may reach velocities over 50 m/sec. Sturzstroms have also been called *rock-fall avalanches* (Varnes 1958) and *rock avalanches* (Evans et al. 1989; Nicoletti and Sorriso-Valvo 1991). Two examples of historic sturzstroms in Switzerland, the Rossberg landslide of 1806 and the Elm landslide of 1881, are discussed in Chapter 1. Hsu (1975) suggested that 5 million m³ is the lower limit of the volume of significant sturzstroms, but Hutchinson (1988) demonstrated that falls in high-porosity, weak European chalk rocks with volumes two orders of magnitude smaller have the same exceptional mobility because of collapse of the pores on impact and consequent high pore-water pressures. Some of Hutchinson's data are reproduced in Figure 3-19.

The motion of sturzstroms probably depends on turbulent grain flow with dispersive stresses arising from momentum transfer between colliding grains. Such a mechanism does not require the presence of a liquid or gaseous pore fluid and can therefore explain lunar and Martian sturzstroms. Van Gassen and Cruden (1989) showed that the motion of the *complex, extremely rapid, dry rock fall–debris flow* that occurred near the town of Frank, Alberta, Canada, in 1903 (see Figure 3-1) could be explained reasonably well by momentum exchange between the moving particles and measured coefficients of friction.

8.2 Topple

A **topple** [Figure 3-19(*b*)] is the forward rotation out of the slope of a mass of soil or rock about a point or axis below the center of gravity of the displaced mass. Toppling is sometimes driven by gravity exerted by material upslope of the displaced mass and sometimes by water or ice in cracks in the mass. Topples may lead to falls or slides of the displaced mass, depending on the geometry of the moving mass, the geometry of the surface of separation, and the orientation and extent of the kinematically active discontinuities. Topples range from extremely slow to extremely rapid, sometimes accelerating throughout the movement.

8.2.1 Modes of Toppling

Flexural toppling was described by Goodman and Bray as

> occurring in rocks with one preferred discontinuity system, oriented to present a rock slope with semi-continuous cantilever beams. . . . Continuous columns break in flexure as they bend forward. . . . Sliding, undermining or erosion of the toe (of the displaced mass) lets the failure begin and it retrogresses backwards with deep, wide tension cracks. The lower portion of the slope is covered with disoriented and disordered blocks. . . . The outward movement of each cantilever produces interlayer sliding (flexural slip) and . . . back-facing scarps (obsequent scarps). . . . It is hard to say where the base of the disturbance lies for the change is gradual. . . . Flexural toppling occurs most notably in slates, phyllites and schists. (Goodman and Bray 1976, 203)

A flexural topple in the proposed classification is a *retrogressive, complex rock topple–rock fall*. Typical examples are shown in Figures 3-20(*a*) and 15-16.

In contrast, *block toppling* occurs

> where the individual columns are divided by widely-spaced joints. The toe of the slope with short columns, receives load from overturning, longer columns above. This thrusts the toe columns forward, permitting further toppling. The base of the disturbed mass is better defined than in the case of flexural toppling; it consists of a stairway generally rising from one layer to

the next. The steps of this stairway are formed by cross-joints. . . . New rock breakage . . . occurs much less markedly than in flexural topples . . . . Thick-bedded sedimentary rocks such as limestones and sandstones, as well as columnar-jointed volcanics exhibit block-toppling. (Goodman and Bray 1976, 203)

Typical examples of block toppling are shown in Figure 15-16.

Chevron topples are block topples in which the dips of the toppled beds are constant and the change of dip is concentrated at the surface of rupture (Cruden et al. 1993). This mode was named from its resemblance to chevron folds (Ramsay 1967, 436). Chevron topples occur on steeper slopes than other block topples. The surface of rupture is often a sliding surface [Figure 3-20(*b*)] forming a *complex rock topple–rock slide*.

Block-flexure toppling is characterized by

pseudo-continuous flexure of long columns through accumulated motions along numerous cross joints. Sliding is distributed along several joint surfaces in the toe [of the displaced mass] while sliding and overturning occur in close association through the rest of the mass. (Goodman and Bray 1976, 204)

Sliding occurs because accumulated overturning steepens the cross joints. There are fewer edge-to-face contacts than in block toppling but still enough to form "a loosened, highly open . . . disturbed zone [displaced mass]. . . . Interbedded sandstone and shale, interbedded chert and shale and thin-bedded limestone exhibit block flexure toppling" (Goodman and Bray 1976, 204). Typical block-flexure topple examples are shown in Figures 3-20(*c*) and 15-16.

8.2.2 Complex and Composite Topples

A *complex rock topple–rock slide* is shown in Figure 3-13(*1*). This cross section of the La Clapière landslide was described by Giraud et al. as follows:

Several distinct movements may be identified as the phenomenon progresses, along with a modification of water flows within the rock mass, due to considerable changes in permeability over time and probably in space as a

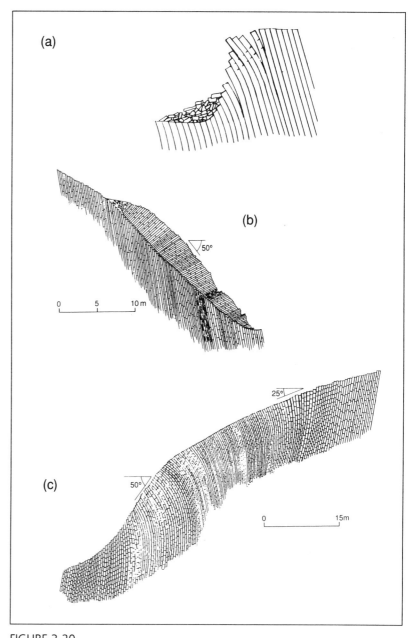

FIGURE 3-20
Three modes of toppling. (*a*) *Flexural:* cracks indicate tension-crack topple; fallen blocks below topple show movement is complex rock topple–rock fall. (*b*) *Chevron:* multiple block topple; hinge surface of chevron may develop into surface of rupture of slide forming complex multiple rock topple–rock slide. (*c*) *Block-flexure.*

result of deformation of the massif. (Giraud et al. 1990, 250)

Goodman and Bray (1976) identified four composite landslide modes in which toppling was caused by earlier sliding. These landslide conditions are discussed in detail in Chapter 15 and are illustrated in Figure 15-17. A *slide head topple* occurs as blocks from the crown of the slide topple onto the head of the displaced mass. According to the naming convention, such a landslide is a *composite rock slide–retrogressive rock topple*.

Rotational sliding of earth or debris above a steeply dipping sedimentary rock mass can cause *slide base toppling* as the sliding induces shear forces in the top of the rock mass (Goodman and Bray 1976). The resulting landslide is, according to the proposed naming convention, a *composite earth slide–advancing rock topple*.

Toppling below the toe of the surface of rupture of a rock slide may be caused by load transmitted from the slide. Such a failure is called a *slide toe topple* (Goodman and Bray 1976). According to the proposed naming convention, it is a *composite rock slide–rock topple*.

The formation of extension cracks in the crown of a landslide may create blocks capable of toppling, or a *tension crack topple* (Goodman and Bray 1976). According to the proposed naming convention, this is a *retrogressive multiple topple*, perhaps forming part of a composite fall or slide. Such failures may also occur in cohesive soils being undercut along streambanks (Figure 3-21).

8.3 Slide

A **slide** is a downslope movement of a soil or rock mass occurring dominantly on surfaces of rupture or on relatively thin zones of intense shear strain. Movement does not initially occur simultaneously over the whole of what eventually becomes the surface of rupture; the volume of displacing material enlarges from an area of local failure. Often the first signs of ground movement are cracks in the original ground surface along which the main scarp of the slide will form. The displaced mass may slide

beyond the toe of the surface of rupture covering the original ground surface of the slope, which then becomes a surface of separation.

8.3.1 Modes of Sliding

Varnes (1978) emphasized the distinction between rotational and translational slides as significant for stability analyses and control methods. That distinction is retained in this discussion. Figure 3-22 shows two rotational slides and three translational slides. Translational slides frequently grade into flows or spreads.

Rotational slides (Figure 3-23) move along a surface of rupture that is curved and concave. If the surface of rupture is circular or cycloidal in profile, kinematics dictates that the displaced mass may move along the surface with little internal deformation. The head of the displaced material may move almost vertically downward, whereas the upper surface of the displaced material tilts backward toward the scarp. If the slide extends for a considerable distance along the slope perpendicular to the direction of motion, the surface of rupture may be roughly cylindrical. The axis of the cylindrical surface is parallel to the axis about which the slide rotates. Rotational slides in soils generally exhibit a ratio of depth of the surface of rupture to length of the surface of rupture, D_r/L_r (see Table 3-4 and Figure 3-5 for definitions of these dimensions), between 0.15 and 0.33 (Skempton and Hutchinson 1969).

Because rotational slides occur most frequently in homogeneous materials, their incidence in fills has been higher than that of other types of movement. Natural materials are seldom uniform, however, and slope movements in these materials commonly follow inhomogeneities and discontinuities (Figure 3-24). Cuts may cause movements that cannot be analyzed by methods used for rotational slides, and other more appropriate methods have been developed. Engineers have concentrated their studies on rotational slides.

The scarp below the crown of a rotational slide may be almost vertical and unsupported. Further movements may cause retrogression of the slide into the crown. Occasionally, the lateral margins of the surface of rupture may be sufficiently high and steep to cause the flanks to move down and into the depletion zone of the slide. Water finding its way into the head of a rotational slide may con-

FIGURE 3-21
Debris topple
(Varnes 1978,
Figure 2.1e).

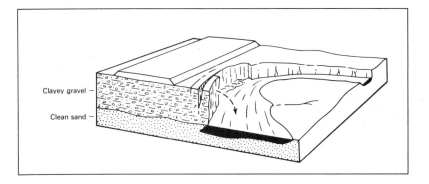

Clayey gravel
Clean sand

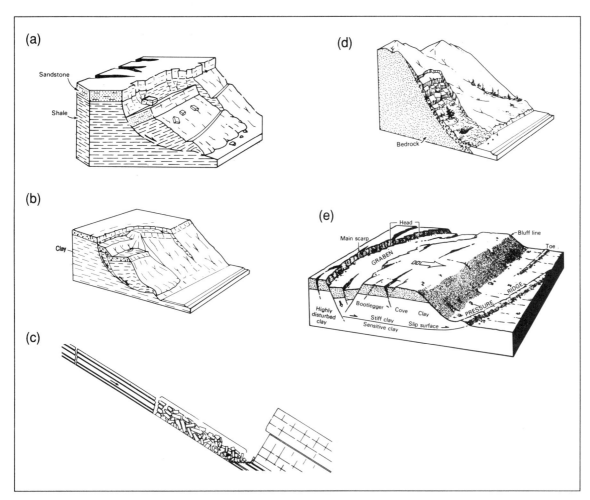

FIGURE 3-22
Examples of rotational and translational slides: (a) rotational rock slide; (b) rotational earth slide; (c) translational rock slide (upper portion is rock block slide); (d) debris slide; (e) translational earth block slide[Varnes 1978, Figures 2.1g, 2.1i, 2.1j2, 2.1k, 2.1l (Hansen 1965)].

tribute to a sag pond in the backward-tilted, displaced mass. This disruption of drainage may keep the displaced material wet and perpetuate the slope movements until a slope of sufficiently low gradient is formed.

In *translational slides* (Figures 3-22, 3-25, and 3-26) the mass displaces along a planar or undulating surface of rupture, sliding out over the original ground surface. Translational slides generally are relatively shallower than rotational slides. Therefore, ratios of D_r/L_r for translational slides in soils are typically less than 0.1 (Skempton and Hutchinson 1969). The surfaces of rupture of translational slides are often broadly channel-shaped in cross section (Hutchinson 1988). Whereas the rotation of a rotational slide tends to restore the displaced mass to equilibrium, translation may continue unchecked if the surface of separation is sufficiently inclined.

As translational sliding continues, the displaced mass may break up, particularly if its velocity or

FIGURE 3-23
Cut through rotational slide of fine-grained, thin-bedded lake deposits, Columbia River valley; beds above surface of rupture have been rotated by slide to dip into slope (Varnes 1978, Figure 2.7).

FIGURE 3-24
Rotational slides
(Varnes 1978,
Figure 2.5).

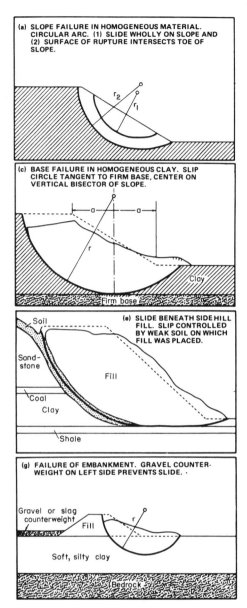

(a) SLOPE FAILURE IN HOMOGENEOUS MATERIAL. CIRCULAR ARC. (1) SLIDE WHOLLY ON SLOPE AND (2) SURFACE OF RUPTURE INTERSECTS TOE OF SLOPE.

(b) SLOPE FAILURE IN NONHOMOGENEOUS MATERIAL. SURFACE OF RUPTURE FOLLOWS DIPPING WEAK BED.

(c) BASE FAILURE IN HOMOGENEOUS CLAY. SLIP CIRCLE TANGENT TO FIRM BASE, CENTER ON VERTICAL BISECTOR OF SLOPE.

(d) BASE FAILURE IN NONHOMOGENEOUS MATERIAL. SURFACE OF RUPTURE FOLLOWS BED OF VERY SOFT CLAY.

(e) SLIDE BENEATH SIDE HILL FILL. SLIP CONTROLLED BY WEAK SOIL ON WHICH FILL WAS PLACED.

(f) FAILURE WITHIN A SIDEHILL FILL.

(g) FAILURE OF EMBANKMENT. GRAVEL COUNTER-WEIGHT ON LEFT SIDE PREVENTS SLIDE.

(h) SLIDE IN FILL INVOLVING UNDERTHRUSTING OF FIRM SURFACE MATERIAL DOWN SLOPE.

FIGURE 3-25
(below)
Translational slide
of colluvium
on inclined
metasiltstone
strata along I-40,
Cocke County,
Tennessee (Varnes
1978, Figure 2.11).

water content increases. The disrupted mass may then flow, becoming a debris flow rather than a slide. Translational sliding often follows discontinuities such as faults, joints, or bedding surfaces or the contact between rock and residual or transported soils.

Translational slides on single discontinuities in rock masses have been called *block slides* (Panet 1969) or *planar slides* (Hoek and Bray 1981). As Hutchinson (1988) pointed out, there is a transition from rock slides of moderate displacement that remain as blocks on the surface of rupture to slides on steeper and longer surfaces that break up into debris or transform into sturzstroms. Where the surface of rupture follows a discontinuity that is parallel to the slope, the toe of the displaced mass may form a wedge that overrides or ploughs into undisplaced material causing folding beyond the toe of the surface of rupture (Walton and Atkinson 1978). *Composite rock slide–rock topples* [Figure 3-13(2)] or *buckles* and confined slides may result.

The surface of rupture may be formed by two discontinuities that cause the contained rock mass to displace down the line of intersection of the discontinuities, forming a *wedge slide* [Figure 3-27(a) and Chapter 15, Figures 15-8 through 15-14]. Similarly shaped displaced masses may be bounded by one discontinuity that forms the main scarp of the slide and another that forms the surface of rupture. The mode of movement depends on the orientations of the free surfaces relative to the discontinuities in the rock masses (Hocking 1976; Cruden 1978, 1984). Stepped rupture surfaces may result if two or more sets of discontinuities, such as bedding surfaces and some joint sets, penetrate the rock masses. As shown in Figure 3-27(b), one set of surfaces forms the risers of the steps and the other forms the treads, creating a *stepped slide* (Kovari and Fritz 1984).

Compound slides are intermediate between rotational and translational slides and their D_r/L_r ratios reflect this position (Skempton and Hutchinson 1969). Surfaces of rupture have steep main scarps that may flatten with depth [Figure 3-22(e)]. The toes of the surfaces of rupture may dip upslope. Displacement along complexly curved surfaces of rupture usually requires internal deformation and shear along surfaces within the displaced material and results in the formation of intermediate scarps. Abrupt decreases in downslope dips of surfaces of rupture may be marked by uphill-facing scarps in displaced masses and the subsidence of blocks of displaced material to form depressed areas, *grabens* [Figure 3-22(e)]. A compound slide often indicates the presence of a weak layer or the boundary between weathered and unweathered material. Such zones control the location of the surface of rupture (Hutchinson 1988). In single compound slides, the width of the graben may be proportional to the depth to the surface of rupture (Cruden et al. 1991).

8.3.2 Complex and Composite Slides

Complex and composite slide movements are common and the literature contains numerous ref-

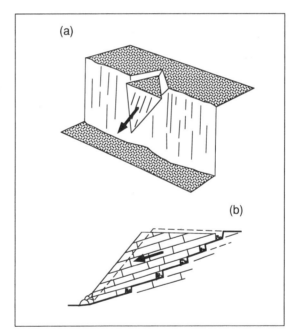

(a)

(b)

FIGURE 3-26
(above)
Shallow translational slide that developed on shaly rock slope in Puente Hills of southern California; slide has low D_r/L_r ratio; note wrinkles on surface [Varnes 1978, Figure 2.33; (Shelton 1966)].
COPYRIGHT JOHN S. SHELTON

FIGURE 3-27
(left)
(a) Wedge and (b) stepped translational slides.

erences to a variety of specialized names. Two of these names, *mudslide* and *flowslide*, are believed to be imprecise and ambiguous, and therefore their use is not recommended.

According to Hutchinson, *mudslides* are

slow-moving, commonly lobate or elongate masses of accumulated debris in a softened clayey matrix which advance chiefly by sliding on discrete, bounding shear surfaces. . . . In long profile, mudslides are generally bilinear with a steeper . . . slope down which debris is fed (by falls, shallow slides and mudslides) to a more gently-inclined slope. . . . Mudslides are especially well-developed on slopes containing stiff, fissured clays, doubtless because of the ease with which such materials break down to provide a good debris supply. (Hutchinson 1988, 12–13)

Evidently these movements begin in either weak rock or earth and retrogress by falls and slides while advancing by sliding. Hutchinson and Bhandari (1971) suggested that the displaced material, fed from above onto the mudslide, acted as an undrained load. Clearly mudslides are composite or perhaps complex both in style of activity and in the breakdown of displaced material into earth or debris. The displaced material is generally moist, though locally it may be wet. A mudslide is therefore often a *retrogressive, composite rock slide-advancing, slow, moist earth slide*. Other mudslide modes may include single earth slides (Brunsden 1984). The current use of *mudslide* for several different landslide modes suggests that more precise terminology should be used where possible.

The term *flowslide* has been used to describe sudden collapses of material that then move considerable distances very rapidly to extremely rapidly. As Hutchinson (1988) and Eckersley (1990) pointed out, at least three phenomena can cause this behavior: (*a*) impact collapse, (*b*) dynamic liquefaction, and (*c*) static liquefaction. Impact-collapse flowslides occur when highly porous, often-saturated weak rocks fall from cliffs, resulting in destruction of the cohesion of the material and the generation of excess fluid pressures within the flowing displaced mass. Clearly these are complex falls in which the second mode of movement is a *debris flow*; if the displaced material is dry debris, the movement may be a sturzstrom, previously defined in Section 8.1.2. This mode of movement can be recognized by both the materials involved and the topography of the flow.

If the structure of the material is destroyed by shocks such as earthquakes, saturated material may liquefy and then flow, sometimes carrying masses of overlying drier material with it. Such a movement is a flow or *liquefaction spread*, which is further defined in Section 8.4.1. Such movements are characteristic of loess, a lightly cemented aeolian

silt. Landslides occurring in loosely dumped anthropogenic materials, both stockpiles and waste dumps, have also been termed *flowslides*. These loose, cohesionless materials contract on shearing and so may generate high pore-water pressures after some sliding (Eckersley 1990). Similar landslides may also occur in rapidly deposited natural silts and fine sands (Hutchinson 1988, 14). Since these movements involve both sliding and then flowage, they may be better described as *complex slide flows*.

Because these separate and distinguishable phenomena are comparatively distinct types of landslides that may be more accurately described by standard descriptors, the use of the term *flowslide* for all these types of movements is redundant, confusing, and potentially ambiguous.

In contrast, one form of compound sliding failure appears to warrant a special term. *Sags* (or *sackungen*) are deformations of the crests and steep slopes of mountain ridges that form scarps and grabens and result in some ridges with double crests and small summit lakes. Material can be displaced tens of meters at individual scarps. The state of activity, however, is generally dormant and may be relict. The term *sag* may be useful to indicate uncertainty about the type of movement visible on a mountain ridge.

Movement is often confined, and small bulges in local slopes are the only evidence of the toes of the displaced material. Detailed subsurface investigations of these features are rare, and classification should await this more detailed exploration. As Hutchinson (1988, 8) demonstrated, the geometry of the scarps (which often face uphill) may be used to suggest types of movement, which may include slides, spreads, and topples. The modes of sagging depend on the lithology of the displaced material and the orientation and strength of the discontinuities in the displacing rock mass. Varnes et al. (1989, 1) distinguished

1. "Massive, strong (although jointed) rocks lying on weak rocks,"
2. "Ridges composed generally of metamorphosed rocks with pronounced foliation, schistosity or cleavage," and
3. "Ridges composed of hard, but fractured, crystalline igneous rocks."

Sags of the first type are usually *spreads* (Radbruch-Hall 1978; Radbruch-Hall et al. 1976), which are discussed in Section 8.4. Sags in foliated metamorphic rocks are often topples, and thus are discussed in Section 8.2, although bedrock flow may also occur (see Section 8.5). Sags in crystalline igneous rocks (Varnes et al. 1989, 22) were modeled by a plasticity solution for "gravity-induced deformation of a slope yielding under the Coulomb criterion." Sags may thus be slides, spreads, or flows, depending on the extent and distribution of plastic flow within the deforming rock mass.

Sags are often associated with glacial features. The absence of Pleistocene snow and ice covers, and the resulting loss of the permafrost and high pore-water pressures these induced, may account for the present inactivity of many sags. Varnes et al. (1990) described an active sag in the mountains of Colorado. Earthquakes, however, can also produce uphill-facing scarps along reactivated normal faults. The surface features of sags require careful investigation before any conclusions can be drawn about the cause and timing of slope movement. Such investigations may be sufficient to allow the identification of sags as other types of landslides.

In many landslides, the displaced material, initially broken by slide movements, subsequently begins to flow (Figure 3-28). This behavior is especially common when fine-grained or weak materials are involved. These landslides have been termed *slump-earth flows*. *Slump* has been used as a synonym for a rotational slide, but it is also used to describe any movement in a fill. It is therefore recommended that this mode of movement be termed a *complex earth slide–earth flow* and that the use of the term *slump* be discontinued.

In permafrost regions, distinctive *retrogressive, complex earth slide–earth flows*, known as *thaw-slumps* (Hutchinson 1988, 21) and *bimodal flows* (McRoberts and Morgenstern 1974), develop on steep earth slopes when icy permafrost thaws and forms flows of very wet mud from a steep main scarp. These special landslide conditions are discussed in more detail in Chapter 25.

8.4 Spread

The term *spread* was introduced to geotechnical engineering by Terzaghi and Peck (1948) to describe sudden movements on water-bearing seams of sand or silt overlain by homogeneous clays or loaded by fills:

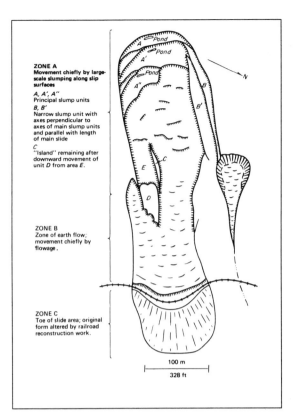

ZONE A
Movement chiefly by large-scale slumping along slip surfaces

A, A', A"
Principal slump units

B, B'
Narrow slump unit with axes perpendicular to axes of main slump units and parallel with length of main slide

C
"Island" remaining after downward movement of unit D from area E.

ZONE B
Zone of earth flow; movement chiefly by flowage.

ZONE C
Toe of slide area; original form altered by railroad reconstruction work.

100 m
328 ft

FIGURE 3-28
Plan of Ames slide near Telluride, Colorado. This enlarging complex earth slide–earth flow occurred in till overlying Mancos shale. Crown of slide retrogressed by multiple rotational slides after main body of displaced material moved. Surface of rupture also widened on left lateral margin [Varnes 1978, Figure 2.10 (modified from Varnes 1949)].

It is characteristic . . . that a gentle clay slope which may have been stable for decades or centuries, moves out suddenly along a broad front. At the same time the terrain in front . . . heaves for a considerable distance from the toe. On investigation, it has invariably been found that the spreading occurred at a considerable distance beneath the toe along the boundary between the clay and an underlying water-bearing stratum or seam of sand or silt. (Terzaghi and Peck 1948, 366)

Recognition of the phenomenon is considerably older. One of the three types of landslides distinguished by Dana (1877, 74) occurs "when a layer of clay or wet sand becomes wet and softened by percolating water and then is pressed out laterally by the weight of the superincumbent layers." An early use of *spread* to describe this phenomenon is by Barlow:

In a landslip [British term for some types of landslide], the spreading of some underlying bed which has become plastic through the percolation of water or for some other cause drags apart the more solid, intractable beds above and produces fissures and fractures transverse to the direction of movement. (Barlow 1888, 786)

Spread is defined here as an extension of a cohesive soil or rock mass combined with a general subsidence of the fractured mass of cohesive material into softer underlying material. The surface of rupture is not a surface of intense shear. Spreads may result from liquefaction or flow (and extrusion) of the softer material. Varnes (1978) distinguished spreads typical of rock, which extended without forming an identifiable surface of rupture from movements in cohesive soils overlying liquefied materials or materials flowing plastically (Figure 3-29). The cohesive materials may also subside, translate, rotate, disintegrate, or liquefy and flow. Clearly these movements are complex, but they are sufficiently common in certain materials and geological situations that the concept of a spread is worth recognizing as a separate type of movement.

8.4.1 Modes of Spreading

In *block spreads*, a thick layer of rock overlies softer materials; the strong upper layer may fracture and separate into strips. The soft underlying material is squeezed into the cracks between the strips, which

may also fill with broken, displaced material [Figure 3-29(a)]. Typical rates of movement are extremely slow.

Such movements may extend many kilometers back from the edges of plateaus and escarpments. The Needles District of Canyonlands National Park, Utah, is an example of a block spread (McGill and Stromquist 1979; Baars 1989). Grabens up to 600 m wide and 100 m deep stretch 20 km along the east side of Cataract Canyon on the Colorado River (Figure 3-30). The grabens extend up to 11 km back from the river. A 450-m-thick sequence of Paleozoic sedimentary rocks has been spread down a regional slope with a 4-degree dip by the flow of an underlying evaporite that is exposed in valley anticlines in the Colorado River and its tributaries (Potter and McGill 1978; Baars 1989). This approximately 60 km³ of displaced material constitutes one of North America's largest landslides.

Liquefaction spreads form in sensitive clays and silts that have lost strength with disturbances that damaged their structure [Figures 3-29(c) and 3-31]. These types of landslides are discussed further in Chapter 24. Movement is translational and often retrogressive, starting at a stream bank or a shoreline and extending away from it. However, if the underlying flowing layer is thick, blocks may

FIGURE 3-29
Rock and earth spreads: (a), (b) rock spreads that have experienced lateral extension without well-defined basal shear surface or zone of plastic flow [Varnes 1978, Figure 2.1m2 (Zaruba and Mencl 1969); Figure 2.1m3 (Ostaficzuk 1973)]; (c) earth spread resulting from liquefaction or plastic flow of subjacent material (Varnes 1978, Figure 2.1o).

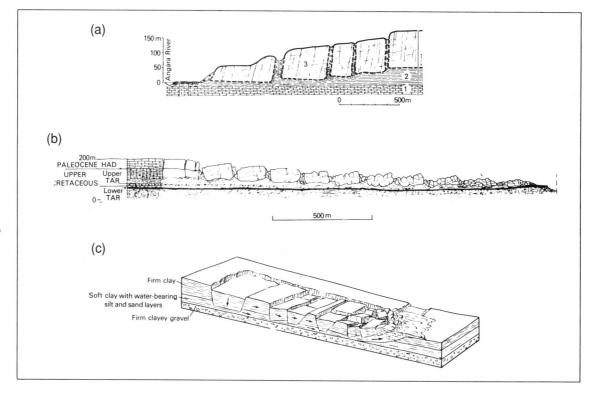

sink into it, forming grabens, and upward flow can take place at the toe of the displaced mass. Movement can begin suddenly and reach very rapid velocities:

> Spreads are the most common ground failure during earthquakes. . . . [They] occur in gentler terrains (commonly between 0.5 per cent [0.3 degrees] and 5 percent [3 degrees]) with lateral movement of a few meters or so. . . . [S]preads involve fracturing and extension of coherent material owing to liquefaction or plastic flow of subjacent material. . . . [S]preads are primarily translational although some associated rotation and subsidence commonly occurs. (Andrus and Youd 1987, 16)

Van Horn (1975) described two movements of more than 8 km² in area on very gentle slopes in flat-lying beds of silty clay, clayey silt, and very fine sand deposited in prehistoric Lake Bonneville, Utah. The displaced material forms ridges parallel to the main scarp of the landslide; distinctive internal structures include gentle folds, shears, and intrusions of liquefied sediment.

8.4.2 Complex Spreads

Major deformations in rock strata were found along many valleys in north-central England during the construction of dams in the late 19th century. These deformations occurred where a nearly horizontal, rigid-jointed cap rock overlaid a thick layer of stiff-fissured clay or clay shale that in turn overlaid a more competent stratum. A bending, or *cambering*, of the rigid strata caused blocks of this stratum to dip toward the valley. This bending of the upper stratum was accompanied by severe deformation and bulging of the softer lower strata in the valley floor.

Hutchinson (1991) defined characteristic features of cambers and valley bulges as follows:

• Marked thinning of the clay substratum as a result of the transfer of clay into the valley bulge;
• Intense folding and distortion in the valley bulge itself as a result of the meeting of the clay masses moving in from beneath each valley side;
• Sympathetic flexuring of the superincumbent capping rocks, producing a valleyward camber, a valley marginal syncline, and an upturn against the flanks of the bulge; and

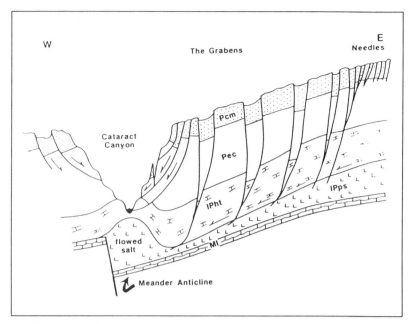

FIGURE 3-30
Geological cross section showing formation of The Grabens in Needles District of Canyonlands National Park, Utah. Colorado River has carved Cataract Canyon to within a few meters of top of Paradox Salt beds (*IPps*), undercutting inclined layers of overlying Honaker Trail (*IPht*), Elephant Canyon (*Pec*), and Cedar Mesa (*Pcm*) formations. These formations have broken up and are moving toward canyon by flowage within salt. Colorado River follows crest of Meander anticline, a salt-intruded fold located above a deep-seated fault zone (Baars 1989).

FIGURE 3-31
Earth spread–earth flow near Greensboro, Florida. Material is flat-lying, partly indurated clayey sand of Hawthorn Formation. Length of slide is 275 m from scarp to edge of trees in foreground. Vertical distance is about 15 m from top to base of scarp and about 20 m from top of scarp to toe. Landslide occurred in April 1948 after a year of unusually heavy rainfall, including 40 cm in the 30 days preceding landslide [Varnes 1978, Figure 2.19 (modified from Jordan 1949)].

• Extension and valleyward toppling of the capping rocks in the camber, resulting in opening of near-vertical joints to form wide-open fissures, termed *gulls*, in valleyward dips of the camber blocks and in the development of dip-and-fault structures between camber blocks as a consequence of their toppling.

The rotation of the dip of the rock blocks produces the slightly arched or convex form popularly called a camber. Rotation is made possible by the extension of "the cap-rock towards the valley producing widened joints (called gulls) often infilled by till" (Hutchinson 1988, 19). The cap rock has spread. The underlying clay exhibits

> a brecciated structure, probably frost-induced, in its upper parts, marked thinning as the valley is approached and intense generally-monoclinal folding in . . . the present valley bottom. . . . The dramatic internal structures appear to be the result principally of valley-ward squeezing and extrusion made possible by the weakening of the clay stratum by multiple freezing and thawing. . . . These cambers and valley bulges are believed to be relict periglacial features. (Hutchinson 1988, 19)

Cambering and valley bulging affected slopes at Empingham, England, that were excavated during the construction of a dam (Horswill and Horton 1976). Figure 3-32(*a*) reproduces a portion of Figure 5 of Horswill and Horton (1976), which shows the details of the structures with a fourfold vertical

exaggeration. Figure 3-32(*b*) is based on a diagram by Hutchinson (1991) that shows the same general section without vertical exaggeration and emphasizes that the displaced materials are found on slopes of less than 5 degrees. According to the proposed naming convention, a camber may be described as a *relict, complex rock spread–rock topple*.

Ward (1948) described as a landslide another complex spread in Britain in which stiff-fissured clays overlaid fine sands but qualified his description as follows:

> So much movement of various types had occurred that it was difficult to trace the movement of the strata from the upper cliff until it arrived in the form of mud on the beach some 180 feet below. . . . The underlying fine sand is in a saturated, quick condition under the blocks when they become detached and they probably flounder forwards and tilt backwards. (Ward 1948, 36)

This description suggests that the clay was being spread by the flow of the sand and thus the movement was a type of *complex earth spread–debris flow*.

8.5 Flow

A **flow** is a spatially continuous movement in which surfaces of shear are short-lived, closely spaced, and usually not preserved. The distribution of velocities in the displacing mass resembles that in a viscous liquid. The lower boundary of the displaced mass may be a surface along which appre-

FIGURE 3-32 Cambering and valley bulging at Empingham, England: (*a*) detailed cross section with 4× vertical exaggeration (modified from Horswill and Horton 1976); and (*b*) generalized cross section drawn without vertical exaggeration (modified from Hutchinson 1991).

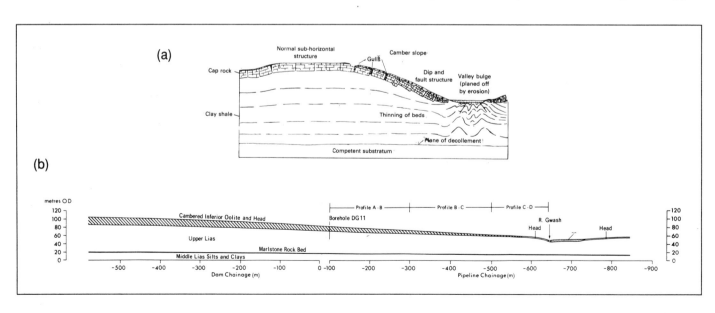

ciable differential movement has taken place or a thick zone of distributed shear (Figure 3-33). Thus there is a gradation from slides to flows depending on water content, mobility, and evolution of the movement. Debris slides may become extremely rapid debris flows or debris avalanches as the displaced material loses cohesion, gains water, or encounters steeper slopes (Figure 3-34).

Varnes (1978, 19–20) used the terms *earth flow* and *slow earth flow* [Figure 3-33(a)] to describe "the somewhat drier and slower earth flows in plastic earth . . . common . . . wherever there is . . . clay or weathered clay-bearing rocks, moderate slopes, and adequate moisture."

Keefer and Johnson (1983) included detailed studies of the movement of earth flows in the San Francisco Bay area (Figure 3-35). They concluded (Keefer and Johnson 1983, 52): "Although some internal deformation occurs within earth flows,

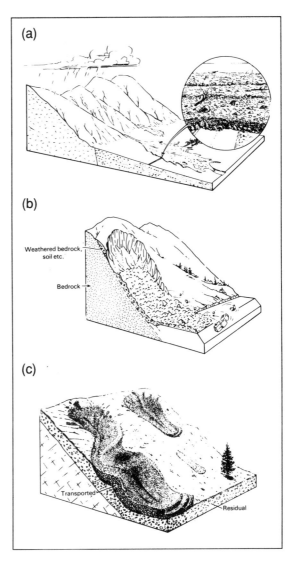

FIGURE 3-34
Channelized debris flows: (a) debris flow, (b) debris avalanche, and (c) block stream (Varnes 1978, Figures 2.1q1, 2.1q3, 2.1q5).

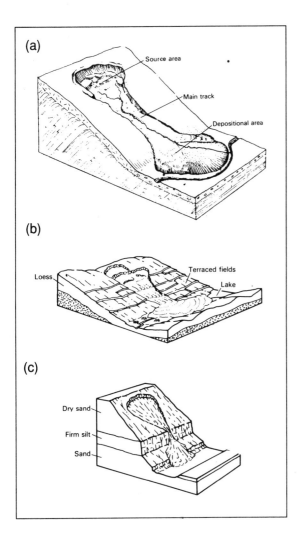

most movement takes place on or immediately adjacent to their boundaries." Their use of *earth flow* thus covers landslide modes from *slow earth flow* through *slow, composite earth slide–earth flow* to *slow earth slide*. When extensive, striated, or slickensided lateral margins or surfaces of rupture are visible, the landslide might well be called an *earth slide*; when the displaced mass is strongly deformed internally, the landslide is probably an *earth flow*. If the same landslide shows both modes of deformation, it is clearly a *composite earth slide–earth flow*.

While defining landslide processes in permafrost regions, McRoberts and Morgenstern (1974) used the term *skin flow* to describe a rapid to very rapid slope movement in which a thin layer, or skin, of thawed soil and vegetation flows or slides

FIGURE 3-33
Examples of flows: (a) slow earth flow [Varnes 1978, Figure 2.1r3 (Zaruba and Mencl 1969)], (b) loess flow, and (c) dry sand flow (Varnes 1978, Figures 2.1r5 and 2.1r4).

FIGURE 3-35
Earth flow
developing from
initial rotational
earth slide near
Berkeley, California
(Varnes 1978,
Figure 2.22).

over the permafrost table, whereas Hutchinson (1988, 12) used the term *active-layer slide*. Seasonal thaw layers, or active layers, up to a meter or so in thickness may contain water originally drawn to the freezing front where it formed segregated ice. Melting of this ice may generate artesian pore-water pressures that greatly reduce the resistance of the active layer to movement. These landslide conditions are discussed in more detail in Chapter 25. Similar shallow failure may also occur in loess materials that become saturated or are subjected to earthquake shaking [Figure 3-33(*b*)].

Open-slope debris flows form their own path down a valley side onto the gentler slopes at the foot. Deposition of levees there may outline a more sinuous channel. The common, small dry flows of granular material may be channelized (Figure 3-36)

FIGURE 3-36
Dry sand flow in
Columbia River
valley; dry sand
from upper terrace
flowed through
notch in cliffs of
more compact
sand and silt below
(Varnes 1978,
Figure 2.24).

or may extend as sheets for some distance across a slope [Figures 3-33(*b*) and 3-37].

Channelized flows follow existing channels (Figures 3-34 and 3-38). As shown in Figure 3-39, debris flows are often of high density, with over 80 percent solids by weight, and may exceed the density of wet concrete (Hutchinson 1988). They can therefore move boulders that are meters in diameter. The mode of flow shown in Figure 3-40 often occurs during torrential runoff following exceptional rainfalls. Soils on steep slopes unprotected by vegetation whose natural cover may have been destroyed by fire are prone to debris flows. Debris may be added to small surface streams by erosion or caving of their banks, increasing the power of the flows. Coarser material may form natural levees, leaving the fines in suspension to move down the channel. Flows can extend many kilometers before they drop their suspended loads upon entering lower-gradient channels. The movement may be in pulses, presumably caused by periodic mobilization of material or by the formation and bursting of dams of debris in the channel.

Pierson and Costa (1987) observed that the term *debris torrent* was misleading and gave two reasons:

> First, mountain torrent or debris torrent is used in European and Japanese literature to mean a very steep channel, not the material that flows in it. . . . Second, the term was coined to differentiate between coarse debris flows occurring in channels and flows occurring on open slopes . . . , a criterion that has no rheologic or other process-specific basis. We suggest that the usage of the term, debris torrent, be discontinued and the more general term, debris flow, be used instead with appropriate descriptive adjectives when specifics are required. (Pierson and Costa 1987, 10)

Debris avalanches are larger, extremely rapid, often open-slope flows [Figure 3-34(*b*)]. The Mt. Huascaran avalanche in Peru (Figure 3-41) involved 50 million m^3 to 100 million m^3 of rock, ice, snow, and soil that traveled at velocities of as much as 100 m/sec. In this case, steam and air cushions were suggested to account for the high velocity and long distance of the debris travel (Varnes 1978, 21). However, the contributions of snow and ice to the movement should also be considered.

According to Varnes, *bedrock flows* include

spatially continuous deformation and surficial as well as deep creep. . . . [They involve] extremely slow and generally nonaccelerating differential movements among relatively intact units. Movements may (1) be along many shear surfaces that are apparently not connected; (2) result in folding, bending, or bulging; or (3) roughly simulate those of viscous fluids in distribution of velocities. (Varnes 1978, Figure 2.1, V)

All the examples given by Varnes (1978) and reproduced as Figure 3-42 show movements that may have been initiated by sliding on the bedding or schistosity of the rock mass. These might all then be classified as complex slides. Further study may define the complex modes of movement to which these examples belong, in which sliding is followed by buckling (Hu and Cruden 1993). Clearly, further examples of bedrock flow should be explored in more detail before they can be more than tentatively classified.

A *lahar* is a debris flow from a volcano. The flow mobilizes the loose accumulations of tephra (airborne solids erupted from the volcano) on the volcano's slopes. Water for the flow may come from the ejection of crater lakes, condensations of erupted steam, the nucleation of water vapor on erupted particles and its precipitation, and the melting of snow and ice accumulated on a sufficiently high volcanic cone (Voight 1990).

9. LANDSLIDE PROCESSES

"The processes involved in slope movements comprise a continuous series of events from cause to effect" (Varnes 1978, 26). In some cases, it may be more economical to repair the effects of a landslide than to remove the cause; a highway on the crest of a slope may be relocated rather than armoring the toe of the slope to prevent further erosion. However, the design of appropriate, cost-effective remedial measures still requires a clear understanding of the processes that are causing the landslide. Although this understanding may require a

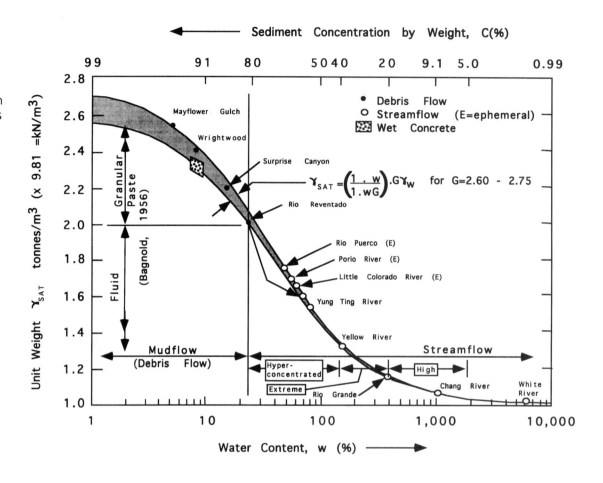

detailed site investigation, a reconnaissance of the landslide as soon as possible after its occurrence can allow important observations of the processes involved. These observations may guide both the site investigation and the remedial measures.

Although Varnes (1978) provided a list of the causes of slides, the aims in this section are less ambitious. The section follows Varnes's distinction that the three broad types of landslide processes are those that

1. Increase shear stresses (Section 9.1),
2. Contribute to low strength (Section 9.2), and
3. Reduce material strength (Section 9.3).

Common landslide triggering mechanisms are discussed at greater length in Chapter 4.

Processes and characteristics that contribute to landslides are summarized in a checklist of landslide causes arranged in four practical groups according to the tools and procedures necessary to begin the investigation (see p. 70). Ground causes

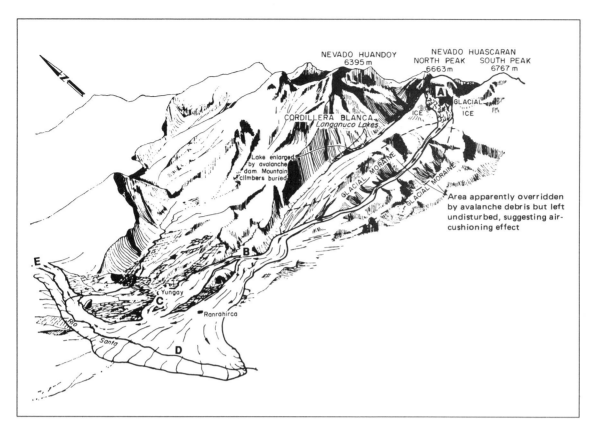

FIGURE 3-41
(top)
May 31, 1970, Huascaran debris avalanche (Peru) originated at *Point A*. Yungay had been protected from January 10, 1962, debris avalanche by ridge up to 240 m high (*Point B*), but portion of later avalanche over-topped protective ridge. Cemetery Hill (*Point C*) was only safe place in Yungay, some 93 people escaping to it before avalanche devastated surrounding area. Moving at an average speed of 320 km/hr, debris arrived at *Point D* on Rio Santa 14.5 km down 15-degree average slope within 3 to 4 min after starting from north peak of Huascaran (*Point A*). Debris flowed 2.5 km up Rio Santa (*Point E*) and continued 160 km downstream to Pacific Ocean, devastating villages and crops on its floodplain [Varnes 1978, Figure 2.27 (modified from Cluff 1971)].

can be identified with the customary tools of site reconnaissance and investigation. Changes in site morphology over time are apparent from the study of surveys, maps, and aerial photographs. Identification of causes of movement requires the collection of data over time from a variety of field instruments, including seismographs, rain gauges, flow gauges, and piezometers. Some changes in material and mass properties with time may, however, be inferred from gradual changes in the mass properties with distance. Anthropogenic causes can be documented by site records, plans, or other observations.

9.1 Increased Shear Stresses

Shear stresses can be increased by processes that lead to removal of lateral support, by the imposition of surcharges, by transitory stresses resulting from explosions or earthquakes, and by uplift or tilting of the land surface.

9.1.1 Removal of Support

The toe of a slope can be removed by erosion, steepening the slope. Typical agents are streams and

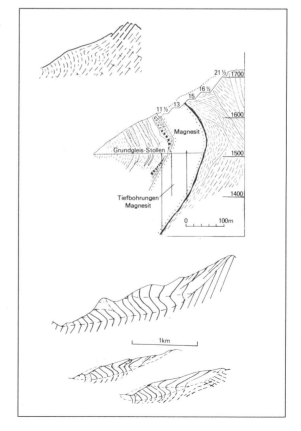

FIGURE 3-42
Examples of rock flows [Varnes 1978, Figure 2.1p1 (Nemčok et al. 1972; Zischinsky 1966)].

Checklist of Landslide Causes

1. Geological causes
   a. Weak materials
   b. Sensitive materials
   c. Weathered materials
   d. Sheared materials
   e. Jointed or fissured materials
   f. Adversely oriented mass discontinuity (bedding, schistosity, etc.)
   g. Adversely oriented structural discontinuity (fault, unconformity, contact, etc.)
   h. Contrast in permeability
   i. Contrast in stiffness (stiff, dense material over plastic materials)
2. Morphological causes
   a. Tectonic or volcanic uplift
   b. Glacial rebound
   c. Fluvial erosion of slope toe
   d. Wave erosion of slope toe
   e. Glacial erosion of slope toe
   f. Erosion of lateral margins
   g. Subterranean erosion (solution, piping)
   h. Deposition loading slope or its crest
   i. Vegetation removal (by forest fire, drought)
3. Physical causes
   a. Intense rainfall
   b. Rapid snow melt
   c. Prolonged exceptional precipitation
   d. Rapid drawdown (of floods and tides)
   e. Earthquake
   f. Volcanic eruption
   g. Thawing
   h. Freeze-and-thaw weathering
   i. Shrink-and-swell weathering
4. Human causes
   a. Excavation of slope or its toe
   b. Loading of slope or its crest
   c. Drawdown (of reservoirs)
   d. Deforestation
   e. Irrigation
   f. Mining
   g. Artificial vibration
   h. Water leakage from utilities

rivers, glaciers, waves and currents, and slope movements. Anthropogenic landslides can be caused by excavations for cuts, quarries, pits, and canals, and by the drawdown of lakes and reservoirs.

Removal of material from the lateral margins of the displaced mass can also cause movement. Material can be removed from below the landslide by solution in karst terrain, by piping (the transport of sediment in groundwater flows), or by mining. In some spreads, the loss of strength of the

material at depth within the displacing mass results in its extrusion or, if the base of the spread has liquefied, in its outward flow. These issues as they relate to landslides in sensitive clay deposits are discussed in Chapter 24.

9.1.2 Imposition of Surcharges

The addition of material can result in increases of both the length and the height of the slope. Water can be added by precipitation, both rain and snow; by the flow of surface and groundwater into the displacing mass; and even by the growth of glaciers. Surcharges can be added by the movement of landslides onto the slope, by volcanic activity, and by the growth of vegetation. Anthropogenic surcharges include construction of fills, stockpiles, and waste dumps; structural weight; and water from leaking canals, irrigation systems, reservoirs, sewers, and septic tanks.

9.1.3 Transitory Stresses

The local stress field within a slope can be greatly changed by transitory stresses from earthquakes and explosions (both anthropogenic and volcanic). Smaller transitory changes in the stress field can result from storms and from human activity such as pile driving and the passage of heavy vehicles.

9.1.4 Uplift or Tilting

Uplift or tilting may be caused by tectonic forces or by volcanic processes. In either case, this type of increased shear stress may be associated with earthquakes, which themselves can trigger landslides (Section 9.1.3). The melting of the extensive Pleistocene ice sheets has caused widespread uplift in temperate and circumpolar regions.

Uplift of an area of the earth's surface generally causes steepening of slopes in the area as drainage responds by increased incision. The cutting of valleys in the uplifted area may cause valley rebound and accompanying fracturing and loosening of valley walls with inward shear along flat-lying discontinuities. The fractures and shears may allow the buildup of pore-water pressures in the loosened mass and eventually lead to landsliding.

9.2 Low Strength

Low strength of the earth or rock materials that make up a landslide may reflect inherent material characteristics or may result from the presence of discontinuities within the soil or rock mass.

9.2.1 Material Characteristics

Materials may be naturally weak or may become weak as a result of common natural processes such as saturation with water. Organic materials and clays have low natural strengths. Rocks that have decomposed to clays by chemical weathering (weathered volcanic tuffs, schists, and serpentinites, for example) develop similar properties.

Besides the nature of the individual particles of which the material is composed, the arrangement of these particles (the fabric of the material) may cause low material strengths. Sensitive materials, which lose strength when disturbed, generally have loose fabrics or textures.

9.2.2 Mass Characteristics

The soil or rock mass may be weakened by discontinuities such as faults, bedding surfaces, foliations, cleavages, joints, fissures, shears, and sheared zones (Chapters 12 and 14). Contrasts in bedded sedimentary sequences—such as stiff, thick beds overlying weak, plastic, thin beds or permeable sands (or sandstones) alternating with weak, impermeable clays (or shales)—are sources of weakness.

9.3 Reduced Shear Strength

Clays are particularly prone to weathering processes and other physicochemical reactions. Hydration of clay minerals results in loss of cohesion, a process often associated with softening of fissured clays. Fissuring of clays may be due to drying or to release of vertical and lateral restraints by erosion or excavation. The exchange of ions within clay minerals with those in the pore water of the clays may lead to substantial changes in the physical properties of some clays. Electrical potentials set up by these chemical reactions or by other processes may attract water to the weathering front.

The effects of extremes of temperature caused by severe weather are not confined to clays. Rocks may disintegrate under cycles of freezing and thawing or thermal expansion and contraction. Dry weather may cause desiccation cracking of weak or weathered rock along preexisting discontinuities, such as bedding planes. Wet weather may dissolve natural rock cements that hold particles together. Saturation with water reduces

effective intergranular pressure and friction and destroys capillary tension.

10. SUMMARY

In the initial reconnaissance of a landslide, the activity and the materials displaced in that type of landslide would be described using terms from Table 3-2, the dimensions defined in Table 3-4 would be estimated, and some preliminary hypotheses would be chosen about the causes of the movements. A simple landslide report form is provided in Figure 3-9; its format would allow the creation of simple data bases suited to much of the data-base management software now available for personal computers. The information collected could be compared with summaries of other landslides (WP/WLI 1991) and used to guide additional investigations and mitigative measures. Further investigation would increase the precision of estimates of the dimensions and increase confidence in the descriptions of activity and material and in the hypotheses about the causes of movement. The new information would then be added to the data base to influence the analysis of new landslides. These data bases could form the foundations of expert systems for landslide mitigation.

REFERENCES

ABBREVIATIONS

IAEG   International Association of Engineering Geology
UNESCO   United Nations Educational, Scientific, and Cultural Organization
WP/WLI   Working Party on the World Landslide Inventory (International Geotechnical Societies and UNESCO)

Abele, G. 1974. Bergstürze in den Alpen: ihre Verbreitung, Morphologie und Folgeerscheinungen. *Wissenschaftliche Alpenvereinshefte*, Heft 25, Munchen.

Andrus, R.D., and T.L. Youd. 1987. *Subsurface Investigation of a Liquefaction Induced Lateral Spread, Thousand Springs Valley, Idaho*. Miscellaneous Paper GL-87-8. U.S. Army Engineer Waterways Experiment Station, Vicksburg, Miss.

Baars, D. 1989. *Canyonlands Country: Geology of Canyonlands and Arches National Parks*. Canon Publishers Ltd., Lawrence, Kans., 140 pp.

Bagnold, R.A. 1956. The Flow of Cohesionless Grains in Fluids. *Proceedings of the Royal Society of London*, Vol. A249, pp. 235–297.

Barlow, W. 1888. On the Horizontal Movements of Rocks and the Relation of These Movements to the Formation of Dykes and Faults and to Denudation and the Thickening of Strata. *Quarterly Journal of the Geological Society*, Vol. 44, pp. 783–796.

Bates, R.L., and J.A. Jackson, eds. 1987. *Glossary of Geology*. American Geological Institute, Falls Church, Va., 788 pp.

Bell, D.H., ed. 1992. *Proc., Sixth International Symposium on Landslides*. A.A. Balkema, Rotterdam, Netherlands, 2 vols., 1495 pp.

Beyer, W. H., ed. 1987. *Handbook of Mathematical Sciences*, 6th ed. CRC Press, Inc., Boca Raton, Florida, 860 pp.

Bishop, A.W. 1973. Stability of Tips and Spoil Heaps. *Quarterly Journal of Engineering Geology*, Vol. 6, No. 4, pp. 335–376.

Bonnard, C., ed. 1988. *Proc., Fifth International Symposium on Landslides*. A.A. Balkema, Rotterdam, Netherlands, 3 vols., 1564 pp.

Brabb, E.E., and B.L. Harrod, eds. 1989. *Landslides: Extent and Economic Significance*: Proc., 28th International Geological Congress: Symposium on Landslides, A.A. Balkema, Rotterdam, Netherlands, 385 pp.

Brown, W.M., D.M. Cruden, and J.S. Denison. 1992. *The Directory of the World Landslide Inventory*. U.S. Geological Survey Open File Report 92-427, 216 pp.

Brunsden, D. 1984. Mudslides. In *Slope Instability* (D. Brunsden and D.B. Prior, eds.), John Wiley and Sons, Chichester, U.K., Chap. 9, pp. 363–418.

Brunsden, D., and D.B. Prior, eds. 1984. *Slope Instability*. John Wiley and Sons, Chichester, U.K., 620 pp.

Canadian Geotechnical Society. 1984. *Proc., Fourth International Symposium on Landslides*. Toronto, Canada, 3 vols., 1484 pp.

Church, H.K. 1981. *Excavation Handbook*. McGraw-Hill, New York, N.Y., 1024 pp.

Cluff, L.S. 1971. Peru Earthquake of May 31, 1970: Engineering Geology Observations. *Bulletin of the Seismological Society of America*, Vol. 61, No. 3, pp. 511–521.

Costa, J.E., and G.F. Wieczorek, eds. 1987. *Debris Flows/Avalanches: Process, Recognition, and Mitigation*. Reviews in Engineering Geology, Vol. 7, Geological Society of America, Boulder, Colo., 239 pp.

Cross, W. 1924. *Historical Sketch of the Landslides of the Gaillard Cut*. Memoir 18. National Academy of Sciences, Washington, D.C., pp. 22–43.

Crozier, M.J. 1984. Field Assessment of Slope Instability. In *Slope Instability* (D. Brunsen and D.B. Prior, eds.), John Wiley & Sons, New York, pp. 103–142.

Crozier, M.J. 1986. *Landslides: Causes, Consequences, and Environment*. Croom Helm, London, 252 pp.

Cruden, D.M. 1978. Discussion of Hocking's Paper. *International Journal of Rock Mechanics and Mining Science*, Vol. 15, No. 4, p. 217.

Cruden, D.M. 1984. More Rapid Analysis of Rock Slopes. *Canadian Geotechnical Journal*, Vol. 21, No. 4, pp. 678–683.

Cruden, D.M. 1986. Discussion of Carter, M., and S.P. Bentley. 1985. The Geometry of Slip Surfaces Beneath Landslides: Predictions from Surface Measurements. *Canadian Geotechnical Journal*, Vol. 23, No. 1, p. 94.

Cruden, D.M. 1991. A Simple Definition of a Landslide. *Bulletin of the International Association of Engineering Geology*, No. 43, pp. 27–29.

Cruden, D.M., Z.Q. Hu, and Z.Y. Lu. 1993. Rock Topples in the Highway Cut West of Clairvivaux Creek, Jasper, Alberta. *Canadian Geotechnical Journal*, Vol. 30, No. 6, pp. 1016–1023.

Cruden, D.M., S. Thomson, and B.A. Hoffman. 1991. Observations of Graben Geometry in Landslides. In *Slope Stability Engineering—Developments and Applications: Proc., International Conference on Slope Stability*, Isle of Wight, 15–18 April (R.J. Chandler, ed.), Thomas Telford Ltd., London, pp. 33–36.

Dana, J.D. 1877. *New Text-book of Geology Designed for Schools and Academies*. Ivison, Blakeman, Taylor, New York, 366 pp.

De Freitas, M.H., and R.J. Watters. 1973. Some Field Examples of Toppling Failure. *Geotechnique*, Vol. 23, No. 4, pp. 495–514.

Eckersley, J.D. 1990. Instrumented Laboratory Flowslides. *Geotechnique*, Vol. 40, No. 3, pp. 489–502.

Eisbacher, G.H., and J.J. Clague. 1984. *Destructive Mass Movements in High Mountains: Hazard and Management*. Paper 84-16. Geological Survey of Canada, Ottawa, Ontario, 230 pp.

Eliot, T.S. 1963. *Collected Poems 1909–1962*. Faber, London, 234 pp.

Engineering News Record. 1971. Contrived Landslide Kills 15 in Japan. Vol. 187, p. 18.

Evans, S.G., J.J. Clague, G.J. Woodsworth, and O. Hungr. 1989. The Pandemonium Creek Rock Avalanche, British Columbia. *Canadian Geotechnical Journal*, Vol. 26, No. 3, pp. 427–446.

Giraud, A., L. Rochet, and P. Antoine. 1990. Processes of Slope Failure in Crystallophyllian Formations. *Engineering Geology*, Vol. 29, No. 3, pp. 241–253.

Goodman, R.E., and J.W. Bray. 1976. Toppling of Rock Slopes. In *Proc., Specialty Conference on Rock Engineering for Foundations and Slopes*, Boulder, Colo., Aug. 15–18, American Society of Civil Engineers, New York, Vol. 2, pp. 201–234.

Haefeli, R. 1965. Creep and Progressive Failure in Snow, Soil, Rock, and Ice. In *Proc., Sixth International Conference on Soil Mechanics and Foundation Engineering*, University of Toronto Press, Vol. 3, pp. 134–148.

Hansen, W.R. 1965. *Effects of the Earthquake of March 27, 1964, at Anchorage, Alaska*. U.S. Geological Survey Professional Paper 542-A, 68 pp.

Harrison, J.V., and N.L. Falcon. 1934. Collapse Structures. *Geological Magazine*, Vol. 71, No. 12, pp. 529–539.

Harrison, J.V., and N.L. Falcon. 1936. Gravity Collapse Structures and Mountain Ranges as Exemplified in South-western Persia. *Quarterly Journal of the Geological Society of London*, Vol. 92, pp. 91–102.

Heim, A. 1932. Bergsturz und Menschenleben. *Beiblatt zur Vierteljahrsschrift der Naturforschenden Gesellschaft in Zurich*, Vol. 77, pp. 1–217. Translated by N. Skermer under the title *Landslides and Human Lives*, BiTech Publishers, Vancouver, British Columbia, 1989, 195 pp.

Hocking, G. 1976. A Method for Distinguishing Between Single and Double Plane Sliding of Tetrahedral Wedges. *International Journal of Rock Mechanics and Mining Science*, Vol. 13, No. 7, pp. 225–226.

Hoek, E., and J.W. Bray. 1981. *Rock Slope Engineering*. Institution of Mining and Metallurgy, London, 358 pp.

Horswill, P., and A. Horton. 1976. Cambering and Valley Bulging in the Gwash Valley at Empingham, Rutland. *Philosophical Transactions of the Royal Society of London*, Series A, Vol. 283, No. 1315, pp. 427–462.

Hsu, K.J. 1975. Catastrophic Debris Streams (Sturzstroms) Generated by Rockfalls. *Bulletin of the Geological Society of America*, Vol. 86, No. 1, pp. 129–140.

Hu, X.Q., and D.M. Cruden. 1993. Buckling Deformation in the Highwood Pass, Alberta. *Canadian Geotechnical Journal*, Vol. 30, No. 2, pp. 276–286.

Huder, J. 1976. *Creep in Bundner Schist*. Norwegian Geotechnical Institute (Laurits Bjerrum Memorial Volume), Oslo, pp. 125–153.

Hungr, O., and S.G. Evans. 1988. Engineering Evaluation of Fragmental Rockfall Hazards. In *Proc., Fifth International Symposium on Landslides* (C. Bonnard, ed.), A.A. Balkema, Rotterdam, Netherlands, Vol. 1, pp. 685–690.

Hutchinson, J.N. 1967. The Free Degradation of London Clay Cliffs. In *Proc., Geotechnical Conference on Shear Strength Properties of Natural Soils and Rocks*, Norwegian Geotechnical Institute, Oslo, Vol. 1, pp. 113–118.

Hutchinson, J.[N.] 1968. Mass Movement. In *Encyclopedia of Geomorphology* (R.W. Fairbridge, ed.), Reinhold, New York, pp. 688–695.

Hutchinson, J.N. 1973. The Response of London Clay Cliffs to Differing Rates of Toe Erosion. *Geologia Applicata e Idrogeologia*, Vol. 8, pp. 221–239.

Hutchinson, J.N. 1983. Methods of Locating Slip Surfaces in Landslides. *Bulletin of the International Association of Engineering Geology*, Vol. 20, No. 3, pp. 235–252.

Hutchinson, J.N. 1988. General Report: Morphological and Geotechnical Parameters of Landslides in Relation to Geology and Hydrogeology. In *Proc., Fifth International Symposium on Landslides* (C. Bonnard, ed.), A.A. Balkema, Rotterdam, Netherlands, Vol. 1, pp. 3–35.

Hutchinson, J.N. 1991. *Periglacial and Slope Processes*. Engineering Geology Special Publication 7, Geological Society of London, pp. 283–331.

Hutchinson, J.N., and R.K. Bhandari. 1971. Undrained Loading, a Fundamental Mechanism of Mudflows and Other Mass Movements. *Geotechnique*, Vol. 21, pp. 353–358.

Hutchinson, J.N., and T.P. Gostelow. 1976. The Development of an Abandoned Cliff in London Clay at Hadleigh, Essex. *Philosophical Transactions of the Royal Society of London*, Series A, Vol. 283, pp. 557–604.

IAEG Commission on Landslides. 1990. Suggested Nomenclature for Landslides. *Bulletin of the International Association of Engineering Geology*, No. 41, pp. 13–16.

Jones, F.O., D.R. Embody, and W.L. Peterson. 1961. *Landslides Along the Columbia River Valley, Northeastern Washington*. U.S. Geological Survey Professional Paper 367, 98 pp.

Jordan, R.H. 1949. A Florida Landslide. *Journal of Geology*, Vol. 57, No. 4, pp. 418–419.

Keefer, D.K., and A.M. Johnson. 1983. *Earthflows: Morphology, Mobilization and Movement*. U.S. Geological Survey Professional Paper 1264, 56 pp.

Kostak, B., and D.M. Cruden. 1990. The Moiré Crack Gauges on the Crown of the Frank Slide. *Canadian Geotechnical Journal*, Vol. 27, No. 6, pp. 835–840.

Kovari, K., and P. Fritz. 1984. Recent Developments in the Analysis and Monitoring of Rock Slopes. In *Proc., Fourth International Symposium on Landslides*, Canadian Geotechnical Society, Toronto, Vol. 1, pp. 1–16.

Kozlovskii, E.A. (ed.). 1988. *Landslides and Mudflows.* UNESCO-UNEP, Moscow, 2 vols., 376 pp.

Krahn, J., and N.R. Morgenstern. 1979. The Ultimate Frictional Resistance of Rock Discontinuities. *International Journal of Rock Mechanics and Mining Science,* Vol. 16, No. 2, pp. 127–133.

Kyunttsel, V.V. 1988. Landslides. In *Landslides and Mudflows* (E.A. Kozlovskii, ed.), United Nations Environment Programme, UNESCO, Moscow, Vol. 1, pp. 35–54.

McCalpin, J. 1984. Preliminary Age Classification of Landslides for Inventory Mapping. In *Proc., 21st Engineering Geology and Soils Engineering Symposium,* University of Idaho, Moscow, pp. 99–120.

McConnell, R.G., and R.W. Brock. 1904. Report on the Great Landslide at Frank, Alberta. In *Annual Report for 1903,* Department of the Interior, Ottawa, Canada, Part 8, 17 pp.

McGill, G.E., and A.W. Stromquist. 1979. The Grabens of Canyonlands National Park, Utah: Geometry, Mechanics, and Kinematics. *Journal of Geophysical Research,* Vol. 84, No. B9, pp. 4547–4563.

McRoberts, E.C., and N.R. Morgenstern. 1974. Stability of Slopes in Frozen Soil, Mackenzie Valley, North West Territories. *Canadian Geotechnical Journal,* Vol. 11, No. 4, pp. 554–573.

Morgenstern, N.R. 1985. Geotechnical Aspects of Environmental Control. In *Proc., 11th International Conference on Soil Mechanics and Foundation Engineering,* A.A. Balkema, Rotterdam, Netherlands, Vol. 1, pp. 155–185.

Mueller, L. 1964. The Rock Slide in the Vaiont Valley. *Rock Mechanics and Engineering Geology,* Vol. 2, No. 3-4, pp. 148–212.

Nemčok, A., J. Pasek, and J. Rybár. 1972. Classification of Landslides and Other Mass Movements. *Rock Mechanics,* Vol. 4, pp. 71–78.

Nicoletti, P.G., and M. Sorriso-Valvo. 1991. Geomorphic Controls of the Shape and Mobility of Rock Avalanches. *Bulletin of the Geological Society of America,* Vol. 103, No. 10, pp. 1365–1373.

Ostaficzuk, S. 1973. Large-Scale Landslides in Northwestern Libya. *Acta Geologica Polonica,* Vol. 23, No. 2, pp. 231–244.

Palmer, L. 1977. Large Landslides of the Columbia River Gorge, Oregon and Washington. In *Landslides* (D.R. Coates, ed.), Reviews in Engineering Geology, Vol. 3, Geological Society of America, Boulder, Colo., pp. 69–83.

Panet, M. 1969. Discussion of K.W. John's Paper (ASCE Proc. Paper 5865, March 1968). *Journal of the Soil Mechanics and Foundation Division,* ASCE, Vol. 95, No. SM2, pp. 685–686.

Pierson, T.C., and J.E. Costa. 1987. A Rheologic Classification of Subaerial Sediment—Water Flows. In *Debris Flows/Avalanches: Process, Recognition, and Mitigation* (J.E. Costa and G.F. Wieczorek, eds.), Reviews in Engineering Geology, Vol. 7, Geological Society of America, Boulder, Colo., pp. 1–12.

Potter, D.B., and G.E. McGill. 1978. Valley Anticlines of the Needles District, Canyonlands National Park, Utah. *Bulletin of the Geological Society of America,* Vol. 89, No. 6, pp. 952–960.

Prior, D.B., N. Stephens, and D.R. Archer. 1968. Composite Mudflows on the Antrim Coast of North-East Ireland. *Geografiska Annaler,* Vol. 50A, No. 2, pp. 65–78.

Radbruch-Hall, D.H. 1978. Gravitational Creep of Rock Masses on Slopes. In *Rockslides and Avalanches* (B. Voight, ed.), Vol. 1: Natural Phenomena, Elsevier, Amsterdam, Netherlands, pp. 607–657.

Radbruch-Hall, D.H., D.J. Varnes, and W.Z. Savage. 1976. Gravitational Spreading of Steep-Sided Ridges ("sackung") in Western United States. *Bulletin of the International Association of Engineering Geology,* No. 14, pp. 23–35.

Ramsay, J.G. 1967. *Folding and Fracturing of Rocks.* McGraw-Hill, New York, 568 pp.

Ritchie, A.M. 1963. Evaluation of Rockfall and Its Control. In *Highway Research Record 17,* HRB, National Research Council, Washington, D.C., pp. 13–28.

Rybár, J., and J. Dobr. 1966. Fold Deformations in the North-Bohemian Coal Basins. *Sbornik geologickych věd,* Rada HIG, No. 5, pp. 133–140.

Shaller, P.J. 1991. Analysis of a Large, Moist Landslide, Lost River Range, Idaho, U.S.A. *Canadian Geotechnical Journal,* Vol. 28, No. 4, pp. 584–600.

Shelton, J.S. 1966. *Geology Illustrated.* W.H. Freeman and Co., San Francisco, 434 pp.

Shreve, R.L. 1968. *The Blackhawk Landslide.* Special Paper 108. Geological Society of America, Boulder, Colo., 47 pp.

Shroder, J.F. 1971. *Landslides of Utah.* Bulletin 90. Utah Geological and Mineralogical Survey, 50 pp.

Skempton, A.W. 1970. First-Time Slides in Over-Consolidated Clays. *Geotechnique,* Vol. 20, No. 3, pp. 320–324.

Skempton, A.W., and J.N. Hutchinson. 1969. Stability of Natural Slopes and Embankment Foundations. In *Proc., Seventh International Conference on Soil Mechanics and Foundation Engineering,* Sociedad Mexicana de Mecána de Suelos, Mexico City, State of the Art Volume, pp. 291–340.

Skempton, A.W., and A.G. Weeks. 1976. The Quaternary History of the Lower Greensand Escarpment and Weald Clay Vale near Sevenoaks, Kent. *Philosophical Transactions of the Royal Society of London,* Series A, Vol. 283, pp. 493–526.

Swaminathan, C.G., ed. 1980. *Proc., Third International Symposium on Landslides*, Sarita Prakashan, New Delhi, India, 3 vols., 927 pp.

Tavenas, F., J.Y. Chagnon, and P. LaRochelle. 1971. The Saint-Jean Vianney Landslide: Observations and Eyewitness Accounts. *Canadian Geotechnical Journal*, Vol. 8, No. 3, pp. 463–478.

Ter-Stepanian, G. 1980. Measuring Displacements of Wooded Landslides with Trilateral Signs. In *Proc., Third International Symposium on Landslides* (C.G. Swaminathan, ed.), Sarita Prakashan, New Delhi, India, Vol. 1, pp. 355–361.

Terzaghi, K. 1950. Mechanism of Landslides. In *Application of Geology to Engineering Practice* (S. Paige, ed.), Geological Society of America, New York, pp. 83–123.

Terzaghi, K., and R.B. Peck. 1948. *Soil Mechanics in Engineering Practice*. John Wiley & Sons, New York, 566 pp.

Thomson, S., and D.W. Hayley. 1975. The Little Smoky Landslide. *Canadian Geotechnical Journal*, Vol. 12, No. 3, pp. 379–392.

Van Gassen, W., and D.M. Cruden. 1989. Momentum Transfer and Friction in the Debris of Rock Avalanches. *Canadian Geotechnical Journal*, Vol. 26, No. 4, pp. 623–628.

Van Horn, R. 1975. Largest Known Landslide of Its Type in the United States—A Failure by Lateral Spreading in Davis County, Utah. *Utah Geology*, Vol. 2, pp. 82–87.

Varnes, D.J. 1958. Landslide Types and Processes. In *Special Report 29: Landslides and Engineering Practice* (E.B. Eckel, ed.), HRB, National Research Council, Washington, D.C., pp. 20–47.

Varnes, D.J. 1978. Slope Movement Types and Processes. In *Special Report 176: Landslides: Analysis and Control* (R.L. Schuster and R.J. Krizek, eds.), TRB, National Research Council, Washington, D.C., pp. 11–33.

Varnes, D.J. 1984. *Landslide Hazard Zonation: A Review of Principles and Practice*. UNESCO, Paris, 63 pp.

Varnes, D.J., D.H. Radbruch-Hall, and W.Z. Savage. 1989. *Topographic and Structural Conditions in Areas of Gravitational Spreading of Ridges in the Western United States*. U.S. Geological Survey Professional Paper 1496, 28 pp.

Varnes, D.J., D.H. Radbruch-Hall, K.L. Varnes, W.K. Smith, and W.Z. Savage. 1990. *Measurements of Ridge-Spreading Movements (Sackungen) at Bald Eagle Mountain, Lake County, Colorado, 1975–1989*. U.S. Geological Survey Open File Report 90-543, 13 pp.

Varnes, H.D. 1949. *Landslide Problems of Southwestern Colorado*. U.S. Geological Survey Circular 31, 13 pp.

Voight, B. 1990. The 1985 Nevado del Ruiz Volcano Catastrophe: Anatomy and Retrospection. *Journal of Volcanology and Geothermal Research*, Vol. 44, No. 1-2, pp. 349–386.

Walton, G., and T. Atkinson. 1978. Some Geotechnical Considerations in the Planning of Surface Coal Mines. *Transactions of the Institution of Mining and Metallurgy*, Vol. 87, pp. A147–A171.

Ward, W.H. 1948. A Coastal Landslip. In *Proc., Second International Conference on Soil Mechanics and Foundation Engineering*, International Society of Soil Mechanics and Foundation Engineering, Rotterdam, Vol. 2, pp. 33–38.

Wieczorek, G.F. 1984. Preparing a Detailed Landslide-Inventory Map for Hazard Evaluation and Reduction. *Bulletin of the Association of Engineering Geologists*, Vol. 21, No. 3, pp. 337–342.

Wilson, S.D. 1970. Observational Data on Ground Movements Related to Slope Instability. *Journal of the Soil Mechanics and Foundations Division*, ASCE, Vol. 96, No. SM4, pp. 1521–1544.

Wilson, S.D., and P.E. Mikkelsen. 1978. Field Instrumentation. In *Special Report 176: Landslides: Analysis and Control* (R.L. Schuster and R.J. Krizek, eds.), TRB, National Research Council, Washington, D.C., pp. 112–138.

WP/WLI. 1990. A Suggested Method for Reporting a Landslide. *Bulletin of the International Association of Engineering Geology*, No. 41, pp. 5–12.

WP/WLI. 1991. A Suggested Method for a Landslide Summary. *Bulletin of the International Association of Engineering Geology*, No. 43, pp. 101–110.

WP/WLI. 1993a. A Suggested Method for Describing the Activity of a Landslide. *Bulletin of the International Association of Engineering Geology*, No. 47, pp. 53–57.

WP/WLI. 1993b. *Multilingual Landslide Glossary*. Bi-Tech Publishers, Richmond, British Columbia, Canada, 59 pp.

Zaruba, Q., and V. Mencl. 1969. *Landslides and Their Control*, 1st ed. Elsevier, Amsterdam, Netherlands, 205 pp.

Zaruba, Q., and V. Mencl. 1982. *Landslides and Their Control*, 2nd ed. Elsevier, Amsterdam, Netherlands, 324 pp.

Zischinsky, U. 1966. On the Deformation of High Slopes. In *Proc., First Congress of the International Society for Rock Mechanics*, Lisbon, Sept. 25–1 Oct., 1966, International Society for Rock Mechanics, Lisbon, Portugal, Vol. 2, pp. 179–185.

Chapter 4

GERALD F. WIECZOREK

LANDSLIDE TRIGGERING MECHANISMS

1. INTRODUCTION

Landslides can have several causes, including geological, morphological, physical, and human (Alexander 1992; Cruden and Varnes, Chap. 3 in this report, p. 70), but only one trigger (Varnes 1978, 26). By definition a trigger is an external stimulus such as intense rainfall, earthquake shaking, volcanic eruption, storm waves, or rapid stream erosion that causes a near-immediate response in the form of a landslide by rapidly increasing the stresses or by reducing the strength of slope materials. In some cases landslides may occur without an apparent attributable trigger because of a variety or combination of causes, such as chemical or physical weathering of materials, that gradually bring the slope to failure. The requisite short time frame of cause and effect is the critical element in the identification of a landslide trigger.

The most common natural landslide triggers are described in this chapter, including intense rainfall, rapid snowmelt, water-level change, volcanic eruption, and earthquake shaking, and examples are provided in which observations or measurements have documented the relationship between triggers and landslides. Some geologic conditions that lead to susceptibility to landsliding caused by these triggers are identified. Human activities that trigger landslides, such as excavation for road cuts and irrigation, are not discussed in this chapter. To the extent possible, examples have been selected that illustrate landslide damage to transportation systems.

2. INTENSE RAINFALL

Storms that produce intense rainfall for periods as short as several hours or have a more moderate intensity lasting several days have triggered abundant landslides in many regions, for example, California (Figures 4-1, 4-2, and 4-3). Well-documented studies that have revealed a close relationship between rainfall intensity and activation of landslides include those from California (Campbell 1975; Ellen et al. 1988), North Carolina (Gryta and Bartholomew 1983; Neary and Swift 1987), Virginia (Kochel 1987; Gryta and Bartholomew 1989; Jacobson et al. 1989), Puerto Rico (Jibson 1989; Simon et al. 1990; Larsen and Torres Sanchez 1992), and Hawaii (Wilson et al. 1992; Ellen et al. 1993).

These studies show that shallow landslides in soils and weathered rock often are generated on steep slopes during the more intense parts of a storm, and thresholds of combined intensity and duration may be necessary to trigger them. In the Santa Monica Mountains of southern California, Campbell (1975) found that rainfall exceeding a threshold of 6.35 mm/hr triggered shallow landslides that led to damaging debris flows (Figure 4-4).

During 1982 intense rainfall lasting for about 32 hr in the San Francisco Bay region of California triggered more than 18,000 predominantly shallow landslides involving soil and weathered rock, which blocked many primary and secondary roads (Ellen et al. 1988). Those landslides whose times

FIGURE 4-1
Landslide blocking
State Highway 1 near
Julia Pfeiffer-Burns
State Park, California:
debris slides and
flows from toe and
flanks of reactivated
landslide. Intense
storm February
28–March 1, 1983,
triggered many
debris flows that
blocked primary
and secondary
roads along Big Sur
coastline.
G.F. WIECZOREK, MARCH 25,
1983

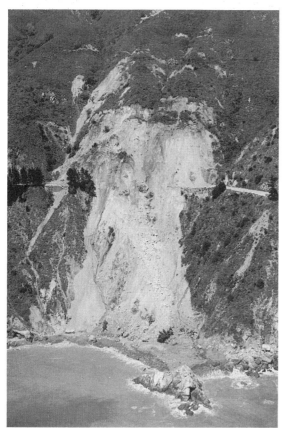

FIGURE 4-3
Landslide blocking
State Highway 1:
excavation of cut of
estimated 6.1 million
m³ in 6-m-wide
benches extending
about 300 m
above roadbed
made this the largest
highway repair job
undertaken in
California history.
Highway reopened
in April 1984
(Works 1984).
CALIFORNIA DEPARTMENT
OF TRANSPORTATION

FIGURE 4-2 *(above left)*
Landslide blocking State Highway 1: May 1, 1983, massive rock slide of 1.2 million m³ incorporated entire
hillside. Exceptionally heavy rainfall during winters of 1981–1982 and 1982–1983 was responsible for raising
groundwater levels and triggering slide. During excavation, groundwater flow of approximately 378,000 L/day
was collected and drained from cut (Works 1984). CALIFORNIA DEPARTMENT OF TRANSPORTATION

FIGURE 4-4
Cumulative rainfall at selected recording gauges in Santa Monica and San Gabriel Mountains, southern California. Known times of debris flows indicated by heavy dots. Steepness of cumulative rainfall line indicates intensity of rainfall (modified from Campbell 1975).

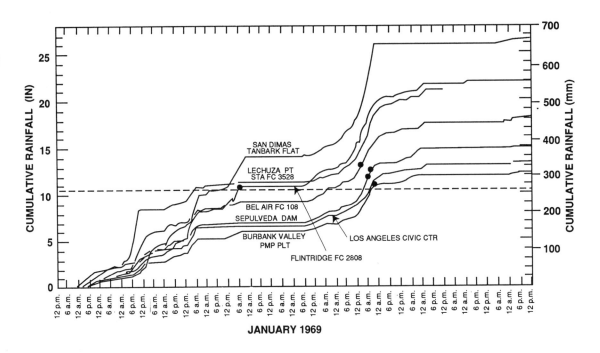

FIGURE 4-5
Rainfall thresholds that triggered abundant landslides in San Francisco Bay region, California. Thresholds for high and low mean annual precipitation (MAP) areas are indicated as curves representing combination of rainfall intensity and duration (modified from Cannon and Ellen 1985).

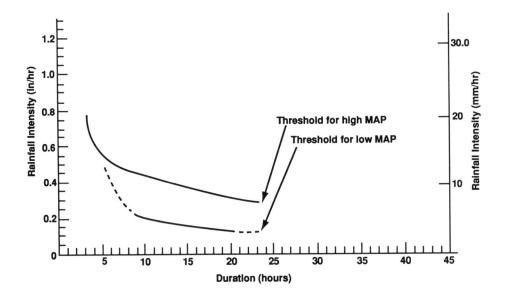

could be well documented were closely associated with periods of most intense precipitation; this documentation permitted identification of landslide-triggering rainfall thresholds based on both rainfall intensity and duration (Figure 4-5) (Cannon and Ellen 1985). Such thresholds are regional, depending on local geologic, geomorphic, and climatologic conditions.

The rapid infiltration of rainfall, causing soil saturation and a temporary rise in pore-water pres-

sures, is generally believed to be the mechanism by which most shallow landslides are generated during storms. With the advent of improved instrumentation and electronic monitoring devices, transient elevated pore pressures have been measured in hillside soils and shallow bedrock during rainstorms associated with abundant shallow landsliding (Figures 4-6 and 4-7) (Sidle 1984; Wilson and Dietrich 1987; Reid et al. 1988; Wilson 1989; Johnson and Sitar 1990; Simon et al. 1990).

Loose or weak soils are especially prone to landslides triggered by intense rainfall. Wildfire may produce a water-repellent (hydrophobic) soil layer below and parallel to the burned surface that, together with loss of vegetative cover, promotes raveling of loose coarse soil grains and fragments at the surface. Increased overland flow and rill formation then lead to small debris flows (Wells 1987). On the lower parts of hill slopes and in stream channels, major storms generate high sediment content in streams (hyperconcentrated flows) or large debris flows (Scott 1971; Wells et al. 1987; Weirich 1989; Florsheim et al. 1991).

Shortly after midnight on January 1, 1934, an intense downpour after more than 12 hr of rainfall resulted in debris flows from several recently burned canyons into the La Cañada Valley of southern California and caused significant property damage and loss of life (Troxell and Peterson 1937). Following an August 1972 wildfire north of Big Sur in central coastal California, storms with intensities of 19 to 22 mm/hr triggered two episodes of debris flows. During the second, more devastating storm on November 15, 1972, large debris flows reached Big Sur about 15 min after intense (22-mm/hr) rain (Johnson 1984). Debris flows blocked California State Highway 1 with mud, boulders, and vegetative debris; the flows partly buried, heavily damaged, or leveled structures and caused one fatality (Jackson 1977).

In arid regions, intense storms can trigger debris flows in thin loose soils on hillsides or in alluvium in stream channels (Woolley 1946; Jahns 1949; Johnson 1984). On September 14, 1974, an intense thunderstorm passed over Eldorado Canyon, Nevada, and although the duration of the rainfall was short (generally less than an hour), the intensities were very high—from 76 to 152 mm/hr for 30 min. The intense rain eroded shallow soils, leaving rills on some of the sparsely vegetated hillsides, and the high runoff scoured unconsolidated alluvium from the larger stream channels. The initial debris-flow surge, heavily laden with sediment and with the consistency of fresh concrete, emerged from the canyon with a high steep front, damaging a marina and killing at least nine people (Glancy and Harmsen 1975).

On June 18, 1982, a very intense thunderstorm occurred over a recently burned steep drainage of the South Fork American River in California between the towns of Kyburz and Strawberry. In

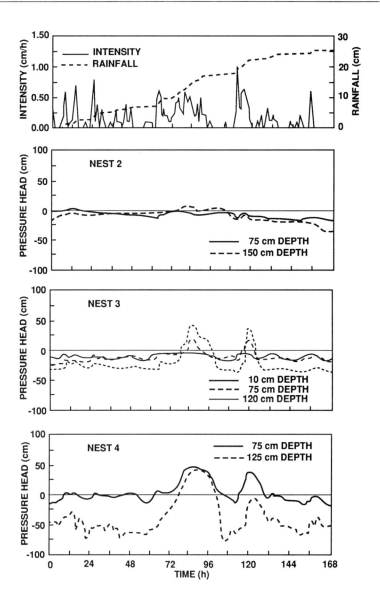

August 1981 a wildfire had removed all vegetation, exposed bare soil, and converted 15 percent of the burned area to a hydrophobic condition; by June 1982 reseeded grasses were establishing themselves because of the wet winter of 1981–1982. In a recording gauge 1.2 km away, rainfall of 46 mm in 6 min, 76 mm in 18 min, and 101 mm in 27 min was measured during the height of the storm. This intense rain triggered a debris flow by sheet and rill erosion from shallow soils and from erosion of alluvium within tributary gullies as well as the main gully. The resulting debris flow and flood closed California State Highway 50 for 5 hr while maintenance crews removed rocky debris from the pavement (Kuehn 1987).

FIGURE 4-6
Response of pore pressure to rainfall in shallow hillside soils of northern California. Positive peaks of pore pressure correspond to periods of high rainfall intensity; negative pore pressures indicate soil tension in partly saturated soil at beginning of storm or during periods between rainfall (modified from Johnson and Sitar 1990).

FIGURE 4-7
(*a*) Acceleration time histories and (*b*) response of pore pressure in liquefied silty sand layer from November 1987 Superstition Hills (California) earthquake. Acceleration time histories were recorded at ground surface and beneath liquefied layer. Piezometers PI, P2, P3, and P5 are in liquefied silty sand layer; P6 is in silt layer that did not liquefy (modified from Holzer et al. 1989).

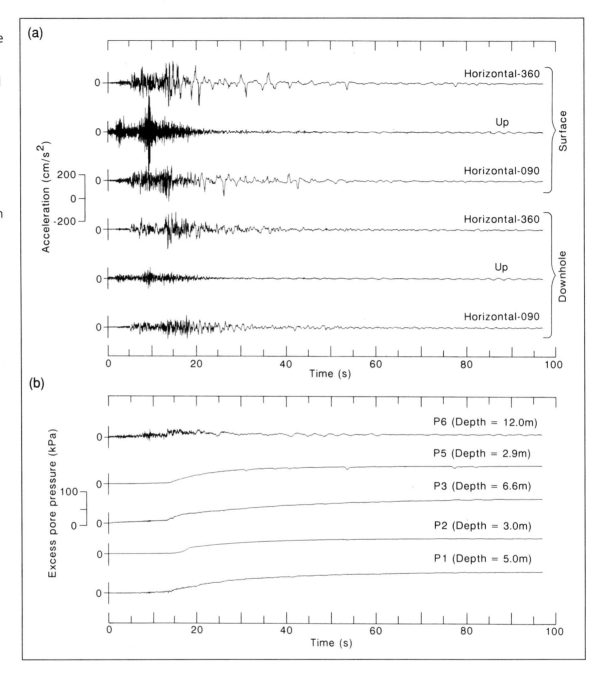

On September 7, 1991, a debris flow triggered by heavy rainfall (63 to 213 mm) within a 24-hr period damaged several houses in a subdivision of North Ogden, Utah. Concentration of runoff from the storm mobilized talus and other debris in tributary channels and scoured material from the main channel into a debris flow, which emerged from the canyon and traveled about 400 m down an alluvial fan before reaching the subdivision (Mulvey and Lowe 1992).

3. RAPID SNOWMELT

Rapid melting of a snowpack caused by sudden warming spells or by rain falling on snow can add water to hillside soils. Horton (1938) examined the infiltration and runoff of melting snow into soil, including the special case of the effects of rain on snow cover. He found that the process of melting may provide a more continuous supply of moisture over a longer time period compared with the usual duration of infiltration from rain. Snowmelt may also recharge shallow fractured bedrock and raise pore-water pressures beneath shallow soils, thus triggering landslides (Mathewson et al. 1990).

Near Wrightwood, California, a steady thaw of a heavy snowpack over a 40-day period in the spring of 1969 triggered mud flows in Heath Creek from saturated debris in steep channels and from steep faces in the toe area of the Wright Mountain landslide (Morton et al. 1979). In Utah during an unusually warm 10-day period from late May to early June 1983, a heavy winter snowpack along the Wasatch Front began to melt rapidly and triggered approximately 150 debris flows and other types of landslides (Pack 1984; Wieczorek et al. 1989). In the Wasatch Front above Farmington, Utah, during the height of this activity (May 28–30, 1983), snowmelt provided the equivalent of approximately 2.1 to 2.6 mm/hr of precipitation; on May 30, 1983, a large debris flow emerged from the canyon of Rudd Creek into the community of Farmington (Vandre 1985).

Rain-on-snow events commonly reduce the water content of the snowpack and add sufficient water to soils to be significant in triggering landslides. In coastal Alaska, Sidle (1984) found that snowmelt before rainfall augmented the piezometric response. In a small watershed of western Oregon, Harr (1981) found that 85 percent of landslides that could be dated accurately were associated with snowmelt during rainfall.

A majority of the documented landslides in the central Sierra Nevada of California in mid-April 1982 and in early and mid-March 1983 occurred during rain-on-snow events (Bergman 1987). Landslides along Stump Springs Road, a major timber-haul route in Sierra National Forest, California, were triggered by a rain-on-snow event that included peak rainfall intensities of 14 to 18 mm/hr supplemented by snowpack losses equivalent to 130 mm of water. Landslide repairs of Stump Springs Road required an estimated $1.3 million along a 23-km section during 1982 and 1983 (DeGraff et al. 1984).

4. WATER-LEVEL CHANGE

The sudden lowering of the water level (rapid drawdown) against a slope can trigger landslides in earth dams, along coastlines, and on the banks of lakes, reservoirs, canals, and rivers. Rapid draw-

down can occur when a river drops following a flood stage, the water level in a reservoir or canal is dropped suddenly, or the sea level drops following a storm tide. Unless pore pressures within the slope adjacent to the falling water level can dissipate quickly, the slope is subjected to higher shear stresses and potential instability (Figure 4-8) (Terzaghi 1943; Lambe and Whitman 1969). In terms of effective stress, Bishop (1954, 1955) introduced a method to estimate the pore-water pressure in terms of reduction of the principal stresses and to analyze slope stability due to the removal of the water load during rapid drawdown.

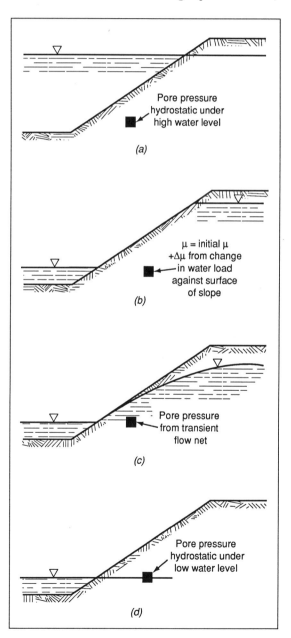

FIGURE 4-8
Response of slope to rapid drawdown: (a) initial equilibrium condition, (b) after drawdown but before consolidation adjustment, (c) after consolidation adjustment, and (d) final equilibrium condition (Lambe and Whitman 1969).

Thick uniform deposits of low-permeability clays and silts are particularly susceptible to landsliding triggered by rapid drawdown. Morgenstern (1963) listed 16 cases in which rapid drawdown triggered landslides in the upstream face of earth dams.

Rapid drawdown triggered four landslides in very low-permeability boulder clay in the Fort Henry and Ardclooney embankments, Ireland. The best documented of these slides occurred after a drawdown of 1.1 m in 10 days; during the last 24 hr the average drawdown rate was 0.35 m/day (Massarsch et al. 1987). In the coastal area of Zeeland, Netherlands, Koppejan et al. (1948) observed that excessive tidal differences of 2.8 to 4.6 m during spring or coinciding with gales triggered wet sand flows. From a few observations, they concluded that movement started during drawdown of the ebb tide between half tide and low water.

Springer et al. (1985) inspected more than 6500 km of the Ohio River system and examined 120 landslide sites in detail. They observed several characteristic types of instabilities, including massive slumps evidently triggered by rapid drops in river level following floods. Other landslides, cohesive wedges of material sliding on thin sand strata, were triggered by recent precipitation that produced high water pressures in tension cracks behind the free face and were not associated with rapid drawdown.

During and following construction of Grand Coulee Dam in Washington State, some 500 landslides were noted between 1941 and 1953 along the shores of Franklin D. Roosevelt Lake. Accurately dated landslides among this sample were most frequent during the filling stage of the reservoir and subsequent to filling during two periods of rapid drawdown (Jones et al. 1961). Even larger drawdowns during the period from 1969 to 1975 were responsible for additional earth slumps, earth spreads, earth flows, and debris flows (Schuster 1979).

Increases in groundwater levels on hill slopes following periods of prolonged above-normal precipitation or during the raising of water levels in rivers, lakes, reservoirs, and canals build up porewater pressure and reduce effective strength of saturated slope materials and can trigger landslides (Figures 4-2 and 4-9). The initial filling of Yellowtail Reservoir, Montana, and also of the Panama Canal were cited by Lane (1967) as examples in which large landslides were triggered by initial raising of the water levels on natural or cut slopes. Rising groundwater levels can also acceler-

FIGURE 4-9
Blucher Valley landslide, approximately 6 km south of Sebastopol, California, began moving after series of 1983 winter storms. Deep-seated (>18 m deep) translational earth block slide on nose of spur ridge moved along bedding planes inclined at only 5 to 8 degrees. Cumulative seasonal rainfall during 1981–1982 and 1982–1983 preceding landslide had been highest recorded historically. Triggering of landslide attributed to high groundwater levels in cracks and grabens (Spittler 1983).

ate landslide movement, as observed at Vaiont Dam, Italy, where a slowly moving landslide rapidly accelerated during the weeks following the initial filling of the reservoir (Lane 1967).

The Mayunmarca landslide of April 25, 1974, blocked the Mantaro River in Peru, and the rising water level behind the dam caused by the landslide resulted in more landslides along the shores of the lake, which destroyed a regional highway (Lee and Duncan 1975). Sudden breaching of the landslide dam and rapid drawdown of the lake level triggered still more landslides along the banks of the lake (R.L. Schuster, personal communication, 1992, U.S. Geological Survey, Denver, Colorado).

There are other examples in which gradually rising groundwater levels caused by irrigation and prolonged or intermittent low- to moderate-intensity rainfall have resulted in landslides. These cases are not cited because the relation of trigger and landsliding is not as closely documented with respect to time as it is for those cases described here, which involve more rapid changes in water levels.

5. VOLCANIC ERUPTION

Deposition of loose volcanic ash on hillsides commonly is followed by accelerated erosion and frequent mud or debris flows triggered by intense rainfall (Kadomura et al. 1983). Irazu, a volcano in central Costa Rica, erupted ash almost continuously from March 1963 through February 1965. Intense rain and high runoff accompanied by sheet and rill erosion of ash-covered slopes triggered more than 90 debris flows in valleys on slopes of this volcano. A large debris flow in the Rio Reventado valley destroyed more than 300 homes and killed more than 20 persons. High runoff and debris flows incised deep channels, resulting in slumping and caving of valley walls and reactivation of landslides, which in turn supplied additional material for debris flows (Waldron 1967).

Following the June 1991 eruption of Mt. Pinatubo in the Philippines, monsoon and typhoon rains triggered many debris flows that originated in thick volcanic-ash deposits (Pierson 1992). Debris flows as deep as 5 m traveled down major channels; during the most rainy periods, three to five debris flows a day were common. Most debris flows were triggered by monsoonal rainstorms with intensities that seldom exceeded 80 to 100 mm over several hours. In addition to disrupting natural drainage patterns, causing lateral migration of river chan-

nels, and inundating agricultural land, debris flows have destroyed most major highway bridges near the volcano (Pierson 1992).

Volcanic eruptions have triggered some of the largest and most catastrophic historic landslides. As a result of the May 18, 1980, eruption of Mount St. Helens, Washington, a massive 2.8-km^3 rock slide–debris avalanche rapidly descended from the north slope of Mount St. Helens and traveled about 22 km down the valley of the North Fork Toutle River (Voight et al. 1983). The avalanche destroyed nine bridges and many kilometers of highways and roads. As a result of rapid melting of snow and ice from the eruption, mud flows surged down several of the valleys that radiated from the mountain. The largest and most destructive of these mud flows entered the valleys of the North Fork and South Fork Toutle River and destroyed or heavily damaged about 200 homes, buried half of the 27-km portion of State Highway 504 and other highways and roads, destroyed 27 km of railway, and destroyed or badly damaged 27 highway and railroad bridges (Figure 4-10) (Schuster 1981).

On November 13, 1985, pyroclastic flows and surges from a relatively small eruption melted snow and ice on the summit of Nevado del Ruiz volcano in Colombia and produced large volumes of meltwater, initiating debris flows in steep channels that swept down and killed more than 23,000 inhabitants of Armero and other areas at or beyond the base of the volcano (Pierson et al. 1990).

6. EARTHQUAKE SHAKING

Strong ground shaking during earthquakes has triggered landslides in many different topographic

FIGURE 4-10
St. Helens Bridge, 75-m steel bridge on State Highway 504 carried about 1/2 km downstream and partially buried by mud flow in 1980.
ROBERT L. SCHUSTER

and geologic settings. Rock falls, soil slides, and rock slides from steep slopes, involving relatively thin or shallow disaggregated soils or rock, or both, have been the most abundant types of landslides triggered by historical earthquakes (Figures 4-11 and 4-12). Earth spreads, earth slumps, earth block slides, and earth avalanches on gentler slopes have also been very abundant in earthquakes (Keefer 1984).

For 40 historic earthquakes, Keefer (1984) determined the maximum distance from epicenter to landslides as a function of magnitude for three general landslide types (Figure 4-13). Using the expected farthest limits of landsliding during

FIGURE 4-11
Rock slide–avalanche onto Sherman Glacier triggered by March 1964 Alaska earthquake: Sherman Glacier on August 26, 1963, showing conditions before earthquake (Post 1967).

FIGURE 4-12
(bottom left)
Rock slide–avalanche onto Sherman Glacier triggered by March 1964 Alaska earthquake: collapse of Shattered Peak in middle distance formed avalanche (Post 1967).

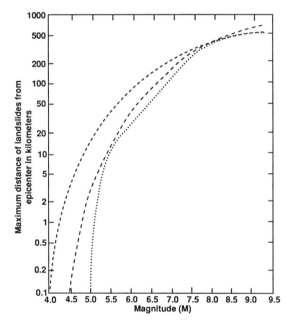

FIGURE 4-13 (*above*)
Maximum distance to landslides from epicenter for earthquakes of different magnitudes. - - - -, disrupted falls and slides; – – –, bound for coherent slides; . . . ., bound for spreads and flows (Keefer 1984).

earthquakes of specific magnitude and location, an outer distance limit for landsliding was prepared for a hypothetical earthquake in the Los Angeles region (Figure 4-14) (Harp and Keefer 1985; Wilson and Keefer 1985). The amount of landslide displacement during an earthquake is a critical factor in hazard assessment; a seismic analysis of earth dams (Newmark 1965) was modified to calculate the displacement of individual landslides on the

basis of records of strong ground shaking (Wilson and Keefer 1983; Jibson 1993).

Landslides involving loose, saturated, cohesionless soils on low to moderate slopes commonly occur as a result of earthquake-induced liquefaction, a process in which shaking temporarily raises pore-water pressures and reduces the strength of the soil (Figure 4-15). Sedimentary environment, age of deposition, geologic history, depth of water

FIGURE 4-14
Map of Los Angeles basin showing predicted limit for coherent landslides from hypothetical **M** 6.5 earthquake with epicenter on northern Newport-Inglewood fault zone (*solid straight line*) (modified from Wilson and Keefer 1985).

FIGURE 4-15
State Highway 1
bridge destroyed
by strong shaking
and liquefaction
of river deposits
at Struve Slough
near Watsonville,
California, during
1989 Loma Prieta
earthquake (Plafker
and Galloway
1989).

table, grain-size distribution, density, and depth determine whether a deposit will liquefy during an earthquake. Generally, cohesionless sediments of Holocene age or younger below the groundwater table are most susceptible to liquefaction (Youd and Perkins 1978).

The May 31, 1970, Richter magnitude (**M**) 7.7 Peru earthquake was the most catastrophic historic earthquake of the Western Hemisphere, causing over 40,000 deaths. The earthquake triggered a huge debris avalanche from the north peak of Huascaran Mountain that buried the town of Yungay and part of the town of Ranrahirca with a loss of more than 18,000 lives. The earthquake also triggered many other landslides within a 30,000-km² area that disrupted communities and temporarily blocked roads; these slides seriously hampered rescue and relief operations and kept the full extent of the disaster unknown until weeks after the earthquake (Plafker et al. 1971).

The **M** 7.5 Guatemala earthquake of February 4, 1976, triggered more than 10,000 landslides, predominantly rock falls and debris slides from steep slopes of Pleistocene pumice deposits (tephras and ash flows) or their residual soils (Harp et al. 1981). Pumice deposits, which stand in steep, near-vertical slopes, lose much of their strength during seismic loading. Strong shaking increases stresses that may break down cohesion in cemented soils or brittle rocks, such as tephra, loess, or sandstone (Sitar and Clough 1983).

On March 5, 1987, two earthquakes (**M** 6.1 and **M** 6.9) 100 km east of Quito, Ecuador, triggered thousands of rock and earth slides, debris avalanches, and debris and mud flows that destroyed nearly 70 km of the Trans-Ecuadorian oil pipeline

and the only highway from Quito to Ecuador's eastern rain forests and oil fields (Crespo et al. 1991; Schuster 1991). Economic losses, principally from landslide-induced damage to the oil pipeline and highway, were estimated to be U.S. $1.5 billion (Nieto and Schuster 1988).

On November 12, 1987, liquefaction of a silt and sand layer during an **M** 6.6 earthquake in Superstition Hills, California, caused sand boils to erupt and resulted in extensive ground cracking indicative of an earth spread. Nearby instrumentation recorded excess pore pressures that began to develop when the peak horizontal acceleration reached 0.21 *g* about 13.6 sec after the earthquake began (Figure 4-7) (Holzer et al. 1989). The pore-pressure buildup was high enough to be the main factor in reducing soil strength and causing the earth spread.

The **M** 7.1 Loma Prieta, California, earthquake of October 17, 1989, triggered an estimated 2,000 to 4,000 rock, earth, and debris falls and slides that blocked a major highway and many secondary roads in the San Francisco–Monterey Bay areas. A debris slide of about 6000 m³ closed the two northbound lanes of California State Highway 17 for 33 days before repairs were completed (Plafker and Galloway 1989; Keefer and Manson in press).

The Loma Prieta earthquake also caused liquefaction and earth spreads between San Francisco and Monterey, including damage to the runways at Oakland International Airport and the Alameda Naval Air Station (Plafker and Galloway 1989; Seed et al. 1990). Numerous earth spreads (about 46) destroyed or disrupted flood-control levees, pipelines, bridge abutments and piers, roads, houses and utilities, and irrigation works in the Monterey Bay area (Plafker and Galloway 1989; Tinsley and Dupre 1993).

7. SUMMARY

Common landslide triggers, including intense rainfall, rapid snowmelt, water-level changes, volcanic eruptions, and strong ground shaking during earthquakes, are probably directly responsible for the majority of landslides worldwide. As illustrated by the foregoing examples, these landslides are responsible for much damage to transportation systems, utilities, and lifelines. These landslide triggers have been well documented, and recent monitoring has provided considerable insight into the mechanics of the triggering processes.

REFERENCES

Alexander, D. 1992. On the Causes of Landslides: Human Activities, Perception, and Natural Processes. *Environmental Geology and Water Sciences*, Vol. 20, No. 3, pp. 165–179.

Bergman, J.A. 1987. Rain-on-snow and Soil Mass Failure in the Sierra Nevada of California. In *Landslide Activity in the Sierra Nevada during 1982 and 1983* (J.V. DeGraff, ed.), Earth Resources Monograph 12, USDA Forest Service, Pacific Southwest Region, San Francisco, pp. 15–26.

Bishop, A.W. 1954. The Use of Pore-Pressure Coefficients in Practice. *Geotechnique*, Vol. 4, No. 4, pp. 148–152.

Bishop, A.W. 1955. The Use of the Slip Circle in the Stability Analysis of Slopes. *Geotechnique* (Proceedings of the Conference on the Stability of Earth Slopes, Stockholm, Sept. 20–25, 1954), Vol. 5, pp. 7–17.

Campbell, R.H. 1975. *Soil Slips, Debris Flows, and Rainstorms in the Santa Monica Mountains and Vicinity, Southern California*. U.S. Geological Survey Professional Paper 851, 51 pp.

Cannon, S.H., and S.D. Ellen. 1985. Rainfall Conditions for Abundant Debris Avalanches, San Francisco Bay Region, California. *California Geology*, Vol. 38, No. 12, pp. 267–272.

Crespo, E., T.D. O'Rourke, and K.J. Nyman. 1991. Effects on Lifelines. In *The March 5, 1987, Ecuador Earthquakes* (R.L. Schuster, ed.), Natural Disaster Studies, Vol. 5, Chap. 6, Committee on Natural Disasters, National Research Council, Washington, D.C., pp. 83–99.

DeGraff, J.V., J. McKean, P.E. Watanabe, and W.F. McCaffery. 1984. Landslide Activity and Groundwater Conditions: Insights from a Road in the Central Sierra Nevada, California. In *Transportation Research Record 965*, TRB, National Research Council, Washington, D.C., pp. 32–37.

Ellen, S.D., G.F. Wieczorek, W.M. Brown III, and D.G. Herd. 1988. Introduction. In *Landslides, Floods, and Marine Effects of the Storm of January 3–5, 1982, in the San Francisco Bay Region, California* (S.D. Ellen and G.F. Wieczorek, eds.), U.S. Geological Survey Professional Paper 1434, pp. 1–5.

Ellen, S.D., R.K. Mark, S.H. Cannon, and D.L. Knifong. 1993. *Map of Debris-flow Hazard in the Honolulu District of Oahu, Hawaii*. U.S. Geological Survey Open-File Report 93-213, 25 pp.

Florsheim, J.L., E.A. Keller, and D.W. Best. 1991. Fluvial Sediment Transport in Response to Moderate Storm Flows Following Chaparral Wildfire, Ventura County, Southern California. *Geological Society of America Bulletin*, Vol. 103, pp. 504–511.

Glancy, P.A., and L. Harmsen. 1975. *A Hydrologic Assessment of the September 14, 1974, Flood in Eldorado Canyon, Nevada*. U.S. Geological Survey Professional Paper 930, 28 pp.

Gryta, J.J., and M.J. Bartholomew. 1983. Debris-Avalanche Type Features in Watauga County, North Carolina. In *Geologic Investigations in the Blue Ridge of Northwestern North Carolina* (S.E. Lewis, ed.), Carolina Geological Society Guidebook, North Carolina Division of Land Resources, Boone, Article 5, 22 pp.

Gryta, J.J., and M.J. Bartholomew. 1989. Factors Influencing the Distribution of Debris Avalanches Associated with the 1969 Hurricane Camille in Nelson County, Virginia. In *Landslide Processes of the Eastern United States and Puerto Rico* (A. P. Schultz and R.W. Jibson, eds.), Geological Society of America Special Paper 236, pp. 15–28.

Harp, E.L., R.C. Wilson, and G.F. Wieczorek. 1981. *Landslides from the February 4, 1976, Guatemala Earthquake*. U.S. Geological Survey Professional Paper 1204-A, 35 pp.

Harp, E.L., and D.K. Keefer. 1985. Landsliding—Predicted Geologic and Seismologic Effects of a Postulated Magnitude 6.5 Earthquake along the Northern Part of the Newport Inglewood Zone. In *Evaluating Earthquake Hazards in the Los Angeles Region: An Earth-Science Perspective* (J.I. Ziony, ed.), U.S. Geological Survey Professional Paper 1360, pp. 436–441.

Harr, R.D. 1981. Some Characteristics and Consequences of Snowmelt during Rainfall in Western Oregon. *Journal of Hydrology*, Vol. 53, pp. 277–304.

Holzer, T.L., T.L. Youd, and T.C. Hanks. 1989. Dynamics of Liquefaction during the 1987 Superstition Hills, California, Earthquake. *Science*, Vol. 244, pp. 56–59.

Horton, R.E. 1938. *Phenomena of the Contact Zone Between the Ground Surface and a Layer of Melting Snow*. Bulletin No. 23, Association Internationale d'Hydrologie Scientifique, Paris, pp. 545–561.

Jackson, L.E., Jr. 1977. Dating and Recurrence Frequency of Prehistoric Mudflows near Big Sur, Monterey County, California. *Journal of Research, U.S. Geological Survey*, Vol. 5, No. 1, pp. 17–32.

Jacobson, R.B., E.D. Cron, and J.P. McGeehin. 1989. Slope Movements Triggered by Heavy Rainfall, November 3–5, 1985, in Virginia and West Virginia, U.S.A. In *Landslide Processes of the Eastern United States and Puerto Rico* (A.P. Schultz and R.W. Jibson, eds.), Geological Society of America Special Paper 236, pp. 1–14.

Jahns, R.H. 1949. Desert Floods. *Engineering and Science,* May, pp. 10–14.

Jibson, R.W. 1989. Debris Flows in Southern Puerto Rico. In *Landslide Processes of the Eastern United States and Puerto Rico* (A.P. Schultz and R.W. Jibson, eds.), Geological Society of America Special Paper 236, pp. 29–56.

Jibson, R.W. 1993. Predicting Earthquake-Induced Landslide Displacements Using Newmark's Sliding Block Analysis. In *Transportation Research Record 1411,* TRB, National Research Council, Washington, D.C., 1993, pp. 9–17.

Johnson, A.M., with contributions by J.R. Rodine. 1984. Debris Flow. In *Slope Instability* (D. Brunsden and D.B. Prior, eds.), John Wiley & Sons Ltd., Chap. 8, pp. 257–361.

Johnson, K.A., and N. Sitar. 1990. Hydrologic Conditions Leading to Debris-Flow Initiation. *Canadian Geotechnical Journal,* Vol. 27, pp. 789–801.

Jones, F.O., D.R. Embody, and W.L. Peterson. 1961. *Landslides Along the Columbia River Valley, Northeastern Washington.* U.S. Geological Survey Professional Paper 367, 98 pp.

Kadomura, H., T. Imagawa, and H. Yamamoto. 1983. Eruption-Induced Rapid Erosion and Mass Movement on Usu Volcano, Hokkaido: Extreme Land Forming Events. *Zeitschrift für Geomorphologie,* Supplement Vol. 46, pp. 123–142.

Keefer, D.K. 1984. Landslides Caused by Earthquakes. *Geological Society of America Bulletin,* Vol. 95, pp. 406–421.

Keefer, D.K., and M.W. Manson. In press. Landslides Produced by the 1989 Loma Prieta, California, Earthquake—Regional Distribution and Characteristics. In *The Loma Prieta Earthquake of October 17, 1989,* U.S. Geological Survey Professional Paper 1551-C, Chap. C, Vol. 2.

Kochel, R.C. 1987. Holocene Debris Flows in Central Virginia. In *Debris Flows/Avalanches: Process, Recognition, and Mitigation* (J.E. Costa and G.F. Wieczorek, eds.), Reviews in Engineering Geology, Vol. 7, Geological Society of America, Boulder, Colo., pp. 139–155.

Koppejan, A.W., B.M. van Wamelen, and L.J.H. Weinberg. 1948. Coastal Flow Slides in the Dutch Province of Zeeland. In *Proc., Second International Conference on Soil Mechanics and Foundation Engineering,* Rotterdam, Netherlands, June 21–30, 1948, Vol. 5, pp. 89–96.

Kuehn, M.H. 1987. The Effects of Exceeding "Probable Maximum Precipitation" on a Severely Burned Watershed in the Sierra Nevada of California. In *Landslide Activity in the Sierra Nevada During 1982 and 1983* (J.V. DeGraff, ed.), Earth Resources Monograph 12, USDA Forest Service, Pacific Southwest Region, San Francisco, pp. 27–40.

Lambe, T.W., and R.V. Whitman. 1969. *Soil Mechanics.* John Wiley & Sons, New York, 553 pp.

Lane, K.S. 1967. Stability of Reservoir Slope. In *Failure and Breakage of Rock* (C. Fairhurst, ed.), Proc., 8th Symposium on Rock Mechanics, American Institute of Mining, Metallurgy and Petroleum Engineering, New York, pp. 321–336.

Larsen, M.C., and A.J. Torres Sanchez. 1992. Landslides Triggered by Hurricane Hugo in Eastern Puerto Rico, September 1989. *Caribbean Journal of Science,* Vol. 28, No. 3-4, pp. 113–125.

Lee, K.L., and J.M. Duncan. 1975. *Landslide of April 25, 1974 on the Mantaro River, Peru.* National Academy of Sciences, Washington, D.C., 72 pp.

Massarsch, K.R., B.B. Broms, M.O. Fives, and A. Nilsson. 1987. The Shannon Embankments—Effects of Sudden Drawdown and Seepage on the Stability. In *Groundwater Effects in Geotechnical Engineering* (E.T. Hanrahan, T.L.L. Orr, and T.F. Widdis, eds.), Proc., 9th European Conference on Soil Mechanics and Foundation Engineering, Dublin, August–September 1987, A.A. Balkema, Boston, pp. 465–468.

Mathewson, C.C., J.R. Keaton, and P.M. Santi. 1990. Role of Bedrock Ground Water in the Initiation of Debris Flows and Sustained Post-storm Stream Discharge. *Bulletin of the Association of Engineering Geologists,* Vol. 27, No. 1, pp. 73–78.

Morgenstern, N. 1963. Stability Charts for Earth Slopes During Rapid Drawdown. *Geotechnique,* Vol. 13, pp. 121–131.

Morton, D.M., R.H. Campbell, A.G. Barrows, Jr., J.E. Kahle, and R.F. Yerkes. 1979. Wright Mountain Mudflow: Spring 1969. In *Landsliding and Mudflows at Wrightwood, San Bernardino County, California,* California Division of Mines and Geology, Special Report 136, Part II, pp. 7–21.

Mulvey, W.E., and M. Lowe. 1992. Cameron Cove Subdivision Debris Flow. *Survey Notes,* Utah Geological Survey, Vol. 25, No. 2, 25 pp.

Neary, D.G., and L.W. Swift, Jr. 1987. Rainfall Thresholds for Triggering a Debris Avalanching Event in the Southern Appalachian Mountains. In *Debris Flows/Avalanches: Process, Recognition, and Mitigation* (J.E. Costa and G.F. Wieczorek, eds.), Reviews in Engineering Geology, Vol. 7, Geological Society of America, Boulder, Colo., pp. 81–92.

Newmark, N.M. 1965. Effects of Earthquakes on Dams and Embankments. *Geotechnique*, Vol. 15, No. 2, pp. 139–160.

Nieto, A.S., and R.L. Schuster. 1988. Mass Wasting and Flooding Induced by the 5 March 1987 Ecuador Earthquakes. *Landslide News* (Japan Landslide Society), No. 2, pp. 1–2.

Pack, R.T. 1984. Debris Flow Initiation in Davis County, Utah, During the Spring Snowmelt Period of 1983. In *Proc., 21st Engineering Geology and Soils Engineering Symposium*, Moscow, Idaho, pp. 59–77.

Pierson, T.C. 1992. Rainfall-Triggered Lahars at Mt. Pinatubo, Philippines, Following the June 1991 Eruption. *Landslide News* (Japan Landslide Society), No. 6, pp. 6–9.

Pierson, T.C., R.J. Janda, J. Thouret, and C.A. Borrero. 1990. Perturbation and Melting of Snow and Ice by the 13 November 1985 Eruption of Nevado del Ruiz, Colombia, and Consequent Mobilization, Flow and Deposition of Lahars. *Journal of Volcanology and Geothermal Research*, Vol. 41, pp. 17–66.

Plafker, G., G.E. Ericksen, and J.F. Concha. 1971. Geological Aspects of the May 31, 1970, Peru Earthquake. *Bulletin of the Seismological Society of America*, Vol. 61, No. 3, pp. 543–578.

Plafker, G., and J.P. Galloway. 1989. *Lessons Learned from the Loma Prieta, California, Earthquake of October 17, 1989.* U.S. Geological Survey Circular 1045, 48 pp.

Post, A. 1967. Effects of the March 1964 Alaska Earthquake on Glaciers. In *The Alaska Earthquake, March 27, 1964: Effects of the Hydrologic Regimen*, U.S. Geological Survey Professional Paper 544-D, pp. D1–D42.

Reid, M.E., H.P. Nielsen, and S.J. Dreiss. 1988. Hydrologic Factors Triggering a Shallow Hillside Failure. *Bulletin of the Association of Engineering Geologists*, Vol. 25, No. 3, pp. 349–361.

Schuster, R.L. 1979. Reservoir-Induced Landslides. *Bulletin of the International Association of Engineering Geology*, Vol. 20, pp. 8–15.

Schuster, R.L. 1981. Effects of the Eruptions on Civil Works and Operations in the Pacific Northwest. In *The 1980 Eruptions of Mount St. Helens, Washington* (P.W. Lipman and D.R. Mullineaux, eds.), U.S. Geological Survey Professional Paper 1250, pp. 701–718.

Schuster, R.L. 1991. Introduction. In *The March 5, 1987, Ecuador Earthquakes—Mass Wasting and Socioeconomic Effects*, Natural Disaster Studies,

Chap. 1, Vol. 5, National Research Council, Washington, D.C., pp. 11–22.

Scott, K.M. 1971. *Origin and Sedimentology of 1969 Debris Flows near Glendora, California.* U.S. Geological Survey Professional Paper 750-C, pp. 242–247.

Seed, R.B., S.E. Dickenson, M.F. Riemer, J.D. Bray, N. Sitar, J.K. Mitchell, I.M. Idriss, R.E. Kayan, A. Kropp, L.F. Harder, and M.S. Power. 1990. *Preliminary Report on the Principal Geotechnical Aspects of the October 17, 1989 Loma Prieta Earthquake.* University of California at Berkeley, Report UCB/EERC-90/05, 137 pp.

Sidle, R.C. 1984. Shallow Groundwater Fluctuations in Unstable Hillslopes of Coastal Alaska. *Zeitschrift für Gletscherkunde und Glazialgeologie*, Vol. 20, pp. 79–95.

Simon, A., M.C. Larsen, and C.R. Hupp. 1990. The Role of Soil Processes in Determining Mechanisms of Slope Failure and Hillside Development in a Humid-Tropical Forest, Eastern Puerto Rico. *Geomorphology*, Vol. 3, pp. 263–286.

Sitar, N., and G.W. Clough. 1983. Seismic Response of Steep Slopes in Cemented Soils. *Journal of Geotechnical Engineering*, ASCE, Vol. 109, No. 2, pp. 210–227.

Spittler, T.E. 1983. Blucher Valley Translational Landslide: Sebastopol, Sonoma County, California. *California Geology*, Aug., pp. 176–180.

Springer, F.M., C.R. Ullrich, and D.J. Hagerty. 1985. Streambank Stability. *Journal of Geotechnical Engineering*, ASCE, Vol. 111, No. 5, pp. 624–640.

Terzaghi, K. 1943. *Theoretical Soil Mechanics.* John Wiley and Sons, Inc., New York, 510 pp.

Tinsley, J.C., and W.R. Dupre. 1993. Liquefaction Hazard Mapping, Depositional Facies, and Lateral Spreading Ground Failure in the Monterey Bay Area, Central California. In *Proc., Fourth U.S./Japan Workshop on Liquefaction, Large Ground Deformation and Their Effects on Lifelines* (T.D. O'Rourke and M. Hamada, eds.), National Center for Earthquake Engineering Research, Honolulu, Technical Report NCEER-92-0019, pp. 71–86.

Troxell, H.C., and J.Q. Peterson. 1937. *Flood in La Cañada Valley, California January 1, 1934.* U.S. Geological Survey Water Supply Paper 796-C, pp. 53–98.

Vandre, B.C. 1985. Rudd Creek Debris Flow. In *Delineation of Landslide, Flash Flood, and Debris Flow Hazards in Utah* (D.S. Bowles, ed.), Proc. of Specialty Conference, Utah Water Research Laboratory, Utah State University, Logan, Aug., pp. 117–131.

Varnes, D.J. 1978. Slope Movement Types and Processes. In *Special Report 176: Landslides: Analysis and Control* (R.L. Schuster and R.J. Krizek, eds.), TRB, National Research Council, Washington, D.C., pp.12–33.

Voight, B., R.J. Janda, H. Glicken, and P.M. Douglass. 1983. Nature and Mechanisms of the Mount St. Helens Rock-slide Avalanche of 18 May 1980. *Geotechnique*, Vol. 33, No. 3, pp. 243–273.

Waldron, E.H. 1967. *Debris Flow and Erosion Control Problems Caused by the Ash Eruptions of Irazu Volcano, Costa Rica.* U.S. Geological Survey Bulletin 1241-I, 37 pp.

Weirich, F.H. 1989. The Generation of Turbidity Currents by Subaerial Debris Flows, California. *Geological Society of America Bulletin*, Vol. 101, pp. 278–291.

Wells, W.G. II. 1987. The Effects of Fire on the Generation of Debris Flows in Southern California. In *Debris Flows/Avalanches: Process, Recognition, and Mitigation* (J.E. Costa and G.F. Wieczorek, eds.), Reviews in Engineering Geology, Vol. 7, Geological Society of America, Boulder, Colo., pp. 105–114.

Wells, W.G. II, P.M. Wohlgemuth, A.G. Campbell, and F.H. Weirich. 1987. Postfire Sediment Movement by Debris Flows in the Santa Ynez Mountains, California. In *Erosion and Sedimentation in the Pacific Rim*, International Association of Hydrological Sciences, Delft, Netherlands, Publication 165, pp. 275–276.

Wieczorek, G.F., E.W. Lips, and S.D. Ellen. 1989. Debris Flows and Hyperconcentrated Floods Along the Wasatch Front, Utah, 1983 and 1984. *Bulletin of the Association of Engineering Geologists*, Vol. 26, No. 2, pp. 191–208.

Wilson, C.J., and W.E. Dietrich. 1987. The Contribution of Bedrock Groundwater Flow to Storm Runoff and High Pore Pressure Development in Hollows. In *Erosion and Sedimentation in the Pacific Rim*, International Association of Hydrological Sciences, Delft, Netherlands, Publication 165, pp. 49–59.

Wilson, R.C. 1989. Rainstorms, Pore Pressures, and Debris Flows: A Theoretical Framework. In *Landslides in a Semi-Arid Environment* (P. Sadler and D.M. Morton, eds.), Inland Geological Society, Riverside, Calif., Vol. 2, pp. 101–117.

Wilson, R.C., and D.K. Keefer. 1983. Dynamic Analysis of a Slope Failure from the 1979 Coyote Lake, California, Earthquake. *Bulletin of the Seismological Society of America*, Vol. 73, pp. 863–877.

Wilson, R.C., and D.K. Keefer. 1985. Predicting Areal Limits of Earthquake Induced Landsliding. In *Evaluating Earthquake Hazards in the Los Angeles Region: An Earth-Science Perspective* (J.I. Ziony, ed.), U.S. Geological Survey Professional Paper 1360, pp. 316–345.

Wilson, R.C., J.D. Torikai, and S.D. Ellen. 1992. *Development of Rainfall Warning Thresholds for Debris Flows in the Honolulu District, Oahu.* U.S. Geological Survey Open-File Report 92-521, 45 pp.

Woolley, R.R. 1946. *Cloudburst Floods in Utah, 1850–1938.* U.S. Geological Survey Water Supply Paper 994, 128 pp.

Works, B. 1984. Geologic Hazard . . . Landslide on State Highway 1, Julia Pfeiffer-Burns State Park, Monterey County. *California Geology*, June, pp. 130–131.

Youd, T.L., and D.M. Perkins. 1978. Mapping Liquefaction-Induced Ground Failure Potential. *Journal of Geotechnical Engineering*, ASCE, Vol. 104, No. GT4, pp. 433–446.

Chapter 5

ROBERT L. SCHUSTER AND
WILLIAM J. KOCKELMAN

PRINCIPLES OF LANDSLIDE HAZARD REDUCTION

1. INTRODUCTION

Careful development of hillside slopes can reduce economic and social losses due to slope failure by avoiding the hazards or by reducing the damage potential. Landslide risk can be reduced by four approaches (Kockelman 1986):

1. Restriction of development in landslide-prone areas;
2. Codes for excavation, grading, landscaping, and construction;
3. Physical measures (drainage, slope-geometry modification, and structures) to prevent or control landslides; and
4. Development of warning systems.

These methods of hazard mitigation, when used with modern technology, can greatly reduce losses due to landslides. Schuster and Leighton (1988) estimated that these methods could reduce landslide losses in California more than 90 percent. Slosson and Krohn (1982) stated that implementation of this methodology has already reduced landslide losses in the city of Los Angeles by 92 to 97 percent.

After a discussion of the aforementioned widely used approaches to landslide hazard reduction, the status of development of landslide insurance programs is reviewed. Although insurance will not reduce overall landslide hazards or costs directly, nonsubsidized landslide insurance does offer promise as a means of distributing landslide costs more widely and of reducing landslide losses for individual property owners. In addition, advice from insurance organizations can positively influence the users of land that is subject to landslide hazards.

Landslides often occur as elements of interrelated multiple natural-hazard processes in which an initial event triggers secondary events or in which two or more natural-hazard processes occur at the same time. Examples are combinations of volcanic eruptions, earthquakes, and landslides. The resulting multiple-hazard problems require a shift in perspective from mitigation of individual hazards, such as landslides, to a broader systems framework that takes into account the characteristics and effects of all the processes involved.

In recent years, risk assessment has become an important factor in landslide hazard reduction. Landslide risk assessment utilizing reliability methods in landslide susceptibility mapping, prediction, and mitigation is discussed in Chapter 6.

Optimal approaches to reduction of landslide hazards generally involve a carefully assembled mix of the above hazard-reduction strategies and techniques. To plan a coordinated and successful reduction program requires input and cooperation from engineers, geologists, planners, landowners and developers, lending organizations, insurance companies, and government entities.

2. PREREQUISITE INFORMATION

Successful landslide hazard-reduction programs in the United States commonly are based on the following factors (U.S. Geological Survey 1982):

1. An adequate base of technical information on the hazards and risks;
2. A technical community able to apply, and enlarge upon, this data base;
3. An able and concerned local government; and
4. A citizenry that realizes the value of and supports a program that promotes the health, safety, and general welfare of the community.

These same concepts apply to other countries, except that most national governments have a stronger role than that of the U.S. government.

The key to a successful landslide hazard-reduction program is awareness and understanding of the landslide problem within the geographic area involved. The work of Varnes (1978) and of Cruden and Varnes (Chap. 3 in this report) in describing and classifying mass movements and in reviewing the principles and practices of zonation as related to landslide hazards (Varnes et al. 1984) has been very helpful in this regard. Recognition and identification of landslides have been discussed in detail by Rib and Liang (1978), Hansen (1989), and Soeters and van Westen (Chap. 8 in this report).

Reliable landslide hazard maps are of significant value in establishing reduction programs. Ideally, these maps indicate where landslides have occurred in the past, the locations of landslide-susceptible areas, and the probability of future occurrences. Brabb (1984) presented examples of various types of landslide maps: inventory, susceptibility, loss evaluation, and risk determination. Of particular interest in reducing the costs of landslide hazard mapping is the use of computer techniques to produce digital maps; in much of the world, the geographic information system (GIS) approach is widely used for producing digital maps and for integrating information to develop, enhance, and complement such maps.

An important element in determining international landslide hazard distribution is the World Landslide Inventory, which is being conducted by the United Nations Educational, Scientific, and Cultural Organization (UNESCO) Working Party on the World Landslide Inventory (WP/WLI)

sponsored by the International Geotechnical Societies (1991). The Working Party, which was formed from the Commission on Landslides of the International Association of Engineering Geology, the Technical Committee on Landslides of the International Society for Soil Mechanics and Foundation Engineering, and representatives of the International Society for Rock Mechanics, assists United Nations agencies in understanding the worldwide distribution of landslides (WP/WLI 1990, 1991, 1993a,b). The five WP/WLI classes of landslide inventory cover a range that includes computer data banks with complete national coverage and systematic data capture (Cruden and Brown 1992). The *Directory of the World Landslide Inventory* (Brown et al. 1992) is a useful worldwide guide to the people and institutions that deal with landslide hazards on a regular basis.

The extent and economic significance of landslides in 136 countries and areas, including land beneath all of the oceans, were reported by Brabb and Harrod (1989). These reports have been an invaluable contribution to the International Decade for Natural Disaster Reduction.

Governmental organizations have reported on landslide hazard-reduction approaches that are of value to others attempting to develop plans for their own areas. For example, the state of Colorado, under the auspices of the Federal Emergency Management Agency (FEMA), published a report (Jochim et al. 1988) that aims to reduce landslide losses by

1. Identifying local governmental resources, plans, and programs that can assist in loss reduction;
2. Determining unmet local needs that must be addressed to reduce losses;
3. Identifying and developing state agency capabilities and initiatives that can deal with unmet local needs;
4. Developing cost-effective projects that reasonably can be expected to reduce landslide losses;
5. Educating state and local officials and emergency-response personnel about landslide hazards and potential methods for loss reduction; and
6. Establishing means to provide a continuing governmental process to reduce losses.

This FEMA report will become part of the overall hazard mitigation plan for Colorado under the auspices of the Colorado Natural Hazards

Mitigation Council. Similar plans based on local needs were prepared for the city of Cincinnati, Ohio (Hamilton County Regional Planning Commission 1976), and Portola Valley, California (Mader et al. 1988). In addition, Hamilton County, Ohio, prepared a report on the duties of the county's Earth Movement Task Force, which provides advice to those wishing to establish similar working groups (Hamilton County 1982). A succinct but comprehensive guidebook for state and local governments interested in reducing landslide losses was sponsored and published by FEMA (Wold and Jochim 1989).

An important aspect of the reduction of landslide hazards is collection and dissemination of landslide information for scientists, engineers, policy makers, and the public. An excellent national example of a public repository of information on landslide hazards is the U.S. Geological Survey's National Landslide Information Center (NLIC) in Golden, Colorado (Brown 1992). The NLIC maintains a multiple-entry data base to foster national and international exchange of landslide information among scientists, engineers, and decision makers. On an international level, the International Union of Forestry Research Organizations (IUFRO) carries out a worldwide exchange of information and assistance on a full spectrum of technical, biological, and economic measures for the control of landslides in mountainous areas.

3. MAJOR POLICY OPTIONS

Alternative management policy options are available to decision makers who are concerned with natural hazards (Petak and Atkisson 1982; Olshansky and Rogers 1987; Olshansky 1990). The three most fundamental options (Rossi et al. 1982) are to

1. Take no action at all,
2. Provide relief and rehabilitation assistance after disasters occur, or
3. Take action to contain or control hazards before serious damage occurs.

Before about 1950, the first two of these options dominated. However, as a result of technical and sociological advances, the concept of prevention of landslide disasters by appropriate land use development or structural retention is becoming increasingly important.

4. APPROACHES

Reduction of landslide hazards in the United States is achieved mainly by

1. Restricting development in landslide-prone areas, a function assisted by mapping landslide susceptibility;
2. Requiring that excavation, grading, landscaping, and construction activities not contribute to slope instability; and
3. Protecting existing development and population (property and structures as well as people and livestock) by physical control measures, such as drainage, slope-geometry modification, and protective barriers, or by monitoring and warning systems.

These techniques, which were discussed by Kockelman (1986), are used individually or in various combinations to reduce or eliminate losses due to existing or potential landslides. The first two methods can be promoted by public legislation. In the United States, such legislation commonly is under the jurisdiction of local governments. However, most other countries with major and continuing landslide losses have incorporated a strong federal or provincial role in dealing with all aspects of landslide hazard-reduction activities to ensure consistent standards of practice and application and to prevent unequal and inadequate performance at provincial, municipal, and private levels (Swanston and Schuster 1989). In the United States the federal government plays a less active role and functions primarily as a source of expertise, research support, and funding of state and local control works.

4.1 Restricting Development in Landslide-Prone Areas

One of the most effective and economical ways to reduce landslide losses is by land use planning to locate developments on stable ground and to dedicate landslide-prone areas to open space or to other low-intensity uses. This procedure, which commonly is known as avoidance, is accomplished by either or both of the following: (a) removing or converting existing development or (b) discouraging or regulating new development in unstable areas.

In the United States, restrictions on land use because of natural hazards generally are imposed and enforced by local governments by means of land use zoning districts and regulations. Some local governments in the United States have adopted ordinances that limit the amount of development in hillside areas (see Section 4.1.3). In many other countries, land use planning that leads to avoidance is a function of the national government. In Japan, which is widely and continually affected by severe landslide problems, land use regulation has not been a common feature of landslide hazard reduction for reasons related to the limited availability of land (Huffman 1986, 96).

4.1.1 Removing or Converting Existing Development

Recurring damage to existing development caused by landslides can be eliminated or reduced by evacuating the area or by converting existing structures or facilities to uses less vulnerable to slope failure. Permanent evacuation of the distressed area commonly requires public acquisition of the land and relocation of the inhabitants and their facilities.

Conversion of existing structures and facilities to uses that are less vulnerable to slope failure may be undertaken by individual property owners, by developers, or, in the case of public properties, by the government. The feasibility of successful conversion depends on the value and criticality of the facilities, their potential for triggering or resisting slope failure, whether they can be successfully retrofitted to resist slope movement, and the level of concern of their owners.

4.1.2 Discouraging New Development

Where feasible, the most effective method of reducing landslide losses is to discourage new development in landslide-prone areas (U.S. Geological Survey 1982). Methods that have been successful in the United States include the following:

- **Public information programs:** Because any program of land use control requires the support of a knowledgeable citizenry, the public must be informed of landslide hazards. Prudent citizens, when properly informed of the existence of hazards, ordinarily will support land use controls that minimize losses due to those hazards.
- **Disclosure of hazards to potential property purchasers:** Governments can discourage develop-

ment in hazardous areas by enacting hazard disclosure laws that alert potential buyers to hazards (Kockelman 1986). For example, Santa Clara County, California, requires sellers of property within the county's landslide, fault-rupture, and flood zones to provide prospective purchasers with written statements of geologic hazard (Santa Clara County Board of Supervisors 1978).
- **Exclusion of public facilities:** Local governments can prohibit construction of public facilities, such as streets and water and sewer systems, in landslide-prone areas.
- **Warning signs:** Warning signs posted by local governments can alert prospective property owners or developers to potential hazards.
- **Tax credits and special assessments:** Tax credits can be applied to properties left undeveloped in hazardous areas. Conversely, special assessments can be levied on landslide-prone properties that lie within especially created assessment districts.
- **Financing policies:** Lending institutions can discourage development in landslide-prone areas by denying loans or by requiring insurance for construction or development in these areas.
- **Insurance costs:** The high cost of nonsubsidized insurance for development in hazardous areas can discourage such development and can encourage land uses that constitute lower risk.
- **Government acquisition:** Government agencies can promote avoidance by acquiring landslide-prone properties by purchase, condemnation, tax foreclosure, dedication, devise (will), or donation. The agencies are then able to control development on these properties for the public interest.
- **Public awareness of legal liabilities:** Property owners and developers can be made aware of liabilities they may have in regard to slope failure.

4.1.3 Regulating Development

To assume that development in landslide-prone areas can be discouraged indefinitely by the nonregulatory methods noted above is unrealistic. Thus, governmental regulation often is needed to prevent or control development of lands subject to landslide hazards. In the United States, restrictions on land use because of natural hazards are generally imposed and enforced by local governments by means of zoning districts and regulations. By means of land development regulations, a local government can prohibit or restrict development in landslide-prone areas. It can zone hazardous areas for open-space uses such as agriculture, grazing, forests, or parks. If development is allowed in areas subject to slope failure, the location or intensity of this development, or both, can be controlled to reduce the risk. Examples of regulations of land use in areas prone to landslide activity are discussed in Sections 4.1.3.1–4.1.3.3.

4.1.3.1 Land Use Zoning Regulations

Land use zoning provides direct benefits by limiting development in landslide-prone areas. Under zoning ordinances enacted and enforced by local governments, land use with the least danger of activating landslides includes parks, woodlands, nonirrigated agriculture, wildlife refuges, and recreation. In addition, these land uses result in relatively small economic losses if landslides do occur. Regulations can include provisions that prohibit specific land uses or operations that might cause slope failure, such as construction of roads or buildings, irrigation systems, storage or disposal of liquid wastes, and operation of off-road vehicles. Zoning regulations can also control the location and density of development in hillside areas.

To assist counties and municipalities in designing land use regulations in hillside areas in the state of Colorado, the Colorado Geological Survey prepared model regulations (Rogers et al. 1974) that permit the following land uses in designated landslide-prone areas:

1. Recreational uses that do not require permanent structures for human habitation, including parks, wildlife and nature preserves, picnic grounds, golf courses, and hunting, fishing, hiking, and skiing areas that do not result in high population concentrations;
2. Low-density agricultural uses, such as forestry, grazing, and truck-crop farming; and
3. Low-density and temporary commercial and industrial uses, such as parking areas and storage yards for portable equipment.

Colorado is currently attempting to set a national precedent in dealing with natural hazards (including landslides) by means of the Colorado Natural Hazards Mitigation Council (1992), an official statewide 300-member group composed of earth scientists, engineers, planners, and local and state policy makers whose goal is to formulate new policies regarding natural hazards in Colorado. A prime strategy of the council is to unify technical experts and policy makers on issue-directed hazard-reduction teams. These teams deal with hazards in areas that are politically responsive to innovative solutions. They prepare statewide plans based on these solutions, and the plans are used to develop policy directions for future state hazard legislation.

4.1.3.2 Subdivision Regulations

Regulating the design of subdivisions (planned local units of land designed with streets, sidewalks, sewerage, etc., in preparation for building homes) is another means of controlling development of landslide-prone areas. Subdivision design and zoning regulations must be based upon geotechnical information.

4.1.3.3 Sewage-Disposal Regulations

Residential sewage-disposal systems that rely on ground absorption (septic-tank systems, leaching fields, and seepage beds and pits) can saturate the surrounding soil and rock and cause slope failure. Thus, the design and installation of these systems must be regulated in landslide-prone areas.

4.1.4 Implementing Avoidance as Landslide Hazard-Reduction Measure

In the United States, implementation of avoidance procedures has met with mixed success in landslide hazard reduction. In some areas, particularly in California, restriction of development in landslide-prone areas has been extensive, and avoidance programs generally have been successful in reducing landslide losses. However, in many states that have landslide-susceptible areas, there are no widely accepted procedures or regulations for considering landslides as part of the land use planning process (Committee on Ground Failure Hazards 1985).

Land use zoning probably has been the most effective means of regulating development. For example, in San Mateo County, California, a landslide-susceptibility map has been in use since 1975 to control the density of development (Brabb et al. 1972). On the basis of this map, the San Mateo County Board of Supervisors (1973) enacted legislation that restricts development in those areas most susceptible to landslides to one dwelling unit per 40 acres (approximately 16 ha) (Figure 5-1). Until 1982, all of the new landslides (mostly slips, slumps, and slides) that occurred in San Mateo County were in areas already mapped as landslides or in areas judged highly susceptible to landsliding. Thus, the zoning procedure was an outstanding success at that point. However, in 1982, under conditions of exceedingly heavy rainfall, thousands of debris flows occurred in areas where few had been observed previously (Brabb 1984). Thus, the 1972 map had accurately predicted the locations of future deep-seated landslides, but was not successful for debris flows. The new debris flows had not been expected because the landslide-susceptibility map was based on interpretation of aerial photographs that showed evidence of only deep-seated landslides.

Another approach to zoning has been used in Fairfax County, Virginia, where maps used for zoning purposes outline various degrees of hazard in different geologic materials (Obermeier 1979). Developers are required to obtain professional engineering advice for sites to be developed in specific geologic materials. The result has been a "drastic reduction in landslides" (Dallaire 1976).

The best examples of removal or conversion of existing development as a tool in the reduction of landslide losses have been those in which develop-

ments have been wholly or partly destroyed by slope failures, and, as a reaction to those losses, a decision has been made to replace the original development with a land use less prone to slope-failure damage. Such efforts commonly have been only partially successful because of the resistance of property owners, developers, or even the communities themselves. An excellent example of such partial success is provided by the city of Anchorage, Alaska, which received heavy damage from soil slides that were triggered by the 1964 Alaska earthquake. As a result of the earthquake, a scientific and engineering task force was established by the federal government to assess the damage, to evaluate future hazards, and to make recommendations that would minimize the impact of any future earthquake or landslide activity.

Especially interesting are land-planning decisions related to the three largest slope failures: the Turnagain Heights, Fourth Avenue, and L Street landslides (Mader et al. 1980). The mixed success of the task force's land use recommendations for these areas is illustrated by the cases discussed in Sections 4.1.4.1–4.1.4.3.

4.1.4.1 Turnagain Heights Landslide
The Turnagain Heights slide was the largest and most spectacular of the 1964 slope failures, covering 53 ha and destroying 75 homes (Figure 5-2). The Alaska State Housing Authority prepared a redevelopment plan for the landslide area calling for park and recreation uses (Mader et al. 1980). However, only the economically least desirable part of the landslide was actually developed as a park. The Anchorage City Council voted against the plan and allowed applications for residential building permits in the landslide area (Selkregg et al. 1970). In 1977 controversy over the issue of rebuilding on parts of the Turnagain Heights landslide led to appointment of the Anchorage Geotechnical Advisory Commission. This commission consistently advised the local government not to allow development on the landslide unless the long-term stability of the slope could be assured. However, a few houses have since been built adjacent to or on the slide. The private property owners believed that compensation at postdisaster values was not sufficient inducement to relocate. In addition, local residents seemed to believe that because a catastrophe had only recently occurred, another would not take place at the same location during their lifetimes.

FIGURE 5-1
Hypothetical property in San Mateo County, California, showing seismic and other geologic constraints. Dwelling units in slope-instability zones are limited to one per 16 ha. Similar lower densities in floodplains and fault-rupture zones are required by San Mateo County Board of Supervisors (1973) (Kockelman and Brabb 1979).

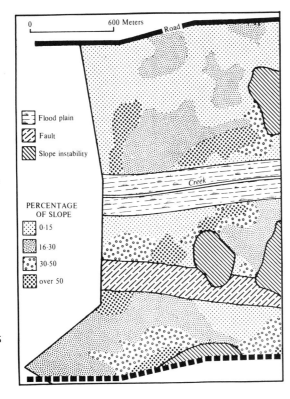

4.1.4.2 Fourth Avenue Landslide

This landslide was a single 15-ha block that moved horizontally about 5 m, destroying a significant part of Anchorage's central business district (Figure 5-3). The task force recommended that future developments in the Fourth Avenue landslide area be limited to parks, parking lots, and light structures no more than two stories high (Hansen et al. 1966). The recommended restrictions were incorporated into an urban renewal plan that was relatively successful because much of the land belonged to the federal government and thus was easily controlled (Mader et al. 1980).

4.1.4.3 L Street Landslide

The study by the task force showed that this 29-ha block slide and considerable adjacent hillside land constituted a significant continuing hazard. Thus, the same land use recommendations were made as for the Fourth Avenue landslide. However, the recommendations were ignored for the L Street area. About one year after the earthquake, the Anchorage City Council decided to rezone the area to permit higher residential densities, and by 1980 extensive new construction, including five new high-rise buildings, had taken place on or adjacent to the 1964 landslide (Mader et al. 1980). As was the case for the Turnagain Heights landslide, the main reason for the partial success in controlling redevelopment of the L Street landslide area was the resistance of property owners to less-intensive redevelopment.

4.2 Developing Excavation, Construction, and Grading Codes

Excavation, construction, and grading codes have been developed to ensure that construction in landslide-prone areas is planned and conducted in a manner that will not impair the stability of hillside slopes. These ordinances commonly

1. Require a permit before slopes are scraped, excavated, filled, or cut;
2. Regulate (including control of design, construction, inspection, and maintenance), minimize, or prohibit excavation and fill;
3. Control disruption of drainage and vegetation; and
4. Provide for proper design, construction, inspection, and maintenance of surface and subsurface drains.

In the United States there is no nationwide uniform code to ensure standardization of the foregoing criteria. Instead, in dealing with stability of hillside slopes on private lands, state and local government agencies apply design and construction criteria that fit the needs of their specific jurisdictions. The federal government usually has not participated directly in the formulation and enforcement of these codes. However, it has influenced the engineering profession by development of the codes used by government agencies in their own construction programs on federal projects. Federal standards for excavation and grading—such as those used by the U.S. Army Corps of Engineers, the Bureau of Reclamation, and the USDA Forest Service, each of which is in charge of major construction activities—often are used by other organizations in both the public and private sectors (Committee on Ground Failure Hazards 1985).

FIGURE 5-2
(top)
Damaged houses on Turnagain Heights landslide, Anchorage, Alaska, triggered by March 27, 1964, Alaska earthquake.
U.S. GEOLOGICAL SURVEY

FIGURE 5-3
(bottom)
Collapse of Fourth Avenue, Anchorage, Alaska, due to landslide triggered by Alaska earthquake of March 27, 1964.
U.S. ARMY

The development of excavation and grading ordinances related to geologic hazards originated in the United States shortly after World War II. At that time the accelerating demand for residential building sites in southern California because of a rapidly expanding population intensified development of hillside and mountain slopes (Scullin 1983, 14). In addition, improved earth-moving technology made development of slope areas economically feasible. The resulting poorly organized development combined with unusually heavy rainfall in southern California in the early 1950s resulted in significant landslide activity and major economic losses (Jahns 1969). As a result, the city of Los Angeles in 1952 adopted the first grading ordinance in the United States. The 1952 code was far from perfect, and during the following 10-year period, hillside developers in southern California faced many difficulties in applying and modifying it. However, this original code formed the basis for all subsequent codes adopted by local governments throughout the United States.

This early work in southern California led the International Conference on Building Officials in 1964 to develop Chapter 70 of the Uniform Building Code, which authorizes local governments to require that developers provide geotechnical reports on sites they intend to develop (Schuster and Leighton 1988). These reports are prepared by registered geotechnical engineers and certified engineering geologists. This code is still in effect (International Conference of Building Officials 1985); it has been adopted directly or used as a model by local governments in many countries.

Heavy rainfall in southern California in 1962 and 1969–1970 resulted in new major landslide losses and, as a result, in improvements to the 1952 code (Schuster and Leighton 1988). The 1962 storms and landslides resulted in more-sophisticated grading-code regulations that required

1. Geotechnical engineering and engineering geology input through the design and construction stages;
2. Definition of responsibilities of the design civil engineer, geotechnical engineer, and engineering geologist; and
3. Adequate subsurface exploration.

The 1969–1970 landslides led to

1. Additional refinement of grading ordinances;
2. A more quantitative approach to the design of slopes, for example, use of soil strength parameters, safety factors, and slope-stability analysis;
3. Emphasis on mud flow–debris flow mitigation; and
4. Proper design of structures above and below natural slopes.

The city of Los Angeles has provided an impressive example of the effective use of excavation and grading codes as deterrents to landslide activity and damage in the development of hillside slopes. The Los Angeles loss-reduction program relies heavily on regulations that require specific evaluations of landslide potential by engineering geologists and geotechnical engineers before construction as well as inspection of grading operations during construction. As noted above, controls on hillside grading and development in Los Angeles were almost nonexistent before 1952. In 1952, following severely damaging winter storms, a grading code was adopted that instituted procedures for safe development of hillsides; these grading regulations were significantly improved in 1963. The benefits resulting from these regulations were illustrated by the distribution of landslide damage in Los Angeles during severe storms in 1968–1969 and 1978. During the storms of 1968–1969, for a comparable number of building sites, the damage to sites developed before institution of grading codes in 1952 was nearly 10 times as great as damage to sites developed after 1963 (Slosson 1969). Similar results were observed after the 1978 storm (Table 5-1).

4.3 Protecting Existing Development

In spite of the avoidance and regulatory techniques presented above, development of hillside and mountain slopes that are subject to slope failure will continue. Thus, land use planning programs should include physical techniques to protect property, structures, and people in landslide-prone areas. These protective measures can be divided into (*a*) physical methods of control of unstable slopes and (*b*) monitoring and warning systems. These measures are introduced here and described in detail in Chapters 10, 11, 17, and 18.

Table 5-1
Relationship Between Modern Grading Codes and Slope Failures for Los Angeles Building Sites from Catastrophic February 1978 Southern California Storm (Slosson and Krohn 1979)

BUILDING CODE IN EFFECT	NO. OF SITES DEVELOPED	NO. OF SITE FAILURES	PERCENTAGE OF SITE FAILURES	DAMAGE COSTS ($)
Pre-1963 (premodern code)	37,000	2,790	7.5	40–49 million
Post-1963 (modern code)	30,000	210	0.7	1–2 million

4.3.1 Physical Controls

The most commonly used physical methods for control of unstable slopes are as follows:

1. **Drainage:** Because of its high stabilization efficiency in relation to cost, drainage of surface water and groundwater is the most widely used, and generally the most successful, slope-stabilization method (Committee on Ground Failure Hazards 1985).
2. **Slope modification:** Stability of a slope can be increased by removing all or part of the landslide mass.
3. **Earth buttresses:** Earth-buttress counterforts placed at the toes of potential slope failures are often successful in preventing failure. In California this is the most common mechanical (as contrasted to hydrologic) method of landslide control (Committee on Ground Failure Hazards 1985).
4. **Restraining structures:** When Methods 1 through 3 will not ensure slope stability by themselves, structural controls, such as retaining walls, piles, caissons, or rock anchors, commonly are used to prevent or control slope movement. In most cases restraining structures are used in conjunction with any or all of the following: drainage, slope modification, and construction of earth counterfort berms. Properly designed restraining structures are useful in preventing or controlling slumps and slips, particularly where lack of space precludes slope modification. However, use of restraining structures should be limited to control of small landslides because they seldom are successful on large ones.

All of these physical control methods have been discussed at length in the landslide literature [e.g., by Veder (1981), Holtz and Schuster (Chap. 17 in this report), and Wyllie and Norrish (Chap. 18 in this report)]. Their principal shortcoming is relatively high cost, which restricts effective usage to those sites for which avoidance is not feasible. Thus, they are most commonly used where landslide costs are high because of high population densities and property values.

4.3.2 Monitoring and Warning Systems

Landslide-prone hillside slopes can be monitored to provide warning of slope movement to downslope residents. Monitoring techniques include field observation and the use of extensometers, tiltmeters, piezometers, electrical fences, and trip wires. Recent innovations in monitoring devices include acoustic instruments, television, guided radar, laser beams, and vibration meters. Data from these devices can be telemetered to central receiving stations.

One of the most significant areas of recent landslide research is the development of real-time warning systems for landslides triggered by major storms. Such a system has been developed for the San Francisco Bay region, California, by the U.S. Geological Survey in cooperation with the National Weather Service. The procedure is based on (*a*) empirical and theoretical relations between rainfall intensity and duration and landslide initiation, (*b*) geologic determination of areas susceptible to landslides, (*c*) real-time monitoring of a regional network of telemetering rain gauges, and (*d*) National Weather Service precipitation forecasts (Keefer et al. 1987). The procedure was used to issue over public television and radio stations the first regional public warnings in the United States during the storms of February 12–21, 1986, which produced 800 mm of rainfall in the San Francisco

Bay region. According to eyewitness accounts of landslide occurrence, the warnings accurately predicted the times of major landslide events. Although analysis after the storms suggested that modifications to and additional development of the system are needed, it can be used as a prototype for systems in other landslide-prone areas.

The Territory of Hong Kong also relies on a rainfall-monitoring system for identifying periods of high landslide potential. This system is maintained by the Geotechnical Control Office (GCO) within the Engineering Department of the Hong Kong government (Geotechnical Control Office 1985). During heavy rainstorms, the GCO operates on an emergency basis to provide advice on remedial measures for landslides.

Landslide monitoring systems also have been developed in other countries, notably Japan, New Zealand, and the alpine countries of Europe. Ancient landslides that may be reactivated by the filling of Clyde Reservoir on the Clutha River on the South Island of New Zealand were being monitored at more than 2,000 points (as of July 1991) by means of piezometers, inclinometers, survey points, and flow measurements; many more monitoring installations are planned (Gillon et al. 1992). In the aftermath of the catastrophic 1987 Val Pola rock avalanche, which dammed the Adda River in northern Italy, the Italian Department of Civil Defense installed an on-line monitoring system on both the unstable slope and the landslide dam (Cambiaghi and Schuster 1989). This system included down-hole inclinometers, Invar-wire extensometers, an acoustic monitoring system, rain gauges, ultrasonic hydrometers, and thermometers. Data obtained by these instruments were transmitted by radio to a computer system operated by the local government; these data provided real-time analysis of applicable risk scenarios for the down-valley populace and for construction crews operating beneath the unstable slopes.

Monitoring and warning systems are installed primarily to protect lives and property, not to prevent landslides. However, these systems often provide warning of slope movement in time to allow the construction of physical control measures that will reduce the immediate or long-term hazard.

5. LANDSLIDE INSURANCE

Although insurance programs are not intended to reduce landslide hazards directly as do the mitiga-

tive measures discussed above, they can reduce the impact of landslide losses on individual property owners by spreading these losses over a larger base (Schuster and Fleming 1986). In addition, the high cost of nonsubsidized insurance for development in landslide-prone areas can discourage such development and encourage lower-risk land uses. The use of insurance as a landslide hazard-reduction technique has the following advantages over other strategies (Olshansky and Rogers 1987; Olshansky 1990):

1. In theory, insurance provides equitable distribution of costs and benefits. If property owners in landslide-prone areas were to pay premiums reflecting their actual risk and if insurance were to fully compensate victims, costs and benefits would be equitably distributed.
2. Landslide insurance encourages hazard reduction if premium rates reflect not only the degree of natural hazard but also the quality of physical control measures.
3. Using insurance to reduce the impact of landslide hazards appeals to those opposed to government regulation because, as compared with the other approaches, it depends more on the private market than on government intervention.

The most successful application of insurance to landslide mitigation has been in New Zealand, where a national insurance program assists homeowners whose dwellings have been damaged by landslides or other natural hazards that could not within reason have been prevented or controlled by the homeowners. A disaster fund, accumulated from a surcharge to the national fire insurance program, reimburses property owners for losses (O'Riordan 1974). This natural-hazard insurance program is an outgrowth of New Zealand's Earthquake and War Damage Act of 1944.

Landslide insurance can be divided into two types, public and private. In the United States, public landslide insurance is available in certain circumstances through the National Flood Insurance Program (NFIP), which was created by the Housing and Urban Development Act of 1968. An amendment to this act extended the application to "mudslides" in 1969. However, the range of phenomena defined by the term *mudslide* was not made clear. As presently worded, the regulations include mudslides that are proximately caused or precipitated by accumulations of surface water or groundwater (Committee on Methodologies for Predicting Mudflow Areas 1982). The insurance on these "water-caused" landslides is provided by private insurance companies but is little used; however, it is underwritten and subsidized by the federal government.

FEMA, which administers the NFIP, has been unable to implement an effective mudslide insurance program, largely because of technical difficulties in defining *mudslide* and in mapping mudslide hazard zones (Olshansky and Rogers 1987). A possible solution to this dilemma would be to add all types of landslides to the NFIP. A bill proposing this solution was introduced in the U.S. House of Representatives in 1981. This bill would have immediately added landslide coverage to the nearly 2 million existing NFIP policies, and a new landslide mapping program would have been undertaken to provide data for the underwriters. However, FEMA opposed it because of its high cost and difficulty in administration. Partly as a result of this opposition, the bill was killed before reaching a vote by the House.

Yelverton (1973) reviewed U.S. experience in landslide insurance as a basis for proposing a national landslide insurance program. However, other than through the NFIP, landslide insurance generally is not available in the United States. Although the concept of private landslide insurance is an appealing one, it has certain drawbacks in practice, and the private sector does not appear to be interested in offering this coverage. The reluctance to provide landslide insurance is of long standing, partially based on several costly and highly publicized landslides along the California coast. In addition, private insurers hesitate to offer landslide coverage because of the problem of "adverse selection," which is the tendency for only those who are in hazardous areas to purchase insurance (Olshansky and Rogers 1987; Olshansky 1990).

Olshansky and Rogers summarized the need and general requirements for landslide insurance in the United States as follows:

Insurance can equitably provide funds to compensate for landslide damage that will inevitably occur even when there are strict land use and grading controls. For insurance to be an effective solution, though, a comprehensive government landslide insurance fund is needed, or alternatively, some other form of government intervention is needed to induce or require private insurers to cover landslides. Controls on building, development, and property maintenance would need to accompany mandatory insurance. Insurance and appropriate government intervention can operate together, each filling a need not served by the other, and each improving the performance of the other in reducing landslides and compensating victims. (Olshansky and Rogers 1987, 992)

The cost of insurance can directly reduce landslide risk by discouraging development in hazardous areas or by encouraging land uses that are less subject to damage. Landslide insurance from private sources is costly for areas known to be susceptible to landslides because losses due to landslides lack the random nature necessary for a sound insurance program. In this respect, landslide areas are comparable with flood areas, and the statement by the American Insurance Association on flood insurance may be applicable:

Flood insurance covering fixed-location properties in areas subject to recurrent floods cannot feasibly be written because of the virtual certainty of loss, its catastrophic nature, and the reluctance or inability of the public to pay the premium charge required to make the insurance self-sustaining. (American Insurance Association 1956)

Sound insurance programs at reasonable rates cannot be made available in known fault-rupture, flood, and landslide areas unless the premium costs are subsidized. Government subsidies to property owners and their mortgagers who suffer damage may lead to development in hazardous areas because the potential loss is indemnified. According to Miller (1977), after national flood insurance became available, lending institutions in 4 of the 15 communities studied reversed earlier restrictions on mortgages in hazardous coastal areas. On the basis of a survey of 1,203 local governments, Burby and French (1981, 84) concluded, "It often appears that the NFIP induces increased flood plain development...." State and local officials interviewed by Kusler (1982, 36, footnote 55) argued that "bank financing would not have been available for much of the new development without flood insurance."

6. MULTIPLE-HAZARD REDUCTION

Typically, landslide hazard reduction is undertaken as an individual exercise. However, landslides often occur as elements of interrelated multiple natural-hazard processes in which an initial event triggers secondary events (Advisory Board on the Built Environment 1983; Advisory Committee on the International Decade for Natural Hazard Reduction 1987). In other cases, two or more natural-hazard processes, not directly related to each other but triggered by a common cause, may occur at the same time in the same or adjacent localities. Examples are the 1964 Alaska earthquake, which triggered tsunamis, local flooding, and many landslides, and the 1980 Mount St. Helens eruption, which led to major landslides, floods, and wildfires that consumed large tracts of timber.

Multiple-hazard problems require a shift of perspective from mitigation of separate hazards, such as landslides, to a broader systems framework that takes into account the characteristics of more than one hazard (Advisory Committee on the International Decade for Natural Hazard Reduction 1987). Therefore, in planning landslide hazard-reduction programs, attention should be paid to possible interrelationships between landslides and other hazards. Building-code requirements in one geographic area may deal individually with landslides, floods, earthquakes, and tornadoes, but the ideal requirement is one that takes into account all of these hazards. For example, a building moved from a floodplain to a hillside to avoid floods may be at increased risk from landslides or earthquakes. In such cases, the failure or loss of the building may be caused by several hazard mechanisms or modes. The planning of mitigation should consider all possible hazard modes. A probabilistic approach for estimating risks associated with multiple hazards is presented in Chapter 6 of this report. A discussion of multiple-hazard mapping—preparation, format, and limitations—was presented in the development planning primer prepared for the Organization of American States (1991, Chap. 6).

7. ELEMENTS OF NATIONAL PROGRAM

As noted earlier, Swanston and Schuster (1989) reviewed landslide hazard management strategies in several countries (Austria, Canada, France, Italy, Japan, New Zealand, Norway, the former Soviet Union, Sweden, and Switzerland) and in Hong Kong, where landslides constitute a major socioeconomic problem. On the basis of the collective experience from these areas and the United States, a successful unified national program of landslide hazard reduction conceivably would include the following key elements (Swanston and Schuster 1989):

1. Identification of a central organization for management of a national landslide loss-reduction program;
2. Establishment of limits of responsibility of federal, state and provincial, municipal, and private entities in dealing with landslide hazards;
3. A national effort to identify and map hazardous areas, define process characteristics, and determine degree of risk;
4. Development of guidelines for application of reduction techniques to identified hazards;
5. Development of minimum standards of application and professional practice (standards should be created by professional societies in collaboration with federal and national governments);
6. Regulation of minimum standards of application and professional practice (in conjunction with professional societies) through periodic review and upgrading of practice guidelines, building codes, and land use practices;
7. Strong support of federal and national government and university research dealing with process mechanics, reduction techniques, and warning systems;
8. Provision of a central clearing house for collection and distribution of publications and guidelines to professionals, agencies, and local governments; and
9. Relief and compensation programs through federal and national and private insurance funds.

Similarly, a comprehensive national program for landslide hazard reduction developed by the U.S. Geological Survey (1982) set forth goals and tasks for making landslide studies, evaluating and mapping the hazard (sometimes called "zonation"), disseminating the information to potential users, and subsequently evaluating the use of the information. Examples of the types and maps to be developed under such a program and lists of typical users and communication techniques were included in the program, which has yet to be actuated except in piecemeal fashion.

REFERENCES

ABBREVIATIONS

UNESCO   United Nations Educational, Scientific, and Cultural Organization

WP/WLI   Working Party on the World Landslide Inventory (International Geotechnical Societies and UNESCO)

Advisory Board on the Built Environment. 1983. *Multiple Hazard Mitigation: Report of a Workshop on Mitigation Strategies for Communities Prone to Multiple Natural Hazards*. National Research Council, National Academy Press, Washington, D.C., 60 pp.

Advisory Committee on the International Decade for Natural Hazard Reduction. 1987. *Confronting Natural Disasters—An International Decade for Natural Hazard Reduction*. National Research Council, National Academy Press, Washington, D.C., 60 pp.

American Insurance Association. 1956. *Studies of Floods and Flood Damages, 1952–55*. New York, N.Y., 296 pp.

Brabb, E.E. 1984. Innovative Approaches to *Landslide Hazard and Risk Mapping*. Proc., Fourth International Symposium on Landslides, Canadian Geotechnical Society, Toronto, Canada, Vol. 1, pp. 307–323.

Brabb, E.E., and B.L. Harrod (eds.). 1989. *Landslides: Extent and Economic Significance*. Proc., 28th International Geological Congress: Symposium on Landslides, Washington, D.C., July 17, A.A. Balkema, Rotterdam, Netherlands, 385 pp.

Brabb, E.E., E.H. Pampeyan, and M.G. Bonilla. 1972. *Landslide Susceptibility in San Mateo County, California*. U.S. Geological Survey Miscellaneous Field Studies Map MF-360.

Brown, W.M. III. 1992. Information for Disaster Reduction: The National Landslide Information Center, U.S. Geological Survey. In *Landslides: Proc., Sixth International Symposium on Landslides* (D.H. Bell, ed.), Christchurch, New Zealand, Feb. 10–14, A.A. Balkema, Rotterdam, Netherlands, Vol. 2, pp. 891–892.

Brown, W.M. III, D.M. Cruden, and J.S. Denison. 1992. *The Directory of the World Landslide Inventory*. U.S. Geological Survey Open-File Report 92-427, 216 pp., Appendix 19 pp.

Burby, R.J., and S.P. French. 1981. Coping with Floods—The Land Use Management Paradox. *Journal of the American Planning Association*, Vol. 47, No. 3, pp. 289–300.

Cambiaghi, A., and R.L. Schuster. 1989. Landslide Damming and Environmental Protection—A Case Study from Northern Italy. In *Proc., Second International Symposium on Environmental Geotechnology* (H.Y. Fang and S. Pamukcu, eds.), Shanghai, May, Envo Publishing Company, Inc., Vol. 1, pp. 469–480.

Colorado Natural Hazards Mitigation Council. 1992. *Annual Report 1991–1992*. Office of Emergency Management, Golden, Colo., 55 pp.

Committee on Ground Failure Hazards. 1985. *Reducing Losses from Landsliding in the United States*. Commission on Engineering and Technical Systems, National Research Council, Washington, D.C., 41 pp.

Committee on Methodologies for Predicting Mudflow Areas. 1982. *Predicting a Methodology for Delineating Mudslide Hazard areas for the National Flood Insurance Program*. National Research Council, Washington, D.C., 35 pp.

Cruden, D.M., and W.M. Brown III. 1992. Progress Towards the World Landslide Inventory. In *Landslides: Proc., Sixth International Symposium on Landslides* (D.H. Bell, ed.), Christchurch, New Zealand, Feb. 10–14, A.A. Balkema, Rotterdam, Netherlands, Vol. 1, pp. 59–64.

Dallaire, G. 1976. Consultants Reviewing Plans for Other Consultants in Fairfax County, Va.; Landslides Greatly Reduced. *Civil Engineering*, ASCE, Vol. 46, No. 9, pp. 77–79.

Geotechnical Control Office. 1985. *Description of the Geotechnical Control Office Engineering Development Department*. Hong Kong, 8 pp.

Gillon, M.D., P.F. Foster, G.T. Proffitt, and A.P. Smits. 1992. Monitoring of the Cromwell Gorge Landslides. In *Landslides: Proc., Sixth International Conference on Landslides* (D.H. Bell, ed.), Christchurch, New Zealand, Feb. 10–14, A.A. Balkema, Rotterdam, Netherlands, Vol. 2, pp. 1135–1140.

Hamilton County. 1982. *Earth Movement Task Force*. Hamilton County, City of Cincinnati, Ohio, 38 pp.

Hamilton County Regional Planning Commission. 1976. *Hillside Development Study—Hillside Development Plan and Strategy*. Regional Planning Commission, Hamilton County, Cincinnati, Ohio, 86 pp.

Hansen, A. 1989. Landslide Hazard Analysis. In *Slope Instability* (S. Brunsden and D.B. Prior, eds.), John Wiley & Sons, New York, pp. 523–595.

Hansen, W.R., E.B. Eckel, W.E. Schaen, R.E. Lyle, W. George, and G. Chance. 1966. *The Alaska Earthquake of March 27, 1964—Investigations and*

Reconstruction Effort. U.S. Geological Survey Professional Paper 541, 111 pp.

Huffman, J. 1986. *Government Liability and Disaster Mitigation—A Comparative Study*. University Press of America, Lanham, Md., 626 pp.

International Conference of Building Officials. 1985. *Uniform Building Code*. Whittier, Calif., 817 pp.

Jahns, R.H. 1969. Seventeen Years of Response by the City of Los Angeles to Geologic Hazards. In *Geologic Hazards and Public Problems* (R.A. Olson and M.M. Wallace, eds.), Conference Proceedings, May, Office of Emergency Preparedness, Washington, D.C., pp. 283–296.

Jochim, C.L., W.P. Rogers, J.O. Truby, R.L. Wold, Jr., G. Weber, and S.P. Brown. 1988. *Colorado Landslide Hazard Mitigation Plan*. Colorado Geological Survey, Department of Natural Resources, Denver, 149 pp.

Keefer, D.K., R.C. Wilson, R.K. Mark, E.E Brabb, W.M. Brown III, S.D. Ellen, E.L. Harp, G.F. Wieczorek, C.S. Alger, and R.S. Zatkin. 1987. Real-Time Landslide Warning during Heavy Rainfall. *Science*, Vol. 238, pp. 921–925.

Kockelman, W.J. 1986. Some Techniques for Reducing Landslide Hazards. *Bulletin of the Association of Engineering Geologists*, Vol. 23, No. 1, pp. 29–52.

Kockelman, W.J., and E.E. Brabb. 1979. Examples of Seismic Zonation in the San Francisco Bay Region. In *Progress on Seismic Zonation in the San Francisco Bay Region* (E.E. Brabb, ed.), U.S. Geological Survey Circular 807, pp. 73–84.

Kusler, J.A. 1982. *Regulation of Flood Hazard Areas to Reduce Flood Losses*. U.S. Water Resources Council, Washington, D.C., Vol. 3, 357 pp.

Mader, G.G., W.E. Spangle, and M.L. Blair. 1980. *Land Use Planning after Earthquakes*. William Spangle and Associates, Inc., Portola Valley, Calif., 24 pp.

Mader, G.G., T.C. Vlasic, and P.A. Gregory. 1988. *Geology and Planning: The Portola Valley Experience*. William Spangle and Associates, Inc., Portola Valley, Calif., 75 pp.

Miller, H.C. 1977. *Coastal Flood Hazards in the National Flood Insurance Program*. National Flood Insurance Program, Department of Housing and Urban Development, Washington, D.C., 58 pp.

Obermeier, S.F. 1979. *Slope Stability Map of Fairfax County, Virginia*. U.S. Geological Survey Miscellaneous Field Studies Map MF-1072.

Olshansky, R.B. 1990. *Landslide Hazard in the United States—Case Studies in Planning and Policy Development*. Garland Publishing, Inc., New York, 176 pp.

Olshansky, R.B., and J.D. Rogers. 1987. Unstable Ground: Landslide Policy in the United States. *Ecology Law Quarterly*, Vol. 13, No. 4, pp. 939–1006.

Organization of American States. 1991. *Primer on Natural Hazard Management in Integrated Regional Development Planning*. Department of Regional Development and Environment, Washington, D.C., 418 pp.

O'Riordan, T. 1974. The New Zealand Natural Hazard Insurance Scheme: Application to North America. In *Natural Hazards—Local, National, Global* (G.F. White, ed.), Oxford University Press, New York, pp. 217–219.

Petak, W.J., and A.A. Atkisson. 1982. *Natural Hazard Risk Assessment and Public Policy—Anticipating the Unexpected*. Springer-Verlag, New York, 489 pp.

Rib, H.T., and T. Liang. 1978. Recognition and Identification. In *Special Report 176: Landslides: Analysis and Control* (R.L. Schuster and R.J. Krizek, eds.), TRB, National Research Council, Washington, D.C., pp. 34–80.

Rogers, W.P., L.R. Ladwig, A.L. Hornbaker, S.D. Schwochow, S.S. Hart, D.C. Shelton, D.L. Sroggs, and J.M. Soule. 1974. Appendix: Model Geologic Hazard Area Control Regulations. In *Guidelines and Criteria for Identification and Land-use Controls of Geologic Hazard and Mineral Resource Areas*, Colorado Geological Survey, Denver, pp. 135–146.

Rossi, P.H., J.D. Wright, and E. Weber-Burdin. 1982. *Natural Hazards and Public Choice—The State and Local Politics of Hazard Mitigation*. Academic Press, New York, 337 pp.

San Mateo County Board of Supervisors. 1973. *Adding a Resource-Management District and Regulations to the County Zoning Ordinance*. Ordinance No. 2229. Redwood City, Calif., 24 pp.

Santa Clara County Board of Supervisors. 1978. *Geological Ordinance NS-1205.35: Santa Clara Code, Section C-12–600 ff*. San Jose, Calif.

Schuster, R.L., and R.W. Fleming. 1986. Economic Losses and Fatalities due to Landslides. *Bulletin of the Association of Engineering Geologists*, Vol. 23, No. 1, pp. 11–28.

Schuster, R.L., and F.B. Leighton. 1988. Regulations in California, U.S.A. In *Landslides and Mudflows* (E.A. Kozlovskii, ed.), UNESCO/UNEP, Moscow, Vol. 2, pp. 116–122.

Scullin, C.M. 1983. *Excavation and Grading Code Administration, Inspection, and Enforcement*. Prentice-Hall, Inc., Englewood Cliffs, N.J., 405 pp.

Selkregg, L., E.B. Crittenden, N. Williams, Jr. 1970. Urban Planning in the Reconstruction. In *The Great Alaska Earthquake of 1964—Human Ecology*, National Academy of Sciences, Washington, D.C., pp. 186–242.

Slosson, J.E. 1969. The Role of Engineering Geology in Urban Planning. In *The Governor's Conference on Environmental Geology*, Colorado Geological Survey Special Publication 1, Denver, pp. 8–15.

Slosson, J.E., and J.P. Krohn. 1979. AEG Building Code Review—Mudflow/Debris Flow Damage; February 1979 Storm—Los Angeles Area. *California Geology*, Vol. 32, No. 1, pp. 8–11.

Slosson, J.E., and J.P. Krohn. 1982. Southern California Landslides of 1978 and 1980. In *Storms, Floods, and Debris Flows in Southern California and Arizona, 1978 and 1980*, Proceedings of a Symposium, National Research Council and Environmental Quality Laboratory, California Institute of Technology, Pasadena, 17–18 Sept. 1980, National Academy Press, Washington, D.C., pp. 291–319.

Swanston, D.N., and R.L. Schuster. 1989. Long-Term Landslide Hazard Mitigation Programs: Structure and Experience from Other Countries. *Bulletin of the Association of Engineering Geologists*, Vol. 26, No. 1, pp. 109–133.

U.S. Geological Survey. 1982. *Goals and Tasks of the Landslide Part of a Ground-Failure Hazards Reduction Program*. U.S. Geological Survey Circular 880, 49 pp.

Varnes, D.J. 1978. Slope-Movement Types and Processes. In *Special Report 176: Landslides: Analysis and Control* (R.L. Schuster and R.J. Krizek, eds.), TRB, National Research Council, Washington, D.C., pp. 11–33.

Varnes, D.J., and International Association of Engineering Geology Commission on Landslides and Other Mass Movements on Slopes. 1984. *Landslide Hazard Zonation—A Review of the Principles and Practice*. UNESCO, Natural Hazards Series, No. 3, 63 pp.

Veder, C. 1981. *Landslides and Their Stabilization*. Springer-Verlag, New York, 247 pp.

Wold, R.L., Jr., and C.L. Jochim. 1989. *Landslide Loss Reduction: A Guide for State and Local Government Planning*. Earthquake Hazards Reduction Series 52. Federal Emergency Management Agency, Washington, D.C., Aug., 50 pp. (Also published as Colorado Geological Survey Special Publication 33, Denver, 50 pp.)

WP/WLI. 1990. A Suggested Method for Reporting a Landslide. *Bulletin of the International Association of Engineering Geology*, No. 41, pp. 5–12.

WP/WLI. 1991. A Suggested Method for a Landslide Summary. *Bulletin of the International Association of Engineering Geology*, No. 43, pp. 101–110.

WP/WLI. 1993a. A Suggested Method for Describing the Activity of a Landslide. *Bulletin of the International Association of Engineering Geology*, No. 47, pp. 53–57.

WP/WLI. 1993b. *Multilingual Landslide Glossary*. BiTech Publishers, Richmond, British Columbia, Canada, 59 pp.

Yelverton, C.A. 1973. Land Failure Insurance. In *Geology, Seismicity and Environmental Impact*, Association of Engineering Geologists, Brentwood, Tenn., pp. 15–20.

Chapter 6

TIEN H. WU,
WILSON H. TANG, AND
HERBERT H. EINSTEIN

LANDSLIDE HAZARD AND RISK ASSESSMENT

1. INTRODUCTION

As a general principle, the choice among different landslide management options should be based on cost. The direct or initial cost, such as cost of construction or removal; social costs; and the costs of lost opportunity and potential failures need to be considered. The choice among management options described in Chapter 5 is made under conditions of uncertainty because future events that may trigger landslides, such as rainstorms and earthquakes, cannot be forecast with certainty. Uncertainty also arises because of insufficient information about site conditions and incomplete understanding of landslide mechanisms. The uncertainties prevent accurate predictions of landslide occurrence or of the performance of physical control measures. In a broader sense, uncertainty should include the probability of success or failure of hazard-reduction measures such as avoidance and codes.

Geotechnical engineers are familiar with risk and decision making under uncertainty. The nature of risk and the need to balance safety with economy in geotechnical design were noted by Casagrande (1965) 30 years ago. A rational decision process used to choose among management options should account for the uncertainties. The concepts of decisions under uncertainty and probabilistic decision models have been well established in business management for over 30 years (e.g., Schlaifer 1959; Raiffa and Schlaifer 1961) and have been successfully applied to engineering problems (Keeney and Raiffa 1976). The basic elements of hazard and risk assessment and decision making as applied to landslide management are summarized in this chapter.

2. DESCRIPTION OF UNCERTAINTY

When there is uncertainty, the conventional approach is to make conservative estimates of the design parameters. In probabilistic analysis, the uncertainty about a variable, called a random variable, is described by a probability density function, $f(x)$ [see Figure 6-1(a)], with mean μ_x and standard deviation σ_x. The probability that the random variable x may have values between a and b is given by the shaded area. The function $f(x)$ may be obtained by fitting to data. When data are insufficient for determination of $f(x)$, it is still possible to obtain reasonable estimates of the mean and standard deviation. Opinions based on experience and judgment can be incorporated as subjective probability. Engineers frequently express their opinions in the form of a best estimate and a range. This can be conveniently described by a subjective probability that has a triangular distribution [see Figure 6-1(b)], where b represents the best estimate and a and c represent the upper and lower limits of the range. Formal methods for evaluating subjective probability were described by Winkler (1969), Brown (1974), Staël von Holstein and Matheson (1979), and Agnew (1985).

Roberds (1990) provided a review of the assessment of subjective probability.

The following sources of uncertainty are commonly encountered in geotechnical engineering. First, future loads and environmental conditions cannot be predicted with certainty. For example, the occurrence of an earthquake of a given magnitude or a given ground acceleration can only be estimated on a probabilistic basis and expressed as the probability that the acceleration will exceed a given value. Pore-water pressure and seepage forces due to future rainstorms may be treated in a similar manner.

The second source of uncertainty concerns site conditions. Spatial variability of geologic materials requires the engineer to make extrapolations from material types observed at boreholes and samples and to make inferences about material types that may exist at points where no observations are made. Such extrapolations involve a large degree of uncertainty. For example, geologic anomalies can be present at a site even though they were not detected during site exploration (Baecher 1979; Halim and Tang 1991). The persistence and location of joints in rock, which are planes of weakness, cannot be accurately determined in site exploration (Baecher et al. 1977). In addition, errors in estimating material properties are introduced when the number of samples is insufficient; when the test method does not accurately measure the property, such as the in situ strength; and when test procedures contain random errors. The above errors were reviewed and the associated uncertainties estimated by Lumb (1975), Tang et al. (1976), Baecher (1979), and Wu (1989).

Analytical models are used to predict performance of geotechnical structures. Models commonly used in landslide analysis include those for stability analysis and seepage. Analytical models contain errors, introduced through inadequacies and simplifications in theory, simplifications in boundary conditions, and approximations in numerical computations. Empirical evidence provides some rough measure of model error. Results of model tests performed to check the predictions of theory have been used to estimate model error.

Although probabilistic methods have been developed to estimate the uncertainties associated with the three sources described above, a fourth

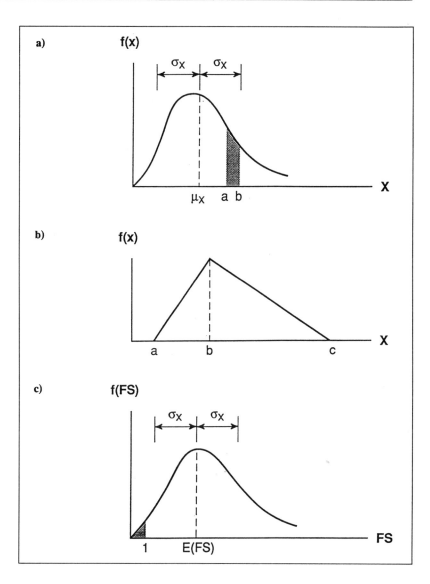

source of uncertainty, that caused by possible omissions, cannot be formally described. Omission refers to the failure by the engineer to consider possible modes of failure or factors that could affect performance. Good engineering practice should avoid omissions, although the probability of omissions is difficult to quantify.

In view of the uncertainties involved in the various stages of geotechnical design, it is frequently necessary to revise estimates of site conditions and foundation performance as more information becomes available. This revision is the essence of the "observational approach" (Terzaghi 1961; Peck 1969). Updating an estimate on the basis of new observations can be modeled via Bayes' theorem:

FIGURE 6-1
Probability
distribution functions.

$$P''(\theta_1|Z) = kL(Z|\theta_1)P'(\theta_1) \qquad (6.1)$$

where

$P'(\theta_1)$ = prior probability, which is the probability that state θ is θ_1 before the new observation;

$P''(\theta_1|Z)$ = posterior probability, which is the probability that state θ is θ_1 given the observed results Z;

$L(Z|\theta_1)$ = likelihood function, which is the probability of observing the results Z given that state θ is θ_1; and

k = normalization constant, which is needed to make the sum of the probabilities over all possible θ's equal to 1.

The state θ may also be used to represent the unknown parameter, such as the mean μ of the probability distribution of a random variable. Moreover, the state θ can be a continuous random variable. In this case, the probabilities $P'(\theta_1)$ and $P''(\theta_1)$ in Equation 6.1 are replaced by the corresponding probability density functions $f'(\theta)$ and $f''(\theta)$, respectively. Bayes' theorem provides a vehicle for combining observational information with professional opinion quantified as subjective probabilities.

3. ESTIMATION OF HAZARD

Available methods for estimating hazard, defined as failure probability P_f, range from reliability analysis to purely empirical estimates. In formal reliability analysis, the performance of a geotechnical system (embankment, slope, etc.) is expressed as a function of controlling factors (precipitation, soil strength, etc.) that are considered to be random variables because their values are not precisely known. As described in Section 2, each random variable x is characterized by a probability density function, $f(x)$, with mean μ_x and standard deviation σ_x. Logically, the mean of an input should represent the engineer's best estimate of the true value without conservatism, whereas the variance σ_x^2 should represent uncertainty about the true value. Thus, the mean and variance reflect the technical expert's judgment about the uncertain variable.

To evaluate reliability, the performance of a geotechnical system may be expressed as a performance function. For example, the performance function that defines the safety of a slope could be the factor of safety, which in turn is a function of random variables that include load and strength. The probability density functions of strength, and so on, are used to derive the probability density function of the performance function, which is then used to calculate the failure probability. Thus far, this has been accomplished only for simple problems because the complexity of the various relations in the performance function makes it difficult to obtain closed-form solutions. Most solutions have used the first-order, second-moment (FOSM) method (Ang and Tang 1975).

In FOSM, Y is the performance variable, such as the safety factor. It is a function G of the random variables $X_1, X_2, \ldots, X_n$, which represent strength, ..., or

$$Y = G(X_1, X_2, \ldots, X_n) \qquad (6.2)$$

To translate means, variances, and correlations of input variables $X_1, X_2, \ldots$ to the mean and variance of the performance function, a simple linear approximation is used to obtain the following relations (Benjamin and Cornell 1970; Ang and Tang 1975):

$$\bar{Y} \cong G(\bar{X}_1, \bar{X}_2, \ldots, \bar{X}_n) \qquad (6.3)$$

$$\text{Var}(Y) \cong \Sigma_i \Sigma_j \left(\frac{\partial G}{\partial X_i} \frac{\partial G}{\partial X_j} \Big|_m \right) \text{Cov}(X_i, X_j) \qquad (6.4)$$

where

X_i = mean or expected value of X_i,
$\text{Cov}(X_i, X_j)$ = covariance of X_i and X_j, and
$\text{Var}(Y)$ = variance of Y.

In Equation 6.4, $|_m$ indicates that the partial derivatives are evaluated at the mean values of X_i, X_j. When the parameters are all mutually independent, Equation 6.4 reduces to

$$\text{Var}(Y) = \Sigma_i \left(\frac{\partial G}{\partial X_i} \Big|_m \right)^2 \text{Var}(X_i) \qquad (6.5)$$

The simple relationships in Equations 6.2 through 6.5 may be used to estimate the mean and variance of a material property, such as strength, because of uncertainties about various input parameters and test conditions. Similarly, the mean and variance of the resistance along a potential slip surface due to uncertainty about the shear strength and other soil properties can be estimated. When the performance variable Y is the safety factor, it is commonly assumed that the safety factor has a log-normal distribution, with mean $E(FS)$ and standard deviation σ_{FS} [see Figure 6-1(c)], as determined from Equations 6.2 through 6.5. It is then possible to calculate the probability that the safety factor is equal to or less than 1, which is the shaded area in Figure 6-1(c).

Instead of the probability of failure, the safety may be expressed by a reliability index (Hasofer and Lind 1974):

$$\beta = \frac{[E(FS) - 1]}{\sigma_{FS}} \qquad (6.6)$$

The numerator of this equation is the distance along the abscissa [Figure 6-1(c)] that measures the difference between the mean safety factor $E(FS)$ and failure at $FS = 1$. This difference is equivalent to a margin of safety. When this difference is divided by σ_{FS}, the margin of safety becomes relative to the uncertainty about the safety factor. Thus, β is a measure of safety while taking into account the magnitude of the uncertainties involved. Clearly, when the uncertainty is large, larger safety factors are necessary to maintain the same level of safety. The reliability index as defined by Equation 6.6 is obtained directly from the two moments, mean and variance, and requires no assumption about probability density functions. The FOSM approach can be extended to problems with performance vectors and to performance functions expressed numerically by finite-difference or finite-element methods (Ditlevsen 1983).

The procedures described above provide probability estimates for many practical problems. However, these probability estimates are only approximate in most cases. When more accurate estimates of the reliability are required, first-order reliability and second-order reliability methods (FORM and SORM) may be applied (Ang and Tang 1984; Madsen et al. 1986). In these methods the partial derivatives in Equations 6.2 through 6.5 are evaluated at the most likely failure point on the surface defined by the performance function. An iterative procedure is required to obtain the failure point and the probability of failure. Finally, Monte Carlo simulations can be used when the system performance assessment becomes too complex for analytical evaluations.

Solutions were obtained for reliability of soil slopes by Tang et al. (1976) and of rock slopes by Einstein et al. (1983) and Scavia et al. (1990). Chowdhury and Tang (1987) presented a review

of probabilistic analysis of slope stability. Methods for evaluating landslides induced by precipitation and by earthquake were reviewed by Ang et al. (1985). Where high pore pressures caused by precipitation may initiate instability, estimates of failure probability should account for the probable occurrence of the critical pore pressure that will cause failure. Examples of estimating pore pressures due to rainfall were given by Wu and Swanston (1980), Bevan (1982), Sangrey et al. (1984), and Suzuki and Matsuo (1991). Failure probability can also be calculated for the various types of protective physical control systems mentioned in Chapter 5 and described further in Chapters 17 and 18.

Where available data are insufficient for statistical analysis, the landslide hazard can be estimated by judgment. Roberds (1991) described a procedure by which slopes along a route were inspected and a probability of failure was assigned to each possible mode of failure on the basis of subjective judgment. Probabilities of failure can also be assigned to slopes on the basis of observed failures. The probabilities of failure or success pertaining to avoidance, codes, and zoning involve social and political considerations and are much more difficult to evaluate. The use of subjective probability based on experience may be the only way to estimate such probabilities [see examples given by Keeney and Raiffa (1976)].

Safety of a geotechnical system may mean satisfactory performance with respect to several modes of failure. For instance, a retaining wall can fail by overturning, sliding, or lacking adequate bearing capacity. Landslides can be induced by precipitation or earthquake. In a broad view, landslides often occur as one mode of multiple failure modes triggered by some event (Advisory Board on the Built Environment 1983; Advisory Committee on the International Decade for Natural Hazard Reduction 1987). Examples are the 1964 Alaska earthquake, which triggered tsunamis, local flooding, and many landslides, and the 1980 Mount St. Helens eruption, with associated landslides, floods, and wildfires.

Multiple modes of failure require a shift of perspective from mitigation of individual modes, such as landslides, to mitigation of an entire system that takes all of the modes into account (see Chapter 5, Section 6). For example, a building moved from a floodplain to a hillside to avoid floods may be at increased hazard from landslides or earthquakes.

The plan of mitigation should consider all possible failure modes. When failure may be caused by landslide or by flood, the failure probability (P_f) is

$$P_f = P(A \cup B) = P(A) + P(B) - P(A \cap B) \quad (6.7)$$

where

A = failure due to landslide,
B = failure due to flooding, and
$\cup$ and $\cap$ = union and intersection between events.

In this simple case, events A and B may be considered independent; hence $P(A \cap B)$ is given by the product of $P(A)$ and $P(B)$. Otherwise the conditional probability $P(A|B)$ or $P(B|A)$ is needed. For complex problems in which the effect of correlations between the respective failure modes may not be easily determined, one can estimate the upper and lower bounds on P_f for system failure (Ang and Tang 1984).

4. DECISION UNDER UNCERTAINTY

In probabilistic decision theory, the choice between available options is dependent on the preferences of the decision maker concerning all possible outcomes of each of the options. A sim-

plified flow diagram that illustrates the decision-making process is shown in Figure 6-2. Step 1 in a geotechnical engineering project is site exploration and soil testing to define material distribution and material properties. The results are used to characterize the site. In Step 2, all possible failure modes are identified. In Step 3, the hazard or probability of failure P_f for each mode is estimated, as described in the previous section. For each mode of failure, the consequence C_f is estimated (Step 4). Then the risk, defined as

$$R = P_f C_f \quad (6.8)$$

is calculated (Step 5). This procedure is repeated for each of the available options in the management strategy, which obviously should include the options of doing nothing (as discussed in Chapter 5) and of additional investigation. The choice of a management strategy is made on the basis of the decision criterion, which should include the potential risk associated with each option and the initial capital cost of the option. For some problems, the criterion may be the expected cost, which is defined as

$$E(C) = C_o + P_f C_f \quad (6.9)$$

where

C_o = initial cost,
P_f = probability of failure, and
C_f = consequence or cost of failure.

In more complicated situations, some of the consequences cannot be directly represented as monetary values. A utility function is used to express the owner's preferences with respect to the possible outcomes. Multiattribute utility makes it possible to include such attributes as delays or loss of service, lost opportunity, social disruptions, and effects on users and nonusers. The use of multiattribute utility in decision analysis was treated by Raiffa and Schlaifer (1961), Keeney and Raiffa (1976), and Ang and Tang (1984).

After the probabilities and consequences have been estimated, methods of decision analysis (Benjamin and Cornell 1970; Ang and Tang 1984) may be used to arrive at management decisions (Step 6). A hypothetical decision tree, such as that shown in Figure 6-3 for landslide mitigation, can be used to identify the alternatives of actions, pos-

FIGURE 6-2
Decision-making
process for landslide
mitigation.

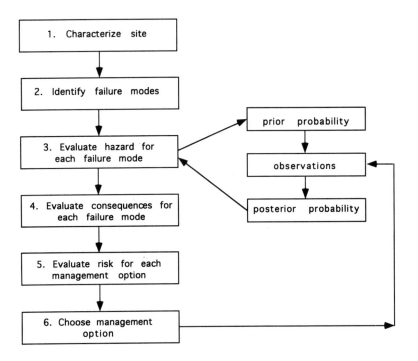

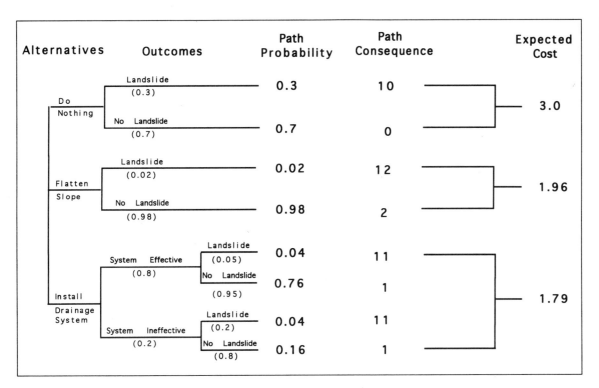

FIGURE 6-3
Decision tree for landslide mitigation.

sible outcomes, and respective consequences or costs for each scenario or path. The probability of each branch of the outcome can be determined either from probabilistic models or subjectively from available information. As given in Equation 6.9, the expected cost of each alternative is the summation of the path probability times the path consequences over the outcomes of all scenarios for that alternative. The alternative with the least expected cost is chosen if the expected value is the criterion for decision. In this example, the optimal solution for this site is the installation of a drainage system. When the probabilities or costs are crude estimates, a sensitivity analysis should be performed to find out if the optimal solution changes with the probabilities and costs.

5. LANDSLIDE HAZARD MAPS

When risk assessment is made over a large area, the results may be expressed in the form of landslide hazard or landslide risk maps (Brabb 1984). Various types of landslide hazard maps are described in Chapter 8. An ideal landslide hazard map should provide information concerning the spatial and temporal probabilities of all anticipated landslide types within the mapped area, and also include information about their types, magnitudes, velocities, and sizes. In this section, the construction of hazard maps is described on the basis of the methodology outlined in the previous section.

A comprehensive mapping procedure for landslide management was proposed by Einstein (1988). Maps containing different types of information are constructed in sequence. According to Einstein, *state-of-nature maps* are those that present data without interpretation. These include geologic and topographic maps, precipitation data, and so forth, as well as the results of site investigation. Construction of such maps corresponds to Step 1 in Figure 6-2. *Danger maps* indicate the possible modes or failure mechanisms, such as debris flows, rock falls, and so forth. These maps follow from Step 2 in Figure 6-2. *Hazard maps*, Step 3 in Figure 6-2, show the probability of failure for various failure modes shown on danger maps. For simple failure modes, the probability that such a failure could occur during a given time interval can be estimated on the basis of the probability distributions of the triggering mechanism (rainstorm, earthquake, etc.), soil or rock properties, slope geometry, and other controlling factors. The results can be shown on a map that delineates zones with different failure probabilities (Viberg 1984; Wu 1992). Alternatively, hazards are expressed qualitatively as high, medium, or low. If the consequence of failure, also called "vulnerability" by Varnes et al. (1984) and van Westen (1993), can be estimated, the risk as defined in Equation 6.6 can be calculated for the construction of a risk map, as shown in Steps 4 and 5 of Figure 6-2. Finally, a *management map* can be used to summarize the management decisions (Step 6 in Figure 6-2).

Most available maps fall into the categories of danger and hazard maps. Danger maps can be constructed by consideration of lithology, rock structure, and hydrology and the relation of these factors to the topography. When data are insufficient for analytical evaluation of failure probability, a rating system may be used to obtain a hazard rating of low, medium, or high. Failure probabilities may be estimated from observed landslide frequencies (Brabb et al. 1972; Wieczorek 1984). These estimates may be based on examination of multiple-date aerial photographs (Canuti 1986), ground observations, or review of historical records. Danger maps provide a spatial distribution of landslides (Wright et al. 1974). A combination of historical and subjective assessment has been used for regions with similar geology, topography, and climate. This approach was used to produce the U.S. landslide susceptibility map by Radbruch-Hall et al. (1982). Hazard can also be estimated by statistical correlation with factors considered to correlate with landslides (Carrara et al. 1991). Fully worked out examples of quantitative landslide risk analysis are rare because of the difficulties in defining both hazard and consequence quantitatively (Einstein 1988; Kienholz 1992). One example of a complete mapping system is the proposed methodology called *Plans d'Exposition aux Risques Naturels* (PER) (Office of the Prime Minister 1985).

6. HISTORIC FAILURE RATE

Historic failure rates can be helpful by providing a broader perspective to the application of hazard prediction methods. Ideally, landslide hazard should be expressed as the probability of failure per time per area. Hazards for other forms of instability can also be included. For slopes undergoing creep movement, the probability of accelerating movement may be appropriate. Examples of historic failure rates include those for the Alpine and Pre-Alpine regions of Switzerland, where "super events," such as the Flims and Sierre landslides,

have return periods of 10^4 years or greater, whereas "major events," such as the landslides at Goldau or Deborence, have return periods of 10^2 to 10^3 years (Einstein 1988). However, there is a general lack of formal statistical evaluation of landslides.

In the design of corrective measures, the engineer may refer to failure rates of geotechnical systems. Although failure statistics have not been formally compiled for geotechnical systems, rough estimates can be made based on data from various studies. Baecher et al. (1980) determined that the historical rate of failure of dams is about 0.0001 per dam-year, or about 0.01 for the average life of a dam, for a wide range of locations and times of construction. The details of failure mechanisms are often imprecise, but it appears that about one-third of the failures were due to overtopping and another one-third to internal erosion, piping, or seepage. These numbers are close to what would be expected from the judgment and experience of the profession. The remaining one-third of the failures are due to slides and other mechanisms.

Results of an extensive survey of embankment and cut-slope failures along British motorways showed that "percent of failure" for embankments and cuts in different geologies ranged between 0 and 0.13 (Perry 1989). The percent of failure was defined as length of failed slope to total length of slope, and the age of the earthworks ranged from 2 to 26 years. An exceptionally low percentage of failure of 0.003 was found for cuts in London clay, and this low percentage was attributed to the flatter design slopes adopted for this well-known material. The statistics reflect a wide range in construction quality and design criteria.

More recent experience by the Ministry of Transportation of Ontario (MTO) with embankments on clay soils similar to Leda clay and New Liskard clay showed that no failures have occurred where the design safety factor is 1.3 for a shear strength measured by the MTO field vane (M.S. Devata, personal communication, 1992, Ministry of Transportation of Ontario, Canada). Since there have been at least 1,000 embankments of this type, it appears that the failure probability is less than 0.001 per embankment during the initial period of several years when modern design and construction methods have been used with a material that has been thoroughly investigated and a geology that is without surprises and anomalies.

Estimates of failure rates have also been expressed by individuals on the basis of their experience. For example, Meyerhof (1970) estimated the overall failure rate to be around 0.001 for earthworks and retaining structures and 0.0001 for foundations over the lifetime of the structure. Lambe's (1985) estimate of the lifetime failure probability for embankments and slopes designed by qualified engineers and built with adequate supervision and monitoring of performance was 0.0001 or less when the design safety factor was 1.5. Besides these judgmental estimates, results of reliability analyses on previously designed structures are also worth noting. The lifetime failure probabilities for foundations of offshore gravity structures (Wu et al. 1989) and of highways (Barker et al. 1991) were estimated to be 0.001 or less. Hence, a lifetime failure probability on the order of 0.001 may be acceptable to the profession. However, one should be cautioned that the consequences of failure in individual cases can vary substantially. Hence, acceptable failure probability can also vary between projects. In this regard, Whitman's (1984) plot of annual failure probability versus consequences of failure for a number of structures and civil engineering projects might provide some indication of acceptable risks (see Figure 6-4).

7. APPLICATIONS

Although it is possible to use probabilistic methods to arrive at decisions on location of transportation and other facilities in which socioeconomic factors are involved [for examples, see work by Keeney and Raiffa (1976)], such methods have not been widely used in landslide hazard reduction by regulation or zoning. Applications of probabilistic methods to design and operations of more limited scope are plentiful. Several examples are given below, beginning with the simple and progressing to the complex. The degree of complexity depends on the nature of the problem.

A scoring system for rock-fall hazards is used by the Oregon Department of Transportation (Pierson 1992) for management of rock slopes along highways. Scores are assigned according to geologic structure, erosion, and so forth. The scores are based largely on observed frequency of rock falls and are proportional to probabilities of failure, given the geologic structure, erosion, and other factors. Scores are also assigned for route conditions, such as sight distance and roadway width. These scores represent consequences, be-

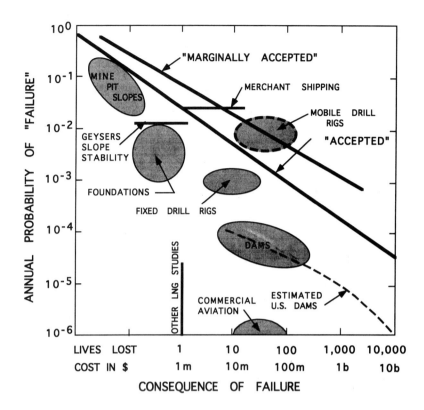

cause short sight distance and narrow roadways are more likely to result in accidents, given a rock fall. Thus, the total score contains the essential elements of risk as defined in Equation 6.8. The ranges in cost of various methods for rock-fall mitigation are estimated. Criteria for choosing slopes for mitigation include a slope's score in the rock hazard rating system (RHRS), the RHRS score relative to cost of mitigation, and others. Although this system does not use the formal method of the decision tree, it contains the essential elements of decision analysis.

The method used by Wagner et al. (1987) to identify hazards of rock and debris slides in Nepal is close to Einstein's mapping procedure. State-of-nature maps consist of a geologic map and a slope map (Maps 1 and 2 in Figure 6-5). A morphostructural map (Map 3 in Figure 6-5) is constructed from the slopes and dips of discontinuities, which are planes of weakness. Where the dip of the slope exceeds that of a discontinuity and the two dips are in the same direction, the slope is considered "structural," and wedge or block failures are possible. Thus, the morphostructural map is analogous to Einstein's danger map. The probability of failure is estimated empirically by weights assigned to the structure, lithology, hydrology, tectonics,

FIGURE 6-4
Risks for selected engineering projects (Whitman 1984).
REPRINTED WITH PERMISSION OF THE AMERICAN SOCIETY OF CIVIL ENGINEERS

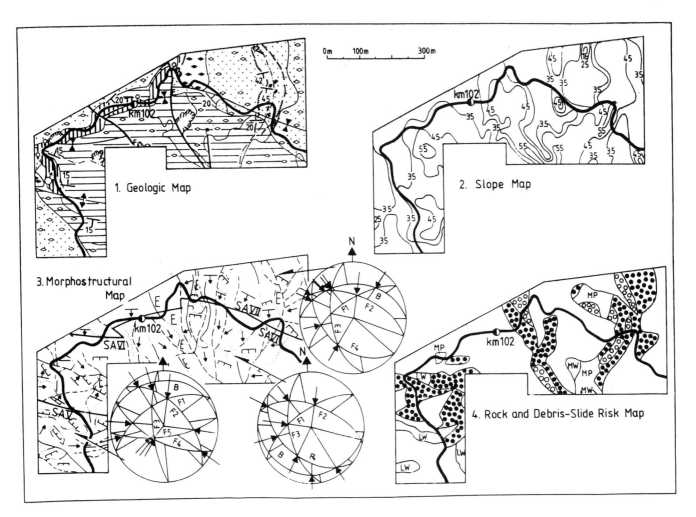

1. Geologic Map   2. Slope Map

3. Morphostructural Map

4. Rock and Debris-Slide Risk Map

1. Geologic Map

    Thick eluvial or colluvial soil

    Rather thin eluvial or colluvial soil
In general sparse outcrops of rock

    Calcareous quartzite with laminae of phyllite;
lithological susceptibility of sliding very high

    Carbonaceous, micaceous, and garnet phyllite;
lithological susceptibility of sliding very high to high

Fault   Anticline and syncline axis

Dip of the rock   Crushed rock

Spring and seepage   Landslide

Groundwater level   Unstable area

2. Slope Map

    Slope contour line (grads)

3. Morphostructural Map

Ridge or crest   Sharp ridge or crest

Rivulet   Limit of slope unit

Nonstructural slope unit

Possible structural slope unit (bed of rock)

Possible structural slope unit (fracture)

4. Rock and Debris Slide Risk Map

Risk of large failures

High risk of planar failure

MP   Medium risk of planar failure

LP   Low risk of planar failure

Risk of medium and small failures

High risk of wedge failure

MW   Medium risk of wedge failure

LW   Low risk of wedge failure

Very low risk of rock and debris slides; possible soil
failure within wet areas

and weathering. The sum of the weights for each of the factors considered is used to obtain a qualitative description of hazards as low, medium, or high. The resulting hazard map, called a risk map, is shown as Map 4 in Figure 6-5. Major rock and debris slides that occurred after completion of the road were located in the three areas of high risk shown on Map 4. This hazard identification procedure has been adopted for selection of routes in mountainous regions (Deoja and Thapa 1989).

A complete event-tree analysis has been used to choose a maintenance program for rock slopes along transportation routes (Roberds 1991). The procedure consists of the following steps:

1. Identify possible failure modes,
2. Estimate probability of failure,
3. Evaluate consequences of failure, and
4. Evaluate effectiveness and cost of maintenance activities.

An example of this method is the choice of maintenance actions for rock slopes on a route with low traffic density. The decision tree for the slope management problem is shown in Figure 6-6. Maintenance activities include the following: M_0, do nothing; M_1, scaling; M_2, installation of rock bolts; . . . , M_7, construction of toe protection. Failure modes are F_1, isolated rock falls; F_2, small individual wedge slides; and so on. The consequences for each failure mode are assessed in terms of four components, namely, C_1, cost of repair; C_2, service disruption; C_3, number of injuries or deaths; and C_4, litigation. The effectiveness of each maintenance activity may be expressed by the mean reduction in the number of failures and by the mean reduction of each component of consequence, as shown in Table 6-1. These mean values are computed from the subjective distributions of the number of slope failures, such as $P(NF_i \mid M_i)$, and those of each component of consequence, such as $P(C_j \mid F_i \mid M_i)$, as shown in Figure 6-6, by Monte Carlo simulation.

By translating the service disruption and injuries, deaths, or both, to equivalent costs of \$20,000 per day and \$100,000 per person, respectively, each

FIGURE 6-5
(opposite page)
Examples of
(1) geologic,
(2) slope, (3)
morphostructural,
and (4) risk maps
from Nepal
(Wagner et al. 1987).

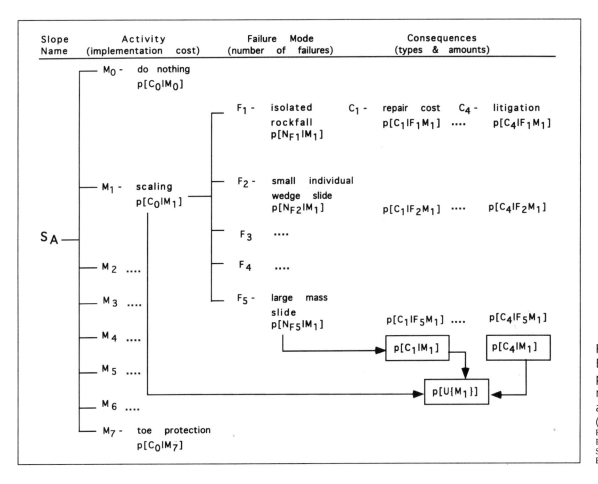

FIGURE 6-6
Evaluation of
preventive
maintenance
activity
(Roberds 1991).
REPRINTED WITH
PERMISSION OF AMERICAN
SOCIETY OF CIVIL
ENGINEERS

Table 6-1
Effectiveness of Maintenance Activity

MAINTENANCE ACTIVITY	REDUCTION IN NO. OF FAILURES	REDUCTION IN CONSEQUENCES[a]			
		C_1	C_2	C_3	C_4
M_1	0.6	0	0	0	0
M_2	0.1	0	0	0	0.2
M_3	0.9	0	0	0	0.2
•					
•					
•					
M_7	0	0.5	0.5	0.8	0.2

[a] Fraction of consequence of M_0.

maintenance activity can be measured by a probability distribution of its overall utility, $P[U\{M_i\}]$. The various maintenance activities can then be compared on the basis of their expected utility or the probability that $U_i > U_j$, where U_i and U_j are the utilities of maintenance activities i and j.

REFERENCES

Advisory Board on the Built Environment. 1983. *Multiple Hazard Mitigation: Report of a Workshop on Mitigation Strategies for Communities Prone to Multiple Natural Hazards.* National Research Council, National Academy Press, Washington, D.C., 60 pp.

Advisory Committee on the International Decade for Natural Hazard Reduction. 1987. *Confronting Natural Disasters—An International Decade for Natural Hazard Reduction.* National Research Council, National Academy Press, Washington, D.C., 60 pp.

Agnew, C.E. 1985. Probability Assessment by Dependent Experts. *Journal of the American Statistical Society,* Vol. 80, No. 390, pp. 343–347.

Ang, A.H-S., and W.H. Tang. 1975. *Probability Concepts in Engineering Planning and Design.* John Wiley and Sons, New York, Vol. 1, 409 pp.

Ang, A.H-S., and W.H. Tang. 1984. *Probability Concepts in Engineering Planning and Design.* John Wiley and Sons, New York, Vol. 2, 562 pp.

Ang, A.H-S., W.H. Tang, Y.K. Wen, and B.C. Yen. 1985. Methods for Engineering Hazard Analysis. In *Proc., PRC-US-Japan Trilateral Symposium/ Workshop on Engineering for Multiple Natural Hazard Mitigation,* Beijing, People's Republic of China, Jan. 7–12, pp. G-1-1 to G-1-18.

Baecher, G.B. 1979. Analyzing Exploration Strategies. In *Site Characterization and Exploration* (C.H.

Dowding, ed.), American Society of Civil Engineers, New York, pp. 220–243.

Baecher, G.B., N.A. Lanney, and H.H. Einstein. 1977. Statistical Description of Rock Properties and Sampling. In *Proc., 18th U.S. Symposium on Rock Mechanics,* Colorado School of Mines Press, Golden, pp. 5C1-1 to 5C1-8.

Baecher, G.B., M.E. Pate, and R. de Neufville. 1980. Risk of Dam Failure in Benefit-Cost Analysis. *Water Resources Research,* Vol. 16, No. 3, pp. 449–456.

Barker, R.M., J.M. Duncan, K.B. Rojiani, P.S.K. Ooi, C.K. Tan, and S.G. Kim. 1991. *NCHRP Report 343: Load Factor Design Criteria for Highway Structure Foundations.* Final Report. TRB, National Research Council, Washington, D.C.

Benjamin, J.R., and C.A. Cornell. 1970. *Probability, Statistics, and Decision for Civil Engineers.* McGraw-Hill, New York, 684 pp.

Bevan, K. 1982. On Subsurface Storm Flow: Predictions with Simple Kinematic Theory for Saturated and Unsaturated Flows. *Water Resources Research,* Vol. 18, pp. 1627–1633.

Brabb, E.E. 1984. Innovative Approaches to Landslide Hazard and Risk Mapping. In *Proc., Fourth International Symposium on Landslides,* Canadian Geotechnical Society, Toronto, Canada, Vol. 1, pp. 307–323.

Brabb, E.E., E.H. Pampeyan, and M.G. Bonilla. 1972. *Landslide Susceptibility in San Mateo County, California.* U.S. Geological Survey Miscellaneous Field Studies Map MF-360.

Brown, R.V. 1974. *Decision Analysis for the Manager.* Holt, Rinehart and Winston, Inc., New York, 618 pp.

Canuti, P. 1986. Slope Stability Mapping in Tuscany, Italy. In *International Geomorphology* (V. Gardiner, ed.), John Wiley and Sons, New York, Part 1, pp. 231–239.

Carrara, A., M. Cardinali, R. Detti, F. Guzzetti, V. Pasqui, and P. Reichenbach. 1991. GIS Techniques and Statistical Models in Evaluating Landslide Hazard. *Earth Surface Processes and Landforms,* Vol. 16, pp. 427–445.

Casagrande, A. 1965. Rule of the Calculated Risk in Earthwork and Foundation Engineering. *Journal of the Soil Mechanics and Foundations Division,* ASCE, Vol. 91, No. SM4, pp. 1–40.

Chowdhury, R.N., and W.H. Tang. 1987. Comparison of Risk Models for Slopes. In *Proc., Fifth International Conference on Applications of Statistics and Probability in Soil and Structural Engineering,* Vancouver, Canada, pp. 863–897.

Deoja, B.B., and B. Thapa. 1989. *Manual on Mountain Risk Engineering.* International Center for Integrated Mountain Development, Kathmandu, Nepal.

Ditlevsen, O. 1983. *Uncertainty Modeling.* McGraw-Hill, New York, 412 pp.

Einstein, H.H. 1988. Landslide Risk Assessment Procedure. In *Proc., Fifth International Symposium on Landslides,* Lausanne, Switzerland, A.A. Balkema, Rotterdam, Netherlands, Vol. 2, pp. 1075–1090.

Einstein, H.H., D. Veneziano, G.B. Baecher, and K.J. O'Reilly. 1983. The Effect of Discontinuity Persistence on Rock Slope Stability. *International Journal of Rock Mechanics and Mining Sciences,* Vol. 20, No. 5, pp. 227–236.

Halim, I.S., and W.H. Tang. 1991. Reliability of Undrained Clay Slope Containing Geologic Anomaly. In *Proc., Sixth International Conference on Application of Statistics and Probability in Civil Engineering,* Mexico City, Mexico, pp. 776–783.

Hasofer, A.M., and N. Lind. 1974. An Exact and Invariant First Order Reliability Format. *Journal of the Engineering Mechanics Division,* ASCE, Vol. 100, No. EM1, pp. 111–121.

Keeney, R.L., and H. Raiffa. 1976. *Decisions with Multiple Objectives.* John Wiley and Sons, New York, 569 pp.

Kienholz H. 1992. Risk Assessment in Mountains. In *Proc., 1 Simposio Internacional sobre Sensores Remotos y Sistemas de Informacion Geografica (SIG) para el Estudio de Riesgos Naturales,* Bogotá, Colombia, Vol. 2, 20 pp.

Lambe, T.W. 1985. Amuay Landslides. In *Proc., 11th International Conference on Soil Mechanics and Foundation Engineering,* Golden Jubilee Volume, A. A. Balkema, Rotterdam, Netherlands, pp. 137–158.

Lumb, P. 1975. Spatial Variability of Soils. In *Proc., Second International Conference on Applications of Statistics and Probability in Soil and Structural Engineering,* Vol. 2, pp. 397–421.

Madsen, H.O., S. Krenk, and N.C. Lind. 1986. *Methods of Structural Safety.* Prentice-Hall, Englewood Cliffs, N.J.

Meyerhof, G.G. 1970. Safety Factors in Soil Mechanics. *Canadian Geotechnical Journal,* Vol. 7, No. 4, pp. 349–355.

Office of the Prime Minister. 1985. *Mise en oeuvre des Plans d'Exposition aux Risques Naturels prévisibles.* Rapport Administratif et Technique Provisoire. Délégation aux risques majeurs, Paris, 14 pp.

Peck, R.B. 1969. Advantages and Limitations of the Observational Method in Applied Soil Mechanics. *Geotechnique,* Vol. 19, No. 2, pp. 171–187.

Perry, J. 1989. *A Survey of Slope Condition on Motorway Earthworks in England and Wales.* Research Report 199. U.K. Transport and Road Research Laboratory, Crowthorne, Berkshire, England, 49 pp.

Pierson, L.A. 1992. Rockfall Hazard Rating System. In *Transportation Research Record 1343,* TRB, National Research Council, Washington, D.C., pp. 6–13.

Radbruch-Hall, D.H., R.B. Colton, W.E. Davies, I. Lucchitta, B.A. Skipp, and D.J. Varnes. 1982. *Landslide Overview Map of the Conterminous United States.* U.S. Geological Survey Professional Paper 1183, 25 pp.

Raiffa, H., and R. Schlaifer. 1961. *Applied Statistical Decision Theory.* MIT Press, Cambridge, Mass., 356 pp.

Roberds, W.J. 1990. Methods for Developing Defensive Subjective Probability Assessments. In *Transportation Research Record 1288,* TRB, National Research Council, Washington, D.C., pp. 183–190.

Roberds, W.J. 1991. Methodology for Optimizing Rock Slope Preventive Maintenance Programs. In *Proc., Geotechnical Engineering Congress,* American Society of Civil Engineers, New York, Vol. 1, pp. 634–645.

Sangrey, D.A., K.O. Harrop-Williams, and J.A. Klaiber. 1984. Predicting Groundwater Response to Precipitation. *Journal of Geotechnical Engineering,* ASCE, Vol. 110, No. 7, pp. 957–975.

Scavia, C., G. Barla, and V. Bernaudo. 1990. Probabilistic Stability Analysis of Rock Toppling Failure in Rock Slopes. *International Journal of Rock Mechanics and Mining Sciences,* Vol. 27, pp. 465–478.

Schlaifer, R. 1959. *Probability and Statistics for Business Decisions.* McGraw-Hill, New York, 732 pp.

Staël von Holstein, C.A., and J.E. Matheson. 1979. *A Manual for Encoding Probability Distributions.* SRI International, Menlo Park, Calif.

Suzuki, H., and M. Matsuo. 1991. Methods for Predicting Slope Failure during Rainfall Using Estimation Model of Field Suction. In *Proc., Sixth International Conference on Applications of Statistics and Probability in Civil Engineering,* Mexico City, Mexico, Vol. 2, pp. 792–799.

Tang, W.H., M.S. Yucemen, and A.H–S. Ang. 1976, Probability-Based Short Term Design of Soil Slopes. *Canadian Geotechnical Journal,* Vol. 13, No. 3, pp. 201–215.

Terzaghi, K. 1961. Past and Future of Applied Soil Mechanics. *Journal of the Boston Society of Civil Engineers,* Vol. 48, No. 2, pp. 110–139.

van Westen, C.J. 1993. *Application of Geographic Information Systems to Landslide Hazard Zonation.* ITC Publication 15. International Institute for Aerospace Survey and Earth Sciences (ITC), Enschede, Netherlands, 245 pp.

Varnes, D.J., and International Association of Engineering Geology Commission on Landslides and Other Mass Movements on Slopes. 1984. *Landslide Hazard Zonation—A Review of the Principles and Practice.* United Nations Economic, Scientific, and Cultural Organization, Natural Hazards Series, No. 3, 63 pp.

Viberg, L. 1984. Landslide Risk Mapping in Soft Clays in Scandinavia and Canada. In *Proc., Fourth International Symposium on Landslides,* Canadian Geotechnical Society, Toronto, Canada, Vol. 1, pp. 325–348.

Wagner, A., R. Olivier, and E. Leite. 1987. Rock and Debris Slide Risk Maps Applied to Low-Volume Roads in Nepal. In *Transportation Research Record 1106,* TRB, National Research Council, Washington, D.C., Vol. 2, pp. 255–267.

Whitman, R.V. 1984. Evaluating Calculated Risk in Geotechnical Engineering. *Journal of Geotechnical Engineering,* ASCE, Vol. 110, No. 2, pp. 145–188.

Wieczorek, G. 1984. Preparing a Detailed Landslide-Inventory Map for Hazard Evaluation and Pre-diction. *Bulletin of the Association of Engineering Geologists,* Vol. 21, pp. 337–342.

Winkler, R.L. 1969. Scoring Rules and the Evaluation of Probability Assessors. *Journal of the American Statistical Association,* Vol. 64, pp. 1071–1078.

Wright, R.H., R.H. Campbell, and T.H. Nilsen. 1974. Preparation and Use of Isopleth Maps of Landslide Deposits. *Geology,* No. 2, pp. 483–485.

Wu, T.H. 1989. Variability of Geologic Materials. In *Art and Science of Geotechnical Engineering* (E.J. Cording, W.J. Hall, J.D. Haltiwanger, A.J. Hendron, and G. Mesri, eds.), Prentice-Hall, Englewood Cliffs, N.J., pp. 221–239.

Wu, T.H. 1992. Prediction and Mapping of Landslide Hazards. In *Proc., Second US-Asia Conference on Engineering for Mitigating Natural Hazards Damage,* Yogyakarta, Indonesia, pp. G10-1 to G10-08.

Wu, T.H., O. Kjekstad, I.M. Lee, and S. Lacasse. 1989. Reliability Analysis of Foundation Stability for Gravity Platforms in the North Sea. *Canadian Geotechnical Journal,* Vol. 26, pp. 359–368.

Wu, T.H., and D.N. Swanston. 1980. Risk of Landslides in Shallow Soils and Its Relation to Clear-cutting in Southeastern Alaska. *Forest Science,* Vol. 26, pp. 495–510.

PART 2

INVESTIGATION

Chapter 7

A. KEITH TURNER AND
VERNE C. MCGUFFEY

ORGANIZATION OF INVESTIGATION PROCESS

1. INTRODUCTION

To investigate is "to observe or inquire into in detail; examine systematically," as defined in *The American Heritage Dictionary* (New College Edition). Investigation and characterization of subsurface conditions form the core of landslide studies.

Geotechnical engineering applications incorporate naturally occurring materials characterized by highly variable physical properties. Although most engineers work with materials that have known properties and undertake designs reflecting these properties, geologists and geotechnical engineers must utilize a structured investigation process to deduce the properties of naturally occurring materials and their geometrical relationships. Successful investigations require the investigator to have sound judgment and the ability to make decisions.

2. FIELD INVESTIGATION OF LANDSLIDES

Field investigation has long been recognized as the central and decisive part of a study of landslides and landslide-prone regions (Philbrick and Cleaves 1958; Sowers and Royster 1978). Landslide investigation supports the adage that a problem is already half solved when one recognizes that a problem exists. Investigation should be directed toward both recognition of actual or potential slope movements and identification of the type and causes of the movement. Both aspects are important in identifying appropriate procedures for the prevention or correction of landslides.

Rib and Liang (1978) suggested that landslide investigations should be designed with reference to four basic guidelines that have evolved over many years of experience:

- Most landslides or potential failures can be predicted if proper investigations are performed in time;
- The cost of preventing landslides is less than the cost of correcting them, except for small landslides that can be handled by normal maintenance procedures;
- Massive landslides that may cost many times the cost of the original facility should be prevented; and
- The occurrence of initial slope movement can lead to additional unstable conditions and movements.

3. DEFINITION OF INVESTIGATION PROCESS

An appropriate investigation process cannot be defined by the rigid application of a set of procedural rules. Because the investigation process is so central to geotechnical applications, it has been discussed in numerous textbooks and papers in professional journals (Burwell and Roberts 1950; Terzaghi and Peck 1967; Kiersch 1969; Peck 1969; Dowding 1979a; Boyce 1982; Clayton et al. 1982;

Fookes et al. 1985). The American Society of Civil Engineers sponsored a specialty conference on the topic and subsequently published a book containing the principal papers and discussions (Dowding 1979a). Clayton et al. (1982) aimed at "improving the quality of site investigation by providing a relatively simple reference book." Their book relates to British conditions but includes numerous examples defining basic site investigation principles that should guide all investigators.

Investigations produce information that forms the basis for design decisions. In a few cases, subsurface conditions at a site are generally conceded to be so complicated that any ordinary and reasonable investigation will yield only a partial and incomplete evaluation. Under such rare circumstances, steps are taken to allow for changes in the facility design or the construction methods as actual conditions are revealed. This flexibility is expensive, however, and in the majority of cases, investigation is expected to yield reasonably accurate predictions of subsurface conditions.

3.1 Investigation Failures

An investigation is inadequate if it fails to reveal information concerning subsurface conditions that is needed to produce a safe and economical design or fails to determine appropriate construction methods. Yet investigations generally should not, and usually will not, remove all uncertainty. Minor unexpected conditions are often found during construction; in fact, such changed conditions are to be expected. Investigations are considered to have failed only when the revealed conditions are found to differ grossly from the predictions. Osterberg (1979) stated that geotechnologists must "take every advantage of every method, tool, and observational opportunity to communicate with personnel involved in order to avoid such failures."

Osterberg (1979) suggested that there are five general reasons for investigation failures:

- General knowledge of geologic processes was not used in planning the exploration program and in evaluating the findings of the investigation;
- The investigator had a preconceived notion of what the site evaluation should be and showed reluctance, or even refused, to consider evidence that contradicted the preconceived idea;

- Not all the available tools were used for site evaluation even though they may have been simple and obvious;
- The investigator failed to properly discuss the goals of the exploration program with all the persons involved; and
- Open and free lines of communication were not set up.

3.2 Site Characterization

A number of authors emphasize the concept of *characterization* during the investigation process. According to Dowding (1979b), characterization of the subsurface includes identification of the geometry of relatively homogeneous zones as well as the constitutive properties of the material within the zones. Constitutive properties are those parameters that allow the prediction of a material's strength, deformation, or permeability in response to changes over time due to stress or other environmental conditions.

Duncan (1979) stated that such characterization serves two distinct purposes:

- Anticipating problems and effects, and
- Quantifying site geometric characteristics or material properties.

Duncan suggested that these purposes interact, the first providing a more or less qualitative definition of the critical issues and problems and the second providing more detailed and quantitative definitions suitable for analysis and design.

3.3 Effect of Economic Factors

Few investigations have sufficient time or money to permit the collection of every pertinent fact; thus critical factors must be identified and assessed on the basis of limited data, relying on the judgment and experience of the investigator. Duncan (1979) quoted Peck (1974) as stating that "even the most experienced practitioner has to form his judgments on the basis of less than perfect data." Subsurface soil and rock conditions are notoriously variable, and reality often differs from expectation. The investigation process must be supported by a logical thought process for appropriate conclusions to be reached.

3.4 Importance of Proper Site Investigation Procedures

The importance of proper site investigation procedures has long been a topic of concern to leading engineers. For example, in reviewing the experience with soil mechanics before World War I, Terzaghi stated:

> Engineers imagined that the future science of foundations would consist of carrying out the following program: Drill a hole in the ground. Send soil samples obtained from the hole through a laboratory with standardized apparatus served by conscientious human automatons. Collect the figures, introduce them into equations, and compute the result. (Terzaghi 1936, 14)

Terzaghi then went on to lament the status of civil engineering education, which he suggested was biased toward the concept that all engineering problems should and could be resolved with a priori assumptions regarding the material properties. Peck (1969) suggested that Terzaghi's great success was due to his use of observation and also to his insistence on full personal responsibility and authority concerning all details of critical investigations.

Underwood suggested that there were two important problem areas in investigation:

> One attitude that has discouraged the writer over the past few years is the apparent hope for some new magical development that will fill in the gaps between a few poorly sampled, widely spaced and often poorly logged borings. . . . Field investigations are often hurriedly and carelessly conducted and the incomplete data is then carefully analyzed by precise (out to 8 digit accuracy) computer techniques which produce impressive but erroneous results which in turn lead to inaccurate design assumptions. (Underwood 1974)

These problems logically lead to the following conclusions about the importance of the proper investigation process:

- New and ever-more-sophisticated equipment will never substitute for a properly designed and adequate sampling program;
- Trained personnel, familiar with the reasons for the investigation, must conduct and supervise the activities in the field;

- The validity of the test results and analyses is based entirely on the quality and extent of the field investigation on which they rely; and
- Overrefinement of analysis does not lead to improved design, which depends entirely on improved investigation.

4. ELEMENTS OF AN INVESTIGATION

Several proposals have been made concerning the design of an ideal investigation. All authors agree that the investigation process should be conducted in an iterative fashion. Clayton et al. (1982) suggested that the ideal investigation should follow a sequence of 11 stages (or events) as defined in Table 7-1. In contrast, Dowding (1979b) suggested that the investigation process should be considered in terms of only three steps, namely,

- Review available information and surface reconnaissance;
- Undertake detailed surface mapping, preliminary borings, initial laboratory testing, and preliminary analysis; and
- Undertake borings to recover specialized samples, geophysical surveys, test excavations (adits, test pits, calyx holes, etc.), and specialized testing.

Dowding further suggested that the results of each step should be integrated with the design process

TABLE 7-1
Ideal Order of Events for Site Investigation
(Clayton et al. 1982)

Event	Description
1	Preliminary desk study or fact-finding survey
2	Aerial photograph interpretation
3	Site walkover survey
4	Preliminary subsurface exploration
5	Soil classification by description and simple testing
6	Detailed subsurface exploration and field testing
7	Physical survey (laboratory testing)
8	Evaluation of data
9	Geotechnical design
10	Field trials
11	Liaison by geotechnical engineer with site staff during project construction

in order to identify the unknowns that should be discovered in the next step or element.

Johnson and DeGraff (1988) suggested that an investigation should include five elements:

- Formulation of the investigation,
- Data collection,
- Data interpretation,
- Application of analysis techniques, and
- Communication of results.

Because landslides are continually changing phenomena, field investigations are not isolated or easily defined activities; they are frequently iterative in their application. New data generate new questions that require more data for resolution.

The critical aspects of a landslide investigation for each of the five investigation elements defined by Johnson and DeGraff (1988) are described briefly in the following sections.

4.1 Formulation of the Investigation

According to Johnson and DeGraff (1988), formulation of the investigation is the element that is most often forgotten or overlooked. This formulation involves two components:

- The identification of the question or questions that the investigation must answer, a clear definition of the purpose of the investigation; and
- Identification of other aspects of the investigation, including its scope, the area and depth to be investigated, and its duration.

Inadequate attention to formulation may cause the investigation to be conducted in an inefficient manner. It may take longer and cost more to complete, and, in some cases, the appropriate information is not obtained at all.

4.1.1 Purpose

Field investigations of landslides may be conducted for two distinct purposes:

- When new facilities are planned, to identify areas that are potentially or currently subject to landsliding; in the case of transportation facilities, this investigation would be conducted during the route-selection phase.
- When a landslide is adjacent to a facility, to define the landslide dimensions, features, and

characteristics and to assess environmental factors that may contribute to the landsliding.

These two purposes require somewhat different approaches.

Unstable areas prone to landsliding usually exhibit symptoms of past movement and incipient failure. During preliminary planning stages, these may be identified by interpretation of aerial photographs or by remote-sensing methods. The potential for landsliding can also be evaluated by a number of numerical mapping and assessment methods. Other cases can only be identified by a detailed field investigation before design. Such investigations can show how to prevent, or at least minimize, future movements, and they can suggest alternate routes that are less subject to landsliding.

Once a landslide has developed, either during construction of a facility or subsequently, the investigation is undertaken to diagnose the factors affecting the movements and to determine what corrective measures are appropriate for preventing or minimizing further movements. Such investigations have much in common with other types of site-investigation programs. However, in many cases these investigations may have to be undertaken with some urgency because the landslide is a threat to property or public safety or is disrupting use of a transportation facility.

4.1.2 Scope

Sowers and Royster (1978) included a rather lengthy checklist of features that should be considered in planning a field investigation of a landslide (see opposite page). It is not expected that any single landslide investigation would involve all the items on this list.

4.1.3 Area

The area of an investigation is controlled by the size of the project and the extent of the topographic and geologic features that are involved in the landslide activity. At sites where there is potential for movement, the area that must be investigated cannot be easily defined in advance. The extent of the investigation can be better defined once a landslide has occurred. However, in either case, the area studied must be considerably larger than that comprising the suspected activity or known movement for two reasons:

Checklist for Planning a Landslide Investigation (Sowers and Royster 1978)

I TOPOGRAPHY
  A. Contour map
    1. Land form
    2. Anomalous patterns (jumbled, scarps, bulges)
  B. Surface drainage
    1. Continuous
    2. Intermittent
  C. Profiles of slope
    1. Correlate with geology (II)
    2. Correlate with contour map (IA)
  D. Topographic changes
    1. Rate of change by time
    2. Correlate with groundwater (III), weather (IV), and vibration (V)

II GEOLOGY
  A. Formations at site
    1. Sequence of formations
    2. Colluvium
      a. Bedrock contact
      b. Residual soil
    3. Formations with bad experience
    4. Rock minerals susceptible to alteration
  B. Structure: three-dimensional geometry
    1. Stratification
    2. Folding
    3. Strike and dip of bedding or foliation
      a. Changes in strike or dip
      b. Relation to slope and slide
    4. Strike and dip of joints with relation to slope
    5. Faults, breccia, and shear zones with relation to slope and slide
  C. Weathering
    1. Character (chemical, mechanical, and solution)
    2. Depth (uniform or variable)

III GROUNDWATER
  A. Piezometric levels within slope
    1. Normal
    2. Perched levels, relation to formations and structure
    3. Artesian pressures, relation to formations and structure
  B. Variations in piezometric levels [correlate with weather (IV), vibration (V), and history of slope changes (VI)]
    1. Response to rainfall
    2. Seasonal fluctuations
    3. Year-to-year changes
    4. Effect of snowmelt
  C. Ground surface indications of subsurface water
    1. Springs
    2. Seeps and damp areas
    3. Vegetation differences
  D. Effect of human activity on groundwater
    1. Groundwater utilization
    2. Groundwater flow restriction
    3. Impoundment and additions to groundwater
    4. Changes in ground cover and infiltration opportunity
    5. Surface water changes
  E. Groundwater chemistry
    1. Dissolved salts and gases
    2. Changes in radioactive gases

IV WEATHER
  A. Precipitation
    1. Form (rain or snow)
    2. Hourly rates
    3. Daily rates
    4. Monthly rates
    5. Annual rates
  B. Temperature
    1. Hourly and daily means
    2. Hourly and daily extremes
    3. Cumulative degree-day deficit (freezing index)
    4. Sudden thaws
  C. Barometric changes

V VIBRATION
  A. Seismicity
    1. Seismic events
    2. Microseismic intensity
    3. Microseismic changes
  B. Human induced
    1. Transport
    2. Blasting
    3. Heavy machinery

VI HISTORY OF SLOPE CHANGES
  A. Natural process
    1. Long-term geologic changes
    2. Erosion
    3. Evidence of past movement
    4. Submergence and emergence
  B. Human activity
    1. Cutting
    2. Filling
    3. Changes in surface water
    4. Changes in groundwater
    5. Changes in vegetative cover, clearing excavation, cultivation, and paving
    6. Flooding and sudden drawdown of reservoirs
  C. Rate of movement
    1. Visual accounts
    2. Evidence in vegetation
    3. Evidence in topography
    4. Photographic evidence
      a. Oblique
      b. Stereo aerial photographs
      c. Aerial photographs
      d. Spectral changes
    5. Instrumental data
      a. Vertical changes, time history
      b. Horizontal changes, time history
      c. Internal strains and tilt, including time history
  D. Correlations of movements
    1. Groundwater [correlate with groundwater (III)]
    2. Weather [correlate with weather (IV)]
    3. Vibration [correlate with vibration (V)]
    4. Human activity [correlate with human-induced vibration (VB)]

- The landslide or potential landslide must be referenced to the stable area surrounding it, and
- Most landslides enlarge with passage of time, and moreover many landslides are much larger than first suspected from the overt indications of activity.

As a rule of thumb, the area studied should be two to three times wider and longer than the area suspected. In some mountainous areas, it is necessary to investigate to the top of the slope or to some major change in lithology or slope angle. The lateral area must encompass sources of groundwater and geologic structures that are aligned with the area of instability.

4.1.4 Depth

The depth of the investigation is even more difficult to define in advance. Borings or other direct techniques should extend deep enough to identify those materials that have not been subject to past movement but that could be involved in future movement and the underlying formations that are likely to remain stable. The boring depth is sometimes revised hourly as field operations proceed. When instrumentation of a landslide yields data on the present depth of activity, planned depths are sometimes found to be insufficient and increases are necessary. The specifications should be flexible enough to allow additional depth of investigation when the data obtained suggest deeper movements. Longitudinal cross sections should be drawn through the center of the landslide depicting possible toe bulges and uphill scarps; circular or elliptical failure surfaces sketched through these limits can suggest the maximum depth of movement. Continuous, thick, hard strata within the slope can limit the depth. However, at least one boring should extend far below the suspected depth of shear: sometimes deep, slow movements are masked by the greater activity at shallower depths. Experience demonstrates that the depth of movement below the ground surface at the center of a landslide is seldom greater than the width of the zone of surface motion.

4.1.5 Duration

Ideally, the investigations should continue over periods of time adequate to evaluate the changing environmental factors and shifting topography. Often the duration of these investigations is constrained by the need for preventive or corrective design.

Since most landslides are influenced by climatic changes, a minimum period for investigation should include one seasonal cycle of weather—one year in most parts of the world. However, because long-term climatic cycles that occur every 11 or 22 years are superimposed on the yearly changes, it could be necessary to continue a landslide investigation for more than two decades. Such a long investigation is almost impossible, however, because of the need to draw conclusions and take corrective action.

Investigations made during a period in which the climatic conditions are less severe than the maximum will prove too optimistic, and those made during a period of bad conditions may appear too pessimistic. The worst climatic conditions that develop during the life of the project control the risk to engineering construction. Experience has indicated that many false conclusions have been reached regarding the causes of landslides and the effectiveness of corrective measures because worsened climatic changes were not considered by the engineers and geologists concerned.

4.2 Data Collection

Data collection involves both office and field studies. Office studies include the discovery and assembly of all existing pertinent information. These data are commonly found in diverse government sources and may include maps, reports, aerial photographs, and historical documents. The appropriate use of such information can materially assist the investigator before and during an initial site visit and guide the planning of the first steps in field data collection.

Field data collection may involve a variety of activities ranging from relatively simple, low-cost reconnaissance studies to sophisticated, frequently expensive specialized instrumentation installations. Investigations are generally most efficient when the simpler and more rapid reconnaissance methods are used initially to obtain a basic understanding of the site and the more expensive and time-consuming sampling methods are employed subsequently where they can be used for maximum benefit. These data collection activities are discussed in more detail in Chapters 8 through 11.

4.3 Data Interpretation

An investigation is incomplete without an interpretation of the data from the office and field studies. In most landslide investigations, data collection and data interpretation go on continuously and interactively. Interpretation of data gathered during initial stages of an investigation will usually suggest the need for additional volumes and types of data and modifications to the investigation process. An efficient investigation process requires a continual review and interpretation of the data as they are gathered.

Data interpretation usually begins with reduction and reorganization of the initial raw data. This activity results in the production of tables, graphs, maps, profiles, and cross sections. For most landslide investigations, spatial and temporal comparisons of the data are of great interest.

4.4 Application of Analysis Techniques

Once data are in manageable form, analysis of the data is usually fairly easy. Analysis may involve graphical techniques, but numerical methods, including both statistical analysis and mathematical modeling approaches, are increasingly being employed. Numerous slope stability analysis procedures are possible. Most involve simplifying assumptions. Slope stability analysis methods are discussed in Chapter 13 for soil slopes and in Chapter 15 for rock slopes.

4.5 Communication of Results

Many sources emphasize the need for clear and precise communication of investigation results (Osterberg 1979; Williams 1984). If the answers obtained by an investigation are not transmitted to those who will use them, the investigation will have served no purpose. Some landslide investigation results are reported to government boards, commissions, or similar entities. Numerous guides are available for authors preparing such documents (California Division of Mines and Geology 1975; Cochrane et al. 1979; Hansen 1991). Litigation may result from some landslides, and some landslide investigations may be developed for such applications. Kiersch (1969) provided guidance for geologists involved as technical (expert) witnesses in such litigation.

5. HIGH-QUALITY INVESTIGATIONS

Clayton et al. (1982) proposed six key factors for improving site investigations. The following are modifications of their factor descriptions that reflect the needs and realities of investigations at landslide sites:

- Insistence on the full use of available documentary evidence in a comprehensive factual survey during the early stages of the investigation process;
- Use of aerial photography, remote-sensing, and possibly numerical map analysis methods in the early stages of an investigation, preferably by trained and experienced personnel;
- Development of a plan of subsurface investigation that is specifically designed for the site and reflects expected geological and environmental subsoil conditions;
- Field supervision of drilling by experienced engineers, who should be aware of the aims of the investigation;
- Frequent revision of the aims and methods of the site investigation as information becomes available and as a result of liaison among geotechnical engineers, designers of the proposed corrective or preventive measures, and, where possible, the contractor who will undertake the work; and
- Close observation by an experienced team of geotechnical engineers during the construction.

6. OVERVIEW OF CHAPTERS 8–11

The major aspects of the investigation process for landslides are defined in the following four chapters.

The use of aerial photographs and other remote-sensor imagery products for landslide mapping is discussed in Chapter 8. Also described is the use of computer-based spatial mapping approaches in performing regional landslide hazard assessments.

In Chapter 9 the initial office and field data collection efforts, including various surface observations and geologic mapping methods, are reviewed. Various surveying methods to supply quantitative data on landslide movements are summarized.

Chapter 10 continues the discussion of field data collection activities for a landslide investigation, covering the entire range of exploration and sampling options to characterize the subsurface conditions. The merits of various geophysical exploration methods as well as the wide range of methods involving in situ testing, borings, test excavations, and sample handling procedures are surveyed.

Chapter 11 completes the data collection process with a discussion of the various field instrumentation options to identify and monitor subsurface movements.

REFERENCES

Boyce, R.C. 1982. An Overview of Site Investigations. *Bulletin of the Association of Engineering Geologists*, Vol. 12, pp. 167-171.

Burwell, E.B., Jr. and G.D. Roberts. 1950. The Geologist in the Engineering Organization. In *Applications of Geology to Engineering Practice* (Berkey Volume), Geological Society of America, Boulder, Colo., pp. 1–9.

California Division of Mines and Geology. 1975. *Guidelines for Geologic/Seismic Considerations in Environmental Impact Reports*. Note 46. Sacramento, Calif.

Clayton, C.R.I., N.E. Simons, and M.C. Matthews. 1982. *Site Investigation: A Handbook for Engineers*. Halsted Press, New York, 423 pp.

Cochrane, W., P. Fenner, and M. Hill. 1979. *Geowriting*. American Geological Institute, Falls Church, Va., 80 pp.

Dowding, C.H., ed. 1979a. *Site Characterization and Exploration*. American Society of Civil Engineers, New York, 395 pp.

Dowding, C.H. 1979b. Perspective and Challenges of Site Characterization. In *Site Characterization and Exploration* (C.H. Dowding, ed.), American Society of Civil Engineers, New York, pp. 10–35.

Duncan, J.M. 1979. Site Characterization for Analysis. In *Site Characterization and Exploration* (C.H. Dowding, ed.), American Society of Civil Engineers, New York, pp. 70–82.

Fookes, P.G., M. Sweeney, C.N.D. Manby, and R.P. Martin. 1985. Geological and Geotechnical Engineering Aspects of Low-cost Roads in Mountainous Terrain. *Engineering Geology*, Vol. 21, pp. 1–152.

Hansen, W.R., ed. 1991. *Suggestions to Authors of Reports of the United States Geological Survey*, 7th ed. U.S. Geological Survey, Washington, D.C., 289 pp.

Johnson, R.B., and J.V. DeGraff. 1988. *Principles of Engineering Geology*. John Wiley and Sons, New York, 497 pp.

Kiersch, G.A. 1969. The Geologist and Legal Cases—Responsibility, Preparation and the Expert Witness. In *Legal Aspects of Geology in Engineering Practice*, Geological Society of America, Boulder, Colo., pp. 1–6.

Osterberg, J.O. 1979. Failures in Exploration Programs. In *Site Characterization and Exploration* (C.H. Dowding, ed.), American Society of Civil Engineers, New York, pp. 3–9.

Peck, R.B. 1969. Advantages and Limitations of the Observational Method in Applied Soil Mechanics: 9th Rankine Lecture. *Geotechnique*, Vol. 19, No. 2, pp. 171–187.

Peck, R.B. 1974. *The Selection of Soil Parameters for the Design of Foundations*. Second Nabor Carillo Lecture. Sociedad Mexicana de Mechanica de Suelos.

Philbrick, S.S., and A.B. Cleaves. 1958. Field and Laboratory Investigations. In *Special Report 29: Landslides and Engineering Practice*, HRB (now TRB), National Research Council, Washington, D.C., pp. 93–111.

Rib, H.T., and Ta Liang. 1978. Recognition and Identification. In *Special Report 176: Landslides: Analysis and Control*, TRB, National Research Council, Washington, D.C., Chap. 3, pp. 34–80.

Sowers, G.F., and D.L. Royster. 1978. Field Investigation. In *Special Report 176: Landslides: Analysis and Control*, TRB, National Research Council, Washington, D.C., Chap. 4, pp. 81–111.

Terzaghi, K. 1936. Relation between Soil Mechanics and Foundation Engineering, Presidential Address. In *Proc., First International Conference on Soil Mechanics and Foundation Engineering*, Harvard University Press, Vol. 3, pp. 13–18.

Terzaghi, K., and R.B. Peck. 1967. *Soil Mechanics in Engineering Practice*, 2nd ed. John Wiley and Sons, New York, 729 pp.

Underwood, L.B. 1974. Exploration and Geologic Prediction for Underground Works. In *Proc., Specialty Conference on Subsurface Exploration for Underground Excavation and Heavy Construction*, Henniker, N.H., American Society of Civil Engineers, New York, pp. 65–83.

Williams, J.W. 1984. Engineering Geology Information for Varied Audiences: The Professional—the Planner—the General Public. *Bulletin of the Association of Engineering Geologists*, Vol. 21, pp. 365–369.

Chapter 8

ROBERT SOETERS AND
CORNELIS J. VAN WESTEN

SLOPE INSTABILITY RECOGNITION, ANALYSIS, AND ZONATION

1. INTRODUCTION

Slope instability processes are the product of local geomorphic, hydrologic, and geologic conditions; the modification of these conditions by geodynamic processes, vegetation, land use practices, and human activities; and the frequency and intensity of precipitation and seismicity.

The engineering approach to landslide studies has focused attention on analysis of individual slope failures and their remedial measures. The techniques used in these studies were in accordance with their required large scale and did not allow for zonation of extensive areas according to their susceptibility to slope instability phenomena. The need for this type of zonation has increased with the understanding that proper planning will decrease considerably the costs of construction and maintenance of engineering structures.

Considering the many terrain factors involved in slope instability, the practice of landslide hazard zonation requires

- A detailed inventory of slope instability processes,
- The study of these processes in relation to their environmental setting,
- The analysis of conditioning and triggering factors, and
- A representation of the spatial distribution of these factors.

Some methodological aspects of slope instability hazard zonation are dealt with in Section 2 of this chapter. The essential role of the earth scientist in modeling the spatial distribution of terrain conditions leading to instability is noted, and a scheme is given for a hierarchical approach to slope instability zonation that is similar to the phases recognized in engineering projects. By following such a systematic approach, the necessary steps to a hazard assessment are defined, taking into consideration both direct and indirect mapping techniques. An overview of current practice is given.

In Section 3 emphasis is on the application of remote-sensing techniques to landslide studies and hazard zonation. A systematic guide is presented for recognition and interpretation of slope movements. The applicability of different remote-sensing data to landslide recognition is evaluated, considering their characteristic spatial, spectral, and temporal resolutions.

The capabilities of a geographic information system (GIS) for analyzing terrain factors that lead to slope instability are highlighted in Section 4. An integration of data collection and analysis techniques is proposed for slope instability zonation at different scales.

The terminology concerning hazards used in this chapter conforms to the definitions proposed by Varnes (1984):

- *Natural hazard* means the probability of occurrence of a potentially damaging phenomenon within a specified period of time and within a given area;

- *Risk* means the expected number of lives lost, persons injured, damage to property, or disruption of economic activity because of a particular natural phenomenon; and
- *Zonation* refers to the division of land in homogeneous areas or domains and the ranking of these areas according to their degrees of actual or potential hazard caused by mass movement.

To determine risk, Varnes gave the following definitions:

- *Vulnerability* means the degree of loss to a given element (or set of elements) at risk resulting from the occurrence of a natural phenomenon of a given magnitude;
- *Element at risk* means the population, properties, economic activities, and so on, at risk in a given area; and
- *Specific risk* means the expected degree of loss due to a particular natural phenomenon.

Landslide hazard is commonly shown on maps that display the spatial distribution of hazard classes (or landslide hazard zonation). Landslide hazard zonation requires a detailed knowledge of the processes that are or have been active in an area and of the factors leading to the occurrence of the potentially damaging phenomenon. This knowledge is considered the domain of earth scientists. Vulnerability analysis requires detailed knowledge of the population density, infrastructure, and economic activities and the effects of a specific damaging phenomenon on these elements at risk. Therefore this part of the analysis is done mainly by persons from disciplines other than the earth sciences, such as specialists in urban planning and social geography, economists, and engineers.

As discussed in Chapter 6, fully developed examples of risk analysis on a quantitative basis are still scarce in the literature (Einstein 1988; Kienholz 1992; Innocienti 1992; Keaton 1994), partly because of the difficulties in defining quantitatively both hazard and vulnerability. Hazard analysis is seldom executed in accordance with the definition given above, since the probability of occurrence of potentially damaging phenomena is extremely difficult to determine for larger areas. The determination of actual probabilities requires analysis of such triggering factors as earthquakes or rainfall, as discussed in Chapter 4, or the applica-

tion of complex models. In most cases, however, there is no clear relationship between these factors and the occurrence of landslides. Therefore, in most hazard maps the legend classes used generally do not give more information than the susceptibility of certain areas to landsliding or relative indications of the degree of hazard, such as high, medium, and low. From the review of many example studies, it appears that susceptibility usually expresses the likelihood that a phenomenon (in this case, a landslide) will occur in an area on the basis of the local terrain conditions; the probability of occurrence, which depends also on the recurrence of triggering factors such as rainfall or seismicity, is not considered. The terms *hazard* and *susceptibility* are frequently used synonymously.

In this chapter only the techniques for recognition and analysis of landslides and methods for *hazard assessment* are treated.

2. PRINCIPLES OF HAZARD ZONATION

An ideal map of slope instability hazard should provide information on the spatial probability, temporal probability, type, magnitude, velocity, runout distance, and retrogression limit of the mass movements predicted in a certain area (Hartlén and Viberg 1988). A reliable landslide inventory defining the type and activity of all landslides, as well as their spatial distribution, is essential before any analysis of the occurrence of landslides and their relationship to environmental conditions is undertaken. The differentiation of slope instability according to type of movement is important, not only because different types of mass movement will occur under different terrain conditions, but also because the impact of slope failures on the environment has to be evaluated according to type of failure.

2.1 General Considerations

Prediction of landslide hazard for areas not currently subject to landsliding is based on the assumption that hazardous phenomena that have occurred in the past can provide useful information for prediction of future occurrences. Therefore, *mapping* these phenomena and the factors thought to be of influence is very important in hazard zonation. In relation to the analysis of the terrain conditions leading to slope instability, two basic methodologies can be recognized:

1. The first mapping methodology is the experience-driven applied-geomorphic approach, by which the earth scientist evaluates direct relationships between landslides and their geomorphic and geologic settings by employing direct observations during a survey of as many existing landslide sites as possible. This is also known as the *direct mapping methodology*.

2. The opposite of this experience-based, or heuristic, approach is the *indirect mapping methodology*, which consists of mapping a large number of parameters considered to potentially affect landsliding and subsequently analyzing (statistically) all these possible contributing factors with respect to the occurrence of slope instability phenomena. In this way the relationships between the terrain conditions and the occurrence of landslides may be identified. On the basis of the results of this analysis, statements are made regarding the conditions under which slope failures occur.

Another useful division of techniques for assessment of slope instability hazard was given by Hartlén and Viberg (1988), who differentiated between *relative-hazard assessment techniques* and *absolute-hazard assessment techniques*. Relative-hazard assessment techniques differentiate the likelihood of occurrence of mass movements for different areas on the map without giving exact values. Absolute-hazard maps display an absolute value for the hazard, such as a factor of safety or a probability of occurrence.

Hazard assessment techniques can also be divided into three main groups (Carrara 1983; Hartlén and Viberg 1988):

1. *White box models*, based on physical models (slope stability and hydrologic models), also referred to as *deterministic models*;
2. *Black box models*, not based on physical models but strictly on statistical analysis; and
3. *Grey box models*, based partly on physical models and partly on statistics.

2.2 Scale-Related Objectives

The development of a clear hierarchical methodology in hazard zonation is necessary to obtain an acceptable cost/benefit ratio and to ensure the practical applicability of the zonation. The working scale for a slope instability analysis is determined by the requirements of the user for whom the survey is executed. Because planners and engineers form the most important user community, the following scales of analysis have been differentiated for landslide hazard zonation (International Association of Engineering Geology 1976):

- National scale (< 1:1 million)
- Regional scale (1:100,000 to 1:500,000)
- Medium scale (1:25,000 to 1:50,000)
- Large scale (1:5,000 to 1:15,000)

The national hazard zonation mapping scale is intended to give a general inventory of problem areas for an entire country that can be used to inform national policy makers and the general public. The level of detail will be low because the assessment is done mostly on the basis of generally applicable rules.

The regional mapping scale is meant for planners in the early phases of regional development projects or for engineers evaluating possible constraints due to instability in the development of large engineering projects and regional development plans. The areas to be investigated are large, on the order of 1000 km² or more, and the required level of map detail is low. The map indicates areas in which mass movements can be a constraint on the development of rural or urban transportation projects. Terrain units with an areal extent of several tens of hectares are outlined and classified according to their susceptibility to occurrence of mass movements.

Medium-scale hazard maps can be used for the determination of hazard zones in areas affected by large engineering structures, roads, and urbanization. The areas to be investigated may cover upward of a few hundreds of square kilometers; yet a considerably higher level of detail is required at this scale. The detail should be such that adjacent slopes in the same lithology are evaluated separately and may obtain different hazard scores depending on their characteristics, such as slope angle or form and type of land use. Within the same terrain unit, distinctions should be made between different slope segments. For example, a concave slope should receive a different rating, when appropriate, than an adjacent straight or convex slope.

Large-scale hazard zonation maps can be used at the level of the site investigation before the design phase of engineering projects. This scale allows evaluation of the variability of a safety factor as a function of variable slope conditions or under the influence of triggering factors. The size of area under study may range up to several tens of square kilometers. The hazard classes on such maps should be absolute, indicating the probability of failure for each grid cell or mapping unit with areas down to one hectare or less clearly defined.

Although the selection of the scale of analysis is usually determined by the intended application of the mapping results, the choice of a mapping technique remains open. This choice depends on type of problem and availability of data, financial resources, and time for the investigation, as well as the professional experience of those involved in the survey.

2.3 Input Data

Slope instability phenomena are related to a large variety of factors involving both the physical environment and human interaction. Thus, assessment of landslide hazard requires knowledge about these factors, ranging from geologic structure to land use. For this reason landslide hazard assessments should preferably involve multidisciplinary teams.

The input data needed to assess landslide hazard at the regional, medium, and large scales are given in Table 8-1. The list is extensive, and only in an ideal case will all types of data be available. However, as will be explained in Section 2.4, the amount and type of data that can be collected will determine the type of hazard analysis that can be applied, ranging from qualitative assessment to complex statistical methods.

The input data needed for landslide hazard analysis can be subdivided into five main groups: geomorphology; topography; engineering geology, geotechnology, or both; land use; and hydrology. Each group may be subdivided to form a sequence of so-called *data layers*. Each data layer may be represented by an individual map containing that one type of data. As discussed in Section 4, when GIS techniques are employed, it is important that each data-layer map be composed of only one type of data element (points, lines, or areas and polygons) and have one or more accompanying tables to define the characteristics of each. Of course, the data layers required by landslide hazard analysis may vary to account for the characteristics of different environments.

In the second column of Table 8-1 the various parameters that are stored in "attribute tables" connected to each map are indicated. In the third column a summary is given of the method by which each data layer is collected, which refers to the three phases of data collection (image interpretation, fieldwork, and laboratory analysis). A number of data layers, such as material sequences, seismic acceleration maps, and water table maps, require the use of specific models in addition to the conventional data collection techniques. Specific algorithms within a GIS may be used to convert topographic elevation values into slope categories or to perform other topographic analyses (see Section 4).

The ratings in the last three columns of Table 8-1 indicate the relative feasibility of collecting certain data types for each of the three scales under consideration. The feasibility of collecting data for a certain scale does not imply that the specific type of data is also useful for that particular scale. A map using terrain mapping units, for example, can be prepared at a 1:10,000 scale but will be of limited use because of its generalized content.

Because of the large areas to be studied and the objectives of a hazard assessment at the regional scale (see Section 2.2), detailed data collection for individual factors (geomorphology, lithology, soils, etc.) is not a cost-effective approach. Data gathered for this scale should be limited to the delineation of homogeneous terrain mapping units, for example, with the use of stereoscopic satellite imagery and the collection of regional tectonic or seismic data.

For the medium scale almost all data layers given in Table 8-1 can be gathered easily with the exception of detailed groundwater and geotechnical information. The data collection at this scale should be focused on the production of detailed multitemporal landslide distribution maps and the various parameters required in statistical analysis.

For large-scale hazard zonation, in which work is carried out in relatively small areas, all of the proposed data layers can be readily collected. Data collection at this scale should relate to the parameters needed for slope stability modeling (for example, material sequences, seismic accelerations, and hydrologic data).

2.4 General Trends

A great deal of research concerning slope instability hazard has been done over the last 30 years.

Table 8-1
Overview of Input Data for Landslide Hazard Analysis

DATA LAYERS FOR SLOPE INSTABILITY HAZARD ZONATION	ACCOMPANYING DATA IN TABLES	METHOD USED	SCALE OF ANALYSIS		
			REGIONAL	MEDIUM	LARGE
GEOMORPHOLOGY					
1. Terrain mapping units	Terrain mapping units	SII + walk-over survey	3	3	3
2. Geomorphological (sub)units	Geomorphological description	API + fieldwork	2	3	3
3. Landslides (recent)	Type, activity, depth, dimension etc.	API + API checklist + fieldwork + field checklist	1	3	3
4. Landslides (older period)	Type, activity, depth, dimension, date, etc.	API + API checklist + landslide archives	1	3	3
TOPOGRAPHY					
5. Digital terrain model	Altitude classes	With GIS from topographic map	2	3	3
6. Slope map	Slope angle classes	With GIS from DTM	2	3	3
7. Slope direction map	Slope direction classes	With GIS from DTM	2	3	3
8. Slope length	Slope length classes	With GIS from DTM	2	3	3
9. Concavities/convexities	Concavity/convexity	With GIS from DTM	1	1	3
ENGINEERING GEOLOGY					
10. Lithologies	Lithology, rock strength, discontinuity spacing	Existing maps + API + fieldwork, field and laboratory testing	2	3	3
11. Material sequences	Material types, depth, USCS classification, grain-size distribution, bulk density, c and φ	Modeling from lithological map + geomorphological map + slope map, field descriptions, field and laboratory testing	1	2	3
12. Structural geological map	Fault type, length, dip, dip direction, fold axis, etc.	SII + API + fieldwork	3	3	3
13. Seismic accelerations	Maximum seismic acceleration	Seismic data + engineering geological data + modeling	3	3	3
LAND USE					
14. Infrastructure (recent)	Road types, railway lines, urban extension, etc.	API + topographical map + fieldwork + classification of satellite imagery	3	3	3
15. Infrastructure (older)	Road types, railway lines, urban extension, etc.	API + topographical map	3	3	3
16. Land use map (recent)	Land use types, tree density, root depth	API + classification of satellite imagery + fieldwork	2	3	3
17. Land use map (older)	Land use types	API	2	3	3
HYDROLOGY					
18. Drainage	Type, order, length	API + topographical maps	3	3	3
19. Catchment areas	Order, size	API + topographical maps	2	3	3
20. Rainfall	Rainfall in time	From meteorological stations	2	3	3
21. Temperature	Temperature in time	From meteorological stations	2	3	3
22. Evapotranspiration	Evapotranspiration in time	From meteorological stations and modeling	2	3	3
23. Water table maps	Depth of water table in time	Field measurements of K_{sat} + hydrological model	1	1	2

NOTE: The last three columns indicate the possibility for data collection for the three scales of analysis: 3 = good, 2 = moderate, and 1 = poor.
Abbreviations used: SII = satellite image interpretation, API = aerial photointerpretation, DTM = digital terrain model, GIS = geographic information system, K_{sat} = saturated conductivity testing.

Initially the investigations were oriented mainly toward solving instability problems at particular sites. Techniques were developed by engineers for the appropriate design of a planned structure as well as the prevention of slope failure. Therefore, research emphasized site investigation techniques and the development of deterministic and probabilistic models. However, the heterogeneity of the natural environment at the regional scale and the large variability in geotechnical properties such as cohesion and internal friction are in sharp contrast to the homogeneity required by deterministic models. This contrast, coupled with the costly and time-consuming site investigation techniques required to obtain property values, makes the engineering approach unsuitable for application over large areas.

In engineering projects, such large areas must often be assessed during early phases of planning and decision making. To solve this problem, several other types of landslide hazard analysis techniques have been developed during the last decades. These techniques provide hazard assessment based on a careful study of the natural conditions of an area and analysis of all the possible parameters involved in slope instability processes. These various methodological approaches, which were reviewed in detail by Hansen (1984) and Varnes (1984), are summarized in the following sections. Some examples are included in this chapter for illustration.

2.4.1 Landslide Inventory

The most straightforward approach to landslide hazard zonation is a *landslide inventory*, based on any or all of the following: aerial photointerpretation, ground survey, and a data base of historical occurrences of landslides in an area. The final product gives the spatial distribution of mass movements, which may be represented on a map either as affected areas to scale or as point symbols (Wieczorek 1984). Such mass movement inventory maps are the basis for most other landslide hazard zonation techniques. They can, however, also be used as an elementary form of hazard map because they display the location of a particular type of slope movement. They provide information only for the period shortly preceding the date that aerial photographs were taken or the field-work was conducted. They provide no insight into temporal changes in mass movement distribution. Many landslides that occurred some time before

photographs were taken may have become undetectable. Therefore a refinement is the construction of *landslide activity maps*, which are based on multitemporal aerial photointerpretation (Canuti et al. 1979). Landslide activity maps are indispensable to the study of effects of temporal variation of a factor, such as land use, on landsliding.

Landslide distribution can also be shown in the form of a *density map*. Wright et al. (1974) presented a method for calculating landslide densities using counting circles. The resulting density values are interpolated and presented by means of *landslide isopleths*. Although the method does not investigate the relationship between mass movements and causal factors, it is useful in presenting landslide densities quantitatively.

2.4.2 Heuristic Approach

In heuristic methods the expert opinion of the geomorphologist making the survey is used to classify the hazard. These methods combine the mapping of mass movements and their geomorphologic setting as the main input factor for hazard determination. Two types of heuristic analysis can be distinguished: geomorphic analysis and qualitative map combination.

2.4.2.1 Geomorphic Analysis

The basis for *geomorphic analysis* was outlined by Kienholz (1977), who developed a method for producing a combined hazard map based on the mapping of "silent witnesses" (*Stumme Zeugen*). The geomorphic method is also known as the *direct mapping method*. The hazard is determined directly in the field by the geomorphologist. The process is based on individual experience and the use of reasoning by analogy. The decision rules are therefore difficult to formulate because they vary from place to place. Examples of this methodology for the appraisal of terrain to determine its susceptibility to slope instability are especially common from Europe, where ample experience exists in geomorphic and engineering geologic mapping (Carrara and Merenda 1974; Kienholz 1977, 1978; Malgot and Mahr 1979; Kienholz et al. 1983, 1988; Ives and Messerli 1981; Rupke et al. 1988). There are many other examples from other regions, however (Hansen 1984; Varnes 1984). The French program that produces 1:25,000-scale ZERMOS maps (Meneroud and Calvino 1976) is probably the best example, but the reproducibility of these maps has been much debated (Antoine 1977). The same is true for the method used by Brunsden and his collaborators (1975), who do not even present a hazard zonation analysis for a project related to a road alignment. Rather, they directly suggest the alignment for the best possible road according to their assessment of the slope stability.

2.4.2.2 Qualitative Map Combination

To overcome the problem of the "hidden rules" in geomorphic mapping, other qualitative methods, based on *qualitative map combination*, have been developed. In qualitative map combination the earth scientist uses the expert knowledge of an individual to assign weighting values to a series of parameter maps. The terrain conditions at a large number of locations are summed according to these weights, leading to hazard values that can be grouped into hazard classes. Stevenson (1977) developed an empirical hazard rating system for an area in Tasmania. On the basis of his expert knowledge of the causal factors of slope instability, he assigned weighting values to different classes on a number of parameter maps. Qualitative map combination has become very popular in slope instability zonation. The problem with this method is in determining the exact weighting of the various parameter maps. Often, insufficient field knowledge of the important factors prevents the proper establishment of the factor weights, leading to unacceptable generalizations.

2.4.3 Statistical Approach

In statistical landslide hazard analysis the combinations of factors that have led to landslides in the past are determined statistically, and quantitative predictions are made for areas currently free of landslides but where similar conditions exist. Two different statistical approaches are used in landslide hazard analysis: bivariate and multivariate.

2.4.3.1 Bivariate Statistical Analysis

In *bivariate statistical analysis* each factor map (for example, slope, geology, land use) is combined with the landslide distribution map, and weighting values based on landslide densities are calculated for each parameter class (for example, slope class, lithologic unit, land use type). Brabb et al. (1972) provided the first example of such an analysis.

They performed a simple combination of a landslide distribution map with a lithologic map and a slope map. Several statistical methods have been applied to calculate weighting values; these have been termed the *landslide susceptibility method* (Brabb 1984; van Westen 1992, 1993), *information value method* (Yin and Yan 1988; Kobashi and Suzuki 1988), and *weight-of-evidence modeling method* (Spiegelhalter 1986). Chung and Fabbri (1993) described several methods, including *Bayesian combination rules*, *certainty factors*, *Dempster-Shafer method*, and *fuzzy logic*.

2.4.3.2 Multivariate Statistical Analysis

Multivariate statistical analysis models for landslide hazard zonation were developed in Italy, mainly by Carrara (1983, 1988) and his colleagues (Carrara et al. 1990, 1991, 1992). In their applications all relevant factors are sampled either on a large-grid basis or in morphometric units. For each of the sampling units, the presence or absence of landslides is also determined. The resulting matrix is then analyzed using multiple regression or discriminant analysis. With these techniques good results can be expected in homogeneous zones or in areas with only a few types of slope instability processes, as shown in the work of Jones et al. (1961) concerning mass movements in terrace deposits. When complex statistics are applied, as was done by Carrara and his collaborators (Carrarra et al. 1990, 1991, 1992), by Neuland (1976), or by Kobashi and Suzuki (1988), a subdivision of the data according to the type of landslide should be made as well. Therefore, large data sets are needed to obtain enough cases to produce reliable results. The use of complex statistics implies laborious efforts in collecting large amounts of data, because these methods do not use selective criteria based on professional experience.

2.4.4 Deterministic Approach

Despite problems related to collection of sufficient and reliable input data, deterministic models are increasingly used in hazard analysis of larger areas, especially with the aid of GIS techniques, which can handle the large number of calculations involved in determination of safety factors over large areas. Deterministic methods are applicable only when the geomorphic and geologic conditions are fairly homogeneous over the entire study area and the landslide types are simple. The advantage of these white box models is that they are based on slope stability models, allowing the calculation of quantitative values of stability (safety factors). The main problem with these methods is their high degree of oversimplification. A deterministic method that is usually applied for translational landslides is the infinite slope model (Ward et al. 1982). These deterministic methods generally require the use of groundwater simulation models (Okimura and Kawatani 1986). Stochastic methods are sometimes used to select input parameters for the deterministic models (Mulder and van Asch 1988; Mulder 1991; Hammond et al. 1992).

2.4.5 Evaluation of Trends in Methodology

As discussed in Section 2.2, not all methods of landslide hazard zonation are equally applicable at each scale of analysis. Some require very detailed input data, which can only be collected for small areas because of the required levels of effort and thus the cost (see Section 2.3). Therefore, suitable methods have to be selected to define the most useful types of analysis for each of the mapping scales while also maintaining an acceptable cost/benefit ratio. Table 8-2 provides an overview of the various methods of landslide hazard analysis and recommendations for their use at the three most relevant scales.

Evaluation of the methodological approaches (see Table 8-2) and the literature on slope instability hazard zonation practices suggests that heuristic methods, described in Section 2.4.2, aim to establish the real causes for slope instability on the basis of scientific and professionally oriented reasoning. However, considering the scale of slope failures and the complexity of the conditions that may lead to slope instability, these direct mapping methods have to be executed on a large scale. Therefore, they are impractical to use over large areas and do not support implementation of a hierarchical approach to hazard zonation. The combination of geomorphic analysis with the application of weights to the contributing parameters, as used by Kienholz (1977; 1978), improves the objectivity and reproducibility of these heuristic methods. This is particularly the case when the weights are based on the contribution of various parameters to slope instability, with the contributions established by simple statistics.

For regional landslide hazard zonations at small scales (1:50,000 to 1:100,000), many ap-

proaches combine indirect mapping methods with more analytical approaches. At these scales a terrain classification can be made using stereo satellite imagery, thereby defining homogeneous lithomorphologic zones or terrain mapping units (Meijerink 1988). These terrain mapping units are further analyzed by photointerpretation and ground surveys. The characteristics of each terrain mapping unit are defined by attributed values, which define their probable values, or range of values, for a suite of parameters. An attribute data base is created in which the characteristics of all the terrain mapping units are defined in a series of tables. Relevant parameters are identified on the basis of an evaluation of slope instability in the area, and these are then used to define hazard categories. These categories are extrapolated to the terrain mapping units throughout the region being mapped according to the presence or absence of these relevant parameters (sometimes referred to as contributing factors) in the attribute data base.

As the project continues, the landslide zonation evolves. The size of the area being studied decreases and the scale of the investigation increases. Additional analytical studies are possible because more time and money become available. Factor maps, displaying the spatial distribution of the most important factors, together with increased analysis of possible contributing parameters based on statistics increase the accuracy of predictions of susceptibility to instability. An adjustment or refinement of the decision rules for the hazard assessment can be obtained by verifying the results of the initial assessment through comparison with the real situation in the field. If necessary, weights assigned to parameters can be adjusted and a new hazard assessment produced. This iterative method becomes necessary when the hazard assessment decision rules are extrapolated over areas with a similar geologic or geomorphic setting but where little ground truth is available because studies have shown that areas with apparently equal conditions may produce weighting values that vary considerably.

In detailed studies of small areas, large amounts of data may become available; thus, simple deterministic or probabilistic models, discussed in Section 2.4.4, become increasingly practical as methods for landslide hazard zonation. They allow the approximation of the variability of the safety factor for slope failure and thus yield information useful to design engineers.

Table 8-2
Analysis Techniques in Relation to Mapping Scales

Type of Analysis	Technique	Characteristics	Required Data Layers[a]	Scale of Use Recommended		
				Regional (1:100,000)	Medium (1:25,000)	Large (1:10,000)
Inventory	Landslide distribution analysis	Analyze distribution and classification of landslides	3	Yes [b]	Yes	Yes
	Landslide activity analysis	Analyze temporal changes in landslide pattern	4,5,14,15,16,17	No	Yes	Yes
	Landslide density analysis	Calculate landslide density in terrain units or as isopleth map	1,2,3	Yes [b]	No	No
Heuristic analysis	Geomorphological analysis	Use in-field expert opinion in zonation	2,3,4	Yes	Yes [c]	Yes [c]
	Qualitative map combination	Use expert-based weight values of parameter maps	2,3,5,6,7,8,9,10, 12,14,16,18	Yes [d]	Yes [c]	No
Statistical analysis	Bivariate statistical analysis	Calculate importance of contributing factor combination	2,3,5,6,7,8,9,10, 12,14,16,18	No	Yes	No
	Multivariate statistical analysis	Calculate prediction formula from data matrix	2,3,5,6,7,8,9,10, 12,14,16,18	No	Yes	No
Deterministic analysis	Safety factor analysis	Apply hydrological and slope stability models	6,11,12,13,16, 20,21,22,23	No	No	Yes [e]

[a] The numbers in this column refer to the input data layers given in Table 8-1.
[b] But only with reliable data on landslide distribution because mapping will be out of an acceptable cost/benefit ratio.
[c] But strongly supported by other more quantitative techniques to obtain an acceptable level of objectivity.
[d] But only if sufficient reliable data exist on the spatial distribution of the landslide controlling factors.
[e] But only under homogeneous terrain conditions, considering the variability of the geotechnical parameters.

2.5 Accuracy and Objectivity

The most important question to be asked in each landslide hazard study relates to its degree of accuracy. The terms *accuracy* and *reliability* are used to indicate whether the hazard map makes a correct distinction between landslide-free and landslide-prone areas.

The accuracy of a landslide prediction depends on a large number of factors, the most important of which are

- Accuracy of the models,
- Accuracy of the input data,
- Experience of the earth scientists, and
- Size of the study area.

Many of these factors are interrelated. The size of the study area determines to a large degree what kind and density of data can be collected (see Table 8-1) and what kind of analysis technique can be applied (see Table 8-2).

Evaluation of the accuracy of a landslide hazard map is generally very difficult. In reality, a hazard prediction can only be verified by observing if failure takes (or has taken) place in time—the so-called "wait and see" procedure. However, this is often not a very useful method, for obvious reasons. There are two possible forms of prediction inaccuracies: landslides may occur in areas that are predicted to be stable, and landslides may actually not occur in areas that are predicted to be unstable. Both cases are undesirable, of course. However, the first case is potentially more serious, because a landslide occurring in an area predicted to be free of landsliding may cause severe damage or loss of life and may lead to lawsuits. The second case may result in additional expenses for unnecessary exploration, for design of complex structures, or for the realignment of facilities from what are in actuality perfectly acceptable areas to areas in which construction is more expensive.

The two possible cases of error in prediction are not equally easy to evaluate. In estimating the magnitude of the first case, in which a landslide occurs in an area predicted to be stable, the investigator is faced with the task of proving the presence of something that does not currently exist.

Accordingly, one of the most frequently used methods for checking the accuracy of hazard maps is the comparison of the final predictive hazard map with a map showing the pattern of existing landslides. A frequency distribution is made relating the hazard scores to identified landslide-prone and non-landslide-prone areas. From this frequency distribution the percentage of mapped landslides found in areas predicted to be stable (non-landslide-prone) can be calculated. This error is then assumed to be the same as the error in predicting landslides in currently landslide-free areas. This method can be refined if multitemporal landslide distribution maps are available. The landslide prediction, based on an older landslide distribution map, can then be checked with a younger landslide distribution to see if newer movements confirm the predictions (Chung et al. in press). The comparison of landslide hazard maps made by different methods (for example, by statistical and by deterministic methods) may also provide a good idea of the accuracy of the prediction.

Related to the problem of assessing the accuracy of hazard maps is the question of their objectivity. The terms *objective* and *subjective* are used to indicate whether the various steps taken in the determination of the degree of hazard are verifiable and reproducible by other researchers or whether they depend on the personal judgment of the earth scientist in charge of the hazard study.

Objectivity in the assessment of landslide hazard does not necessarily result in an accurate hazard map. For example, if a very simple but verifiable model is used or if only a few parameters are taken into account, the procedure may be highly objective but will produce an inaccurate map. On the other hand, subjective studies, such as detailed geomorphic slope stability analyses, when made by experienced geomorphologists may result in very accurate hazard maps. Yet such a good, but subjective, assessment may have a relatively low objectivity because its reproducibility will be low. This means that the same evaluation made by another expert will probably yield other results, which can have clearly undesirable legal effects.

The degree of objectivity of a hazard study depends on the techniques used in data collection and the methods used in data analysis. The use of objective analysis techniques, such as statistical analysis or deterministic analysis, may still lead to subjective results, depending on the amount of

subjectivity that is required for creating the parameter maps. Studies were conducted by Dunoyer and van Westen (1994) to assess the degree of subjectivity involved in the interpretation of landslides from large-scale aerial photographs (at a scale of 1:10,000) by a group of 12 photointerpreters, several of whom had had considerable experience and some who had local knowledge. These comparisons have shown that differences between interpretations can be large (Dunoyer and van Westen 1994). These findings confirm other similar investigations on the subjectivity of photointerpretation in slope instability mapping (Fookes et al. 1991; Carrara et al. 1992).

Many of the input maps used in landslide hazard analysis are based on aerial photointerpretation and will therefore include a large degree of subjectivity. Even data concerning factors that are obtained by means of precise measurements, such as soil strength, may have a high degree of subjectivity in the resulting parameter maps because the individual sample values, representing the conditions at the sampled location, have to be linked to mapped units on a material map produced by photointerpretation and fieldwork in order to provide a regional distribution of the sampled property.

For each type of data collection and analysis, different levels of objectivity and accuracy may be encountered at the various hierarchical levels corresponding to the various scales of hazard analysis. The demand for higher levels of objectivity has led several researchers to replace the subjective expert's opinion on the causative factors related to slope failure with statistical analysis of all terrain conditions observed in areas with slope failures (Carrara et al. 1978; Neuland 1976). Although the objectivity of such an approach is guaranteed, doubts may exist as to the accuracy of the assessment, especially when the experience and skill required in the data collection and the labor required to complete the extensive data sheets are considered (Figure 8-1).

Because of the limitations inherent in the data collection and analysis techniques and the restrictions imposed by the scale of mapping, a landslide hazard survey will always retain a certain degree of subjectivity, which does not necessarily imply inaccuracy. The objectivity and reproducibility of the hazard assessment can be improved considerably by interpretation of sequential imagery, by use of clear and if possible quantitative descriptions of

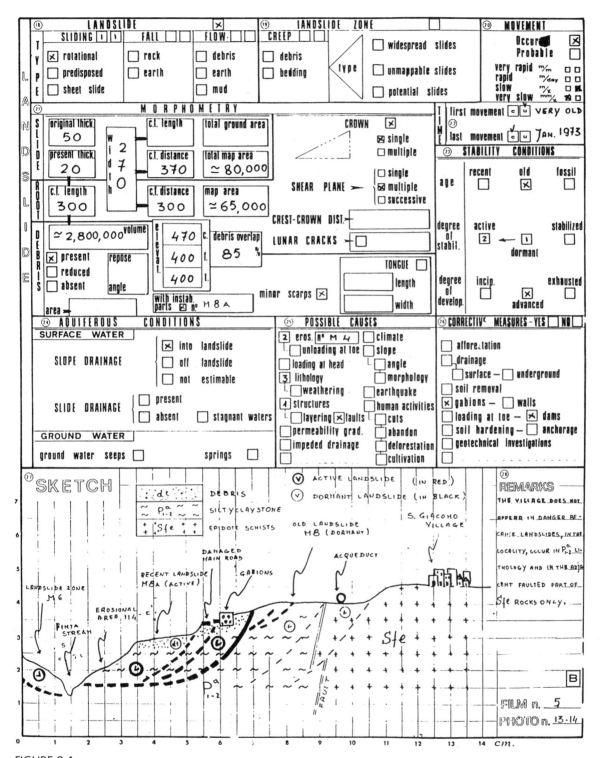

FIGURE 8-1
Page from data form used by Carrara and collaborators
(Carrara and Merenda 1974).

the factors considered, and by application of well-defined analytical procedures and decision rules. The most important aspect, however, remains the experience of the interpreter with regard to both the various factors involved in slope instability hazard surveys and the specific conditions of the study area.

3. REMOTE SENSING IN SLOPE INSTABILITY STUDIES

Because landslides directly affect the ground surface, remote-sensing techniques are well suited to slope instability studies. The term *remote sensing* is used here in its widest sense, including aerial photography and imagery obtained by satellites or any other remote-sensing technique. Remote sensing is particularly useful when stereo images are used because they depict in the stereo model the typical morphologic features of landslides, which often can provide diagnostic information concerning the type of movement (Crozier 1973). Also, the overall terrain conditions, which are critical in determining the susceptibility of a site to slope instability, can profitably be interpreted from remote-sensing data.

The value of photointerpretation of aerial photographs for identifying slope instability has been reported by many investigators. Rib and Liang (1978) discussed these photointerpretation techniques in considerable detail. Mollard (1977) also demonstrated the utility of aerial photography in examining landslides. Several basic textbooks on slope instability refer to the importance of aerial photographs in the study of landslides (Brunsden and Prior 1984). Scientists at the University of Bari in Italy have successfully used aerial photographs to evaluate both active landslides (Guerricchio and Melidoro 1981) and historic movements (Cotecchia et al. 1986). However, during the past two decades, considerable development in remote sensing has occurred; thus an overview is presented here of the types of images available and their relevant characteristics for landslide investigations.

3.1 Remote-Sensing Products

There has been little development in aerial photography during the past two decades. Panchromatic black-and-white and color film are available to cover the visible part of the electromagnetic spectrum, and black-and-white infrared and false-color infrared film extend the imagery sensitivity into the reflective near-infrared regions. The spatial resolution of these films is excellent, and the aerial photographs are normally taken so as to provide stereoscopic coverage, producing a three-dimensional picture of the terrain that gives detailed morphologic information. However, the spectral resolution is much less than that provided by multispectral data imagery sources because the photography integrates the broad spectral band into a single picture. The organization of an aerial photographic mission is time-consuming, and in some locations the number of days during the year with climatic conditions suitable for acceptable aerial photography may be very limited. Thus temporal resolution, that is, the number of images of the same area over time, of aerial photography can be much less than that provided by satellite imagery. On the other hand, constraints imposed by orbital mechanics restrict satellite imagery to a fixed schedule of viewing opportunities, and these may not coincide with optimum weather conditions at a landslide site. In this regard, aerial photography may have more flexibility in scheduling, but at some considerable economic cost.

The application of satellite data has increased enormously in the past decade. Table 8-3 compares the specifications of resolution for LANDSAT and SPOT satellite images. After the initial low-spatial-resolution images of the LANDSAT MSS (which were about 60 by 80 m), LANDSAT now offers thematic mapper (TM) images with a spatial resolution of 30 m and excellent spectral resolution. LANDSAT TM provides six bands to cover the entire visible, near-infrared, and middle-infrared portions of the spectrum, with one additional band providing a lower resolution of the thermal infrared. LANDSAT satellite orbits are arranged to provide good coverage of a large portion of the earth's surface. The satellite passes over each location every 18 days, offering a theoretical temporal resolution of 18 days, although weather conditions are a serious limiting factor in this respect. Clouds frequently hamper the acquisition of data from the ground surface. The degree of weather interference naturally varies with climate regions. The weakest point of the LANDSAT system is the lack of an adequate stereovision capability. Theoretically a stereomate of a LANDSAT TM image can be produced with the help of a good digital elevation

Table 8-3
Comparison of Specifications of Different Multispectral Remote-Sensing Products

	LANDSAT MSS	LANDSAT TM	SPOT MULTISPECTRAL	PANCHROMATIC
No. of spectral bands	4	7	3	1
Spectral resolution (μm)	0.5-1.1	0.45-2.35 10.4-12.5	0.5-0.9	0.5-0.7
Spatial resolution (m)	80	30[a]	20	10
Swath width (km)	185	185	2 × 60	2 × 60
Stereo	No	No	Yes	Yes
Temporal resolution	18 days	18 days	26 days, 5 days off nadir	26 days, 5 days off nadir

[a] 120 m in thermal infrared band.

model (DEM), but this remains a relatively unattractive option because very detailed DEMs are not currently available for most locations.

It should be noted that the terms *digital elevation model* (DEM) and *digital terrain model* (DTM) are frequently used interchangeably. However, some authors prefer to use DEM to refer to values that merely provide topographic elevation values, usually on a regular or gridded basis. They prefer to restrict DTM to those situations in which a more complete description of the terrain is provided, for example, by slope or geomorphic classification. Since the applications concerning the creation of stereoscopic images require only elevation values, the term *DEM* will be used in this chapter.

The French SPOT satellite is equipped with two sensor systems that cover adjacent paths, each with a swath width of 60 km. The sensors have an off-nadir viewing capability, offering the possibility of producing images with good stereoscopic vision. The option of viewing sideways also provides for potentially higher temporal resolution because the satellite can observe a location not directly under its orbital path. SPOT senses the terrain in a single wide panchromatic band and in three narrower spectral bands corresponding to the green, red, and near-infrared portions of the spectrum (see Table 8-3). The spatial resolution in the panchromatic mode is 10 m, whereas the three spectral bands have a spatial resolution of 20 m. The system lacks spectral bands in the middle-infrared and far-infrared (thermal) portions of the spectrum.

Radar satellite images, available from the European ERS-1 and the Japanese JERS satellites, offer all-weather viewing capability because radar systems can penetrate clouds. Theoretically this type of imagery can yield detailed information on surface roughness and micromorphology. However, the currently applied radar wavelengths and viewing angles have not been very appropriate for applications in mountainous terrain. Initial results of research with radar interferometry are promising, indicating that detailed terrain models with an accuracy of less than 1 m can be created. Such resolution suggests the possibility of monitoring landslide activity.

New commercial satellites with 1-m panchromatic and 3-m multispectral image resolutions have recently been announced with launch dates in 1996 and 1997. Not only will these satellites provide much higher spatial resolution than the present LANDSAT or SPOT satellites, but they will also provide greater spectral resolution. Only a few simulated image products have been produced to suggest the capabilities of these new satellites. At the present time it is impossible to predict accurately how these new imagery sources will affect landslide mapping efforts, but readers of this report are encouraged to be aware of, and to investigate, the potential of new developments in the rapidly changing satellite image collection field.

3.2 Landslide Interpretation from Remote-Sensing Images

Landslide information extracted from remote-sensing images is mainly related to the morphology, vegetation, and drainage conditions of the slope. Slope morphology is best studied by examination of a stereoscopic model. The study of variations in tone and texture or of pattern, shape, and lineaments has to be related to the expected ground conditions or landforms associated with slope instability processes (see Figure 8-2).

The interpretation of slope movements from remote-sensing images is based on recognition or identification of elements associated with slope movements and interpretation of their significance to the slope instability process. The implication is that a particular type of slope failure is seldom recognized directly but is interpreted to exist by analysis of a certain number of elements pertaining to slope instability features that are observed on the remotely sensed images.

As a consequence, the categorization of slope movements obtained by interpretation of aerial photographs is not as detailed as the classifications of Cruden (see Chapter 3 in this report) or those of other authors (Hansen 1984; Crozier 1986; Hutchinson 1988). These classifications include field evidence in their considerations. Experience has shown that photointerpretation of landslides has to use a simpler classification. Local adaptations to existing classifications can be justified to prevent ambiguities and therefore misclassifications. Table 8-4 shows a checklist constructed on one recent project to provide a systematic characterization of slope failures as observed on aerial photographs.

The types of slope movements considered were based on local knowledge in this specific region in the Colombian Cordillera (van Westen 1992, 1993). Table 8-4 also gives an indication of the type of information that can be obtained by experienced photointerpreters using aerial photographs at a scale of 1:20,000. The degree of landslide activity, as classified by aerial photointerpretation, is determined in this region by the freshness of the features related to the landsliding. The morphology of older landslides, showing a degradation of their morphologic forms and usually already overgrown by vegetation, is classified as stable.

Table 8-5 is a summary of the terrain features frequently associated with slope movements, the relationship of these features to landslides, and their characterization on aerial photographs. These elements are used to develop an interpretation and classification of slope failure according to the characteristics in Table 8-6. The interpretation of these various types of mass movements is discussed in the following sections.

3.2.1 Falls and Topples

Falls and *topples* are always related to very steep slopes, mostly those steeper than 50 degrees, where bedrock is directly exposed. Falls are mainly con-

FIGURE 8-2
Faint tonal and textural differences characterizing surficial mud slides in marly clays. Slope failures can be readily interpreted when stereogram is viewed with pocket stereoscope. Although these mass movements are significant because they frequently damage roads, their size precludes them from being interpreted on smaller-scale images (original photograph scale 1:17,000, Basilicata, Italy).

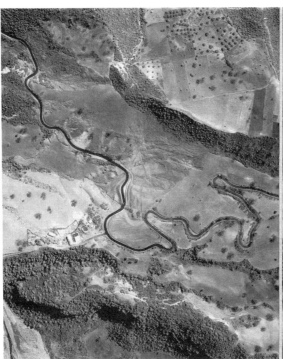 

Table 8-4
Interpretation of Landslides in Colombia Using Aerial Photographs and GIS (Van Westen 1993)

Code	Type	Subtype	Activity	Depth	Vegetation	Body
1	Slide	Rotational	Stable	Surficial	Bare	Landslide scar
2	Lateral spread	Translational	Active	Deep	Low	Runout body
3	Flow	Complex			High/dense	
4	Debris avalanche	Unknown				

Note: Landslide delineations were digitized and stored in georeferenced image data base, and digital codes, representing the ID number and landslide information, were stored in attribute data base.

Table 8-5
Morphologic, Vegetation, and Drainage Features Characteristic of Slope Instability Processes and Their Photographic Characteristics

Terrain Features	Relation to Slope Instability	Photographic Characteristics
Morphology		
Concave/convex slope features	Landslide niche and associated deposit	Concave/convex anomalies in stereo model
Steplike morphology	Retrogressive sliding	Steplike appearance of slope
Semicircular backscarp and steps	Head part of slide with outcrop of failure plane	Light-toned scarp, associated with small, slightly curved lineaments
Back-tilting of slope facets	Rotational movement of slide blocks	Oval or elongated depressions with imperfect drainage conditions
Hummocky and irregular slope morphology	Microrelief associated with shallow movements or small retrogressive slide blocks	Coarse surface texture, contrasting with smooth surroundings
Infilled valleys with slight convex bottom, where V-shaped valleys are normal	Mass movement deposit of flow-type form	Anomaly in valley morphology, often with lobate form and flow pattern on body
Vegetation		
Vegetational clearances on steep scarps, coinciding with morphological steps	Absence of vegetation on headscarp or on steps in slide body	Light-toned elongated areas at crown of mass movement or on body
Irregular linear clearances along slope	Slip surface of translational slides and track of flows and avalanches	Denuded areas showing light tones, often with linear pattern in direction of movement
Disrupted, disordered, and partly dead vegetation	Slide blocks and differential movements in body	Irregular, sometimes mottled grey tones
Differential vegetation associated with changing drainage conditions	Stagnated drainage on back-tilting blocks, seepage at frontal lobe, and differential conditions on body	Tonal differences displayed in pattern associated with morphological anomalies in stereo model
Drainage		
Areas with stagnated drainage	Landslide niche, back-tilting landslide blocks, and hummocky internal relief on landslide body	Tonal differences with darker tones associated with wetter areas
Excessively drained areas	Outbulging landslide body (with differential vegetation and some soil erosion)	Light-toned zones in association with convex relief forms
Seepage and spring levels	Springs along frontal lobe and at places where failure plane outcrops	Dark patches sometimes in slightly curved pattern and enhanced by differential vegetation
Interruption of drainage lines	Drainage anomaly caused by head scarp	Drainage line abruptly broken off on slope by steeper relief
Anomalous drainage pattern	Streams curving around frontal lobe or streams on both sides of body	Curved drainage pattern upstream with sedimentation or meandering in (asymmetric) valley

Table 8-6
Image Characteristics of Mass Movement Types

Type of Movement	Characterization Based on Morphological, Vegetational, and Drainage Aspects Visible on Stereo Images	
Fall and topple	Morphology:	Distinct rock wall or free face in association with scree slopes (20 to 30 degrees) and dejection cones; jointed rock wall (>50 degrees) with fall chutes
	Vegetation:	Linear scars in vegetation along frequent rock-fall paths; vegetation density low on active scree slopes
	Drainage:	No specific characteristics
Sturzstrom	Morphology:	Extremely large (concave) scars on mountain, with downslid blocks of almost geological dimensions; rough, hummocky depositional forms, sometimes with lobate front
	Vegetation:	Highly irregular/chaotic vegetational conditions on accumulative part, absent on sturzstrom scar
	Drainage:	Irregular disordered surface drainage, frequent damming of valley and lake formed behind body
Rotational slide	Morphology:	Abrupt changes in slope morphology characterized by concave (niche) and convex (runout lobe) forms; often steplike slopes; semilunar crown and lobate frontal part; back-tilting slope facets, scarps, hummocky morphology on depositional part; D/L ratio 0.3 to 0.1; slope 20 to 40 degrees
	Vegetation:	Clear vegetational contrast with surroundings, absence of land use indicative for activity; differential vegetation according to drainage conditions
	Drainage:	Contrast with nonfailed slopes; bad surface drainage or ponding in niches or back-tilting areas; seepage in frontal part of runout lobe
Compound slide	Morphology:	Concave and convex slope morphology; concavity often associated with linear grabenlike depression; no clear runout but gentle convex or bulging frontal part; back-tilting facets associated with (small) antithetic faults; D/L ratio 0.3 to 0.1, relatively broad in size
	Vegetation:	As with rotational slides, although slide mass will be less disturbed
	Drainage:	Imperfect or disturbed surface drainage ponding in depressions and in rear part of slide
Translational slide	Morphology:	Joint controlled crown in rock slides, smooth planar slip surface; relatively shallow, certainly in surface material over bedrock; D/L ratio <0.1 and large width; runout hummocky, rather chaotic relief, with block size decreasing with larger distance
	Vegetation:	Source area and transportational path denuded, often with lineations in transportation direction; differential vegetation on body, in rock slides; no land use on body
	Drainage:	Absence of ponding below crown, disordered or absent surface drainage on body; streams deflected or blocked by frontal lobe
Lateral spread	Morphology:	Irregular arrangement of large blocks tilting in various directions; block size decreases with distance and morphology becomes more chaotic; large cracks and linear depressions separating blocks; movement can originate on very gentle slopes (<10 degrees)
	Vegetation:	Differential vegetation enhancing separation of blocks; considerable contrast with unaffected areas
	Drainage:	Disrupted surface drainage; frontal part of movement is closing off valley, causing obstruction and asymmetric valley profile
Mudslide	Morphology:	Shallow concave niche with flat lobate accumulative part, clearly wider than transportation path; irregular morphology contrasting with surrounding areas; D/L ratio 0.05 to 0.01; slope 15 to 25 degrees
	Vegetation:	Clear vegetational contrast when fresh; otherwise differential vegetation enhances morphological features
	Drainage:	No major drainage anomalies beside local problems with surface drainage
Earth flow	Morphology:	One large or several smaller concavities, with hummocky relief in source area; main scars and several small scars resemble slide type of failure; path following stream channel and body is infilling valley, contrasting with V-shaped valleys; lobate convex frontal part; irregular micromorphology with pattern related to flow structures; slope > 25 degrees; D/L ratio very small

continued on next page

Table 8-6 *(continued)*

TYPE OF MOVEMENT	CHARACTERIZATION BASED ON MORPHOLOGICAL, VEGETATIONAL, AND DRAINAGE ASPECTS VISIBLE ON STEREO IMAGES	
Earth flow, *continued*	Vegetation:	Vegetation on scar and body strongly contrasting with surroundings, land use absent if active; linear pattern in direction of flow
	Drainage:	Ponding frequent in concave upper part of flow; parallel drainage channels on both sides of body in valley; deflected or blocked drainage by frontal lobe
Flowslide	Morphology:	Large bowl-shaped source area with steplike or hummocky internal relief; relatively great width; body displays clear flow structures with lobate convex frontal part (as earth flow); frequently associated with cliffs (weak rock) or terrace edges
	Vegetation:	Vegetational pattern enhancing morphology of scarps and blocks in source area; highly disturbed and differential vegetation on body
	Drainage:	As with earth flows, ponding or disturbed drainage at rear part and deflected or blocked drainage by frontal lobe
Debris avalanche	Morphology:	Relatively small, shallow niches on steep slopes (>35 degrees) with clear linear path; body frequently absent (eroded away by stream)
	Vegetation:	Niche and path are denuded or covered by secondary vegetation
	Drainage:	Shallow linear gully can originate on path of debris avalanche
Debris flow	Morphology:	Large amount of small concavities (associated with drainage system) or one major scar characterizing source area; almost complete destruction along path, sometimes marked by depositional levees; flattish desolate plain, exhibiting vague flow structures in body
	Vegetation:	Absence of vegetation everywhere; recovery will take many years
	Drainage:	Disturbed on body; original streams blocked or deflected by body

trolled by rock discontinuities (joints and fractures), giving the rock slope a rough appearance; these discontinuities are expressed in the image by a coarse texture. Toppling is favored by the presence of a steeply inclined joint set with a strike aligned approximately parallel to the slope face. Therefore, fine lineaments at the crest that are oriented parallel to the free face may be related to open joints behind toppling blocks. The accumulation of talus at the foot of the slope or the occurrence of coarse scree on slopes below the rock face is associated with rough micromorphology and results in a relatively coarse textural appearance in the image.

Talus accumulations and colluvial slopes formed by fall processes typically have slopes between 25 and 35 degrees. Scattered trees or bushes are the most frequent vegetation on these slopes. The density of this vegetation is indicative of the degree of slope-movement activity. At specific places where falls occur more frequently, chutes are eroded in the rock wall and talus cones are formed at its base. Linear patterns, also visible in the vegetation, are indicative of the paths along which the blocks are falling.

Large rock falls may create large, rapidly moving rock or debris avalanches, or *sturzstroms* (see Chapter 3, Section 8.1.2). These failures are associated with major morphologic anomalies and scars on mountain slopes (see Figure 8-3). The accumulation of these materials may spread a considerable distance from the source area, often creating rather chaotic landforms in which enormous blocks form an extremely irregular, rough morphology. This chaotic morphology is enhanced in the image by the very irregular vegetational pattern. Lobate convex forms are sometimes associated with the front of the mass. The drainage pattern in the whole area is generally seriously disturbed by these large, complex landslide deposits. Surface drainage can be blocked by the accumulated mass, creating lakes, or rivers are deflected, finding their way around the mass. Abrupt changes in the width and pattern of the river and clearly asymmetric valley slopes at the location of the accumulative mass are other characteristics. It is quite often observed that the deflection of river channels by larger mass movements induces slope instability features on the opposite valley side caused by the resulting erosion and undercutting of these slopes.

FIGURE 8-3

Major complex rock fall (*sturzstrom*) or rock slide occurring in limestones overlying sequence of mudstones. All morphologic features associated with such gigantic mass movements are clearly defined: large back scarp and huge block that has slid (*left side*); runout of material onto slope underlain by fine pelitic materials (*right side*). Tonal changes around larger block (*lower part of stereogram*) are associated with soft clays squeezed out of slope by impact of block. Block field begins within stereogram and continues for almost a kilometer farther eastward. Disturbed drainage conditions at interface of limestones and fine pelitic sequence are causing ongoing slope instability, mainly in form of flow-type features, as can be observed by grey tones and elongated patterns along contact (original photograph scale 1:18,000, Province of Malaga, Spain).

FIGURE 8-4
Translational slides controlled by dip slopes in sequence of very friable sandstones alternating with siltstones and mudstones. Joint-controlled back scarp, linear patterns, and micromorphology of area where the sandstones have slid away (*lower part*) are diagnostic for landslides. Somewhat comparable morphology and change in land use (*upper part*) are indicative of landslides that have rapidly transformed into earth flow because of higher clay content (orginal photograph scale 1:17,000, Basilicata, Italy).

3.2.2 Slides

A *rotational slide* is mainly associated with slopes ranging from 20 to 40 degrees and is recognized by a characteristic slope morphology. The crown of the slide, with its frequently semilunar shape, initiates an abrupt change in the slope. Concave and convex slope forms are related to the landslide niche and the deposit, which are directly connected to each other. These landslides generally have a depth-to-length (D/L) ratio on the order of 0.3 to 0.1. Successive or retrogressive sliding results in a steplike morphology. The lower frontal part (or toe) of the landslide has a generally convex lobate form. The rotational movement often results in back-tilting of slope facets. The overall micromorphology of these landslides is irregular, resulting in textural and tonal variations on the aerial photograph. The differential drainage conditions on these landslides and their disturbed vegetational conditions enhance textural and tonal variations. When the scale of the image is appropriate, these variations form a characteristic pattern in association with slide scars and back-tilted blocks. Poor surface drainage, or even ponding, occurs on the main landslide mass and behind back-tilting blocks. Wet zones and springs are characteristic along the toe of the slide. These wetter conditions and their distinctive vegetation influence the tone on the photographs. The vegetation on a slide shows a disturbed and chaotic aspect. The absence of cultivation or differences in land use in comparison with those in the surrounding area are often remarkable and also indicative of the activity of the movements.

In a *translational slide* the failure surface usually reflects a weak layer or preexisting structural discontinuity (see Figure 8-4). This characteristic has clear consequences for the morphologic aspects of the mass movement. In the first place, the D/L ratio for translational slides is many times smaller than that for rotational slides, whereas the width of the zone of movement in translational slides is greater than that for most rotational slides. When the failure is controlled by the interface between surficial materials and bedrock, the movement will be shallow and the displacement may extend over a considerable distance. Such slope failures are also commonly relatively wide. Flowage features in the runout material are frequently observed, especially when the coherence of the material is low and strong rainfall is the triggering mechanism. The source area and the path along which the material moved are denuded of vegetation, resulting in a clear tonal contrast with the surroundings. Lineaments parallel to the direction of the movement are common. Vegetation conditions are chaotic on the displaced mass, and most land use activities will be absent when the movement is recent (only a few years old). Also, the drainage conditions are disordered on the displaced material, although the typical poorly drained areas associated with rotational slides are normally absent.

A *compound slide* is a form that is transitional between a rotational and a translational slide. From the point of view of the photointerpreter, these slides are in many respects similar to rotational slides but their upper portions often contain grabenlike depressions and have a less pronounced runout. Their D/L ratio is normally smaller than that for rotational slides, whereas their width is generally greater.

A *rock slide* is also characterized by a small D/L ratio (usually less then 0.1) and a large width. Joints and fractures provide structural control of the failure surface and at the crown of the slide, and these joints and fractures are often distinct on the photograph. The morphology in the runout area is very rough, and decreasing block size with increasing distance is characteristic. Enormous slabs occur close to the source area, whereas chaotic and irregular "block fields" occur at a greater distance. Vegetation is absent in the source area and along the path. On the slide mass, vegetation is chaotic and in patches. Drainage conditions are normally good because most of the drainage will be internal. Springs may be found at the toe of the slide, and the front of the mass can obstruct local streams.

In the category of complex and composite slides, *mudslides* can be differentiated by photointerpretation. Mudslides are generally shallow mass movements occurring on slopes of between 15 and 25 degrees composed of fine clayey materials. The clearly differentiated source area, transportation path, and accumulative zone are diagnostic features of mudslides. The morphology of mudslides is characterized by a clear concave niche from which the material was derived, comparable with the landslide scar of shallow slides. The transportation path is often represented by a more-or-less straight channel originated by failure due to undrained loading. Mudslide runout deposits are spread over a much wider area than the width of the source area or the transportation path, where the material was confined as in a channel. The tongue of the mudslide displays a lobate form. The dilation of the material and the flatness of the lobe are characteristic and relate to the fluid nature of the material during movement. The D/L ratio for mudslides is on the order of 0.05 to 0.01, much smaller than that for rotational slides.

The term *flowslide* has been used to describe a sudden collapse of material that then moves a considerable distance very rapidly to extremely rapidly. Hutchinson (1988) pointed out that at least three phenomena can cause this behavior:

1. Collapse of weak rocks, such as chalk, along cliffs;
2. Destruction of the normal structure of saturated material by shocks such as earthquakes; and
3. Movement of loosely dumped materials in waste piles or in rapidly deposited, loosely compacted silts and fine sands.

Depending on the material in which the failure occurs, the size of the failure, and the place from which the movements are derived, the overall morphology of flowslides can resemble large rock avalanches (sturzstroms), translational slides produced by failure along a weak horizon, or liquefaction spreads, which are described in the following section. For these reasons, Cruden and Varnes (see Chapter 3 in this report) suggest that the term *flowslide* is redundant, confusing, and potentially ambiguous. They suggest that these different kinds of landslides be described with more appropriate terms.

Nevertheless, the sudden collapse of loose, saturated, almost cohesionless soils or weak rocks occurring on moderate to gentle slopes or even in almost flat terrain produces a distinctive pattern that can be readily evaluated by photointerpretation. The area from which the landslide is derived

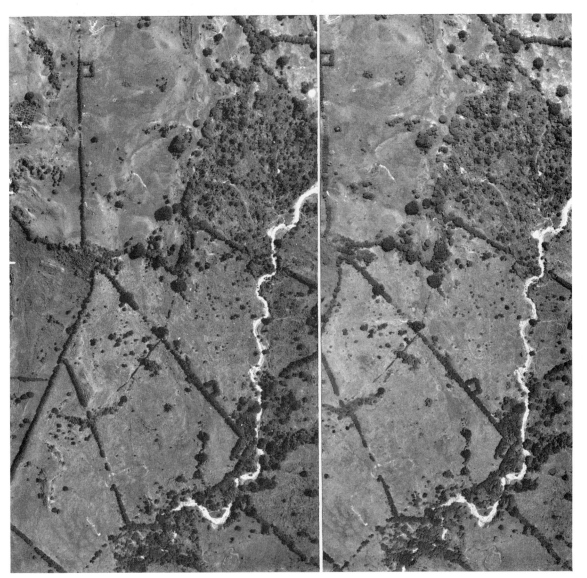

FIGURE 8-5
Highly unstable zone with numerous rotational landslides, mud slides, and earth flows. Well-cemented but strongly jointed volcanoclastic sequence, dipping gently to southwest (*north is up*), overlies series of almost unconsolidated sandstones and mudstones and claystones deposited in shallow marine to coastal environment. Steplike morphology in volcanoclastic material, sometimes showing slight back-tilting, is characteristic for rotational slides. Hummocky morphology with varying grey tones and faint linear elements is indicative of more flow-type movements on slopes near creek (original photograph scale approx. 1:10,000, Antioquia, Colombia).

is often an extensive flat concave zone with an irregular hummocky or undulating micromorphology. Within this area the drainage conditions are completely disturbed and ponded water is frequently encountered. These conditions are in strong contrast to the surroundings, which usually show a smooth topography with mostly complete internal drainage. A relatively narrow opening or neck indicates the place through which the landslide movement occurred. The transportation path, which varies in length according to the slope of the area and the fluidity of the mass, is clearly recognized on the images by a tonal contrast and lineaments parallel to the flow. The accumulative mass has a flat, slightly convex lobate form. Flow structures are clearly visible in both the micro-

morphology and the vegetation because of differential drainage conditions (see Figure 8-5).

3.2.3 Spreads

Spreads are mass movements occurring on gentle to moderate slopes where a slow plastic deformation or liquefaction occurs in a subsurface horizon overlain by a more coherent surface layer. This upper layer is broken up by the movements of the underlying material and moves and slides outward on the underlying layer. The areal extent of the movement is often considerable (up to several square kilometers), and the limits of the movement at the surface can be diffuse and difficult to distinguish both on aerial photographs and on the ground. Linear features corresponding to cracks and tilting of blocks of surface material are visible on remote-sensing images, forming indicators of the initial movements. The presence of these cracks is often enhanced by vegetation differences. The morphologic anomalies increase in the middle portions of the landslide. The surface material breaks up into irregular blocks that are chaotically disposed. These chaotic conditions are reflected by the morphology, the drainage, and the vegetation conditions. Bulging of the lower slopes, which usually display a typical convex form, indicates extrusion of the unstable subsurface material. Poor drainage conditions and seepage horizons are characteristic for this zone, causing tonal differences in the photographs. Lateral spreads often result in overall drainage anomalies because the movements may narrow or block valleys and deflect streams. These drainage anomalies usually result in increased stream erosion at the location where the spread blocks the valley, which in turn results in development of numerous local rotational slides of considerably smaller size than the lateral spread.

3.2.4 Flows

Flows comprise a large range of slope failures, including relatively slow-moving earth flows, extremely fast debris avalanches, devastating debris flows induced by the failure of the barrier forming a natural or artificial lake, or the equally devastating lahars caused by volcanic activity.

Earth flows often originate as one of various types of mass movements. The coherence within the initial landslide mass is lost because of the initial failure, and the mass continues as a viscous flow down the slope. Water contained within the mass may contribute to this flowage. Earth flows may continue long distances, following stream channels and reaching main valleys where they may obstruct the drainage. The source area of earth flows often has the aspect of a zone with complex mass movements, landslides coming from various directions and showing generally a clear retrogressive progression. The transportation path is distinct, following first the maximum slope and then a stream channel. The earth-flow material exhibits morphologic features that are often comparable with those of glaciers or lava flows, with cracks (lineaments on the aerial photograph) parallel to the movement and transverse cracks at places where the slope and

flow velocity increase. The transverse section of an earth flow shows a slightly convex ground profile, which may be readily visible in the three-dimensional photoimage because of the exaggerated stereoscopic relief. Earth flows infilling valleys create a clear morphologic anomaly, contrasting with the V-shaped valleys in mountainous areas (see Figure 8-6). The frontal portions of earth flows have clearly lobate convex forms. The source area is generally devoid of any vegetation, whereas the vegetation on the earth flow, if any, looks patchy because of differential surface drainage conditions in the material. The drainage conditions in the source area are disturbed, and local ponding can occur. Two small streams normally develop in a valley subjected to an earth flow, one on each side of the flow. These form an easily recognizable drainage anomaly, as does the deflection of the stream channel around the frontal lobe.

Debris avalanches are extremely fast and sometimes relatively small slope failures on straight steep slopes with inclinations generally greater than 35 degrees. They are characterized by a concave niche from which a long, narrow, light-toned tail originates. The linear character remains visible on aerial photographs even when secondary vegetation has invaded the area affected by the debris avalanche. Debris avalanches are most common on steep slopes that are at almost their maximum angle of stability. They are especially common where the slope equilibrium has been disturbed by a vegetation or land use change or by engineering work such as road construction. They are often triggered by earthquakes.

Debris flows can be caused by a large number of factors, but in all cases considerable amounts of loose material are suddenly moved by an excessive amount of water and transported in an extremely fast and destructive flow through a valley. Depending on the origin of the debris flow, the morphologic characteristics of the source area may vary. The zone can be characterized by a large number of surficial debris slides. Figure 8-7 shows a debris flow in Thailand. Extremely intense rainfall triggered a large number of superficial landslides in weathered granitic rocks, and these flowed together to form a devastating debris flow. However, debris flows may also originate from a single slope failure or be caused by the failure of a dam. Extremely large volumes of debris-flow deposits caused by massive glacial-lake outburst floods

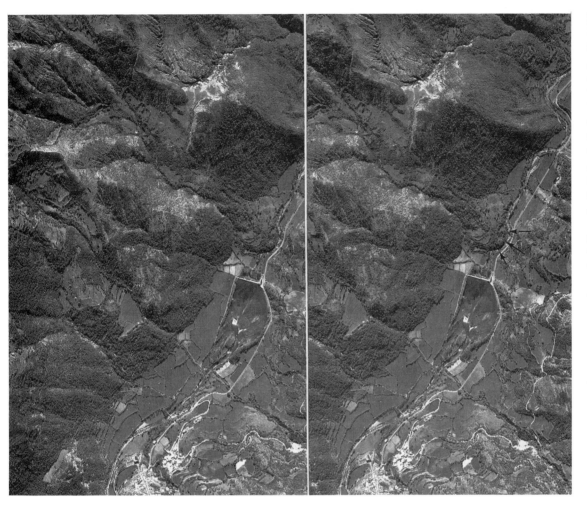

FIGURE 8-6
Major earth flow
within small valley
and almost blocking
main valley. Convex
form of earth-flow
lobe is characteristic
of earth flows and
contrasts with
concave forms of
alluvial fans. River is
pressed against
opposite valley side
and may cause
undercutting and
slope failures on
that side. Earth
flow occurred in
prehistoric time
following a rock fall
in conglomerates
(*just beyond upper
left part*). Impact of
rock fall triggered
earth flow in
weathered slates
underlying
conglomerates
(original photograph
scale 1:22,000,
Province of Lleida,
Spain).

have been identified in the northwestern portions of the United States and elsewhere. *Lahars* are debris flows originated from loose pyroclastic deposits on volcanic slopes. They display morphologic forms similar to those of other debris flows.

Common to all debris flows are the marks left behind by the devastating flow. Some debris flows have such large dimensions that these marks are clearly recognizable even on small-scale images and for a considerable time after the event occurs. The appearance of the depositional mass varies with the type of material transported but generally consists of a desolate, flat area engulfing isolated small higher vegetated areas that correspond to an older topographic surface. Large blocks of rock floating in the mud may create an irregular micromorphology, which is recognizable on large-scale photographs by a rough or coarse texture. Flow structures are often absent in this chaotically deposited mass. Drainage conditions on the flow

mass are disturbed, and the mass itself deflects or obstructs streams in the area of deposition.

Bedrock flows, or deep-seated creep, generally do not affect the morphologic conditions sufficiently to be interpreted in a preliminary photointerpretation. These flows can only be mapped with good knowledge of the local conditions. Once the characteristic features for creep are known in an area, this knowledge may successfully be extrapolated on the aerial photographs. Creep can create an irregular micromorphology, sometimes reflected in the drainage and vegetation, that causes a contrast with the zones not affected by creep. The surrounding areas typically show very smooth forms with subdued photographic grey tones in comparison with the creep-affected areas. Bulging of slopes is associated with deep-seated creep. When sagging develops, it is generally accompanied by elongated depressions along the slope and back-tilting slope facets (see Chapter 3 in this report).

FIGURE 8-7
SPOT multispectral
satellite image with
relatively good
spatial resolution
can be used
profitably in
assessment of areas
affected by slope
movements.
Example shows area
in Thailand where
large number of
shallow debris slides
generated large
debris flow that
traveled through
valleys, damaging
agricultural fields
and houses.

3.2.5 Cartographic Aspects

Mass movements may be represented cartographically at two levels of generalization. In those cases in which the area affected by the slide is too small to represent the real outlines of the slope movement at the scale of the map, mass movements are represented by a symbol defining the type of movement, thus differentiating falls and topples, slides, and flows. At larger map scales, or for larger landslides, the outlines of the mass movement may be shown on the map (see Figure 8-8). In these situations the map usually also shows all the valuable information related to the slope movement processes, such as scarps, cracks, steps in slopes, back-tilted blocks, and depressions. Indirect evidence of instability, such as seepage horizons, stagnated drainage, or ponding, may also be mapped. In large or complex mass-movement areas, simple symbols are commonly combined to represent morphologic details. The degree of activity may be indicated by using solid and dashed lines or by using symbols of different colors, for example, red for active areas and black for inactive areas. However, such interpretations must be used with caution because the differentiation between active and inactive landslides, especially when using satellite imagery, is based only on the degree of freshness of the morphologic features associated with the movement. In large or complex slides, small active movements are frequently indicative of ongoing activity within the larger mass.

When an inventory map of landslides is prepared for a GIS-based hazard analysis, the landslides are delineated on the image and labeled with an identification number and a digital code defining the landslide type, subtype, activity, and depth. Standard landslide classifications and uniform photointerpretation methods should be used as much as possible. However, experience has shown that photointerpretation of landslides may require slight modifications to these classifications in order to produce an unambiguous and consistent interpretation.

3.3 Mapping Terrain Parameters from Remote-Sensing Images

As discussed in Section 2.3, a landslide hazard assessment should not be based only on the production of a landslide inventory map. A complete landslide hazard assessment also requires an analysis of the factors leading to instability and the classification of the terrain into susceptibility classes for slope failures. These susceptibility classes are defined to reflect the presence and intensity of slope instability causative factors. The interpretation of either satellite images or aerial photographs plays a crucial role in the evaluation of the many factors that must be taken into account for landslide hazard analysis and their display as parameter maps (see Table 8-1). The interpretation process may also be used to perform terrain analysis, thereby producing a single map defining the area characteristics in terms of homogeneous map units. Accordingly, remote-sensing methods can be applied to obtain information concerning landslide hazard assessment factors or parameters by two distinctive processes:

1. *Preparation of individual thematic maps:* This obviously highly desirable method of evaluating landslide hazards involves the representation of the various factors potentially affecting slope instability (such as geomorphology, slope angle, length, convexity, land use, and lithology). However, at relatively detailed scales, such as 1:50,000 or larger, preparing the many individual maps requires a large amount of time for photointerpretation, fieldwork, map creation, and the subsequent digitization of these maps if GIS techniques are to be used (see Section 4 in this chapter). Furthermore, the digitization of identical boundary lines shown on different maps must be conducted with much care and frequently with several editing steps in order for them to coincide exactly. If these lines do not coincide, the overlaying of factor maps produces a large number of small areas containing spurious hazard assessments. For these reasons, this method is most appropriate for the assessment of relatively small areas.

2. *Terrain classification approach:* A terrain classification divides the landscape into homogeneous zones or natural divisions by using the interrela-

tionships among geology, geomorphology, and soils. Because reliable quantitative data on geology, geomorphology, soils, and so forth, are frequently scarce during the early stages of planning for development and engineering projects, terrain classification may be used during these stages to transform the available earth science data into information reflecting applications such as slope stability. Terrain classifications reduce the seemingly infinite variations of the terrain into a manageable number of classes. They also allow landslide hazard analysis to proceed with the creation and digitization of only a single additional map. This process is thus especially attractive during the earlier regional assessment stages.

3.3.1 Geomorphic Mapping

The production of individual thematic maps will normally require the use of specialized mapping procedures. One broad class of such mapping methods is called geomorphic mapping. Many different geomorphic mapping systems have been proposed, either for universal application or for specific areas or regions such as mountainous terrain. Overviews of conventional medium-scale and large-scale geomorphic mapping systems were presented by Demek and Embleton (1978) and van Zuidam (1986). The use of several different systems in prac-

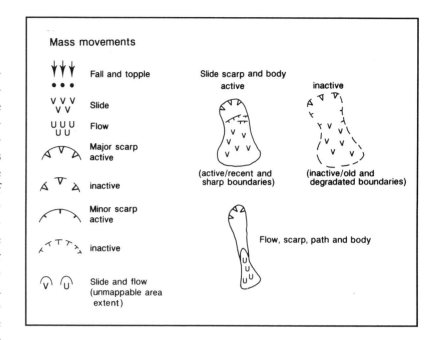

FIGURE 8-8
Example of legend for slope instability map. By use of such symbols, landslide type and activity may be differentiated. Morphologic features and drainage conditions associated with mass movements can also be indicated by symbols (modified from Sissakian et al. 1983).

tice suggests that no universally accepted system is adequate for mapping in different environments, in contrast to the case with soil mapping. In the conventional geomorphic mapping systems for scales of 1:25,000 and larger, various symbols, lines, colors, and hatchings are used to represent the morphometry, morphography, drainage, genesis, chronology, and materials of the landscape features or the processes forming them. These systems differ in the importance assigned to each feature and in the method of representation. They all combine the different types of geomorphic data onto one map sheet. Such a system is not amenable to computer-based representations using GIS methods. Thus, the construction of detailed geomorphic maps suitable for use in a GIS representation requires a different, and much more complicated, mapping method (Dikau 1992; van Westen 1993).

3.3.2 Terrain Classification Systems

Terrain classification methods were developed to replace the mapping of individual landscape parameters on multiple maps with a single mapping unit that can be shown on a single map. Many different terrain classification systems have been developed over the years. The principal systems were compared by van Zuidam (1986) and Cooke and Doornkamp (1990). These systems differ in the way they utilize or depend on geomorphic, analytic, morphometric, physiographic, biogeographic, or lithologic-geologic parameters. Most terrain classification systems have a rigid hierarchical structure, which may hinder their flexible use, or they are based on a single parameter or a limited set of parameters.

To overcome these problems, a terrain classification system based on the delineation of *terrain mapping units* (TMUs) was developed by Meijerink (1988). A TMU is defined as a unit that groups zones of interrelated landforms, lithology, and soil. It is a natural division of the terrain that can be distinguished on stereo SPOT imagery or small-scale aerial photographs and verified on the ground. The units are differentiated on the basis of photomorphic properties in the stereo model. Meijerink's TMU system does not have a strict hierarchical structure. The user can construct the legend according to the important parameters encountered in the study area and the purpose of the study. An individual TMU differs from other adjoining TMUs either because the landforms are evidently different or because the phenomena associated with the landform, such as the nature of the weathered zone, the lithology, or the type of soil, are different. The TMU approach has been used successfully in various geomorphic and engineering geologic applications, such as highway planning (Akinyede 1990).

In conventional thematic mapping, a TMU can be considered as a legend unit. In terms of GIS techniques, a TMU may be described as the geographic location of entities (polygons) that relate to a unique set of attributes (terrain conditions). These are linked to the geographic TMU polygons by attribute tables in a data base (see Figure 8-9). TMUs allow the grouping of the following interrelated landscape variables:

- Geomorphic origin and physiography,
- Lithology,
- Morphometry, and
- Soil geography.

3.4 Image Resolution and Interpretability

The interpretation of landslides from remote-sensing sources requires knowledge of the distinctive features associated with slope movements and of the image characteristics associated with these features. An adequate interpretation depends on image characteristics. The interpretability of features in an image is influenced by the contrast that exists between features and their background. For the image interpretation of landslides, this contrast results from the spectral or spatial differences that exist between the landslide and its surroundings. These are affected by

1. The period elapsed since the failure, because erosional processes and vegetation recovery tend to obscure the features created on the land surface by landslide movements, and
2. The severity with which the landsliding affects the morphology, drainage, and vegetation conditions.

The spatial resolution of the remote-sensing images provides the primary control of the interpretability of slope instability phenomena and thus the applicability of any type of remote-sensing data

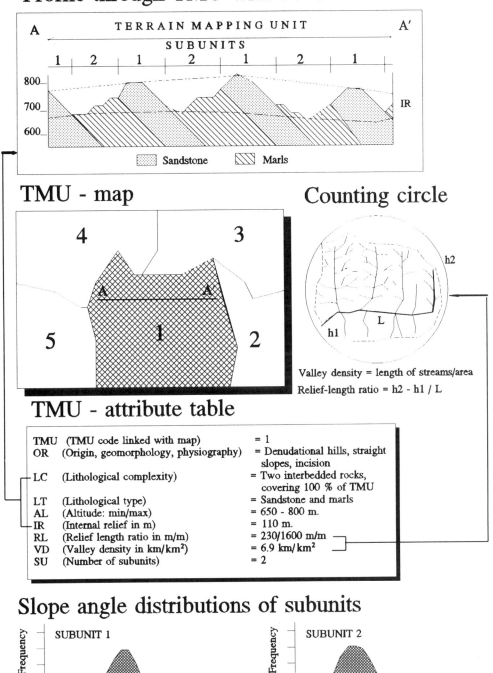

Profile through TMU with subunits

A   TERRAIN MAPPING UNIT   A'

SUBUNITS

1   2   1   2   1   2   1

800
700
600

IR

▨ Sandstone   ▨ Marls

TMU - map

4   3

A   A

5   1   2

Counting circle

h2

L

h1

Valley density = length of streams/area
Relief-length ratio = h2 - h1 / L

TMU - attribute table

TMU	(TMU code linked with map)	= 1
OR	(Origin, geomorphology, physiography)	= Denudational hills, straight slopes, incision
LC	(Lithological complexity)	= Two interbedded rocks, covering 100 % of TMU
LT	(Lithological type)	= Sandstone and marls
AL	(Altitude: min/max)	= 650 - 800 m.
IR	(Internal relief in m)	= 110 m.
RL	(Relief length ratio in m/m)	= 230/1600 m/m
VD	(Valley density in km/km²)	= 6.9 km/km²
SU	(Number of subunits)	= 2

Slope angle distributions of subunits

SUBUNIT 1

Frequency

Slope angle

SUBUNIT 2

Frequency

Slope angle

FIGURE 8-9

Mapping by TMU approach (Meijerink 1988). Simplified example of TMU map together with parameters stored in attribute table. Subdivision of units is shown according to lithologic type, determination of internal relief (*IR*) and valley density (*VD*) with help of counting circle, and computation of slope angle distributions.

for landslide studies. The relationship between image resolution and the size of the features necessary to identify or characterize the slope movement is obviously critical. If the resolution is too low, the features cannot be recognized or identified.

Comparison of the spatial resolution of photography and nonphotographic remote sensing requires the use of the concept of a ground resolution cell (GRC), first introduced by Rengers et al. (1992). In nonphotographic remote sensing the GRC is the size of a scene element, the dimensions on the ground of the basic elements (or pixels) of the image.

At any given scale, aerial photography provides a higher resolution, and therefore a smaller GRC size, than remote-sensor imagery at the same scale (see Figure 8-10). According to Naithani (1990), aerial photography provides a GRC with a size equal to 0.4 times the value of the GRC for nonphotographic remote-sensing imagery. Strandberg (1967) suggested that the following formula be used to relate the GRC and the photographic scale:

$$GRC = \frac{S}{1000R}$$

where

   GRC = ground resolution cell (m),
      S = scale number of image (i.e., denominator of scale ratio), and
      R = resolution of photographic system (line pairs/mm).

The resolution of aerial photography systems is on the order of 40 linepairs/mm for conventional aerial photographic cameras and films with extreme contrast.

A certain minimum number of pixels is needed to recognize a feature in an image. The actual minimum number of pixels varies according to the grey-tone contrast between the feature and its background. Although it is difficult to give precise data on the minimum number of pixels required, experience with visual interpretation of remote-sensing imagery suggests that the values shown in Table 8-7 are appropriate.

When the required minimum number of pixels recommended in Table 8-7 is multiplied by the size of the GRC, it will indicate the minimum size of landslide that is likely to be identified. Several other factors may also influence the minimum number of pixels necessary for satisfactory identification of landslides. These include factors related to the skill of the individual interpreter, including professional experience and local knowledge related to type and occurrence of landslides. These factors together define the reference level of the interpreter. A high reference level is very important for an adequate interpretation, as demonstrated by Fookes et al. (1991), who compared the photointerpretations made by five recognized professionals of an unknown area before a large landslide.

The implications of Table 8-7 are illustrated by Figures 8-11 and 8-12, which were derived from large-scale aerial photographs. These photographs were digitized with a raster size corresponding to a GRC of 0.3 m. The individual photographs in Figures 8-11 and 8-12 were then created by artificially aggregating and averaging these pixels to reflect GRC sizes of 1, 3, 10, and 30 m.

Table 8-7
Number of Ground Resolution Cells Needed To Identify and Interpret Object of Varying Contrast in Relation to Its Background

	No. of GRCs	
	For Identification	For Interpretation
Extreme contrast: white or black object against variable grey-tone background	20–30	40–50
High contrast: dark or light object in grey-tone background	80–100	120–140
Low contrast: grey feature with grey-tone background	1,000–1,200	1,600–2,000

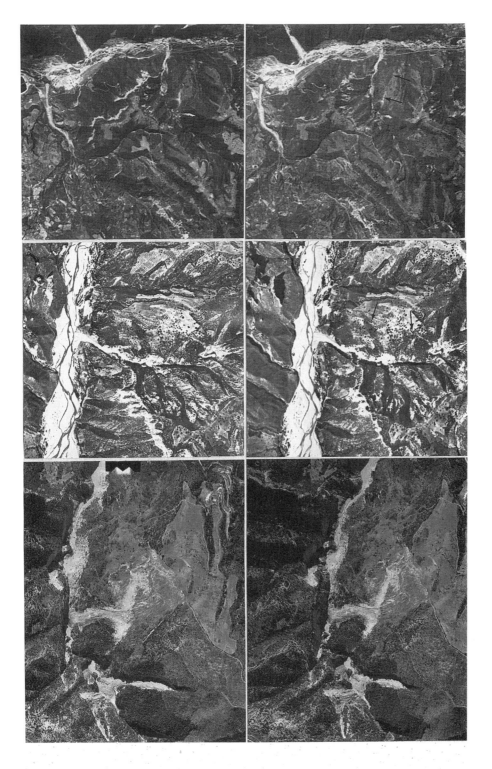

FIGURE 8-10
Comparison of interpretability of huge complex landslide in Sant Arcangelo Basin (Basilicata, Italy) shown on stereo SPOT image (scale approx. 1:70,000), medium-scale aerial photographs (1:33,000), and large-scale aerial photographs (1:17,000). Flight line on medium-scale photographs is north-south, and viewing direction on SPOT and flight line of large-scale photographs is east-west.

Figure 8-11 shows photographs with varying GRC sizes of a landslide in the Spanish Pyrenees (Sissakian et al. 1983). The landslide scar with its shadow, which provides high contrast, is observed in the upper part of the image and a depositional area of landslide debris is observed in the central lower part. This area of accumulation is recognizable by a characteristic surface texture due to an irregular microrelief (formed by low-contrast features) and by the surrounding band of higher vegetation recognizable as such in those pictures with small GRC sizes. The photographs in Figure 8-11

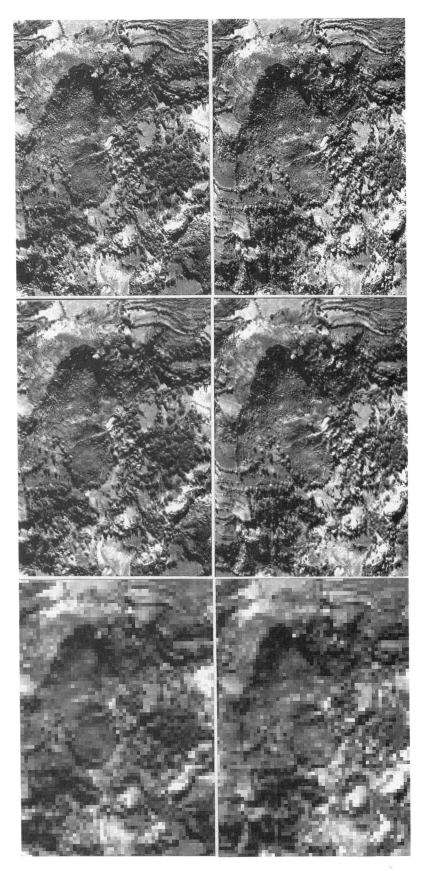

are arranged to form stereopairs. They demonstrate the enormous advantage of stereoscopic viewing in the interpretation of mass movements. The values in Table 8-8 give only a general indication of the order of magnitude of the minimum object size that may be recognized or identified on the basis of shape and pattern. Tonal or spectral aspects are considered only in terms of contrast of the feature against the background.

For evaluation of the suitability of remote-sensing images in landslide inventory mapping, the size of individual slope failures in relation to the GRC is of crucial importance. Although sizes of landslides may vary enormously, even within a single type of slope failure, some useful information

FIGURE 8-11 *(left)*
Area in Spanish Pyrenees showing landslide scar in shadow, thereby providing high contrast, and depositional area of landslide debris *(lower center)*. These illustrations, which form stereopairs, are derived from digitized large-scale aerial photograph. Artificially aggregated pixels represent ground resolution cell (GRC) sizes of 1, 3, and 10 m. Although image corresponding to 10-m GRC resolution already shows serious degradation, landslide is still clearly interpreted when viewed stereoscopically.

FIGURE 8-12 *(below)*
Same area as that shown in Figure 8-11 but with 30-m GRC resolution. At this image scale, with such a large pixel, no satisfactory stereoscopic image can be obtained.

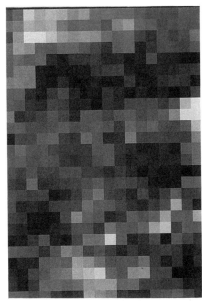

Table 8-8
Minimum Object Size Needed for Landslide Identification or Interpretation

| | | SIZE (m²) NEEDED FOR | | | |
| | | HIGH CONTRAST | | LOW CONTRAST | |
	GRC SIZE (m)	IDENTIFICATION	INTERPRETATION	IDENTIFICATION	INTERPRETATION
LANDSAT MSS	80	160,000	288,000	7,040,000	11,520,000
LANDSAT TM	30	22,500	40,500	990,000	1,620,000
SPOT Multispectral	20	10,000	18,000	440,000	720,000
SPOT Panchromatic	10	2,500	4,500	110,000	180,000
Aerial photographs					
1:50,000	1	25	45	1,100	1,800
1:15,000	0.3	6.5	11.5	300	450

NOTE: The values given depend on the conditions of contrast between the elements of the slide and the background. The data for aerial photographs are somewhat optimistic since optimal photographic conditions and processing are assumed.

can be found in the literature. Crozier (1973) described the morphometric analysis of landslides and provided average values for several types of movements. For a total of almost 400 slope failures, Carrara et al. (1977) computed a mean crown-to-tip distance of 262 m and an average total area involved in a failure of 42 000 m². This total map area per failure approximately corresponds to 20×20 pixels on a SPOT Panchromatic image (having a resolution of 10 m), or 10×10 pixels on SPOT multispectral images (which have a resolution of 20 m). According to Table 8-7, this number of pixels would be sufficient to identify a landslide displaying high contrast but is insufficient for a proper analysis of the elements pertaining to the failure, thus making it impossible to establish the characteristics and type of landslide. Cleaves (1961) gave mean values of landslide area and size dimensions, also based on a large number of observations, which are even smaller than those of Crozier and Carrara. He concluded that 1:15,000 is the most appropriate photographic scale for analysis of landslides.

Experience at the International Institute for Aerospace Survey and Earth Sciences with the use of photointerpretation techniques in support of landslide hazard investigations in various climatic zones and for a considerable variety of terrain conditions suggests that a scale of 1:15,000 appears to be the optimum scale for aerial photographs, whereas a scale of 1:25,000 should be considered the smallest useful scale for analyzing slope instability phenomena with aerial photographs. A slope failure may be recognized on smaller-scale photography provided that the failure is large enough and the photographic contrast is sufficient. However,

such interpretations will lack the analytical information that would enable the interpreter to make conclusions concerning landslide types or causes. Furthermore, many slope movements will not be identifiable on these smaller-scale photographs.

Nevertheless, these smaller-scale aerial photographs are useful for some aspects of landslide hazard assessment, especially for analyzing the overall geologic and geomorphic settings that tend to result in slope failures. Scanvic and Girault (1989) and Scanvic (1990) reached similar conclusions. These authors describe a study in which the applicability of SPOT satellite imagery to slope instability mapping near La Paz, Bolivia, was evaluated. The authors concluded that SPOT yielded excellent data complementary to large-scale photographs.

Thus it appears that small-scale photographs are useful in determining the regional spatial distribution of variables affecting landsliding, whereas large-scale aerial photographs support landslide inventory and analysis activities, including interpretation of possible causal factors. These two types of activities suggest that landslide hazard assessment when larger areas must be evaluated can be most efficiently conducted with small-scale photography, and large-scale photography would be used only in pilot areas to establish the relation between landslide and terrain condition. Furthermore, good-quality, relatively small-scale aerial photography used during the reconnaissance stages of a project can be enlarged and used for more detailed subsequent studies, making an additional flight to obtain new photographs unnecessary. Table 8-9 gives results from a comparative study of

Table 8-9
Relative Suitability of Different Scales of Aerial Photographs for Different Elements in Slope Instability Mapping (modified from Sissakian et al. 1983)

SUBJECT	SIZE (m)	PHOTOSCALE 1:20,000	1:10,000	1:5,000
Recognition of instability phenomena	< 20	0	0	2
	20–75	0→1	1→2	3
	> 75	1→2	2	3
Recognition of activity of unstable areas	< 20	0	0	1
	20–75	0	0→1	2
	> 75	1	1→2	3
Recognition of instability elements (cracks, steps, depressions, etc.)	< 10	0	0	0
	10–75	0	0→1	1→2
	> 75	1	2	3

NOTE: 0 = less adequate, 1 = limited use, 2 = useful, 3 = very useful.

the interpretability of slope instability features on aerial photographs at their original scale and at three levels of enlargement (Sissakian et al. 1983).

3.5 Spectral and Temporal Resolution of Remote-Sensing Data

Vegetation and soil moisture conditions produce distinctive spectral responses in the infrared portions of the spectrum. For example, healthy vegetation produces infrared reflectance values that are very different from those of stressed, or unhealthy, vegetation. Strong differences in the infrared responses may be detected even when the vegetation appears normal in the visible portion of the spectrum. In a similar fashion, slight changes in soil moisture conditions are readily detected in the infrared portions of the spectrum. Landslides frequently produce subtle changes in the health and vigor of vegetation and may also cause increases in soil moisture caused by disruption of subsurface water movements. Thus, remote-sensing systems that are sensitive to the infrared part of the spectrum are most effective in landslide inventory studies. The use of infrared-sensitive film, and false-color infrared film in particular, is highly recommended for landslide studies in view of the capability of these films to register small anomalies in vegetation or drainage conditions. Optimal differences in vegetation conditions may be expected in either the very early or very late stages of the growing season. Differences in drainage conditions are optimal shortly after the first rainstorms of the rainy season in many tropical regions or shortly after the

spring snowmelt period in cold and temperate climates.

Satellite imagery offers more detailed spectral information than is usually available in photographs because the satellite sensors are designed to obtain the reflected electromagnetic radiation in various wavelength (spectral) bands. Black-and-white images of individual spectral bands may be displayed, but more commonly these are combined to form color composites, of which the *false-color composite*, comparable with false-color infrared photography, is the most common. Digital processing of the spectral data offers the possibility for detailed analysis of the obtained reflectance values and enhancement of small spectral variations that seem to be correlated with slope instability features.

The size of areas with anomalous drainage conditions or disturbed vegetation causing an anomalous spectral response is often too small to allow the interpretation of individual instability features on the basis of spectral criteria. However, spectral interpretation of satellite data has been used successfully in slope instability studies when this spectral information is used in conjunction with other data related to slope failures. Together these multiple information sources provide converging evidence for slope movements. Practical applications of spectral information from satellite imagery are also possible when, on the basis of terrain evidence, a direct relationship is known between slope instability and vegetation or drainage anomalies. McKean and coworkers (1991) demonstrated that spectral vegetation indices can be used in mapping spatial patterns of grass senescence that were found to be related to soil thickness and slope instability. In another case landslides in a homogeneous forested area exposed differences in understory vegetation and soils, thereby altering the spectral characteristics.

In general, it can be stated that spectral information can be used in the same way as spatial data to delineate terrain variables that are correlated or assumed to be related to slope movements. In the case of landslides, these terrain variables are mostly related to vegetation and drainage conditions. In special cases where the landslide conditions produce high contrast or in the case where the landslide has unusually large dimensions, or in both cases, the feature itself may be identified on the basis of spectral information. However, seldom will this type of information alone be sufficient for analyzing the type of failure.

Satellite systems orbiting the earth also provide the opportunity to obtain data from the same areas on a regular schedule, allowing for the monitoring of processes over time. Images obtained shortly after a period of slope instability will show high contrast between the zones affected by slope instability and the stable surroundings, resulting in clearly detectable spatial and spectral changes. These changes allow the interpreter to develop a slope instability impact assessment, such as that shown in Figure 8-7, which shows an area in Thailand affected by debris flows following an exceptionally heavy rainstorm. The interpretation of sequential images allows for the correlation of climatic or seismic events with the occurrence and intensity of slope movements. Finally, the comparison of imagery obtained at different times may indicate the activity of the slope processes in an area. However, it must be noted that even 20 years' worth of satellite images is still a rather small amount to obtain a good idea of the activity of slope instability processes because they are mainly triggered by low-frequency spasmodic events. Furthermore, adverse weather conditions or certain system limitations are additional limiting factors to the acquisition of satellite data at the most appropriate times and serve to restrain the full achievement of the available temporal resolution.

3.6 Applicability of Satellite Remote Sensing

On the basis of the foregoing discussion, it may be concluded that currently available satellite remote sensing has limited application to the *direct mapping* of slope instability features. The spatial resolution of these systems is insufficient to allow the identification of landslide features smaller than about 100 m, even when conditions favor a strong contrast between the landslide and the background areas. If contrast conditions are less favorable, identification may be limited to features greater than 400 m. Such dimensions are greater than those of many economically significant landslides.

The lack of stereoscopic imagery, except in the case of the SPOT satellite, further limits the applicability of much of the currently available satellite imagery. Stereoscopic imaging capability allows for the visualization of the land surface in three dimensions. Such three-dimensional information is necessary for the interpretation of the characteristic and diagnostic morphologic features of slope failures. Therefore, accepting the limitations related to the spatial resolution, stereoscopic SPOT satellite images may sometimes be used for small-scale regional hazard zonation studies. LANDSAT thematic mapper images could also be used for the same purpose, but only when stereomates, which provide a means for stereoscopic viewing, are made with the help of a detailed DEM. Such detailed DEM data are not often available.

In contrast, satellite images are valuable tools for *indirect mapping* methods. These methods require information concerning the spatial distribution of landslide-controlling variables, such as a particular geomorphic condition, a specific lithology, or a particular type of land use. These may be mapped rapidly and reliably over large areas with the use of satellite images. In practice, this mapping process implies the use of a combination of satellite imagery and large-scale photography. Large-scale aerial photography is used for the initial landslide inventory mapping and analytical stages of slope instability assessment. These findings are then extrapolated and used to assist in the interpretation of the smaller-scale satellite imagery. In the interpretation of small-scale images, the local reference level of the interpreter is of great influence. The reference level of the interpreter is greatly improved when local large-scale information is combined with regional small-scale synoptic data. The value of stereoscopic SPOT satellite imagery can hardly be overestimated in these applications, as was demonstrated by Scanvic and Girault (1989) and Scanvic (1990) with a case study of landslide vulnerability mapping near La Paz, Bolivia. The combined use of aerial photographs and satellite imagery in slope instability studies has also been emphasized by Tonnayopas (1988) and especially by van Westen (1992, 1993). The efficiency of integrating satellite images, aerial photographic interpretations, and field data was further improved by van Westen through the application of GIS techniques.

The potential of radar imagery for landslide hazard zonation requires further investigation. The results with radar interferometry are promising, and terrain roughness classification methods appear encouraging (Slaney and Singhroy 1991). Evans (1992) provided an overview of the applicability of synthetic aperture radar (SAR) to the study of geologic processes. However, problems remain with the use of radar imagery for landslide evaluations. Figure 8-13 is an ERS-1 satellite radar image for an area in southern Italy characterized by intensive slope processes and a local topographic relief that does not exceed 100 m. The geometric distortions due to foreshortening, a characteristic of the radar systems, and speckling due to the surface reflection characteristics resulted in a poor image.

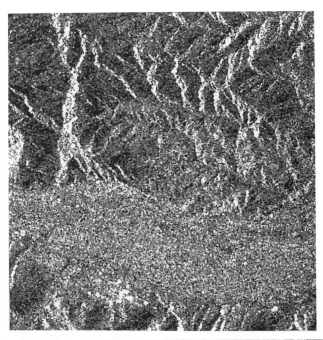

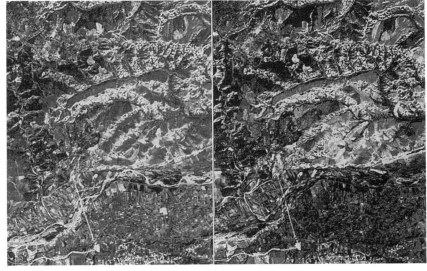

FIGURE 8-13
Comparison of ERS-1 radar image of area in Basilicata, southern Italy, with SPOT image of same zone. Internal topographic relief is less than 100 m; nevertheless, radar image already shows considerable geometric distortion of slopes (foreshortening). Speckling is characteristic of radar images.

The application of thermal infrared (TIR) remote-sensing imagery for slope instability studies is still in an early research phase. The spatial resolution of the thermal band of the LANDSAT thematic mapper is far too coarse for landslide investigations. Yet a higher spatial resolution apparently would markedly degrade the thermal resolution of the detectors from their current sensitivity of 0.1°C. A number of materials identification experiments have been performed with airborne TIR sensors (Lasky 1980; Bison et al. 1990). The lower altitudes of these airborne sensors compared with the LANDSAT TM thermal sensor allow them to achieve increased spatial resolution while maintaining the thermal resolution.

Bison et al. (1990) conducted some promising thermal inertia mapping research on a small area in Italy that is subject to landsliding. Thermal inertia provides a measure of the ease with which an object changes temperature. In natural soil materials the presence of water greatly reduces the rate at which the soil changes temperature in response to diurnal heating and cooling cycles. Thus, comparison of thermal images taken at different times within the daily cycle allows the determination of relative thermal inertia values. If other factors are the same, wetter soils will show greater relative thermal inertia values. Bison et al. evaluated TIR images collected immediately before and after sunrise in the autumn to produce maps of relative thermal inertia. Effects of vegetation and shadowing were identified and removed. Detectors installed in the ground registered variations in temperature of the soils on a slope. These variations could be correlated with variations in the soil moisture content and with patterns visible in the thermal imagery. However, no threshold values were established for the soil moisture content in relation to the occurrence of slope failures.

The term *small-format aerial photographs* refers to all photographs taken from airborne platforms and that have a negative size smaller than the conventional 23- by 23-cm aerial photographs. Useful images have been collected at a variety of negative sizes, including the common 35-mm film size. Small-format oblique aerial photographs may be obtained with a hand-held camera from helicopters, light aircraft, and even ultralight aircraft. These methods allow for almost real-time synoptic information on landslides to be obtained. Such information is extremely useful for the evaluation of large, complex slope failures. To make precise measurements of objects from these photographs, techniques of nonconventional photogrammetry must be employed. These techniques are becoming more promising because software programs have been developed that allow for detailed quantitative work with a minimum of ground control points (V. Kaufmann, personal communication, 1993, Technical University, Graz, Austria).

4. GEOGRAPHIC INFORMATION SYSTEMS IN HAZARD ZONATION

The occurrence of slope failures depends generally on complex interactions among a large number of partially interrelated factors. Analysis of landslide hazard requires evaluation of the relationships between various terrain conditions and landslide occurrences. An experienced earth scientist has the capability to mentally assess the overall slope conditions and to extract the critical parameters. However, an objective procedure is often desired to quantitatively support the slope instability assessment. This procedure requires evaluation of the spatially varying terrain conditions as well as the spatial representation of the landslides. A geographic information system (GIS) allows for the storage and manipulation of information concerning the different terrain factors as distinct data layers and thus provides an excellent tool for slope instability hazard zonation.

4.1 Geographic Information Systems

A GIS is defined as a "powerful set of tools for collecting, storing, retrieving at will, transforming, and displaying spatial data from the real world for a particular set of purposes" (Burrough 1986). The first experimental computerized GIS was developed as early as the 1960s, but the real boom came in the 1980s with the increasing availability of inexpensive personal computers (PCs). For an introduction to GIS, the reader is referred to textbooks such as those by Burrough (1986) or Aronoff (1989). Generally a GIS consists of the following components:

1. Data input and verification,
2. Data storage and data-base manipulation,
3. Data transformation and analysis, and
4. Data output and presentation.

Currently there are many different systems on the market, ranging from public domain software for PCs to very expensive systems for mainframe computers. In general, the systems differ with respect to

- Type of data structure (vector versus raster);
- Data compression technique (Quadtrees, run-length coding);
- Dimension (two-dimensional versus three-dimensional);
- Mainframe, minicomputer, and microcomputer hardware; and
- User interface (pop-up menus, mouse-driven, help options, etc.).

The advantages of the use of GIS as compared with conventional spatial analysis techniques are treated extensively by Burrough (1986) and Aronoff (1989). An ideal GIS for landslide hazard zonation combines conventional GIS procedures with image-processing capabilities and a relational data base. Since frequent map overlaying, modeling, and integration with remote-sensing images (scanned aerial photographs and satellite images) are required, a raster system is preferred. The system should be able to perform spatial analysis on multiple-input maps and connected attribute data tables. Necessary GIS functions include map overlay, reclassification, and a variety of other spatial functions incorporating logical, arithmetic, conditional, and neighborhood operations. In many cases landslide modeling requires the iterative application of similar analyses using different parameters. Therefore, the GIS should allow for the use of batch files and macros to assist in performing these iterations. Since most data sets required for landslide hazard zonation projects are still relatively small, mostly less than 100 megabytes, they can be readily accommodated by inexpensive PC-based GIS applications.

The advantages of GIS for assessing landslide hazard include the following:

1. A much larger variety of hazard analysis techniques becomes attainable. Because of the speed of calculation, complex techniques requiring a large number of map overlays and table calculations become feasible.
2. It is possible to improve models by evaluating their results and adjusting the input variables. Users can achieve the optimum results by a process of trial and error, running the models several times, whereas it is difficult to use these models even once in the conventional manner. Therefore, more accurate results can be expected.
3. In the course of a landslide hazard assessment project, the input maps derived from field observations can be updated rapidly when new data are collected. Also, after completion of the project, the data can be used by others in an effective manner.

The disadvantages of GIS for assessing landslide hazard include the following:

1. A large amount of time is needed for data entry. Digitizing is especially time-consuming.
2. There is a danger in placing too much emphasis on data analysis as such at the expense of data collection and manipulation based on professional experience. A large number of different techniques of analysis are theoretically possible, but often the necessary data are missing. In other words, the tools are available but cannot be used because of the lack or uncertainty of input data.

4.2 Examples from the Literature

The first applications of a simple, self-programmed, prototype GIS for analyzing landslide hazard zonation date from the late 1970s. Newman et al. (1978) reported on the feasibility of producing landslide susceptibility maps using computers. Carrara et al. (1978) reported results of multivariate analysis applied on grid cells with a ground resolution of 200 by 200 m and using approximately 25 variables. Huma and Radulescu (1978) reported an example from Romania that provided a qualitative hazard analysis by including the factors of mass movement occurrence, geology, structural geologic conditions, hydrologic conditions, vegetation, slope angle, and slope aspect. Radbruch-Hall et al. (1979) wrote their own software to produce small-scale (1:7,500,000) maps of the United States. Each map contained about 6 million pixels, which showed hazards, unfavorable geologic conditions, and areas in which construction or land development may exacerbate existing hazards. These maps were made by qualitative overlay of several input maps.

During the 1980s the use of GIS for slope instability mapping increased sharply because of the development of a great variety of commercial systems, such as Arc/Info (Environmental Systems Research Institute 1992) and Intergraph Corporation's MGE (Intergraph Corporation 1993). The increasing power and availability of PCs led to the development of several GIS applications that would work on these computers, including Tydac Corporation's SPANS and IDRISI (Eastman 1992a, 1992b).

The majority of case studies presented in the literature concerning the use of GIS methods for slope instability investigations deal with qualitative hazard zonation. The importance of geomorphic input data is stressed in the methods of Kienholz et al. (1988), who used detailed aerial photointerpretation in conjunction with a GIS for qualitative mountain hazard analysis. They state that because of the lack of good models and geotechnical input data, the use of a relatively simple model based on geomorphology seemed to be the most realistic method. Most examples of qualitative hazard analysis with GIS are recent (Stakenborg 1986; Bertozzi et al. 1992; Kingsbury et al. 1992; Mani and Gerber 1992; van Westen and Alzate-Bonilla 1990). Many examples are presented in the proceedings of specialty conferences, such as those edited by Alzate (1992) and by Goodchild et al. (1993).

Examples of landslide susceptibility analysis utilizing GIS techniques have been reported by U.S. Geological Survey (USGS) personnel in Menlo Park, California (Brabb 1984, 1987; Brabb et al. 1989). These studies extended earlier studies and took into account additional factors besides landslide activity, geology, and slope. Other examples of quantitative statistical analysis of landslide cause or potential with GIS are rather scarce (Choubey and Litoria 1990; Lopez and Zinck 1991; van Westen 1993). This lack of examples is strange, since one of the strong advantages of using a GIS is the capability to test the importance of each factor, or combination of factors, and assign quantitative weighting values.

Recent examples of multivariate statistical analysis using GIS have been presented by Carrara and his team from Italy. Their work initially used large rectangular grid cells as the basis for analysis (Carrara et al. 1978; Carrara 1983, 1988). Later studies evolved toward the use of morphometric units (Carrara et al. 1990, 1991, 1992). The method itself has not undergone major changes. The statistical model is built up in a "training area" where the spatial distribution of landslides is (or

should be) well known (Carrara 1988). In the next step the model is extended to the entire study area, or "target area," on the basis of the assumption that the factors that cause slope failure in the target area are the same as those in the training area.

Another example of multivariate analysis of landsliding using a GIS was presented by Bernknopf et al. (1988), who applied multiple regression analysis to a data set using presence or absence of landslides as the dependent variable and the factors used in a slope stability model (soil depth, soil strength, slope angle) as independent variables. Water table and cohesion data were not considered, however. The resulting regression function allows the computation of landslide probability for each pixel.

Deterministic modeling of landslide hazard using GIS has become popular. Most examples deal with infinite slope models, since they are simple to use for each pixel separately (Brass et al. 1989; Murphy and Vita-Finzi 1991; van Westen 1993). Hammond et al. (1992) presented methods in which the variability of the factor of safety is calculated from selected input variables utilizing Monte Carlo techniques. This implies a large number of repeated calculations, which are readily supported by use of a GIS.

Another useful application of GIS has been the prediction of rock slides. The prediction is made for a series of pixels by comparing discontinuity measurements within structurally homogeneous regions with slope and aspect values for each pixel (Wentworth et al. 1987; Wagner et al. 1988). The method is feasible only in structurally simple areas, however.

A relatively new development in the use of GIS for slope instability assessment is the application of so-called "neighborhood analysis." Most of the conventional GIS techniques are based on map overlaying, which allows only for the spatial comparison of different maps at common pixel locations. In contrast, neighborhood operations permit evaluation of the neighboring pixels surrounding a central pixel. The process is repeated for a sequence of central pixels, the analysis neighborhood, or window, moving around the map. Neighborhood functions are used to compute, or determine, such morphometric and hydrologic features as slope angle, slope aspect, downslope and cross-slope convexity, ridge and valley lines, catchment areas, stream ordering, and the contributing areas for each pixel in the map area by evaluating the data contained in a gridded DEM. Zevenbergen and Thorne (1987) presented a method for the automatic extraction of slope angle, slope aspect, and downslope and cross-slope convexity. An overview of the algorithms applied in the extraction of morphometric parameters from DEMs was given by Gardner et al. (1990).

The potential value of DEMs for dynamic slope stability analysis was stressed by Pike (1988) and Wadge (1988). Carrara automatically identified from a detailed DEM the homogeneous units he used as the basis for a multivariate analysis. The morphometric and hydrologic parameters used in that analysis were also extracted automatically (Carrara et al. 1990; Carrara 1988). Niemann and Howes (1991) performed a statistical analysis based on automatically extracted morphometric parameters (slope angle, slope aspect, downslope and cross-slope convexity, and drainage area), which they grouped into homogeneous units using cluster analysis. Various authors (Okimura and Kawatani 1986; Brass et al. 1989) have used neighborhood analysis in the modeling of groundwater tables over time, and used these values as one of the input factors in infinite slope modeling. A simple type of neighborhood analysis was applied by van Dijke and van Westen (1990) to model the runout distances for rock-fall blocks. Ellen et al. (1993) developed a dynamic model for simulating the runout distance of debris flows with a GIS.

A recent development in the use of GIS for slope instability zonation is the application of expert systems. Pearson et al. (1991) developed an expert system in connection with a GIS in order to remove the constraint that users should have considerable experience with GIS. A prototype interface between a GIS (Arc/Info) and an expert system (Nexpert Object) was developed and applied for translational landslide hazard zonation in an area in Cyprus. The question remains, however, as to whether the rules used in this expert system apply only to this specific area or are universally applicable.

4.3 GIS-Based Landslide Hazard Zonation Techniques

The most useful techniques for the application of GIS landslide hazard zonation are presented in the following discussion. A brief description of the various landslide hazard analysis techniques was

given in Section 2.4. Each technique described here is shown schematically in a simplified flow-chart. An overview is given of the required input data (as discussed in Section 2.3), and the various steps required in using GIS techniques are mentioned briefly. A recommendation is also provided regarding the most appropriate working scale (see also Section 2.2).

4.3.1 Landslide Inventory

The input consists of a field-checked photointerpretation map of landslides for which recent, relatively large-scale aerial photographs have been used combined with a table containing landslide parameters obtained from a checklist. GIS can perform an important task in transferring the digitized photointerpretation to the topographic base map projection using a series of control points and camera information.

The GIS procedure is as follows:

- Digitize the mass movement phenomena, each with its own unique label and a six-digit code containing information on the landslide type, subtype, activity, depth, and site vegetation and whether the unit is a landslide scarp or body;
- Recode the landslide map showing the parameters for landslide type or subtype into maps that display only a single type or process.

In this technique, the GIS is used only to store the information and to display maps in different forms (e.g., only the scarps, only the slides, or only the active slides). Although the actual analysis is very simple, the use of GIS provides a great advantage for this method. The user can select specific combinations of mass movement parameters and obtain better insight into the spatial distribution of the various landslide types. The method is represented schematically in Figure 8-14.

The code for mass movement activity given to each mass movement phenomenon can also be used in combination with mass movement distribution maps from earlier dates to analyze mass movement activity. Depending on the type of terrain being studied, time intervals of 5 to 20 years may be selected. This method of interval analysis provides estimates of the numbers or percentages of reactivated, new, or stabilized landslides.

Mass movement information can also be presented as a percentage cover within mapping

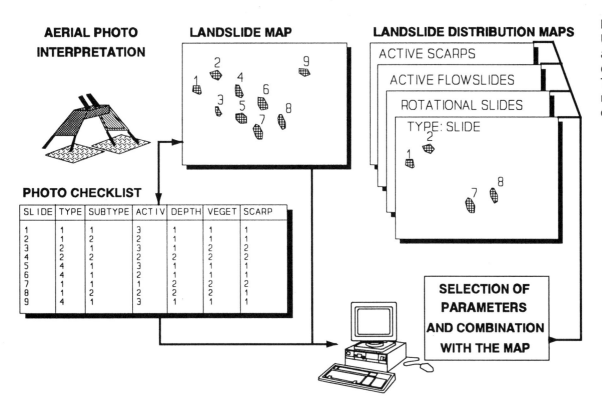

FIGURE 8-14
Use of GIS for
analysis of landslide
distribution. See
Table 8-4 for code
numbers in photo
checklist.

units. These mapping units may be TMUs, geomorphic units, geologic units, or any other appropriate map unit. This method may also be used to test the importance of each individual parameter for predicting the occurrence of mass movements. The required input data consist of a mass movement distribution map and a land-unit map. If the method is used to test the importance of specific parameter classes, the user decides, on the basis of his or her field experience, which individual parameter maps, or combination of parameter maps, will be used.

The following GIS procedures are used for mass movement density analysis:

- Calculation of a bit map (presence or absence) for the specific movement type for which the analysis is carried out;
- Combination of the selected parameter map with the bit map through a process called "map crossing," which spatially correlates the conditions on the two maps; and
- Calculation of the area percentage per parameter class occupied by landslides.

With a small modification, the number of landslides can be calculated instead of the areal density. In this case a bit map is not made, and the mass

movement map itself, in which each polygon has a unique code, is overlaid by the parameter map. The method is represented schematically in Figure 8-15.

Isopleth mapping is a special form of mass movement density mapping. The method uses a large circular counting filter that calculates the landslide density for each circle center automatically. The resulting values for the circle centers are interpolated and contours of equal density are drawn. The scale of the pixels and the size of the filter used define the values in the resulting density map.

The method is most appropriate at medium or large scales. At the regional scale the construction of a mass movement distribution map is very time-consuming and too detailed for procedures of general regional zoning. Nevertheless, when possible, it is advisable to prepare such a map also for the regional scale, although with less detail.

4.3.2 Heuristic Analysis

As explained in Section 2.4.2, when a heuristic approach is used, the hazard map is made by the mapping geomorphologist using site-specific knowledge obtained through photointerpretation and fieldwork. The map can be made either directly in the field or by recoding a geomorphic map. The criteria on which hazard class designa-

FIGURE 8-15
Use of GIS for
analysis of landslide
density.

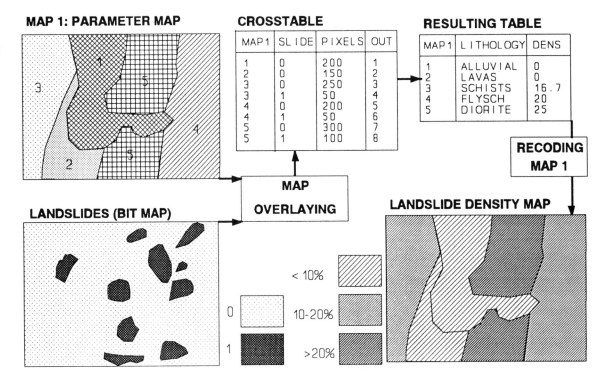

tions are based are not formalized in generally applicable rules and may vary from polygon to polygon. GIS can be used in this type of work as a drawing tool, allowing rapid recoding of units and correction of units that were coded erroneously. GIS is not used as a tool for the analysis of the important parameters related to the occurrence of mass movements. The method can be applied at regional, medium, or large scales in a relatively short time period. It does not require the digitizing of many different maps. However, the detailed fieldwork requires a considerable amount of time.

If the analysis is done by combining several parameter maps, qualitative weighting values are assigned to each class of parameter map, and each parameter map receives a different weight. The earth scientist decides which maps will be utilized and which weighting values will be assigned on the basis of field knowledge of the causal factors (see Figure 8-16).

The following GIS procedures are used:

- Classification of each parameter map into a number of relevant classes;
- Assignment of weighting values to each of the parameter classes (e.g., on a scale of 1 to 10);
- Assignment of weighting values to each of the parameter maps; and

- Calculation of weights for each pixel and classification in a few hazard classes.

The method is applicable on all three scales. Each scale has its own requirements as to the required detail of the input maps.

4.3.3 Statistical Analysis

In statistical methods, overlaying of parameter maps and calculation of landslide densities form the core of the analysis. If bivariate techniques are chosen, the importance of each parameter or specific combinations of parameters can be analyzed individually. Several methods exist for calculating weighting values (see Section 2.4.3). Most are based on the relationship between the landslide density per parameter class compared with the landslide density over the entire area. Each method has its specific rules for data integration required to produce the total hazard map.

The weighting values can also be used to design decision rules, which are based on the experience of the earth scientist. It is possible to combine various parameter maps into a map of homogeneous units, which is then combined or overlaid with the

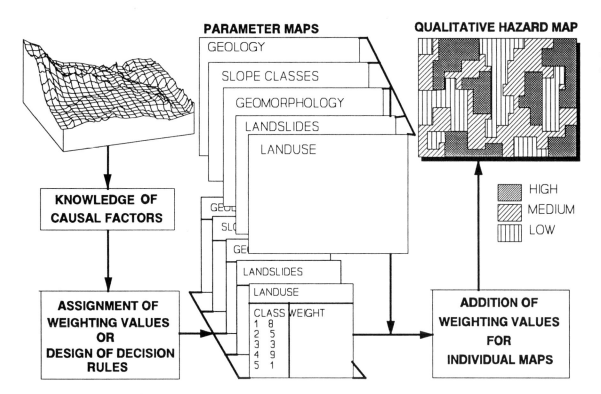

PARAMETER MAPS

QUALITATIVE HAZARD MAP

FIGURE 8-16
Use of GIS for
qualitative map
combination.

landslide map to produce a landslide density for each unique combination of input parameters.

GIS is very suitable for use with this method, especially with macro commands for repetitive calculations involving a large number of map combinations and manipulation of attribute data. It should be stressed that the selection of parameters also has an important subjective element in this method. However, the user can test the importance of individual parameter maps and decide on the final input maps in an iterative manner. The following GIS procedures are used (see Figure 8-17):

- Classification of each parameter map into a number of relevant classes;
- Combination of the selected parameter maps with the landslide map by the process known as map crossing to produce cross-tabulations defining the spatial correlations between the parameter maps and the landslide map;
- Calculation of weighting values based on the cross-tabulation data; and
- Assignment of weighting values to the various parameter maps or design of decision rules to be applied to the maps and classification of the resulting scores in a few hazard classes.

Multivariate statistical analyses of important factors related to landslide occurrence give the relative contribution of each of these factors to the total hazard within a defined land unit. The analyses are based on the presence or absence of mass movement phenomena within these land units, which may be catchment areas, interpreted geomorphic units, or other kinds of terrain units.

Several multivariate methods have been proposed in the literature. Most of these, such as discriminant analysis or multiple regression, require the use of external statistical packages. GIS techniques are used to sample parameters for each land unit. However, with a PC-based GIS, the large volume of data may become a problem. The method requires a landslide distribution map and a land-unit map. A large number of parameters are used, sometimes up to 50. The following GIS procedures are used (see Figure 8-18):

- Determination of the list of factors that will be included in the analysis. Because many input maps (such as geology) are of an alphanumerical type, they must be converted to numerical maps. These maps can be converted to presence or absence values for each land unit or presented as percentage cover, or the parameter

FIGURE 8-17
Use of GIS for
bivariate statistical
analysis.

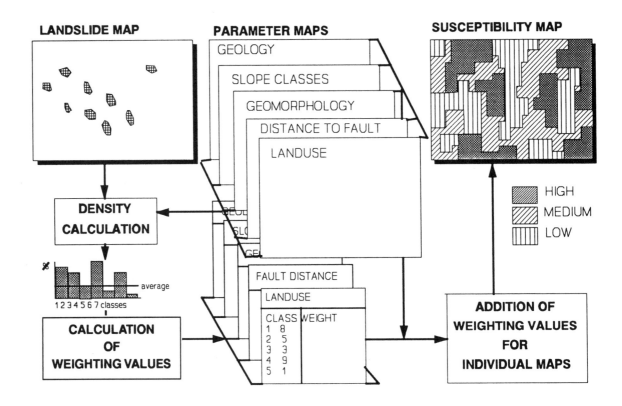

LANDSLIDE MAP

PARAMETER MAPS

GEOLOGY

SLOPE CLASSES

GEOMORPHOLOGY

DISTANCE TO FAULT

LANDUSE

SUSCEPTIBILITY MAP

HIGH

MEDIUM

LOW

**DENSITY
CALCULATION**

%

average

1 2 3 4 5 6 7 classes

**CALCULATION
OF
WEIGHTING VALUES**

FAULT DISTANCE

LANDUSE

CLASS	WEIGHT
1	8
2	5
3	3
4	9
5	1

**ADDITION OF
WEIGHTING VALUES
FOR
INDIVIDUAL MAPS**

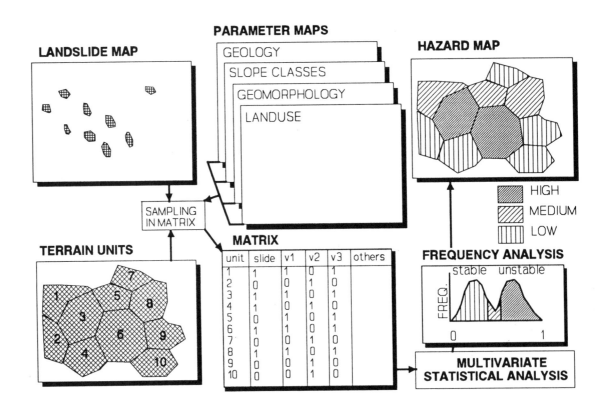

LANDSLIDE MAP

PARAMETER MAPS

GEOLOGY

SLOPE CLASSES

GEOMORPHOLOGY

LANDUSE

HAZARD MAP

HIGH

MEDIUM

LOW

SAMPLING
IN MATRIX

TERRAIN UNITS

MATRIX

unit	slide	v1	v2	v3	others
1	1	1	0	1	
2	0	0	1	0	
3	1	0	1	1	
4	1	0	1	0	
5	0	1	0	1	
6	1	1	0	1	
7	1	0	1	0	
8	1	0	1	0	
9	0	0	0	1	
10	0	0	1	1	

FREQUENCY ANALYSIS

stable   unstable

FREQ.

0   1

**MULTIVARIATE
STATISTICAL ANALYSIS**

FIGURE 8-18
Use of GIS for
multivariate
statistical analysis.

classes can be ranked according to increasing mass movement density. By combining the parameter maps with the land-unit map, a large matrix is created.

- Combination of the land-unit map with the mass movement map by map overlaying and dividing the stable and the unstable units into two groups.
- Exportation of the matrix to a statistical package for subsequent analysis.
- Importation of the results for each land unit into the GIS and recoding of the land units. The frequency distribution of stable and unstable classified units is checked to see whether the two groups are separated correctly.
- Classification of the map into a few hazard classes.

Although the statistical techniques can be applied at different scales, their use becomes quite restricted at the regional scale, where an accurate input map of landslide occurrences may not be available and where most of the important parameters cannot be collected with appropriate accuracy. At large scales different factors will have to be used, such as water-table depth, soil layer sequences, and thicknesses. These data are very difficult to obtain even for relatively small areas. Therefore the medium scale is considered most appropriate for this technique.

4.3.4 Deterministic Analysis

The methods described thus far give no information on the stability of a slope as expressed in terms of its factor of safety. For such information, slope stability models are necessary. These models require input data on soil layer thickness, soil strength, depth below the terrain surface to the potential sliding surfaces, slope angle, and pore pressure conditions to be expected on the slip surfaces. The following parameter maps must be available in order to use such models:

- A material map showing the distribution both at ground surface and in the vertical profile with accompanying data on soil characteristics,
- A groundwater level map based on a groundwater model or on field measurements, and
- A detailed slope-angle map derived from a very detailed DEM.

Several approaches allow for the application of GIS in deterministic modeling (see Figure 8-19):

- The use of an infinite slope model, which calculates the safety factor for each pixel;
- Selection of a number of profiles from the DEM and the other parameter maps, which are exported to external slope stability models; and
- Sampling of data at predefined grid points and exportation of these data to a three-dimensional slope stability model.

The result is a map showing the average safety factor for a given magnitude of groundwater depth and seismic acceleration. The variability of the input data can be used to calculate the probability of failure in connection with the return period of triggering events. Generally the resulting safety factors and probabilities should not be used as absolute values unless the analysis is done in a small area where all parameters are well known. Normally they are only indicative and can be used to test different scenarios of slip surfaces and groundwater depths. The method is applicable only at large scales and over small areas. At regional and medium scales, the required detailed input data, especially concerning groundwater levels, soil profile, and geotechnical descriptions, usually cannot be provided.

4.4 Phases of Landslide Hazard Analysis Using GIS

A GIS-supported landslide hazard analysis project requires a number of unique phases, which are distinctly different from those required by a conventional landslide hazard analysis project. An overview of the 12 phases is given in Table 8-10. There is a logical order to these phases, although some may overlap considerably. Phases 7 to 11 are carried out using the computer. Data-base design (Phase 4) occurs before the computer work starts and even before the fieldwork because it determines the way in which the input data are collected in the field.

Table 8-10 also indicates the relative amount of time spent on each phase at each of the three scales of analysis. The time amounts are expressed as a percentage of the time spent on the entire process and are estimated on the basis of experience. Absolute time estimates are not given, since

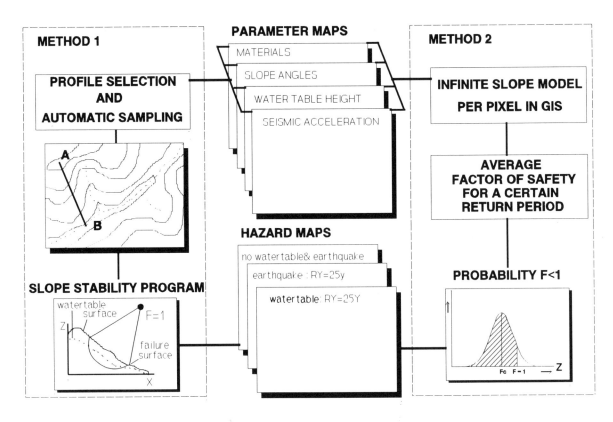

FIGURE 8-19
Use of GIS in deterministic analysis.

Table 8-10
Percentage of Time Spent in Various Phases of Landslide Hazard Assessment Projects at Different Scales Using GIS and
Conventional Methods

Phase	Regional Scale		Medium Scale		Large Scale	
	Conventional Methods	GIS-Based Methods	Conventional Methods	GIS-Based Methods	Conventional Methods	GIS-Based Methods
1 Choice of scale and methods	<5	<5	<5	<5	<1	<5
2 Collection of existing data	<5	<5	<5	<5	8	8
3 Image interpretation	50	50	30	30	10	20
4 Data-base design	0	<5	0	<5	0	<5
5 Fieldwork	<5	<5	7	7	10	20
6 Laboratory analysis	0	0	0	<5	0	10
7 Data entry	0	20	0	30	0	15
8 Data validation	0	<5	0	5	0	5
9 Data manipulation	0	<5	0	5	0	5
10 Data analysis	30	10	48	10	61	10
11 Error analysis	0	<5	0	<5	0	<5
12 Final map production	10	<5	10	<5	10	<5

these depend on too many variable factors, such as the amount of available input data, the size of the study area, and the experience of the investigator or investigators.

The percentage of time needed for image interpretation using the GIS approach decreases from the regional scale to the large scale, and fieldwork and laboratory analysis tasks become more important. Data entry requires the most time at the medium scale because of the large number of parameter maps that have to be digitized. Because analysis is based on only one basic data layer of TMUs, the time needed for data entry on the regional scale is much lower.

Working with a GIS considerably increases the time needed for the preanalysis phases, mainly because of the tedious job of hand-digitizing input maps. Time needed for data analysis, however, is not more than 10 percent in the GIS approach versus almost 50 percent using conventional techniques. Many of the analysis techniques are almost impossible to execute without a GIS. Working with GIS considerably reduces the time needed to produce the final maps, which are no longer drawn by hand.

REFERENCES

Akinyede, J.O. 1990. *Highway Cost Modelling and Route Selection Using a Geotechnical Information System*. Ph.D. thesis. Technical University of Delft, Delft, Netherlands, 221 pp.

Alzate, J.B. (ed.). 1992. *Proc., 1er Simposio Internacional sobre Sensores Remotos y Sistemas de Informacion Geografica (SIG) para el Estudio de Riesgos Naturales*, Bogotá, Colombia, Instituto Geografico Agustin Codazzi, Bogotá, 686 pp.

Antoine, P. 1977. Réflexions sur la Cartographie ZERMOS et Bilan des Expériences en Cours. *Bulletin, Bureau de Recherches Géologiques et Minières* (deuxieme série), Section III, Vol. 1/2, pp. 9–20.

Aronoff, S. 1989. *Geographical Information Systems: A Management Perspective*. WDL Publications, Ottawa, Canada, 294 pp.

Bernknopf, R.L., R.H. Campbell, D.S. Brookshire, and C.D. Shapiro. 1988. A Probabilistic Approach to Landslide Hazard Mapping in Cincinnati, Ohio, with Applications for Economic Evaluation. *Bulletin of the Association of Engineering Geologists*, Vol. 25, No. 1, pp. 39–56.

Bertozzi, R., M. Locatelli, and G. Vianello. 1992. Model for the Correlation Between Landuse Dynamics and Hydrogeological Risk. In *Proc., Interpraevent 1992*, Bern, Switzerland, Vol. 3, pp. 131–144.

Bison, P., E. Grinzato, A. Pasuto, and S. Silvano. 1990. Thermal IR Remote Sensing in Landslides Survey. In *Proc., Sixth International Congress, International Association of Engineering Geology*, Amsterdam (D.G. Price, ed.), A.A. Balkema, Rotterdam, Netherlands, pp. 873–884.

Brabb, E.E. 1984. Innovative Approaches to Landslide Hazard and Risk Mapping. In *Proc., Fourth International Symposium on Landslides*, Canadian Geotechnical Society, Toronto, Canada, Vol. 1, pp. 307–324.

Brabb, E.E. 1987. Analyzing and Portraying Geologic and Cartographic Information for Landuse Planning, Emergency Response and Decision Making in San Mateo County, California. In *Proc., GIS'87*, San Francisco, Calif., American Society of Photogrammetry and Remote Sensing, Falls Church, Va., pp. 362–374.

Brabb, E.E., E.H. Pampeyan, and M.G. Bonilla. 1972. *Landslide Susceptibility in San Mateo County, California*. Misc. Field Studies Map MF360 (scale 1:62,500). U.S. Geological Survey, Reston, Va.

Brabb, E.E., F. Guzzetti, R. Mark, and R.W. Simpson. 1989. The Extent of Landsliding in Northern New Mexico and Similar Semi-Arid Regions. In *Landslides in a Semi-Arid Environment* (P.M. Sadler and D.M. Morton, eds.), Inland Geological Society, University of California, Riverside, Vol. 2, pp. 163–173.

Brass, A., G. Wadge, and A.J. Reading. 1989. Designing a Geographical Information System for the Prediction of Landsliding Potential in the West Indies. In *Proc., Economic Geology and Geotechnics of Active Tectonic Regions*, University College, University of London, April, 13 pp.

Brunsden, D., J.C. Doornkamp, P.G. Fookes, D.K.C. Jones, and J.M.H. Kelly. 1975. Large Scale Geomorphological Mapping and Highway Engineering Design. *Quarterly Journal of Engineering Geology*, Vol. 8, pp. 227–253.

Brunsden, D., and D.B. Prior (eds.). 1984. *Slope Instability*. John Wiley and Sons, New York, 620 pp.

Burrough, P.A. 1986. *Principles of Geographical Information Systems and Land Resources Assessment*. Clarendon Press, Oxford, England, 194 pp.

Canuti, P., F. Frascati, C.A. Garzonio, and C. Rodolfi. 1979. *Dinamica Morfologica di un Ambiente Soggetto a Fenomeni Franosi e ad Intensa Attiva Agricola*. Publ. No. 142. Consiglio Nazionale delle Ricerche, Perugia, Italy, pp. 81–102.

Carrara, A. 1983. Multivariate Models for Landslide Hazard Evaluation. *Mathematical Geology*, Vol. 15, No. 3, pp. 403–427.

Carrara, A. 1988. Landslide Hazard Mapping by Statistical Methods: A "Black Box" Approach. In *Workshop on Natural Disasters in European Mediterranean Countries*, Perugia, Italy, Consiglio Nazionale delle Ricerche, Perugia, pp. 205–224.

Carrara, A., and L. Merenda. 1974. Metodologia per un Censimento Degli Eventi Franoso in Calabria. *Geologia Applicata e Idrogeologica*, Vol. 9, pp. 237–255.

Carrara, A., C.E. Pugliese, and L. Merenda. 1977. Computer Based Data Bank and Statistical Analysis of Slope Instability Phenomena. *Zeitschrift für Geomorphologie N.F.*, Vol. 21, No. 2, pp. 187–222.

Carrara, A., E. Catalano, M. Sorriso-Valvo, C. Realli, and I. Ossi. 1978. Digital Terrain Analysis for Land Evaluation. *Geologia Applicata e Idrogeologica*, Vol. 13, pp. 69–127.

Carrara, A., M. Cardinali, R. Detti, F. Guzzetti, V. Pasqui, and P. Reichenbach. 1990. Geographical Information Systems and Multivariate Models in Landslide Hazard Evaluation. In *ALPS 90 Alpine Landslide Practical Seminar, Sixth International Conference and Field Workshop on Landslides*, Aug. 31–Sept.12, Milan, Italy, Universitá degli Studi de Milano, pp. 17–28.

Carrara, A., M. Cardinali, R. Detti, F. Guzzetti, V. Pasqui, and P. Reichenbach. 1991. GIS Techniques and Statistical Models in Evaluating Landslide Hazard. *Earth Surface Processes and Landforms*, Vol. 16, No. 5, pp. 427–445.

Carrara, A., M. Cardinali, and F. Guzzetti. 1992. Uncertainty in Assessing Landslide Hazard and Risk. *ITC Journal*, No. 2, pp. 172–183.

Choubey, V.D., and P.K. Litoria. 1990. Terrain Classification and Land Hazard Mapping in Kalsi-Chakrata Area (Garhwal Himalaya), India. *ITC Journal*, No. 1, pp. 58–66.

Chung, C.J., and A.G. Fabbri. 1993. The Representation of Geoscience Information for Data Integration. *Nonrenewable Resources*, Vol. 2, No. 3, pp. 122–139.

Chung, C.J., A.G. Fabbri, and C.J. van Westen. In press. Multivariate Statistical Analysis in Landslide Hazard Zonation. In *Proc., International Workshop on GIS in Natural Hazard Assessment*, Perugia, Italy, Sept. 21–23, Consiglio Nazionale delle Ricerche, Perugia.

Cleaves, A.B. 1961. *Landslide Investigations: A Field Handbook for Use in Highway Location and Design*. Bureau of Public Roads, U.S. Department of Commerce, 67 pp.

Cooke, R.U., and J.C. Doornkamp. 1990. *Geomorphology in Environmental Management*. Clarendon Press, Oxford, England, 410 pp.

Cotecchia, V., A. Guerricchio, and G. Melidoro. 1986. The Geomorphogenetic Crisis Triggered by the 1783 Earthquake in Calabria (Southern Italy). In *Proc., International Symposium on Engineering Geology Problems in Seismic Areas*, University of Bari, Italy (also published in *Geologia Applicata e Idrogeologica*, Vol. 21, pp. 245–304).

Crozier, M.J. 1973. Techniques for the Morphometric Analysis of Landslips. *Zeitschrift für Geomorphologie N.F.*, Vol. 17, No. 1, pp. 78–101.

Crozier, M.J. 1986. *Landslides: Causes, Consequences and Environment*. Croom Helm, England, 245 pp.

Demek, J., and C. Embleton (eds.). 1978. *Guide to Medium-Scale Geomorphological Mapping*. IGU Commission on Geomorphological Survey and Mapping, E. Schweizerbart'sche Verlagsbuchhandlung, Stuttgart, Germany, 348 pp.

Dikau, R. 1992. Aspects of Constructing a Digital Geomorphological Base Map. *Geologisches Jahrbuch Reihe A*, Heft 122, pp. 357-370.

Dunoyer, M., and C.J. van Westen. 1994. Assessing Uncertainty in Interpreting Landslides from Airphotos. *ITC Journal*, No. 3.

Eastman, J.R. 1992a. *User's Guide; IDRISI Version 4.0*. Graduate School of Geography, Clark University, Worcester, Mass., 178 pp.

Eastman, J.R. 1992b. *Technical Reference; IDRISI Version 4.0*. Graduate School of Geography, Clark University, Worcester, Mass., 213 pp.

Einstein, H.H. 1988. Special Lecture: Landslide Risk Assessment Procedure. In *Proc., Fifth International Symposium on Landslides*, Lausanne (C. Bonnard, ed.), A.A. Balkema, Rotterdam, Netherlands, Vol. 2, pp. 1075–1090.

Ellen, S.D., R.K. Mark, S.H. Cannon, and D.L. Knifong. 1993. *Map of Debris-Flow Hazard in the Honolulu District of Oahu, Hawaii*. Open-File Report 93-213. U.S. Geological Survey, Reston, Va., 25 pp.

Environmental Systems Research Institute (ESRI). 1992. *Understanding GIS; The Arc/Info Method*. Redlands, Calif., 500 pp.

Evans, D.L. 1992. Geologic Process Studies Using Synthetic Aperture Radar (SAR) Data. *Episodes*, Vol. 15, No. 1, pp. 21–31.

Fookes, P.G., S.G. Dale, and J.M. Land. 1991. Some Observations on a Comparative Aerial Photography Interpretation of a Landslipped Area. *Quarterly Journal of Engineering Geology*, Vol. 24, pp. 249–265.

Gardner, T.W., K. Conners-Sasowski, and R.L. Day.

1990. Automatic Extraction of Geomorphometric Properties from Digital Elevation Data. *Zeitschrift für Geomorphologie N.F.*, Suppl. Vol. 60, pp. 57–68.

Goodchild, M.F., B.O. Parks, and L.T. Steyaert (eds.). 1993. *Environmental Modeling with GIS*. Oxford University Press, New York, 488 pp.

Guerricchio, A., and G. Melidoro. 1981. Movimenti di Massa Pseudo-Tettonici nell' Apennino dell' Italia Meridionale. *Geologia Applicata e Idrogeologica*, Vol. 16, pp. 251–294.

Hammond, C.J., R.W. Prellwitz, and S.M. Miller. 1992. Landslide Hazard Assessment Using Monte Carlo Simulation. In *Proc., Sixth International Symposium on Landslides* (D.H. Bell, ed.), Christchurch, New Zealand, A.A. Balkema, Rotterdam, Netherlands, Vol. 2, pp. 959–964.

Hansen, A. 1984. Landslide Hazard Analysis. In *Slope Instability* (D. Brunsden and D.B. Prior, eds.), John Wiley and Sons, New York, pp. 523–602.

Hartlén, J., and L. Viberg. 1988. General Report: Evaluation of Landslide Hazard. In *Proc., Fifth International Symposium on Landslides*, Lausanne (C. Bonnard, ed.), A.A. Balkema, Rotterdam, Netherlands, Vol. 2, pp. 1037–1057.

Huma, I., and D. Radulescu. 1978. Automatic Production of Thematic Maps of Slope Instability. *Bulletin of the International Association of Engineering Geology*, No. 17, pp. 95–99.

Hutchinson, J.N. 1988. Morphological and Geotechnical Parameters of Landslides in Relation to Geology and Hydrogeology. In *Proc., Fifth International Symposium on Landslides* (C. Bonnard, ed.), Lausanne, A.A. Balkema, Rotterdam, Netherlands, Vol. 1, pp. 3–35.

Innocienti, L. 1992. *Data Integration for Landslide Risk Mapping in Urban Areas Using GIS*. M.Sc. thesis. International Institute for Aerospace Survey and Earth Sciences (ITC), Enschede, Netherlands, 88 pp.

Intergraph Corporation. 1993. *MGE Technical Reference Manuals*. Huntsville, Ala., various pagings.

International Association of Engineering Geology (IAEG). 1976. *Engineering Geological Maps: A Guide to Their Preparation*. UNESCO Press, Paris, 79 pp.

Ives, J.D., and B. Messerli. 1981. Mountain Hazards Mapping in Nepal: Introduction to an Applied Mountain Research Project. *Mountain Research and Development*, Vol. 1, No. 3–4, pp. 223–230.

Jones, F.O., D.R. Embody, and W.C. Peterson. 1961. *Landslides along the Columbia River Valley, Northeastern Washington*. Professional Paper 367. U.S. Geological Survey, Reston, Va., 98 pp.

Keaton, J.R. 1994. Risk-Based Probabilistic Approach to Site Selection. *Bulletin of the Association of Engineering Geologists*, Vol. 31, No. 2, pp. 217–229.

Kienholz, H. 1977. Kombinierte Geomorphologische Gefahrenkarte 1:10,000 von Grindelwald. *Catena*, Vol. 3, pp. 265–294.

Kienholz, H. 1978. Maps of Geomorphology and Natural Hazards of Grindelwald, Switzerland, scale 1:10,000. *Arctic and Alpine Research*, Vol. 10, pp. 169–184.

Kienholz, H. 1992. Risk Assessment in Mountains. In *Proc., 1er Simposio Internacional sobre Sensores Remotes y Sistemas de Informacion Geografica (SIG) para el Estudio de Riesgos Naturales*, Bogotá, Colombia (J.B. Alzate, ed.), Instituto Geografico Agustin Codazzi, Bogotá, Vol. 2, 20 pp.

Kienholz, H., M. Bichsel, M. Grunder, and P. Mool. 1983. *Kathmandu-Kakani Area, Nepal: Mountain Hazards and Slope Stability Map*. Mountain Hazards Mapping Project Map No. 4 (scale 1:10,000). United Nations University, Tokyo, Japan.

Kienholz, H., P. Mani, and M. Kläy. 1988. Rigi Nordlene: Beurteilung der Naturgefahren und Waldbauliche Prioritätenfestlegung. In *Proc., Interpraevent 1988*, Graz, Austria, Vol. 1, pp. 161–174.

Kingsbury, P.A., W.J. Hastie, and A.J. Harrington. 1992. Regional Landslip Hazard Assessment Using a Geographical Information System. In *Proc., Sixth International Symposium on Landslides* (D.H. Bell, ed.), Christchurch, New Zealand, A.A. Balkema, Rotterdam, Netherlands, Vol. 2, pp. 995–999.

Kobashi, S., and M. Suzuki. 1988. Hazard Index for the Judgment of Slope Stability in the Rokko Mountain Region. In *Proc., Interpraevent 1988*, Graz, Austria, Vol. 1, pp. 223–233.

Lasky, L.R. 1980. *The Thermal Inertia of Engineering Geologic Units*. M.Sc. thesis. Colorado School of Mines, Golden, 223 pp.

Lopez, H.J., and J.A. Zinck. 1991. GIS-Assisted Modelling of Soil-Induced Mass Movement Hazards: A Case Study of the Upper Coello River Basin, Tolima, Colombia. *ITC Journal*, No. 4, pp. 202–220.

Malgot, J., and T. Mahr. 1979. Engineering Geological Mapping of the West Carpathian Landslide Areas. *Bulletin of the International Association of Engineering Geology*, No. 19, pp. 116–121.

Mani, P., and B. Gerber. 1992. Geographische Informationssysteme in der Analyse von Naturgefahren. In *Proc., Interpraevent 1992*, Bern, Switzerland, Vol. 3, pp. 97–108.

McKean, D.F., H.W. Calkins, and D.J. Peuquet.

1991. Remote Sensing and Landslide Hazard Assessment. *Photogrammetric Engineering and Remote Sensing,* Vol. 57, No. 9, pp. 1185–1193.

Meijerink, A.M.J. 1988. Data Acquisition and Data Capture Through Terrain Mapping Units. *ITC Journal,* No. 1, pp. 23–44.

Meneroud, J.P., and A. Calvino. 1976. *Carte ZER-MOS, Zones Exposées à des Risques liés aux Mouvements du Sol et du Sous-Sol à 1:25,000, Region de la Moyenne Vesubie (Alpes Maritimes).* Bureau de Recherches Géologiques et Minières, Orléans, France, 11 pp.

Mollard, J.D. 1977. Regional Landslide Types in Canada. In *Reviews in Engineering Geology,* Vol. 3 (D.R. Coates, ed.), Geological Society of America, Boulder, Colo., pp. 29–56.

Mulder, H.F.H.M. 1991. *Assessment of Landslide Hazard.* Ph.D. thesis. University of Utrecht, Netherlands, 150 pp.

Mulder, H.F.H.M., and T.W.J. van Asch. 1988. A Stochastical Approach to Landslide Hazard Determination in a Forested Area. In *Proc., Fifth International Symposium on Landslides,* Lausanne (C. Bonnard, ed.), A.A. Balkema, Rotterdam, Netherlands, Vol. 2, pp. 1207–1210.

Murphy, W., and C. Vita-Finzi. 1991. Landslides and Seismicity: An Application of Remote Sensing. In *Proc., Eighth Thematic Conference on Geological Remote Sensing,* Denver, Colo., Environmental Research Institute of Michigan, Ann Arbor, Vol. 2, pp. 771–784.

Naithani, K.K. 1990. Can Satellite Images Replace Aerial Photographs? A Photogrammetrist's View. *ITC Journal,* No. 1, pp. 29–31.

Neuland, H. 1976. A Prediction Model of Landslips. *Catena,* Vol. 3, pp. 215–230.

Newman, E.B., A.R. Paradis, and E.E. Brabb. 1978. *Feasibility and Cost of Using a Computer to Prepare Landslide Susceptibility Maps of the San Francisco Bay Region, California.* Bulletin 1443. U.S. Geological Survey, Reston, Va., 29 pp.

Niemann, K.O., and D.E. Howes. 1991. Applicability of Digital Terrain Models for Slope Stability Assessment. *ITC Journal,* No. 3, pp. 127–137.

Okimura, T., and T. Kawatani. 1986. Mapping of the Potential Surface-Failure Sites on Granite Mountain Slopes. In *International Geomorphology* (J. Gardiner, ed.), John Wiley and Sons, New York, Part 1, pp. 121–138.

Pearson, E., G. Wadge, and A.P. Wislocki. 1991. An Integrated Expert System/GIS Approach to Modelling and Mapping Natural Hazards. In *Proc., European Conference on GIS (EGIS),* Egis

Foundation, University of Utrecht, Netherlands, Session 26, pp. 763–771.

Pike, R.J. 1988. The Geometric Signature: Quantifying Landslide-Terrain Types from Digital Elevation Models. *Mathematical Geology,* Vol. 20, No. 5, pp. 491–511.

Radbruch-Hall, D.H., K. Edwards, and R.M. Batson. 1979. Experimental Engineering Geological Maps of the Conterminous United States Prepared Using Computer Techniques. *Bulletin of the International Association of Engineering Geology,* No. 19, pp. 358–363.

Rengers, N., R. Soeters, and C.J. van Westen. 1992. Remote Sensing and GIS Applied to Mountain Hazard Mapping. *Episodes,* Vol. 15, No. 1, pp. 36–45.

Rib, H.T., and T. Liang. 1978. Recognition and Identification. In *Special Report 176: Landslides: Analysis and Control* (R.L. Schuster and R.J. Krizek, eds.), TRB, National Research Council, Washington, D.C., Chap. 3, pp. 34–80.

Rupke, J., E. Cammeraat, A.C. Seijmonsbergen, and C.J. van Westen. 1988. Engineering Geomorphology of the Widentobel Catchment, Appenzell and Sankt Gallen, Switzerland: A Geomorphological Inventory System Applied to Geotechnical Appraisal of Slope Stability. *Engineering Geology,* Vol. 26, pp. 33–68.

Scanvic, J.Y. 1990. Mapping the Vulnerability of Ground to Landslides: Potential Use of SPOT Stereoscopic Data for La Paz, Bolivia. In *Proc., 23rd ERIM Symposium,* Bangkok, Environmental Research Institute of Michigan, Ann Arbor, pp. 703–708.

Scanvic, J.Y., and F. Girault. 1989. Imageri SPOT-1 et Inventaire des Mouvements de Terrain: L'Exemple de La Paz (Bolivie). *Revue Photointerpretation,* No. 2, Fasc. 1.

Sissakian, V., R. Soeters, and N. Rengers. 1983. Engineering Geological Mapping from Aerial Photographs: The Influence of Photoscale on Map Quality and the Use of Stereo-Orthophotographs. *ITC Journal,* No. 2, pp. 109–118.

Slaney, R., and V. Singhroy. 1991. SAR for Non-Renewable Resources Applications. In *Proc., IUGS Meeting on Remote Sensing in Global Geoscience Processes,* Boulder, Colo., International Union of Geological Sciences, Trondheim, Norway.

Spiegelhalter, D.J. 1986. Uncertainty in Expert Systems. In *Artificial Intelligence and Statistics* (W.A. Gale, ed.), Addison-Wesley, Reading, Mass., pp. 17–55.

Stakenborg, J.H.T. 1986. Digitizing Alpine Morphology: A Digital Terrain Model Based on a Geomorphological Map for Computer-Assisted Applied Mapping. *ITC Journal*, No. 4, pp. 299–306.

Stevenson, P.C. 1977. An Empirical Method for the Evaluation of Relative Landslide Risk. *Bulletin of the International Association of Engineering Geology*, No. 16, pp. 69–72.

Strandberg, C.A. 1967. *Aerial Discovery Manual.* John Wiley and Sons, New York, 249 pp.

Tonnayopas, F. 1988. *Approche Cartographique des Instabilités de Versants à Partir de Photographies Aériennes et d'Images Satellitaires.* Thèse d'études supérieures mémoire GDTA, Paris, France.

van Dijke, J.J., and C.J. van Westen. 1990. Rockfall Hazard: A Geomorphological Application of Neighborhood Analysis with ILWIS. *ITC Journal*, No. 1, pp. 40–44.

van Westen, C.J. 1992. Medium Scale Landslide Hazard Analysis Using a PC Based GIS: A Case Study from Chinchina, Colombia. In *Proc., 1er Simposio Internacional sobre Sensores Remotes y Sistemas de Informacion Geografica (SIG) para el Estudio de Riesgos Naturales*, Bogotá, Colombia (J.B. Alzate, ed.), Instituto Geografico Agustin Codazzi, Bogotá, Vol. 2, 20 pp.

van Westen, C.J. 1993. *Application of Geographic Information Systems to Landslide Hazard Zonation.* ITC Publication No. 15. International Institute for Aerospace Survey and Earth Sciences (ITC), Enschede, Netherlands, 245 pp.

van Westen, C.J., and J.B. Alzate–Bonilla. 1990. Mountain Hazard Analysis Using a PC Based GIS. In *Proc., Sixth International Congress, International Association of Engineering Geology* (D.G. Price, ed.), Amsterdam, A.A. Balkema, Rotterdam, Netherlands, pp. 265–271.

van Zuidam, R.A. 1986. *Terrain Classification.* ITC Textbook. International Institute for Aerospace Survey and Earth Sciences (ITC), Enschede, Netherlands.

Varnes, D.J. 1984. *Landslide Hazard Zonation: A Review of Principles and Practice.* UNESCO Press, Paris, 63 pp.

Wadge, G. 1988. The Potential of GIS for Modelling of Gravity Flows and Slope Instabilities. *International Journal on GIS*, Vol. 2, No. 2, pp. 143–152.

Wagner, A., R. Olivier, and E. Leite. 1988. Rock and Debris Slide Risk Maps Applied to Low-Volume Roads in Nepal. In *Transportation Research Record 1106*, TRB, National Research Council, Washington, D.C., pp. 255–267.

Ward, T.J., L. Ruh-Ming, and D.B. Simons. 1982. Mapping Landslide Hazard in Forest Watershed. *Journal of the Geotechnical Engineering Division*, ASCE, Vol. 108, No. GT2, pp. 319–324.

Wentworth, C.M., S.D. Ellen, and R.K. Mark. 1987. Improved Analysis of Regional Engineering Geology Using GIS. In *Proc., GIS'87*, 26–30 October, San Francisco, Calif., American Society of Photogrammetry and Remote Sensing, Falls Church, Va., pp. 636–649.

Wieczorek, G.F. 1984. Preparing a Detailed Landslide Inventory Map for Hazard Evaluation and Reduction. *Bulletin of the Association of Engineering Geologists*, Vol. 21, No. 3, pp. 337–342.

Wright, R.H., R.H. Campbell, and T.H. Nilsen. 1974. Preparation and Use of Isopleth Maps of Landslide Deposits. *Geology*, Vol. 2, pp. 483–485.

Yin, K.L., and T.Z. Yan. 1988. Statistical Prediction Model for Slope Instability of Metamorphosed Rocks. In *Proc., Fifth International Symposium on Landslides*, Lausanne (C. Bonnard, ed.), A.A. Balkema, Rotterdam, Netherlands, Vol. 2, pp. 1269–1272.

Zevenbergen, L.W., and C.R. Thorne. 1987. Quantitative Analysis of Land Surface Topography. *Earth Surface Processes and Landforms*, Vol. 12, pp. 47–56.

Chapter 9

JEFFREY R. KEATON AND
JEROME V. DEGRAFF

SURFACE OBSERVATION AND GEOLOGIC MAPPING

1. INTRODUCTION

By assessing causes of and factors contributing to slope movement, surface observation and geologic mapping of slopes provide the basis for subsurface investigations and engineering analyses that follow. Accurate interpretation of the surface features of a landslide can be used to evaluate the mode of movement, judge the direction and rate of movement, and estimate the geometry of the slip surface. Surface observation and geologic mapping should be done on slopes with active or inactive landslides and on slopes with no evidence of past landslides to provide a basis for evaluation of the likelihood of new or renewed slope movement. Information obtained through surface observation and geologic mapping of a particular site extends and utilizes knowledge of landslide types and processes discussed in Chapter 3 and the recognition and identification procedures described in Chapter 8. The results of geologic mapping provide the basis for planning the subsurface investigations described in this chapter and for locating the instrumentation discussed in Chapter 11. The geologist should remain involved in the project during the subsurface investigation to aid in the correlation of surface and subsurface data. During the design of slope stabilization measures, discussed in Chapters 17 and 18, the geologist should be available to answer questions about the geology, and during construction the geologist should be on site to compare conditions encountered with those predicted.

Basic tools and techniques needed in surface observation and geologic mapping include access to existing information, use of topographic maps and aerial photographs, use of aircraft for aerial reconnaissance, access to the field site, use of limited hardware and simple instruments, and the ability to observe and interpret geologic features caused by and related to slope instability. Some of the techniques that are described in this chapter are relatively new and have not received widespread exposure. Those techniques that are well known will be mentioned in the context of the chapter but not discussed in detail. A review of more sophisticated survey technology applicable to landslide evaluations is also included in this chapter.

1.1 Duties of the Geologist

Surface observation and geologic mapping are useful for an existing project that has developed an unstable slope and for a proposed project that has the potential for slope movement. The usual sequence of events is similar for both projects; the sense of urgency and the time available to respond are different.

The geologist is notified that the project has been authorized and the area is described as well as possible by a knowledgeable person, perhaps someone who witnessed or discovered a slope movement. Geologic and topographic maps and aerial photographs are examined if available to provide an initial understanding of the general character of the site.

An aerial reconnaissance is made if possible, providing a very useful perspective of the geology; the nature of the slope movement if it has occurred; the general configuration of the landscape, including vegetation, geomorphology, and surface water features; and access to the site. Proximity to utilities and other nongeologic features of importance also can be observed readily from the air. The most practical aircraft for geologic reconnaissance is a high-wing, single-engine airplane. Helicopters are excellent reconnaissance aircraft; however, they are much more expensive than fixed-wing aircraft and generally are less available.

The geologist begins the actual investigation with a ground-based geologic reconnaissance. The relationship of the topographic map and aerial photographs to the actual landscape is recognized and geologic mapping begins. If slope movement has occurred, particular attention is paid to features such as ground cracks, closed topographic depressions, tilted trees, and seeps and springs. Other geologic features, such as bedrock exposures, surface water drainage patterns, surficial deposits, and geologic structure, are also mapped and recorded. A special effort should be made to photograph important features. Photographs are a visual supplement to the geologic map and the field notes produced during the reconnaissance. Photographs can record information that may become less evident over time and can facilitate communication with specialists unfamiliar with technical landslide terminology.

Early in the surface observation and geologic mapping effort at sites with active slope movements, reconnaissance instrumentation should be deployed across selected ground cracks and within the body of the landslide. Instrumentation measurements may be repeated several times during the geologic mapping to provide early information on the rate and nature of slope movement. Shallow groundwater information acquired by simple instrumentation is especially valuable.

Topographic profile and geologic information should be collected to produce geologic cross sections at important locations. Topographic maps usually provide adequate detail for cross sections in less important locations. Conceptualization of geologic conditions follows field data collection; however, much of the conceptualization is developed as part of the "multiple working hypothesis" approach used in geology (Chamberlin 1965). The

formal geologic map is prepared from field notes, and a verbal report consisting of pertinent findings, observations, and recommendations for locations and numbers of borings and test pits may be given to the design engineer. A written report is presented to those responsible for planning the subsurface investigation, and the geologist assists in coordinating the subsurface investigation results with the surface observations and geologic maps.

1.2 Active Slides Versus Stable Slopes

Investigations of stable slopes, even those with inactive landslide deposits, may be more methodical than investigations of active slide areas because such analyses are not performed under conditions of urgency. On active slides the investigative tools and techniques used for stable slope areas are supplemented with other techniques. The immediate information needed by a design team investigating an active landslide includes the boundaries of the slope movement, the rate and direction of movement, and the probable causes of movement. The engineering geologist is well equipped to collect and interpret this kind of information rapidly. Reconnaissance instrumentation for monitoring deformations and pore pressures should be deployed in the early stages of an investigation of an active slide to provide an early and long record. Experience on the part of the geologist is needed because of the lack of time for methodical investigation and because of possible hazards such as open ground cracks, falling rock, or debris flows.

2. WORK REQUIRED BEFORE FIELD VISITATION

Efficient surface observation and geologic mapping must be planned in the office before the site is visited. The area of interest must be identified, and available geologic and geotechnical information, aerial photographs, and topographic maps must be collected and reviewed.

2.1 Area of Interest

The area of interest includes the slope with the active landslide or the potential for slope movement as well as adjacent regions that could be contributing to causes of movement. Adjacent land uses, such as agricultural irrigation, may be important factors. Regional geologic conditions could be directing groundwater from adjacent recharge areas into the area subject to slope movement. Some landslide types are capable of traveling relatively far from their sites of origination. When such landslide types are anticipated, it is prudent to consider the adjacent areas upslope from the project site where such landslides might originate as well as the adjacent areas downslope that might be affected by landslides generated at the project site.

The area of interest must be defined to permit searching for available geologic and geotechnical information, aerial photographs, and topographic maps and for planning aerial and ground-based geologic reconnaissance.

2.2 Geologic and Geotechnical Information

Regional geologic and tectonic information provides an understanding of the geologic context, which will be helpful in anticipating those factors that will be important in controlling slope stability. Regional maps generally are made at scales of 1:100,000 or less and are usually published by government agencies.

Local geologic and tectonic information provides a useful basis for an understanding of the general nature of the site geology. The rock types, surficial deposits, ages, stratigraphic relationships, and structural features are better portrayed on maps at scales larger than 1:100,000. (In the United States, common scales for local geologic maps are 1:62,500 and 1:24,000; in many other countries, common scales are 1:50,000 and 1:10,000.) Unpublished theses from local universities may include larger-scale geologic maps to supplement published maps.

Groundwater conditions are particularly important in slope stability evaluations. Preliminary information often can be collected from regional and local geologic maps. Recharge and discharge areas may be discernible from a knowledge of the regional climate and preliminary analysis of the terrain, including the distribution of rock types, faults and fractures, and springs and marshes. Isohyetal maps and other maps representing climatic information are available from government agencies. Some geologic maps provide reasonably detailed information on surficial deposits, including landslide deposits. Other geologic maps are made specifically to portray bedrock relationships; these maps often disregard the surficial deposits and show bedrock relationships as if surficial materials were not present.

Some countries have published soil survey reports that may be helpful in understanding the weathering products of some geologic formations. The U.S. Department of Agriculture (USDA), for example, includes the Natural Resources Conservation Service (formerly the Soil Conservation Service), an agency responsible for mapping soils in agricultural areas. The USDA Forest Service maps soils on national forests, which cover extensive areas in the western United States and Alaska. Soil surveys usually are restricted to the upper 1.5 m of the soil profile. Nonetheless, this information can be particularly useful because of the level of detail (Hasan 1994). Soil moisture, seeps, springs, and marshy areas are important in agricultural soil surveys; therefore, these features are well documented in the published surveys. Furthermore, the maps in soil survey reports are on an aerial-photographic base. These photographs, although not stereoscopic in the published soil surveys, are helpful in several ways, as discussed in Section 2.3. Soil surveys usually include some limited data on the engineering properties of selected soils. This information can serve as the basis for estimating initial values for preliminary slope stability calculations (Hasan 1994).

Areas of slope instability often have recurring problems. A report may have been prepared before construction of a highway project, but a slope movement may occur after the project has been completed. In such cases, a report on the specific geology or geotechnology, or both, prepared by an agency or consultant may be available for the subject area. Such reports are excellent sources of pertinent information and should be utilized to the greatest extent possible.

2.3 Aerial Photographs

Aerial photographs, which are discussed in Chapter 8, may be available from government agencies or commercial aerial-photography companies, usually for specific projects or purposes, possibly for making topographic maps. Although older photographs are commonly in black and white, since the late 1960s they have usually been in natural color. The older black-and-white photographs were produced at scales that are of marginal usefulness for detailed slope mapping. Scales of 1:15,000 or larger are required before the photographs have the detail needed for mapping. Some features are lost even at a scale of 1:5,000 in natural color photographs (see Chapter 8). Aerial photographs at scales between 1:24,000 and 1:12,000 are available in the United States for areas covered by the Natural Resources Conservation Service soil surveys and for federal land administered by the Forest Service, Bureau of Land Management, and National Park Service.

For most modern highway projects, including some low-volume roads, project-specific aerial photographs are used in planning and design. Often the photographs are used to prepare topographic strip maps of the right-of-way. Such photographs are particularly valuable as a base for geologic mapping. Items used in working with aerial photographs are shown in Figure 9-1. Acetate sheets (such as those used for overhead projector transparencies) may be taped over the photographs and special extra-fine-point permanent-ink pens (overhead projector pens) used to make notations in the field. To help keep the photographs clean and dry in the field, suitably sized plastic food storage bags with tightly fitting closures (self-sealing bags) are particularly useful and inexpensive. Nonprescription reading glasses are an alternative to a pocket stereoscope; the glasses may be worn like regular glasses, even over prescription lenses.

When current aerial photography is available, it is often possible to obtain photographs taken for the same area 5, 10, or even 50 years earlier. Older photographs may show evidence of past landslide movement at or near the area of interest. In areas with extensive urban development, the original topography will be visible on aerial photographs taken before urbanization. These photographs can provide valuable insight into conditions affecting slope stability in these areas.

Landslide features, particularly scarps and fissures, can be accentuated by low-sun-angle illumination. Data and procedures described in a current solar ephemeris permit calculation of the optimum time of day and day of year for morning and afternoon low-sun-angle illumination at any location where the latitude and longitude are known. Figure 9-2 shows a plot and data table for the position of the sun referenced to the Thousand Peaks landslide area in northern Utah on August 23, 1990, the date when photographs were taken for a pipeline investigation. A detail of the highlighted

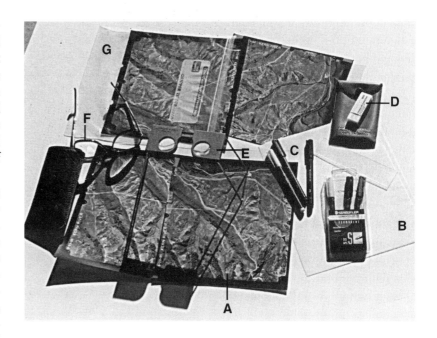

and shadowed scarps visible on one of the photographs is shown in Figure 9-3.

2.4 Topographic Maps

Existing topographic maps provide information for assessing site access and general conditions. However, these maps usually are not at a scale suitable for the level of detail needed in most landslide investigations. Furthermore, existing maps most likely were made before recent slope movement, and thus the features of active landslides may not be visible.

Project-specific topographic maps can be tailored to suit the needs of the project. These maps usually are produced by photogrammetric methods from stereoscopic aerial photographs. Therefore, the photographs and maps are useful in surface field-investigation activities. Current technology allows for production of orthogonally rectified aerial photographs. The best base map can be realized by superimposing the topographic contours on an orthophotographic base.

3. ON-SITE ENGINEERING-GEOLOGIC INVESTIGATIONS

On-site engineering-geologic investigations of areas of active and potential slope movements consist of reconnaissance observations, engineering-geologic mapping, and reconnaissance instrumentation.

FIGURE 9-1 Items useful in working with aerial photographs: A, contact prints of stereoscopic aerial photographs; B, acetate sheets (overhead projector films) to be taped to aerial photographs; C, extra-fine-point permanent-ink pens for writing on acetate sheets (overhead projector pens); D, eraser with emulsion for erasing permanent ink from acetate sheets; E, pocket stereoscope; F, nonprescription reading glasses, which can be used as pocket stereoscope; G, self-sealing plastic food storage bags to keep photographs clean and dry in field.

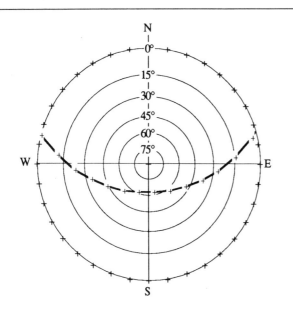

```
THESE SUNANGLE DATA ARE FOR:
THOUSAND PEAKS LANDSLIDE AREA - UTAH
NORTH LATITUDE:  40.98 DEGREES  ( 40° 59'  0")
WEST LONGITUDE: 111.08 DEGREES  (111°  5'  0")
ON 23 AUGUST 1990
SUNRISE AT  6:47  AT AZIMUTH  74.9
SUNSET  AT  8:06  AT AZIMUTH 284.8
```

MOUNTAIN DAYLIGHT TIME (HR)	SUN AZIMUTH (DEG)	ANGLE OF SUN ABOVE HORIZON (DEG)	COORDINATES OF LOWER HEMISPHERE EQUAL ANGLE PROJECTION (UNIT RADIUS) X	Y
6	67.1	-8.4		
SUNRISE	74.9	0.0	0.966	0.260
7	77.1	2.4	0.935	0.214
8	86.8	13.6	0.786	0.044
9	96.8	24.9	0.634	-0.076
10	108.1	35.9	0.485	-0.159
11	122.2	46.1	0.341	-0.215
12	141.2	54.6	0.200	-0.249
1	166.9	59.6	0.061	-0.265
2	196.1	59.3	-0.076	-0.264
3	221.1	53.8	-0.215	-0.246
4	239.4	45.0	-0.356	-0.210
5	253.1	34.7	-0.502	-0.152
6	264.2	23.6	-0.651	-0.066
7	274.1	12.2	-0.804	0.058
8	283.8	1.1	-0.953	0.234
SUNSET	284.8	0.0	-0.967	0.255
9	293.9	-9.7		

The results of the earlier office-based investigations should be well understood before this field investigation is begun.

3.1 Reconnaissance Observations

Aerial reconnaissance is the preferred type of initial surface investigation. The perspective obtained from the air is valuable in understanding the relationships among landslide features, surficial and bedrock materials, landforms, vegetation, and surface water features. In addition, some aspects of logistics, such as roads and trails, may be observed best from the air at a close distance.

General geologic features and landforms can be noted on topographic maps, aerial photographs, or both during the aerial reconnaissance. Features such as bedrock exposures, vigorous vegetation marking shallow groundwater or springs, breaks in slope angle, terraces, ground cracks, tilted trees, and areas of eroded bare slopes should be noted for subsequent examination on the ground. A thorough understanding of these types of features often leads to an interpretation of the causes of or factors contributing to slope movement.

Reconnaissance observations also can be of value in assessing physical access to the site. The location of roads and trails is important, not only for the geologist, but also as access for subsurface investigation equipment. In some locations, utilities may cross the site. Even in relatively remote areas, buried communications cables and pipes carrying water for stock could be present. To the extent possible, these utilities should be identified and located during the reconnaissance or subsequent engineering-geologic mapping. Overhead electric-power lines can represent a constraint to drilling equipment. A nearby canal or pond for watering stock can be a source of water for the drilling operation.

Property ownership and land use should be investigated. Permission may be needed to cross

FIGURE 9-2
Position of sun calculated on hourly basis from sunrise to sunset on August 23, 1990, at Thousand Peaks landslide area in northern Utah. Sun position is calculated using solar ephemeris with reference to local horizon at point of interest. Input data required to calculate sun position are latitude and longitude, number of hours of difference between local time and Greenwich mean time, and Greenwich hour angle and declination from ephemeris for day of interest (day after for areas west of Greenwich and day before for areas east of Greenwich). Tabulated data of sun azimuth and sun angle are converted to coordinates of lower hemisphere equal angle projection (Wulff net) for plotting on stereographic projection.

property for access to a specific site. In addition, a project, such as a pipeline, can have a right-of-way much narrower than a specific landslide or potential slope movement, requiring permission from adjacent property owners for access to conduct investigations of the pertinent slopes. Adjacent land use can influence the performance of a slope. For example, agricultural irrigation can transmit water to places that would otherwise be dry. Nearby buried water pipelines could be threatened by slope movement, or they could leak and contribute to the potential for slope movement.

3.2 Engineering-Geologic Mapping

The purpose of engineering-geologic mapping is to document surface conditions to provide a basis for projecting subsurface conditions and to assist the design engineer in understanding key factors that must be accommodated in construction. A geologic map is an artistic depiction of the geology of a site. An engineering-geologic map must not only be interpretative, it should also be documentary. For example, an area mapped as sandstone should be exposed sandstone, not colluvial deposits containing abundant fragments of sandstone. Notes should be made regarding uncertainties, with an attempt to quantify them if possible. If the uncertainties cannot be quantified, notes should be made that outline the reasons.

Engineering-geologic mapping should characterize a site or region in terms meaningful for and useful to the design engineer. Areas of similar geologic characteristics are designated structural domains for rock slope engineering (Piteau and Peckover 1978). Boundaries of structural domains usually coincide with major geologic features, such as faults and bedrock formation contacts. Boundaries of straight construction slope segments that have similar orientations are determined by the design engineer and superimposed on an engineering-geologic map that shows structural domain boundaries to form design sectors (Piteau and Peckover 1978). Within each design sector, representative rock defect data are selected for use in design calculations (see Chapter 15).

Engineering-geologic mapping for landslide investigations should focus on landslide features, as shown in Figure 9-4. In addition, surficial deposits and bedrock exposures must be carefully documented. Surface-water features are important in

interpreting causes of landslides and potential for slope movement. An interpretive result of engineering-geologic mapping will be classification of slopes in terms of stability, as shown in Figure 9-5.

3.2.1 Landslide Surface Features

Features on the ground surface provide the key to understanding the details of landslide processes and causes. Landslide types and processes are discussed in Chapter 3 and common landslide triggers are described in Chapter 4. Landslide deposits can be classified by age and type of movement according to the features observed on the ground surface. Therefore, particular attention must be paid to these features, and detailed notes and sketches should be made.

The boundaries of landslide deposits may appear gradational, but a boundary may actually be a zone of subparallel cracks and bulges that mark a shear zone. With continuing displacement, the cracks and bulges may give way to a single continuous crack along which lateral shearing occurs. Figure 9-6 illustrates this evolution at the Aspen Grove landslide in central Utah, showing conditions (a) in August 1983 and (b) in August 1984.

An active landslide may affect existing structures, utilities, and other artificial features in ways that provide insight into the processes and cause of

FIGURE 9-3 *(above)* Part of aerial photograph of Thousand Peaks landslide area in northern Utah taken on August 23, 1990, at about 8:00 a.m. local time. Nominal negative scale of photograph is 1:12,000. Scarps and ground cracks caused by movement of landslides in 1983 are enhanced by highlights or shadows caused by low angle of sun.
COURTESY OF KERN RIVER GAS TRANSMISSION COMPANY

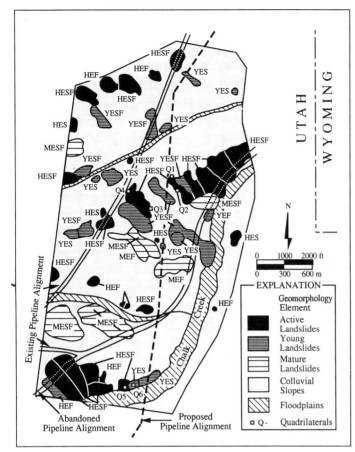

FIGURE 9-4 Engineering-geologic map of Thousand Peaks landslide area in northern Utah (modified from Keaton et al. 1992). See Table 9-2 for definition of landslide abbreviations.
USED WITH PERMISSION OF AMERICAN SOCIETY OF CIVIL ENGINEERS

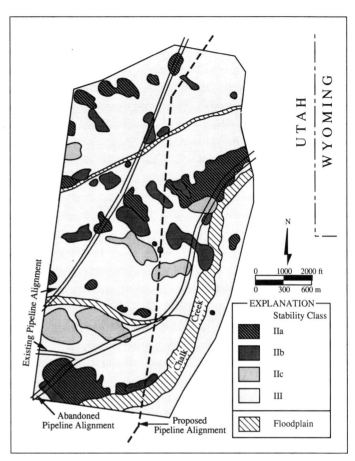

FIGURE 9-5 Stability classification of Thousand Peaks landslide area in northern Utah (modified from Keaton et al. 1992). See Table 9-6 for definition of stability classes.
USED WITH PERMISSION OFAMERICAN SOCIETY OF CIVIL ENGINEERS

the feature. Cracks in pavement, foundations, and other brittle materials can support inferences about the stress produced by movement of the landslide. The timing of breakage of water lines, electrical cables, and similar utilities can suggest the sequence of deformation before field observations or supplement observations of continuing movement. Measuring the tilt of structures assumed to be vertical or horizontal before movement can give an idea of the amount of displacement on certain parts of the landslide.

Internal features, such as those described in Chapters 3 and 8, should be documented by the geologist when and where observed so they may be used to interpret the subsurface conditions. The geometry and nature of the sliding surface are among the most important of the subsurface conditions in landslide evaluations. Surface measurements can be employed to estimate the shape of

the slip surface (Carter and Bentley 1985). A series of lines is projected through stations used to construct a topographic cross section from the main scarp to the toe of the landslide. By graphical representation, the lines define the probable slip surface. Hutchinson (1983) noted several other techniques using surface observations to infer the slip surface and related subsurface movement. Seismic-refraction and electrical-resistivity techniques (Carroll et al. 1968; Miller et al. 1980; Cummings and Clark 1988; Palmer and Weisgarber 1988) and acoustic-emission and electromagnetic methods (McCann and Forster 1990) have been used in landslide investigations. It probably will be necessary to interpret the results obtained along a number of geophysical lines and to incorporate surface observations of ground cracks and bedrock exposures. Landslide deposits commonly are extremely variable, resulting in severe

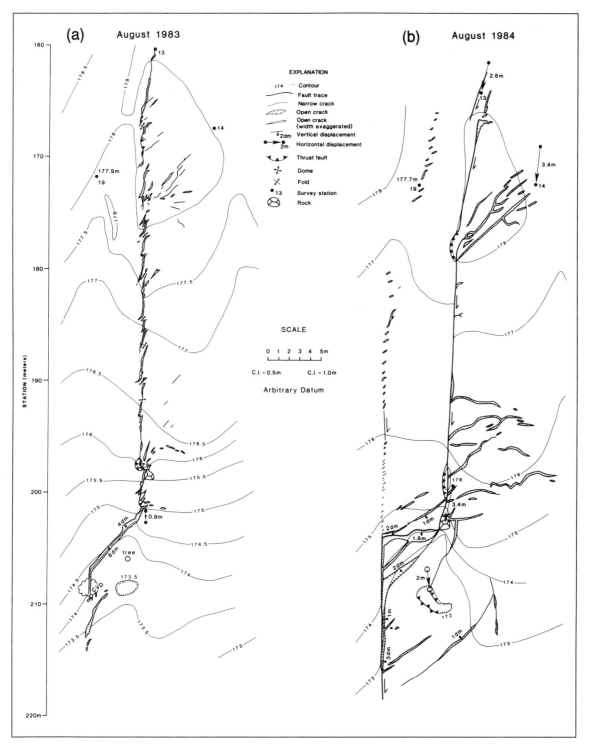

FIGURE 9-6
Plane-table maps of part of right flank of Aspen Grove landslide, Utah, (a) in 1983 and (b) in 1984 (Fleming and Johnson 1989). Scale is for reference to specific landslide features. Mapping was done in field at scale of 1:200. Note evolution of en echelon cracks into through-going slip planes.
REPRINTED WITH PERMISSION OF ELSEVIER SCIENCE PUBLISHERS

energy attenuation and complex arrival times in seismic-refraction surveys and complicated patterns in electrical-resistivity soundings.

Landslide features become modified with age. Active landslides have sharp, well-defined surface features, whereas landslides that have been stable for tens of thousands of years have features that are subdued and poorly defined. The changes of landslide features from sharp and well-defined to subdued and poorly defined were incorporated into an age classification by McCalpin (1984), as shown in Table 9-1, for the Rocky Mountains of western

North America. The key features are the main scarp, lateral flanks, internal morphology, vegetation, and toe relationships. The rate of change of landslide features in climates other than that of the Rocky Mountains has not been documented, and the estimated age of most recent movement shown in Table 9-1 may not be valid for other climates. However, by intuition, the general sequence of changes must occur in all climates.

A classification system based on activity, degree of certainty of identification of the slide boundaries, and the dominant type of slide movement was developed by Wieczorek (1984). McCalpin's age classification and Wieczorek's certainty and type of movement classification were combined for this chapter into the Unified Landslide Classification System, shown in Table 9-2. Changes in typical features for different ages of landslides are shown in Figure 9-7. An example of a map made with this system is presented in Figure 9-8.

The reader is cautioned that these classifications are different from the system proposed by the UNESCO Working Party on the World Landslide Inventory (WP/WLI 1990, 1991, 1993a, b) that is described at length in Chapter 3 of this report. A major source of confusion may result because all classification systems use similar, even identical, terminology with different meanings. The basis of all terms used must always be clearly defined.

3.2.2 Surficial Deposits

Surficial deposits must be mapped in terms that will be meaningful to the design engineer. The Genesis-Lithology-Qualifier (GLQ) System of engineering-geologic mapping symbols (Galster 1977; Keaton 1984; Compton 1985) provides a useful method for accomplishing this; it is proposed here that the name of this system be changed to the Unified Engineering Geology Mapping System.

Table 9-1
Age Classification of Most Recent Activity for Landslides in Rocky Mountain–Type Climate (modified from McCalpin 1984)

ACTIVITY STATE	MAIN SCARP	LATERAL FLANKS	INTERNAL MORPHOLOGY	VEGETATION	TOE RELATIONSHIPS	ESTIMATED AGE (YEARS)
Active, reactivated, or suspended; dormant-historic	Sharp; unvegetated	Sharp; unvegetated; streams at edge	Undrained depressions; hummocky topography; angular blocks separated by scarps	Absent or sparse on lateral and internal scarps; trees tilted and/or bent	Main valley stream pushed by landslide; floodplain covered by debris; lake may be present	< 100 (historic)
Dormant-young	Sharp; partly vegetated	Sharp; partly vegetated; small tributaries to lateral streams	Undrained and drained depressions; hummocky topography; internal cracks vegetated	Younger or different type or density than adjacent terrain; older tree trunks may be bent	Same as for active class but toe may be modified by modern stream	100 to 5,000 (Late Holocene)
Dormant-mature	Smooth; vegetated	Smooth; vegetated; tributaries extend onto body of slide	Smooth, rolling topography; disturbed internal drainage network	Different type or density than adjacent terrain but same age	Terraces covered by slide debris; modern stream not constricted but wider upstream floodplain	5,000 to 10,000 (Early Holocene)
Dormant-old or relict	Dissected; vegetated	Vague lateral margins; no lateral drainage	Smooth, undulating topography; normal stream pattern	Same age, type, and density as adjacent terrain	Terraces cut into slide debris; uniform modern floodplain	> 10,000 (Late Pleistocene)

NOTE: See Chapter 3 for definitions of terms. Activity states dormant-stabilized and dormant-abandoned may have features of any age classification; the stabilized and abandoned states must be interpreted from other conditions.

Table 9-2
Unified Landslide Classification System (modified from Wieczorek 1984)

AGE OF MOST RECENT ACTIVITY[a]		DOMINANT MATERIAL[b]		DOMINANT TYPE OF SLOPE MOVEMENT[b]	
SYMBOL	DEFINITION	SYMBOL	DEFINITION	SYMBOL	DEFINITION
A	Active	R	Rock	L	Fall
R	Reactivated	S	Soil	T	Topple
S	Suspended	E	Earth	S	Slide
H	Dormant-historic	D	Debris	P	Spread
Y	Dormant-young			F	Flow
M	Dormant-mature				
O	Dormant-old				
T	Stabilized				
B	Abandoned				
L	Relict				

NOTE: See Chapter 3 for further definitions of terms. Landslides classified using this system are designated by one symbol from each group in the sequence activity-material-type. For example, MDS signifies a mature debris slide, HEF signifies a historic earth flow, and ARLS signifies an active rock fall that translated into a slide.
[a] Based on activity state in Table 3-2 and age classification in Table 9-1.
[b] Based on material and type in Table 3-2.

This system consists of a series of letters to indicate how the material was deposited (genesis) and its basic grain size (lithology). If additional information is needed to improve the understanding of the geology, qualifying features may be indicated by additional letters. The basic elements of the Unified Engineering Geology Mapping System are shown in Table 9-3.

3.2.3 Bedrock

Bedrock consists of the rock material itself and the discontinuities that cut through it. Bedrock must be mapped in terms that will be meaningful to the design engineer. The Unified Engineering Geology Mapping System, introduced in Section 3.2.2 and described in Table 9-3, and the Unified Rock Classification System (Williamson 1984), shown in Table 9-4, provide useful methods for accomplishing this purpose. The Unified Engineering Geology Mapping System uses a version of the conventional geologic shorthand consisting of two capital letters to denote bedrock type. The elements of the Unified Rock Classification System are degree of weathering, estimated strength, and estimated density.

In the Unified Rock Classification System, degrees of weathering are designated States 1 through 5. The degree of weathering is determined by examining intact rock fragments with the unaided eye and by simple strength tests. Rocks that do not reveal staining under hand lens examination are considered to be fresh (State 1). Rocks that do not appear stained to the unaided eye but do reveal stained areas under examination with a hand lens are considered to be slightly weathered (State 2). Rocks that are stained but cannot be broken by hand are considered to be moderately weathered (State 3). Rocks that can be broken by hand into gravel and larger fragments of rock in a soil matrix are considered to be severely weathered (State 4). Rocks that can be completely disaggregated into mineral grains are considered to be completely weathered (State 5).

In the Unified Rock Classification System, strength is also designated as one of five states. Strength is estimated with the aid of a ball-peen or geological hammer. A rock from which the ball-peen hammer rebounds without leaving a mark is much stronger than concrete and is considered to have very high strength (State 1). A rock that reacts elastically and on which a ragged pit is produced is stronger than concrete and is considered to have high strength (State 2). A rock that can be dented by a ball-peen hammer has an unconfined compressive strength approximately the same as concrete and is considered moderately strong (State 3). A rock that reacts plastically and on which a dent is produced surrounded by a sheared crater is not as strong as concrete and is considered to have low strength (State 4). A rock that can be broken by hand has very low strength (State 5).

(a)

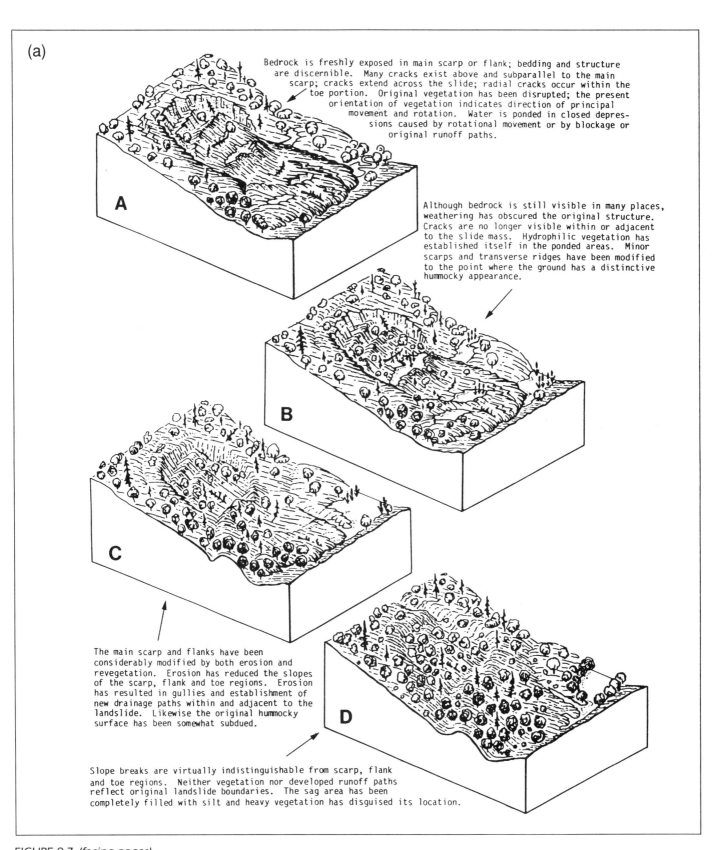

Bedrock is freshly exposed in main scarp or flank; bedding and structure are discernible. Many cracks exist above and subparallel to the main scarp; cracks extend across the slide; radial cracks occur within the toe portion. Original vegetation has been disrupted; the present orientation of vegetation indicates direction of principal movement and rotation. Water is ponded in closed depressions caused by rotational movement or by blockage or original runoff paths.

Although bedrock is still visible in many places, weathering has obscured the original structure. Cracks are no longer visible within or adjacent to the slide mass. Hydrophilic vegetation has established itself in the ponded areas. Minor scarps and transverse ridges have been modified to the point where the ground has a distinctive hummocky appearance.

The main scarp and flanks have been considerably modified by both erosion and revegetation. Erosion has reduced the slopes of the scarp, flank and toe regions. Erosion has resulted in gullies and establishment of new drainage paths within and adjacent to the landslide. Likewise the original hummocky surface has been somewhat subdued.

Slope breaks are virtually indistinguishable from scarp, flank and toe regions. Neither vegetation nor developed runoff paths reflect original landslide boundaries. The sag area has been completely filled with silt and heavy vegetation has disguised its location.

FIGURE 9-7 *(facing pages)*
Block diagrams of morphologic changes with time of idealized landslide *(a)* in humid climate (Wieczorek 1984) and *(b)* in arid or semiarid climate (modified from McCalpin 1984): A, active or recently active (dormant-historic) landslide features are sharply

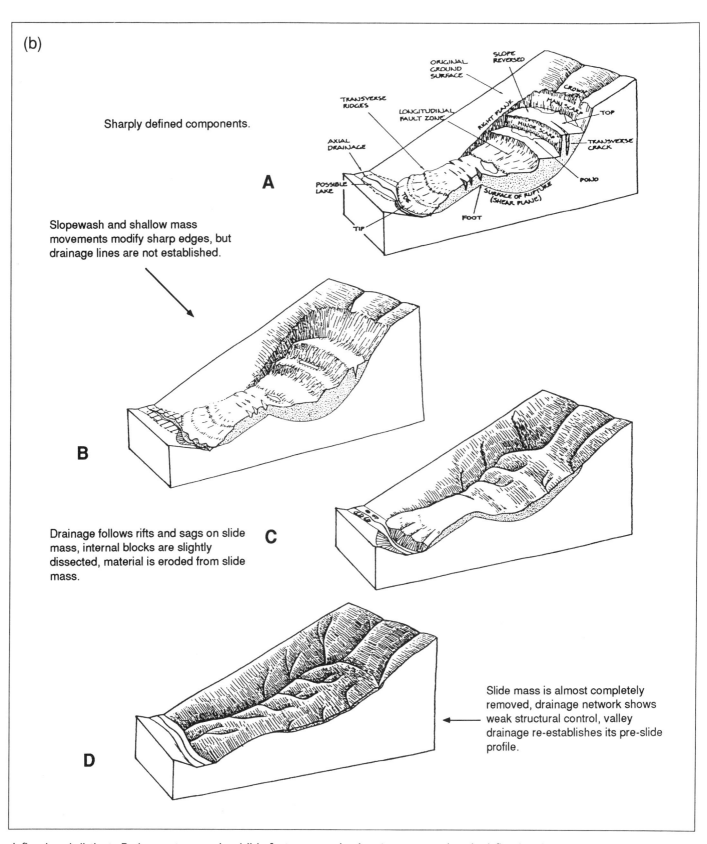

(b)

Sharply defined components.

Slopewash and shallow mass movements modify sharp edges, but drainage lines are not established.

Drainage follows rifts and sags on slide mass, internal blocks are slightly dissected, material is eroded from slide mass.

Slide mass is almost completely removed, drainage network shows weak structural control, valley drainage re-establishes its pre-slide profile.

A labels: ORIGINAL GROUND SURFACE, SLOPE REVERSED, CROWN, TRANSVERSE RIDGES, LONGITUDINAL FAULT ZONE, MAIN SCARP, TOP, RIGHT FLANK, MINOR SCARP, TRANSVERSE CRACK, AXIAL DRAINAGE, POSSIBLE LAKE, POND, SURFACE OF RUPTURE (SHEAR PLANE), FOOT, TIP

defined and distinct; *B, dormant-young landslide features remain clear but are not sharply defined owing to slope wash and shallow mass movements on steep scarps; C, dormant-mature landslide features are modified by surface drainage, internal erosion and deposition, and vegetation; D, dormant-old landslide features are weak and often subtle.*

Table 9-3
Basic Elements of Unified Engineering Geology Mapping System (modified from Keaton 1984 and Compton 1985)

	Symbol	Definition	Symbol	Definition	Symbol	Definition
Surficial Deposits						
Genetic	A	Alluvial	C	Colluvial	E	Eolian
	F	Fill	G	Glacial	L	Lacustrine
	M	Marine	R	Residual	V	Volcanic
Lithologic	c	Clay	m	Silt	s	Sand
	g	Gravel	k	Cobbles	b	Boulders
	r	Rock rubble	t	Trash or debris	e	Erratic blocks
	p	Peat	o	Organic material	d	Diatomaceous earth
Qualifier (deposits)						
Alluvial	(f)	Fan morphology	(fp)	Floodplain	(te)	Terrace
	(p)	Pediment	(df)	Debris fan		
Colluvial	(sw)	Slope wash	(ta)	Talus	(cr)	Creep deposits
Eolian	(d)	Dune morphology	(l)	Loess		
Fill	(u)	Uncompacted	(e)	Engineered		
Glacial	(t)	Till	(m)	Moraine	(o)	Outwash
	(es)	Esker	(k)	Kame	(ic)	Ice contact
Lacustrine and marine	(b)	Beach	(de)	Delta	(ma)	Marsh
	(tc)	Tide channel				
Residual	(sa)	Saprolite	(bh)	B horizon	(kh)	Calcic horizon
Volcanic	(af)	Air fall	(pf)	Pyroclastic flow	(s)	Surge
	(pc)	Pyroclastic cone	(l)	Lahar		
Bedrock Materials						
Sedimentary	SS	Sandstone	ST	Siltstone	CS	Claystone
	CG	Conglomerate	LS	Limestone	SH	Shale
Igneous	GR	Granite	AN	Andesite	BA	Basalt
	SY	Syenite	RH	Rhyolite	DI	Diorite
Metamorphic	QT	Quartzite	SC	Schist	GN	Gneiss
	SL	Slate	MA	Marble	SE	Serpentine

NOTE: Surficial deposits are designated by a composite symbol: Ab(c), where A is a genetic symbol, b is a lithologic symbol, and (c) is a qualifier symbol; for example, Csmg(sw) signifies colluvial slope wash composed of silty sand and gravel. This system was formerly called the Genesis-Lithology-Qualifier (GLQ) System because of the symbol for surficial deposits. Bedrock materials are designated by two capital letters; for example, LS signifies limestone bedrock. See Table 9-4 for additional bedrock classifications.

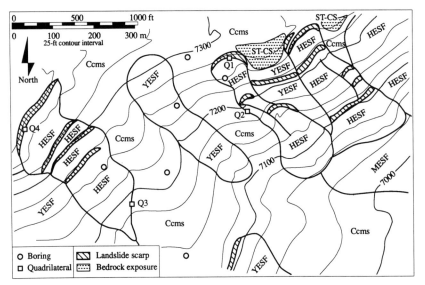

Rocks in strength States 4 and 5 should be treated as soil rather than rock in the engineering sense.

The estimates of rock density utilized by the Unified Rock Classification System can be determined rapidly for rock samples using Archimedes' principle, a spring-loaded "fish" scale, and a bucket of water. A rock sample is suspended on a string from the scale and weighed in air; the weight of the string is neglected. The same sample is then sub-

FIGURE 9-8 *(left)*
Detail of map of part of Thousand Peaks landslide area in northern Utah (scale 1:4,800). Landslides defined by Unified Landslide Classification System (Table 9-2); materials defined using Unified Engineering Geology Mapping System (Table 9-3). *Ccms* stands for colluvial deposits composed of silty and sandy clay; *ST-CS* stands for siltstone-claystone bedrock.

Table 9-4
Unified Rock Classification System (modified from Williamson 1984; Geological Society Engineering Working Party 1977; and Hoek and Bray 1977)

STATE	DEFINITION	DESCRIPTION
DEGREE OF WEATHERING		
1	None	No visible sign of rock material weathering
2	Slight	Discoloration on major discontinuity surfaces; rock material may be discolored and somewhat weaker than fresh rock
3	Moderate	Less than half of the rock material is present either as a continuous framework or as corestones
4	Severe	Most of rock material is decomposed, disintegrated to a soil, or both; original mass structure is largely intact
5	Complete	All rock material is converted to a soil; mass structure and material fabric are destroyed; a large change in volume has occurred, but soil has not been transported significantly
ESTIMATED STRENGTH		
1	Very high	Geological hammer rebounds; can be chipped with heavy hammer blows; unconfined compressive strength: $q_u > 100$ MPa
2	High	Geological hammer makes pits; cannot be scratched with knife blade; unconfined compressive strength: $50 \leq q_u < 100$ MPa
3	Moderate	Geological hammer makes dents; can be scratched with knife blade; unconfined compressive strength: $20 \leq q_u < 50$ MPa (range of concrete)
4	Low	Geological hammer makes craters; can be cut with knife blade; unconfined compressive strength: $5 \leq q_u < 20$ MPa
5	Very low	Moldable by hand; can be gouged with knife blade; unconfined compressive strength: $q_u < 5$ MPa (behaves like soil)
ESTIMATED DENSITY		
1	Very high	$D > 25$ kN/m^3
2	High	$23.5 \leq D < 25$ kN/m^3
3	Moderate	$22 \leq D < 23.5$ kN/m^3 (range of concrete)
4	Low	$20.5 \leq D < 22$ kN/m^3
5	Very low	$D < 20.5$ kN/m^3 (behaves like soil)

NOTE: Bedrock materials are designated by a composite symbol: AAbcd, where AA is rock type from bedrock materials section of Table 9-3, b is weathering, c is strength, and d is density. For example, SS324 signifies moderately weathered, high strength, low density sandstone.

merged in water and weighed again. The unit weight, or density, of the rock is computed by

$$\gamma_r = \frac{W_a}{W_a - W_w} \times D_w \qquad (9.1)$$

where

γ_r = density of rock sample,
D_w = unit weight of water,
W_a = weight of rock sample in air, and
W_w = weight of rock sample in water.

Rock defects or discontinuities, such as bedding planes, joints, and faults, are very important in rock-slope engineering. The most representative bedrock observations can be made on cut slopes rather than on natural exposures, on which the characteristics of bedding planes, joints, faults, and fractures often are masked by surface processes. The aspects of discontinuities that are important to rock-slope engineering are orientation, continuity (length), aperture, roughness, infilling materials, and water condition, as summarized in Table 9-5. These parameters are used in many rock mass classifications, as described by Bieniawski (1989).

Discontinuity orientation is expressed as strike and dip or dip direction and dip magnitude (Compton 1985; Hoek and Bray 1977). Strike and dip are measured with a Brunton compass, and dip direction and dip magnitude are measured with a Clar compass. These compasses are shown in Figure 9-9 and the measurement techniques are shown in Figure 9-10. More complete discussion of the Brunton compass was given by Compton (1985), and Hoek and Bray (1977) described the use of the Clar compass. The use of rock-structure data in rock-slope engineering is discussed in Chapter 14. Continuity of rock defects is an expression of the length or persistence of a joint or other feature across a rock exposure. A single long or through-going joint can contribute more to instability than a number of parallel but short joints.

FIGURE 9-9
Geologic compasses: *Right,* Brunton compass, also known as pocket transit; *left,* Clar compass, also known as Breithupt-Kässel rock structure compass. See Figure 9-10 for use of compasses; refer to Chapter 14 for analysis of rock-structure data.

Table 9-5
Rock Defect Data (modified from Bieniawski 1989)

	SYMBOL	DEFINITION
Defect type	J	Joint
	F	Fault
	B	Bedding
	T	Fracture
	S	Schistosity/foliation
Length	1	Very short: $L < 1$ m
	2	Short: $1 \leq L < 3$ m
	3	Moderate: $3 \leq L < 10$ m
	4	Long: $10 \leq L < 20$ m
	5	Very long: $L > 20$ m
Aperture	1	Closed:
	2	Small: $A < 2$ mm
	3	Moderate: $2 \leq A < 20$ mm
	4	Large: $20 \leq A < 100$ mm
	5	Very large: $A > 100$ mm
Roughness	1	Very rough: irregular, jagged
	2	Rough: ridges, ripples
	3	Moderate: undulations, minor steps
	4	Smooth: planar, minor undulations
	5	Very smooth: slickensided, polished
Infilling	1	None
	2	Hard infilling < 5 mm thick
	3	Hard infilling $\geq$ 5 mm thick
	4	Soft infilling < 5 mm thick
	5	Soft infilling $\geq$ 5 mm thick
Water Condition	1	Dry
	2	Damp
	3	Wet
	4	Dripping
	5	Flowing

NOTE: These data usually are recorded on a data collection table. Orientation is measured directly as strike and dip or dip direction and dip magnitude (see text). See Chapter 14 for discussion of rock mass properties.

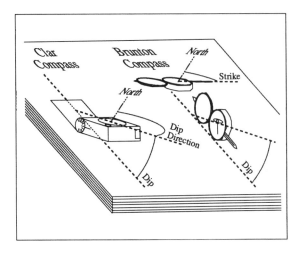

FIGURE 9-10
Use of geologic compasses. Brunton compass is used to measure strike and dip of discontinuity planes; two measurements are required, one using magnetic needle and one using inclinometer. Clar compass is used to measure dip direction and dip of discontinuity planes; one measurement provides both values. Hinge of Clar compass is color-coded inclinometer. If reference mark on inclinometer is on black dip angle, black end of magnetic needle indicates dip direction; if reference mark is on red dip angle, red end of needle indicates dip direction.

Aperture of rock defects is an expression of the openness or lack of contact of rock surfaces across the defects. Roughness is a measure of the irregularities on the defect surfaces. Roughness is important because of its contribution to the coefficient of friction and resistance to sliding. Joint roughness can be estimated using a fractal-dimension procedure (Carr and Warriner 1987) and shadow profilometry (Maerz et al. 1990). Infilling materials, such as clay, can contribute significantly to instability, as can the abundance of water along the defects.

Harp and Noble (1993) developed an engineering rock classification to evaluate seismic rock-fall susceptibility using a modification of the rock mass quality designation (Q system) of Barton et al. (1974). The rock mass quality classification for rock-fall susceptibility is

$$Q = \frac{115 - 3.3\,Jv}{Jn} \cdot \frac{Jr}{Ja} \cdot \frac{1}{AF} \qquad (9.2)$$

where

- Jv = total number of joints per cubic meter;
- Jn = joint set number, ranging from 0.5 for no or few joints to 20 for crushed rock;
- Jr = joint roughness number, ranging from 0.5 for slickensided planar joints to 4 for discontinuous joints;
- Ja = joint alteration number, ranging from 0.75 for tightly healed, hard, nonsoftening joints to 4 for low friction clay filling the joints; and
- AF = aperture factor, modified from the stress reduction factor of Barton et al. (1974), ranging from 1 for all tight joints to 15 for many joints open more than 20 cm.

Harp and Noble (1993) described a seismic rock-fall susceptibility rating as follows:

Rock Mass Quality	Category	Rating
$Q \le 1.41$	A	Highly susceptible
$1.41 < Q \le 2.83$	B	Susceptible
$2.83 < Q \le 3.87$	C	Moderately stable
$3.87 < Q$	D	Mostly stable

3.2.4 Surface-Water Features

Surface-water features of importance in slope-stability evaluations consist of streams and lakes as well as springs, seeps, marshes, and closed or nearly closed topographic depressions. Streams on some landslides may make abrupt changes in direction or gradient. Springs and seeps near the crest of a slope can supply recharge zones that provide groundwater to the unstable or potentially unstable slope. Springs and seeps near the base of a slope indicate discharge zones that can be helpful in projecting piezometric surfaces in the slope. Localized closed depressions on slopes usually are zones of groundwater recharge, particularly if ground cracks are present in or adjacent to them.

Culverts and other artificial features diverting surface water or affecting subsurface flow should be identified. Culverts may divert water from a greater area than would normally contribute surface flow to a landslide. Poorly constructed ditches provide depressions for water to pond and recharge groundwater. The compacted material forming the road prism or related retaining structures may act as barriers to downslope movement of groundwater, which may be closer to the ground surface in the slope immediately above these features.

3.2.5 Field-Developed Cross Sections

The topographic profile of a geologic cross section can be obtained directly from a topographic base map or it can be measured in the field during geologic data collection. The field-developed profile can be measured using several methods. The conventional method is to use surveying instruments and a two-person crew to collect topographic data along the desired line of section. This method does not produce a profile in the field for plotting geologic data.

Alternative single- or two-person methods utilize a 50-m-long tape measure, 2-m-long folding rulers, a hand level, and a Brunton compass (Williamson et al. 1981; Koler and Neal 1989). The tape measure is stretched on the ground surface in the line of profile, and its orientation is measured with the compass. The folding ruler is used to position the hand level at a known height above the end of the tape measure. The hand level is used to sight to the tape measure on the ground or to another folding ruler held by a member of the field crew, as shown in Figures 9-11(a) and (b), respectively. The slope distance and hand level height are recorded and can be plotted in the field. Pertinent geologic information along the tape measure, such as landslide ground cracks and

FIGURE 9-11
Methods of
surveying for
developing cross
sections in the field:
(a) single-person
method using hand
level, tape, and
folding ruler; (b)
two-person method
using hand level
and folding rulers;
and (c) incremental
measurements of
slope profile using a
circle level, or
Slope-a-Scope
method (Lips and
Keaton 1988). (See
text for details.)

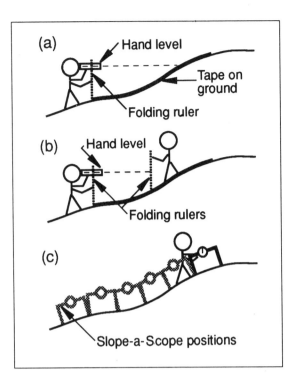

bedrock exposures, is recorded along with the topographic data. An optical range finder can be used in lieu of a tape measure.

A third single-person method uses a rigid two-dimensional frame equipped with a circle level—the Slope-a-Scope (Lips and Keaton 1988). The slope is measured in inclined increments equal to the length of the crossbar of the frame (commonly 1 or 2 m), as shown in Figure 9-11(c). The inclination and increment length are noted and can be plotted in the field. Geologic information is recorded while the profile data are being collected.

3.2.6 Classes of Slope Stability

Slopes range from apparently stable segments to actively moving landslides. Surface observation and geologic mapping provide the basis for interpreting degree of slope stability. A stability classification for slopes developed on the basis of ideas by Crozier (1984) is presented in Table 9-6. This classification is based on recurrence of movement, analogy to stable or unstable slopes, and results of stress analyses. The reader is once again cautioned that this classification differs from the classification system of the UNESCO Working Party on the World Landslide Inventory (WP/WLI 1990, 1991, 1993a, b) defined in Chapter 3. Active, reactivated, or suspended landslides in the Unified

Landslide Classification System (Tables 9-1 and 9-2) would be mapped as Ia, Ib, or Ic in the stability classification in Table 9-6. Dormant-historic, -young, -mature, and -old landslides would be mapped as IIa, IIb, IIc, and IId, respectively. Slopes that do not show evidence of prior slope movement but appear potentially unstable would be mapped as III and those that appear stable would be mapped as IV.

Other classifications of landslide hazards exist and have been summarized by Varnes (1984) and Hansen (1984). Such classifications are generalized and best suited for regional application rather than for specific landslide sites. Many classification schemes are based on multivariate regression of numerous attributes, including such factors as rock type, which must be assigned an ordinal rank value for inclusion in regression analyses (Jade and Sarkar 1993; Anbalagan 1992; Pachauri and Pant 1992).

3.3 Reconnaissance Instrumentation and Surveying

Reconnaissance instrumentation and surveying are intended to provide early quantitative information regarding landslide movement and piezometric level. Formal field instrumentation is described in Chapter 11. Reconnaissance instrumentation must be simple and easy to install. The instrumentation discussed in this section is restricted to open-standpipe piezometers. Other reconnaissance techniques, such as reference bench marks, aerial and terrestrial photography, and quadrilaterals, are directly related to surveying, which is discussed in Section 4.

Preliminary information on shallow piezometric surfaces can be collected at the reconnaissance level if the soil is relatively soft. The device used is an open-standpipe piezometer constructed of conventional 1.25-cm-diameter galvanized or black iron pipe. It is convenient to use three or four 1-m-long sections that can be connected by threaded couplings (Figure 9-12). A bolt is placed in the bottom end of the first pipe section and driven into the ground using a capped, 30-cm-long section of larger-diameter pipe that will slide over the piezometer pipe. A coupling or cap small enough to fit inside the driver pipe should be used on the piezometer pipe to prevent damage to the threads. Subsequent sections of piezometer pipe are

Table 9-6
Stability Classification of Slopes and Landslides (modified from Crozier 1984)

	CLASS	DESCRIPTION
I UNSTABLE SLOPES		
	Ia	Active landslides; material is currently moving, and landslide features are fresh and well defined
	Ib	Reactivated landslides; material is currently moving and represents renewed landslide activity; some landslide features are fresh and well defined; others may appear older
	Ic	Suspended landslides; slopes with evidence of landslide activity within the past year; landslide features are fresh and well defined
II SLOPES WITH INACTIVE LANDSLIDES		
	IIa	Dormant-historic landslides; slopes with evidence of previous landslide activity that have undergone most recent movement within the preceding 100 years (approximately historic time)
	IIb	Dormant-young landslides; slopes with evidence of previous landslide activity that have undergone most recent movement during an estimated period of 100 to 5,000 years before present (Late Holocene)
	IIc	Dormant-mature landslides; slopes with evidence of previous landslide activity that have undergone most recent movement during an estimated period of 5,000 to 10,000 years before present (Early Holocene)
	IId	Dormant-old landslides; slopes with evidence of previous landslide activity that have undergone most recent movement more than 10,000 years before present (Late Pleistocene)
III POTENTIALLY UNSTABLE SLOPES		
		Slopes that show no evidence of previous landslide activity but that are considered likely to develop landslides in the future; landslide potential is indicated by analysis or comparison with other slopes
IV APPARENTLY STABLE SLOPES		
	IVa	Stabilized landslides; slopes with evidence of previous landslide activity but that have been modified by artificial means to a stable state
	IVb	Abandoned landslides; slopes with evidence of previous landslide activity but that are stable because external forces causing movement are no longer active
	IVc	Relict landslides; slopes with evidence of previous landslide activity that clearly occurred under geomorphological or climatic conditions not currently present
	IVd	Stable slopes; slopes that show no evidence of previous landslide activity and that by analysis or comparison with other slopes are considered stable

NOTE: See Chapter 3 for definition of terms.

connected to the initial section as it is driven into the ground. The piezometer may be driven to a depth of 2.5 to 3.5 m with 0.5 m of pipe protruding above the ground surface. The piezometer is then pulled upward about 2 cm, creating a gap between the bolt and the bottom of the piezometer pipe to allow free entry of groundwater.

4. SURVEYS OF LANDSLIDE SITES

The topography at a landslide site often provides the first indications of potential instability and the degree to which the area has undergone landslide activity. As discussed in Chapters 7 and 8, initial reconnaissance studies frequently utilize existing maps or aerial photography to provide information concerning topography. In the case of larger landslides, these existing sources can be supplemented by larger-scale aerial photographs and topographic maps produced from them by photogrammetric methods to provide an overall view of the site conditions. However, because considerable topographic detail is required to locate many critical landslide elements, which in many environments may be masked by vegetation, detailed ground surveys generally must be included as a major component of landslide investigations.

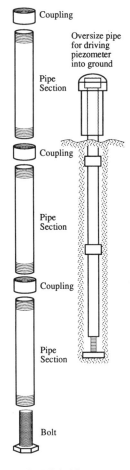

FIGURE 9-12
Reconnaissance-level open-standpipe piezometer. Sections of galvanized steel or black iron pipe connected with couplings are driven into ground with aid of oversized pipe section and cap. Open bottom of piezometer pipe is protected from clogging with soil by insertion of loose-fitting bolt. After piezometer is driven to desired depth (usually about 3 m), it is pulled upward about 2 cm to allow water to enter freely around bolt.

In this section, information provided in Chapters 4 and 5 of a previous landslide report (Sowers and Royster 1978; Wilson and Mikkelsen 1978) is expanded. Basic survey methods have been well described in numerous textbooks (e.g., Moffitt and Bouchard 1975); however, information on recent advances in survey techniques was obtained from appropriate experts (D. Little, personal communication, 1994, Virginia Department of Transportation, Richmond).

4.1 Purposes of Ground Surveys

In an active landslide area, surface movements are normally monitored to determine the extent of landslide activity and the rate of movement (Merriam 1960; Franklin and Denton 1973). Such monitoring requires the establishment of an accurate three-dimensional reference system, including identifiable reference stations, which can be established only by ground-based surveying. Therefore, ground surveys are required to

1. Establish the ground control for photogrammetric mapping and instrumentation,
2. Obtain topographic details where the ground surface is obscured by vegetation (these details are particularly important because of the accuracy required in mapping landslides), and
3. Establish a frame of reference against which movements of the ground surface can be compared.

Terzaghi stated:

> If a landslide comes as a surprise to the eyewitness, it would be more accurate to say that the observers failed to detect the phenomena which preceded the slide. (Terzaghi 1950, 110)

The implication is that the smallest possible movements should be measured at the earliest possible time. In the following sections each of these three requirements is discussed in turn.

4.1.1 Ground Control

The first requirement of ground surveys is a system of local bench marks that will remain stable during the course of the investigation and as far into the future as movements will be observed. These bench marks must be located far enough outside the suspected zone of movement that they will not be affected by any movements. Ultimately, the bench marks should be related to a geographic reference, such as control monuments of federal and state survey systems or latitude and longitude. The Global Positioning System can be useful for locating bench marks, particularly in remote areas. However, for convenience, a subsystem of local bench marks should be established close enough to the zone of movement so that they can be used as ready references for continuing surveys. At least two monuments of position and elevation should be established on each side of the zone of suspected movement. As indicated in Figure 9-13, these monuments should be as close as possible to the movement zone but not influenced by future enlargement of the landslide. Experience suggests that the distance from a bench mark to the closest point of known movement should be at least 25 percent of the width of the landslide zone. In areas with previous landslides, the minimum distance may be greater. In mountainous areas, adequate outcrops of bedrock sometimes can be found uphill or downhill from the landslide; in areas of thick soil, deep-seated bench marks may be necessary.

The bench marks should be tied together by triangulation and precise leveling traverses. With sufficient bench marks, movement by any one can be detected by changes in the control network. Intermediate or temporary bench marks are some-

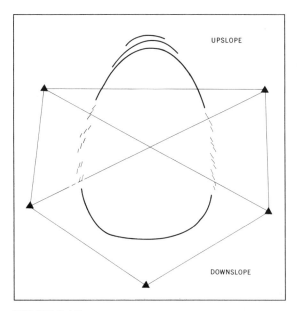

FIGURE 9-13
Bench marks and triangulation leveling network (Sowers and Royster 1978).

times established closer to the zone of movement for use in the more frequent surveys of the landslide area. However, the locations of these bench marks should be determined relative to the permanent monument grid each time they are used.

The direct surface measurements described above supplement and serve as a basis for the more sophisticated instrument systems that are used to determine the at-depth movements (see Chapter 11). Surveying must be used to accurately define the surface location and elevation of the reference points of these instrumentation systems each time observations are made.

4.1.2 Topographic Details

Aerial photographs may not provide sufficiently accurate or detailed topographic information for landslide studies because vegetation obscures the ground surface or because the important landslide features cannot be identified. Therefore, detailed on-site mapping is necessary to define major features, such as scarps, cracks, bulges, and areas of disrupted topography (Figure 9-14).

Because of the changing nature of landslides, surface surveys conducted after aerial photographs have been taken may not correspond directly to

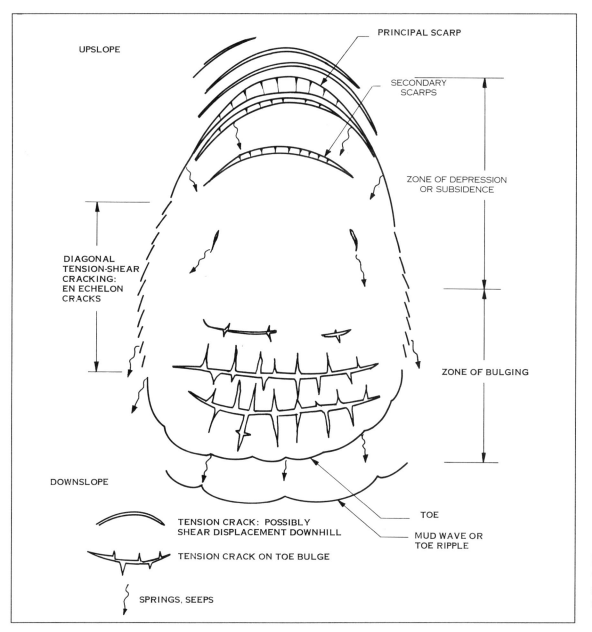

FIGURE 9-14
Cracks, bulges, scarps, and springs (Sowers and Royster 1978).

the features shown on the photographs. It may not be possible to obtain a precise correlation between the surface topography determined on the ground over a period of days or weeks and that obtained at a single instant from aerial surveys. Differences between ground surveys and aerial surveys should be expected, and these differences are particularly useful in understanding landslide deformation. Topographic maps should include the accurate representation of landslide features belonging to two classes: (*a*) cracks and bulges and (*b*) springs and seeps.

4.1.2.1 Cracks and Bulges

Although many cracks and bulges, as well as other minor topographic details, can be identified in aerial photographs, their full extent seldom can be determined unless the photographs are taken with an unusually high degree of resolution in vegetation-free areas (see Chapter 8). Therefore, independent surveys of cracks and bulges should be made by surface methods. Developing cracks, particularly the ends, often are obscured by grass, leaves, and root mats; these cracks should be carefully uncovered so that their total extent can be mapped. Hidden cracks can be identified by subtle changes in leaf mold patterns, torn shrubs, and distorted trees and tree-root systems. Boulder alignments or sliding trajectories should be noted. Cracks should be staked on both sides, and all stakes should be referenced to the movement monitoring system because the entire crack system shifts with continuing landslide movement.

4.1.2.2 Springs and Seeps

Springs and seeps are the ultimate areas of discharge for water-bearing strata and cracks and thus are indicators of the water flow paths that influence soil and rock stability. Because seeps often follow cracks that have been opened by soil or rock movement, they can sometimes be traced to sources uphill. The points of disappearance of surface runoff into cracks and fissures should be mapped also. Seeps, springs, and points of water loss change with rainfall, snowmelt, and ground movement. Thus, meaningful data on their location and shifts cannot be obtained by a single survey or at regular intervals of observation. Instead, they should be located during and shortly after periods of intense rainfall or snowmelt and after episodes of significant movement.

4.1.3 Movement Grids and Traverses

The continuing movement of a landslide can be measured by a system of grids or traverses across the landslide area (Figures 9-15 and 9-16). Typically, a series of lines more or less perpendicular to the axis of the landslide and spaced 15 to 30 m apart with stakes at intervals of 15 to 30 m should be maintained and referenced to the control bench marks. Grids should be laid out so that the reference points are aligned with trajectories of maximum slope or apparent continuing movement. In addition, where soil and rock weaknesses cause secondary movements that are skewed to the major landslide, intermediate points should be established. For small landslides or widely spaced areas of suspected movements, single traverse lines of reference often are used (Figure 9-16).

Line-of-sight monuments can be established with end monuments on stable ground and intermediate points on the moving mass in areas where vegetation does not obscure visibility. The monuments are established in a line and smooth or polished metal plates are fixed on top of the intermediate points. A line perpendicular to the line of sight is scribed onto each plate. A transit is located on or over one of the stable monuments and the other stable monument is used to set the horizontal angle of the instrument. Each of the intermediate points is sighted, and a line is scribed on its plate. Movement from one measurement to the next is simply measured at each plate with the aid of a machinist's rule. Scribed lines are visible in Figure 9-17.

Appropriate location flags or markers should be placed nearby so that the staked points can be found despite severe movement. The elevation and coordinates of each point should be determined on the traverse or reference grids by periodic surveys. In areas where highly irregular topography suggests rapid differences in movement from one point to another, reference points should be spaced more closely regardless of any predetermined grid pattern. Such closely spaced stakes help to define the lateral limits of the landslide as well as the direction of movement of localized tongues within the landslide. This is particularly important in the later stages of movement because secondary movements often develop as a result of weakening of the displacing materials. Depending on the rate of movement, these grid points should be checked at

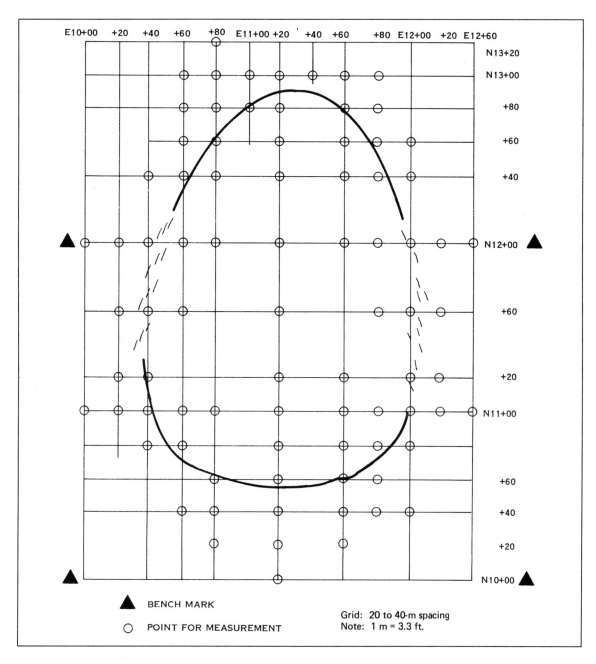

FIGURE 9-15
Observation grid
(Sowers and Royster
1978).

intervals ranging from a few days to several months. In addition, they should be observed after periods of unusual environmental conditions, such as snowmelt, high rainfall, or marked temperature changes. In this way any relation between landslide movement and climatic changes can be established.

4.2 Surveying Methods

Surveying is an integral part of the broader surface observations and geologic mapping activities during landslide investigations. The mechanics of conducting an accurate survey should not be allowed to obstruct general observations of topographic features. As discussed in the following section, general observations often provide important information and should be incorporated into the surveyor's notes and reports.

Conventional surveying methods commonly are adapted to the needs of landslide investigators. Table 9-7 is a summary of the characteristics, advantages, and limitations of several survey methods. Conventional survey methods, which are discussed further in Section 4.2.2, are sometimes supple-

FIGURE 9-16
Observation
traverses for rough
topography or less
important landslides
(Sowers and Royster
1978).

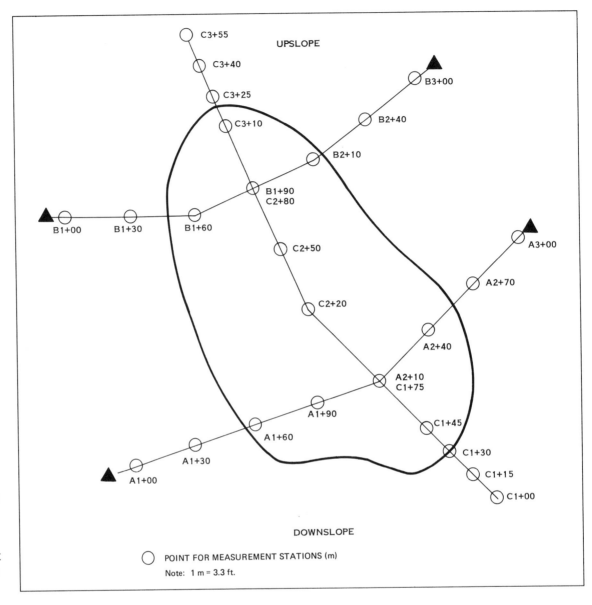

FIGURE 9-16
Observation
traverses for rough
topography or less
important landslides
(Sowers and Royster
1978).

FIGURE 9-17
(below)
Smooth metal plate
on line-of-sight
monument with
scribed lines
showing movement
between successive
observations.

mented by other, more specialized survey methods, which are discussed further in Section 4.2.3.

4.2.1 General Observations

When surface cracking is not apparent, detection of small surface movements requires a trained observer. Horizontal stretching results in localized instability of individual rocks that can be detected by small gaps between the soil and the rocks or exposure of different coloring or surface weathering patterns. Trees whose trunks are inclined at the base but change to vertical orientation a meter or so above the ground (J-shaped trees) typically indicate snow creep, which affects only small-diameter

Table 9-7
Comparison of Ground Survey Methods (modified from Wilson and Mikkelsen 1978 and Cording et al. 1975)

METHOD	RANGE	ACCURACY	ADVANTAGES	LIMITATIONS	RELIABILITY
Compass and pace	Variable	Low; provides only approximate values	Rapid; requires no equipment; may be useful in establishing overall dimensions of landslide	Reasonable initial estimates on uniform terrain without obstacles	Varies with experience of personnel and roughness of terrain
Compass, hand level and tape	Variable	Moderate; provides approximate values	Rapid; moderate accuracy; produces a map or section	Reasonable estimates on rough terrain	Varies with experience of personnel and roughness of terrain
Plane table and alidade	5 to 500 m	1:100	Relatively rapid; moderate accuracy; produces a map	Awkward in rough or steep terrain	Moderate; errors may be due to instrument, drafting, and instability of plane table
Transit and stadia	5 to 500 m	1:200 to 1:500	Relatively rapid; moderate accuracy	Complex corrections needed for measurements along steeply inclined directions	Moderate to good; may be lower in vegetated and rough areas
Transit and subtense bar	50 to 500 m	1:800 at 500 m; 1:8,000 at 50 m	Relatively rapid; most accurate of optical methods; can measure inclined distances	Careful orientation of subtense bar required; may be difficult to use in rough terrain	Good
Direct measurement by tape or chain Ordinary survey	Variable	1:5,000 to 1:10,000	Simple and inexpensive; provides direct observation	Requires clear, relatively flat surface between measured points and stable reference monuments	Excellent
Precise survey	Variable	1:20,000 to 1:20,0000	Relatively simple and inexpensive; provides direct observation	Corrections for temperature and slope must be applied and standard chain tension used	Excellent
Electronic distance measurement (EDM)	20 to 3000 m	1:50,000 to 1:30,0000	Precise, long-range, and rapid; usable over rough terrain	Accuracy is influenced by atmospheric conditions; accuracy over shorter distances (30 to 90 m) is less for most instruments	Good
Total station	1.5 to 3000 m	5 mm minimum error; 1:300,000 over longer distances	Precise and rapid; reduces computational effort and errors; provides digital data; usable over rough terrain	Accuracy may be influenced by atmospheric conditions	Good to excellent
Global Positioning System (GPS)	1.5 m to 40 km	100 m with a single receiver; 1:300,000 with two or more receivers recording four or more satellites; 0.3-cm minimum error	Precise absolute horizontal position; possible precise vertical position	Accurate measurements require 45 to 60 min continuous operation; elevations require receiver dedicated to bench mark; tree limbs over antenna interfere with reception; possible backscatter	Excellent

continued on next page

Table 9-7 (continued)

METHOD	RANGE	ACCURACY	ADVANTAGES	LIMITATIONS	RELIABILITY
Optical leveling Ordinary (second or third order)	Variable	3 to 6 mm (vertical)	Simple and fast, especially with modern self-leveling instruments	Limited precision; requires good bench marks	Excellent
Precise (first order)	Variable	0.6 to 1.2 mm (vertical)	More precise than ordinary leveling	Requires good bench marks and reference points; careful adherence to standard procedures	Excellent
Offsets from baseline Theodolite and scale	0 to 1.5 m	0.6 to 1.5 mm	Simple; provides direct observation	Requires baseline unaffected by ground movements and good monuments; accuracy can be improved by using a target with a vernier and repeating sighting from opposite end of baseline	Excellent
Laser and photocell detector	0 to 1.5 mm	1.5 mm	More rapid than theodolite-and-scale method	Seriously affected by atmospheric conditions	Good
Triangulation	Variable	0.6 to 12 mm	Usable when direct measurements are impossible; useful in tying into points outside immediate area	Requires precise measurement of base distance and angles; requires good reference monuments	Good
Terrestrial photogrammetry	Variable	1:5,000 to 1:50,000	Records hundreds of potential movements at one time for determining overall displacement pattern	Limited by weather conditions	Good

trees, rather than old landslide movement. Trees with straight trunks that are inclined in different directions within an area ("jackstrawed" trees or "drunken" forests) indicate recent movement. Trees with completely curved trunks (C-shaped trees) are found on slopes with long histories of continuing minor movement. Observations of trees should not be used as the sole basis for identifying landslide movement because other phenomena may produce similar effects (DeGraff and Agard 1984).

Cracks covered with leaves, surface litter, or duff can be detected by an experienced observer walking over the area and noting firmness of foot support. Livestock and other animals may avoid grazing or browsing in an active landslide area because of uncertain support or hidden fissures. Small openings on the downhill sides of structures or next to tree trunks may indicate incipient move-

ment. Overly taut or excessively sagging utility lines, misalignment of fence posts or utility poles, or distress to pavement are excellent indicators of ground movements. Such movements, when accurately monitored, serve as important tools in assessing the potential hazard to transportation facilities, nearby structures, and the public.

4.2.2 Conventional Surveying

The principles of surveying have been described extensively in numerous textbooks (e.g., Moffitt and Bouchard 1975) and thus will not be discussed here. One of the basic operations of surveying is the determination of the distance between two points on the surface of the earth. In surveys of areas of limited extent, which is the case for most landslides, the distance between two points at different elevations is reduced to an equivalent horizontal distance either by the procedure used to make the measurement or by computation of the horizontal distance from the measured slope distance and inclination. Vertical distances are independently computed and recorded. In the case of landslides, both vertical and horizontal distances, and rates of change in these distances, provide important information needed to evaluate landslide mechanisms.

As noted in Table 9-7, many surveying techniques may be used to determine distances and inclinations for computing horizontal and vertical distances. Some simple methods, such as pacing to determine approximate distances, are of limited use in most landslide investigations.

4.2.2.1 Tacheometry

A series of methods has been developed for indirectly measuring distances using optical surveying instruments in conjunction with measuring bars or rods. These methods may be referred to collectively as *tacheometry* (Moffitt and Bouchard 1975, 14). They include plane table and alidade surveys, transit and stadia surveys, and transit and subtense bar surveys (see Table 9-7).

These measurements are performed rapidly and comparatively well over rough, uneven ground. They may be sufficiently accurate for many landslide investigations, but their accuracy is considerably less than that of other techniques involving direct distance measurement by tapes or chains or the determination of distances by electronic means. Thus, tacheometry is often used to provide rapid supplementary data concerning the locations of intermediate points of landslide features, particularly when field-developed maps are needed for geologic descriptions. A somewhat more accurate tacheometry measurement involves the use of a transit and a short horizontal baseline, referred to as a subtense bar. Distances are determined by using a precise transit to measure the small horizontal angle subtended by the bar.

Tacheometric techniques remain a viable and economical surveying solution for landslide investigations in many locations, and the required surveying equipment is commonly available. However, whenever possible, tacheometric methods should be replaced with the more accurate and rapid distance measurements provided by electronic distance measurement (EDM) equipment or by the even more modern total station

surveying methods. These approaches are discussed further in Sections 4.2.2.3 and 4.2.2.4.

4.2.2.2 Transit and Tape Methods

Optical instrument surveys and tape measurements are commonly used to determine lateral and vertical positions of points accurately. Bench marks and transit stations located on stable ground provide the basis from which subsequent movements of monuments can be determined optically and by tape measurement. As shown in Figure 9-18, transit lines can be established so that the vertical and horizontal displacements at the center and toe of the landslide can be observed. Lateral motions can be detected by transit and tape measurements from each monument. When a tension crack has opened above the top of a landslide, simple daily measurements across the crack can be made between two markers, such as stakes or pieces of concrete-reinforcing steel driven into the ground. In many cases the outer limit of the ground movements is not known, and establishing instrument setups on stable ground may be a problem.

Various techniques and accuracies achieved in optical leveling, offset measurements from transit lines, chaining distances, and triangulation have been discussed extensively in the literature (Gould and Dunnicliff 1971; British Geotechnical Society 1974), particularly for dams, embankments, and buildings. Although conventional surveys, particularly higher-order surveys, can define the area of movement, more accurate measurements may be required in many cases.

4.2.2.3 Electronic Distance Measurement Equipment

Electronic distance measurement (EDM) devices have proved particularly suitable for rugged terrain; they are more accurate and much faster than ordinary surveying techniques and require fewer personnel (Dallaire 1974). Lightweight EDM instruments can be used efficiently under ideal conditions for distances as short as 20 m and as long as 3 km; errors are as small as 3 mm (St. John and Thomas 1970; Kern and Company Ltd. 1974). Larger instruments that employ light waves or microwaves can be used at much longer distances. The accuracy of EDMs is influenced by weather and atmospheric conditions; comparative readings with three different instruments were described by Penman and Charles (1974).

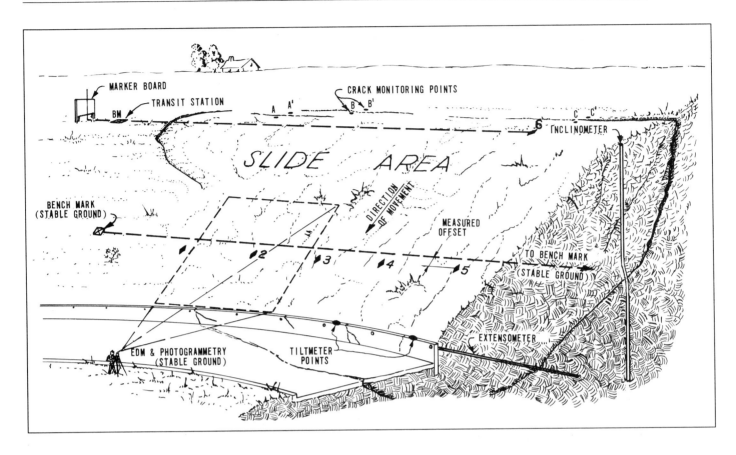

FIGURE 9-18
Movement
measurements
in typical
landslide area
(Wilson and
Mikkelsen 1978).

EDM equipment can be used to monitor large landslides with large movements and provide a rapid way to survey many points on the mass from a single, readily accessible location. An example of such an installation involves a reactivated ancient landslide in the state of Washington along the Columbia River where the active landslide is more than 0.6 km wide and 5 km long. Movements vary from 1 to 9 m per year, and the rates of movement depend on the time of year and rainfall intensity and duration. A permanent station, readily accessible all year, has been set up on the side of the river opposite the landslide and distance readings are taken monthly to 14 points located on the landslide and 2 points located on stable ground outside the landslide boundaries. The distances involved vary from about 1.5 to 6 km. Figure 9-19(a) shows the movements (changes in distance) during a one-year period (1972–1973) for two selected points at this Columbia River landslide based on monthly readings recorded by an EDM instrument. At the end of the year, the points were resurveyed by triangulation. The discrepancy is about 10 cm, which, although larger than anticipated, is quite satisfactory considering the to-

tal movements. Figure 9-19(b) shows recorded changes for two points believed to be on stable ground. The variation of monthly readings is seen to be 60 mm and the variation is no greater for a 4765-m distance than for a 1844-m distance.

4.2.2.4 Total Station Equipment

Total station survey systems have improved surveying practices to a great extent. These devices can measure vertical and horizontal positions within a three-dimensional coordinate framework having x-, y-, and z-axes (east, north, and elevation). They have proved particularly suitable for rugged terrain, and building on the experiences gained with the earlier EDM systems, they perform even more accurately and faster than these earlier systems and require fewer personnel.

Current total station survey systems allow the surveyor the flexibility not only of measuring the horizontal and vertical positions of any point with a high degree of accuracy, but also of recording all readings into hand-held or instrument-located data collection devices. Field data can be downloaded from these devices to computers, where the data can be processed, printed in formal

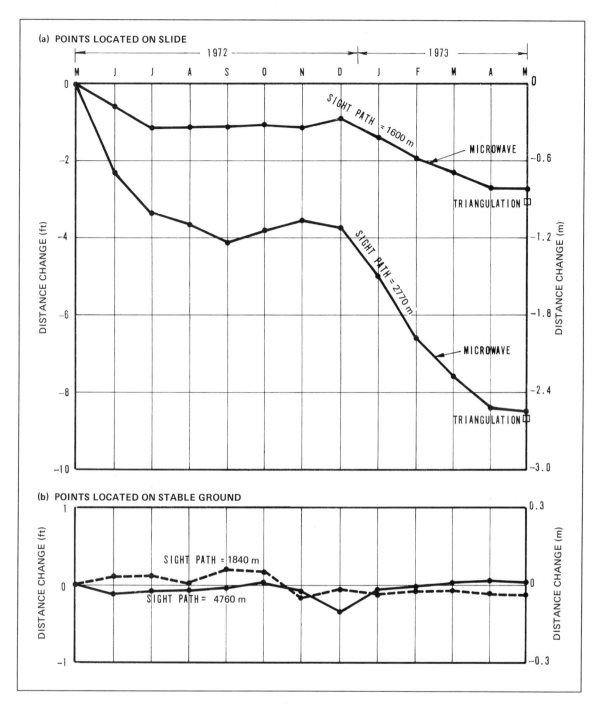

(a) POINTS LOCATED ON SLIDE

(b) POINTS LOCATED ON STABLE GROUND

FIGURE 9-19
Microwave
measurements of
landslide
movements along
Columbia River,
Washington State
(Wilson and
Mikkelsen 1978).

records, and plotted as maps. In some cases data reduction and checking can be performed in the field while the survey continues, allowing inaccuracies to be identified and corrected immediately and saving remobilization costs.

Field-developed maps can be produced with total station equipment, a plane table, and a three-person crew. A geologist positions the retroprism pole on features of significance. The surveyor

determines the position of the points, and another person plots the points on the plane table. Portable radios facilitate communication of geologic information to be recorded on the map.

Extreme care must be taken concerning the calibration and adjustment of all equipment involved in total station surveys. The following equipment must be checked carefully to confirm calibration or proper adjustment:

- Total station,
- Tripods,
- Retroprisms,
- Prism poles, and
- Optical plummet tribrachs.

As with EDM equipment, the accuracy of total station surveys is influenced by weather and atmospheric conditions. Lightweight total stations can be used efficiently for distances as short as 1.5 m and as long as 3 km. Errors may be as small as 30 mm over a 3-m distance; however, these instruments generally have a minimum resolvable distance accuracy on the order of 5 mm. Thus for short distances, the possible error in the distance measurement will always be at least 5 mm. For many landslide investigations, this accuracy is much better than the many potential sources of error in locating and reestablishing reference points.

Results of repeated surveys can be plotted effectively as direction and cumulative magnitude of movement, as shown in Figure 9-20. The plot for

FIGURE 9-20
Plots of direction and cumulative magnitude of monitor point movement. Point A: moving point with consistent direction and uniform trend of magnitude. Point B: stable point with random direction and small incremental magnitude that is caused by error of surveying technique.

Point A shows consistent direction of movement and uniform, incremental increase in magnitude of the displacement vector. Point B shows random direction and small incremental magnitude of cumulative movement; it is a stable point and documents the survey's accuracy.

4.2.3 Other Surveying Techniques

In some landslide investigations the conventional surveying approaches discussed in the previous sections are supplemented by other, more specialized surveying techniques. These techniques include the use of lasers, methods involving aerial and terrestrial photogrammetry, techniques for monitoring internal deformation and the growth and movement of surface cracks, and the use of specialized equipment such as tiltmeters. In addition, videography and digital image analysis and the current Global Positioning System (GPS) are discussed in the following subsections. The use of specialized slope movement monitoring devices is discussed in Chapter 11.

These supplementary surveying methods continue to evolve rapidly. Some already are in fairly extensive use (e.g., EDM equipment and lasers), whereas others are in limited use or are in the developmental or experimental stage (e.g., terrestrial photogrammetry and videography). They will undoubtedly find increasing use in field measurements in the future.

4.2.3.1 Lasers

Laser instruments already are widely used for setting alignments, and they are well suited for setting a reference line for offset measurements to surface monuments. Laser beams are also used with some EDM instruments. It should be possible to measure offsets with errors no greater than 3 to 6 mm (Gould and Dunnicliff 1971).

Laser total station instruments are available in hand-held, monopod, and tripod formats (Laser Technology, Inc. 1992). These instruments are particularly well suited for rapid topographic profiling in locations too steep or dangerous for other methods. The laser total station is equipped with a fluxgate compass and tilt-angle sensor, in addition to a serial port for communicating data to other computer applications. The range of the laser is up to about 460 m to a target that is 20 percent reflective; to a retroprism the range is 12 200 m. Accuracies of

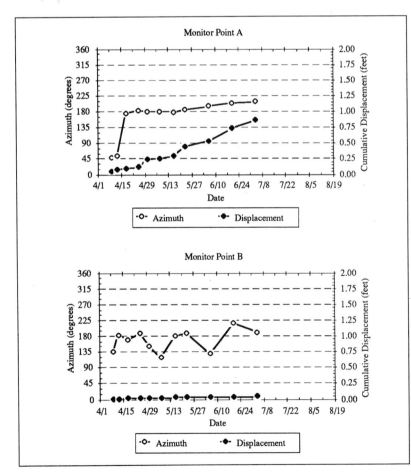

0.3 degree azimuth, 0.2 degree vertical angle, and 9.4 cm range can be obtained from units that weigh less than 3 kg and operate up to 8 hr on a Nicad battery.

4.2.3.2 Aerial and Terrestrial Photography and Photogrammetry

Important features of active landslides cannot be documented on aerial photographs taken before the slope movement; therefore, photographs taken at the beginning of an investigation of an active landslide permit documentation of current conditions for navigation as well as geologic interpretation. Photographs taken with a small-format (35-mm) hand-held camera from a fixed-wing aircraft during aerial reconnaissance can provide an excellent base for use on the ground (Wracker 1973). Stereoscopic photographs can be taken in this way in vertical or near-vertical and oblique orientations. Commercial rapid processing of color film is available in many cities. Thus, an aerial reconnaissance can be made and prints received for use in the field within a single day. An alternative to hand-held photography from a fixed-wing aircraft would be to contract with an aerial photography company to produce stereoscopic photographs for subsequent use on the project.

Prominent targets visible from the air or from a distance on the ground can be used as references for aerial or terrestrial photographs taken at some interval of days, weeks, or months. Suitable targets are circles or squares 30 cm to 1 m across with opposite quadrants painted red and white, as shown in Figure 9-21. If such targets are to be deployed from aircraft onto landslide surfaces, both sides of the targets should be painted so that it will not matter which side faces up. If a single photograph includes a stable bench mark and one or more landslide bench marks, the direction and magnitude of deformation can be estimated with the accuracy of the photographs. Oblique photographs, of course, show more complicated distortion than do vertical aerial photographs.

Terrestrial photographs can serve as bench marks for qualitative changes in landslides. Several examples of photographs documenting crack propagation in a large landslide may be found in a paper by Fleming and Johnson (1989). Simple photographic bench marks involve taking a photograph next to some prominent feature that can be found readily on subsequent visits to the

site. The feature should be identified in the initial photograph for targeting in later photographs to ensure comparable views. Showing some object or measure, such as a folding ruler, for scale in the photograph is recommended.

For greater precision the location of the photographic bench mark can be established for later use by driving a metal or plastic stake into the ground. The height of the camera above the stake should be noted along with the azimuth and inclination of the view. A tripod can be used to simplify establishing these measurements and setting up for later photographs. More precision is obtained by using the same camera and lens combinations. Terrestrial photographs are more usable when later views are taken at a similar illumination (time of day) as the initial ones. This ensures similar contrast and visibility of the features among different photographs. Malde (1973) offered additional ideas for using terrestrial photography for geologic bench marks.

Photogrammetric measurements of ground geometry can be made from oblique photographs obtained at the ground surface. For example, two or more permanent photography sites that overlook a landslide area can be used to document landslide movement through successive sets of stereoscopic photographs. Phototheodolites are used to take successive stereophotographs from these fixed stations; movements are identified in a stereocomparator, and accuracies of 6 to 9 mm have been reported (Moore 1973). Although the data reduction may be more complex than that for conventional aerial photogrammetric mapping, the technique is useful for determining movement of any selected points provided that they can be seen in the photograph. Terrestrial photogrammetry has been used in some cases to measure changing dam deflections (Moore 1973) and to monitor rock slopes in open-pit mines (Ross-Brown and Atkinson 1972), but no landslide measurements using this method have been found in the literature.

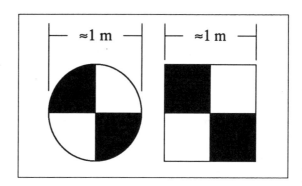

FIGURE 9-21
Simple reference targets for monitoring landslide displacements. Targets can be made of plywood or sheet metal and painted with bright colors such as red and white. Targets can be deployed rapidly for photographic documentation. It is wise to place two or three targets outside landslide boundary to document absolute landslide displacements.

FIGURE 9-22
(*below*)
Quadrilateral
configuration for
determining
displacement,
strain, and tilt
across landslide
flank. Initially
deployed at
positions *A, B, C,*
and *D*, quadrilateral
deforms to positions
A, B', C', and *D* at
second observation.
Quadrilateral
defines four
triangles; *ABC,*
ACD, ABD, and
BCD, which deform
to *AB'C', AC'D,*
AB'D, and *B'C'D.*
Each triangle is
analyzed as
conventional
three-point problem
for orientation.
Changes in position
of *B, C,* and *D* with
respect to *A* are
used to calculate
displacements.
Changes in leg
length and triangle
shape and area are
used to calculate
strains and tilts.

4.2.3.3 Quadrilaterals

Quadrilaterals provide a convenient method for measuring slope deformation (Baum et al. 1988). A quadrilateral consists of an array of four stakes initially in a nearly square configuration. Quadrilaterals that span the lateral flank of a landslide or are located within the body of a slide deform with landslide movement. Subsequent measurements of quadrilaterals are compared with initial and previous measurements; they are used to compute displacements across landslide boundaries and strains and tilts within the body of the landslide. The geometry of a quadrilateral straddling a landslide boundary is illustrated in Figure 9-22.

The measurements required for quadrilateral deformation analyses consist of the relative elevations of the four stakes, the six distances between pairs of stakes, and the azimuth of one leg of the quadrilateral. Computer programs for this purpose were written in BASIC by Johnson and Baum (1987). The measurements of the quadrilaterals are used to define four triangles. The shape and tilt of each triangle are computed, and changes are used to define landslide deformation.

The accuracy of displacement, strain, and tilt estimates is limited by measurement error. Quadrilaterals measured by tape and hand level require large movements before the differences are significant. Laser geodimeters or tape extensometers provide the most precise measurements but may exceed the intentions of reconnaissance instrumentation. Tops of steel rods can be machined to allow repeated positioning of tape measures (Figure 9-23), and a cap can be machined to fit over the steel rod with a groove so that the tape can be positioned for repeated measurements (Figure 9-24). A special sleeve-type anvil is required to allow the

FIGURE 9-23
Machined pin on steel rod used in quadrilateral.

FIGURE 9-24
Machined cap to fit over quadrilateral rods for repeated-measurement accuracy.

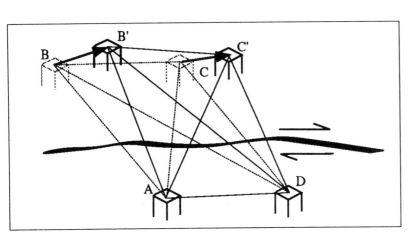

steel rods to be driven into the ground without damage to the machined pins. A tensioned steel tape corrected for temperature (Moffitt and Bouchard 1975) provides a suitable level of precision in distance measurements. Elevation differences can be determined precisely with a water level or manometer (Figure 9-25). Care must be exercised in selecting quadrilateral locations so that the relief within the quadrilateral does not prevent subsequent measurements.

4.2.3.4 Crack Monitoring

Monitoring the system of cracks found on the surface of a moving mass often is of critical importance for landslide investigation programs. Crack monitoring includes crack mapping and crack measurement or gauging.

1. *Crack Mapping:* Most earth movements are accompanied by cracking and bulging of the ground (Figure 9-14). Survey points, in addition to the predefined grid or traverse points (Figures 9-15 and 9-16), should be set on the more prominent of these features and in areas beyond them. Repeated, detailed mapping of areas of cracking serves to document the evolution of the cracks, which provides the basis for interpretation of the stresses responsible for them. An example of repeated, detailed mapping of an area of landslide cracks by plane table and alidade is shown in Figure 9-6.

2. *Crack Measurement:* Instrumentation of landslides is discussed in detail in Chapter 11. However, certain reconnaissance and supplementary measurements should be a part of the survey program. When geologic mapping is considered adequate to describe the area affected by a landslide, simple qualitative measurements can provide knowledge of the activity. Movements on cracks, particularly those uphill and downhill from well-defined zones of movement, indicate possible increasing size associated with many landslides. Therefore, it is desirable to monitor the change in width as well as the change in elevation across the cracks. This can be done easily by direct measurement from quadrilaterals straddling the cracks or pairs of markers set on opposite sides of the cracks.

Quadrilateral measurement techniques were described in the previous section. Crack width changes also may be measured directly by taping between stakes set on opposite sides of the crack. Crude, simple gauges can be constructed in the field

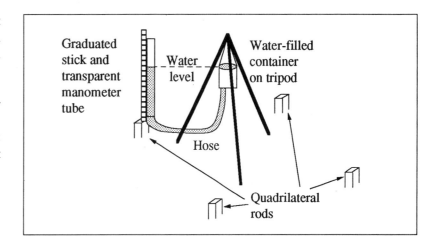

to provide accurate and continuing indications of crack movement. Figure 9-26 shows one such device, which consists of two vertical reinforcing steel rods and a heavy-duty elastic rubber strap or band stretched between the two rods (Nasser 1986). Initial measurements of length, bearing, and inclination of the band provide a basis for comparison with subsequent measurements. Reduction of the field measurements provides values of lateral and vertical movements along the crack. Vertical offsets on cracks and scarps also may be obtained from direct measurement. If total station surveying equipment is available, similar measurements can be made readily and referenced to stable bench marks.

FIGURE 9-25
Details of manometer for use in determining elevation of quadrilateral rods.

FIGURE 9-26
Crack measurement with rubber-band extensometer (modified from Nasser 1986).

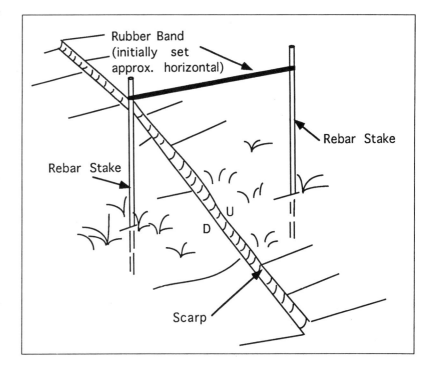

However, such sophisticated surveying methods are not always justified.

In areas of cracked pavement or jointed rock, the change in crack or joint width can be determined by scribing marks or bonding washers onto the pavement or rock surface on opposite sides of the crack or joint and simply measuring between the pairs of marks. Small-scale quadrilaterals 5 to 10 cm across may be installed to monitor displacements, strains, and tilts, as described in the previous section. On pavement or rock surfaces the quadrilaterals can consist of stainless steel pins with washers fixed by epoxy into small-diameter drilled holes. Another simple device is a small hardwood wedge lightly forced into an open crack and marked at the level of the pavement or rock surface. If the crack opens, the wedge will fall deeper into the crack. Such a wedge indicator may not be helpful if the crack closes or is subject to shear movement.

Cracks often are visible in the early stages of landslide deformation. In places where the cracks are not likely to be destroyed rapidly, such as on rock faces or paved surfaces in urban areas, a simple technique for monitoring crack propagation rate is to mark the end of the crack and observe the crack over a period of time, such as days or weeks. The end may be marked by placing a straight-edged piece of cardboard or similar material across it so that it is just covered and spraying paint so that a straight-edged mark is left on the rock face or paved surface (Figure 9-27). The monitoring process can be repeated using different colors of paint and the distance between successive marks can be measured. In active traffic areas where repaving may destroy monitoring locations, attempts to monitor crack propagation may not provide acceptable results.

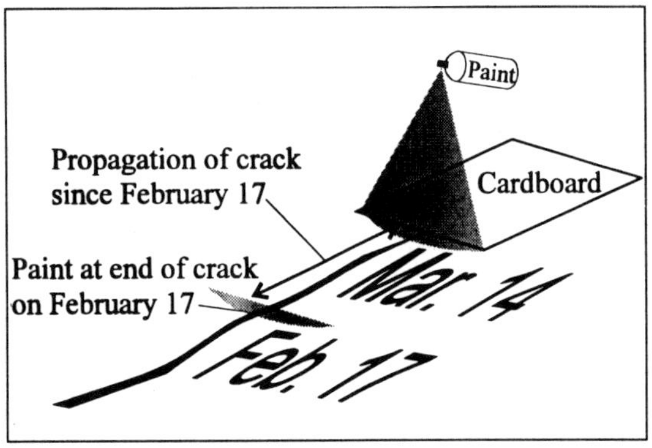

FIGURE 9-27
Paint at end of crack to monitor propagation. Straight-edged material such as cardboard is positioned at end of crack on landslide mass, small paint mark is made, and date is recorded. Marks made in subsequent observations allow crack propagation to be documented. Paint marks have greatest utility if made on paved surfaces or bedrock exposures. Broad-tipped pens or markers could be used to mark cracks in lieu of paint.

Monitoring of cracks usually provides information on movement within the upper half to two-thirds of the landslide mass. The lower half to one-third of the landslide may exhibit little cracking because of compression and overriding of material within this portion. Movement in this area must be detected in reference to fixed objects adjacent to or within the path of this overriding. Photographic points that show the toe of the landslide in relation to fixed features such as trees, rock outcrops, buildings, utility poles, or roads are one means for this monitoring. The photographic point must be established in a location that permits a clear view of the lower part of the landslide and the reference features. Establishing the photographic point should follow the procedures described for terrestrial photography in Section 4.2.3.2.

Another way to monitor movement at the toe of the landslide is to establish one or more rows of targets or stakes parallel to and within the expected path of the landslide. The first target or stake should be placed as near as possible to the toe. Succeeding targets should be placed a set distance apart. For example, a slow movement feature might be monitored by placing five targets or stakes at 1-m intervals. It is important to mark the individual stakes with paint or some other indelible substance. As the landslide movement overrides or covers the first stakes, determination of the amount of movement depends on knowing the number of stakes buried and where the end stake is located in the line. The number of parallel lines of targets or stakes depends on how broad the toe of the landslide is and the configuration of the slope below it.

4.2.3.5 Tiltmeters

Tiltmeters can be used to detect tilt (rotation) of a surface point, but such devices have had fairly limited use in landslide investigations. They have been used mostly to monitor slope movements in open-pit mines and highway and railway cuts, but they may be used in any area where the failure mode of a mass of soil or rock can be expected to contain a rotational component. One type of tiltmeter is shown in Figure 9-28, and sample tiltmeter data from a mine slope are shown in Figure 9-29. Tiltmeters use the same types of servo-accelerometers as those used with the more sensitive inclinometers described in Chapter 11. The prime advantages of tiltmeters are their light weight, simple operation,

FIGURE 9-28
Portable tiltmeter
(Wilson and
Mikkelsen 1978).

compactness, and relatively low cost. Disadvantages of portable tiltmeters in reconnaissance instrumentation are their power requirements and their digital data output.

Alternative devices to detect tilt on landslides are quadrilaterals and existing linear features on the slide, such as utility poles and trees. Quadrilaterals were described in previous subsections of this chapter and provide excellent information that can be used to compute tilt. Trees or structures can constitute crude tiltmeter bench marks within landslides or on slopes expected to move. A line is marked along which the inclination of the structure is measured. Placing a Brunton compass or similar device

capable of measuring orientation along the line provides the initial measurement of its inclination. By comparing the initial measurement with later ones along the same line, any tilting related to displacement on the landslide can be detected. Trees and poles should be measured parallel to their lengths at points 0, 90, 180, and 270 degrees around their circumferences. This measurement provides better control of the direction of tilting when the tree or pole could tip in any direction. The uneven surface of the tree can be made smoother for more accurate measurement by nailing wood strips parallel to its length (Figure 9-30). These strips also serve to mark the alignment for later remeasurement. The orientation of the wood strips can also be measured with a Clar compass.

4.2.3.6 Videography and Digital Image Analysis

Videography and digital image analysis are emerging technologies with promise for application in surface observation and geologic mapping of landslides. Videography is a computer-aided procedure in which multiple videotape recordings of the same area taken with different spectral bands or lens filters are registered to each other or to a geographic reference system and processed to create a single interpreted image. Bartz et al. (1994) successfully de-

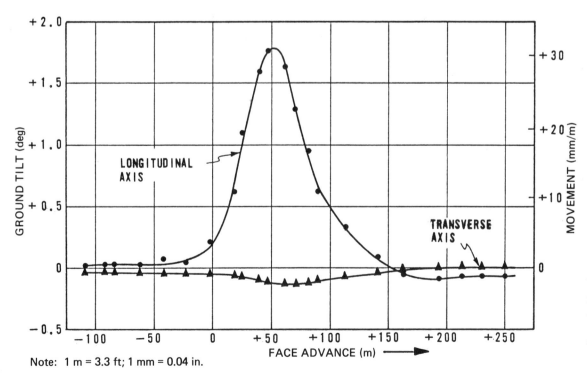

Note: 1 m = 3.3 ft; 1 mm = 0.04 in.

FIGURE 9-29
Tilt at ground
surface due to
advance of longwall
mining face
(Wilson and
Mikkelsen 1978).

FIGURE 9-30
Tree trunk with
wood strips tacked
at 90-degree
intervals to provide
reference lines
for successive
measurement of tilt.
Trend and plunge of
strips are measured
on successive field
observations to
document direction,
amount, and rate of
tilting.

lineated riparian vegetation types using multispectral airborne videography. They collected three bands of data—green (centered at 0.55 μm), red (centered at 0.65 μm), and near-infrared (centered at 0.85 μm)—and digitized selected frames of the imagery using special electronic computer hardware known as a frame-grabbing and display board. Multispectral images were created by superimposing the three individual bands, and continuous sections of their study area were prepared as mosaics of the individual three-band images. The computer is "trained" to interpret specific combinations of image characteristics as specific geographic features by having the computer operator classify areas of the feature. Areas of ponded water, for example, would have a distinctive set of characteristics on the different spectral bands.

Bartz et al. (1994) concluded that multispectral airborne videography can be used to collect inexpensive and timely baseline information. A similar procedure based on multispectral aerial videography was used by Shoemaker et al. (1994) for jurisdictional delineation of wetlands. Videography has not yet been applied to landslide mapping, but the potential value is great for rapid classification of a landslide area once features such as ground cracks and seeps have been identified.

Digital image analysis is similar to videography, except that the analysis is based on a single image rather than on multiple superimposed images. Terrestrial digital image analysis is an emerging technology that is related in some respects to satellite image analysis. Multispectral remote sensing of landslide-susceptible areas was addressed by Jadkowski (1987). Vandre (1993) applied digital image analysis to determine grain size distribution of cobble- and boulder-size particles; he also has used this technology on landslide evaluations in Utah (B. Vandre, personal communication, 1994, AGRA Earth & Environmental, Inc., Salt Lake City, Utah). The basis for digital image analysis is rapid estimates of size based on pixel dimensions and a suitable scale visible in the image. A conventional video camera is used to record images in the field. The videotape is replayed through a computer monitor using a card with a special video jack (RCA format). Selected frames of the video are grabbed for display and analysis. Quantitative estimates of landslide movement can be obtained from digital image analysis of repeated terrestrial photography if the photographs are digitized and the features displaced or deformed by landslide movement are visible in the repeated photographs.

4.2.3.7 Global Positioning System Surveying

Global Positioning System (GPS) surveying has developed rapidly in the past few years. This technique appears to provide precise surveying technology anywhere in the world, but it has some severe limitations. The basis for GPS surveying is a network of satellites (26 in 1994) deployed by the U.S. Department of Defense. These satellites make up a positioning system called the Navigation Satellite Timing and Ranging (NAVSTAR) GPS. The positioning system in each satellite transmits at two frequencies with three modulations (Reilly 1992). The atomic clocks in the satellites generate a fundamental frequency that is multiplied by two different factors to generate two carrier frequencies. Modulated onto these carrier frequencies are pseudo-random-noise codes to protect military use for precise navigation. Civilian users do not have access to the precise military navigation code. The satellites also transmit messages containing clock correction coefficients, satellite ephemeris parameters, and other data, including almanac data, which give orbit information for all of the satellites in the NAVSTAR GPS "constellation."

A single GPS receiver can be used to determine the position of an unknown point with an accuracy of about 100 m. This position is in an Earth-centered, Cartesian-coordinate system that can be expressed in terms of latitude, longitude, and height on Earth's reference ellipsoid. Much better accuracy in horizontal position can be achieved with at least two GPS receivers simultaneously recording at least four satellites. Horizontal accuracies of 1:300,000 are routine (Reilly 1992).

GPS surveying is not well suited for determining ground-surface elevation. The best way to determine elevation from a GPS survey is to have at least one GPS receiver dedicated to a bench mark with accurate known elevation. The bench mark receivers are used to calibrate the satellite signals for determining the elevation of the roving receivers. Most of the error in elevation is related to setting up the receiver antenna, and an accuracy of 0.3 cm is thought to be reasonable for careful GPS measurements (Reilly 1992).

Accurate measurements with GPS receivers require positions to be operated continuously for 45 to 60 min. Tree limbs and other material blocking the antenna can cause problems with reception of satellite transmissions. Troublesome backscatter of transmissions can be caused by reflective surfaces such as buildings, pavement, and buried pipelines. GPS surveying is a powerful tool, but for most landslide investigations, the exact geographic position of the landslide is not needed, only accurate relative positions of points on and adjacent to it.

4.3 Representation of Topographic Data

Survey data must be displayed and analyzed in order to be useful to landslide investigators. The four most common representation methods are topographic maps, profiles, displacement vectors and trajectories, and strain ellipses.

4.3.1 Topographic Maps

Generally, topographic information obtained by photogrammetric methods is correlated with ground survey controls and detailed topographic information to establish two or more maps of the landslide area. The first map encompasses the landslide (or suspected landslide) plus the surrounding area, including the topography extending uphill and downhill beyond major changes in slope or lithology. Topography should be developed on each side for a distance of approximately twice the width of the moving area (or more when the zone of potential movement is not well defined). Typical scales for such mapping of large landslides may be 1:2,500 to 1:10,000. A portion of such a map developed for a landslide investigation near Vail, Colorado, is reproduced in Figure 9-31 (Casals 1986).

The second topographic map is more detailed and encompasses the observed landslide area plus all of the uphill and downhill cracks and seeps associated with the landslide. Typically, the detailed map extends beyond the boundaries of the landslide uphill and downhill for a distance of half the length of the landslide or to significantly flatter slopes. The detailed topography should extend beyond the limits of the landslide laterally at least half the width of the landslide area. Contour intervals in such detailed topography should be as close as 0.5 m for landslides that do not have too great a degree of vertical relief. The horizontal scale is typically 1:2,500 or larger. A portion of such a landslide map prepared as part of field investigations of a landslide in western Colorado (Umstot 1988) is shown in Figure 9-32.

Topographic data commonly are collected in the field electronically and compiled in the office digitally. Many topographic maps for landslide investigations are plotted by computer. Data from successive surveys can be compared with the aid of the computer, and plots of changes in topographic contour-line position can be produced. Such comparative contour-line plots provide useful information on changes in topography of landslide masses but require complete resurveying of landslide sites. An example of a map showing changes in elevation of the surface of the Thistle landslide (Duncan et al. 1985) is shown in Figure 9-33.

4.3.2 Profiles

In addition to a topographic map, profiles of the landslide area are prepared (Figures 9-34 and 9-35). The most useful profiles are perpendicular to the steepest slope of the landslide area. Where the movement definitely is not perpendicular to the steepest slope, two sets of profiles are necessary: one

FIGURE 9-31 Portion of topographic map prepared by photogrammetric methods showing region around head of landslide near Vail, Colorado (Casals 1986). Original map scale was 1:2,400. Elevations are in feet; major contours are at intervals of 25 ft (8 m), with supplementary contours at 5-ft (1.5-m) intervals.

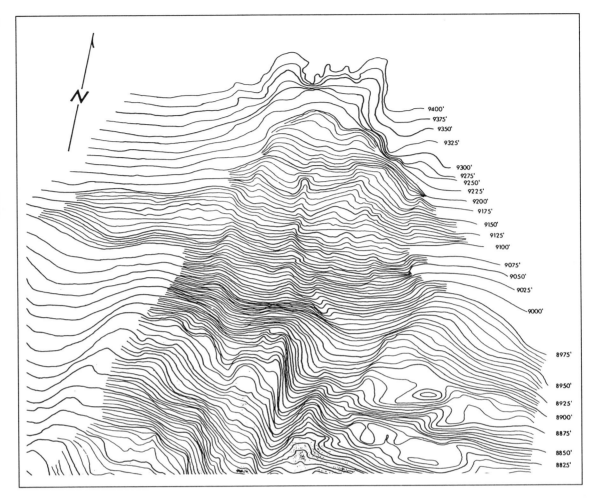

set should be parallel to the direction of movement and the other parallel to the steepest ground-surface slope. For small landslides, three profiles may be sufficient; these should be at the center and quarter points of the landslide width (or somewhat closer to the edge of the slide than the quarter points). For very large landslides, the longitudinal profiles should be obtained at spacings of 30 to 60 m. It is particularly important that the profiles be selected so as to depict the worst and less critical combinations of slope and movement within the landslide area. It is usually desirable to have at least one additional profile in the stable ground 15 to 30 m beyond the limits of the landslide area on each side so that the effect on the movement of ground-surface slope alone can be determined.

Each longitudinal profile of the landslide is usually plotted separately. Successive profiles can be shown on the same drawing to illustrate changing site topography. The original topography should be estimated from old maps and shown for com-

parison where possible. However, it is difficult to reference old maps precisely to the more detailed topography obtained during landslide investigations. Differences between existing topography and prelandslide topography may represent survey mismatches as well as deformation of the site ground surface. Adjustments of the prelandslide profile from old maps to the profile from new surveys may in some cases be made by comparing old and new topography beyond the limits of observed landslide movements.

4.3.3 Displacement Vectors and Trajectories

Survey grid points and other critical points are plotted on the more detailed topographic map of the landslide area. Both the topography and the depicted grid points should be referenced to the same geographic reference system, whether it is an arbitrary reference convenient for the landslide site or

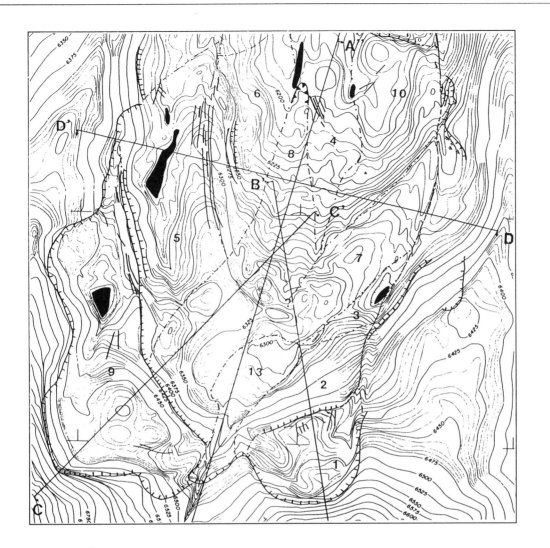

FIGURE 9-32 Portion of detailed map showing topographic and landslide features developed during landslide investigation in western Colorado (Umstot 1988). Original map scale was 1:2,400 and contour interval was 5 ft (1.5 m).

FIGURE 9-33 (below) Elevation changes within Thistle landslide, Utah (modified from Keaton 1989 from Duncan et al. 1985).

true geographic position on the Earth's surface. From the consecutive readings on the survey grids and traverses, the horizontal and vertical displacements of the ground surface can be determined. If the movements are large, the subsequent positions of the reference points can be plotted on the topographic map. However, if the movements are small, the successive positions of the monuments may be plotted separately to a larger scale depicting vectors of movement. The vector map may show reference-point locations and displacement vectors on a map with the landslide outline (Figure 9-36), or a topographic base map may be used if the information can be shown clearly. Although the initial positions of the points are shown in their proper scale relations, the displacement vectors may be plotted to a larger scale; this difference in scale should be noted. Elevations at successive dates can be entered beside the grid points.

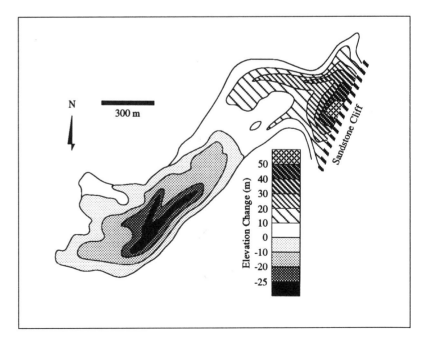

FIGURE 9-34
Landslide contours
and profile locations
(Sowers and Royster
1978).

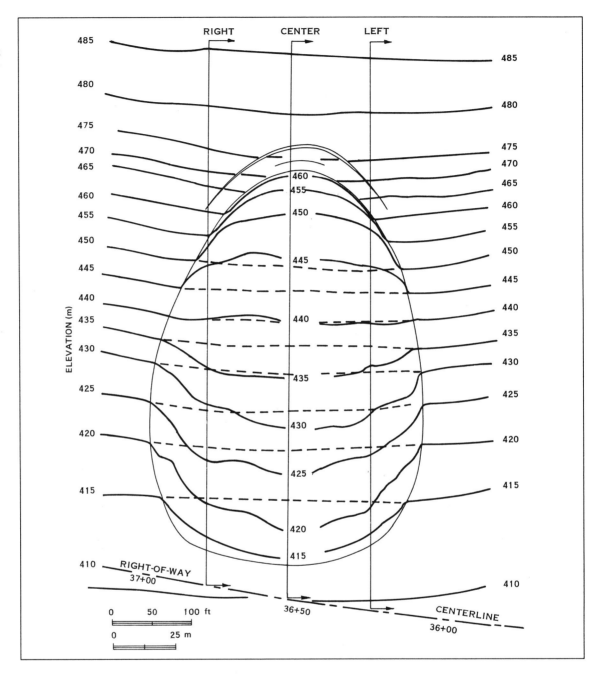

Because topography changes significantly with continuing movement, the dates of the surveys should be noted on the maps. Furthermore, if a significant period of time has elapsed between the dates of the surveys that establish the topography and the surveys that establish the movement grid, the elevations of the points on the grids will not necessarily correspond to those on the topographic map.

Displacement-vector data also can be displayed separately from maps showing the distribution of

the grid points. An example of a plot of direction and cumulative magnitude of displacement is shown in Figure 9-20.

4.3.4 Strain Vectors and Ellipses

Quadrilateral monitoring (Section 4.2.3.3) provides a means for calculating strain from changes in the length of chords, called *stretch* (Baum et al. 1988), between points on each quadrilateral. The directions and magnitudes of maximum and min-

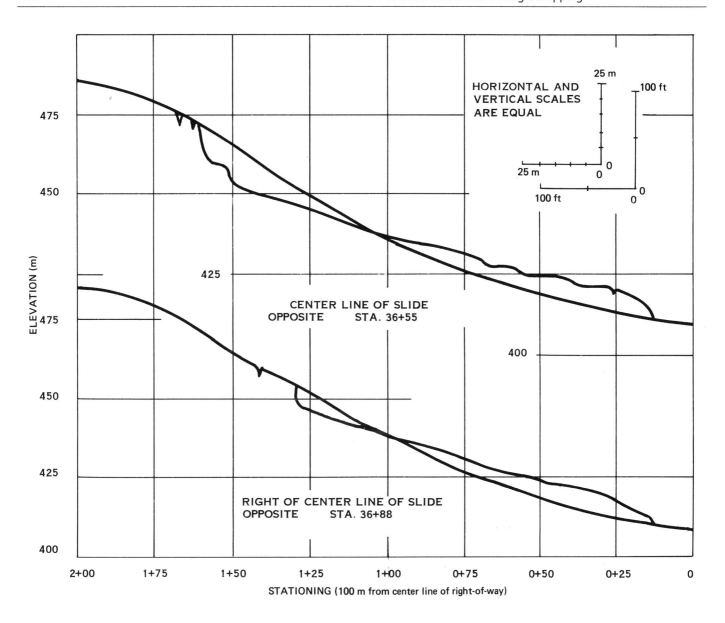

FIGURE 9-35
Landslide ground-
surface profiles
(Sowers and Royster
1978).

imum principal stretches can be determined and plotted in the form of ellipses on the topographic base map or landslide outline map. Similarly, the directions and magnitudes of area strain and finite shear strain (Baum et al. 1988) can be determined and plotted on maps to represent internal deformation rather than displacement.

5. INTERPRETATION AND DATA PRESENTATION

The geologic data collected in the field must be interpreted and presented in a form that communicates useful and relevant information to non-geologists, usually engineers. The primary form of

geologic data presentation is the geologic map. The map and explanation must be carefully crafted to present relevant information accurately and clearly. Geologic sections are used to illustrate subsurface relationships interpreted from surface observations. Sections can be connected to portray three-dimensional relationships. Selected photographs of critical features enhance the report user's understanding. The geologic report discussing methods, findings, conclusions, and recommendations must be carefully written to communicate relevant information without using unexplained geologic terminology.

The concept of multiple working hypotheses is fundamental to geologic interpretation and is de-

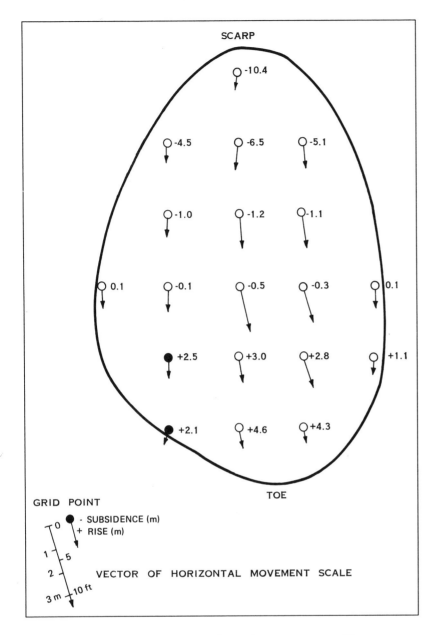

FIGURE 9-36
Movement vectors showing displacements since beginning of measurement (Sowers and Royster 1978).

scribed briefly in the next section. The measurements obtained from reconnaissance instrumentation also are discussed, followed by elaboration on the geologic maps, sections, and report.

5.1 Importance of Multiple Working Hypotheses

The multiple-working-hypothesis method proposed by Chamberlin (1965) has become the conventional geologic approach to scientific investigation, although some believe the name should no longer be applied to the method (Johnson 1990). This method consists of continuous evaluation of alter-

native explanations for the data that have been collected. As more and more observations are made, the hypotheses continue to be tested; those hypotheses that can no longer be supported by the data are rejected. Ultimately the goal is to explain all of the facts with a unique model. This ideal goal may be difficult to achieve, and after some investigation, two or more viable alternative explanations may remain. This situation identifies the essence of the frequent difficulties in communication between geologists and engineers. Engineers try to base their analyses and design on the concept that a unique solution exists and can be identified.

It is essential that uncertainties in geologic interpretation be explained in terms relevant to engineers. Often with several viable alternative explanations, one appears most likely. This situation should be explained clearly, and the likelihood should be quantified if possible. If the geologist does not offer this kind of assistance to the engineer responsible for design, the engineer may be forced to make geologic decisions or to ignore geology altogether (Keaton 1990). The engineer responsible for design, on the other hand, should seek relevant information from the geologist. The engineer and the geologist must work together as a team to accurately characterize a site and formulate a design based on that characterization.

Much geologic analysis is based on the concept that current processes can be used to interpret the geologic record; in other words, the present is the key to the past. For example, a stream transporting sand and gravel can be interpreted to mean that gravelly sandstone bedrock was deposited in a fluvial environment. The engineering-geologic corollary to this concept states that processes that were active in the recent geologic past should be expected to remain active in the near future; in other words, the recent past is the key to the near future. For example, subdued hummocky terrain may indicate that landslides that were active in prehistoric time have the potential to become active again.

5.2 Analysis of Reconnaissance Instrumentation

Information developed from reconnaissance instrumentation can be valuable in understanding the cause, rate, and direction of slope movement. These factors in turn may be helpful in planning the subsurface investigation and interpreting the results. Measurements made with the instrumentation should be summarized as basic data and then interpreted in a geologic context.

5.3 Engineering-Geologic Maps and Explanations

Engineering-geologic information is most useful when plotted on a topographic base map. An orthophotograph with superimposed topographic contours (Figure 9-37) provides a most useful method for rapidly communicating relevant geologic data to nongeologists if the photograph documents current conditions. The scale of the base map controls the detail that can be shown. Engineering-geologic mapping should be completed at the scale that will be used for design drawings if possible. Detailed geologic mapping, particularly of complex areas, can be accomplished efficiently with the aid of a plane table and alidade, as was done by Fleming and Johnson (1989) for the Aspen Grove landslide in Utah (Figure 9-6).

Three levels of detail may be warranted in landslide assessments: areal assessment, regional evaluation, or site investigation. In urban areas, Leighton (1976) called these levels *regional, tract,* and *site*. Areal landslide maps show the general distribution and shape of large landslide deposits. Good examples are a landslide map of Utah (Figure 9-38) by Harty (1991) at a scale of 1:500,000 and a preliminary map of landslide deposits in San Mateo County, California, by Brabb and Pampeyan (1972) at a scale of 1:62,500. Areal landslide maps provide some basis for characterizing large-area suitability; feasibility and design decisions, however, require greater detail.

Increased detail is achieved at the regional evaluation level. Information on the landslide types and their features would be documented on maps for regional evaluation. Baum and Fleming (1989) mapped landslides in central Utah at a scale of 1:12,000 (Figure 9-39). Wieczorek (1982) mapped landslides near La Honda, California, at 1:4,800, and Baum and Fleming (1989) showed the Aspen Grove landslide in Utah at about 1:4,000 (Figure 9-40); these are reasonable scales for site investigations. Fleming and Johnson (1989) mapped the Aspen Grove landslide at 1:200 using a plane table; a portion of this map is shown in Figure 9-41 at 1:833. A smaller part of the Aspen Grove land-

FIGURE 9-37
Part of orthotopographic map of Thousand Peaks landslide area, northern Utah. Photographs at nominal negative scale of 1:12,000 were adjusted to a map scale of 1:4,800. Topographic contours are at 5-ft (1.5-m) intervals. Compare with Figure 9-8.
COURTESY OF KERN RIVER GAS TRANSMISSION COMPANY

FIGURE 9-38
Example of landslide map for areal assessment, scale 1:500,000 (Harty 1991). Original map is colored and distinguishes between deep-seated, shallow, and undifferentiated landslides compiled from different sources; dots represent landslides smaller than 600 m across, and patches depict landslides larger than 600 m across. Area shown includes Ephraim Canyon, central Utah; increasing detail is shown in Figures 9-39, 9-40, 9-41, and 9-6.

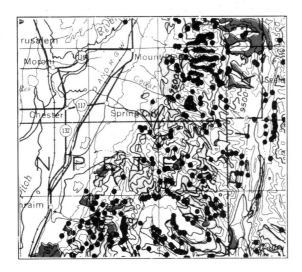

slide mapped in 1983 and 1984 is shown in Figure 9-6 at 1:333.

Regional evaluation maps give sufficient detail for establishing feasibility and some preliminary design. Areas of exposed bedrock should be mapped as bedrock; where surficial deposits are present at the surface, even if bedrock is at a shallow depth, these deposits should be shown. The Unified Engineering Geology Mapping System, described in Section 3.2.2, provides a method for indicating the estimated thickness of surficial deposits overlying bedrock.

The site investigation map is utilized for designing a project and therefore the greatest possible level of detail should be presented. Not only should the information on the bedrock and surficial deposits contained in a regional evaluation map be retained, but greater detail should also be included.

Geologic map explanations should contain brief but complete descriptions of the units mapped and the symbols used. Map units may be correlated with geologic formation names and ages. The locations of sections and reconnaissance instrumentation should also be indicated on the map and included in the explanation.

Comparison of mapped landslides with other physical characteristics of the area may provide insight on conditions influencing stability. A landslide map is essentially a map on which points within an area have been classified as either related to a landslide or not. There is little difficulty in comparing landslide maps with geologic maps where points within an area have been classified according to what geologic material is present. It is more difficult to make a meaningful comparison

with contoured data such as elevation on a topographic map or isohyetal lines on a precipitation map. A landslide map can be transformed into contoured data by isopleth mapping (Wright et al. 1974; DeGraff and Canuti 1988), as shown in Figure 9-42. The percentage of a standard area that is underlain by landslide deposits is determined on a grid superimposed on the landslide map. The percentage values determined for all the points on the grid are contoured to show isopleths of landslide deposits.

The isopleth landslide map provides a way to represent landslide activity in an area and improve the understanding of conditions leading to instability (DeGraff 1985), as shown in Figure 9-43. It facilitates comparison of large and small landslides that may be present in different parts of the area. Isopleth maps for different landslide types in an area can be constructed to assess whether their distribution is controlled by different sets of factors. Locations with especially high percentages of landslides may merit more detailed investigation.

5.4 Engineering-Geologic Sections

Engineering-geologic sections should be located so as to provide information that will be needed for planning subsurface investigations and stability analyses. Commonly these positions are perpendicular to topographic contours and aligned longitudinally down active or dormant landslide deposits. The projected shape of sliding surfaces, geologic contacts, and piezometric surfaces should be shown along with the topographic profile (Figure 9-44). Geologic sections should be constructed at the same horizontal and vertical scales. Exaggerated vertical scales do not permit direct measurement of angular relationships and can be misinterpreted. Furthermore, they do not aid in communicating geologic information to nongeologists.

5.5 Three-Dimensional Representations

Communication of geologic information can be enhanced by diagrams of interconnected geologic sections. These interconnected sections (called *fence diagrams*) provide three-dimensional or pseudo-three-dimensional representation of the surface and subsurface conditions (Figure 9-45). *Block diagrams* are created by constructing perpendicular sections and sketching the surface geology in an isometric projection. (Figure 9-7 shows

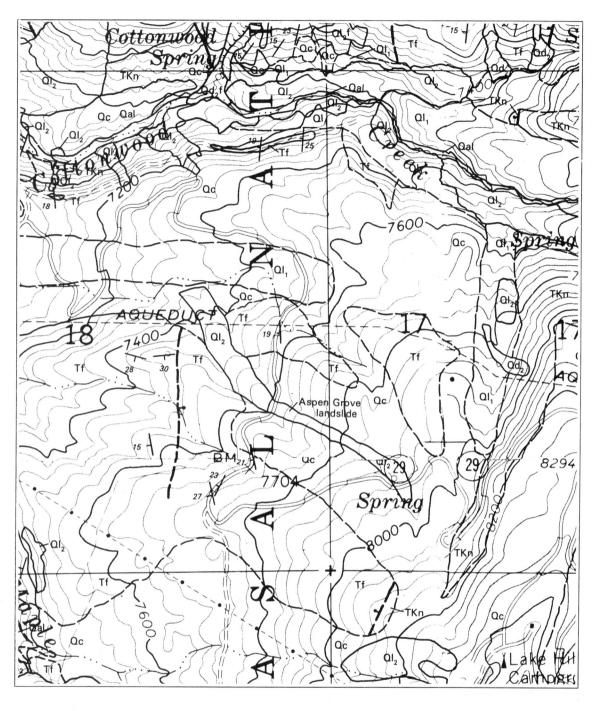

FIGURE 9-39 Example of landslide map for regional assessment, scale 1:12,000 (Baum and Fleming 1989). Aspen Grove landslide, Utah, is identified near center; increasing detail is shown in Figures 9-40, 9-41, and 9-6; less detail is shown in Figure 9-38. Abbreviated explanation: Ql_1, pre-1983 landslide; Ql_2, landslide active in 1983 to 1986; Qd_2, debris flow active in 1983 to 1986; Qal, alluvial deposits; Qc, colluvial deposits; Tf, Eocene-Paleocene Flagstaff limestone; TKn, Paleocene–Upper Cretaceous North Horn Formation.

schematic block diagrams of landslide deposits of differing ages.) Scale models constructed from thin sheets of styrofoam (foam-core boards) are the equivalent of orthogonal block diagrams (Figure 9-46). Computer-generated sections, fence diagrams, and block diagrams can be produced routinely with the use of commercially available computer programs. These diagrams can be particularly helpful in communicating geologic information to nongeologists. They are most useful when data from subsurface investigations has been integrated into them.

5.6 Engineering-Geologic Reports

Engineering-geologic reports most likely will be used by nongeologists; therefore, geologic terminology must be minimized and explained in non-

FIGURE 9-40
Example of landslide
map for site
assessment (Aspen
Grove landslide,
Utah), scale about
1:4,000 (modified
from Baum et al.
1988). More detail
of part of landslide
is shown in Figures
9-41 and 9-6; less
detail is shown in
Figures 9-38 and
9-39.

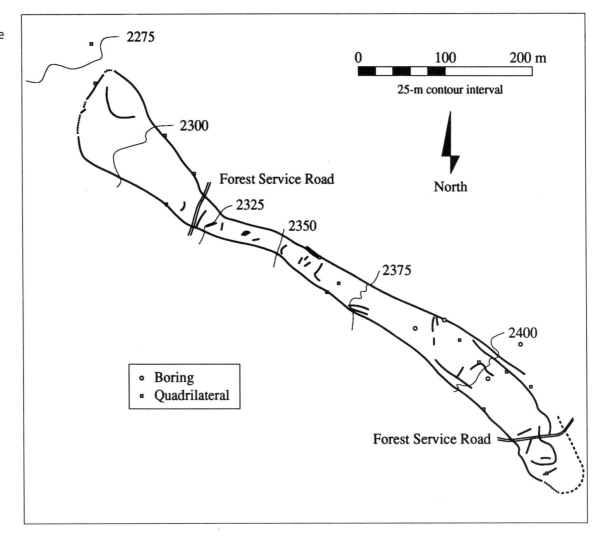

geologic terms. For example, "fossiliferous silty limestone" should be used rather than "biomicrite." Engineering geology essentially is geology that is relevant to an engineering project; geologic information that is not relevant to the project should be omitted or placed in an appendix to the report.

Engineering-geologic descriptions must be quantitative to the greatest extent possible. Phrases such as "moderately weathered," "moderately strong," and "moderately dense" are not meaningful unless they are quantified as in the Unified Rock Classification System (Table 9-4). The basis for geological opinions and recommendations should be clearly stated. As additional information becomes available, such as the results of the subsurface investigation, the basis for opinions can be evaluated, and the opinions can be verified or modified.

6. GUIDANCE FOR SUBSURFACE INVESTIGATIONS

The geologist often is the first technical specialist at a site or the first to spend a substantial amount of time there. Thus, valuable information can be collected regarding logistical issues, such as access to the site for drilling and other equipment, locations and types of utilities in the vicinity, and existing land uses.

The results of the surface observation and engineering-geologic mapping should provide detailed and quantitative information on

- Distribution, modes, and rates of slope movement;
- Positions and geometries of slip surfaces;
- Position of groundwater table and piezometric surfaces;

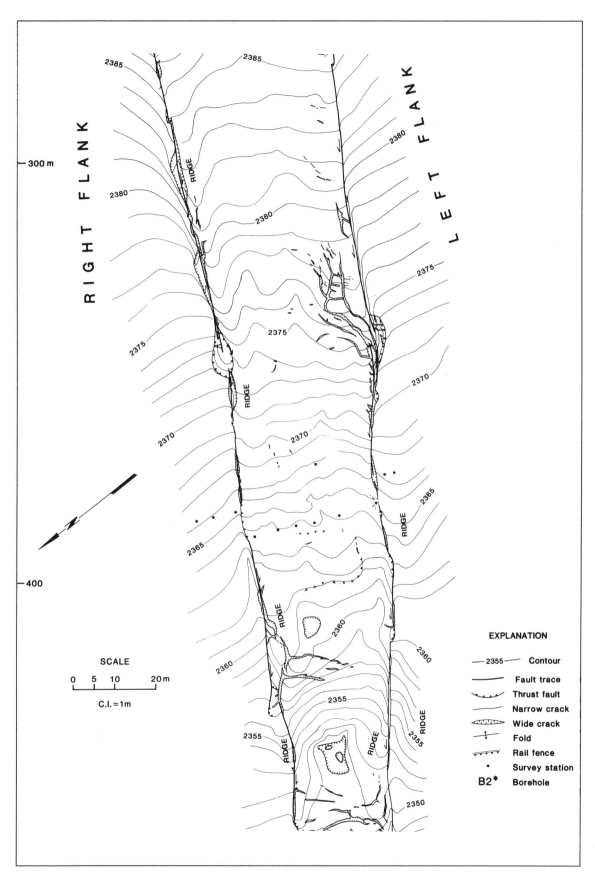

FIGURE 9-41
Example of landslide
map for site
assessment (part of
Aspen Grove
landslide, Utah).
Fleming and
Johnson (1989)
originally mapped
this landslide at
scale of 1:200, but
original illustration
was published at
scale of about
1:833. More detail
of part of this
landslide is shown
in Figure 9-6 at
scale of 1:300; less
detail is shown in
Figures 9-38, 9-39,
and 9-40.

EXPLANATION

—2355— Contour
——— Fault trace
⌣⌣⌣⌣ Thrust fault
⌒⌒⌒⌒ Narrow crack
▨▨▨▨ Wide crack
⊥⊥⊥ Fold
⊤⊤⊤⊤ Rail fence
• Survey station
B2* Borehole

SCALE

0 5 10   20 m

C.I. = 1 m

FIGURE 9-42
Steps in constructing
isopleth map of
landslide distribution
(DeGraff and Canuti
1988).

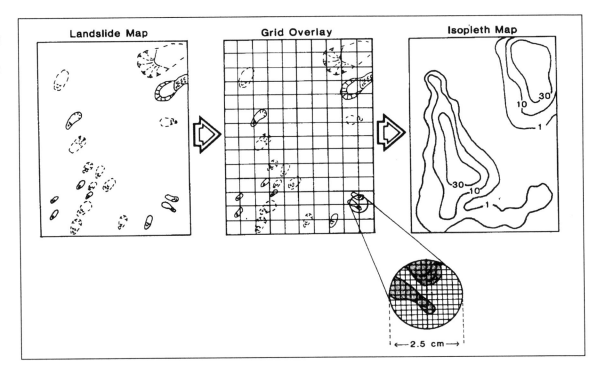

FIGURE 9-43
(below)
Example of isopleth
map of landslide
distribution
(DeGraff 1985).

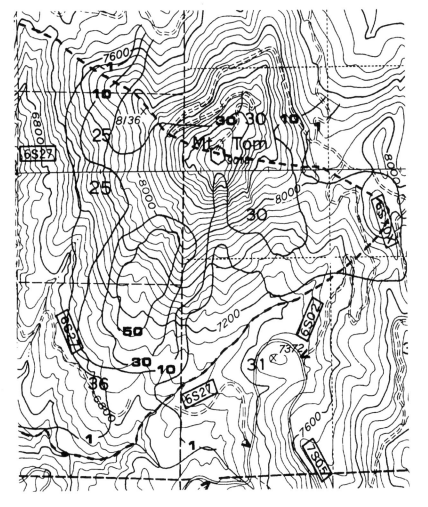

- Potentially difficult drilling or excavation conditions, or both;
- Recommendations for locations of surface geophysics, borings, and test pits; and
- Possible locations and schemes for stabilizing structures.

Of particular importance are the geologist's recommendations concerning the best locations for subsurface investigations. Engineers may plan an investigation program of regularly spaced borings, test pits, or both, hoping to collect sufficient data to adequately characterize the site. However, geology is not random; therefore, borings and test pits at appropriately selected locations are more likely to provide information needed to test geologic-model hypotheses. It must be recognized that additional borings may be needed to collect sufficient samples for laboratory testing.

7. CORRELATION OF SURFACE AND SUBSURFACE DATA

The geologist should observe and log subsurface exposures and samples to provide a basis for correlating surface mapping with subsurface information. Subsurface data generally permit more specific geologic interpretation than do surface data. The geologic interpretation of the site based

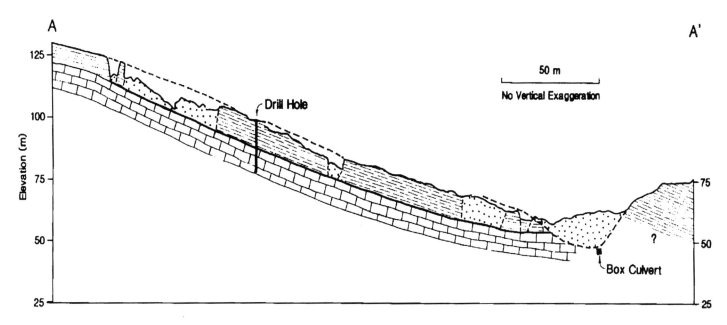

on surface mapping must be evaluated in the context of the subsurface data. If appropriate, revisions should be made to the engineering-geologic map and sections. The reevaluation of the geologic interpretation of the site may reveal the need for supplemental subsurface data at locations not initially selected for investigation.

The basic goal of the surface and subsurface investigations is accurate, quantitative characterization of the site. Accurate site characterization permits selection of the best geologic-geotechnical model of the slope system for use in stability analyses. The best model permits the most appropriate design for the project.

8. ADDITIONAL FIELD INFORMATION

Additional field information may be needed to permit prediction of the performance of the slope and the potential risk of damage to the project, adjacent facilities, or both. Long-term monitoring of slope deformation and piezometric levels may be appropriate. The reconnaissance instrumentation can be converted to permanent instrumentation by replacing the rapidly deployed devices with more stable ones.

Adjacent land use strongly influences the consequences of damage. High-value or high-occupancy property adjacent to a slope movement may require a more conservative design and more extensive field monitoring to provide a basis for legal defense in the event of litigation on behalf of

the adjacent property. Some adjacent land uses, such as agricultural irrigation, can contribute to stresses causing slope movement at the project site. Repeated surface observations and instrumentation can permit early detection of the beginning of a problem. Alternatively, such observations can provide a legal basis for justifying compensation from the adjacent land owner for part of the damages at the site.

The potential for excessive precipitation, strong earthquake shaking, or both, to occur at a site

FIGURE 9-44
Example of engineering-geologic section of landslide deposit (Jibson 1992).

FIGURE 9-45
(*below*)
Example of fence diagram showing subsurface conditions in landslide area (Sowers and Royster 1978).

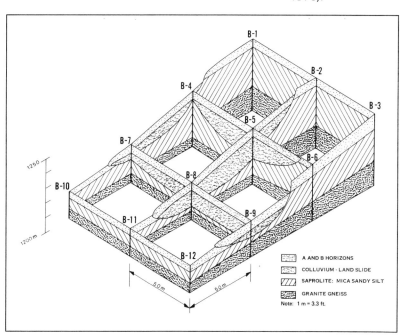

FIGURE 9-46
Example of styrofoam (foam-core) model of Thistle landslide in central Utah made by printing copies of topographic map at scale of 1:2,400 and fixing them onto sheet of styrofoam with spray adhesive. Thickness of each sheet of styrofoam was equal to vertical rise of 40 ft (12 m), and model was made with no vertical exaggeration. Equivalent sun illumination is early to mid-afternoon.

should be recognized early in the investigation. It may be appropriate to consider installing precipitation gauges, strong-motion accelerographs, or both, to collect information that may be of great value in improving design of slope systems. If slope movement occurs in response to precipitation or an earthquake, the intensity or magnitude of the event at the slope will be known. Otherwise, the intensity or magnitude of the event will have to be projected from the nearest recording station, which may be a substantial distance from the site.

9. COST OF SURFACE INVESTIGATIONS

The pre-field investigation involves identifying, collecting, and reviewing available geologic and topographic information and aerial photographs. Usually no more than a few days of a geologist's time is required for this task. Up to several hundred dollars may be needed to purchase maps and available aerial photographs, depending upon the size of the project. It may require six to eight weeks to obtain aerial photographs from government agencies in the United States and Canada; elsewhere it may take much longer. Where aerial photography is produced by the air force of a particular country, its availability requires overcoming major obstacles. However, many smaller countries in the British Commonwealth produce their topographic

maps based on aerial photography taken by the Directorate of Overseas Surveys of the Government of the United Kingdom. This is commonly black-and-white photography at a scale of 1:20,000. A project-specific aerial-photograph mission may be warranted for some projects; the cost of such missions varies widely depending upon the size and location of the project, but generally costs at least several thousand dollars.

The amount of time required for a geologist to complete surface observation and geologic mapping depends on the size of the site, the ease of access, the ruggedness of the site, and the degree of geologic complexity. Under most circumstances, however, one to two weeks should be sufficient. The reconnaissance instrumentation usually consists of relatively inexpensive wooden or steel stakes and small-diameter iron pipe. The equipment needed to conduct the monitoring is simple and is probably owned by most geologists so that rental or purchase would not be needed.

The amount of involvement by the geologist during the subsurface investigation can be extensive or minor, depending on the organization of the effort. Often the geologist is assigned to the drilling or excavation equipment to make field observations. In these cases the geologist can be involved for several weeks or even months. If the geologist is not directly involved in the subsurface-data collection, approximately one to two weeks of effort may be required to examine the samples and correlate the subsurface and surface data.

It may be appropriate for the geologist to assist in installing instrumentation and to be involved in long-term monitoring efforts. The geologist can make repeated surface observations at the time the instruments are being read. Long-term monitoring can extend over a period of years, with readings being taken at intervals of weeks or months. The amount of time required for each series of readings varies with the number of instrument locations and the difficulty of access. Generally, a few days to one week is sufficient to read the instruments at a typical site.

Correlation and analysis of surface and subsurface geologic data are commonly done as the data are being collected. It may be appropriate for the geologist to synthesize the data at the end of a major effort. Typically, a few days to one week is sufficient time for such correlation and analysis.

REFERENCES

ABBREVIATIONS

UNESCO United Nations Educational, Scientific, and Cultural Organization

WP/WLI Working Party on the World Landslide Inventory (International Geotechnical Societies and UNESCO)

Anbalagan, R. 1992. Landslide Hazard Evaluation and Zonation Mapping in Mountainous Terrain. *Engineering Geology*, Vol. 32, No. 4, pp. 269–278.

Barton, N., R. Lien, and J. Lunde. 1974. *Engineering Classification of Rock Masses for the Design of Tunnel Support*. Norwegian Geotechnical Institute, Oslo, Norway, 48 pp.

Bartz, K.L., J.L. Kershner, R.D. Ramsey, and C.M.U. Neale. 1994. Delineating Riparian Cover Types Using Multispectral, Airborne Videography. In *Proc., 14th Biennial Workshop on Color Photography and Videography in Resource Monitoring*, May 25–28, 1993, Utah State University, Logan, American Society of Photogrammetry and Remote Sensing, Bethesda, Md.

Baum, R.L., and R.W. Fleming. 1989. *Landslides and Debris Flows in Ephraim Canyon, Central Utah*. U.S. Geological Survey Bulletin 1842-C, pp. C1–C12 (scale 1:12,000).

Baum, R.L., A.M. Johnson, and R.W. Fleming. 1988. *Measurement of Slope Deformation using Quadrilaterals*. U.S. Geological Survey Bulletin 1842, 23 pp.

Bieniawski, Z.T. 1989. *Engineering Rock Mass Classifications*. Wiley Interscience, New York, 251 pp.

Brabb, E.E., and E.H. Pampeyan. 1972. *Preliminary Map of Landslide in San Mateo County, California*. U.S. Geological Survey Miscellaneous Field Studies Map MF-344 (scale 1:62,500).

British Geotechnical Society. 1974. *Field Instrumentation in Geotechnical Engineering*. John Wiley & Sons, New York, 720 pp.

Carr, J.R., and J.B. Warriner. 1987. Rock Mass Classification Using Fractal Dimension. In *Rock Mechanics: Proceedings of the 28th U.S. Symposium* (I.W. Farmer et al., eds.), A.A. Balkema, Rotterdam, Netherlands, pp. 73–80.

Carroll, R.D., F.T. Lee, J.H. Scott, and C.S. Robinson. 1968. *Seismic-Refraction Studies of the Loveland Basin Landslide, Colorado*. U.S. Geological Survey Professional Paper 600-B, pp. B83–B87.

Carter, M., and S.P. Bentley. 1985. The Geometry of Slip Surfaces Beneath Landslides: Predictions from Surface Measurements. *Canadian Geotechnical Journal*, Vol. 22, pp. 234–238.

Casals, J.F. 1986. *Analysis of a Landslide Along Interstate 70 Near Vail, Colorado*. Master of Engineering Report ER-3172. Colorado School of Mines, Golden, 121 pp.

Chamberlin, T.C. 1965. The Method of Multiple Working Hypotheses. *Science*, Vol. 148, pp. 754–759 (reprinted from *Science*, Feb. 7, 1890).

Compton, R.R. 1985. *Geology in the Field*. John Wiley & Sons, New York, 398 pp.

Cording, E.J., A.J. Hendron, Jr., W.H. Hansmire, J.W. Mahar, H.H. MacPherson, R.A. Jones, and T.D. O'Rourke. 1975. *Methods for Geotechnical Observation and Instrumentation in Tunneling*. Department of Civil Engineering, University of Illinois, Urbana-Champaign, and National Science Foundation, Washington D.C., Vol. 2, pp. 293–566.

Crozier, M.J. 1984. Field Assessment of Slope Instability. In *Slope Instability* (D. Brunsen and D.B. Prior, eds.), John Wiley & Sons, New York, pp. 103–142.

Cummings, D., and B.R. Clark. 1988. Use of Seismic Refraction and Electrical Resistivity Surveys in Landslide Investigations. *Bulletin of the Association of Engineering Geologists*, Vol. 25, No. 4, pp. 459–464.

Dallaire, E.E.. 1974. Electronic Distance Measuring Revolution Well Under Way. *Civil Engineering*, Vol. 44, No. 10, pp. 66–71.

DeGraff, J.V. 1985. Using Isopleth Maps of Landslide Deposits as a Tool in Timber Sale Planning. *Bulletin of the Association of Engineering Geologists*, Vol. 22, No. 4, pp. 445–453.

DeGraff, J.V., and S.S. Agard. 1984. Defining Geologic Hazards for Natural Resources Management Using Tree-Ring Analysis. *Environmental Geology and Water Sciences*, Vol. 6, pp. 147–155.

DeGraff, J.V., and P. Canuti. 1988. Using Isopleth Mapping to Evaluate Landslide Activity in Relation to Agricultural Practices. *Bulletin of the International Association of Engineering Geology*, Vol. 38, pp. 61–71.

Duncan, J.M., R.W. Fleming, and F.D. Patton. 1985. *Report of the Thistle Slide Committee to the State of Utah Department of Natural Resources, Division of Water Rights*. Salt Lake City, Utah, 62 pp.

Fleming, R.W., and A.M. Johnson. 1989. Structures Associated with Strike-Slip Faults That Bound Landslide Elements. *Engineering Geology*, Vol. 27 (Richard H. Jahns Memorial Volume), Nos. 1–4, pp. 39–114.

Franklin, J.A., and P.E. Denton. 1973. The Monitoring of Rock Slopes. *Quarterly Journal of Engineering Geology*, Vol. 6, No. 3, pp. 259–286.

Galster, R.W. 1977. A System of Engineering Geology Mapping Symbols. *Bulletin of the Association of Engineering Geologists*, Vol. 14, No. 1, pp. 39–47.

Geological Society Engineering Working Party. 1977. The Description of Rock Masses for Engineering Purposes. *Quarterly Journal of Engineering Geology*, Vol. 10, No. 4, pp. 355–388.

Gould, J.P., and C.J. Dunnicliff. 1971. Accuracy of Field Deformation Measurements. In *Proc., Fourth Pan-American Conference on Soil Mechanics and Foundation Engineering*, San Juan, Puerto Rico, American Society of Civil Engineers, New York, Vol. 1, pp. 313–366.

Hansen, A. 1984. Landslide Hazard Analysis. In *Slope Instability* (D. Brunsden and D.B. Prior, eds.), John Wiley & Sons, New York, pp. 523–602.

Harp, E.L., and M.A. Noble. 1993. An Engineering Rock Classification to Evaluate Seismic Rock-Fall Susceptibility and Its Application to the Wasatch Front. *Bulletin of the Association of Engineering Geologists*, Vol. 30, No. 3, pp. 293–319.

Harty, K.M. 1991. *Landslide Map of Utah*. Utah Geological Survey Map 133, 28 pp. (scale 1:500,000).

Hasan, S.E. 1994. Use of Soil Survey Reports in Geotechnical Projects. *Bulletin of the Association of Engineering Geologists*, Vol. 31, No. 3, pp. 367–376.

Hoek, E., and J.W. Bray. 1977. *Rock Slope Engineering*. Institute of Mining and Metallurgy, London, 402 pp.

Hutchinson, J.N. 1983. Methods of Locating Slip Surfaces in Landslides. *Bulletin of the Association of Engineering Geologists*, Vol. 20, No. 3, pp. 235–252.

Jade, S., and S. Sarkar. 1993. Statistical Models for Slope Instability Classification. *Engineering Geology*, Vol. 36, No. 1/2, pp. 91–98.

Jadkowski, M.A. 1987. *Multispectral Remote Sensing of Landslide Susceptible Areas*. Ph.D. thesis. Utah State University, Department of Civil and Environmental Engineering, Logan, 260 pp.

Jibson, R.W. 1992. The Mameyes, Puerto Rico, Landslide Disaster of October 7, 1985. In *Landslides/Landslide Mitigation* (J.E. Slosson et al., eds.), Reviews in Engineering Geology, Geological Society of America, Boulder, Colo., Vol. 9, pp. 37–54.

Johnson, A.M., and R.L. Baum. 1987. *BASIC Programs for Computing Displacements, Strains, and Tilts from Quadrilateral Measurements*. U.S. Geological Survey Open-File Report 87-343. 19 pp.

Johnson, J.G. 1990. Method of Multiple Working Hypotheses: A Chimera. *Geology*, Vol. 18, No. 1, pp. 44–45.

Keaton, J.R. 1984. Genesis-Lithology-Qualifier (GLQ) System of Engineering Geology Mapping Symbols. *Bulletin of the Association of Engineering Geologists*, Vol. 21, No. 3, pp. 355–364.

Keaton, J.R. 1989. The Thistle Landslide, Utah County, Utah. In *Engineering Geology of Western United States Urban Centers* (J.R. Keaton and R. Morris, eds.), Field Trip Guidebook T181, American Geophysical Union, Washington, D.C., pp. 59–63.

Keaton, J.R. 1990. Construction Geology: As-built Conditions. In *The Heritage of Engineering Geology; The First Hundred Years* (G.A. Kiersch, ed.), Centennial Special Volume 3, Geological Society of America, Boulder, Colo., pp. 429–466.

Keaton, J.R., R.M. Robison, and J.D.J. Bott. 1992. Landslide Hazard Analysis for Pipeline Design, Northeast Utah. In *Stability and Performance of Slopes and Embankments* (R.B. Seed and R.W. Boulanger, eds.), Geotechnical Special Publication 31, American Society of Civil Engineers, New York, N.Y., Vol. 1, pp. 192–204.

Kern and Company, Ltd. 1974. *The Mekometer, Kern ME3000, Electro-Optical Precision Distance Meter*. Tech. Information 10. Kern and Co., Aarau, Switzerland, 4 pp.

Koler, T.E., and K.G. Neal. 1989. Chestershire and Backdrop Timber Sales: Case Histories of the Practice of Engineering Geology in the Olympic National Forest. In *Engineering Geology in Washington* (R.W. Galster, ed.), Washington Division of Geology and Earth Resources Bulletin 78, Vol. 2, pp. 933–944.

Laser Technology, Inc. 1992. *Criterion Series Specifications*. Laser Technology, Inc., Englewood, Colo., 2 pp.

Leighton, F.B. 1976. Urban Landslides: Targets for Land-Use Planning in California. In *Urban Geomorphology* (D.R. Coates, ed.), Geological Society of America Special Paper 174, pp. 37–60.

Lips, E.W., and J.R. Keaton. 1988. Slope-a-Scope: A Convenient Tool for Rapid Topographic Profiling. In *Abstracts and Program*, Association of Engineering Geologists 31st Annual Meeting, Kansas City, p. 51.

Maerz, N.H., J.A. Franklin, and C.P. Bennet. 1990. Joint Roughness Measurement Using Shadow Profilometry. *International Journal of Rock Mechanics, Mineral Science, and Geomechanics Abstracts*, Vol. 27, No. 5, pp. 329–343.

Malde, H.E. 1973. Geologic Bench Marks by Terrestrial Photography. *Journal of Research, U.S. Geological Survey*, Vol. 1, pp. 193–206.

McCalpin, J. 1984. Preliminary Age Classification of Landslides for Inventory Mapping. In *Proc.,*

21st Engineering Geology and Soils Engineering Symposium, University of Idaho, Moscow, pp. 99–120.

McCann, D.M., and A. Forster. 1990. Reconnaissance Geophysical Methods in Landslide Investigations. *Engineering Geology*, Vol. 29, No. 1, pp. 59–78.

Merriam, R. 1960. Portuguese Bend Landslide, Palos Verdes Hills, California. *Journal of Geology*, Vol. 68, No. 2, pp. 140–153.

Miller, C.H., A.L. Ramirez, and T.F. Bullard. 1980. *Seismic Properties Investigation of the Springer Ranch Landslide, Powder River Basin, Wyoming.* U.S. Geological Survey Professional Paper 1170-C, 7 pp.

Moffitt, F.H., and H. Bouchard. 1975. *Surveying*, 6th ed. Harper & Row, New York, 879 pp.

Moore, J.F.A. 1973. The Photogrammetric Measurement of Constructional Displacements of a Rock-fill Dam. *Photogrammetric Record*, Vol. 7, No. 42, pp. 628–648.

Nasser, K.H. 1986. *Slope Stability Analysis of the Wolcott Landslide, Eagle County, Colorado.* Master of Engineering Report ER-3110. Colorado School of Mines, Golden, 111 pp.

Pachauri, A.K., and M. Pant. 1992. Landslide Hazard Mapping Based on Geological Attributes. *Engineering Geology*, Vol. 32, No. 1/2, pp. 81–100.

Palmer, D.F., and S.L. Weisgarber. 1988. Geophysical Survey of the Stumpy Basin Landslide, Ohio. *Bulletin of the Association of Engineering Geologists*, Vol. 25, No. 3, pp. 363–370.

Penman, A.D.M., and J.A. Charles. 1974. Measuring Movements of Embankment Dams. In *Field Instrumentation in Geotechnical Engineering*, British Geotechnical Society; John Wiley & Sons, New York, pp. 341–358.

Piteau, D.R., and R.J. Peckover. 1978. Engineering of Rock Slopes. In *Special Report 176: Landslides: Analysis and Control* (R.L. Schuster and R.J. Krizek, eds.), TRB, National Research Council, Washington, D.C., Chap. 9, pp. 192–228.

Reilly, J.P. 1992. *Practical Surveying with GPS.* P.O.B. Publishing Company, Canton, Mich., 184 pp.

Ross-Brown, D.M., and K.B. Atkinson. 1972. Terrestrial Photography in Open Pits: Description and Use of the Phototheodolite in Mine Surveying. *Transactions, Institute of Mining and Metallurgy*, London, Vol. 81, pp. 205–213.

Shoemaker, J.A., T.B. Hardy, and C.M.U. Neale. 1994. Jurisdictional Delineation of Wetlands with Multispectral Aerial Videography. In *Proc., 14th Biennial Workshop on Color Photography and Videography in Resource Monitoring*, May 25–28,

1993, Utah State University, Logan, American Society of Photogrammetry and Remote Sensing, Bethesda, Md.

Sowers, G.F., and D.L. Royster. 1978. Field Investigation. In *Special Report 176: Landslides: Analysis and Control* (R.L. Schuster and R.J. Krizek, eds.), TRB, National Research Council, Washington, D.C., Chap. 4, pp. 81–111.

St. John, C.M., and T.L. Thomas. 1970. The N.P.L. Mekometer and Its Application in Mine Surveying and Rock Mechanics. *Transactions, Institute of Mining and Metallurgy*, London, Vol. 79, Sec. A, pp. A31–A36.

Terzaghi, K. 1950. The Mechanism of Landslides. In *Application of Geology to Engineering Practice* (S. Paige, ed.), Geological Society of America, Boulder, Colo., Berkey Volume, pp. 83–123.

Umstot, D. 1988. *Field Studies and Modeling Analysis of the Roan Creek Landslide, Garfield County, Colorado.* Master of Engineering Report ER-3420. Colorado School of Mines, Golden, 186 pp.

Vandre, B.C. 1993. What Is Coarse and Durable Mine Waste Rock? In *Proc., 29th Symposium on Engineering Geology and Geotechnical Engineering* (G.M. Norris and L.E. Meeker, eds.), Reno, Nevada, March 22–24, pp. 22–33.

Varnes, D.J. 1984. *Landslide Hazard Zonation: A Review of Principles and Practice.* United Nations Educational, Scientific and Cultural Organization, Paris, France, 63 pp.

Wieczorek, G.F. 1982. *Map Showing Recently-Active and Dormant Landslides Near La Honda, Central Santa Cruz Mountains, California.* U.S. Geological Survey Miscellaneous Field Studies Map MF-1422 (scale 1:4,800).

Wieczorek, G.F. 1984. Preparing a Detailed Landslide-Inventory Map for Hazard Evaluation and Reduction. *Bulletin of the Association of Engineering Geologists*, Vol. 21, No. 3, pp. 337–342.

Williamson, D.A. 1984. Unified Rock Classification System. *Bulletin of the Association of Engineering Geologists*, Vol. 21, No. 3, pp. 345–354.

Williamson, D.A., K.G. Neal, and D.A. Larson. 1981. *The Field-Developed Cross-Section—A Systematic Method of Presenting Three-Dimensional Subsurface Information.* Olympic National Forest Geotechnical Section, USDA Forest Service, Ft. Lewis, Wash., 35 pp.

Wilson, S.D., and P.E. Mikkelsen. 1978. Field Instrumentation. In *Special Report 176: Landslides: Analysis and Control* (R.L. Schuster and R.J. Krizek, eds.), TRB, National Research Council, Washington, D.C., Chap. 5, pp. 112–138.

WP/WLI. 1990. A Suggested Method for Reporting a Landslide. *Bulletin of the International Association of Engineering Geology*, No. 41, pp. 5–12.

WP/WLI. 1991. A Suggested Method for a Landslide Summary. *Bulletin of the International Association of Engineering Geology*, No. 43, pp. 101–110.

WP/WLI. 1993a. A Suggested Method for Describing the Activity of a Landslide. *Bulletin of the International Association of Engineering Geology*, No. 47, pp. 53–57.

WP/WLI. 1993b. *Multilingual Landslide Glossary*. Bi-Tech Publishers, Richmond, British Columbia, Canada, 59 pp.

Wracker, D.A. 1973. Small Format Aerial Photography. *Mining Engineering*, Vol. 25, No. 11, pp. 47–48.

Wright, R.H., R.H. Campbell, and T.H. Nilsen. 1974. Preparation and Use of Isopleth Maps of Landslide Deposits. *Geology*, Vol. 2, No. 10, pp. 483–485.

Chapter 10

VERNE C. MCGUFFEY,
VICTOR A. MODEER, JR.,
AND A. KEITH TURNER

SUBSURFACE EXPLORATION

1. INTRODUCTION

Slope instability reflects soil, rock, and groundwater conditions that are hidden beneath the ground surface. Although geologic structures and strength properties of earth materials can often be inferred from surface investigations, subsurface investigations are also required to obtain definitive data and samples. Subsurface explorations exhibit a wide range in cost. In order to save time and money, subsurface exploration programs should be undertaken following site reconnaissance and surficial investigation programs (see Chapters 7 and 9) and before the selection of instrumentation (see Chapter 11). Subsurface investigation follows an iterative process that incorporates new procedures and adjustments as information is discovered and tested against multiple working hypotheses and proposed mitigation strategies.

Selection of exploration methods and development of a plan for the subsurface exploration program are based on considerations of study objectives, size of the landslide area, geologic conditions, surface conditions, access to the area, and limitations of budget and time. Available information concerning the site, including any plans for construction or remedial treatment, should be used to support this selection and planning process. A subsurface exploration program should provide information that allows for qualification and quantification of pertinent material properties. The exploration program should provide values for the undisturbed and residual shear strength or friction angle of all geologic deposits, pore pressures in water-bearing strata, depth to controlling features, and probable vertical and lateral limits of sliding. Interpretation of such data identifies and quantifies potential solutions for landslide movements.

1.1 Classification of Subsurface Exploration Methods

Subsurface exploration methods may be classed as *direct methods* and *indirect methods* (Hunt 1984). Direct methods, such as test borings and the excavation of test pits, allow the examination of materials, usually with the removal of samples. Indirect methods, such as geophysical surveys and use of the cone penetrometer, provide a measure of material properties that, by correlation with other data, allows the estimation of material type. Exploration methods may be further classified into the following key categories:

- Reconnaissance methods,
- Surface-based geophysical methods,
- Test and core borings,
- Borehole logging, and
- Field testing, including specialized sampling from test pits, adits, and shafts.

1.2 Definition of Appropriate Exploration Program

Decisions regarding the type and location of subsurface explorations are dependent on the information needed to quantify the various working hypotheses. Some rules of thumb that may be helpful in deciding on a reasonable approach to a subsurface exploration program are as follows:

- Reconnaissance methods involve low-cost techniques requiring a minimum of equipment. They provide both direct and indirect data.
- Surface geophysical explorations provide only indirect data but are relatively inexpensive and can cover a large area in a very short time.
- Borings constitute the most common subsurface explorations. They include a wide variety of techniques and can vary from relatively routine and low-cost approaches to highly specialized and expensive methods. Because borings generally are used to provide samples, they provide direct data. Samples obtained by different techniques vary considerably in their utility; in many cases samples obtained from borings produce inaccurate values for material properties because of their relatively small volumes.
- Field tests range from relatively inexpensive penetration tests that can be performed as part of exploratory boring programs to expensive specialized test pits. Results obtained from field tests provide confirmation of strength property estimates obtained in laboratory tests.
- Test pits provide direct data and the potential for collecting large samples or performing in situ field tests to obtain landslide information not available from other sources. These pits can usually reach only shallow depths; they become extremely costly as the depth increases.
- Geophysical or other methods for logging test borings often provide valuable information at a modest additional cost.
- Specialized sampling and investigations requiring the construction of adits and shafts are extremely expensive and time consuming. Adits and shafts may be hazardous in landslide areas because of the nature and inherent instability of earth materials; accordingly, they are used only rarely during landslide investigations.

1.3 Safety Considerations

The safety of the subsurface exploration team should be evaluated before the site is occupied. Explorations of landslides are often located in difficult terrain, and excavations may require temporary falsework to protect personnel. The landslide may still be active and thus may pose a constant risk to the workers and equipment. The possibility of having loose or unstable material upslope of the exploration crew should be considered, and precautions, such as building protective cages or setting up manual or automated movement-warning devices, should be taken. In some situations, it may be advisable for crews to work in shifts around the clock to reduce the duration of such safety measures.

1.4 Supervision by Geologist or Geotechnical Engineer

On-site supervision by a knowledgeable experienced geologist or geotechnical engineer is critical to the success of most subsurface investigation programs to ensure that the intent of all specifications is preserved and that the field activities are properly executed so that the desired results can be achieved. The chief functions of the supervision are to

- Enforce all technical and legal contract specifications;
- Maintain liaison with the designer of the exploration program;
- Select and approve modifications to the program specifications as new or unanticipated conditions are revealed (such as the addition or deletion of borings, changes in depths of borings, changes in the types, depths, or intervals of sampling, etc.);
- Ensure that complete and reliable field reports are developed; and
- Identify all geologic conditions accurately.

Lack of such a knowledgeable on-site decision maker during the exploration program can lead to large additional expenses if site revisitation becomes necessary to obtain additional required information. In some instances, without such expertise available, serious mistakes can be made during the exploration program that will aggravate the landslide conditions.

1.5 Sources of Information

Numerous sources of information are available for more specific guidance concerning the importance of subsurface investigation or proper procedures for planning and conducting subsurface explorations on landslides. Several basic engineering geology and geotechnical engineering textbooks contain chapters on subsurface exploration techniques (Schultz and Cleaves 1955; Krynine and Judd 1957; Hunt 1984; Johnson and DeGraff 1988). Such sources may be supplemented by government manuals prepared by agencies for training and guidance of their personnel (USBR 1974; NAVFAC 1982). In addition, numerous guides to suitable sampling and exploration practices have been prepared by professional societies, standards-setting organizations, manufacturers of exploration equipment, and commercial publishers (Hvorslev 1949; ASTM 1951; Mohr 1962; Merritt 1974; Lowe and Zaccheo 1975; Broms 1980; Hunt 1984).

Research literature concerning subsurface exploration procedures and the advantages and disadvantages of several techniques includes discussions of the applicability of geophysical exploration methods to engineering investigations by Griffiths and King (1969), Saayman (1978), van Zijl (1978), Hunt (1984), and Johnson and DeGraff (1988). Methods of logging boreholes have been described by Deere (1963), Myung and Baltosser (1972), Underwood (1974), and Knill (1975), and in a report prepared by the Association of Engineering Geologists (AEG 1978). The use of penetrometers and the evaluation of penetrometer data have been discussed by Krynine and Judd (1957), Sanglerat (1972), Alperstein and Leifer (1976), and Schmertmann (1978). Additional references to specific applications of various subsurface exploration techniques are given in subsequent sections of this chapter.

2. PLANNING SUBSURFACE INVESTIGATIONS

The initial planning of a subsurface investigation program incorporates information concerning terrain features, site accessibility, and anticipated geologic conditions to define the areal extent of the investigation; types of investigative procedures; test boring locations, spacings, and depths; and required types of samples and sampling frequencies.

Previously conducted surface investigations (see Chapter 9) will often suggest possible modes of landsliding. The subsurface exploration program must be designed to resolve the remaining uncertainties and to define the operative landslide mode (or modes) from among the various

hypotheses. A successful subsurface exploration program will identify the controlling subsurface deposits and quantify all variables that might control landslide activity according to the various alternative hypotheses using an iterative process that must be continuously modified to answer the critical design questions. The subsurface exploration program must define the spatial relationships and provide quantitative information on the density, shear strength, and permeability of each of the subsurface layers. The necessary parameters required by design solutions for the landslide should be quantified, including definition of properties for the very strong as well as for the very weak materials. Instrumentation may be needed to quantify the ranges of water pressure that can be expected in each of the important geologic deposits. The subsurface exploration program must be coordinated and integrated with the instrumentation program (see Chapter 11) so that the parameters that cannot be quantified by using conventional exploration techniques can be defined by the instrumentation.

Alternative exploration strategies and their required equipment and techniques should be identified on the basis of the initial site evaluations, both those in the office and from initial field inspection. This information should clearly identify the anticipated conditions, thereby allowing the investigator to select appropriate equipment, such as a Christensen core barrel, a borehole camera, or undisturbed-sampling tools. Careful attention must be given to alternative methods for sealing high artesian water pressures if they are encountered.

The preliminary layout, spacing, and depth of borings will depend on the prior site information. As a minimum, there should be borings near the top, middle, and bottom of a potential landslide, with as many profiles of borings as appear to be required to define the subsurface conditions. Philbrick and Cleaves (1958) suggested that a profile of borings be developed along the centerline of the landslide and that the first boring be placed between the midpoint and the scarp or head of the landslide. The area outside the landslide perimeter should also be explored to provide comparative data on the stable and unstable portions of the slope. Such information may also be needed to provide data on possible further expansion of the landslide or possible design of remedial measures.

2.1 Area of Investigation

The area of the investigation is determined partly by the size and type of an affected transportation project and partly by the extent and type of topographic and geologic features believed to affect the landslide activity. At sites where there is potential for future landslide movements, the area to be investigated cannot be easily defined in advance. A grid of borings should be placed within the suspected area to delineate the landslide (Figure 10-1). Once a landslide has occurred, the area of investigation can be better defined (Figure 10-2). However, in either case the area studied must be considerably larger than that comprising the suspected activity or known movement for three reasons:

- The landslide or potential landslide must be referenced to the stable area surrounding it,
- Most landslides enlarge with passage of time, and
- Many landslides are much larger than first suspected from the overt indications of activity.

As a rule of thumb, the area to be studied should be two to three times wider and longer than the area suspected. In some mountainous areas, it is necessary to investigate to the top of the slope or to some major change in lithology or slope angle. The lateral area must encompass sources of groundwater and geologic structures that affect the landslide stability.

2.2 Depth of Investigation

The depth of investigation is even more difficult to define in advance. Initial estimates of investigation depths can be made by applying various rules of thumb, including the following:

- The depth of movement at the center of the slide is rarely greater than the width of the zone of surface movement.
- The maximum depth of the failure surface is often approximately equal to the distance from the break in the orginal ground surface slope to the most uphill crack or scarp (McGuffey 1991).

Longitudinal cross sections drawn along the landslide centerline may also be helpful in defining initial investigation depths. Circular or elliptical failure surfaces connecting possible toe bulges and

uphill scarps can be sketched onto these cross sections; these surfaces may suggest possible maximum depths for movement. Continuous thick, hard strata within the slope may limit depths of movement. However, at least one boring should extend far below the suspected failure surface; deep, slow movements often are masked by more rapid movements at shallower depths.

Borings or other direct investigative techniques should extend deep enough (*a*) to identify materials that have not been subjected to movements in the past but that might be involved in future movements and (*b*) to clearly identify underlying stable materials. Boring depths are sometimes revised repeatedly as field investigation proceeds. Later, when field instrumentation has been installed and has begun to yield data, the existing or planned boring depths may be found to be insufficient, and increases in these depths may become necessary. The exploration program specifications should be flexible enough to allow for additional depths of investigation when the data obtained suggest deeper movements.

2.3 Duration of Investigation

Since most landslides are affected by climate changes, a minimum period for investigation should include one seasonal cycle of weather, which is one year in most parts of the world. Longer-term climate cycles, such as several years with periods of wetter and drier weather, are common, however; thus landslide investigations often require a monitoring phase lasting for many years or even several decades. In practical terms, such a long-term assessment often is impossible because there is usually a need to draw conclusions and make decisions concerning corrective action much more quickly.

Experience has shown that false conclusions have often been reached on the causes of landslides and the effectiveness of corrective measures because the effects of severe climate conditions were not adequately considered by the engineers and geologists. The worst climate conditions possible during the life of a project are likely to control the risk to the project of landsliding. Investigations made during climate conditions that are less severe can prove to be too optimistic, and those made during a particularly severe climate cycle may be too pessimistic.

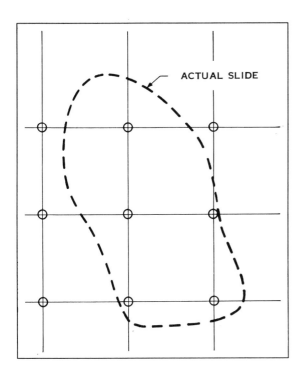

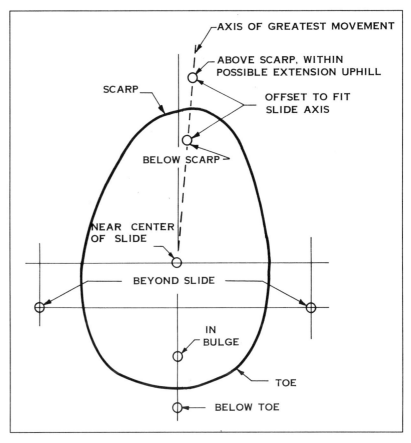

3. RECONNAISSANCE METHODS

Reconnaissance exploration methods range from relatively inexpensive ones involving the use of hand tools, soundings, and penetration tests to relatively expensive ones involving excavations. Reconnaissance exploration usually entails appraisal of conditions over a sizable area. The characteristics of commonly used reconnaissance procedures are summarized in Table 10-1.

Reconnaissance procedures should define the general characteristics of the earth materials suspected of involvement in landsliding; they should also identify and direct special attention to those locations at which significant or unusual problems may arise. Problems that should be rated as significant include any landslide features or conditions that may influence the selection of stabilization measures or suggest further growth in the severity or scale of landsliding.

Reconnaissance explorations emphasize relatively quick and economical methods for assessing earth material properties. They must also allow for consideration of the full range of engineering problems resulting from landsliding as well as potential mitigation solutions. Some detailed field methods may be required to analyze certain important problems, but in general the required information should be easily and rapidly obtained. Terrain evaluation and the application of geologic and pedo-

logic principles, as discussed in Chapters 8 and 9, form important components of most reconnaissance explorations. Throughout the reconnaissance exploratory process, the investigator must maintain a perspective between the level of detail required to identify potentially significant landslide stability conditions and the amount of information required to subsequently analyze and resolve these problems. Reconnaissance exploration methods are divided into three broad categories according to the principal tools and methods used: portable hand tools and soundings, shallow test pits and trenches, and penetration tests.

3.1 Portable Hand Tools and Soundings

Hand tools and soundings provide low-cost, rapid preliminary evaluations of subsurface conditions. These methods involve strictly manual hand tools and lightweight power tools, some of which may be mounted on small vehicles. *Soundings* are conducted by means of metal bars driven to shallow depths, either by blows from a hammer or by application of deadweights. Soundings may provide

FIGURE 10-1
(above left)
Grid of borings
in suspected
landslide area
before movement
commences.

FIGURE 10-2
(above right)
Location of
borings in a
known landslide.

Table 10-1
Reconnaissance Methods

CATEGORY	APPLICATIONS	LIMITATIONS
Soundings	Provide information on thicknesses and depths to shallow bedrock by using metal probes pushed through soft shallow soils	No samples obtained; strata are not identified
Portable hand tools Retractable-plug samplers	Provide subsurface profile; locate buried objects (such as utility lines), boulders, and soil-bedrock interface	Has limited penetration in clay materials
Hand augers	Provide continuous profiling in granular soils above the groundwater table and in clayey soils of firm or greater consistency above and below groundwater table	Samples are disturbed; cannot penetrate below groundwater table in granular soils; penetration in strong soils very difficult
Shallow test pits and trenches	Provide visual examination of strata, groundwater conditions, soil-rock interface, discontinuities, and rupture surfaces	Limited depth when lightweight machinery used; safety issues often critical; expensive or impossible to use below groundwater table
Penetration tests	Are fast and efficient methods of determining continuous penetration resistance for all but strongest of soils	Samples are not recovered; cannot penetrate strong soils or rock

some measure of penetration resistance, but no samples are obtained. *Hand augers* or *post-hole diggers* may be used to obtain disturbed soil samples at moderate depths, depending on the soil properties. A 2.5-cm diameter *retractable-plug sampler* consists of a hollow tube of the specified diameter that can be driven by hammering. It can collect samples up to 1 m long before the tube is filled and the sampler must be removed from the hole and the sample extracted. Under favorable circumstances in soft clays, these samplers can be driven to several tens of meters, but usually they can only be used to much shallower depths.

3.2 Shallow Test Pits and Trenches

Shallow test pits can be excavated with hand tools. Usually, however, mechanical equipment such as backhoes can excavate shallow test pits more efficiently and economically and may be required to excavate deeper pits or long trenches. The sides of the excavation should be sampled, logged, and photographed in detail to provide a three-dimensional picture of the materials and geologic structures (Hatheway 1982). In addition, these excavations allow the undertaking of large-scale

in-place field tests (see Section 7.2). Such large-scale tests may overcome the primary limitation of laboratory tests, namely, their inability to integrate the variations in soil properties because of their small sample sizes (Sowers and Royster 1978). When the observations and tests are complete, the excavation should be filled in or, in some cases, incorporated into the remedial design, such as by serving as a drainage outlet.

3.3 Penetration Tests

Penetration tests involve the use of exploratory drilling equipment; nevertheless, they are frequently conducted as part of early reconnaissance investigations as well as during subsequent, more detailed investigations. Penetration tests provide an extremely valuable and relatively low-cost supplement to the data obtained by direct borehole sampling and logging procedures. Conducted during the advancement of boreholes, the tests measure the resistance of the materials to the advancement of standard probes (Sanglerat 1972; Schmertmann 1978). Subsurface conditions are inferred from correlations of soil properties with resistance values. Details of the two main types of penetration test, the *standard penetration test* (SPT) and the *continuous cone penetrometer test* (CPT) are discussed in Section 7.1.1.

4. SURFACE-BASED GEOPHYSICAL METHODS

Surface-based geophysical exploration often provides the quickest and most economical means of obtaining general information on subsurface conditions over relatively large and rugged areas (Griffiths and King 1969; Johnson and DeGraff 1988). These methods supply indirect information concerning landsliding because they involve the investigation and mapping of the distribution of physical properties of earth materials, such as the speed of transmission of elastic waves or the ability to transmit electrical currents, which are not directly related to landsliding processes. These geophysical techniques do not replace test borings, samples, test pits, or trenches. Rather, they supplement these more expensive subsurface exploratory methods, assist in correlation of information among widely spaced exploration locations, and greatly reduce the time, cost, and environmental problems associated with large-scale drilling programs.

There are two broad categories of geophysical exploration: surface-based exploration and borehole logging methods. Borehole logging and seismic methods are discussed in Section 6; the latter involve the direct measurement of wave velocities within and between boreholes (see Section 6.6). Table 10-2 summarizes the characteristics of the most commonly used surface-based geophysical exploration methods, including

- Electrical and electromagnetic surveys,
- Seismic surveys,
- Microgravity surveys, and
- Ground-penetrating radar surveys.

4.1 Electrical and Electromagnetic Surveys

Many geophysical exploration methods use electrical and electromagnetic principles (Griffiths and King 1969; Dobrin 1976; Hunt 1984; Johnson and DeGraff 1988). Some operate by measuring the conductance of electrical currents through earth materials, whereas others induce currents in earth materials by electromagnetic fields or measure the variations of such fields caused by variations in earth conductivity. Sources of electrical energy may occur naturally, they may exist as a consequence of human activities (for example, radio transmissions), or they may be specifically generated during a geophysical survey. Electrical conductance of earth materials—or the inverse value, resistivity—may be determined by airborne, surface, or borehole surveys. Details of electrical surveying techniques have been widely discussed (Scharon 1951; Van Nostrand and Cook 1966; Donaldson 1975; Dobrin 1976).

4.1.1 Resistivity Methods

Surface-based measurement of the electrical resistivity of earth materials involves the introduction of an electrical current into the ground and the measurement of the materials' resistance to the current. There are several variations to the resistivity survey method. All introduce a controlled electrical current into the earth materials through two *current electrodes*. The resistance of the materials to the current is measured by the potential difference between two *potential electrodes* placed within the field created by the current electrodes.

The presence of moisture and dissolved salts within the pores of the soil or rock largely controls the apparent conductivity of the earth materials, and hence the inverse, apparent resistivity. The flow of current through earth materials is by ionic conduction, which depends on the salinity of the fluid present, porosity, and percent saturation. For example, a dense granite with few voids and little moisture will demonstrate high resistance, whereas a moist clay will show low resistance. Sometimes a rupture surface in a landslide will be detected as a zone of low resistance because of the concentration of moisture along the surface. However, these conditions may show considerable seasonal variation. Seasonal fluctuations in resistivity of as much as 200 percent have been reported (Brooke 1973).

Commonly used electrode configurations include the *Wenner array*, the *Lee-partitioning array*, and the *Schlumberger array* (Figure 10-3). The depth of investigation of a resistivity survey is proportional to the spacing of the current electrodes. However, the penetration of the electrical current into the ground depends on both the resistance of individual layers and their distribution. As a consequence, a weakness of electrical methods is

Table 10-2
Surface-Based Geophysical Methods

Type of Survey	Applications	Limitations
Electrical and electromagnetic		
Electrical resistivity	Locates boundaries between clean granular and clay strata, groundwater table, and soil-rock interface	Difficult to interpret and subject to correctness of the hypothesized subsurface conditions; does not provide engineering strength properties
Electromagnetic conductivity profiling	Locates boundaries between clean granular and clay strata, groundwater table, and rock-mass quality; offers even more rapid reconnaissance than electrical resistivity	Difficult to interpret and subject to correctness of hypothesized subsurface conditions; does not provide engineering strength properties
Seismic		
Seismic refraction profiling	Determines depths to strata and their characteristic seismic velocities	May be unreliable unless strata are thicker than a minimum thickness, velocities increase with depth, and boundaries are regular. Information is indirect and represents average values
Direct seismic (uphole, downhole, and crosshole surveys)	Obtains velocities for particular strata, their dynamic properties, and rock-mass quality	Data are indirect and represent averages; may be affected by mass characteristics
Microgravity	Extremely precise; locates small volumes of low-density materials utilizing very sensitive instruments	Use of expensive and sensitive instruments in rugged terrain typical of many landslides may be impractical; requires precise leveling and elevation data; results must be corrected for local topographic features; requires detailed information on topography and material variations; not recommended for most landslide investigations
Ground-penetrating radar	Provides a subsurface profile; locates buried objects (such as utility lines), boulders, and soil-bedrock interface	Has limited penetration in clay materials

that no simple proportionality exists between the electrode spacing and the depth of investigation.

Resistivity surveys can be conducted to provide vertical or horizontal profiling. In vertical profiling, the center of the electrode spread is kept fixed at a desired location, and the spacings of the electrodes are increased. Because increased spacings result in increased depths of investigation, this procedure is called *sounding*. In contrast, horizontal profiling, sometimes referred to as *electrical mapping*, employs a constant electrode spacing with the array moving so as to center at a series of desired map locations. Usually the spacing to be used for the horizontal profiling is selected following analysis of several vertical soundings.

The Wenner array uses four electrodes spaced equally at a distance x in a straight line on the ground surface. A known current is passed into the ground between the outer two electrodes, and the difference in electrical potential generated by the resistance to the current flow is measured between the two inner electrodes. The x spacing between the electrodes is increased by a fixed amount, and the measurements are repeated.

In the Schlumberger array, the total spacing X is varied while the spacing x of the potential electrodes is kept constant (within certain limits). Both the Wenner and Schlumberger arrays assume laterally uniform materials. Lateral variability is expected in landslides, and knowledge of such variations is most valuable. The Lee-partitioning array allows for determination of lateral variations. A third potential electrode placed centrally between the first two potential electrodes permits detection of nonuniform lateral resistivity variations (Johnson and DeGraff 1988). The Wenner array and its Lee-partitioning variation are the most commonly used electrical resistivity surveying techniques.

The major advantages of the resistivity surveying techniques lie in the portability and simplicity

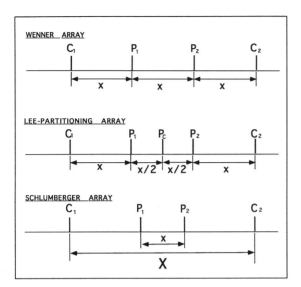

WENNER ARRAY

LEE-PARTITIONING ARRAY

SCHLUMBERGER ARRAY

FIGURE 10-3
Common electrode
configurations
used in resistivity
surveys: C = current
electrodes; P =
potential electrodes.

of the instrument. Large areas can be covered relatively rapidly at small cost. The major disadvantage is that the interpretation of the measurements is neither simple nor unique, especially in areas where the strata are not horizontal, the structures are complex, the layers are nonuniform, or contrasts in material resistivities are not great (van Zijl 1978). Such weak contrasts can occur where very-low-moisture colluvium overlies dense parent-rock materials.

4.1.2 Electromagnetic Conductivity Profiling

The electromagnetic (EM) conductivity measurement technique utilizes an instrument having two coils: a transmitter and a receiver. The transmitter coil uses an alternating electrical current of a specified frequency to produce an associated magnetic field, which in turn induces an electrical current in the ground. This current then creates a secondary current in the receiver coil. The coils are connected by a reference cable, which is monitored to provide a value of the apparent conductivity of the earth materials between the two coils. The *skin depth*, generally regarded as the depth below which no appreciable contribution to the apparent conductivity is made, is affected by coil spacing and frequency of the alternating-current signals as well as the conductivity of the earth materials (Kaufman and Keller 1983). Different instruments use different coil spacings and frequencies. Smaller coil spacings coupled with high frequencies give high resolutions but shallow skin depth, whereas large

coil spacings and low frequencies provide greater depths of exploration with lower spatial resolution. For average earth-material conductivity values, commercial EM instruments provide depths of exploration ranging from less than 10 m to about 80 m.

The data collection procedures for the EM technique are relatively simple. Because it is desirable to repeat the profiles with different coil spacings, frequencies, or both, field survey lines should be clearly marked with regularly spaced measurement stations. Horizontal and vertical control for all such stations is necessary. In addition to varying the coil spacings and frequencies, some systems allow the measurement of different components of the electromagnetic dipole. For example, when the coils are held parallel to the ground, the vertical dipole is measured; when the coils are held in a vertical plane, the horizontal dipole is measured. Because these different orientations typically provide different penetration depths, both should be collected where possible (Kaufman and Keller 1983).

4.1.3 Applications of Electrical and EM Surveys to Landslides

Resistivity surveys have been used successfully to map the limits of landslide masses. Both vertical sounding and horizontal profiling methods have been used. Because landsliding results in the disruption of earth materials and the development of an irregular rupture surface, large contrasts in electrical resistivity are frequently associated with these movements. Subsurface water may accumulate at some locations and drain from others. Where test borings or the presence of springs permits the recognition of water table conditions, the extent of such saturated and reduced-stability zones may be mapped by resistivity methods. Case histories describing the use of resistivity surveys for landslides have been reported by Trantina (1963), Takada (1968), Moore (1972), Brooke (1973), and Bogoslovsky and Ogilvy (1977).

As with resistivity surveys and soundings, the apparent conductivity of earth materials measured by EM techniques is largely a reflection of the presence and salinity of groundwater. Therefore, EM techniques are essentially as applicable to landslide investigations as resistivity surveys. EM methods offer the potential for still more rapid reconnaissance of suspect areas, however.

4.2 Seismic Surveys

Seismic surveys include refraction, reflection, and acoustic techniques (Johnson and DeGraff 1988). All are based on the fact that the elastic properties of earth materials determine the velocities of waves propagating through them (Aikas et al. 1983). Shock waves generated by some energy source propagate through the body of the earth as *body waves* and along the near surface of the earth as *surface waves* (Dobrin 1976). These waves follow multiple paths from source to point of detection. Initially, as *direct waves*, they follow a direct path, and the measurement of the elapsed time of travel and the distance permits the computation of the wave velocity through the material. At greater distances, the waves encounter different materials and are *refracted* and *reflected* at the velocity interfaces (Figure 10-4). Complex sequences of wave disturbance patterns are thus detected at most locations following the release of a single energy pulse from a discrete source.

There are numerous types of both body and surface waves. Surface waves are not used in most geophysical engineering explorations but have important roles in more sophisticated applications (Dobrin 1976); they will therefore not be discussed further here. Body waves include both compressional and shear waves. Compressional body waves travel the fastest of all the waves, and consequently are called *primary* or *P-waves* because they arrive first at a detector or receiver. Compressional waves are used for most engineering seismic applications. They are the dominant body wave generated by explosives in a shallow borehole, hammer blows to a metal plate on the surface, or the dropping of a heavy weight. Compressional wave velocities for many types of earth materials have been reported by Redpath (1973), Dobrin (1976), and Telford et al. (1976).

Shear waves arrive at a detector following the P-waves but ahead of surface waves. For this reason they are often referred to as *secondary* or *S-waves* (Johnson and DeGraff 1988). Shear waves travel only through solids. Because the usual sources of shock-wave energy do not generate strong shear waves, specialized techniques are required. Measurement and analysis of shear-wave velocities can supply important information to landslide investigations. Compressional-wave velocity (V_p) is frequently affected by the degree of saturation, whereas shear-wave velocity (V_s) is unaffected. Consequently the ratio V_s/V_p is often of value in determining degree of saturation (Johnson 1976). Shear-wave velocities and the V_s/V_p ratio also offer a superior method for in situ estimation of dynamic elastic moduli, such as the Young's modulus, shear modulus, and Poisson's ratio (Griffiths and King 1969; Johnson and DeGraff 1988).

4.2.1 Seismographs

Seismographs are used to record shock-wave travel times between a source and a receiver, or geophone, over a series of selected distances. Seismographs may be either single channel or multichannel.

In most seismic work involving landslides, a multichannel seismograph system is used, which includes a number of detectors or geophones that are placed on the surface at varying distances from the shock source, amplifiers that enhance the signals, and a recording oscillograph that produces a time-based record of the signals received from all the detectors [Figure 10-5(*a*)]. Multichannel seismographs allow more sophisticated data filtering, recording, and processing of an entire series of

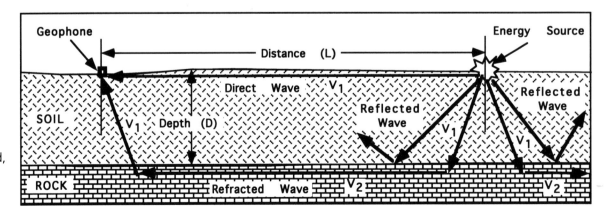

FIGURE 10-4
Refracted, reflected, and direct shock waves (Weaver 1978).

records from a single source. They are more complex and expensive than the single-channel seismographs. With a single-channel seismograph, the energy source must be repeated for different geophone distances until a suitable range of travel times has been collected [Figure 10-5(b)]. This procedure requires more time and the use of multiple energy events. An important capability of any seismograph is enhancement of the signal-to-noise ratio, usually by adding the effects of multiple sequential energy sources (such as hammer blows) to produce a single record. Weak signals are thus enhanced and random noise events are partially cancelled.

4.2.2 Seismic Refraction

The seismic refraction method is based on measurement of the time required for shock compression waves generated by hammer impact or by detonation of an explosive at or just below the ground surface to pass from one point to another through the earth. Some of the waves are deflected or refracted by the more rigid, deeper formations and return to the surface where their times of arrival are recorded.

When a shock wave from the explosion or hammer impact reaches each geophone, it appears on the recording as a pronounced change in the trace and is termed the *first arrival*. The time of first arrival at each geophone may be used to compute the depth to successively more rigid strata. Methods of analysis have been described by Dobrin (1976), Bullock (1978), and Johnson and DeGraff (1988). If the velocity interfaces causing the wave refraction are not parallel to the surface, apparent rather than true velocities will be obtained. It is imperative that seismic refraction profiles be run in both forward and reverse directions so that the proper computations can be made (Redpath 1973; Dobrin 1976).

4.2.3 Seismic Reflection

Reflected shock waves have many advantages over refracted waves in accurately calculating depths. However, the seismic reflection method has not found widespread use in shallow engineering investigations (Hunter et al. 1984). The reason for this lack of use is the difficulty in recognizing shallow-depth reflected-wave arrivals when intermixed

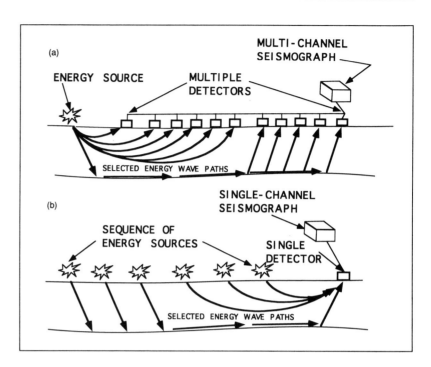

with a series of refracted body-wave and direct surface-wave arrivals. Some procedural changes, coupled with new instruments, digital data recording methods, and computer-aided data analysis, have now made the use of reflection data more feasible for shallow-depth computations (Hunter et al. 1984; Johnson and DeGraff 1988).

4.2.4 Applications of Seismic Surveys to Landslides

As with electrical resistivity surveys, seismic surveys offer several advantages in landslide investigation: the environment is not disturbed, the equipment is portable, and large areas can be covered at relatively small cost. However, interpretation of seismic measurements is also conjectural where the geology is complex and velocities of the various materials are not in sharp contrast. Landslides can cause erratic changes in density, and therefore wave transmission may occur in complex patterns that often are difficult to quantify. However, the limits of sliding activity are often identifiable by changes in signal response when the edges of the landslide are reached.

The refraction method has been used most frequently to determine depths and geometries of landslide rupture surfaces and hence landslide volumes. Case histories of such applications for landslides in a variety of soil and rock types have been

FIGURE 10-5
(a) Multichannel and (b) single-channel refraction seismic surveys (modified from Saayman 1978).

reported by Trantina (1963), Carroll et al. (1972), Brooke (1973), Murphy and Rubin (1974), and Bogoslovsky and Ogilvy (1977). Reductions in shear-wave velocities have been reported in some landslides. If shear-wave velocities can be determined at a landslide site, the V_s/V_p ratios for disturbed and undisturbed materials are of considerable value in defining the rupture surface and seasonally monitoring the water table and degree of saturation (Johnson and DeGraff 1988).

Even before observable landslide movements occur, compressional wave velocities may be progressively reduced by the opening of cracks in the rock mass. Periodic wave-velocity measurements in areas surrounding open-pit mines have identified changes in these velocities and hence have permitted the monitoring of the initiation and progression of tension fracturing that can lead to subsequent slope failure (Lacy 1963; Dechman and Ouderhoven 1976). Attenuation of seismic energy is also typically greater in landslide materials. The increased attenuation is related to reduction of shock-wave velocity and reflects disturbance of the material. Equipment that can record geophone outputs can be used to measure any progressive attenuation of energy. Such progressive attenuation may over time indicate potential slope failure (Tamaki and Ohba 1971).

4.2.5 Subaudible Rock Noise Monitoring

The detection of subaudible rock noise (SARN), also known as acoustic emissions, has been attempted at a number of landslide locations with varied success (Kennedy and Niermeyer 1971; McCauley 1976; Jurich 1985; Jurich and Miller 1987). The method relies on the detection of low-intensity noises emitted by movements of earth masses within the landslide. SARN technology developed from microseismic monitoring of rock bursts within mines (Obert and Duvall 1942, 1957; Hardy 1981). SARN data are gathered on landslides by using receivers attached to wave guides. Metal rods or pipes driven into the ground, metal borehole casings, or metal strips attached to plastic casings are suitable wave guides. Improved instrumentation has allowed the three-dimensional location of acoustic emission sources under favorable circumstances (Hardy 1981).

McCauley (1976) and Jurich (1985) both emphasized that the noise rate, not the number of events, is the significant measurement. SARN monitoring of landslides near heavily traveled transportation facilities may be difficult, however (Jurich and Miller 1987). Kennedy and Niermeyer (1971) used SARN to successfully predict a major slope failure in an open-pit mine.

4.3 Microgravity Surveys

Gravity surveys have been used to detect major subsurface geologic structures. In comparatively recent times, extremely precise gravity surveys, termed *microgravity surveys*, have been applied to selected engineering applications (Greenfield 1979). These surveys utilize very sensitive instruments (gravimeters) that permit measurement of the gravitational attraction at a given location to within 0.01 milligal (a milligal corresponds to 0.001 of the value of the gravitational constant). This precision corresponds to a 24-cm change in thickness for a stratum having a density contrast of 1 g/cm^3 with its surroundings. Thus, in theory, such precise microgravity surveys can detect areas of low density in relatively large landslides. Colluvium or landslide debris is usually less dense than bedrock, so large bodies of loosened rock or soil can be identified where the density contrast is sufficiently great.

However, practical use of microgravity surveys in most landslide investigations is doubtful. Gravity measurements require precise leveling and elevation determinations. The recorded values reflect topographic conditions and must be corrected to remove the influence of local topographic features. Such corrections require detailed information on the topography and material compositional variations. Gravimeters are expensive and sensitive instruments, and their use and transport across rugged landslide surfaces may be difficult. Accordingly, the use of microgravity surveys is not recommended for most landslide investigations.

4.4 Ground-Penetrating Radar Surveys

Ground-penetrating radar (GPR) has experienced rapid development (Moffatt 1974; Morey 1974; Arcone 1989; Doolittle and Rebertus 1989). The method appears to offer important potential for rapid subsurface profiling. GPR instruments constantly emit and receive signals as they are moved across the ground surface. Energy emitted by GPR instruments in the radio portion of the electro-magnetic spectrum is reflected back to the receiver. Some applications permit the use of airborne equipment.

Common earth materials differ in degree of transparency to radar energy and consequently exhibit different degrees of radar penetration (Cook 1974). A chief limitation to the use of current GPR equipment for most landslide investigations is the extremely poor transmission of radar energy through clay soils and shales. The presence of even minor amounts of clay limits effective GPR penetration to only a few meters at best.

5. TEST AND CORE BORINGS

Exploratory borings form a critical component of subsurface exploration. They are commonly undertaken to satisfy several objectives, including

- Identification of the subsurface distribution of materials with distinctive properties, including the presence and geometry of layers of material (referred to by geologists as *stratigraphy*);
- Determination of parametric data on the characteristics of each layer by
  - Retrieving representative samples and conducting laboratory tests on these samples to provide data concerning moisture content, gradation, plasticity, shear strength, and other properties as required;
  - Conducting in situ field tests, such as penetration tests, as the borehole is advanced; and
  - Performing geophysical and visual borehole logging activities; and
- Acquisition of groundwater data by observing conditions encountered while drilling and by converting exploratory holes into groundwater monitoring wells to provide such long-term data as free-water levels, artesian pressures, flow rates, and water samples.

In order to obtain the desired information, it is usually necessary to pay particular attention to how the borehole is advanced and how the material is removed from the borehole, how the sides of the hole are stabilized to prevent mixing and contamination of samples, and how the fluid pressure in the hole is controlled to prevent collapse of the hole, excessive deformation to the materials sur-rounding the hole, or disturbance of the material to be sampled. Exploratory borings must always be undertaken using methods that minimize any changes in the nature of the strata being sampled and tested.

5.1 Borehole Drilling Methods

Creation of a borehole in either soil or rock involves three stages:

- Fragmentation of the earth materials,
- Removal of the material fragments from the hole, and
- Stabilization of the hole walls to prevent their collapse.

Various methods are available to meet each of these requirements; accordingly, the undertaking of exploratory borings requires careful selection of the most appropriate and economical techniques to achieve the best possible exploration under given site conditions. Hunt (1984) provided summaries of the advantages and limitations of the common methods.

5.1.1 Fragmentation of Materials

Several methods may be used to fragment or disaggregate materials encountered by the borehole as it is advanced. The commonly used methods in soil include

- *Washing* or *jetting*: the use of powerful water jets in cemented materials and the circulation of water in loose sands, soft clays, and organic soils;
- *Chopping*: the repeated dropping and twisting of the bit to break up the materials; and
- *Augering*: the use of cutting or grinding bits to excavate moderately strong soils and weak rocks.

The commonly used methods in rock include

- *Percussion drilling*: the use of repeated impacts of a hardened probe to disaggregate rock or cemented soils; and
- *Core drilling*: the use of an abrasive cutting head to form an annular ring around the circumference of the hole while preserving a central rock core.

Only a few drilling methods are recommended for subsurface investigation of landslides. Where undisturbed samples are required, hollow-stem augers should be used in soils and core drills should be used in rock. If only disturbed samples are required in soils, continuous-flight augers of various diameters, including quite large diameters in some cases, may be used.

Rock cores are collected by using core bits and any of a variety of core barrels. A core bit consists of a hollow ring of durable cutting teeth (usually composed of tungsten carbide or diamonds) designed to fragment the annular area around the circumference of the hole. The resulting central rock core is preserved in a core barrel that can be retrieved periodically. Although core drilling may not always provide undisturbed samples of rock, other methods of drilling boreholes in soil or rock do not provide enough accurate information concerning the subsurface conditions or useful samples for determining physical material properties and so are not recommended for use in most landslide investigations.

5.1.2 Removal of Material Fragments

Once fragmented, the materials can be removed from the hole by

- *Dry methods*, including the use of continuous-flight and hollow-stem augers; and
- *Circulating fluids*, which may be high-pressure air, relatively clean water circulated within a casing, or mud slurry. Mud slurry may also stabilize the borehole.

5.1.3 Stabilization of Hole

Some form of stabilization is often needed to prevent collapse of the walls of a borehole, even while the hole is being extended. Usually no stabilization is needed in strong soils above the groundwater table or in relatively competent rocks. Two methods of stabilization are common: *casing* and *mud slurry*. Hollow-stem augers provide their own casing.

Casing is used in sands and gravels above the groundwater table and in most soils below the groundwater table. It may also be required in very soft soils. Casing is available in several standard sizes and is composed of a variety of metal or plastic materials. Polyvinylchloride (PVC) plastic is the most commonly used plastic casing. Metal casing is commonly used for support of side walls during drilling, whereas PVC casing is commonly used in permanent or long-term observation wells. Metal casing can be driven ahead of the hole by use of a hammer; plastic casing is usually installed within a hollow-stem auger before it is withdrawn or in holes that are marginally stable and remain open long enough to allow installation of the casing.

Casing is often necessary, yet the use of any type of casing potentially has a number of disadvantages:

- Installation by driving may be slow in strong soils;
- Recovery of the casing often is impossible;
- Sampling to locate changes in strata may not be possible if the casing has already been driven past the point where such changes occur;
- Obstacles, such as boulders, cannot be penetrated by the casing, and further extensions of the hole require the use of a smaller casing that fits within the first casing;
- Driving of casings in gravels requires chopping to break up the gravel particles;
- If sufficient drilling fluid (either wash water or drilling mud) is not present within the casing, loose or granular soils below the water table tend to flow into the casing because of liquefaction effects, causing plugging of the casing and loosening of soils below the casing; and
- The presence of the casing prevents the use of several borehole remote-sensing and logging devices (see Section 6) and may interfere with the accuracy of penetration tests.

Mud slurry can be formed by mixing commercial bentonite pellets with water at the drill site or by the natural mixing of clayey soils and circulating water in the hole. Use of mud slurry is a fast and efficient method of stabilizing boreholes in many situations. However, it also has some disadvantages:

- Mud cakes the hole walls and prevents the use of some visual logging devices, such as borehole cameras;
- Mud may penetrate the hole walls and contaminate some samples;
- Mud-caked walls may interfere with accurate water-level determinations;
- Mud slurry may not prevent hole closure in soft soils; and

- Large pumps may be required to circulate the slurry, especially when hole depths exceed 10 m, and excessive pump wear may occur if care is not taken to remove sand-sized particles by cycling the slurry into settling ponds.

Drilling fluid provides dual functions: it cleans and lubricates the drilling bit and it stabilizes the hole. Especially in soils and weak rocks, the fluid must be pumped at a carefully monitored pressure to prevent creation of cavities around the hole or undue disturbance of materials immediately below the hole that are to be sampled subsequently.

5.2 Observations During Drilling

Observations made during drilling often provide very important indications that can help evaluate subsurface conditions. Some examples of such observations are the following:

- The rate of rock core advance and the change of pressure on the bit can help to identify changes in the strata.
- In addition to notations on gain or loss of water during drilling, it is useful to measure the rise or fall of the drilling fluid in the hole. These measurements can sometimes provide an estimate of the piezometric elevation and the flow rate through the layer being investigated.
- Measurements of the depth of drilling water at the beginning and end of the day and at the beginning and end of any work stoppage can be used to quantify permeabilities and static water tables. The use of drilling mud may mask critical observations on groundwater and artesian pressures; therefore, drilling mud should be used only where absolutely necessary in landslide investigations.

The field instrumentation plan should be coordinated with the subsurface investigation plan since many of the instruments must be installed through drill holes.

5.3 Test Borings and Sampling in Soils

A large variety of soil boring techniques and sampler types are available. Detailed explanations of the many variations are beyond the scope of this report but are readily available from numerous sources (Hvorslev 1949; ASTM 1951; USBR 1974;

Broms 1980; NAVFAC 1982; Hunt 1984). Such manuals and guidance documents offer detailed recommendations concerning the choice of equipment and field procedures. Two broad categories of soil samples may be obtained from boreholes: disturbed and undisturbed samples.

Disturbed soil samples are primarily used for soil classification tests and must contain all the soil constituents, although the soil structure is not preserved. Disturbed soil samples are usually collected using split-barrel samplers following the procedure recommended by ASTM D1586, Penetration Test and Split-Barrel Sampling of Soils. Some common samplers used to collect disturbed soil samples are described in Table 10-3.

Undisturbed soil samples usually do not entirely represent truly undisturbed in situ soil conditions because the process of sampling inevitably introduces some disturbance into the soil structure. However, undisturbed samples are collected in ways that minimize the degree of such disturbance. These samples are taken primarily for laboratory strength and compressibility tests and for the determination, as closely as possible, of in situ soil properties. Undisturbed soil samples are collected using thin-wall tube samplers in soft to firm clays and coring samplers in other types of soils (Hunt 1984). ASTM D1587, Thin-Walled Tube Sampling of Soils, defines recommended procedures. In all cases, undisturbed soil samples should be collected so as to satisfy the following criteria:

- They should contain no visible distortion of stratification or softening, cracking, or modification of material conditions by drying or freezing;
- The length of the recovered sample should exceed 95 percent of the length of the sampled interval; and
- The annular cross-sectional area of the sampler should be less than 15 percent of the total area of the sampler; in other words, the walls of the sampling device should be as thin as possible.

Some common samplers used to recover undisturbed soil samples are described in Table 10-4. Usually, samples are collected at 1- or 2-m intervals (in North America at standard 5-ft intervals) or at changes in strata. Continuous sampling is frequently desired in landslide investigations. Hutchinson (1983) discussed methods of locating rupture surfaces in landslides.

Table 10-3
Common Samplers To Collect Disturbed Soil Samples (NAVFAC 1982)

SAMPLER	TYPICAL DIMENSIONS	SOILS THAT GIVE BEST RESULTS	METHOD OF PENETRATION	CAUSE OF DISTURBANCE OR LOW RECOVERY	REMARKS
Split barrel	Standard is 50 mm outside diameter (OD) and 35 mm inside diameter (ID); penetrometer available up to 100 mm OD and 89 mm ID	All fine-grained soils that allow sampler to be driven; gravels invalidate drive data	Hammer driven	Vibration	SPT is made using standard penetrometer and hammer (see text); undisturbed samples obtained by using liners, but some sample disturbance is likely
Retractable plug	Tubes 150 mm long and 25 mm OD; maximum of six tubes can be filled during a single penetration	Silts, clays, fine and loose sands	Hammer driven	Improper soil types for sampler; vibration	Lightweight, highly portable units can be hand carried; some sample disturbance is likely
Continuous-helical-flight auger	Diameters range 76 to 406 mm; penetrations to depths exceeding 15 m	Most soils above water table; will not penetrate hard soils or those containing cobbles or boulders	Rotation	Hard soils, cobbles, boulders	Rapid method of determining soil profile; bag samples can be obtained; log and sample depths must account for lag time between penetration of bit and arrival of sample at surface
Hollow-stem auger	Generally 150 to 200 mm OD with 75 to 100 mm ID hollow stem	Same as flight auger	Rotation	Same as flight auger	Special type of flight auger with hollow center through which undisturbed samples or SPT can be taken
Disc auger	Up to 1070 mm diameter; usually has maximum penetration depth of 8 m	Same as flight auger	Rotation	Same as flight auger	Rapid method of determining soil profile; bag samples can be obtained
Bucket auger	Up to 1220 mm diameter common; larger sizes available; with extensions, depths over 24 m are possible	Most soils above water table; can dig harder soils than above types and can penetrate soils with cobbles and boulders if equipped with a rock bucket	Rotation	Soil too hard to dig	Several bucket types available, including those with ripper teeth and chopping tools; progress is slow when extensions are used

Table 10-4
Common Samplers To Collect Undisturbed Soil Samples (NAVFAC 1982)

SAMPLER	TYPICAL DIMENSIONS	SOILS THAT GIVE BEST RESULTS	METHOD OF PENETRATION	CAUSE OF DISTURBANCE OR LOW RECOVERY	REMARKS
Shelby tube	76 mm OD and 73 mm ID most common; available from 50 to 127 mm OD; 760-mm sampler length standard	Cohesive fine-grained or soft soils; gravelly soils will crimp tube	Pressing with fast, smooth stroke; can be carefully hammer driven	Erratic pressure applied during sampling, hammering, gravel particles, crimping of tube edge, improper soil types for sampler	Simplest device for undisturbed samples; boring should be clean before sampler is lowered; little waste area in sampler; not suitable for hard, dense, or gravelly soils
Stationary piston	76 mm OD most common; available from 50 to 127 mm OD; 760-mm sampler length standard	Soft to medium clays and fine silts; not for sandy soils	Pressing with continuous, steady stroke	Erratic pressure during sampling, allowing piston rod to move during press, improper soil types for sampler	Piston at end of sampler prevents entry of fluid and contaminating material; requires heavy drill rig with hydraulic drill head; samples generally less disturbed compared with Shelby tube; not suitable for hard, dense, or gravelly soil; no positive control over specific recovery ratio
Hydraulic piston (Osterberg)	76 mm OD is most common; available from 50 to 101 mm OD; 910-mm sampler length standard	Silts and clays, some sandy soils	Hydraulic or compressed air pressure	Inadequate clamping of drill rods, erratic pressure	Needs only standard drill rods; requires adequate hydraulic or air capacity to activate sampler; samples generally less disturbed compared with Shelby tube; not suitable for hard, dense, or gravelly soil; not possible to limit length of push or amounts of sample penetration
Denison	89 to 177 mm OD, producing samples 60 to 160 mm; 610-mm sampler length standard	Stiff to hard clay, silt, and sands with some cementation, soft rock	Rotation and hydraulic pressure	Improper operation of sampler; poor drilling procedures	Inner tube face projects beyond outer tube, which rotates; amount of projection can be adjusted; generally takes good samples; not suitable for loose sands and soft clays
Pitcher sampler	105 mm OD; uses 76-mm diameter Shelby tubes; sample length 610 mm	Same as Denison	Same as Denison	Same as Denison	Differs from Denison in that inner tube projection is spring controlled; often ineffective in cohesionless soils

Ensuring the best possible quality of soil samples involves experience and skill. In the collection of soil samples from boreholes, the following guidelines should be considered:

- Whenever there is danger of erosion or collapse of the borehole walls, commonly referred to as *caving*, a viscous drilling fluid or a borehole casing, or both, must be used while the borehole is being advanced.
- When samples are being collected above the water table, the borehole should be kept dry whenever possible.
- When samples are being collected below the water table, the borehole should be maintained full of water or drilling fluid during drilling, cleaning of the borehole, sampling and sample withdrawal, and removal of cleanout tools. If continuous samples are required, the casing should remain full of water or drilling fluid for the entire drilling and sampling operation.
- A single sampling tube should not be used to obtain an undisturbed sample of a soft or loose soil found directly below a stiff or compact soil. The driving of the sampling tube should be stopped as soon as a sudden decrease in resistence is observed.

5.4 Test Borings and Sampling in Rock

Rock cores are collected by using core bits and any of a variety of core barrels. A core bit consists of a hollow ring of durable cutting teeth (usually composed of tungsten carbide or diamonds) designed to fragment the annular area around the circumference of the hole. The resulting central rock core is preserved in a core barrel, which can be retrieved periodically. There are many types of rock coring bits and core barrels (USBR 1974; Hunt 1984; Johnson and DeGraff 1988).

Deere (1963) defined the standard numerical method of describing the degree of fracturing of rock masses from drilled rock cores, termed the *rock quality designation* (RQD). RQD is computed as the summation of the lengths of all rock core fragments more than 10 cm long divided by the total length of rock core drilled.

Rock cores may not always provide undisturbed samples of rock. The quality and degree of core recovery are a function of many factors, including speed of drill rotation, bit pressure and rate of advance, drilling fluid pressure, and core blockage within the core barrel. Especially when weak zones are encountered, the core may be damaged or destroyed by the jamming of rock fragments within the core barrel that results in grinding of the sample as the drilling continues. Such occurrences are common during landslide investigations.

Skilled and experienced drilling personnel are required to obtain the best possible results. Double and triple tube core barrels have been designed to minimize such problems. They should be used whenever possible to provide the best quality of samples.

6. BOREHOLE LOGGING

Direct sampling of subsurface materials by means of borings provides much important information, but detection of critical conditions, such as thin sand seams or lenses, rupture surfaces, or rock fractures, is often a problem because even with continuous sampling or coring, sample recovery usually does not completely represent subsurface conditions. Alternatively, the borehole may be logged by a sensing device that measures the desired soil and rock characteristics at closely spaced intervals as it is lowered down or pulled up the borehole. Direct visual observations of subsurface conditions are sometimes possible with borehole cameras. A number of different material properties can be determined by different logging devices, including self-potential, electrical resistivity, nuclear radiation, density (based on nuclear absorption), water content (based on hydrogen ion reaction), and sound wave or impulse response.

Borehole logging produces a graph of each property plotted as a function of depth. Most logging devices provide measurements within about 15 to 30 cm around the hole. Thus, effects of drilling muds and borehole casing installation must be considered in selecting logging procedures and interpreting results. Other geophysical methods, such as downhole and crosshole seismic surveys and downhole and crosshole resistivity surveys, are effective tools for special applications such as finding voids in rock.

Several devices can be used to log boreholes and provide continuous in situ high-resolution measurements that are more repesentative of subsurface conditions than samples from boreholes. An adequate assessment of subsurface conditions may require the use of a suite of logging methods because each responds to a different property of soil, rock, or fluid. Some techniques allow for measurements from inside a plastic or metal borehole casing, and some allow measurements in both unsaturated and saturated zones.

Borehole logging measurements can be correlated with known geologic strata and conditions in one borehole and then used to correlate and identify similar strata in nearby boreholes. Thin layers, not readily detected in soil or core samples, can often be resolved by logging. By providing high-resolution data that are independent of subjective interpretations of soil or rock type, logging can improve the correlation of strata and subsurface conditions between borings. For example, it is difficult to compare samples obtained from two different holes to determine whether soils or rocks that have similar classification characteristics represent the same stratum. However, by comparing continuous borehole logs, one can match the patterns of the different properties; similar patterns suggest similar stratification. Thus, although borehole logging may have limited engineering significance by itself in one hole, it is a significant tool for boring interpretation and correlation when used in adjacent holes.

Table 10-5 summarizes some of the more commonly used borehole logging methods, their applications, the conditions under which they may be used, and some of their limitations. The logging methods are grouped into six classes, which are described in the following subsections:

- Caliper logging, which measures the diameter of an uncased borehole;
- Electric logging, which measures electrical conductance or resistivity of the subsurface materials;
- Nuclear logging, which measures the radioactivity of subsurface materials and hence their lithology, bulk density, and moisture content;
- Remote sensing;
- Thermal profiling; and
- Seismic methods.

6.1 Caliper Logging

A caliper log measures the diameter of the uncased borehole. The mechanical caliper device is lowered to the bottom of the hole, the caliper arms are spread, and the device measures the hole diameter continuously as it is raised. The caliper device is connected to a recording device on the surface. The profile of the borehole diameter produced by the caliper logging device is needed to interpret the results obtained by many other logging techniques. Changes in borehole diameter differentiate between hard and soft rock and may identify swelling zones or locations of possible borehole shearing.

6.2 Electric Logging

Electric logging methods include several devices for measuring apparent resistivities of earth materials and *self potentials*, which are naturally occurring small electrical currents generated within earth materials. These logging methods are analogous to the surface-based electrical resistivity methods discussed in Section 4.1.1.

6.2.1 Induction Log

The induction log is similar to the electromagnetic (EM) conductivity profiling measurements described in Section 4.1.2. It measures the electrical conductivity of the soil or rock surrounding open or PVC-cased boreholes above or below the water table. The induction log can be used to identify lithology and provide stratigraphic correlation between boreholes. Variations in electrical conductivity with depth are related to changes in the specific conductance of pore fluids, which in turn indicate changes in clay content, permeability, degree of fracturing, or contaminants in the fluids.

The induction log has a radius of investigation of about 1 m, much greater than that for other logging methods. Thus, the induction log is almost unaffected by mud on the borehole walls or construction conditions. It is a good indicator of general soil and rock conditions surrounding the borehole. The induction log provides data that are similar to those provided by the resistivity log, but because the induction log does not require electrical contact with the earth materials, it can measure in both the unsaturated vadose zone and the saturated zone and can be used in PVC-cased boreholes.

6.2.2 Resistivity Log

The resistivity log measures the apparent resistivity of soil or rock immediately surrounding a borehole. Because this device requires electrical contact with

Table 10-5
Borehole Logging Methods

CATEGORY	PARAMETER MEASURED	CASING	SATURATED	UNSATURATED	RADIUS OF INVESTIGATION	EFFECT OF BOREHOLE DIAMETER AND MUD	APPLICATIONS	LIMITATIONS
Caliper logging	Borehole diameter	No	Yes	Yes	At immediate borehole wall	NA	Used to continuously measure and record borehole diameter	Requires an uncased hole
Electric logging								
Induction	Electrical conductivity	Yes	Yes	Yes	75 cm	Negligible	Provides continuous measure of conductivity for materials surrounding borehole	Information has lower resolution than resistivity log but can evaluate unsaturated zone and PVC-cased boreholes
Resistivity	Electrical resistivity	No	Yes	No	30 to 150 cm	Significant to minimal, depending on probe	Provides continuous measure of resistivity from which material types can be deduced when compared with borehole material logs	Generally information provided is only semi-quantitative; requires borehole log and is restricted to saturated zone and uncased borehole
Spontaneous-potential	Electrical potentials from mineral reactions and groundwater flow	No	Yes	No	Immediately adjacent to borehole wall	Significant	Identifies lithology; oxidation-reduction reaction zones, and subsurface flows	Provides ambiguous data that require considerable interpretation; can only valuate saturated zone in uncased borehole
Nuclear logging								
Natural-gamma	Natural-gamma radiation	Yes	Yes	Yes	15 to 30 cm	Moderate	Determines presence and integrity of clay and shale formations	May be affected by presence of mud coatings on borehole walls
Gamma-gamma	Material density	Yes	Yes	Yes	15 cm	Significant	Provides continuous measurement of material density	Provides only material density measurements; health and safety regulations may influence operational costs

Method	Parameter measured					Depth/location	Effect	Description	Remarks
Neutron-neutron	Moisture content (above water table); porosity (below water table)	Yes	Yes	Yes	Yes	15 to 30 cm	Moderate	Provides continuous measurement of natural moisture content; locates rupture zones when used in combination with gamma logging	Provides only in situ moisture values; health and safety regulations may influence operational costs
Remote sensing Borehole cameras	Visual images of fractures and structures in borehole walls	Yes	No	Yes	Yes	At immediate borehole wall	Significant	Special videocamera obtains continuous image of borehole walls; software can be used to interpret dips of fractures	Requires uncased hole; images are affected by water quality in hole
Ultrasonic acoustic	Continuous images of borehole wall showing fractures	Yes	No	Yes	Yes	At immediate borehole wall	Significant to moderate	Provides continuous image of borehole wall showing fractures and other discontinuities	Requires uncased hole; images are much less clear than those obtained by borehole cameras
Thermal profiling	Temperature	Yes	No	Yes	No	Within borehole	NA	Determines zone of water inflow into borehole	Requires uncased hole
Seismic methods Uphole survey	Material dynamic properties	Yes	No	Yes	Yes	NA	NA	Determines dynamic properties and rock-mass quality of materials surrounding borehole	Requires uncased and mud-filled hole
Downhole survey	Material dynamic properties	Yes	No	Yes	Yes	NA	NA	Determines dynamic properties and rock-mass quality of materials surrounding borehole	Requires uncased and mud-filled hole
Crosshole survey	Material dynamic properties	Yes	No	Yes	Yes	NA	NA	Determines dynamic properties and rock-mass quality of selected stratum	Requires array of uncased holes

NOTE: NA = not applicable.

the borehole wall, it can only be run in uncased boreholes filled with water or drilling fluid.

Resistivity logging devices are produced with various electrode spacings. Short probes have spacings of about 45 cm between electrodes. They produce high-resolution logs showing the presence of thin layers, but the measurements are made only in the zone immediately surrounding the borehole. Long probes have electrode spacings of about 1.6 m and provide resolutions and penetrations that are similar to those achieved by induction logs.

6.2.3 Spontaneous-Potential Log

The spontaneous-potential log measures the natural potential (in millivolts) developed between the borehole fluid and the surrounding rock materials. It can only be run in uncased boreholes within the saturated zone, and its radius of investigation is highly variable.

The observed spontaneous potential is composed of two components, the first being developed by the electrochemical reactions among dissimilar materials and the second being a result of the movement of ionized water through permeable materials. The measurements are only semiquantitative and are subject to considerable noise from the electrodes, local hydrogeologic conditions, and the borehole fluids. However, these measurements may, under favorable circumstances, yield information concerning the lithology, oxidation-reduction conditions, and subsurface fluid flow.

6.3 Nuclear Logging

Nuclear logging includes three techniques that are closely similar and, because they support each other, are often used together. Gamma-gamma and neutron-neutron nuclear probes have been used to monitor changes in moving slopes and to successfully locate a rupture zone in a relatively uniform deposit (Cotecchia 1978). These methods may suffer from restrictions and cost escalation because of health and environmental regulations surrounding the use of radioactive materials. Liability insurance against loss of a nuclear source probe in a drill hole may not be available. Lack of such insurance may prevent the use of these logging techniques for many projects.

6.3.1 Natural-Gamma Log

The natural-gamma log records the amount of gamma radiation naturally emitted from earth materials. This log is used chiefly to identify lithologies in either uncased or cased boreholes both above and below the water table. It has a radius of investigation of about 15 to 30 cm. Natural-gamma emissions come chiefly from potassium-40 and daughter products of the uranium and thorium decay series. Because clays and shales concentrate these elements through processes of ion exchange and adsorption, the natural-gamma activity of clay and shale-bearing formations is much higher than that for clean sands, sandstones, or limestones. The natural-gamma log is therefore useful in determining the presence and integrity of clays and shales.

6.3.2 Gamma-Gamma (Density) Log

Gamma-gamma logging provides a continuous measurement of material in situ density. This log can be used in uncased and cased holes above and below the water table. Gamma-gamma logging uses an active probe containing both a radiation source and a detector. The probe measures the backscatter of gamma rays emitted by the probe averaged over the distance between the source and the detector. The radius of investigation is usually no more than 15 cm; thus this log is more likely to be affected by variations in borehole diameter, mud coatings on the borehole walls, and other well-construction factors.

6.3.3 Neutron-Neutron (Porosity) Log

Neutron-neutron logging provides a continuous measurement of natural moisture content above the water table and porosity below the water table. This logging can be run in cased or uncased holes above or below the water table. The neutron-neutron log uses an active probe that measures the backscatter of gamma rays and neutrons resulting from bombardment by fast neutrons generated by a source within the probe. The backscatter is a function of the hydrogen content of the materials, which is correlated with the natural water content, porosity, or both. However, this estimation is highly dependent on the clay or shale content of the material. The radius of investigation is typically about 15 cm, rising to about 30 cm in very porous formations, so borehole diameter fluctuations and similar construction factors can influence the results obtained. However, these influences are less severe than for the gamma-gamma log.

6.4 Remote Sensing

Several methods of remotely sensing borehole conditions have been used. Both film cameras and television (video) cameras have been used for a number of years. With video-recording devices and special software, the television camera images can be analyzed to determine the strike and dip of fractures or other planar features intersected by the borehole. Ultrasonic acoustic logging devices provide a three-dimensional view of the borehole wall showing fractures and other discontinuities. These methods all require an uncased hole.

6.5 Thermal Profiling

A thermal profile down a borehole can be obtained using an accurate thermocouple at the end of a wire attached to a high-resolution ohmmeter. The temperature is obtained continuously as the thermocouple is lowered into the hole. When continuous recording equipment is not available, readings at intervals of 30 cm have proved satisfactory. Differences in thermal conductivity from different geologic deposits allow inferences to be drawn concerning their characteristics. Many profiles of the same hole are needed because information from the thermal profile is often masked by surface effects; for example, daily and seasonal changes affect the upper 10 m of a borehole profile.

The hole must be kept open with drilling mud or some type of casing (such as a plastic pipe) so that the temperature in the hole can stabilize. Multiple readings over a long period are desirable to account for surface variations. It is known that the temperature within the earth surface increases approximately 1°C in 30 m below the uppermost 10-m depth. Large-scale regional variations in this value exist, reflecting the presence of local sources of magmatic heat within the earth's crust, but every location has a characteristic average thermal gradient. Changes from this average gradient reflect the presence of flowing water or changes in thermal conductivity that relate to changes in the geologic materials.

A near-constant thermal signature identifies impervious strata or water-bearing strata in which groundwater flow is taking place at very slow rates. Sharp changes in the thermal signatures identify strata supplying significant water volumes. Thermal profiling was successfully used by one of the authors of this chapter, McGuffey, to identify subsurface zones of flowing water on State Route 22 at Berlin, New York. This technique was used to identify layers in which subsurface horizontal drains could be installed to improve the factor of safety and stop landslide movement.

6.6 Seismic Methods

Seismic methods that employ seismic sources or sensors in boreholes may be used in conjunction with seismic geophysical surveying techniques to obtain data on the dynamic properties of earth materials and to evaluate rock-mass quality (NAVFAC 1982; Hunt 1984). The borehole techniques include *uphole surveys*, *downhole surveys*, and *crosshole surveys* (Ballard 1976; Auld 1977; Dobecki 1979).

Uphole surveys utilize a sequence of energy sources that are set off at successively decreasing depths in an uncased, mud-filled borehole, starting at the bottom of the hole. Geophones are placed in an array on the surface (Figure 10-6). They should be set on rock where possible if the intent of the survey is to measure rock-mass quality.

Downhole surveys locate the energy source on the surface adjacent to the uncased, mud-filled borehole, and the detectors are incorporated into a sonde that is raised or lowered in the borehole to obtain either a continuous or a discrete set of velocity measurements of the earth materials surrounding the hole (Figure 10-7).

FIGURE 10-6
Uphole direct seismic survey method.

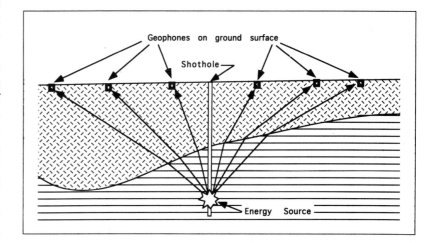

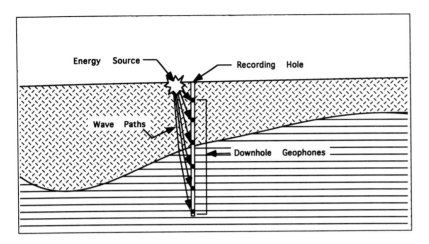

FIGURE 10-7
Downhole direct seismic survey method.

Crosshole surveys utilize an array of boreholes; usually four observational holes are placed around a single shothole. The energy source is placed at a given depth in the central shothole, and the detectors are placed at the same depth or in the same geologic stratum in the surrounding observational holes (Figure 10-8).

Uphole and downhole surveys substantially reduce the influence of reflection and refraction of shock waves. Travel-time velocities may be readily and accurately determined for layers of interest. Crosshole surveys eliminate the influence of surrounding layers and allow the direct determination of the velocity for a single layer of interest.

7. FIELD TESTING

FIGURE 10-8
Crosshole direct seismic survey method.

Evaluation of the stability of a slope requires measurement of the physical strength properties of the materials that make up the slope. Of course, it is usually more convenient to obtain such measures by laboratory tests of undisturbed samples secured from the site, but obtaining representative samples is often difficult for the following reasons:

- Discontinuous samples with relatively small diameters can miss thin critical strata that control landsliding (such as the slickensided surface of movement of a preexisting landslide);
- Distortion, disturbance, and changes in moisture and stress are always associated with taking a sample out of the ground, handling it, transporting it, and preparing it for laboratory testing; and
- The process of creating a hole to obtain samples often changes the nature of the samples. For example, bentonite seams may be washed out by the drilling fluid when drilling through bedrock or clay and silt may be washed out when drilling through bouldery till. It is often desirable to run direct field tests, such as penetration tests (see Section 7.1.1) or vane shear tests (see Section 7.1.5), to compensate for or reduce the influence of these effects.

For the foregoing reasons, some field testing is often recommended to provide in situ estimates of strength values and properties of materials involved in landsliding. Field tests are always more expensive to perform than laboratory tests, are usually more complex to interpret, and often are not as readily reproducible. However, their results are invaluable in confirming the validity of the laboratory test results. When field and laboratory test results do not agree, the investigator must determine the reasons for such discrepancies and their significance.

Field tests range from relatively economical and routine borehole tests to more expensive in situ tests involving test pits or the collection of large block samples and, even more rarely, to the use of exploratory adits and shafts.

7.1 Borehole Tests

Certain physical tests have been devised that can be performed in boreholes drilled for identifying the soil strata and for securing the small-diameter samples. Borehole tests are among the most economical of field tests. Many test procedures are widely adopted. Although borehole tests suffer from the limited volume of material tested, they do allow the soil to be tested without the disturbance produced

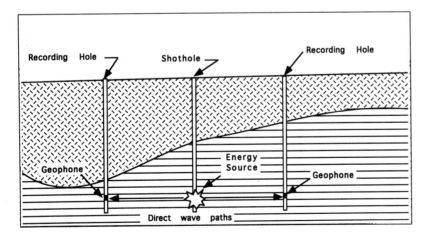

by removing the sample from the ground, taking it to the laboratory, and preparing it for testing. However, it should be noted that some disturbance is caused by stress relief in the borehole walls, and complete changes in soil characteristics often occur as a result of the progression of the hole and the use of drilling fluids. Borehole test measurements may be obtained by a variety of procedures and equipment, including an assortment of penetration tests, use of a dilatometer, and borehole dilation, borehole shear, and vane shear tests.

7.1.1 Penetration Tests

Penetration tests are conducted during the creation of a borehole. They involve measurement of the resistance of materials to the advancement of standard probes (Sanglerat 1972; Schmertmann 1978). Subsurface conditions may be inferred by correlating soil properties with resistance values. Penetration tests provide an extremely valuable and relatively low-cost supplement to the data obtained by direct borehole sampling and logging procedures. The two main forms of penetration test are the *standard penetration test* (SPT), which is conducted in conjunction with split-barrel sampling, and the *continuous cone penetrometer test* (CPT), of which there are numerous varieties.

The SPT as defined by ASTM D1586 (which is in non-SI units) utilizes a standard 50-mm (2-in.) outside-diameter, 37-mm (1.5-in.) inside-diameter split-barrel sampler (Figure 10-9). The sampler is 660 mm (26 in.) long and is driven 450 mm (18 in.) into the undisturbed soil at the bottom of a borehole in three 150-mm (6-in.) increments by blows of a 63.5-kg (140-lb) hammer dropping in free fall from a height of 760 mm (30 in.). The

number of blows required to advance the sampler 150 mm is recorded. The SPT can be conducted at any desired depth below the surface within an advancing test boring. Usually the first 150-mm advance is disregarded because it is likely to be affected by disturbance at the bottom of the previously developed hole. Therefore the SPT includes gaps in the penetration resistance record. The total number of blows for the second and third 150-mm advances is used to define N, the *standard penetration resistance*, which is recorded as the number of blows per 300 mm (blows per foot).

Penetration resistance encompasses both hard and soft seams in the 300-mm distance and is composed of both end resistance and shaft friction. Although some investigators have attempted to drive such sampling tubes as far as 1.8 m, counting blows for each 150-mm increment, the accumulating skin friction and the buildup of soil resistance within the samples usually produce resistances that increase with each successive increment until the sample is withdrawn and the borehole cleaned out. The continuing increases in the values mask useful correlations.

Penetration resistance is generally correlated empirically with soil properties measured by either laboratory tests or field tests of the same material. In this way, large numbers of low-cost penetration tests supplement the more limited information obtained by more expensive laboratory tests. End resistance is more important in granular soils and may be correlated with values of the angle of internal friction, whereas shaft friction is more important in cohesive soils and may be correlated with their consistency. Although many empirical relationships between resistance and soil properties have been published (Terzaghi and Peck 1967; Sowers and

FIGURE 10-9
Split-tube sampler
for standard
penetration test.

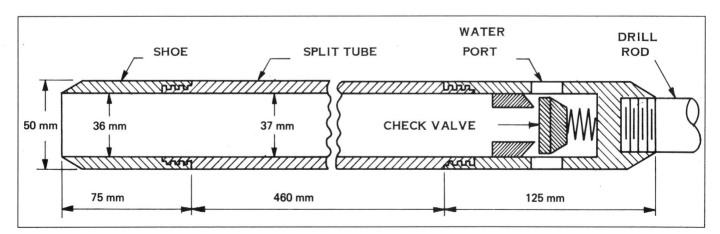

Sowers 1970), these should not be used indiscriminately. Instead, a new correlation should be established from the data obtained at the site in question, or the data should be used to verify the accuracy of the published relations (Terzaghi and Peck 1967).

The CPT is defined by ASTM D3441. Rods with conical tips are forced into the ground while the required force is continuously recorded (Figure 10-10). No samples are recovered. The force may be dynamic, in which the rods are driven by blows from a hammer; static, in which rods are pushed into the ground as deadweights are applied; or quasi-static, in which the rods are pushed by hydraulic pistons reacting against a vehicle or other

machinery. The quasi-static method is the most commonly used and is often referred to as the *static cone penetrometer test* (Alperstein and Leifer 1976). In relatively permeable soils, pore-pressure effects around the cone tip during penetration at standard rates are negligible, and the CPT values are correlatable with fully drained soil-strength properties. In homogeneous plastic clays, the CPT values approximate fully undrained behavior.

There are numerous types of cones. The simplest types, including the *Dutch cone*, consist of a small-diameter rod ending in a cone tip with a 60-degree angle at its point and an effective cross-sectional area of 10 cm² (Figure 10-11). The Dutch cone

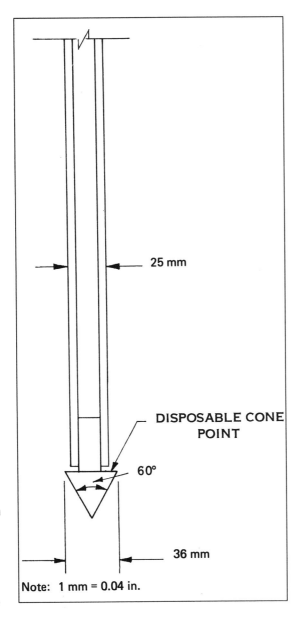

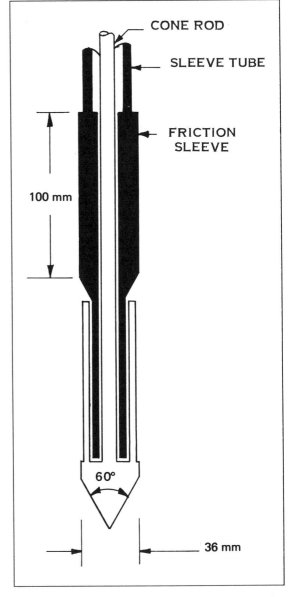

FIGURE 10-10
(right) Drive cone on 25-mm drill rod.

FIGURE 10-11
(far right) Dutch cone with friction sleeve for static test.

measures a combined penetration resistance from both shaft friction f_s and end-cone resistance q_c. More advanced types of cones incorporate a friction jacket above the standard cone tip, permitting separation of the values for f_s and q_c (Figure 10-12). Recently, several refinements have been incorporated into experimental cone penetrometers, including pore-pressure measuring devices (piezocone), geophysical measurements by electrical-resistivity receivers (Figure 10-13) or geophones, and geochemical monitoring devices to measure subsurface contaminants.

Cone penetrometers typically can penetrate as much as a few meters into granular soils such as sands. However, significant amounts of gravel cause the values to be erratic and not readily interpreted. The electric-resistivity cone penetrometer and the piezocone have been used and correlated to estimate shear-strength parameters in clay soils. The static cone test often is not applicable to landslides because of the difficult terrain and hard materials overlying the critical layers. Dynamic cone tests appear to have greater flexibility for difficult landslide situations, but few recorded uses have been published. The cone penetrometer is an extremely valuable supplement to the more direct boring and sampling technique. It helps identify changes and stratification and pinpoints weak materials that should be investigated in greater detail by more direct exploration methods.

7.1.2 Dilatometer Test

The flat dilatometer, developed by Marchetti (1980), is essentially a penetration device capable of obtaining an estimate of lateral soil stiffness (Borden et al. 1986). As shown in Figure 10-14, the dilatometer uses a thin circular membrane with a diameter of 60 mm, which is mounted on the side of a flat blade approximately 95 mm wide and 14 mm thick. When inactive, the membrane is flush with the surrounding flat surface of the blade. The blade is pushed or driven into the ground, and when it is at the desired test depth, the membrane is inflated by means of pressurized gas controlled by a monitoring unit at the surface. Initial readings are taken of the pressure required to just move the membrane (the A pressure), which is related to the lateral stresses existing in the ground. Further readings are taken of the pressure required to move the membrane center an additional 1 mm (0.04 in.) into the soil (the B pressure), which is related to the soil stiffness.

In the usual field testing procedure, the pressure required to drive the dilatometer into the ground is recorded as a form of penetration test. At predetermined intervals (usually about every 20 cm) the penetration is halted, the membrane is inflated, the A and B pressures are determined, the membrane is vented to deflate it, and the dilatometer is advanced to the next depth. The dilatometer is directly inserted into the ground and does not require a borehole, in contrast to the older pressuremeter discussed in Section 7.1.3, which does require a borehole. The dilatometer is thus suitable for rapid profiling of subsurface conditions.

Much testing has been conducted with the dilatometer in a variety of transportation-related applications (Bullock 1983; Borden et al. 1986; Mayne and Frost 1988). Studies by Borden et al.

FIGURE 10-12 Operation of Dutch cone penetrometer and example of typical results (NAVFAC 1982).

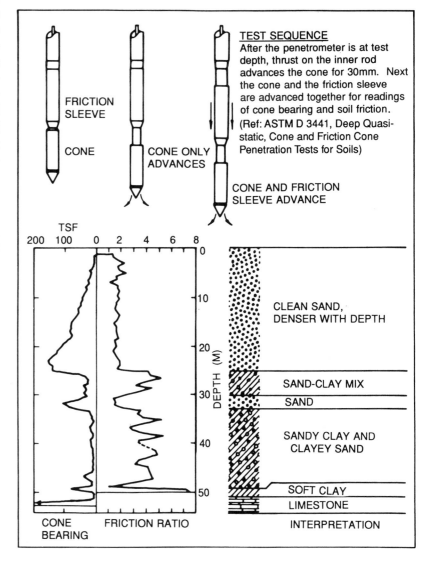

FIGURE 10-13
Fugro electric
friction cone
penetrometer tip
(Hunt 1984).
REPRINTED WITH PERMISSION
OF McGRAW-HILL

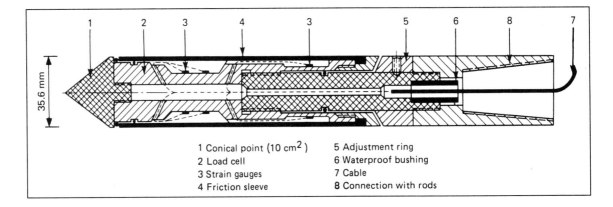

1 Conical point (10 cm^2)   5 Adjustment ring
2 Load cell   6 Waterproof bushing
3 Strain gauges   7 Cable
4 Friction sleeve   8 Connection with rods

(1986) in North Carolina indicated that dilatometer data significantly overestimated soil compressibility. Mayne and Frost (1988) conducted a sequence of dilatometer tests near Washington, D.C., over a 3-year period. They correlated dilatometer test data with standard soil borings and cone penetrometer tests and concluded that the dilatometer offers an expedient and cost-effective method for profiling subsurface conditions, providing reasonable interpretations of soil properties in a variety of geological environments. They noted that the dilatometer cannot be easily inserted in very dense or gravelly soils.

7.1.3 Borehole Dilation Test

FIGURE 10-14
Dilatometer and
control unit
(Borden et al. 1986).

A number of field tests have been developed that are based on the resistance of a cylindrical borehole to dilation from applied internal pressure. The best known of these uses the Menard pressuremeter

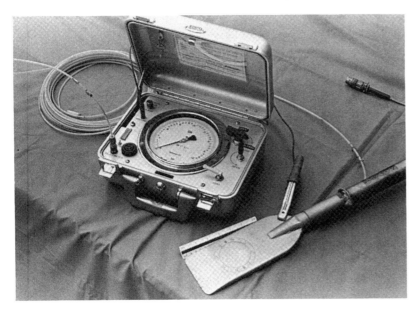

(Menard 1975; Winter 1986). As shown in Figure 10-15, a membrane in close contact with the soil surrounding a borehole is inflated, and the stress-strain characteristics are measured with time. These data have been correlated with shear-strength and consolidation parameters in certain soil systems (Figure 10-16). In the Menard device, the end effects in the measuring cylinder are minimized by means of similar cylindrical rubber tubes above and below; these tubes are inflated with the same pressure as the test cylinder, thereby providing a two-dimensional stress configuration rather than a three-dimensional one, which has a more complicated elliptical zone of strain. Other similar devices omit the end tubes and rely on theoretical interpretation of the elliptical zone of stress and strain. Still others employ mechanical sleeves and strain sensors to measure pressure and displacement.

The instruments are quite delicate and cannot easily be used in very stiff or stony materials. The cost of acquiring equipment and running tests with these units is quite high, and the interpretation of the results is difficult. It appears that good results are obtained primarily in softer clay deposits. Although it is claimed that these devices can provide the user with all of the necessary soil properties to evaluate shear and consolidation, the interpretation is largely empirical and certainly open to question in variable materials.

7.1.4 Borehole Shear Test

The borehole shear test measures the shear strength of the soil in an annular zone surrounding the boring. The device consists of an expandable plug with serrations on its outer surface to grip the soil walls of the borehole when pressure of known magnitude is applied internally by a hydraulic system. The soil is then sheared by pulling the device upward through the hole. If several such tests (essentially undrained direct shear) are made on the same stratum at varying internal pressures, a Mohr failure envelope can be obtained.

The test is of limited utility because it shears the soil in a different direction than that involved in the landslide process. Therefore, if the soil has anisotropic properties (usually the axis of weakness is parallel to the greatest extent of the surface of shear movement and more or less perpendicular to the direction of shear in the borehole shear test), the results may be misleading. There is usually some soil smear in the walls of the borehole; thus, the soil involved in the test may be partially disturbed. The size of the device is such that it integrates the effect of soil irregularities over a cylindrical surface with a diameter of 76 mm and a length of about 300 mm. This test is usually not useful in landslide investigations because the parameters measured (undrained shear) do not control landslide performance and thus have limited application.

7.1.5 Vane Shear Test

Vertical blades at the end of a thin rod produce a vertical cylindrical surface of shear when rotated (Figure 10-17). The torque required to initiate continual rotation is a measure of the peak undrained strength of the soil. The torque required to maintain rotation after several revolutions measures the residual undrained or disturbed strength. To minimize end effects, the length of the vane should be at least twice its width. The blades should be sufficiently thin that there is a minimum soil disturbance due to displacement and sufficiently thick that they do not bend under load.

In very soft soils, the vane and its torque rod are forced into the soil to each level to be tested. A reference test using the torque rod without the vane is required so that the torque necessary to overcome rod friction can be subtracted from the total torque measured when the soil is tested. In firm soils or at great depths, the test is made in undisturbed soil 300 to 760 mm below the bottom of a borehole; hence, the resistance of the torque rod in the hole generally is negligible, but friction in the casing may be of concern. Numerous procedures and forms of equipment, ranging from simple torque wrenches to elaborate torque meters that apply a uniform angular strain rate, have been utilized (Hvorslev 1949).

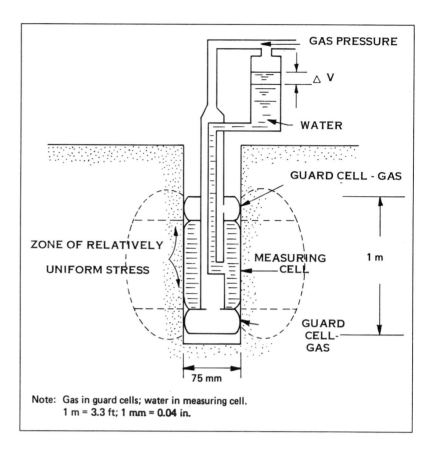

Note: Gas in guard cells; water in measuring cell.
1 m = 3.3 ft; 1 mm = 0.04 in.

FIGURE 10-15 Menard pressuremeter for borehole dilation test (not to scale).

Vane data should be correlated with other shear data for use in analysis and design. Caution should be exercised in interpreting the peak strength; in some cases, the strength measured in a vane test has been found to be as much as 30 percent greater than that measured by other methods. The results of field vane shear tests, like the results of borehole shear tests, have limited utility in solving landslide problems because they reflect undrained shear val-

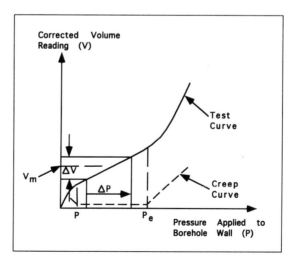

FIGURE 10-16 Typical pressuremeter test curve (Winter 1986).

ues for undisturbed materials, which may not control landslide performance.

7.2 Large-Scale Tests and Sampling in Pits, Adits, or Shafts

When test-boring programs cannot quantify the subsurface characteristics of the slide system adequately, it may be necessary to choose large excavations that allow the direct examination of in situ geologic conditions. Large excavations range from test pits and test trenches, large-diameter holes that may be made relatively cheaply and rapidly in soils, and finally to shafts and adits, which are costly but valuable for investigating complex rock-mass conditions. Excavation of a large hole deep enough into a landslide to obtain required data sometimes poses a safety risk to workers. The risks can be reduced by increasing the lateral support provided by shoring or by using remote-sampling and inspection tools so that the hole does not have to be occupied.

As previously stated, one of the major limitations of laboratory tests is their inability to include variations in the soil, particularly in zones with weak or hard spots. This limitation can be overcome by large-scale, in-place tests performed in pits or trenches excavated to the questionable strata or zones of slickensiding. Although the range of stresses and particularly the range of groundwater pressures that can be evaluated by such tests are limited, these tests permit large volumes of soil to be evaluated under the conditions within the total mass without the problems of sampling disturbance and exposure inherent in small-scale sampling and laboratory testing.

Creating a large test pit in an active landslide requires extensive temporary support for worker safety and therefore can be expected to be expensive. Data obtained from laboratory tests performed on block or tube samples of critical soil layers collected from test pits may give better quantification of the parameters than field tests in the test pits. However, some authors believe that large-scale field tests are important sources of information (Sowers and Royster 1978).

7.2.1 Plate-Load Test

The plate-load test is the oldest form of in-place test (Sowers and Sowers 1970). It has only limited value in determining the strength of soils involved in the instability of large earth masses because there are many different interpretations of such tests, all yielding different values for shear-strength parameters.

A pit is excavated to expose the surface of the stratum in question, and a rigid square or circular plate is placed on the ground. The pit should be as wide as possible but no wider than about two-thirds

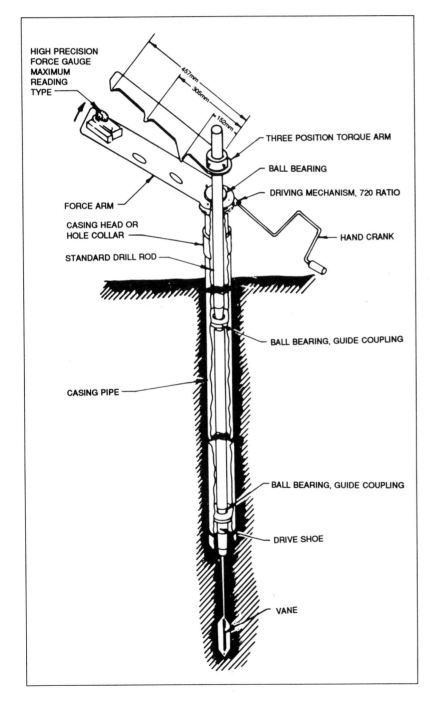

FIGURE 10-17 Borehole vane shear testing device (NAVFAC 1982).

the thickness of the stratum whose strength is being evaluated. Loading is incremental so that at least 10 successively greater loads are applied before the plate shears the soil beneath it. The results can be interpreted in terms of soil-bearing capacity to give the shear strength of the soil along a curvilinear surface that may crudely approximate the rupture surface of a landslide (Sowers and Royster 1978). However, use of the plate-load test as the principal method for determining the strength of soils involved in a landslide is not recommended.

7.2.2 Large-Scale Direct-Shear Test

A large-scale direct-shear test can be performed in a pit at the level of the suspected weak stratum; in the case of an existing landslide, the test may be performed on the actual rupture surface of the soil or rock. The pit is excavated to the level of the stratum or rupture surface to be evaluated. The pit should be large enough and be carefully excavated more deeply around the sides so that the material to be tested, left in the form of an exposed block or stump in the center, will not be disturbed by engineers and technicians working around it. The size of the block is dictated by the geologist's or geotechnical engineer's evaluation of the variation in the material strength properties. It should be large enough to be representative of the stratum as a whole and not just its weaker or stronger elements.

A double box is placed around the block. If there is a definite plane of weakness, the sides of the box should be perpendicular to that plane, and the plane should lie between the top and bottom halves of the box, as shown in Figure 10-18. Good contact is necessary between the sides of the block of material and the sides of the box. Such contact can be achieved by careful trimming of relatively fine-grained or cohesive soils. In other materials, plaster can be poured to fill the spaces between the block and the enclosing box.

A normal load is applied to the block by means of a plate just slightly smaller than the dimensions of the box, which can be fitted inside the upper half of the box. The normal load can be applied by jacking against a piece of heavy machinery above the pit or against a heavy steel beam anchored to the ground by earth anchors located sufficiently far from the test to prevent their having any influence on the tested block. The bottom half of the box is anchored securely by packed soil, concrete, or plas-

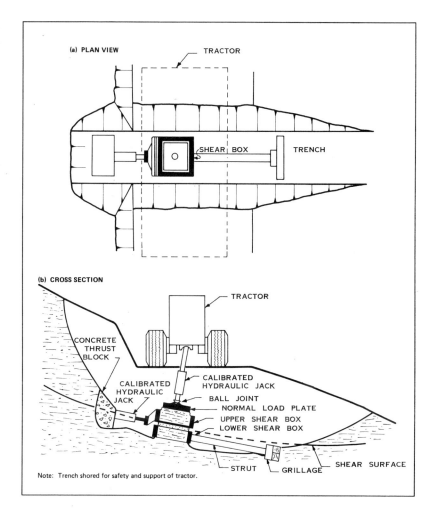

ter in the bottom of the pit. The top half of the box is then jacked sideways by a calibrated system so that the lateral load and the amount of lateral movement can be measured.

The direction of lateral jacking should parallel the direction of suspected or actual landslide movement. The same surface can be tested at several different increasing normal loads if each test stage is stopped as soon as peak strength or significant movement is observed. Such a test sequence determines the average shear stress required to produce rupture on a single predetermined plane. If determinations of the peak strengths for a variety of normal loads are required, separate tests should be performed on fresh sample blocks for each normal load. Such tests require a large test trench, or multiple test pits, and are much more time-consuming and expensive to perform.

Large-scale direct-shear tests simulate the shear strength of the earth materials along an actual rupture surface. If the samples are large enough, they

FIGURE 10-18
Direct-shear test of strength along rupture surface.

integrate the effects of both hard and soft zones. Unfortunately, it is difficult to include the effects of changing water pressure. Sowers and Royster (1978) stated that their experience with large-scale direct-shear tests indicated that these tests do provide meaningful shear-strength data, particularly if the tests are conducted during the wet season. Sowers and Royster also suggested that these large-scale field tests be correlated with the results of the much cheaper standard laboratory tests in order to further extend the data. They concluded that a more reliable evaluation of shear strength is provided by a combination of judiciously selected large-scale field tests and laboratory shear tests than by laboratory tests alone.

Sowers and Royster noted that standard equipment for making such large-scale tests does not exist. They suggested that suitable tests be conducted with materials fabricated in the field to fit the sample sizes and test-pit dimensions. Steel angles, channels, and plates can be fabricated at reasonable cost to create the shear box. Loads can be provided and measured by means of calibrated hydraulic jacks or jacks with load cells, and the movements can be measured with micrometer dial gauges. The lack of precision offered by such equipment is more than compensated for by its realistic representation of field conditions.

7.2.3 Collection of Large Block Samples

Laboratory determination of soil strength and stiffness requires high-quality samples. As described in Chapter 19, such determinations are particularly valuable when dealing with landslides involving residual and tropical soils (Geological Society of London 1990). Highest-quality intact samples are obtained by block sampling in test pits and trenches. Stepped excavation of the test pit or trench provides convenient benches for cutting and trimming samples.

Figure 10-19 illustrates the method of preparing and removing hand-carved block samples. Normally, block samples should be at least six times larger than the largest particle present but not larger than a 200-mm cube. Large samples tend to be too heavy for handling and transport without risk of damage. Soils with sensitive structures should be encased in protective boxes immediately upon trimming. Their exposed faces should be coated with sealants before undercutting and removal are attempted.

The boxed samples should be packed with damp sawdust or plastic foam to support the sample from shocks during transit (Dearman and Turk 1985).

An alternative method for taking intact samples from soils free of gravel-sized particles is with the core cutter (Figure 10-20). This procedure may be used for friable soils with little cohesion or for soils with a very sensitive grain structure. The cutter should not be forced into the ground. Rather, a roughly cylindrical block should be cut from the ground so as to have about 10 mm in excess to be pared off as the core cutter is jacked down steadily. A California bearing ratio (CBR) mold fitted with a suitable cutting shoe may be used with this method to obtain an intact sample.

7.2.4 Adits and Shafts

Exploration using adits and shafts is extremely expensive compared with the cost of more usual methods of subsurface investigation, but occasionally it may be justified during the investigation of old landslides where dams or transportation facilities will alter the topography or groundwater conditions. The usual purpose of this type of exploration is to obtain very detailed information on joint and fracture spacing and orientation, to collect high-quality samples of failure material or infillings, or to intercept groundwater.

An adit must be carefully designed to perform safely and to provide subsurface information. In landslide investigations, an adit is typically designed to penetrate the landslide base. If tunneling through weak materials is involved, additional subsurface evaluation may be required to provide design parameters for the adit bracing. In most cases, data from existing test borings and other landslide investigations will provide adequate information without a specific investigation to determine adit support parameters. Adits should be constructed to drain by gravity if groundwater is expected. Sloping the adit to provide drainage also facilitates muck removal.

Shafts are adits that are oriented vertically or nearly vertically. Bracing design and construction techniques are similar to those for adits and are derived from the mining industry. Shafts smaller in diameter than 1 m are normally drilled. If the shaft is drilled through colluvium, casing is used with cutouts to allow for observation of wall materials (Wyllie 1992).

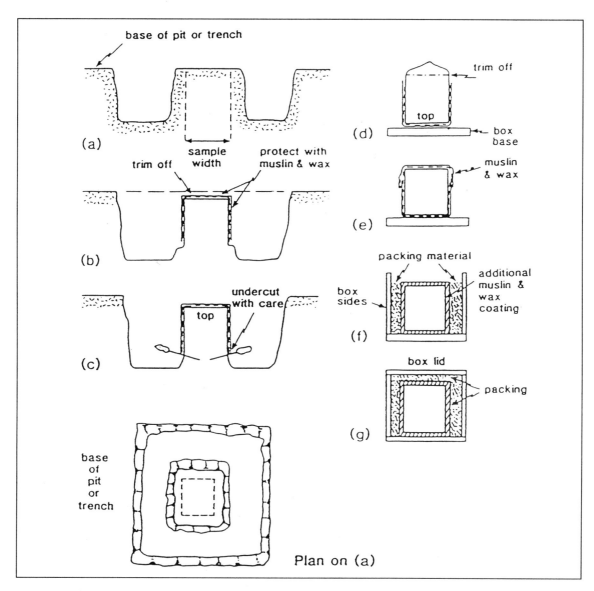

FIGURE 10-19
Stages in taking and protecting a block sample (Geological Society of London 1990): (a) isolate block sample by excavating a trench around it; (b) trim top and sides of block to size and cover with muslin and wax; (c) carefully undercut a slightly oversized sample and remove it; (d) invert sample onto base of shipping box and trim excess from sample; (e) cover trimmed sample with muslin and wax; (f) add more muslin and wax coatings as necessary, construct shipping box sides, and surround sample with cushioning packing materials; and (g) add more packing material and close shipping box.

High-quality cores can be taken at any angle within the adit or shaft (Krynine and Judd 1957) to test for natural and residual shear strength. In situ loading and direct-shear strength tests can be performed in adits and shafts, although the confined space makes such tests even more expensive to perform than those in test pits. Such tests are common in dam-site studies (Krynine and Judd 1957) but are very uncommon in landslide investigations.

8. TREATMENT OF SAMPLES AND CORES

A primary responsibility of the field investigation team is the accurate labeling and location of all samples. It is self-evident that samples that are mislabeled or mislocated in the field have no value and may in fact be the subject of much fruitless analysis if such mislabeling or mislocation problems are undiscovered.

However, the investigation team has responsibilities that go beyond the accurate identification of all samples. Samples are relatively small representatives of the much larger volumes of soil and rock materials that make up the landslide. Samples can be easily damaged, and in most situations, no laboratory test procedures can compensate for careless handling of samples in the field or during shipment to the laboratory. Common problems are subjecting soil samples to excessive shocks and blows; allowing soil samples to dry; allowing some rock samples, especially claystones and shales, to slake; and exposing samples to heat or freezing.

FIGURE 10-20
Jacking undisturbed
core-cutter sampler
(Geological Society
of London 1990):
(*a*) overall view of
equipment; (*b*)
core-cutter assembly
resting on exposed
block sample; (*c*)
rough trimming of
block to about 10
mm larger than
core-cutter
diameter; (*d*) jacking
of core-cutter down
onto sample,
allowing excess
material along sides
to be shaved off;
(*e*) final sample
within cutter
assembly, allowing
it to be carefully
undercut and
removed.

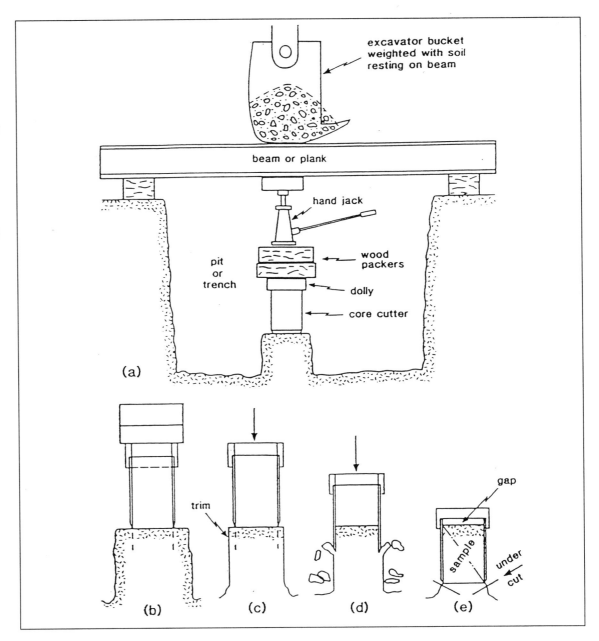

8.1 Preservation, Shipment, and Storage

All samples should be described and logged immediately upon recovery from the borehole. Sampling devices should be dismantled carefully to avoid unnecessary shocks, especially to soil samples. Rock cores often need careful handling to prevent additional breaking. RQD values for rock cores should be determined immediately. Soil samples especially must be protected from drying because consistency of cohesive soils changes with drying and details of stratification may be lost.

Samples must be quickly preserved and prepared for shipment and storage. Many types of soil samples require protection and sealing to prevent drying. Even disturbed samples or samples from augered holes, which are merely mixtures of materials, should be handled in ways that preserve their field character to the maximum extent possible.

Split-barrel soil samples should be placed in wide-mouthed jars that are sufficiently large so that samples will not be pushed, squeezed, or otherwise deformed when placed in the jars. The jars should be sealed with appropriate caps and wax should be applied over the cap to prevent any possible moisture loss.

Thin-wall tube samples are normally retained within the tubes in the field. A small amount of sample is removed from one end of the tube and a nonshrinking wax is used to seal the end, which is then further capped, taped, and labeled. The tube is inverted and the other end is similarly sealed. The tubes should be shipped upright if possible and be protected with padding to prevent them from striking together.

Rock cores are usually placed in special wooden *core boxes* that contain wooden spacers to separate the core runs. Rough handling of core boxes may result in significant damage to the cores.

8.2 Extrusion of Undisturbed Samples

Undisturbed soil samples are usually shipped to the laboratory in their sampling tubes, sealed and labeled as described in Section 8.1. Before an attempt is made to remove, or extrude, the samples from their tubes, the tubes may be subjected to X-ray examination to determine the conditions of the sample, details of the profile, degree of disturbance, and presence of anomalies. New York State Department of Transportation practice now requires that every tube sample be X-rayed before being extruded. This practice has greatly improved the quality of sampling and the testing results.

The samples are extruded from their tubes in the laboratory. A few practitioners extrude the sample in the field, cut off short 15-cm sections, wrap them in aluminum foil, and surround the foil covers with wax in a carton. This reduces some transport problems but also requires additional field and laboratory sample handling that may result in disturbance of some weak soils.

Samples should always be extruded in the same direction that they entered the tube in the field, or sample quality may be compromised. In the case of strongly cohesive soils, wall friction may become excessive, and extrusion may only be possible after the sample tubes have been cut into short lengths.

9. INVESTIGATION OF GROUNDWATER CONDITIONS

Groundwater is a major cause or triggering mechanism of many landslides (Chapter 4). Identification and detection of groundwater conditions should therefore form a critical aspect of subsur-

face investigations in areas of suspected or active landslides. Groundwater conditions within landslide masses are frequently complex, a result of the disruption of rock and soil strata and structures by the landslide movements.

Surface observations and geologic mapping (Chapter 9) should identify surface water features of importance to landslide investigations. Springs, seeps, marshes, and topographic depressions within and adjacent to landslides are helpful in identifying the position of the upper boundary of a saturated zone, the water table, within portions of the landslide. Landslide movements may cause disruption, damming, or diversion of surface-water flows causing increased saturation of the landslide mass.

Movements of landslide materials may also disrupt or block subsurface movement and drainage of groundwater from a slope. Investigations should record the presence and flow volumes of springs and seeps near the toe of a landslide. Such discharges are usually beneficial to the stability of a landslide since they represent drainage of groundwater from the unstable mass. In many situations, the reduction or cessation of such discharges should be considered a warning of potential increased instability.

Many landslides involve complex interrelationships between surface-water and groundwater conditions. Details concerning methods of groundwater flow analysis for complex subsurface conditions typical of most landslides are beyond the scope of this report and may be found in standard texts on groundwater hydrogeology (Davis and DeWeist 1966; Cherry and Freeze 1979). Simplistic assumptions concerning groundwater conditions, such as the presence of a single unsaturated vadose zone overlying a saturated zone with a water table defining the boundary between these zones, are unrealistic for most landslide situations and lead to much confusion when field observations are interpreted.

Landslide movements often result in the juxtaposition of materials having low and high permeabilities. Such permeability contrasts lead to the formation of isolated shallow zones of saturation, known as *perched water tables*, and isolated zones of saturated permeable materials containing groundwater under elevated pressures. Significant vertical flow gradients exist within the groundwater at many landslides caused by the complex variations and geometries of the landslide materials and by the considerable topographic relief. As a consequence, groundwater pressure measurements at different depths at a common location may yield dramatically different values.

The interpretation of groundwater pressure values obtained from within and around landslides is complex. Increases in pressure with depth indicate the potential for upward flow, whereas decreases in pressure with depth indicate the potential for downward flow. However, poor hydraulic connectivity between zones often prevents or restricts such flows from occurring. Groundwater conditions within many landslides are best considered as complex, multiple, partially connected flow systems. The degree of independence or isolation of these flow systems may change over time as landslide movements continue or as a result of seasonal changes.

Each landslide has unique relationships, so generalizations must be used with caution. However, investigators can often be guided by the following assumptions and recommendations:

- Surface observations are essential in determining the effect of groundwater on landslide instability.
- Periodic or seasonal influx of surface water to groundwater will not be detected unless groundwater observations are conducted over extended time periods.
- Landslide movements may open cracks and develop depressions at the head of a landslide that increase the rate of infiltration of surface water into the landslide mass. Water infiltration at the head of a landslide generally results in increased instability.
- Ponding of surface water anywhere on the landslide may cause increased infiltration of water into the landslide and should be investigated.
- Disruption of surface water channels and culverts may also result in increased infiltration of surface water into the landslide.
- Landslide movements may result in blockage of permeable zones that were previously freely draining. Such blockage may cause a local rise in the groundwater table and increased saturation and instability of the landslide materials. Subsurface observations should therefore be directed to establishing groundwater conditions in the undisturbed areas surrounding the landslide.
- Low-permeability soils, which are commonly involved in landslides, have slow response times to changes in groundwater conditions and pres-

sures. Long-term groundwater monitoring is required in these soils.
- Accurate detection of groundwater in rock formations is often difficult because shale or claystone layers, intermittent fractures, and fracture infilling may occlude groundwater detection by boring or excavation.
- Borings should never be the only method of groundwater investigation; nevertheless they are a critical component of the overall investigation.

The accurate definition of groundwater conditions within and adjacent to a landslide is often very difficult. As noted in earlier sections of this chapter, geophysical field and borehole logging methods are potentially useful techniques for identifying groundwater conditions, but these methods have limitations. As discussed in Section 4.1, electromagnetic conductivity and resistivity profile surveys reflect the presence and salinity of groundwater. Where test borings or the presence of springs or seeps indicates shallow groundwater conditions, the extent of such saturated zones may be further mapped by these methods. When exploratory boreholes are logged by electric or nuclear logging devices (Sections 6.2 and 6.3), some semiquantitative information on soil moisture conditions may be obtained. Thermal profiles of boreholes (Section 6.5) have been successful in identifying subsurface zones of flowing water in landslides. Recently developed experimental cone penetrometers have included a pore-pressure measuring device (the piezocone) that can assist in locating zones of higher groundwater flow.

Direct measurement of groundwater pressure conditions is necessary and can be accomplished by a monitoring well, or *piezometer*, an instrument used to measure groundwater pressure at a point in the subsurface. There are three main types of piezometers (Table 10-6), which consist of three basic components: the tip, the standpipe, and seals. The tip includes a perforated section of casing, well screen, porous element, or other similar feature that allows interaction with the groundwater. In fine-grained or unstable materials, so common in landslides, the tip installation may be surrounded by a filter material. The standpipe is a watertight pipe or measurement conduit of the smallest practical diameter that is attached to the tip and extends to the surface of the ground. A seal, or multiple seals, consisting of grout, bentonite slurry,

Table 10-6
Types of Piezometers

Instrument	Advantages	Disadvantages
Open standpipe piezometer or wellpoint	Simple and reliable, with a long record of experience; no elaborate recording terminal point required	Slow response time; liable to freeze in cold weather
Pressure sensor piezometer		
Pneumatic	Location and elevation of terminal recording point are independent of measuring-tip location; can exhibit a rapid response	Humid air must be prevented from entering measurement tubing
Electrical	Location and elevation of terminal recording point are independent of measuring-tip location; can exhibit a rapid response and high sensitivity; suitable for automatic readout	Expensive; temperature correction may be necessary; errors may arise due to drift over longer observation periods

or other impermeable material must be placed between the standpipe and the boring walls to isolate the zone being monitored by the tip.

The open standpipe piezometer is the simplest and most common type used in landslides. A common version of the open standpipe piezometer (the so-called Casagrande type) uses a porous tip, commonly constructed of ceramic, that can be driven to the desired depth. These tips may be damaged when driven into the ground, so metal tips have been developed. The pores in the tip are about 50 μm, making it possible to use them in direct contact with fine-grained soils. For certain landslide investigations and for long-term monitoring of landslides, more sophisticated pneumatic or electrical piezometer installations may be warranted.

Problems arise because the installation of piezometers in landslides is often difficult, the observations are often ambiguous because of installation errors or malfunctions of the piezometers, and the complexity of groundwater conditions within the landslide precludes simple conceptualizations. The following recommendations are made:

• It is necessary to accurately detect and monitor groundwater in most soils and rock formations through the installation of monitoring wells (piezometers).
• As many exploratory borings as possible should be converted to operate as piezometers; if these installations are to accurately detect groundwater, drilling techniques that conform to the

level of care accorded groundwater monitoring wells should be used.
• Drilling activities aimed solely at producing maximum rates of advance during development of a borehole in soil or rock can often occlude a seam containing groundwater. Modified drilling techniques, such as maintaining a high rate of return water flow and permitting solids to settle out of the drilling fluid before its reintroduction into the drill hole, can help avert such problems.
• The groundwater level should be measured at the depth at which water is first encountered in a borehole as well as at the level at which it stabilizes after drilling. The borehole should be kept open with a perforated casing until stabilization occurs.

The continuing movements within a landslide may destroy piezometers, so in many cases only the cheapest and simplest piezometer installations can be justified or are attempted. Thus open standpipe piezometers are the most commonly used but may be unsatisfactory because of their inherent limitations.

All piezometers are potentially subject to error, major sources of which include gas bubbles and tube blockage (Table 10-7). The magnitude of errors can be controlled to some degree by proper piezometer selection, installation, and deairing techniques. A common problem is not establishing the measuring tip at the appropriate elevation within the landslide materials so that it measures critical higher-pressure groundwater zones; as a re-

Table 10-7
Sources of Error and Corrective Methods for Piezometer Measurements

PIEZOMETER TYPE	SOURCE OF ERROR	CORRECTIVE METHOD
Open standpipe	Piezometer partially filled with liquid such as oil, kerosene, or antifreeze to prevent freezing or reduce corrosion; position of interface between liquids may be uncertain due to leakage or evaporation	Avoid use of such liquids whenever possible; carefully monitor leakage or evaporative loss potentials
Open standpipe	Piezometer completely filled with liquid other than water to prevent freezing or corrosion; differences in surface tension at interface between liquid and groundwater may cause appreciable and misleading changes in pressure determinations	Avoid using liquids that fill the entire piezometer so that liquid interface can be determined within piezometer tube
Open standpipe	Air or gas bubbles caught by internal roughness or joints in open tube may cause stabilized water level to rise above true groundwater level	Use pipe diameter large enough and interior smooth enough to permit gas bubbles to rise; avoid downward-protruding edges of joints
Pneumatic pressure sensor	Gas bubbles entrapped within liquid cause increased time lag in piezometer response, but gas entrapped above gauge does not affect equalized pressures	Provide gas trap and outlet valve and flushing facilities; use materials that do not promote electrolysis and development of gas
All types	Gas bubbles in soil near intake cause increased lag time in piezometer response due to decreased effective permeability and changes in gas volumes	Use well points or screens constructed of materials that do not cause electrolysis with soils and having pores or holes large enough to permit escape of any gas bubbles

sult, values may not represent actual landslide instability conditions.

Open standpipe piezometers will freeze if subjected to sufficiently cold weather, and use of antifreeze to counteract this problem introduces additional difficulties and potential error in the observations. The open systems are also prone to error due to gas bubbles within the standpipe or in the surrounding filter pack or tip. In fine-grained soils, open standpipe systems may be slow to respond to changes in groundwater levels.

Closed systems, such as the pneumatic and electrical piezometers, overcome several of these limitations. Electrical piezometers also allow for automatic and remote recording. Because pneumatic and electrical piezometers are more expensive and will become inoperative just as quickly as open standpipe piezometers when landslide movements damage them, they are most often used to monitor landslides that pose major risks.

The most common problem in subsurface investigation of landslides is the selection of an adequate

number of piezometers and their proper installation. Water elevations in open or screened boreholes extending through the landslide mass, or large sections of it, generally do not provide meaningful data. Nested installations consisting of several piezometers with measuring tips at different elevations provide much more valuable information. In some situations, it may be suggested that several piezometers be installed in a single borehole with impervious seals separating the measuring zones. However, such installations require considerable skill for the proper placement and development of the impervious seals to ensure that they do not allow leakage between measuring zones. Such installations become even more difficult when landslide deposits are unstable or moving, creating problems in maintaining a stable borehole configuration. Also, if measurements are required in zones that are closely spaced, perhaps 3 m or less, adequate seals cannot be expected to form. For these reasons, it is generally best to create piezometer nests by installing individual instruments at different depths in different, but closely spaced, bore-holes. This procedure increases the total length of boreholes to be drilled but may be cheaper overall since each piezometer can be installed more quickly.

More details of piezometer installations and operation are discussed in Chapter 11.

10. DATA PRESENTATION

Subsurface exploration of landslides is frequently conducted under less-than-ideal environmental or working conditions. Information describing these conditions should be recorded in great detail at the time of observations or measurements because such knowledge may materially influence the methods of analysis of the resulting data and the degree of confidence that can be placed in them. Information that may be important to subsequent interpretation of the data are details on

- Tools and techniques used;
- Samples obtained;
- Weather conditions preceding and during the exploration program;
- Impressions of the investigators regarding operation of the equipment (i.e., the "feel of the tools");
- Each geophysical exploration, such as seismic and acoustic measurements, test borings, and test pit, adit, or shaft results;
- Field instrumentation installed; and
- Field tests associated with the exploration.

These details of the subsurface exploration program should be incorporated into the overall geotechnical report for the project. Appropriate interpretive techniques should be identified, and all data should be correlated in a consistent manner. The investigator should highlight any anomalies or areas of conflicting evidence exposed in the exploration program.

During any investigation, considerable data will be accumulated. The variations in these data should be studied in three dimensions because the mass of soil or rock involved in a landslide is three-dimensional. When examined spatially, the variations may exhibit random or definitive patterns, lending support to some hypotheses concerning landslide mechanisms and eliminating others. The data must therefore be compiled into a geotechnical report, which is the primary product of the

subsurface investigation program. The basic elements of a geotechnical report include a location plan, interpretive maps, geologic cross sections, interpretations resulting from any geophysical methods, and logs of all boreholes, trenches, or test pits. When observations have been conducted over a period of time, the geotechnical report should include a separate section discussing such time-based observations and correlations.

10.1 Location Plan

All exploration efforts, including boreholes, test pits, field tests, geophysical profiles, and any other exploration activities, should be indicated accurately on a location plan drawn to an appropriate scale (Figure 10-21). For most studies, a single plan should be sufficient to locate all the activities. However, at the site of very extensive investigations, the data may be arranged on more than one plan or enlargements of selected areas may be added to provide better readability. The base map for the site should include topographic contours of sufficient resolution and other local landmarks to allow future investigators to relocate themselves on the site. The plan may include the local geology, or this may be shown on a separate map. A smaller-scale regional map clearly identifying the site location should also be included. Many reports lack an accurate description of the site location. Such information is useful when tying the detailed site information to regional geologic conditions.

10.2 Interpretive Maps

Some subsurface observational data vary with geographic position and are best shown in the form of maps. If the variations appear to be systematic, such as lines of equal strength or lines of equal groundwater elevation, they can be drawn as contours. Contouring is frequently an interpretive process, with different individuals producing different contour maps from the same data. These interpretive maps preferably should be plotted on the same base map used for the location plan map or as overlays that can be superimposed on the base map. Several interpretive maps may be desirable.

10.3 Geologic Cross Sections

Cross sections of the data are particularly useful (Figure 10-22). Data from many different exploration methods may be combined or interpreted to

FIGURE 10-21
Typical landslide
location plan
(Jurich and Miller
1987).

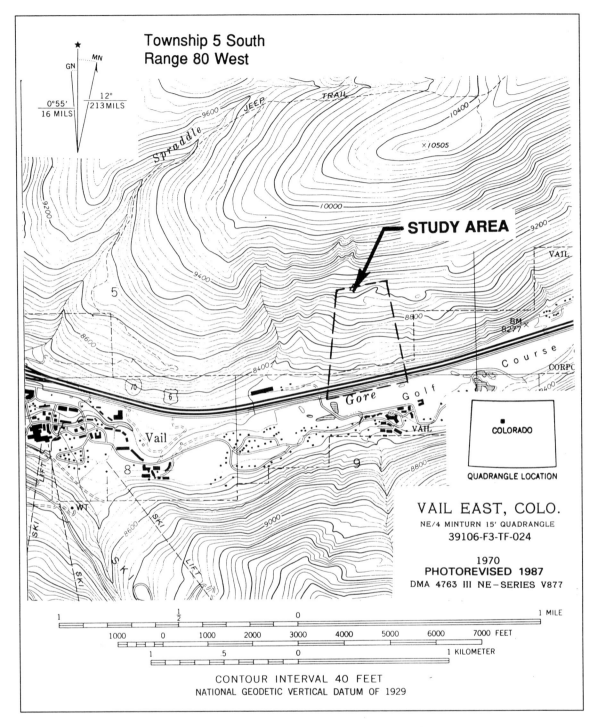

form the basis for typical geologic cross sections that illustrate the significant geologic conditions affecting the landslide. Cross sections are also useful in evaluating the stability of the landslide mass. For this purpose, cross sections are required parallel to the direction of maximum slope, to the maximum water-pressure gradient, or to the observed trajectories of movement. Geologic conditions may re-

quire cross sections in other orientations, however, in order to further define important geologic controls on the landslide movements. Fence diagrams, or three-dimensional sections, are helpful for sites with complex geology (for example, see Figure 9-45). Recent advances in computer-generated spatial data-handling and visualization capabilities (Figure 10-23) have led to the experimental use of

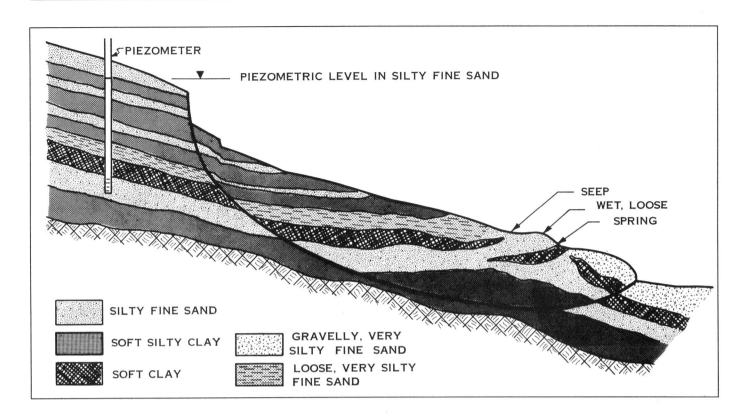

SILTY FINE SAND

SOFT SILTY CLAY

SOFT CLAY

GRAVELLY, VERY
SILTY FINE SAND

LOOSE, VERY SILTY
FINE SAND

true three-dimensional digital models of landslide masses (Chiarle and Powers 1994). Such analyses have been found helpful in the limited number of cases to which they have been applied and may become a more commonly used analysis and reporting tool as the required computer programs become more widely disseminated.

10.4 Geophysical Survey Interpretations

The geotechnical report should include adequate documentation concerning the methods used to interpret the geophysical data. The geophysical surveys and geophysical logging methods described previously can be used not only as a means of defining the stratification of the materials within the landslide, but also as a direct measure of certain physical properties of the materials. Although some of these properties may not necessarily be those needed to evaluate landslide stability, empirical correlations among the properties can sometimes be helpful in establishing the stability of the landslide.

For example, the seismic refraction technique yields the compression-wave velocity of the materials. If the density of the materials is known or can be reasonably estimated, the compression-wave ve-

locity can be used to compute the dynamic modulus of elasticity and hence to estimate the rigidity of the mass. The compression-wave velocity is reduced by cracks or microfissures in a rock mass, and thus it is an important indicator of the degree of fracturing in a rock mass.

At best, however, geophysical methods are only indirect supplements to the more direct means of evaluating the qualities of the soil and rock materials. Field data collection using geophysical methods is relatively simple and rapid. However, interpretation of the data frequently incorporates a considerable number of simplifying assumptions, and quality of the interpretation often depends on the skill and experience of the person making the interpretation.

10.5 Borehole and Excavation Logs

The results of test and core borings, test pits, and other reconnaissance methods should be presented in logs that include all pertinent information. Boring logs should be prepared to provide complete documentation on the drilling, sampling, and coring operations, as well as detailed descriptions of the materials, groundwater conditions, and other aspects of the subsurface environ-

FIGURE 10-22
Typical geologic cross section through center of landslide.

FIGURE 10-23
Computer-generated
three-dimensional
visualization
of Slumgullion
landslide,
Hinsdale County,
Colorado
(Chiarle and
Powers 1994).
DIGITAL ELEVATION MODEL
BASED ON 1990 AERIAL
PHOTOGRAPHY, U.S.
GEOLOGICAL SURVEY

FIGURE 10-24
(opposite page)
Typical borehole
report log.
COURTESY OF ILLINOIS
DEPARTMENT OF
TRANSPORTATION

ment encountered by the borehole. Because logs frequently provide the basis for analysis and design, clear and precise presentation of all data and complete descriptions are important.

Two sets of logs are usually generated: *field logs* and *report logs*, each serving different purposes. A field log records all the basic data and information regarding the boring operation. It is designed to describe each sample in detail and provide all information needed by the engineer to establish the validity of the data obtained. It does not contain all the information necessary for design.

A report log records the data necessary for design or analysis. It is prepared after laboratory testing of samples is completed and contains laboratory identification test results and notations concerning the various tests performed. The report log may incorporate changes in the material descriptions from those in the earlier field logs so that the descriptions agree with the gradation and plasticity test results from laboratory testing (Figure 10-24).

10.6 Time-Based Observations and Correlations

The significance of the different factors involved in a landslide can frequently be found from empirical correlations between the movement and observed forces or environmental factors. For example, a time-based graph of both landslide movement and rainfall, or accumulated rainfall and snowmelt, may show a visual relationship (Figure 10-25). Similar time-based relations can often be observed between such environmental factors as groundwater levels or construction activities and the initiation, movement, or reactivation of landslides. In some cases, plots of the observed phe-

Illinois Department of Transportation

Bridge Foundation Boring Log

Sh. 1 of 1 Sh.

PROJECT _____ BRIDGE __ FA Route 14 (Ill.3) @ __ Date ____ 4-14-89 ____

ROUTE __ FA 14 __ ____ Chester City Park ____ **Bored By** __ Larry Ford ____

SEC. __ 121-I __ STA. __ 1130+00± to 1135+00± __ **Checked By** __ Robert Nebelsick ____

COUNTY __ Randolph __

Boring No. __ 8 __
Station __ 1132+33 __
Offset __ 122' Lt.℄ __

Surface Water El. _____

Groundwater El. at Completion _____

After _____ Hours _____

Elevation	N	Qu t/s.f.	w (%)	(description)	Elevation	N	Qu t/s.f.	w (%)
Ground Surface 574.2 — 0								
				BROWN AND GRAY VERY WEATHERED SHALE W/WEATHERED LIMESTONE	— 3 7 10	S/5 2.7	22	
	1 3 4	S/10 1.1	29		-25 8 21 26	S/5 3.7	18	
BROWN AND GRAY VERY SILTY CLAY				547.7				
-5	2 3 5	S/15 1.4	28	GRAY SHALE & LIMESTONE SHELVES	— 50/ .3	-	-	
				545.7±				
	1 3 4	B 1.2	28	GRAY SHALE (W/CALCARIOUS STREAKS)	-30 50/ .3	-	-	
-10	2 4 3	S/10 0.9	28	542.0± LIMESTONE SHELVES & GRAY SHALE 540.7±	— 50/ .3	-	-	
	1 3 3	B 1.0	28	GRAY SHALE 539.5	-35 50/ .2	-	-	
560.2±				END OF BORING				
-15	2 3 4	B 1.3	27					
BROWN AND GRAY SILT	2 6 9	S 1.3	23		-40			
553.7	-20 4 7 10	S 1.3 S/5 2.3	21 25					
					-45			

N-Standard Penetration Test- Blows per foot to drive 2″ O.D. Split Spoon Sampler 12″ with 140 No. hammer falling 30″.

Qu-Unconfined Compressive Strength - t/sf

w - Water Content - percentage of oven dry weight-%.

Type failure:
B - Bulge Failure
S - Shear Failure
E - Estimated Value
P - Penetrometer

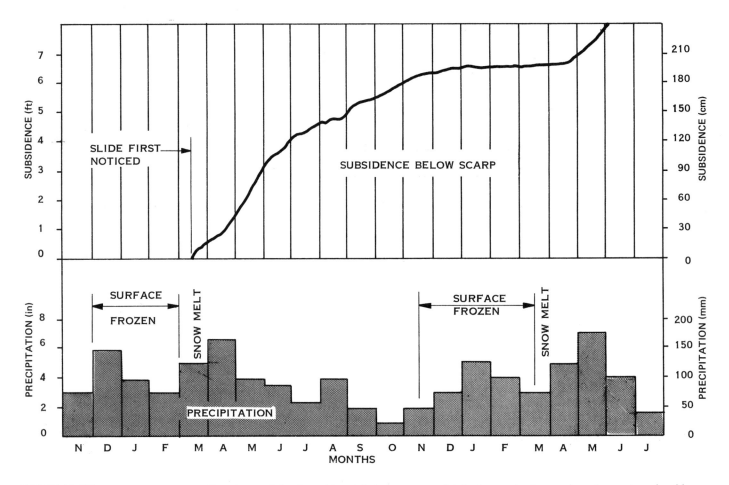

FIGURE 10-25
Correlation of
landslide movement
with precipitation.

nomena as functions of the logarithm of time are instructive.

11. SUMMARY

A technically sound solution to a landslide problem can be derived only from technically sound data. A properly conducted subsurface investigation program leading to a comprehensive geotechnical report forms the basis for subsequent stability analyses, which in turn form the basis for remedial action plans. The subsurface investigation program identifies potential problem areas and defines the features involved in an existing landslide. Subsurface investigation techniques include a considerable variety of methods, ranging from the relatively inexpensive methods utilized in the earlier reconnaissance stages to the more expensive and specialized sampling and field testing methods used to provide estimates of physical-strength properties of the landslide materials. Unfortunately, numerous cases of inadequate subsurface investigations have been documented in

which the geologists and engineers involved have jumped to erroneous conclusions on the basis of incomplete data. Such situations invariably lead to increased difficulties and costs.

REFERENCES

ABBREVIATIONS

AEG   Association of Engineering Geologists
ASTM   American Society for Testing and Materials
NAVFAC   Naval Facilities Engineering Command
USBR   Bureau of Reclamation, U.S. Department of
          the Interior

AEG. 1978. A Guide to Core Logging for Rock Engineering: Core Logging Committee Report, South African Section. *Bulletin of the Association of Engineering Geologists*, Vol. 15, pp. 295–328.

Aikas, K., P. Loven, and P. Sakka. 1983. Determination of Rock Mass Modulus of Deformation by Hammer Seismograph. *Bulletin of the International Association of Engineering Geology*, No. 26-7, pp. 131–133.

Alperstein, R., and S.A. Leifer. 1976. Site Investigation with Static Cone Penetrometer. *Journal of the Geotechnical Engineering Division*, ASCE, Vol. 102, No. GT5, pp. 539–555.

Arcone, S. 1989. Some Airborne Applications of Subsurface Radar. In *Transportation Research Record 1192*, TRB, National Research Council, Washington, D.C., pp. 111–119.

ASTM. 1951. *Symposium on Surface and Subsurface Reconnaissance.* Special Technical Publication 122. Philadelphia, Pa., 228 pp.

Auld, B. 1977. Cross-hole and Down-hole V_s by Mechanical Impulse. *Journal of the Geotechnical Engineering Division*, ASCE, Vol. 103, No. GT12, pp. 1381–1398.

Ballard, R.F., Jr. 1976. Method for Crosshole Seismic Testing. *Journal of the Geotechnical Engineering Division*, ASCE, Vol. 102, No. GT12, pp. 1261–1273.

Bogoslovsky, V.A., and A.A. Ogilvy. 1977. Geophysical Methods for the Investigation of Landslides. *Geophysics*, Vol. 42, No. 3, pp. 562–571.

Borden, R.H., R.E. Saliba, and W.M. Lowder. 1986. Compressibility of Compacted Fills Evaluated by the Dilatometer. In *Transportation Research Record 1089*, TRB, National Research Council, Washington, D.C., pp. 1–10.

Broms, B.B. 1980. Soil Sampling in Europe: State of the Art. *Journal of the Geotechnical Engineering Division*, ASCE, Vol. 106, No. GT1, pp. 65–98.

Brooke, J.P. 1973. Geophysical Investigation of a Landslide near San Jose, California. *Geoexploration*, Vol. 11, No. 2, pp. 61–73.

Bullock, P.J. 1983. *The Dilatometer: Current Test Procedures and Data Interpretation.* Master's thesis. University of Florida, Gainesville.

Bullock, S.J. 1978. The Case for Using Multichannel Seismic Refraction Equipment for Site Investigation. *Bulletin of the Association of Engineering Geologists*, Vol. 15, No. 1, pp. 19–36.

Carroll, R.D., J.H. Scott, and F.T. Lee. 1972. Seismic Refraction Studies. In *Geological, Geophysical and Engineering Investigations of the Loveland Basin Landslide, Clear Creek County, Colorado, 1963–65*, U.S. Geological Survey Professional Paper 673-C, pp. 17–19.

Cherry, J.A., and R.A. Freeze. 1979. *Groundwater.* Prentice-Hall, Englewood Cliffs, N.J., 604 pp.

Chiarle, M., and P. Powers. 1994. *Surficial Horizontal Displacements on Slumgullion Landslide, Hinsdale County, Colorado, 1985 to 1990 (Determined by Direct Visual Comparison).* U.S. Geological Survey Open-File Report 94-233, 34 pp.

Cook, J.C. 1974. Status of Ground Probing Radar and Some Recent Experience. In *Proc., Specialty Conference on Subsurface Exploration for Underground Excavation and Heavy Construction, Henniker, New Hampshire*, American Society of Civil Engineers, New York, pp. 175–194.

Cotecchia, V. 1978. Systematic Reconnaissance Mapping and Registration of Slope Movements. *Bulletin of the International Association of Engineering Geology*, No. 17, pp. 5–37.

Davis, S.N., and R.J.M. DeWeist. 1966. *Hydrogeology.* John Wiley & Sons, Inc., New York, 463 pp.

Dearman, W.R., and N. Turk. 1985. Sampling and Testing of Residual Soils in the United Kingdom. In *Sampling and Testing of Residual Soils* (E.W. Brand and H.B. Phillipson, eds.), Scorpion Press, Hong Kong, pp. 175–181.

Dechman, G.H., and M.S. Ouderhoven. 1976. *Velocity-Based Method for Slope Failure Detection.* Report of Investigations 8194. Bureau of Mines, U.S. Department of the Interior, 19 pp.

Deere, D.U. 1963. Technical Description of Rock Cores for Engineering Purposes. *Rock Mechanics and Engineering Geology*, Vol. 1, pp. 18–22.

Dobecki, T.L. 1979. Measurements of In Situ Dynamic Properties in Relation to Geologic Conditions. In *Geology in the Siting of Nuclear Power Plants*, Reviews in Engineering Geology No. 4, Geological Society of America, Boulder, Colo., pp. 201–225.

Dobrin, M.B. 1976. *Introduction to Geophysical Prospecting*, 3rd ed. McGraw-Hill, New York, 630 pp.

Donaldson, P.R. 1975. Geotechnical Parameters and Their Relationships to Rock Resistivity. In *Proc., 13th Annual Engineering Geology and Soils Engineering Symposium*, University of Idaho, Moscow, pp. 169–185.

Doolittle, J.A., and R.A. Rebertus. 1989. Ground-Penetrating Radar as a Means of Quality Control for Soil Surveys. In *Transportation Research Record 1192*, TRB, National Research Council, Washington, D.C., pp.103–110.

Geological Society of London. 1990. Tropical Residual Soils: Report of the Geological Society Engineering Group Working Party on Tropical Residual Soils. *Quarterly Journal of Engineering Geology*, Vol. 23, No. 1, pp. 1–101.

Greenfield, R.J. 1979. Review of Geophysical Approaches to the Detection of Karst. *Bulletin of the Association of Engineering Geologists*, Vol. 16, pp. 393–408.

Griffiths, D.H., and R.F. King. 1969. *Applied Geophysics for Engineers and Geologists.* Pergamon Press, London, 223 pp.

Hardy, H.R., Jr. 1981. *Applications of Acoustic Emission Techniques to Rock and Rock Structures: A State-of-the-Art Review.* Special Technical Publication 750. ASTM, Philadelphia, Pa., pp. 4–92.

Hatheway, A.W. 1982. Trench, Shaft, and Tunnel Mapping. *Bulletin of the Association of Engineering Geologists,* Vol. 19, No. 2, pp. 173–180.

Hunt, R.E. 1984. *Geotechnical Engineering Investigation Manual.* McGraw-Hill, New York, 983 pp.

Hunter, J.A., S.E. Pullan, R.A. Burns, R.M. Gagne, and R.L. Good. 1984. Shallow Seismic Reflection Mapping of the Overburden-Bedrock Interface with the Engineering Seismograph—Some Simple Techniques. *Geophysics,* Vol. 49, No. 8, pp. 1381–1385.

Hutchinson, J.N. 1983. Methods of Locating Slip Surfaces in Landslides. *Bulletin of the Association of Engineering Geologists,* Vol. 20, No. 3, pp. 235–252.

Hvorslev, M.J. 1949. *Subsurface Exploration and Sampling of Soil for Civil Engineering Purposes.* U.S. Army Corps of Engineers, Waterways Experiment Station, Vicksburg, Miss., 521 pp.

Johnson, R.B., and J.V. DeGraff. 1988. *Principles of Engineering Geology.* John Wiley & Sons, Inc., New York, 497 pp.

Johnson, W.E. 1976. Seismic Detection of Water Saturation in Unconsolidated Material. In *Proc., 14th Annual Engineering Geology and Soils Engineering Symposium,* Boise State University, Boise, Idaho, pp. 221–231.

Jurich, D.M. 1985. *Analysis of Soil Slopes Utilizing Acoustic Emissions.* Master of Engineering Report ER-3138. Colorado School of Mines, Golden, 168 pp.

Jurich, D.M., and R.J. Miller. 1987. Acoustic Monitoring of Landslides. In *Transportation Research Record 1119,* TRB, National Research Council, Washington, D.C., pp. 30–38.

Kaufman, A.A., and G.V. Keller. 1983. *Frequency and Transient Soundings.* Elsevier Science Publishers, BV, New York, 685 pp.

Kennedy, B.A., and K.E. Niermeyer. 1971. Slope Monitoring Systems Used in the Prediction of a Major Slope Failure at the Chuquicamata Mine, Chile. In *Planning Open Pit Mines,* South African Institute of Mining and Metallurgy, pp. 215–225.

Knill, J.L. 1975. *Suggested Methods for Geophysical Logging of Boreholes and Measurements of In Situ Seismic Characteristics.* Draft Report. Committee on Standardization of Laboratory and Field Tests, International Society for Rock Mechanics, Department of Geology, Imperial College, London.

Krynine, D.P., and W.R. Judd. 1957. *Principles of Engineering Geology and Geotechnics.* McGraw-Hill, New York, 730 pp.

Lacy, W.C. 1963. Quantitizing Geological Parameters for the Prediction of Stable Slopes. *Transactions of the Society of Mining Engineers of the American Institute of Mining, Metallurgical and Petroleum Engineers,* Vol. 226, pp. 272–276.

Lowe, J., III, and P.F. Zaccheo. 1975. Subsurface Explorations and Sampling. In *Foundation Engineering Handbook,* Van Nostrand-Reinhold Company, New York, pp. 1–66.

Marchetti, S. 1980. In Situ Tests by Flat Dilatometer. *Journal of the Geotechnical Engineering Division,* ASCE, Vol. 106, No. GT3, pp. 299–321.

Mayne, P.W., and D.D. Frost. 1988. Dilatometer Experience in Washington, D.C., and Vicinity. In *Transportation Research Record 1169,* TRB, National Research Council, Washington, D.C., pp. 16–23.

McCauley, M.L. 1976. Microsonic Detection of Landslides. In *Transportation Research Record 581,* TRB, National Research Council, Washington, D.C., pp. 25–30.

McGuffey, V.C. 1991. Clues to Landslide Identification and Investigation. In *Geologic Complexities in the Highway Environment* (R.H. Ficies, ed.), *Proc., 42nd Annual Highway Geology Symposium,* Albany, New York, New York State Department of Transportation, Albany, pp. 187–192.

Menard, H. 1975. Interpretation and Application of Pressuremeter Test Results. *Sols-Soils,* No. 26.

Merritt, A.H. 1974. Exploration Methods. In *Proc., Specialty Conference on Subsurface Exploration for Underground Excavation and Heavy Construction,* Henniker, New Hampshire, American Society of Civil Engineers, New York, pp. 56–64.

Moffatt, B.T. 1974. Subsurface Video Pulse Radars. In *Proc., Specialty Conference on Subsurface Exploration for Underground Excavation and Heavy Construction,* Henniker, New Hampshire, American Society of Civil Engineers, New York, pp. 195–212.

Mohr, H.A. 1962. *Exploration of Soil Conditions and Sampling Operations.* Harvard University Press, Cambridge, Mass., 79 pp.

Moore, R.W. 1972. Electrical Resistivity Investigations. In *Geological, Geophysical and Engineering Investigations of the Loveland Basin Landslide, Clear Creek County, Colorado, 1963–65,* U.S. Geological Survey Professional Paper 673-B, pp. 11–15.

Morey, R.M. 1974. Continuous Subsurface Profiling by Impulse Radar. In *Proc., Specialty Conference on Subsurface Exploration for Underground*

Excavation and Heavy Construction, Henniker, New Hampshire, American Society of Civil Engineers, New York, pp. 213–232.

Murphy, V.J., and D.I. Rubin. 1974. Seismic Survey Investigations of Landslides. In *Proc., Second International Congress of the International Association of Engineering Geologists*, Sao Paulo, Brazil, Associação Brasileira de Geologica de Engenharia, Paper V-26, Vol. 2, 3 pp.

Myung, J.T., and R.W. Baltosser. 1972. Fracture Evaluation by the Borehole Logging Method. In *Stability of Rock Slopes*, American Society of Civil Engineers, New York, pp. 31–56.

NAVFAC. 1982. *Design Manual: Soil Mechanics*. DM-7.1. U.S. Department of Defense, 360 pp.

Obert, L., and W.I. Duvall. 1942. *Use of Subaudible Rock Noises for the Prediction of Rock Bursts, II.* Report of Investigations 3654. Bureau of Mines, U.S. Department of the Interior, 22 pp.

Obert, L., and W.I. Duvall. 1957. *Microseismic Method of Determining the Stability of Underground Openings*. Bulletin 573. Bureau of Mines, U.S. Department of the Interior, 18 pp.

Philbrick, S.S., and A.D. Cleaves. 1958. Field and Laboratory Investigations. In *Special Report 29: Landslides and Engineering Practice* (E.B. Eckel, ed.), HRB, National Research Council, Washington, D.C., pp. 93–111.

Redpath, B.B. 1973. *Seismic Refraction Exploration for Engineering Site Investigations*. Technical Report E-73-4. U.S. Army Corps of Engineers, Exploration and Excavation Research Laboratory, Waterways Experiment Station, Vicksburg, Miss., 55 pp.

Saayman, C.M. 1978. Fundamentals of the Seismic Method. *Bulletin of the Association of Engineering Geologists*, Vol. 15, No. 1, pp. 1–18.

Sanglerat, G. 1972. *The Penetrometer and Soils Exploration*. Elsevier, Amsterdam, 464 pp.

Scharon, H.L. 1951. *Electrical Resistivity Geophysical Method as Applied to Engineering Problems*. Special Technical Publication 122. ASTM, Philadelphia, Pa., pp. 104–114.

Schmertmann, J.H. 1978. *Guidelines for CPT Performance and Design*. FHWA Report 78–209. U.S. Department of Transportation, 145 pp.

Schultz, J.R., and A.B. Cleaves. 1955. *Geology in Engineering*. John Wiley & Sons, Inc., New York, 592 pp.

Sowers, G.B., and G.F. Sowers. 1970. *Introductory Soil Mechanics and Foundations*, 4th ed. Macmillan, New York, 621 pp.

Sowers, G.F., and D.L. Royster. 1978. Field Investigation. In *Special Report 176: Landslides: Analysis and Control* (R.L. Schuster and R.J. Krizek, eds.), TRB, National Research Council, Washington, D.C., pp. 81–111.

Takada, Y. 1968. A Geophysical Study of Landslides. *Bulletin of the Disaster Prevention Research Institute* (Kyoto University, Japan), Vol. 18, Pt. 2, No. 137, pp. 37–58.

Tamaki, I., and Y. Ohba. 1971. Shallow Reflection Method and Its Use for Landslide Investigation. In *Proc., Fourth Asian Regional Conference on Soil Mechanics and Foundation Engineering*, Bangkok, Thailand, Vol. 1, pp. 227–233.

Telford, W.M., L.P. Geldaret, R.E. Sheriff, and D.A. Keys. 1976. *Applied Geophysics*. Cambridge University Press, Cambridge, England, 859 pp.

Terzaghi, K., and R.B. Peck. 1967. *Soil Mechanics in Engineering Practice*, 2nd ed. John Wiley & Sons, Inc., New York, 729 pp.

Trantina, J.A. 1963. *Investigation of Landslides by Seismic and Electrical Resistivity Methods*. Special Technical Publication 322. ASTM, Philadelphia, Pa., pp. 120–133.

Underwood, L.B. 1974. Exploration and Geologic Prediction for Underground Works. In *Proc., Specialty Conference on Subsurface Exploration for Underground Excavation and Heavy Construction*, Henniker, New Hampshire, American Society of Civil Engineers, New York, pp. 65–83.

USBR. 1974. *Earth Manual*, 2nd ed. Bureau of Reclamation, U.S. Department of the Interior, Denver, Colo., 810 pp.

Van Nostrand, R.G., and K.L. Cook. 1966. *Interpretation of Resistivity Data*. U.S. Geological Survey Professional Paper 499, 310 pp.

van Zijl, J.S.V. 1978. On the Uses and Abuses of the Electrical Resistivity Method. *Bulletin of the Association of Engineering Geologists*, Vol. 15, No. 1, pp. 85–112.

Weaver, C.M. 1978. Case History: Seismic Survey for Smelter Plant Foundations, Pietersburg, Transvaal. *Bulletin of the Association of Engineering Geologists*, Vol. 15, No. 1, pp. 51–62.

Winter, E. 1986. The Pressuremeter in Geotechnical Practice. In *Transportation Research Record 1089*, TRB, National Research Council, Washington, D.C., pp. 1–10.

Wyllie, D.C. 1992. *Foundations on Rock*. Chapman and Hall, London, 331 pp.

Chapter 11

P. ERIK MIKKELSEN

FIELD INSTRUMENTATION

1. INTRODUCTION

Landslides often present the ultimate measurement challenge, in part because of their initial lack of definition and the sheer scale of the problems. Intuitively, there is a desire to quantify the extent of a potential or real disaster, but what should be measured? The answer can be complicated and should be addressed by drawing on the best available information concerning topography, geology, groundwater, and material properties. The measurement problem usually requires information ranging from a coarse scale down to a fine scale and involving a number of instrumentation techniques. The ultimate goal is to select the most sensitive measurement parameters, the ones that will change significantly at the onset of the landslide event. However, because of physical limitations and economic constraints, all parameters cannot be measured with equal ease and success.

Geotechnical instruments have evolved tremendously during the past quarter-century. Early versions of current instruments were developed and used on various projects beginning in the 1960s. Early reports concerning instrumentation that have retained their value include those by Shannon et al. (1962), Dunnicliff (1971), Gould and Dunnicliff (1971), and Peck (1972). The comprehensive report on geotechnical field instrumentation prepared by the British Geotechnical Society (1974) contains several important papers (Burland and Moore 1974; Dibiagio 1974; Green 1974; Penman and Charles 1974; Vaughn 1974).

Transportation Research Record 482 (Transportation Research Board 1974) includes several documented case histories of geotechnical instrumentation applications. Considerable interest continued during the 1970s concerning the most appropriate methods for using instrumentation (Mikkelsen and Bestwick 1976; Wilson and Mikkelsen 1977). Chapter 5 in *Landslides: Analysis and Control* (Wilson and Mikkelsen 1978) was devoted to landslide instrumentation.

By the late 1980s, geotechnical instrumentation had evolved to the point that a major textbook was devoted to the subject (Dunnicliff 1988), and the Transportation Research Board published a second *Transportation Research Record* containing reports of new applications and methods (Green and Mikkelsen 1988; Hinckley et al. 1988). Other case histories were published during the same period (Patton 1983, 1984). In recent years the full impact of new instruments and computer automation in data collection and data reduction has been reported in numerous papers (e.g., Mikkelsen 1989, 1992; Dunnicliff and Davidson 1991; Tatchell 1991; Gillon et al. 1992; Proffitt et al. 1992; Kane and Beck 1994; O'Connor and Wade 1994). Experiences with large-scale instrumentation and automation of data collection for monitoring slope stability at the huge open-pit mines of Syncrude, Canada, Ltd., in northern Alberta were described by McKenna et al. (1994) and McKenna (1995).

The focus of this chapter is primarily field instrumentation and measurement techniques associated with permanent borehole installations that serve to measure subsurface behavior. Typical instruments are piezometers for measuring groundwater pressures and inclinometers and extensometers for ground-displacement measurements. Excluded from this discussion are borehole testing and logging techniques that help define structure, lithology, and material properties; these are described in Chapter 10. Surface surveying and other surface measurement technologies, such as seismic monitoring, remote sensing, and aerial photointerpretation, are discussed in Chapters 8, 9, and 10, and so are also not treated in detail in this chapter.

Field instrumentation is most often used on landslides that have already exhibited some movement. In such cases, the overall characteristics of the landslide frequently may be readily observed and noted. However, small movements of a soil or rock mass before or even during incipient failure are usually not visually evident. Thus, the value of the information that can be obtained at the ground surface is limited. Instrumentation can provide valuable information on incipient as well as fully developed landslides. In this respect, use of instrumentation is not intended to replace field observations and other investigative procedures. Instead, it augments other data by providing supplementary information and by warning of impending major movements. Typical situations for which various instruments have been used are the following:

- Determination of the depth and shape of the sliding mass in a developed landslide so that calculations can be made to define the appropriate strength parameters at failure and design remedial treatments,
- Determination of absolute lateral and vertical movements within a sliding mass,
- Determination of the rate of sliding (velocity) and thus provision of a warning of impending danger,
- Monitoring of the activity of marginally stable natural or cut slopes and identification of effects of construction activity or precipitation,
- Monitoring of groundwater levels or pore pressures normally associated with landslide activity so that effective stress analyses may be performed,
- Provision of remote digital readout or information to a remote alarm system that would warn of possible danger, and
- Monitoring and evaluation of the effectiveness of various control measures.

In the last situation, savings are often realized in remedial treatment by a planned and monitored sequence of construction. For example, drainage might be initially installed and its effect monitored to determine whether planned mechanical stabilization is actually necessary.

2. PLANNING AND DESIGN

Adequate planning is required before a specific landslide is instrumented. The plan should proceed through the following stages:

- Determination of types of measurements required;
- Selection of the specific types of instruments best suited to make the required measurements;
- Definition of the location and depth of instrumentation and number of instruments;
- Development of the data acquisition techniques; and
- Decision as to the management and presentation of the data acquired.

Initially the planning process requires development of ideas on the causes of the landslide and the probable limits of the depth and outer boundaries of the movements. Reconnaissance of the area, study of the geology and groundwater seepage, review of rainfall records, and observation of topographic features, especially recent topographic changes, will often provide clues. Unfortunately, no two landslides are alike in all details, and experience alone without the application of judgment may lead to erroneous concepts.

An instrumentation system is a waste of time and money if the instruments do not extend below the zone of movement, were installed at the wrong locations, or are unsuitable. Instruments must collect data accurately, reliably, efficiently, and in a timely manner; otherwise, corrective treatment may be started too late, resulting in even greater economic losses than would have been experienced without it.

2.1 Types of Measurements Required

Landslides by definition involve movement. The magnitude, rate, and distribution of this movement are generally the most important measurements required. Measurements of pore-water pressures within the landslide are important in resolving many landslide problems. Such measurements are especially important when landslides occur in layered systems in which excess hydrostatic pressures may exist between layers.

As discussed in Chapter 9, if the depth of sliding is readily apparent from visual observations, surface measurements may be sufficient for obtaining the rate of movement. Surface measurements should extend beyond the uppermost limit of visual movement so that any ground movements that might occur before visible surface cracks develop can be monitored. Movement of the ground surface should be measured vertically and horizontally at various locations within the landslide area. Vertical offsets, widening of cracks, and heaving of the toe should be monitored. The direction of movement can often be inferred from the pattern of cracking, particularly by matching the irregular edges of the cracks.

If the depth and thickness of the zone of movement are not apparent, inclinometers or similar devices that can detect the movement with depth must be used. Pore pressures at or near the sliding surface must be measured to support effective stress analysis or to assess the adequacy of drainage measures. Rapid-response piezometers are advantageous, particularly in impervious soils.

2.2 Selection of Instrument Types

Many types and models of instruments are available for measuring the changing conditions in a landslide; they vary in degree of sophistication, particularly with regard to readout capabilities. Instruments have been developed to measure vertical and horizontal earth movements, pore-water pressures, in situ stresses and strains, dynamic responses, and other parameters. However, in most landslides measurements of horizontal movement of the soil or rock and of pore-water pressures are of primary concern. Instruments commonly used for these purposes are described in this chapter.

The types of instruments, layout, and monitoring schedules are usually determined by the specific needs of a project. Several basic requirements, however, should be thoroughly evaluated for any system. Instruments should be reliable, rugged, and capable of functioning for long periods of time without repair or replacement. They must also be capable of responding rapidly and precisely to changes so that a true picture of events can be maintained at all times. High sensitivity is usually a prerequisite when hazard warning is required or landslide performance is monitored during construction, since it is often the rate of change rather than the absolute value that provides the key to proper interpretation.

The location of instruments requires a thorough understanding of geologic and subsurface conditions if meaningful data are to be obtained. Location is particularly crucial for pore-pressure recording devices that are intended to measure pressures in specific zones of weakness or potential instability. Since most measurements are relative, a stable base or datum must be provided so that absolute movements can be determined.

If the movement is known to be along a well-defined shear surface, such as a bedding surface or fracture, simple probe pipes will suffice to determine the depth. If the movement is large and rapid, accuracy is not an essential requirement and even relatively crude inclinometers may suffice. When the rate of movement is small and the depth and distribution are not known, more precise instrumentation is required. Carefully installed precision inclinometers are best in such instances, although there may be cases in which extensometers or strain meters can be used to advantage.

As discussed in Chapter 10, the Casagrande type of piezometer is the most useful general purpose pore-pressure measuring device, but it may give too slow an initial response in fine-grained soils. Therefore, pneumatic or vibrating-wire pressure-sensor types (transducers) may be preferable. High-air-entry, low-flow piezometers may warrant consideration in clays or clay-shales in which permeabilities are low or suction may be present because of unloading.

3. SURFACE MEASUREMENT

Instrumentation that may be used for surface measurement includes conventional surveying equipment, portable tiltmeters, and recently developed differential Global Positioning System (DGPS) location methods.

3.1 Conventional Surveying Equipment

The application of conventional surveying methods to landslide movement measurement and monitoring is discussed at length in Chapter 9. In an active landslide area, surface movements are normally monitored to determine the extent of landslide activity and the rate of movement. Both optical instrument surveys and electronic distance measurement (EDM) devices are used to determine lateral and vertical movements. Bench marks located on stable ground provide the basis for determination of subsequent movements of established survey points, or hubs, by optical surveys, EDM equipment, or tape measurement. Recent advances in surveying equipment, including the so-called total stations (see Chapter 9, Section 4.2.2.4), increase the automation and productivity of these measurement procedures, especially on the often-rough ground surface found on landslides. Survey lines can be established so that vertical and horizontal displacements at the center and toe of the landslide can be observed. Lateral motions can be detected by theodolite and distance measurements from each hub. When a tension crack has opened above a landslide, simple daily measurements across the crack can be made between two hubs driven into the ground. In many cases, the outer limit of the ground movement is unknown, and establishing instrument setups on stable ground may be a problem.

3.2 Tiltmeters

Portable tiltmeters (Figure 11-1) can be used to detect tilt (rotation) of a surface point, but such devices have had relatively limited use. Their most common application is to monitor slope movements in open-pit mines and highway and railway cuts, but they may be used in any area where the failure mode of a mass of soil or rock can be expected to include a rotational component, such as a landslide area. Tiltmeters utilize either elec-

trolytic sensors or servo-accelerometers. Electrolytic sensors provide greater sensitivity, but servo-accelerometers have greater range. The prime advantages of these instruments are their light weight, simple operation, compactness, and relatively low cost. Stationary electrolytic sensors with automatic data acquisition are gaining wider usage in landslide monitoring.

3.3 Differential Global Positioning System

The DGPS is emerging as a candidate surveying method for monitoring surface movement at landslide sites. A base station at a known location is used to provide corrections and refinements to the computed locations of one or several roving stations (Gilbert 1995). All stations use the same global positioning satellite network. The DGPS relates observations at unknown roving station locations to simultaneous observations at the base stations placed at a known and fixed location. As the satellite signals are monitored, errors may suggest that the base station is moving. Since it is known that the base station is not moving, these fluctuations can be used as corrections to the observations of the unknown roving stations to achieve their accurate location. All measurements are relative to the base station position. As long as that position is defined reasonably precisely, the other measurements will be internally consistent. If its location is very poorly defined, additional errors may arise. An assumed value of the true location (proper latitude and longitude) can be used without affecting the accuracy of the distances within the survey.

The accuracy of the DGPS is now approaching that of conventional surveys. Under favorable conditions, an accuracy of better than 1 cm is possible (Kleusberg 1992; Gilbert 1995). The historically large capital investments for such systems are now dropping (Gilbert 1994), and it appears that these systems already may be economical for monitoring larger landslide areas. However, the accuracy of the DGPS may be degraded in areas where the ground surface is obscured by tree cover or in times of poor weather conditions. Nevertheless, the enhanced productivity of the DGPS approach suggests that it will soon be economically practical for many landslide monitoring applications.

FIGURE 11-1
Portable tiltmeter
(Wilson and
Mikkelsen 1978).

4. GROUND-DISPLACEMENT MEASUREMENT

4.1 Inclinometers

Before some of the sophisticated types of inclinometers are described, one of the simplest subsurface measuring devices deserves mention: the *borehole probe pipe*, also called the "slip indicator" or "poor boy." It typically consists of a 25-mm diameter semirigid plastic tube, which is inserted into a borehole. Metal rods of increasing length are lowered inside the tubing in turn, and the rod length that is just unable to pass a given depth gives a measure of the curvature of the tubing at that point. Use of the borehole probe pipe is an acceptable technique for the location of slip surfaces in unstable slopes. This type of measurement can easily be performed in the riser pipe of an observation well or open piezometer. If there are several shear planes or if the shear zone is thick, a section of rod hung on a thin wire can be left at the bottom of the hole. Subsequently, the rod is pulled up to detect the lower limit of movement.

The development of inclinometers has been the most important contribution to the analysis and detection of landslide movements in the past two decades. Although used most extensively to monitor landslides, the inclinometer has also gained widespread use as a monitoring instrument for dams, bulkheads, and other earth-retaining structures and in various areas of research. Two broad classes of inclinometers may be used: probe inclinometers and in-place inclinometers. Both classes use similar sensors.

4.1.1 Inclinometer Sensors

One basic type of inclinometer sensor is currently in use in the United States, the accelerometer type, which uses a closed-loop, servo-accelerometer circuit. Each sensor measures inclination in one plane, and the use of two sensors per unit (biaxial sensors) is common (Mikkelsen and Bestwick 1976). A few older, obsolete types are also still in use but are gradually being replaced or abandoned because of lower accuracy and efficiency. Most inclinometers rely on commercially available, sensitive servo-accelerometers.

The servo-accelerometer is composed of a pendulous *proof mass* that is free to swing within a magnetic field (Figure 11-2). The proof mass is provided with a coil or *torque motor*, which allows a linear force to be applied to the proof mass in response to current passed through the coil. A special *position sensor* detects movement of the proof mass from the vertical position. A signal is then generated and converted by a restorer circuit, or servo, into a current through the coil, which balances the proof mass in its original position. In this manner, the current developed by the servo becomes an exact measure of the inertial force and thus of the transducer's inclination. The proof mass usually consists of a flexure unit that operates on the cantilever principle. Temperature has only a negligible effect on readings taken with accelerometer transducers. Accelerometer systems have a resolution of approximately 10 arc-sec in ranges from 30 to 50 degrees. A typical example of such measurements is shown in Figure 11-3(*b*).

4.1.2 Probe Inclinometer

The probe inclinometer commonly used today was developed from a device built in 1952 by S.D. Wilson at Harvard University (Green and Mikkelsen 1988). The same basic concepts have been incorporated into instruments now manufactured by several firms. Such an inclinometer consists of a probe containing a pendulum-actuated transducer, which

FIGURE 11-2
Servo-accelerometer for inclinometer measurements (Wilson and Mikkelsen 1978).

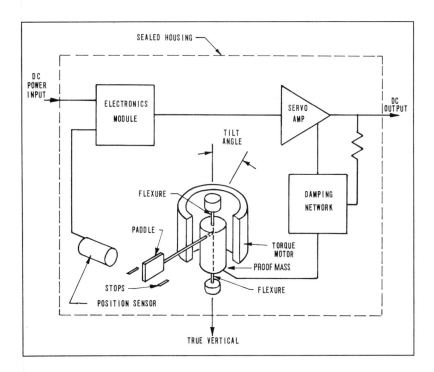

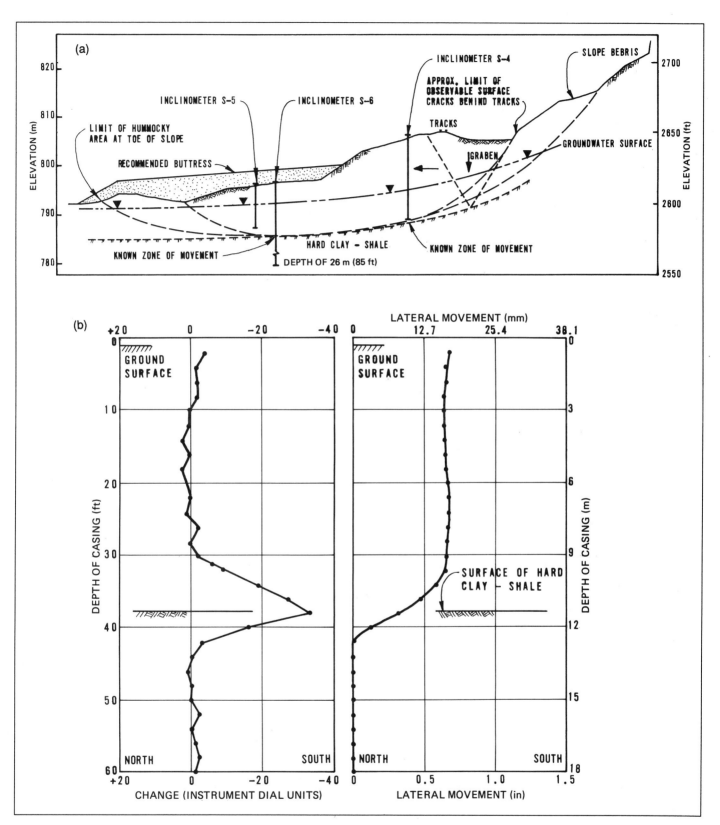

FIGURE 11-3
Inclinometer location and data, Fort Benton landslide, Montana (Wilson and Mikkelsen 1978):
(a) section through Fort Benton landslide; (b) movement of inclinometer S-5 at toe of slope.

is fitted with wheels and lowered by an electrical cable down a plastic casing that is grooved to control alignment. The cable is connected to a readout unit, and data can be recorded manually or automatically (Figure 11-4).

A probe inclinometer system has four main components:

1. A guide casing is permanently installed in a near-vertical borehole in the ground. The casing may be made of plastic, steel, or aluminum. Circular sections generally have longitudinal slots or grooves for orientation of the sensor unit, but square sections are used in some types.
2. A probe sensor unit is commonly mounted in a carriage designed for operation in the guide casing.
3. A control cable raises and lowers the sensor unit in the casing and transmits electrical signals to the surface. For accurate depth control of the sensor unit, the cable is usually graduated or lowered on a separate surveyor's steel tape.
4. A portable control and readout unit at the surface supplies power, receives electric signals, displays readings, and can often store and process data.

The probe inclinometer instrument sensor unit (Figures 11-4 and 11-5) is lowered and raised on an accurately marked cable, its wheels or guides fol-lowing the oriented, longitudinal slots of the casing. Two sets of grooves in the casing allow the inclinometer probe to be oriented in either of two planes set at 90 degrees to each other. At each installation one set of grooves is designated the A orientation plane, and the other set the B orientation plane. The response to slope changes in the casing is monitored and recorded at the surface. Readings are taken at fixed increments, usually equal to the length of the probe, throughout the entire depth.

The wheels of the inclinometer probe provide points of measurement between which the inclination of the instrument is measured. If the reading interval is greater or less than the probe wheel base, correspondence between the position and orientation of the instrument will be only approximate in determining the total lateral displacement profile. Optimum accuracy is achieved only if the distance between each reading interval equals the distance between the upper and the lower wheels of the probe. Gould and Dunnicliff (1971) reported that readings at depth intervals as large as 1.5 m result in poor accuracy.

Instruments differ primarily in the type of sensor used, the accuracy with which the sensor detects the inclination, and the method of alignment and depth control within the borehole. Probe inclinometers can measure changes in inclination on the order of 1.3 to 2.5 mm over a 33-m length of casing (a precision of 1:10,000).

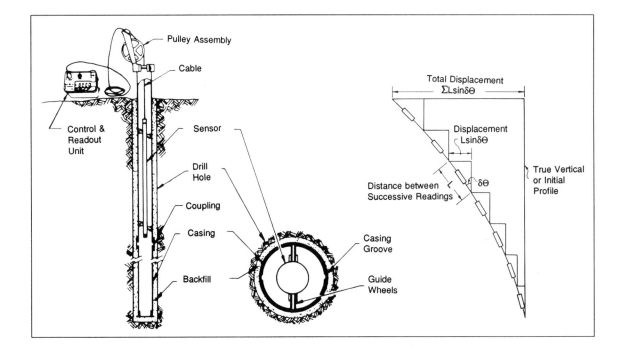

FIGURE 11-4
Principle of inclinometer operation (Wilson and Mikkelsen 1978; Green and Mikkelsen 1988).

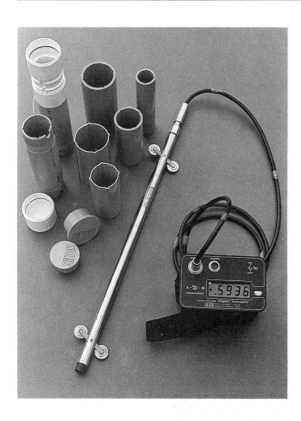

4.1.3 In-Place Inclinometer

The in-place inclinometer (Figure 11-6) employs a series of servo-accelerometer or electrolytic sensors available for both one- and two-axis measurement (Cording et al. 1975). The sealed sensor packages are spaced along a standard grooved inclinometer casing by a series of rods. The rods and sensors are linked by universal joints so that they can deflect freely as the soil and the casing move. The sensors are aligned and secured in the casing by spring-loaded wheels, which fit the casing grooves. They are positioned at intervals along the borehole axis and may be more closely spaced to increase resolution of the displacement profile in zones of expected movement. Use of the grooved casing and guide wheels allows removal of the instrument for maintenance, adjustments, and salvage. Since the casing is a standard inclinometer casing, a probe inclinometer can be used in it after the fixed unit has been removed.

Readings are obtained by determining the change in tilt of the sensor and multiplying by the gauge length or spacing between sensors. The result is the relative displacement of each sensor; these relative displacements can be summed to determine the total displacement at each sensor. Any number of sensors at any spacing can be used in

one borehole. The maximum deflection range is ±30 degrees; the precision is reported to be ±0.01 mm in 1 m, but this accuracy is not usually attained because of long-term drift. A more realistic maximum accuracy is about 0.04 mm in 1 m, giving a precision of 1:25,000, or about 2.5 times the precision of a probe inclinometer. Monitor and alarm consoles and telemetry systems are available for use with in-place inclinometers.

Since fixed units often consist of a number of sensors, they are more expensive and complex than

FIGURE 11-5 *(left)*
Probe inclinometer with manual readout, inclinometer casing, and couplings (Green and Mikkelsen 1988).

FIGURE 11-6 *(below)*
Installation and detail of multiposition in-place inclinometer (Wilson and Mikkelsen 1978).

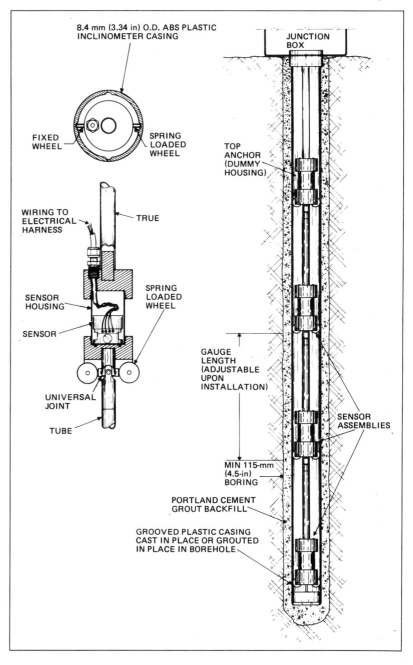

probe inclinometers. However, in-place inclinometers have major advantages over probe inclinometers, including enhanced accuracy and continuous recording capabilities. Fixed units can be monitored remotely by a datalogger or connected to alarm systems. Problems common to probe inclinometers, such as tracking inaccuracies and repeatable positioning (placement error), are eliminated. If the fixed unit is removed for any reason, the overall accuracy will, for the duration, be reduced to no more than that of a probe unit, which can be used in its place. With probe inclinometer units, sensor drift effects are canceled by taking readings in opposite directions. Fixed accelerometer sensors may also drift, and such effects cannot be so easily compensated for. Better long-term performance and resolution are gained by using electrolytic sensor technology. On some projects, an extra inclinometer casing is used so that a probe inclinometer can be used to check readings of the fixed unit.

Because of its potential accuracy (an achievable precision of 1:25,000), an in-place inclinometer can be used to measure small movements in rock and to measure velocity in the hourly time domain. Its adaptability to various remote-monitoring systems allows a continuous record of displacement. By design, most systems are not intended to obtain a continuous profile of deformation. Rather, they measure critical movements of a few sections within the borehole. This instrument, therefore, is not necessarily a replacement for the probe inclinometer unit, but rather has its own purpose and is worthy of consideration for continuous monitoring of movements during critical stages of the landslide.

4.1.4 Principle of Operation

The primary application of the inclinometer to landslides is readily apparent: the definition of slip surfaces or zones of movement relative to stable zones by detection of any change in inclination of the casing from its original near-vertical position. Readings taken at regular incremental depths inside the casing allow determination of the change in slope at various points; integration of the slope changes between any two points yields the relative deflection between those points. The integration is normally performed from the bottom of the boring because the bottom is assumed to be fixed in position and inclination. Repeating such measurements periodically provides data allowing the distribution of lateral movements to be determined as a function of depth below the ground surface and as a function of time (Figure 11-7).

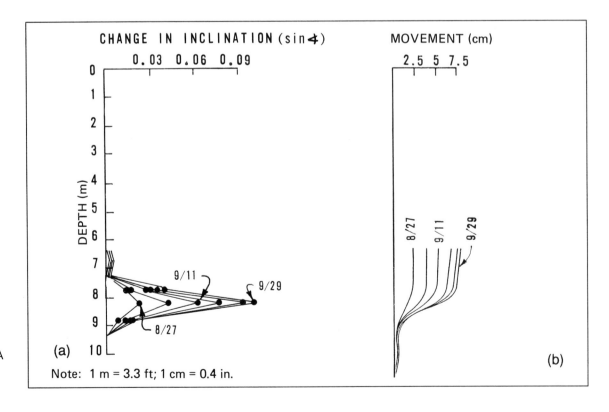

FIGURE 11-7
Measurement of landslide movements with inclinometer: (*a*) observational data, Well A, and (*b*) computed movements, Well A (Wilson and Mikkelsen 1978).

Inclinometers measure the slope of the casing in two mutually perpendicular planes, designated A and B. Thus, horizontal components of movement, both transverse and parallel to any assumed direction of sliding, can be computed from the inclinometer measurements (Figure 11-8). Deviations from this ideal occur because of limitations and compromises in manufacturing and because of installation circumstances. Instruments of different manufacture cannot generally be interchanged without causing discernible effects on the results obtained [Figure 11-8(*a*)]. In the interest of accuracy, the interchange of instruments (probes and cables) of the same manufacture should also be avoided.

Although the principle of operation has remained unchanged, inclinometers have undergone rapid development to improve reliability, provide accuracy, reduce weight and bulk of the instrument, lessen data acquisition and reduction time, and improve versatility of operations under adverse conditions. Use of automatic data recording, personal computers, and data-base software has become standard practice.

4.1.5 Casing Installation

In landslide applications, inclinometer casings are normally installed in exploration drill holes. They must extend through soil and rock suspected of movement and well into materials that, in the best judgment at the time, are assumed to be stable. Some installations extend to depths of more than 200 m (Imrie and Bourne 1981). The annular space between the drill hole and the outer wall of the casing is generally grouted or backfilled with sand. The success of the casing installation depends on the experience and the skill of the personnel; soil, rock, and groundwater conditions; depth of installation; and accessibility to the area.

The accuracy of the inclinometer observations is generally limited not by the sensitivity of the inclinometer sensor, but by the requirement that successive readings be made with the same orientation of the instrument at the same point in the casing. Because the casing must provide a reliable orientation and a continuous tracking for the sensor unit, proper casing installation is of paramount importance.

Since the displacements are referenced with respect to the bottom of the casing, the casing bottom must be extended about 6 m or more into material that will not undergo lateral displacement. If any doubt arises as to the stability of the casing bottom, movement of the casing top should be checked by precise surveying methods. The casing for the inclinometer is usually installed in 1.5- and 3.0-m lengths, which are joined with couplings that are either riveted or cemented, or both, to ensure a firm connection (Figure 11-9). Each coupling represents a possible source of entry for grout or mud, which can seep into the casing and be deposited along the internal tracking system. For this reason, couplings should be sealed with tape or glue. The bottom section of the casing is closed with an aluminum, plastic, or wooden plug, which also should be sealed. If the drill hole is filled with water or drilling mud, the inclinometer casing must be filled with water to overcome buoyancy.

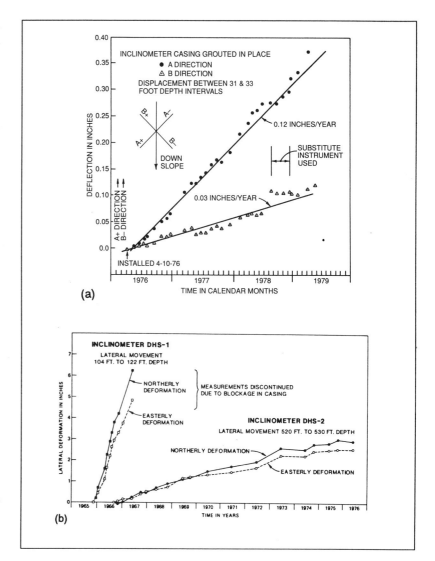

(a)

(b)

FIGURE 11-8
Typical landslide deformation data from inclinometers (Green and Mikkelsen 1988): (*a*) creep movements on thin shear zone in reactivated ancient landslide; (*b*) deformations measured on deep shear zones on a landslide in Canada.

FIGURE 11-9
Details of
inclinometer casing
(Wilson and
Mikkelsen 1978).

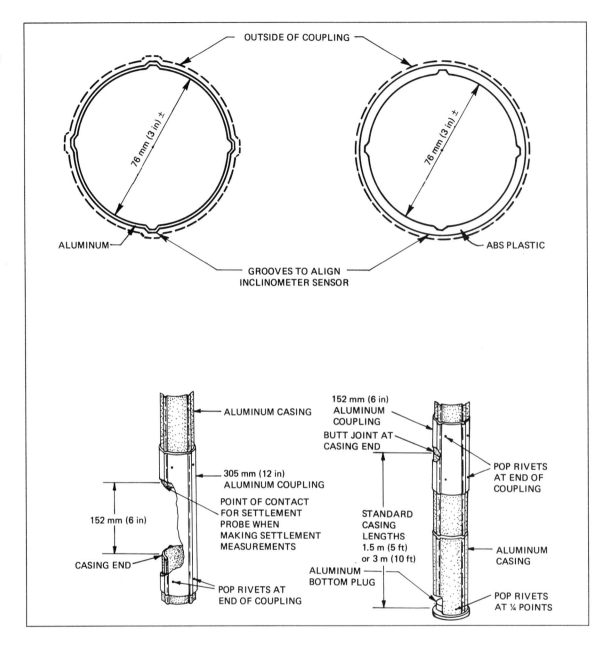

FIGURE 11-9
Details of inclinometer casing (Wilson and Mikkelsen 1978).

Sometimes, extra weight, such as sand bags or a drill stem, may be needed.

The annular space between the boring wall and the inclinometer casing may be backfilled with sand, pea gravel, or grout. The selection of backfill depends mostly on soil, rock, and groundwater conditions (i.e., whether the borehole is dry, wet, caving, or stable without mud). The type of drilling technique (e.g., rotary, hollow-stem, cased holes) is also important. Poorly backfilled casings can introduce scatter into initial measurements. If maximum precision is required, great care should be taken in the selection of the proper backfilling

material. Grouting is generally preferred but may not always be possible, particularly in pervious geologic materials such as talus. Grouting may be facilitated by use of a plastic tube with a diameter of 20 to 25 mm firmly attached to the casing bottom through which grout is pumped from the ground surface until the entire hole is filled. In a small-clearance drill hole, grout can be placed (tremied) through drill rods via a one-way valve at the inside bottom of the casing.

In the as-manufactured casing, the grooves may have some spiral. During installation, the casing can become even more twisted or spiraled so that, at some depth, the casing grooves may not have the same orientation as at the ground surface. Spiraling as great as 18 degrees in a 24-m section of plastic casing has been reported (Green 1974); 1 degree per 3-m section is not uncommon. Because significant error in the assumed direction of movement may result, deep inclinometer casings should be checked after installation by using spiral indicators available from the inclinometer manufacturer. For a particular installation, groove spiraling is generally a systematic error occurring for each section in the same amount and the same direction. In any event, spiraling, whether it is due to manufacture or to torque of the casing during installation, can be measured and is thus not really an error in inclinometer measurement. Spiraling is important only when a determination of the true direction of the movement is required; in such cases, the amount of spiraling should be measured and the results adjusted accordingly.

Small irregularities in the tracking surface of the casing can lead to errors in observation, especially if careful repetition of the depth to each previous reading is not exercised. If plastic casing has been stored in the sun before installation, each section may be locally warped (so that opposing grooves are not parallel), and large measurement errors can occur with minor variations at the depths at which the readings are taken (Gould and Dunnicliff 1971).

Aluminum casings should be used with some caution because severe corrosion may occur if the casing is exposed to alkaline soil, corrosive groundwater, or grout. Epoxy coatings help to minimize this problem. Burland and Moore (1974) recommended cleaning the casing with a stiff brush before taking readings. Frequent flushing with water is also helpful. The top of the casing is generally capped with a tight-fitting plug to prevent intrusion of debris. In addition, the inclinometer casing should be protected and padlocked at the ground surface. When a protective casing or monument covers are selected for use around the top, the need to place a casing-mounted pulley or similar device during inclinometer measurements should be kept in mind.

4.1.6 Observations

Because all inclinometer readings are referenced to an original set of measurements, extreme care must be taken to ensure that the initial set of observations is reliable. Measurements of the original profile should be established by a minimum of a double set of data. If any set of readings deviates from the previous or anticipated pattern, the inclinometer should be checked and the readings repeated. A successful record of measurements generally requires at least two trained technicians, who should be allowed the proper amount of time for setting up, recording observations, and undertaking maintenance of the equipment. Use of the same technicians and instrument for all measurements on a particular project is highly desirable.

When an inclinometer installation is surveyed for the first time with a probe inclinometer to provide the initial set of data, a fixed reference for the probe should be selected so that each time a survey is repeated the same set of grooves is used for alignment orientation and the probe will have the same orientation in the casing. For the typical biaxial inclinometer (Figure 11-5), it is generally recommended that the A component (or sensor) be oriented so that it will register the principal direction of anticipated deformation as a positive change. For example, in an area suspected of landslide activity, the first set of readings is taken by orienting the fixed wheels of the probe in the casing groove closest to the downhill position.

Probe inclinometer measurements generally are recorded as the algebraic sums or differences of pairs of readings. At selected measurement depths, a series of readings in one vertical plane are taken and then repeated with the instrument turned through 180 degrees. Computing the differences of these readings minimizes errors contributed by irregularities in the casing and instrument calibration. One excellent check on the reliability of each measurement is to compute the algebraic sum of each pair of readings. These algebraic sums should be approximately constant except for those readings made while one set of wheels is influenced by a casing joint. Differences between any sum of two readings and the observed constant usually indicate that an error has occurred or that opposite sides of the casing wall are not parallel. This check allows errors resulting from mistaken transcription, faulty equipment, or improper technique to be identified. Thus, the algebraic sum of pairs of readings should be computed on site in the field, so that corrective actions may be taken to resolve transcription errors. However, if the algebraic sum does not remain nearly constant, the sensor unit, readout, and casing should be rechecked before further use.

4.1.7 Data Reduction

Manual computation and reduction of inclinometer data is a tedious and time-consuming operation. A single movement profile for a 33-m casing can involve more than 200 separate computations. Because of the amount of effort involved in taking readings and subsequently computing displacements, a successful measurement program depends primarily on organization and discipline. Field readings must be transposed to discernible measurement, preferably in the form of summary plots indicating successive movement profiles, as soon as possible after the field observation. Successive movement profiles also provide perhaps the best check of instrument reliability. If a record of successive movement profiles is established, the consistency of new measurements can be referenced to previous readings and judged in light of anticipated materials behavior. Because of occasional error in

the accumulation of field readings, the data reduction for a particular casing must be performed with the knowledge of both former movement and expected ground behavior.

Manual data recording is adequate for many small projects. A special data sheet is recommended, which can be used to transcribe the data into a computer. Once this is done, the computer can handle the data with any desired degree of sophistication. It must be recognized, however, that in addition to the time delay, this method is subject to two potential sources of error. The first is in recording the data manually, and the second is in transcribing the data. Figure 11-10 shows a special field data sheet formerly used for recording inclinometer data that were subsequently transcribed for computer processing. Major improvements have been made in inclinometer data acquisition and processing during the past few years.

Personal computers may be used as data collection systems that eliminate hand recording of data and prompt the observer to fill in the right data. Most inclinometer manufacturers now provide units that can support automated data collection. These units can record, store, file, and reduce inclinometer data as collection proceeds in the field, including the correction of systematic errors and incorporation of information defining the spiraling of the casing. One commonly used product is the Digitilt recorder, processor, and printer (RPP) unit, shown in Figure 11-11. This unit can tabulate results, graph them on a built-in electrostatic printer, and store the data on magnetic tape. The RPP can also operate as a computer terminal and send data via telephone modem to a remote computer for additional processing and plotting.

Software for reducing and plotting the results has also greatly improved and now provides rapid, simple operations (Figure 11-12). Typically such software includes the following features:

- Utilization of simple pulldown menus so that there is very little, if any, need to refer to the instruction manual;
- Allowance for all data sets for each inclinometer, representing readings collected at different times, to be stored in a unique data file; and
- Generation of multiple plots by many types of printers, with color plots possible if appropriate equipment is available.

FIGURE 11-10
Simple field sheet formerly used to transcribe inclinometer data before tabulation on punched cards (Green and Mikkelsen 1988).

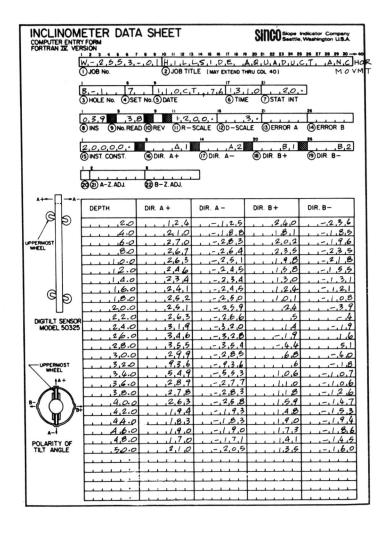

Some recently developed software systems incorporate data-base building and management capabilities. Some software allows linkage of inclinometer data with other data automatically collected from other instruments and telemetered from the landslide site. Such software allows the investigator to readily make a coherent assessment of multiple data sources. Routines are included in many systems for performing time-series analyses, correcting systematic errors, and producing illustrations suitable for inclusion in reports.

When combined with the new data acquisition capabilities, such computer programs can in just a few minutes complete the data analysis, a process that previously required hours. Thus, more time is made available to analyze and make systematic error corrections to the data. Since the mid-1970s, it has been possible to make systematic error corrections to inclinometer data. Because of the speed and convenience of current software, corrections for sensor drift and sensor azimuth alignment changes caused by rotation of the casing can be easily made to improve the results of inclinometer surveys.

4.1.8 Evaluation and Interpretation of Data

All too often, confusing or unexplainable results appear because of instrumentation problems.

These problems are common to all inclinometers; therefore, they are discussed in some detail so that, even without a detailed understanding of the inner workings of the instruments, these anomalies, the limits of accuracy, and the meaning of the results can be recognized. In simple terms, the person closest to the job should be able to answer the following basic questions:

- Is the landslide active?
- How fast is the landslide moving?
- How deep is the landslide?

By means of repeated surveys, inclinometers measure changes in the orientation of a borehole over certain depth intervals during a period of time. A probe inclinometer records this change in orientation at any depth within the limitations of the cable on which the probe is lowered (i.e., weight, strength, and elongation characteristics). Once the active zone has been detected from successive sets of data, the rate of deformation can be determined by plotting the change versus time. Usually, the major zone of movement is only a few meters thick; hence, the sum of the changes over a few consecutive intervals will often be representative of the magnitude and rate of movement of the entire landslide.

The most commonly encountered errors that affect the quality of inclinometer data are the so-called "zero-shift" error and rotation errors due to spiraling of the casing (Figure 11-13). When readings are obtained in a slightly inclined casing, a correction for the apparent drift in the servoaccelerometer readings is necessary. Rotation produces a more irregular pattern in the displacement-versus-depth profile. Both of these effects may mask shear movements occurring at discrete zones and should therefore be evaluated and compensated for during data processing.

To obtain the greatest observation accuracy, the original installation should be as close to vertical as practicable. The error in inclinometer surveys is proportional to the product of the casing inclination and angular changes in the sensor alignment. As shown in Figure 11-14, angular changes in sensor alignment on the order of 1 degree within inclined casings, corresponding to a sensor rotation value (sin a) of about 0.02, may produce errors of several centimeters per 30 m of casing. Sensor-alignment change occurs from time to time because

FIGURE 11-11
Digitilt inclinometer used with recorder, processor, and printer (RPP) unit (Green and Mikkelsen 1988).

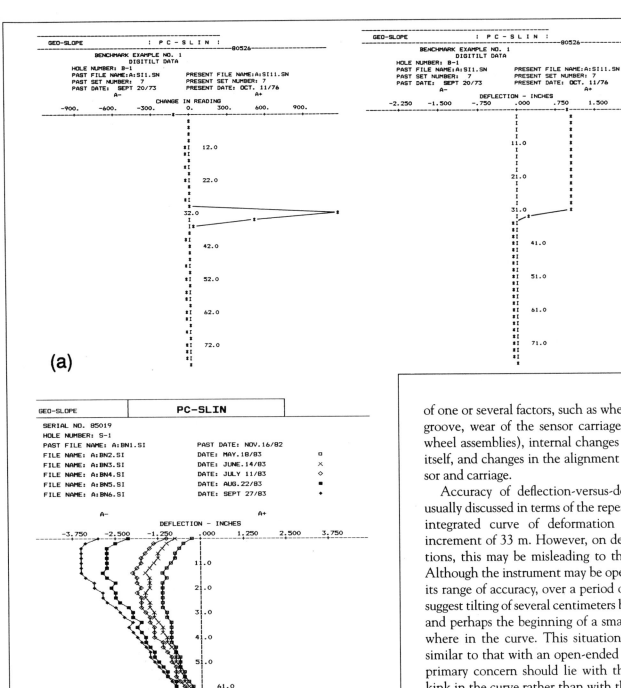

FIGURE 11-12
Graphical computer output of inclinometer data: (*a*) single data set, (*b*) multiple data set (Green and Mikkelsen 1988).

of one or several factors, such as wheel play in the groove, wear of the sensor carriage (particularly wheel assemblies), internal changes in the sensor itself, and changes in the alignment between sensor and carriage.

Accuracy of deflection-versus-depth plots is usually discussed in terms of the repeatability of an integrated curve of deformation for a depth increment of 33 m. However, on deeper installations, this may be misleading to the interpreter. Although the instrument may be operating within its range of accuracy, over a period of time it may suggest tilting of several centimeters back and forth and perhaps the beginning of a small kink somewhere in the curve. This situation is somewhat similar to that with an open-ended traverse. The primary concern should lie with the developing kink in the curve rather than with the overall tilt, which is often related to instrument error at the time of measurement. Again, it must be remembered that the greatest accuracy of the inclinometer lies in its ability to measure change in inclination at a specific depth rather than to survey an exact displacement profile.

4.1.9 Maintenance

Inclinometers are special instruments; consequently, the number of suppliers is limited. Repair or replacement of an inclinometer can be expensive and time-consuming and can result in loss of important data. The best insurance against damage is careful use and systematic maintenance. The sensor unit should be checked frequently during operation and its wheel fixtures and bearings tightened and replaced as necessary. After each casing has been read, the guide wheels should be cleaned and oiled. Most important, the electric readout should be protected against water at all times. The introduction of a few drops of water into the readout circuitry can cause the galvanometer in a pendulum inclinometer to drift or can induce variations in the numbers on the digital voltage display of an accelerometer instrument. If readings are made too rapidly or if an automatic recorder is used, these variations or drift may not be detected and erroneous measurements may result. When the data are subsequently analyzed, the cause of such results cannot be identified.

4.2 Extensometers

Extensometers and strain meters measure the increase or decrease in the length of a wire or rod connecting two points that are anchored in the borehole and whose distance apart is approximately known. One commercially available device is shown in Figure 11-15. When gauge lengths are on the order of a meter or less, these devices are often referred to as strain meters rather than as extensometers. When they are used as extensometers, measurement accuracy and repeatability depend on the type of sensing element and its range of travel and also on the type of connecting wire or rod and the methods used to control the tension. Deadweights are best for maintaining constant tension in wires; if deadweights cannot be used, constant tension springs are acceptable, although there may be some hysteresis. The simplest measurements involve manual readings using a scale or micrometer. Alternatively, linear displacement transducers (potentiometers or vibrating wire) are often used as sensors. These transducers can be monitored by connecting and reading measurements obtained with relatively simple, battery-operated voltmeters. Sensitivity is on the order of 0.1 percent of the range of travel, but repeatability and accuracy may be no better than 0.55 mm, depending on the type of anchor and connecting member. Nevertheless, these parameters are usually sufficient for landslide applications.

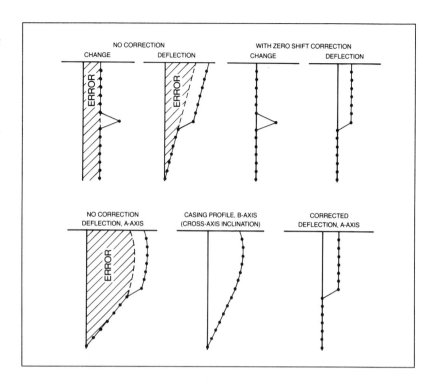

FIGURE 11-13
Typical zero-shift and rotation errors (Green and Mikkelsen 1988).

4.2.1 Strain Meters

When the above devices are used as strain meters, the repeatability and accuracy are essentially the same as the sensitivity. Thus, for a grouted-in-place 3-m-long extensometer assembly composed of an invar rod with a range of 25 mm, unit strains as low as 0.0001 may be detected with relatively inexpensive instrumentation. Horizontal stretching of embankments has been observed by installing anchors at various positions at a given elevation and attaching horizontal wires to deadweights on the downstream face, as was done at Oroville Dam (Wilson 1967). Care is required to ensure that the wires (or rods) do not get pinched if localized vertical shear movements occur. Other applications are described by Dutro and Dickinson (1974), Heinz (1975), and Gillon et al. (1992).

4.2.2 Rod and Wire Extensometers

Wire extensometers, especially those with long distances between the anchor and the sensor, are particularly responsive to changes in wire tension caused by friction along the wire and to hysteresis effects in the sensor or constant tension spring. The sensitivity of these instruments is controlled by these factors. For example, a typical installation might be constructed with a 16-gauge stainless-

FIGURE 11-14
Measurement error
as a result of casing
inclination and
sensor rotation
(Wilson and
Mikkelsen 1978).

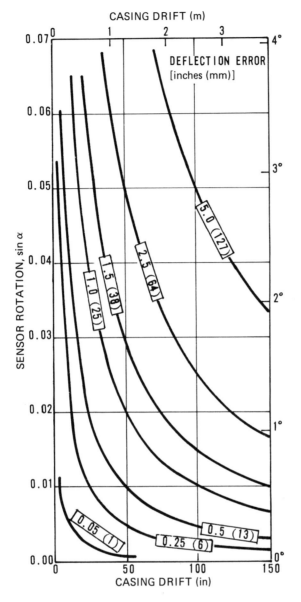

NOTE: CASING DRIFT OVER ANY DEPTH
INCREMENT IS MEASURED AS HORIZONTAL
DEVIATION FROM TRUE VERTICAL AXIS IN
TWO MUTUALLY PERPENDICULAR PLANES.
DEFLECTION ERROR IN ONE PLANE IS RESULT
OF DRIFT IN OTHER PERPENDICULAR PLANE.

should be given to the rate of change in length, because any increase in this rate may be an indication of impending failure.

Extensometer readings are also especially sensitive to temperature changes. If the connecting rods or wires are made of steel, any increase in temperature will result in an increase in their length and thus a reduction in the actual extensometer reading. For example, when the extensometer described in the previous paragraph is subjected to an average change in temperature of 12°C, the same wire will change in length by 3.0 mm. This change in length is about 20 times greater than that detected by the basic sensitivity of the instrument. Daily and seasonal temperature variations are likely to show similar variations in the gauge readings. Not only does the length of the wire change with temperature, but the ground itself expands and contracts as its temperature varies. At a hillside stability project in Montana, rock outcrops were found to expand and contract seasonally by as much as 1.5 mm. Electric lead wires also change their electric resistance with temperature and will cause erroneous readings unless such a change is properly taken into account.

4.2.3 Slope Extensometers

When landslide movements are concentrated along relatively thin shear zones, inclinometers are only capable of measuring movements that are less than 2 to 5 cm. Larger movements exceed the diameter of the inclinometer casing, and the inclinometer becomes useless owing to the inability to access the probe. In such cases, extensometers involving cables and anchors provide a means for continuing the monitoring of the movement. Sometimes extensometer cables and anchors can be installed within the damaged inclinometer casing.

Extensometer cables do not provide accurate measurements until the displacements are greater than the thickness of the shear zone. Before the inclinometer casing becomes crushed, a special detailed survey of the shear zone with the inclinometer probe may be able to determine the thickness of the shear zone, and future slope extensometer results can be corrected accordingly. Alternatively, a completely new subhorizontal extensometer can be installed across the shear zone to provide maximum sensitivity and minimize measurement error.

steel wire 33 m long that is subjected to a pull of 67 N. If the sensor requires 1.11 N to actuate it, the minimum change in length of the wire that can be detected by the sensor is 0.15 mm, which is the effective sensitivity of the extensometer. Up-to-date plots should be kept of changes in length from each anchor to the sensor and of computed changes between anchors. Particular attention

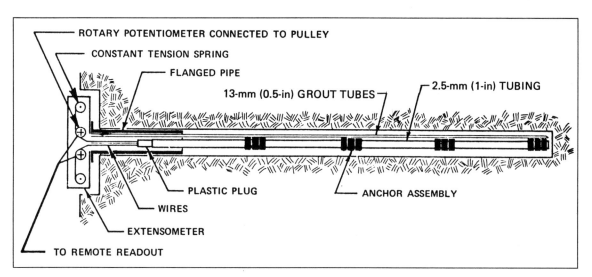

FIGURE 11-15
Extensometer
(Wilson and
Mikkelsen 1978).

Two types of special extensometers that deserve some mention are designed to measure large movements. The first was invented in Europe in the 1970s for monitoring gross landslide and glacier movements. It was adopted for use in the United States to monitor a natural gas pipeline that moved several feet where it crossed a landslide at Douglas Pass, northwestern Colorado.

The second type, a gravity-tensioned slope extensometer, was developed by the Bureau of Reclamation for use on the Big Bull Elk landslide in Montana (L.R. Carpenter, personal communication, 1995, Bureau of Reclamation, U.S. Department of the Interior). The site is located on the Bighorn Reservoir 16 km upstream from Yellowtail Dam and 72 km southeast of Billings. During initial filling of the reservoir, the landslide slumped 3 to 6 m at its contact with a limestone cliff. Measurement points were installed along two lines established on the landslide. These lines were labeled A and B. Manual readings are taken annually between a point scribed on the cliff and reinforcing bars placed at three points on the landslide.

Determination of the Line A displacements was automated in 1985 by establishing a line parallel to and 2 m uphill from the manually read points. The line was divided into sections 30, 50, and 98 m long. A 10.2-cm diameter polyvinyl chloride (PVC) pipe was buried below the ground surface to protect a surveying tape, which was supported by pulleys placed in the pipe at 17-m intervals or at points where the pipe bent to follow a major slope change. Bellows constructed of rubberized fabric connected the individual sections of pipe at such bends. The bellows allowed both linear and angular

displacements of pipe sections and kept the inside of the pipe free of debris. A 30-m extensometer was established in the first section of the line by anchoring the upper end of the surveying tape to an eyebolt installed in the cliff face. This extensometer was tensioned by attaching a weight suspended in a vertical pipe to the lower end of the surveying tape by means of a pulley system installed at the top of the vertical pipe. Two other extensometers were installed in the lower two sections of the line. They were constructed in a similar fashion to the uppermost extensometer. The upper ends of each of these extensometers were anchored by attaching their surveying tape to the axle of the pulley system of the weight-tensioning system of the preceding extensometer.

The displacement of each of these extensometers is determined by monitoring the distance between the lower end of the tape and the position of the lower end of the protecting pipe. This distance is measured by attaching the body of a linear potentiometer to the lower end of the pipe and the sliding potentiometer rod to the tape. The potentiometers exhibit a 5-kΩ electrical change for each 305-mm change in length. The three potentiometer values are monitored by a solar-powered automated data analysis system (ADAS) that transmits the values via a radio link to a personal computer located in the control room at the Yellowtail Dam. These values and other data collected at the dam are relayed through the Geostationary Operational Environmental Satellite (GOES) to a computer at the Bureau of Reclamation office in Denver, Colorado.

5. GROUNDWATER MONITORING

Groundwater levels and pore pressures in a land-slide area can be measured by a variety of commercially available piezometers. The selection of the best type for a particular installation involves several considerations, which are discussed in Chapter 10 (Section 9 and Tables 10-6 and 10-7).

5.1 Open Standpipe Piezometers

The most common water-level recording technique, despite the availability of more sophisticated methods, is observation of the water level in an uncased borehole or observation well. A particular disadvantage of this system is that perched water tables or artesian pressures occurring in specific strata may be interconnected by the borehole

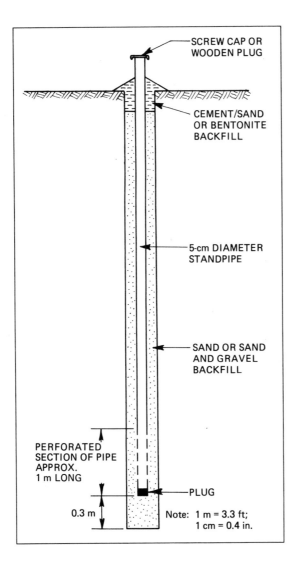

SCREW CAP OR WOODEN PLUG

CEMENT/SAND OR BENTONITE BACKFILL

5-cm DIAMETER STANDPIPE

SAND OR SAND AND GRAVEL BACKFILL

PERFORATED SECTION OF PIPE APPROX. 1 m LONG

PLUG

0.3 m

Note: 1 m = 3.3 ft; 1 cm = 0.4 in.

FIGURE 11-16
Open standpipe piezometer (Wilson and Mikkelsen 1978).

so that the recorded water level may be of little significance for further analysis. Therefore, at all installations, boreholes between different water-bearing strata should be properly sealed, and each water-bearing zone should be separately monitored. Not only will better measurements result, but unnecessary cross-contamination of groundwater supplies will be avoided and most local regulations concerning well construction will be satisfied.

Sealed, open standpipe piezometers vary mainly in diameter of standpipe and type and volume of collecting chamber. The simplest type (Figure 11-16) is merely a cased well sealed above the monitoring portion. The depth to water is measured directly by means of a small probe. In this case, the static head represents the average conditions over the entire zone being monitored. This measured head may be higher or lower than the free water table, but in the case of moderately impervious soils, it may be subject to a large time lag before a stable water level is observed (Hvorslev 1951). The simplicity, ruggedness, and overall reliability of the open standpipe piezometer dictate its use in many installations. However, it is often a poor choice of instrument for measuring water levels in impervious soils because of the inherent time lag, and in partially saturated soils the significance of the measured head may be difficult to evaluate.

The time lag may be reduced by decreasing the diameter of the riser pipe and increasing the diameter of the inflow screen or porous filter (Hvorslev 1951). The Casagrande type of standpipe piezometer (Figure 11-17) consists of a porous stone tip embedded in sand in a sealed-off portion of a boring and connected with a 1-cm diameter plastic riser tube (Shannon et al. 1962). When properly installed, this type of piezometer has proven successful for many materials, particularly in the long term, because it is self-deairing and its nonmetallic construction resists corrosion. The reliability of unproven piezometers is usually evaluated on the basis of how well their results agree with those of adjacent Casagrande piezometers. However, so that commercial water-level probes can be inserted, larger 1- to 2-cm diameter risers, which increase the time lag, are commonly used with the Casagrande-type piezometers. If rapid piezometer responses are a major concern in a landslide, pressure sensor piezometers should be used.

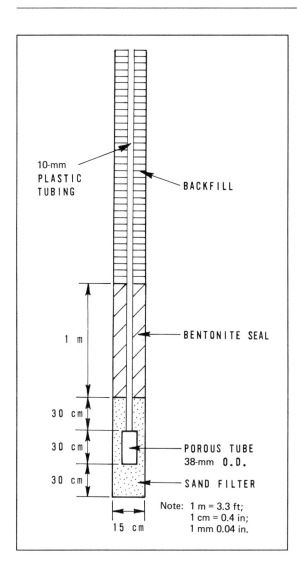

10-mm PLASTIC TUBING

BACKFILL

1 m

BENTONITE SEAL

30 cm

30 cm

POROUS TUBE 38-mm O.D.

30 cm

SAND FILTER

Note: 1 m = 3.3 ft;
1 cm = 0.4 in;
1 mm 0.04 in.

15 cm

FIGURE 11-17
Casagrande
borehole piezometer
(Wilson and
Mikkelsen 1978).

slope face, at any convenient location and elevation. In the pneumatic piezometer shown in Figure 11-18, flow of air through the outlet tube is established as soon as the inlet-tube pressure equals the pore-water pressure. Pneumatic piezometers have the following advantages: (a) negligible time lag because of the small volume change required to operate the valve, (b) simplicity of operation, (c) capability of purging the lines, and (d) long-term stability. Their main disadvantage is the absence of a deairing facility.

Electrical piezometers have a diaphragm that is deflected by the pore pressure acting against one face. The deflection of the diaphragm is proportional to the pressure and is measured by means of various types of electric transducers, the most common being either vibrating-wire or strain-gauge pressure transducers. Such devices have a negligible time lag and are extremely sensitive. However, they cannot be deaired nor can the sensitive electric transducers normally be recalibrated in situ (Dibiagio 1974). For short-term observations at installations where data transmission is over limited distances, standard resistance strain-gauge pressure transducers may be used. Because they are affected by the environment and have poor long-term stability, resistance strain-gauge pressure transducers are generally not recommended for installations in which reliable readings are required over an extended period of time. A typical vibrating-wire electrical piezometer is shown in Figure 11-19. These generally exhibit better long-term stability characteristics. However, not all vibrating-wire types of piezometers have provided consistent long-term stability records.

Although somewhat expensive, the multipoint piezometer offers major advantages for monitoring groundwater in deep landslides. Invented in the mid-1970s and initially used on the Downie landslide located between Mica and Revelstoke on the Columbia River in British Columbia (Patton 1979, 1983, 1984), this system has since become a seasoned and reliable commercial product.

Integrated data gathering and monitoring systems that take full advantage of the latest in microtechnology and single-wire transmission of digital data were launched about a decade ago. These systems, such as the MP System with modular subsurface data acquisition (MOSDAX) developed by Westbay (Patton et al. 1991), are revolutionizing the methods of obtaining and eval-

5.2 Pressure Sensor Piezometers

Electrical vibrating-wire and pneumatic sensors dominate the bulk of the geotechnical piezometer instrumentation market, with the old standby, the Casagrande-style standpipe, as the backup. Considerable improvement in these instruments has occurred during the last decade, and several new commercial products have become available. Each type has distinctly favorable and unfavorable attributes relative to short- and long-term monitoring. Both the electrical vibrating-wire and pneumatic sensors (or transducers) have become less expensive and easier to use.

The pneumatic piezometer consists of a sealed tip containing a pressure-sensitive valve (Penman 1961). The valve opens or closes the connection between two tubes that lead to the surface, or the

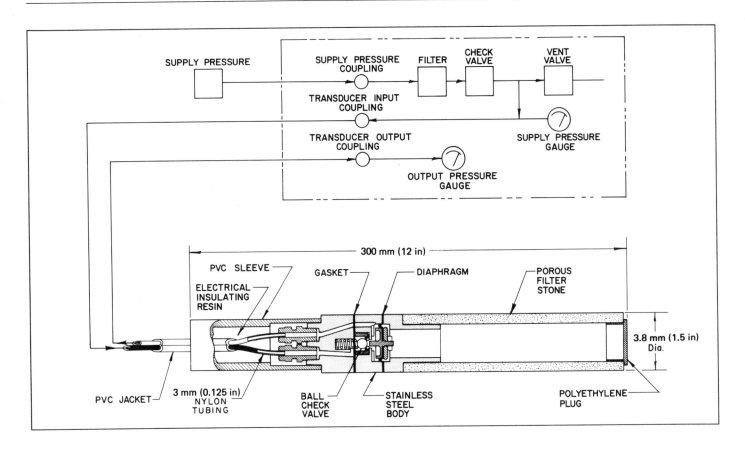

FIGURE 11-18
Pneumatic
piezometer
(Wilson and
Mikkelsen 1978).

uating groundwater data in complex hydrogeological regimes at reservoirs, dams, and landslide areas.

The MP System facilitates the measurement of groundwater or gas pressure at any number of points spaced as close as 1.5 m apart in a single borehole. Sections of 48- × 38-mm PVC pipe are combined with special valved couplings and packer units to isolate multiple pressure-measuring ports. If desired, ports can serve as monitoring sites or be opened to pump and sample groundwater, or a permeability test may be run from these locations. Installations have extended to depths of more than 300 m. MOSDAX is a pressure-sensing module that can be used to probe any individual port in turn, provided manual data collection is acceptable. Automated data collection is possible, even at unattended sites, when dataloggers are combined with multiple MOSDAX units attached to selected ports.

5.3 Importance of Piezometer Sealing

Piezometers in landslides are commonly installed in boreholes advanced into soil or rock. The depth of the monitoring location for each piezometer is selected on the basis of the drill log and identification of materials encountered in the borehole, the estimated position of the water table, and the estimated location of the sliding surface.

The simplest open standpipe piezometer is constructed by placing a tube within the drilled hole that extends to the ground surface. The monitoring portion of the piezometer installation should be pervious to allow observation of the groundwater pressure. This monitoring section may be constructed of a porous material, or the appropriate portion of the tube can be slotted or constructed from a material with a sufficiently fine screen so that free access of groundwater is permitted but movement of soil into the piezometer is prevented. The piezometer tube assembly is centered in the drilled hole, and a known volume of clean sand is placed around the piezometer monitoring section to create a sand filter between the soil and the piezometer sensor.

Construction of an impervious barrier above the piezometer tip and sand pocket is essential. Casagrande, pneumatic, and electric sensors placed within open standpipe piezometer assemblies in boreholes can be sealed in a similar manner. One

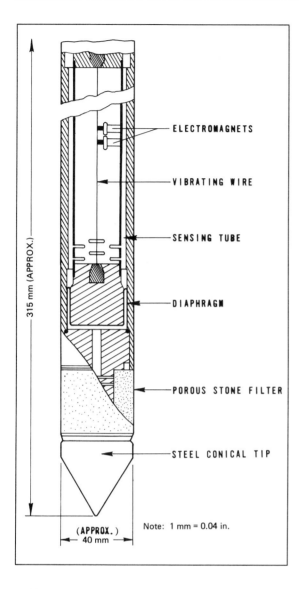

ELECTROMAGNETS

VIBRATING WIRE

SENSING TUBE

DIAPHRAGM

POROUS STONE FILTER

STEEL CONICAL TIP

315 mm (APPROX.)

(APPROX.)
40 mm

Note: 1 mm = 0.04 in.

FIGURE 11-19
Electrical
piezometer
(Wilson and
Mikkelsen 1978).

cement-bentonite grout. Even if the permeability of the grout is significantly higher than that of the surrounding soil, little error will result because of the relatively large grouted length. In many cases, a sand filter need not be included because the piezometer tip can be grouted directly with little error.

Installing only one, or at most two, piezometers in a single borehole has generally been found preferable because bentonite seals are sometimes difficult to construct and may leak slightly despite precautions. However, special-purpose, multiple-point piezometers (with up to four tips) have been used successfully by Vaughan (1969) and merit consideration.

In soft and medium-stiff soils, it may be preferable to push the piezometer directly into the ground. Flush-coupled heavy steel pipe is required to allow for the driving of the probe, and the piezometer tip must be robust enough to resist damage and should be designed to minimize disturbance around the tip (Mikkelsen and Bestwick 1976). Casagrande, pneumatic, or, less commonly, electric sensors may be used. These driven piezometers are self-sealing and rapidly installed. If pneumatic sensors are used, these piezometers have a rapid response. Since the piezometer tip cannot be deaired after installation, it should be soaked in deaired water beforehand and kept in the water until it is driven into place. A low-air-entry filter tip can be placed in a saturated sand pocket, but a high-air-entry tip should be pushed into the soil beyond the base of the hole or surrounded by a porous grout such as plaster of paris. This procedure will ensure more rapid pressure equalization, which may be of considerable value in obtaining reliable pore-pressure measurements soon after installation in low-permeability clay. Both electrical and pneumatic piezometers should be checked for malfunction before and during installation, and particular care is needed in pushing electrical piezometers to prevent overpressuring them.

5.4 Presentation of Data

Piezometric heads should be plotted on time graphs showing amounts of rainfall and other data that may influence the pore pressure. If drainage has been installed, the quantity of seepage should also be recorded and plotted. If possible, the response of each piezometer should be checked periodically to determine its recovery rate.

well-established procedure is to drop balls of soft bentonite into the space between the piezometer tube and the borehole wall and then to tamp those balls around the piezometer tube by using an annular hammer. Prepared bentonite in chip or pellet form is available commercially; it has a specific gravity that is sufficiently great to allow it to sink through the water in the piezometer hole so that hammering is not necessary. Alternatively, a cement-bentonite grout can be tremied into the hole above the sand filter. Such a seal can be pumped through a 1.3- to 2.0-cm diameter pipe adjacent to the piezometer tube. An alternative method of sealing piezometers in boreholes within fine-grained materials was described by Vaughan (1969). Instead of bentonite being placed above a sand filter, the borehole is completely grouted with

6. DATA RECORDING AND TRANSMISSION

Clearly, the evolution of automated data recording in the field has been the greatest innovation in geotechnical instrumentation during the past decade. The accuracy and speed of data transfer from field to office are increased drastically by electronic recording (Mikkelsen 1992). Methods range in sophistication from data keyed into general-purpose electronic notebooks to fully automated data analysis systems (ADAS). At the heart of this technology are the continuously evolving microchips, which seem to get smaller and faster and have more memory every few months. The versatility, speed, power, portability, and convenience of personal computers provide capabilities that were only dreamed of 10 years ago. Today, some data evaluations can even be conducted in the field with devices that are very simple to use. The day of the slide rule is long gone, and so is the manual recording of numerical readings, with date, time, and comments, for many applications.

Successful instrumentation and data-gathering automation require a blending of expertise from diverse areas, including civil engineering, electronics, communications, instrumentation, and computers (Dunnicliff and Davidson 1991). Much of the necessary synergism was encouraged by those involved in dam instrumentation, in which increased sensitivity to dam safety issues coupled with dramatically improved computer hardware and software offered potential solutions to reductions in monitoring staff and other budgetary constraints. Some of the instruments and much of the experience in the automation of field instrumentation networks developed by those involved in dam monitoring can be applied to other geotechnical applications, including landslide investigations. On the basis of experience with dam instrumentation, Dunnicliff and Davidson (1991) reviewed methods for automating data collection from various classes of geotechnical instruments and the ease with which such automation can be achieved (Table 11-1).

Automated data recording and transmission can be undertaken in several ways, each reflecting different levels of automation, investment, and sophistication. Three broad approaches can be defined:

- Readout boxes with memory,
- Multichannel dataloggers, and
- Integrated computer-based ADAS.

6.1 Readout Boxes with Memory

A variety of dedicated readout boxes containing large volumes of digital memory are appearing on the market from several manufacturers. These readout boxes are lightweight and weather- and splashproof, have power for at least one work day, and can communicate with a personal computer. Currently available readout boxes are able to process signals from inclinometers and vibrating-wire electrical and pneumatic piezometer sensors. Typically, these readout boxes are carried into the field and manually connected to a series of instruments in a sequential fashion. The data received from each instrument are temporarily stored in the memory of the readout box. At the end of a suitable period, usually no more than a day, the stored data are offloaded from the readout box into a personal computer, with which an appropriate data base is maintained. Expertise in programming is not necessary for this method of electronic recording. An infrequent user is usually guided through the proper recording procedure by menu displays. Some data-entry mistakes are even allowed without the necessity of restarting the reading session.

For example, preparations for a typical field session of reading a series of inclinometers at a landslide would start in the office by connecting the readout box to a personal computer with a serial interface cable and bringing up the data-base management program. A single data-base file would contain information concerning all the inclinometers and all the readings for one project, including special notes and installation parameters. From this file, any of the inclinometer installation parameters and sets of previous readings would be obtained. In an on-going data acquisition program, the installation parameters for each inclinometer would be loaded into the readout-unit memory, but modifications in the parameters would perhaps be required. Deletion of previously transferred data sets is usually necessary to free up memory in the readout box before initiation of a new field reading session. Only a few minutes would be required to accomplish these tasks before the readout box was ready for the field.

In the field, the reading of the individual inclinometers would proceed as usual; the readout box would be connected to the inclinometer probe and activated by selecting the menu option for collection of readings. Then, by scrolling through a list

Table 11-1
Automated Data Collection from Various Geotechnical Instruments (Dunnicliff and Davidson 1991)

INSTRUMENT[a]	CATEGORY[b]	AUTOMATION METHOD	MANUAL BACKUP
Piezometers			
Open standpipe	2	Pressure transducer down standpipe	Usually straightforward, by measuring water surface
Twin-tube hydraulic	2	Pressure transducers connected to lines	Straightforward, by existing pressure gauges
Pneumatic	3	Pressure actuating and measurement system	Straightforward, with manually operated readout
Vibrating-wire	2	Frequency counter	Straightforward, with manually operated readout
Electrical resistance strain gauge	2	Strain gauge completion circuitry	Straightforward, with manually operated readout
4- to 20-mA pressure transmitter	1,2	Electrical current readout	Straightforward, with manually operated readout
Deformation gauges			
Tiltmeter[c]	1,2	—[c]	Straightforward, with manually operated readout
Probe extensometer	4	N/A	N/A
Embankment and borehole extensometer[c]	1,2	—[c]	Straightforward, with manually operated readout
Inclinometer, probe type	4	N/A	N/A
Plumb line and inverted pendulum[c]	3,4	Infrared or light sensor line and position transducers	Straightforward, with manually operated readout
Liquid level gauge[c]	1,2	Pressure transducer	Straightforward, with manually operated readout
Earth pressure cells			
Pneumatic	3	Pressure actuating and measurement system	Straightforward, with manually operated readout
Electrical resistance strain gauge	2	Strain gauge completion circuity	Straightforward, with manually operated readout
Load cells and strain gauges			
Electrical resistance strain gauge	2	Strain gauge completion circuitry	Straightforward, with manually operated readout
Vibrating-wire	2	Frequency counter	Straightforward, with manually operated readout
Hydraulic load cell	2	Pressure transducer	Straightforward, with manually operated readout
Temperature			
Thermistor	1,2	Resistance readout	Straightforward, with manually operated readout
Thermocouple	2	Low-level voltage readout	Straightforward, with manually operated readout
Miscellaneous			
Surface water elevations, rain gauges	1,2	—[c]	Straightforward, with manually operated readout
Weir for measuring water flow	1,2	—[c]	Straightforward, with manually operated readout
Seismograph 4	N/A		Seismograph chart record

[a] For a definition of instument terms, see work by Dunnicliff (1988).
[b] There are four categories in order of increasing difficulty, specialization, and cost of automation:
 Category 1, Straightforward: Automation can be done by numerous manufacturers of laboratory and industrial ADAS equipment.
 Category 2, More Specialized Requirements: Problems exist in thermally unconditioned environments or with inaccessible power; requires non-standard signal conditioning, low-level voltage measurement, and wide common mode voltage range.
 Category 3, Most Specialized Requirements: Problems exist in hostile outdoor environment; radio network communication required if hard-wire communication unavailable; sensors are difficult to automate (e.g., plumb lines or pneumatic piezometers); involves sensors that communicate via serial or parallel interface.
 Category 4, Usually Not Practical To Automate: Technical complexity or cost outweighs benefits; significant reliability problems exist in hostile environment; customized automation hardware required or impractical to automate.
[c] Automation method and difficulty depend on type of transducer used.

on the display, the user would select the desired inclinometer. The time and date would be logged automatically, and the stored inclinometer parameters would be retrieved from memory. These parameters would be quickly confirmed by the operator, and the reading session would begin. The readings would be recorded at each depth by pressing a remote hand switch. After a series of field readings was completed, the data would be checked for errors in the field and simple calculations performed. Subsequently, in the office, the new readings would be uploaded into the personal computer, the data base updated, and new displacement plots produced.

6.2 Multichannel Dataloggers

Manufacturers of geotechnical instruments have produced a number of sensor-dedicated dataloggers in the past, but it is only recently that more intelligent field-hardened units that can handle several signal types have emerged. In the early 1980s, Campbell Scientific, a manufacturer of weather and environmental data collection units, produced their initial CR10 instrument, the first significant new data collection unit to be widely utilized by the geotechnical industry in North America (Figure 11-20). The CR10 still forms the heart of most of the small unattended datalogger systems produced by several suppliers.

Because they must remain in the field continuously for extended periods, dataloggers have to be even more rugged than readout boxes, which are only carried into the field during actual reading sessions. Dataloggers have to withstand whatever field and weather conditions are experienced for

days, weeks, and months. Sufficient battery power and a great deal of memory are necessary requirements. Continuing improvement in low-power electronics permits dataloggers to be operated unattended from a month to up to a year. Use of solar power for battery recharging can extend unattended use for several years. Dataloggers can monitor from one to a few dozen sensors at a remote site. The sensors can be polled according to a preprogrammed sequence or activated by a triggering event or by manual command.

6.3 Integrated Computer-Based ADAS

A datalogger may form the heart of an ADAS, but much more powerful and flexible systems are built around personal computers. A computer-based ADAS is generally capable of dealing with a variety of transducer types and instrument-monitoring needs. The use of an ADAS leads to some cost reductions and easier system operation because a single computer can replace several dataloggers. The resulting integrated system is more robust. Computer-based systems generally provide powerful communications capabilities that allow the ADAS to be accessed by several remote users, including the system instrumentation manufacturer for monitoring and evaluation of operational diagnostic tests.

The heart of an ADAS is the central network monitor, which houses the host computer and communications hardware allowing receipt of data from one or several sites. Data are acquired at sites by local monitoring units (LMUs), which are directly connected to the central network monitor, or by remote monitoring units (RMUs), which transmit data to the central network monitor by telemetry (usually radio) links. The individual RMU or LMU systems contain modules for the multiplexing (switching), signal conditioning, power excitation, control, and communications functions required to acquire data from each transducer. RMU and LMU systems can be designed to convert raw data into desired engineering units, to collect data readings according to a preprogrammed sequence, or to be polled from the central network module.

All of the above ADAS components can be located at a remote site, for example, a single landslide or group of nearby landslides. A variety of telephone, microwave, radio, or even satellite data

FIGURE 11-20
Campbell Scientific
CR10 datalogger.

transmission links have been used to connect remote ADAS installations to a central office computer system. Dunnicliff and Davidson (1991) provided many important recommendations concerning the specification, design, and maintenance of ADAS installations for dam projects, but some further comments are warranted concerning their application to landslide investigations. Economic considerations are most important. These systems were developed mainly for large dam projects and remain relatively costly for many landslide investigation and monitoring activities. Nevertheless, as noted in the case histories presented later in this chapter, they are already in use at several landslide locations, usually those that exhibit considerable potential for having significant economic impacts or pose large safety hazards. The cost of such systems continues to fall, in line with the rapidly declining costs of most electronic components. As the novelty of these automated systems passes and their cost continues to diminish, their increased use at landslide sites can be expected.

Dunnicliff and Davidson (1991) noted that the initial expectation that automation would cut labor costs does not appear realistic. Compared with manual retrieval of readings, automation does not eliminate the need for personnel. Introduction of a new and more sophisticated technology, the need for maintenance of additional components, and processing and evaluating an increased volume of data generally require the same or a greater number of people. However, better and more timely data, and even real-time continuous data, are advantageous because a much better understanding of the landslide problem is provided.

Automated data collection technology is new to most geotechnical engineers; only a few specialists and instrument suppliers have the expertise to carry out successful programs. The technology is better known in other application sectors, such as hydrological and meteorological monitoring, and by the telecommunications industry. Geotechnical practitioners should seek assistance from those in other fields experienced with this technology before attempting to use it for the first time in geotechnical applications.

7. EXAMPLE PROJECTS

Chapter 5 in *Landslides: Analysis and Control* (Wilson and Mikkelsen 1978) contained several brief case histories of landslide instrumentation that had been reported before 1978. Although now somewhat dated, these case histories include several classic examples. These earlier landslide instrumentation applications included those at a rock cut along Interstate-94 in Minneapolis, Minnesota (Wilson, 1970, 1974); in the Potrero Tunnel of the Southern Pacific Railroad near San Francisco, California (Wilson, 1970; Smith and Forsyth, 1971); in the stabilization of glacial till slopes along Interstate-5 in downtown Seattle, Washington (Wilson, 1970; Palladino and Peck, 1972); and at the Fort Benton landslide along the Missouri River in western Montana (Wilson and Hilts 1971). Other early landslide instrumentation case studies were reported by Durr (1974), Muñoz and Gano (1974), and Tice and Sams (1974).

Five recent landslide monitoring projects that feature a variety of instrumentation issues have been selected for inclusion in this chapter. The instrument programs developed, the purposes of the investigations, the measurements achieved, and the lessons learned are described for each case history.

7.1 Downie Slide near Revelstoke Dam and Reservoir, British Columbia, Canada

Investigations in 1956 of alternative dam sites on the Columbia River north of Revelstoke, British Columbia, led to discovery of a large prehistoric rock slide on the west side of the river (Piteau et al. 1978). The volume of this landslide is approximately 1.5 km³; today its toe extends into Revelstoke Reservoir 66 km upstream from the 160-m-high concrete gravity Revelstoke Dam.

The Downie landslide was first studied in 1965–1966 as part of the design for Mica Dam, located 70 km upstream from the landslide (Imrie 1983). The investigation was undertaken to determine how a blockage of the Columbia River caused by reactivation of the landslide might affect the stability of the downstream toe of the dam. The early investigation consisted of geologic mapping, borings to study the rock layers, and installation of instruments to measure slide movement and groundwater conditions. Five drill holes (three on and two off the landslide) were involved in which core was recovered using double-tube core barrels. Inclinometer casings were placed in the drill holes on the landslide, and measurements were taken. Other early instrumentation included two surface

surveys with 21 monuments, strong-motion seismometer measurements, rock noise surveys, and
water-level measurements in the three leaky inclinometer casings. The results of the study, described
by Piteau et al. (1978), were helpful in answering
questions concerning the stability of Mica Dam but
were inconclusive with regard to the nature and
depth of the slide (Patton 1983).

Beginning in 1973, a new study was undertaken
to evaluate the stability of the landslide and to determine the effect on the landslide of the future
reservoir created by the proposed Revelstoke Dam,
which would raise the water level 70 m into the
toe of the landslide (Imrie and Bourne 1981). This
new study proceeded in several stages using instrumentation, stability analysis, and geologic observations. The instrumentation program eventually
involved 41 drill holes instrumented with 113
piezometers (both single- and multiple-position)
and 10 inclinometer casings (Patton 1983; Moore

1989). Additional instrumentation included 45
surface survey monuments, 2 meteorological stations and 3 snow-course stations, 7 streamflow and
adit-flow gauging stations, and 1 seismometer. The
combination of surface surveys measured by electronic distance measurement (EDM) and inclinometers cemented vertically into holes that
extended as much as 250 m into the landslide
monitored movements as small as 2 mm/year
(Imrie 1983). The locations of the drainage adits
and many of the instrument stations are shown in
Figure 11-21.

During reservoir filling in 1983, a continuous
surveillance system monitored a microseismic unit,
three in-place inclinometer sensors, three extensometers, two automatic flume recorders, and four
fluid-pressure transducers (Patton 1983). Some of
the instrument readings were telemetered to
Revelstoke and Vancouver, British Columbia, to
provide constant information on landslide move-

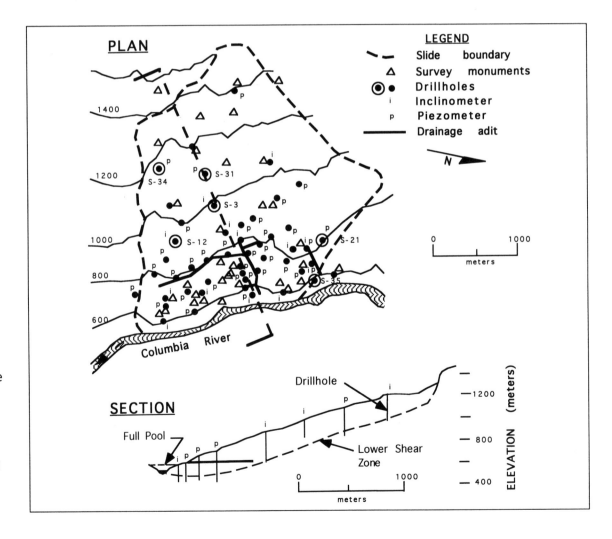

FIGURE 11-21
Plan view and
section for Downie
slide indicating
locations of
drainage adits,
boreholes,
instrumentation,
and survey targets
(modified from
Imrie and Bourne
1981).

ment, seismic activity, and groundwater flow. The telemetry allowed the instruments to serve as a warning system that would allow suitable action, such as lowering the reservoir, to be taken quickly (Imrie 1983).

The Downie slide has been stabilized by means of an extensive drainage system (Imrie et al. 1992). Some 2430 m of exploratory and drainage adits was excavated in and adjacent to the landslide mass in three separate stages between 1974 and 1981. Observations made during initial stages were used to lay out subsequent plans. About 12 200 m of drainholes was drilled from the adits, and a widespread lowering of water pressures at the basal shear zone was achieved (Patton 1983; Moore 1989).

Largely on the basis of the Downie slide studies, Patton (1984) arrived at the following conclusions regarding the collection and use of piezometric data:

• Dormant slides can be affected by maximum long-term piezometric levels and therefore are affected by long-term climatic changes;
• Slope stability analyses should generally consider three classes of piezometric observations: (a) short-term measurements, (b) seasonal or storm-related maxima over a several-year period of record, and (c) estimated long-term maximum piezometric levels;
• Long-term piezometric records are almost always unavailable for slope stability analyses, but they can be estimated for a period to coincide with the expected life of the project using available piezometric records as a basis; and
• Piezometric levels used for slope stability analyses should, as a minimum, be based on the maximum levels recorded over a period of several years.

After 10 to 15 years of use, the monitoring system has begun to decay. Several instruments have become blocked or otherwise inoperable and will have to be replaced (Imrie et al. 1992). The original ADAS, first implemented in 1983 to gather real-time data from key piezometers, inclinometers, extensometers, seismometers, and flow meters (Tatchell 1991), was no longer state of the art and was replaced, beginning in 1991, by advanced telemetry, including hardware and software components (Imrie et al. 1992).

7.2 Cromwell Gorge Landslides, Clyde Dam Project, New Zealand

An extensive network of instruments has been installed to monitor 13 large landslides in Cromwell Gorge of the Clutha River, central Otago, New Zealand (Gillon et al. 1992; Proffitt et al. 1992). The landslides, shown on the map in Figure 11-22, border Lake Dunstan, the reservoir formed by the Clyde Dam, a 100-m-high concrete gravity dam owned by the Electricity Corporation of New Zealand (Gillon and Hancox 1992). Approximately 25 percent of the shoreline of Lake Dunstan is bordered by extensive, deep-seated landslides formed mainly in schist bedrock, schist debris, and colluvium. The total area of landsliding along the shores of the lake is about 20 km², and the landslides are composed of about 1.6 km³ of material.

By mid-1991, more than 2,000 individual points were monitored on or adjacent to these landslides, with the expectation that this monitoring would increase to more than 3,500 points by the completion of lake filling in mid-1992 (Table 11-2). The instruments are distributed over all the landslides along an 18-km stretch of the Clutha River and in more than 16 km of tunnels (Gillon et al. 1992). The elements being measured are deformation, water pressure, water flow rate, and weather. In general, the monitoring techniques are conventional. The challenge is in obtaining data from the great number of monitoring points spread over a large area in steep terrain.

7.2.1 Deformation Measurement

The main deformation-detection instruments being used at Cromwell Gorge are (Gillon et al. 1992)

• Inclinometers for detecting the location and magnitude of small lateral movements;
• Geodetic deformation survey networks for determining the areal extent and magnitude of larger, longer-term movements; and
• Multipoint borehole extensometers installed from tunnels across specific geologic features, such as landslide failure surfaces.

Lower-accuracy techniques include installation of rod extensometers along tunnel walls and across surface scarps and surveying between points in

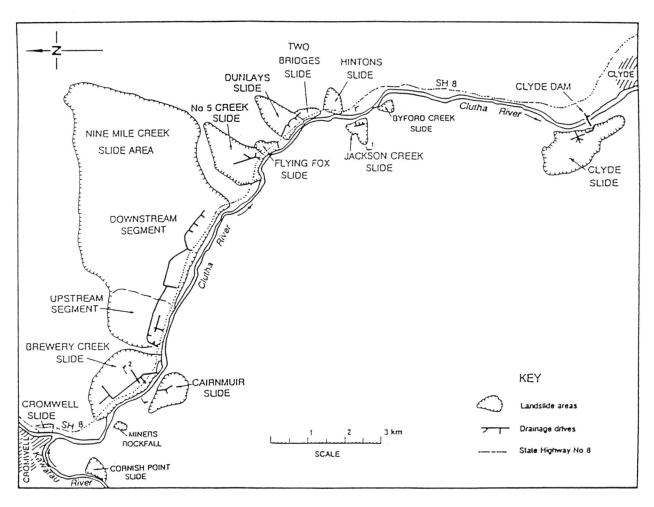

FIGURE 11-22
Map of Cromwell
Gorge landslide
area, New Zealand,
showing locations
and extent of
drainage adits
(modified from
Gillon and Hancox
1992).
REPRINTED FROM D.H. BELL
(ED.), LANDSLIDES/
GLISSEMENTS DE TERRAIN,
PROCEEDINGS OF THE
SIXTH INTERNATIONAL
SYMPOSIUM,
CHRISTCHURCH, 10–14
FEBRUARY, 1992, VOLUME 1,
1992–1994, 1800 PP., 3 VOLS.,
A.A. BALKEMA, OLD POST
ROAD, BROOKFIELD,
VERMONT 05036

tunnels with EDM equipment. The satellite-based GPS was used on a trial basis on one landslide with relative accuracies similar to those of conventional survey methods.

7.2.1.1 Inclinometers

By mid-1991 there were 180 inclinometer installations. About 150 of these, with an average depth of 90 m and a maximum depth of 300 m, were being regularly monitored (Gillon et al. 1992). During filling of the Lake Dunstan reservoir, inclinometers were monitored twice weekly or weekly at sites showing unusual activity or where construction work could cause instability.

Early in the investigation phase, drilling problems and the practice of combining piezometers with inclinometers caused several poor installations. In general, combining standpipe piezometers with inclinometers tends to create a zone of poor bond at each piezometer sand filter, which may reduce data quality for the inclinometer. However, a recent development has been the use of a Westbay

packer, casing components, and pumping port at the bottom of a 70-mm casing to provide an efficient and accurate inclinometer-piezometer. The packer provides a good base bond, and the hole was deliberately drilled deeper to ensure that the piezometer filter avoids any zone of deformation.

7.2.1.2 Multipoint Borehole Extensometers

Before the filling of the Lake Dunstan reservoir, approximately 35 multipoint borehole extensometers were installed from tunnels to intersect basal slide surfaces. Designed and manufactured in house, the extensometers use four untensioned, fully grouted fiberglass rods up to 50 m long. Readings are obtained either manually with a dial gauge or electronically using linear potentiometers linked to a datalogger. These extensometers are accurate to 0.1 mm.

7.2.1.3 Surface Extensometers

Two types of extensometers for measuring surface deformation have been installed across head scarps

Table 11-2
Instruments and Monitoring Points on Cromwell Gorge Landslides (modified from Gillon et al. 1992)

INSTRUMENTS AND POINTS	JULY 1991	FUTURE
Standpipe piezometers	760	800
Westbay multipoint piezometer ports	260	400
Inclinometers	180	200
Multipoint borehole extensometers	20	140
Drive leveling points	30	300
Surface monuments	250	280
Surface leveling points	250	350
Drainage flow meters	350	1,000
Weather and miscellaneous monitoring devices	120	120
Total	2,220	3,590

and other surface expressions of slide movement (Gillon et al. 1992). For small movements, rod extensometers were buried in special casings at a depth of 1 m. The rod material is a carbon-fiber composite with a near-zero coefficient of thermal expansion.

For measurement of larger surface movements, a wire extensometer was developed. The 6-mm wire, mounted on 2-m pylons to prevent animal damage, consists of a kevlar core protected against ultraviolet light by a nylon sleeve. The wire runs across an instrumented pulley and is tensioned by weights.

7.2.2 Groundwater-Level Instrumentation

7.2.2.1 Standpipe Piezometers
Standpipe piezometers are the most common instrument on the Clyde Project (Gillon et al. 1992). As of mid-1991 there were 760 standpipe piezometers in about 360 boreholes. Typically, they were installed in cored 96-mm (HQ) diameter boreholes at depths of 100 to 300 m. Multiple standpipe piezometers, normally two or three but sometimes up to five, were installed in each borehole. Most standpipe piezometers are monitored manually at frequencies ranging from daily to quarterly, with weekly being typical.

7.2.2.2 Westbay Multipoint Piezometers
By mid-1991 Westbay multipoint piezometers had been installed in 13 holes, with more installations planned (Gillon et al. 1992). Placement of these instruments began about midway through construction as part of the move to improve data quality and quantity. They have been restricted to

critical locations to provide data where complex groundwater conditions exist on high-hazard landslides and to confirm the response of landslides to drainage works.

7.2.3 Drainage Flows

As of mid-1991, drainage flows were measured at about 350 locations, principally at individual drain holes. Ultrasonic pipe-flow meters linked to dataloggers are used when a continuous record is required from individual drain holes.

7.2.4 Data Processing

The use of several software packages, including regional data bases, has facilitated the storage, selective retrieval, filtering, manipulation, quantitative assessment, and analysis of a large and rapidly accruing data base from the Clyde Project landslide instrumentation. The following software packages are being utilized (Proffitt et al. 1992):

- TECHBASE, a relational data base and modeling software package for storage, management, manipulation, statistical evaluation, and output of engineering, scientific, and topographic data;
- TIDEDA, a data base for storing, manipulating, and plotting time-dependent data, including water levels, flows, weather, and surveying information; and
- GTILT, a data-base system used for analysis and plotting of inclinometer surveys. Several types of deflection plots are available in GTILT; however, on the Clyde project graphical output has generally been restricted to cumulative deflection and time-series plots (Figure 11-23).

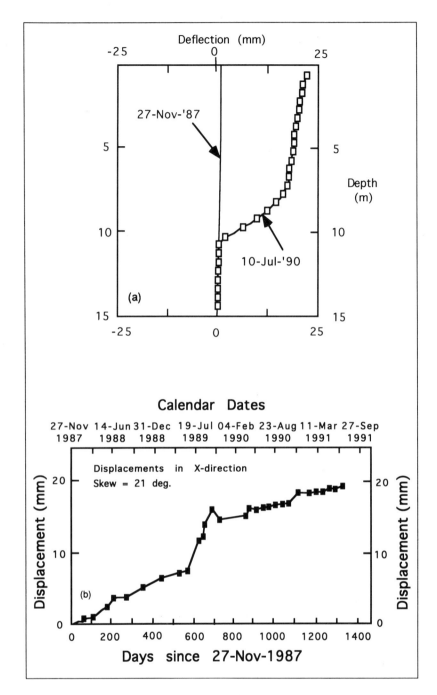

per year. Episodes of major movement correlate with critical rainfall events, each of which generally includes a period of prolonged and high-intensity rainfall of at least 24 hr duration. Catastrophic failure of the 215-m-wide by 580-m-long mass occurred most recently on March 23, 1993, and resulted in closure of US-101 for approximately 2 weeks (Squier Associates, Inc. 1994).

Evaluation of the geologic setting, depth to the failure surface, and high groundwater pressures led to the conclusion that only drainage alternatives would provide cost-effective, long-term stabilization of the landslide. Geotechnical exploration focused on the following key issues (Squier Associates, Inc. 1994):

• Further definition of the artesian zone;
• Measurement of piezometric pressures in the landslide mass, the artesian zone, and the underlying bedrock;
• Relationship of the unconfined water levels to the confined aquifer;
• Relationship of piezometric pressures to precipitation;
• Permeability and variability in the landslide mass and in bedrock;
• Source of artesian pressures; and
• Quantity of groundwater.

Evaluation of these issues has been assisted by drilling of additional exploratory borings and installation of multiple vibrating-wire piezometers in the boreholes, continuous monitoring and evaluation of piezometric pressures and rainfall, and stressing the groundwater system using newly installed wells and pump tests.

The landslide is now being monitored by an instrumentation system incorporating automated monitoring and recording of a number of instruments. Continuous monitoring of one inclinometer, piezometric pressures, and rainfall is being accomplished by an ADAS that allows monitoring of 26 piezometers simultaneously. The locations of borings, test wells, and instrument sites are shown in Figure 11-24.

Six inclinometers were installed by the Oregon Department of Transportation (ODOT) during March through May 1993. Then during early April 1994, ODOT installed (*a*) vibrating-wire transducers in eight open standpipe piezometers, (*b*) a continuous-reading slope inclinometer in one hole,

7.3 Arizona Inn Landslide, Oregon

The Arizona Inn landslide (Figure 11-24) on US-101 in the Klamath Mountains of southwestern Oregon has a history of persistent creep and recurring major movements since highway construction in the 1930s. The landslide mass consists primarily of fine-grained, intensely sheared, mylonitized sandstone and mudstone. The landslide area receives an average of 1900 to 2000 mm of rainfall

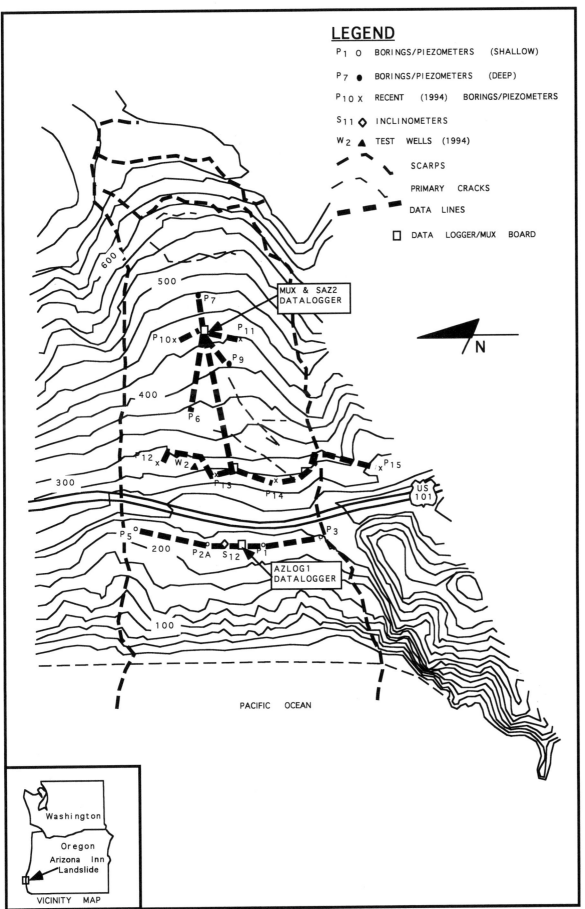

LEGEND

P_1 o   BORINGS/PIEZOMETERS   (SHALLOW)

P_7 ●   BORINGS/PIEZOMETERS   (DEEP)

P_{10} X   RECENT   (1994)   BORINGS/PIEZOMETERS

S_{11} ◇   INCLINOMETERS

W_2 ▲   TEST WELLS (1994)

    SCARPS

    PRIMARY CRACKS

    DATA LINES

□   DATA LOGGER/MUX BOARD

MUX & SAZ2 DATALOGGER

AZLOG1 DATALOGGER

US 101

PACIFIC OCEAN

Washington

Oregon

Arizona Inn Landslide

VICINITY MAP

FIGURE 11-23 *(opposite page)* Landslide in Cromwell Gorge: (*a*) cumulative deflection in Direction A for Inclinometer RD-6 between November 27, 1987, and July 10, 1990 (July 1990 data include corrections for zero shift and rotations) and (*b*) time-series plot (Proffitt et al. 1992).

FIGURE 11-24 Locations of borings, test wells, and instrument sites on Arizona Inn landslide, Oregon (Squier Associates, Inc. 1994).

and (*c*) an on-site tipping-bucket rain gauge. All of these instruments were connected to the ADAS using two dataloggers. The continuous monitoring system was programmed and placed in operation on April 10, 1994.

Subsequently, 16 vibrating-wire piezometers were connected to a datalogger after the addition of two multiplexing boards. Another datalogger records hourly data for four piezometers and a continuous-reading inclinometer. New software was added; it has three primary routines that monitor and record data from 20 piezometers and from rain gauges as well as system information. The routine includes (Squier Associates, Inc. 1994) daily maximum and minimum piezometric head values, complete hourly data recording of all instruments, and user-selectable reading frequencies during special test periods. Summary plots have been produced for the inclinometers, piezometers, and rain gauges. Zones of shear showed rates of movement of about 2.5 to 7.5 cm/year, with an

average rate of about 4 cm/year. Figure 11-25 is a time plot of piezometric readings for the upper part of the landslide.

7.4 Last Chance Grade Landslide, California

The Last Chance Grade landslide, located along US-101 in far northwestern California, is approximately 520 m wide and 460 m long. The landslide, underlain by interbedded shale, sandstone, and conglomerate, is very active; at the road elevation, it moves on the order of 1.5 to 3.0 m/year and affects approximately 235 m of the roadway. Investigation of the Last Chance Grade landslide by the California Department of Transportation (Caltrans) included drilling, borehole logging, geologic mapping, and developing recommendations for stabilizing the highway (Kane and Beck 1994).

Three alternatives are being considered for highway stabilization. The first is to realign the

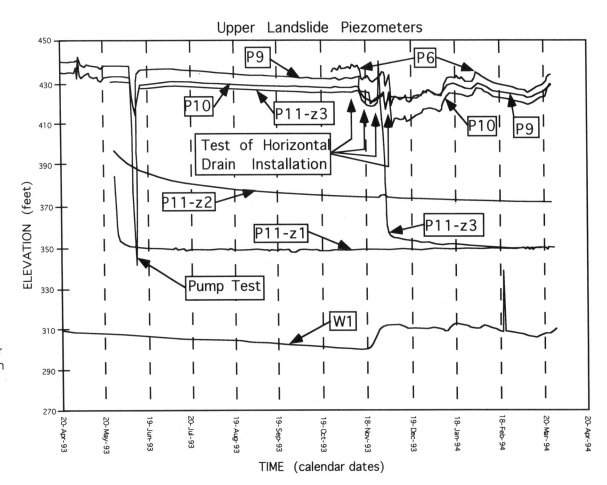

FIGURE 11-25
Time plot of piezometric readings for upper part of Arizona Inn landslide, Oregon; symbols indicate individual piezometers (modified from Squier Associates, Inc. 1994).

roadway in a tunnel excavated behind the sliding surface. The second alternative is to realign the roadway slightly and stabilize the landslide material below the roadway with a soldier-pile and tieback wall and the material above the roadway with several rows of slope reinforcement. The third choice is to realign the highway in a cut excavated behind the sliding surface. Preconstruction determination of the location of the sliding surface will be essential in selecting one of these measures.

Caltrans commonly uses inclinometers to measure landslide movement. The primary disadvantage of the ordinary inclinometer system is the necessity of visiting the site for each set of instrument readings. Because the Last Chance Grade landslide is located far from any Caltrans geotechnical engineering facilities, an alternative method was sought that could be used for denoting zones of landslide movement without actually having to visit the site. For this purpose, as part of the field investigation, while inclinometer casings were installed in two boreholes on the landslide, a coaxial cable was fixed to the outside of one of the casings to allow comparison of time domain reflectometry (TDR) data with standard inclinometer results (Kane and Beck 1994).

TDR is a technique that uses characteristics of returned electrical pulses to determine the amount of strain, or the existence of a rupture, in a coaxial cable. Although developed for cable testing, TDR is finding use in geotechnical applications, especially mining (O'Connor and Wade 1994). Most commonly, a cable tester connected to a coaxial cable installed in a borehole emits a stepped voltage impulse (Figure 11-26). As rock-mass movements deform the cable, its capacitance is changed, and this causes changes in the reflected waveform of the voltage pulse. The time delay between a transmitted pulse and the reflection from a cable deformity determines the damage location (Dowding et al. 1989).

Different wave reflections are received for different cable deformations. The length and amplitude of the reflection indicate the severity of damage. A cable in shear reflects a voltage spike that increases in magnitude in direct proportion to the amount of shear deformation. A distinct spike occurs just before failure. After failure, a permanent reflection is recorded. If the cable is in tension, the wave reflection is a subtle, troughlike voltage signal that in-

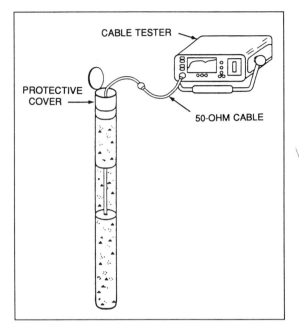

FIGURE 11-26 Diagrammatic sketch of TDR cable installed in borehole (Dowding and Huang 1994)
REPRINTED WITH PERMISSION OF AMERICAN SOCIETY OF CIVIL ENGINEERS

creases in length as the cable is deformed. At tension failure, a small necking trough is visible in the TDR signal, which is distinguishable from a shear-failure reflection (Dowding et al. 1989). The cable tester can be connected to a datalogger, which records and stores reflections. The datalogger controls the cable tester and supplies power during measurements. The data can be collected automatically from a remote location using a computer with a telemetering system (Dowding and Huang 1994).

At the Last Chance Grade landslide, an 82-m-long coaxial cable (type RG 59, commonly used for videocassette recording) was attached to the outer, upslope side of an inclinometer casing by means of nylon ties spaced every 1.5 m. Because the inclinometer hole was positioned on the actively traveled roadway of US-101, the cable was continued beyond the highway shoulder in a shallow groove in the bituminous pavement, thus allowing TDR readings to be obtained without interference by traffic. To further explore the possibility of remotely collecting the data and monitoring the landslide, the TDR system was connected to a datalogger operated by a telemetry system that could be activated using a cellular telephone connection. With this arrangement, the TDR data could be directly loaded into a Caltrans computer located in Sacramento, more than 500 km away, and no one would have to visit the landslide site.

At 148 days after installation, the reflection signature for the TDR cable indicated two notable

FIGURE 11-27
TDR output for
borehole on Last
Chance Grade
landslide, California;
note spikes where
cable enters
inclinometer hole
and in vicinity of
surface of sliding
(modified from Kane
and Beck 1994).

spikes (Figure 11-27). The upper spike, for a depth of 8.6 m, indicated a crimp in the cable where it was squeezed against the lip of the casing. At a depth of approximately 40 m, roughly the same depth as the inclinometer deflections, a trough was visible in the reflection signature, which was interpreted as indicating distress of the cable at the sliding surface (Kane and Beck 1994).

The results of this study indicate that in certain cases TDR may be used instead of, or in parallel with, inclinometers for monitoring landslide movement and that the technology exists to remotely access the TDR installation. The advantages of a remotely accessed system are as follows:

- There is no need to visit the site,
- Debris does not have to be cleaned from the inclinometer casing before TDR readings are taken,
- Cable is inexpensive and readily available, and
- The ability exists to monitor up to 512 holes with one cable tester and multiplexer hookup.

Remote data collection results in cost savings. It also means that hazardous areas can be closely monitored by frequent sampling to determine incipient movements. Another benefit of this technology is that it can be used as part of an early warning system to alert transportation officials of a possible catastrophic highway failure (Kane and Beck 1994).

7.5 Golden Bypass Landslide, Golden, Colorado

In the spring of 1991, during construction of a highway bypass around the city of Golden, Colorado, forming a part of Colorado State Highway 93, an initial minor landslide failure began in one wall of an excavation in Pierre shale (Highland and Brown 1993). Slow movement and enlargement of this earth flow continued until 1994, when the landslide was stabilized by means of a Brazilian-concept tieback panel wall, installation of horizontal drains, and removal of 75 000 m³ of overburden near the landslide headwall (Ed Belknap, personal communication, 1995, Colorado Department of Transportation, Denver, Colorado).

Seven of the tieback anchor cables were equipped with load cells. In addition, five 12.5-mm-diameter piezometers with vibrating-wire trans-

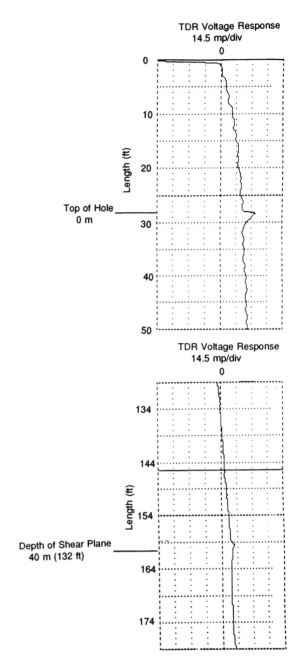

ducers were installed to measure groundwater elevations at the landslide. Automated monitoring of the load cells and piezometers was desired, along with two-way, remote telemetry between the landslide and Colorado Department of Transportation (CDOT) headquarters in Denver, 30 km to the east. Neither power nor communication facilities were available at the site.

As a solution to the problem, an integrated monitoring and telemetry system was installed to remotely automate the load cell and piezometer measurements and provide direct, two-way radio communication (Dennis E. Gunderson, personal communication, 1995, Geomation, Inc., Golden, Colorado). A network monitor station (NMS) was established at CDOT headquarters. The NMS, an IBM-compatible personal computer running GEONET software, provides the operator interface with the system for programming, alarm announcement and acknowledgment, real-time monitoring, supervisory control, and data logging. An RS-232 cable from a serial port in the NMS connects to a radio gateway unit, which is connected to a high-gain directional antenna mounted on a nearby signal tower. The gateway provides a link between the radio communication from the Golden landslide site and the NMS.

Three measurement and control units (MCUs), which are battery-powered, solar-recharged, and radio-linked, were installed at the landslide (Figure 11-28). MCU1, located at the top of the landslide, automatically monitors three piezometers. MCU2 and MCU3 are located near the bottom of the landslide and automatically monitor (a) two load cells and (b) five load cells and two piezometers, respectively. All monitoring and telemetry is user-programmable and is currently set at 66-hr intervals. MCU2 and MCU3, both equipped with

standard whip antennas, communicate with the networked gateway unit through MCU1, which is equipped with a high-gain directional antenna.

Each MCU has its own microprocessor and is capable of multitasking functions, including

- Automated monitoring of any standard measurement signal (analog or digital);
- Time-and-date stamping of all measurements in the field at the point of capture;
- Conditioning and converting physical-parameter signals to desired engineering units;
- Peer-to-peer communication;
- Acting as a repeater and performing store-and-forward functions for other network nodes;
- Enacting direct or remote-control functions; and
- Accomplishing data reduction in the field, if desired.

The user, at any time without system disruption, can initiate system reconfiguration diagnostics and measurement verification and can enact remote control functions from the NMS.

8. PROBLEMS PREVENTING FURTHER DEVELOPMENT OF NEW INSTRUMENTATION TECHNOLOGY

Instrumentation and associated monitoring techniques have made significant advances during the past decade. The general proliferation of miniature electronics has revolutionized many aspects of data acquisition, processing, and communications, and has profoundly benefited both industry and the public. Personal computers have entered offices and homes, and people have come to expect the speed and convenience of information without delay. However, such progress in the geotechnical field, and in the transportation industry generally, has been less dramatic. Four main reasons govern the slow-paced and uneven gains in geotechnical monitoring relative to other data management technologies.

First, the geotechnical instrumentation industry is relatively small and highly specialized. Its practice and progress are easily influenced by economic conditions, levels of government spending, and existing procurement practices. In the United States, a highly competitive and shrinking heavy construction market has also shrunk the market for instrument suppliers.

FIGURE 11-28
MCU and solar power supply attached to monitoring well at Golden Bypass landslide.
COURTESY OF GEOMATION, INC.

Second, as overall levels of support for research and development continue to dwindle, less funds are available for government-sponsored instrument development. Property owners concerned with the stabilization of a specific landslide can rarely justify supporting the development and testing of new instruments to further the state of the art of geotechnical knowledge.

Third, geotechnical engineers have frequently underestimated the costs and requirements of a field installation and operation of geotechnical instrumentation, leading to the failure of several monitoring programs. Past bad experiences engender mistrust and avoidance of further instrumentation by geotechnical engineers.

Last, because of current contracting practices and a highly litigious business environment, especially in the United States, designers of all types of transportation facilities have tended to abdicate their responsibilities for requesting and implementing instrument monitoring at landslides. These responsibilities have been allowed to devolve onto general contractors, who may or may not be particularly concerned about what information instruments can provide. Most general contractors are interested in project production efficiency, and instruments are often merely obstacles in the field that should be minimized. However, instruments allow the monitoring of project performance and safety. To place the contractor in charge of the primary source of such monitoring and safety information raises a potential conflict of interest. If instrumentation selection and operation are restored to the chief engineer and those at the project site, landslide instrumentation will more likely become part of broadly expected standard practice during landslide investigations.

REFERENCES

British Geotechnical Society. 1974. *Field Instrumentation in Geotechnical Engineering*. John Wiley & Sons, Inc., New York, 720 pp.

Burland, J.B., and J.F.A. Moore. 1974. The Measurement of Ground Displacement Around Deep Excavations. In *Field Instrumentation in Geotechnical Engineering*, John Wiley & Sons, Inc., New York, pp. 70–84.

Cording, E.J., A.J. Hendron, Jr., W.H. Hansmire, J.W. Mahar, H.H. MacPherson, R.A. Jones, and T.D. O'Rourke. 1975. *Methods for Geotechnical Observations and Instrumentation in Tunneling*. Department of Civil Engineering, University of Illinois, Urbana-Champaign; National Science Foundation, Vol. 2, pp. 293–566.

Dibiagio, E. 1974. Discussion: Session 2—Equipment. In *Field Instrumentation in Geotechnical Engineering*, John Wiley & Sons, Inc., New York, pp. 565–566.

Dowding, C.H., M.B. Su, and K. O'Connor. 1989. Measurement of Rock Mass Deformation with Grouted Coaxial Cables. *Rock Mechanics and Engineering*, Vol. 22, pp. 1–23.

Dowding, C.H., and F.C. Huang. 1994. Early Detection of Rock Movement with Time Domain Reflectometry. *Journal of Geotechnical Engineering*, ASCE, Vol. 120, No. 8, pp. 1413–1427.

Dunnicliff, C.J. 1971. Equipment for Field Deformation Measurements. In *Proc., Fourth Pan-American Conference on Soil Mechanics and Foundation Engineering*, San Juan, American Society of Civil Engineers, New York, Vol. 2, pp. 319–332.

Dunnicliff, C.J. 1988. *Geotechnical Instrumentation for Monitoring Field Performance*. John Wiley & Sons, Inc., New York, 577 pp.

Dunnicliff, C.J., and D. Davidson. 1991. Lessons Learned from Automating Instrumentation at Some U.S. Dams. *Geotechnical News*, Vol. 9, No. 2, pp. 46–51.

Durr, D.L. 1974. An Embankment Saved by Instrumentation. In *Transportation Research Record 482*, TRB, National Research Council, Washington, D.C., pp. 43–50.

Dutro, H.B., and R.D. Dickinson. 1974. Slope Instrumentation Using Multiple-Position Borehole Extensometers. In *Transportation Research Record 482*, TRB, National Research Council, Washington, D.C., pp. 9–17.

Gilbert, C. 1994. Evaluating the Cost of GPS Data Collection. *Earth Observation Magazine*, Vol. 3, No. 7 (July), pp. 42–43.

Gilbert, C. 1995. How to Do Differential GPS Without Known Coordinates for Your Base Station. *Earth Observation Magazine*, Vol. 4, No. 4 (April), pp. 50–51.

Gillon, M.D., P.F. Foster, G.T. Proffitt, and A.P. Smits. 1992. Monitoring of the Cromwell Gorge Landslides. In *Proc., Sixth International Symposium on Landslides*, Christchurch, New Zealand, A.A. Balkema, Rotterdam, Netherlands, Vol. 2, pp. 1135–1140.

Gillon, M.D., and G.T. Hancox. 1992. Cromwell Gorge Landslides—A General Overview. In *Proc., Sixth International Symposium on Landslides*,

Christchurch, New Zealand, A.A. Balkema, Rotterdam, Netherlands, Vol. 1, pp. 83–102.

Gould, J.P., and C.J. Dunnicliff. 1971. Accuracy of Field Deformation Measurements. In *Proc., Fourth Pan-American Conference on Soil Mechanics and Foundation Engineering*, San Juan, American Society of Civil Engineers, New York, Vol. 1, pp. 313–366.

Green, G.E. 1974. Principles and Performance of Two Inclinometers for Measuring Horizontal Ground Movements. In *Field Instrumentation in Geotechnical Engineering*, John Wiley & Sons, Inc., New York, pp. 166–179.

Green, G.E., and P.E. Mikkelsen. 1988. Deformation Measurements with Inclinometers. In *Transportation Research Record 1169*, TRB, National Research Council, Washington, D.C., pp. 1–15.

Heinz, R.A. 1975. In Situ Soils Measuring Devices. *Civil Engineering*, Vol. 45, No. 10, pp. 62–65.

Highland, L.M., and W.M. Brown III. 1993. The Golden Bypass Landslide, Golden, Colorado. *Earthquakes & Volcanoes*, Vol. 24, No. 1, pp. 4–14.

Hinckley, A.L., B.D. Tanner, and E.C. Campbell. 1988. Automation of Monitoring of Geotechnical Instrumentation. In *Transportation Research Record 1169*, TRB, National Research Council, Washington, D.C., pp. 26–29.

Hvorslev, M.J. 1951. *Time Lag and Soil Permeability in Groundwater Observations*. Bulletin 36. U.S. Army Engineer Waterways Experiment Station, Vicksburg, Miss., 50 pp.

Imrie, A.S. 1983. Taming the Downie Slide. *Canadian Geographic*, Vol. 103, No. 3, pp. 47–51.

Imrie, A.S., and D.R. Bourne. 1981. Engineering Geology of the Mica and Revelstoke Dams. In *Field Guides to Geology and Mineral Deposits*, Geological Association of Canada, Calgary, pp. 393–401.

Imrie, A.S, D.P. Moore, and E.G. Energen. 1992. Performance and Maintenance of the Drainage System at Downie Slide. In *Proc., Sixth International Symposium on Landslides*, Christchurch, New Zealand, A.A. Balkema, Rotterdam, Netherlands, Vol. 1, pp. 751–757.

Kane, W.F., and T.J. Beck. 1994. Development of a Time Domain Reflectometry System to Monitor Landslide Activity. In *Proc., 45th Highway Geology Symposium*, Portland, Oreg., August 17–19, Geology Department, Portland State University, Oreg., pp. 163–173.

Kleusberg, A. 1992. Precise Differential Positioning and Surveying. *GPS World*, Vol. 3, No. 7 (July/August), pp. 50–52.

McKenna, G. 1995. Advances in Slope-Monitoring Instrumentation at Syncrude Canada Ltd.—A Followup. *Geotechnical News*, Vol. 13, No. 1, pp. 42–44.

McKenna, G., G. Livingstone, and T. Lord. 1994. Advances in Slope-Monitoring Instrumentation at Syncrude Canada Ltd. *Geotechnical News*, Vol. 12, No. 3, pp. 66–69.

Mikkelsen, P.E. 1989. Geotechnical Monitoring During Construction of Two Highway Embankments on Soft Ground in Tacoma. In *Engineering Geology of Washington* (R.W. Galster, ed.), Bulletin 78, Washington Division of Geology and Earth Resources, Olympia, Vol. 2, pp. 819–829.

Mikkelsen, P.E. 1992. Instrument Innovations for Monitoring Natural Hazards. *Geotechnical News*, Vol. 10, No. 1, pp. 46–49.

Mikkelsen, P.E., and L.K. Bestwick. 1976. Instrumentation and Performance: Urban Arterial Embankments on Soft Foundation Soil. In *Proc., 14th Annual Symposium on Engineering Geology and Soils Engineering*, Boise, Idaho, Idaho Department of Transportation, Boise, pp. 1–18.

Moore, D.P. 1989. Stabilization of Downie Slide and Dutchman's Ridge. In *Proc., 12th International Conference on Soil Mechanics and Foundation Engineering*, Rio de Janeiro, A.A. Balkema, Rotterdam, Netherlands, Vol. 5, pp. 3063–3065.

Muñoz, A., Jr., and D. Gano. 1974. The Role of Field Instrumentation in Correction of the Fountain Slide. In *Transportation Research Record 482*, TRB, National Research Council, Washington, D.C., pp. 1–8.

O'Connor, K.M., and L.V. Wade. 1994. Applications of Time Domain Reflectometry in the Mining Industry. In *Proc., Symposium and Workshop on Time Domain Reflectometry in Environmental, Infrastructure, and Mining Applications*, Evanston, Ill., Sept., Special Publication 19-94, Bureau of Mines, U.S. Department of the Interior, pp. 494–506.

Palladino, D.J., and R.B. Peck. 1972. Slope Failures in an Overconsolidated Clay, Seattle, Washington. *Geotechnique*, Vol. 22, No. 4, pp. 563–595.

Patton, F.D. 1979. Groundwater Instrumentation for Mining Projects. In *Mine Drainage: Proc., First International Mine Drainage Symposium*, Denver, Colo., Miller Freeman Publications Inc., San Francisco, May, Chap. 5, pp. 123–153.

Patton, F.D. 1983. The Role of Instrumentation in the Analysis of the Stability of Rock Slopes. In *Proc., International Symposium on Field Measurements in Geomechanics*, Zurich, Sept. 5–8, pp. 719–748.

Patton, F.D. 1984. Climate, Groundwater Pressures and Stability Analyses of Landslides. In *Proc., Fourth International Symposium on Landslides*, Canadian Geotechnical Society, Toronto, Vol. 3, pp. 43–60.

Patton, F.D., W.H. Beck, and D. Larssen. 1991. A Modular Subsurface Data Acquisition System (MOSDAX) for Real-time Multi-level Groundwater Monitoring. In *Field Measurements in Geotechnics: Proc., Third International Symposium on Field Measurements in Geomechanics* (G. Sørum, ed.), Oslo, Norway, Sept. 9–11, A.A. Balkema, Rotterdam, Netherlands, Vol. 1, pp. 339–348.

Peck, R.B. 1972. Observation and Instrumentation: Some Elementary Considerations. *Highway Focus*, Vol. 4, No. 2, pp. 1–5.

Penman, A.D.M. 1961. A Study of the Response Time of Various Types of Piezometer. In *Pore Pressure and Suction in Soils*, Butterworths, London, pp. 53–58.

Penman, A.D.M., and J.A. Charles. 1974. Measuring Movements of Embankment Dams. In *Field Instrumentation in Geotechnical Engineering*, John Wiley & Sons, Inc., New York, pp. 341–358.

Piteau, D.R., F.H. Mylrea, and I.G. Blown. 1978. Downie Slide, Columbia River, British Columbia, Canada. In *Rockslides and Avalanches* (Barry Voight, ed.), Elsevier Scientific Publishing Company, New York, Vol. 1, pp. 365–392.

Proffitt, G.T., G.J. Fairless, G.G. Grocott, and P.A. Manning. 1992. Computer Applications for Landslide Studies. In *Proc., Sixth International Symposium on Landslides*, Christchurch, New Zealand, A.A. Balkema, Rotterdam, Netherlands, Vol. 2, pp. 1157–1162.

Shannon, W.L., S.D. Wilson, and R.H. Meese. 1962. Field Problems: Field Measurements. In *Foundation Engineering* (G.A. Leonards, ed.), McGraw-Hill, New York, pp. 102–1080.

Smith, T.W., and R.A. Forsyth. 1971. Potrero Hill Slide and Correction. *Journal of the Soil Mechanics and Foundations Division*, ASCE, Vol. 97, No. SM3, pp. 541–564.

Squier Associates, Inc. 1994. *Phase 1—Arizona Inn Landslide Geotechnical Investigation, Curry, County, Oregon*. Geotechnical Services Unit, Oregon Department of Transportation, Salem, 209 pp.

Tatchell, G.E. 1991. Automatic Data Acquisition Systems for Monitoring Dams and Landslides. In *Field Measurements in Geotechnics: Proc., Third International Symposium on Field Measurements in Geomechanics* (G. Sørum, ed.), Oslo, Norway, Sept. 9–11, A.A. Balkema, Rotterdam, Netherlands, Vol. 1, pp. 349–360.

Tice, J.A., and C.E. Sams. 1974. Experiences with Landslide Instrumentation in the Southeast. In *Transportation Research Record 482*, TRB, National Research Council, Washington, D.C., pp. 18–29.

Transportation Research Board. 1974. *Landslide Instrumentation: Transportation Research Record 482*, TRB, National Research Council, Washington, D.C., 51 pp.

Vaughn, P.R. 1969. A Note on Sealing Piezometers in Boreholes. *Geotechnique*, Vol. 19, No. 3, pp. 405–413.

Vaughan, P.R. 1974. The Measurement of Pore Pressures with Piezometer. In *Field Instrumentation in Geotechnical Engineering*, John Wiley & Sons, Inc., New York, pp. 411–422.

Wilson, S.D. 1967. Investigation of Embankment Performance. *Journal of the Soil Mechanics and Foundations Division*, ASCE, Vol. 93, No. SM4, pp. 135–156.

Wilson, S.D. 1970. Observational Data on Ground Movements Related to Slope Instability. *Journal of the Soil Mechanics and Foundations Division*, ASCE, Vol. 96, No. SM5, pp. 1521–1544.

Wilson, S.D. 1974. Landslide Instrumentation of the Minneapolis Freeway. In *Transportation Research Record 482*, TRB, National Research Council, Washington, D.C., pp. 30–42.

Wilson, S.D., and D.E. Hilts. 1971. Application of Instrumentation to Highway Stability Problems. In *Proc., Joint ASCE-ASME National Transportation Engineering Meeting*, Seattle, American Society of Civil Engineers, New York, and American Society of Mechanical Engineers, New York.

Wilson, S.D., and P.E. Mikkelsen. 1977. *Foundation Instrumentation—Inclinometers* (reference manual for 45-min videotape). FHWA, U.S. Department of Transportation.

Wilson, S.D., and P.E. Mikkelsen. 1978. Field Instrumentation. In *Special Report 176: Landslides: Analysis and Control* (R.L. Schuster and R.J. Krizek, eds.), TRB, National Research Council, Washington, D.C., Chap. 5, pp. 112–138.

PART 3

STRENGTH AND STABILITY ANALYSIS

Chapter 12

TIEN H. WU

SOIL STRENGTH PROPERTIES AND THEIR MEASUREMENT

1. INTRODUCTION

Methods of limiting equilibrium are frequently used to analyze the stability of a soil mass (see Chapter 13). In such analyses, the shear strength of the material is assumed to be fully developed along the rupture surface at failure. In this chapter the basic principles that govern shear strength and the methods that may be used for its measurement are outlined. Brief descriptions of the properties of some common soils are provided.

2. GENERAL PRINCIPLES

The basic principles in the description of strength properties are the failure criterion and the effective stress principle. When a failure criterion derived from test data is used to estimate in situ strength, appropriate attention should be given to possible differences between the stress state of the test and that of the in situ soil when subjected to the expected load.

2.1 Failure Criterion

The Mohr-Coulomb criterion is widely used to define failure; it states that the shear strength (s) is

$$s = c + \sigma \tan\phi \qquad (12.1)$$

where

σ = normal stress on rupture surface,

c = cohesion, and

ϕ = angle of internal friction.

In terms of principal stresses, the Mohr-Coulomb criterion becomes

$$\sigma_1 = \sigma_3 \tan^2\left[\left(\frac{\pi}{4}\right)+\left(\frac{\phi}{2}\right)\right]+2c\tan\left[\left(\frac{\pi}{4}\right)+\left(\frac{\phi}{2}\right)\right] \quad (12.2)$$

where σ_1 is the major principal stress and σ_3 is the minor principal stress.

A more general formulation, which combines failure with stress-strain behavior, is the yield surface (Drucker et al. 1955) and the critical state (Schofield and Wroth 1968). The yield surface is especially useful when evaluation of deformation is required. For limiting equilibrium analysis, the Mohr-Coulomb criterion is still the most convenient failure criterion.

2.2 Effective Stress Versus Total Stress Analysis

Because the shear strength of soils is strongly influenced by drainage conditions during loading, those conditions must be properly accounted for in the use of shear strength in design. A fundamental principle in soil engineering is the use of effective stress σ', which was first defined by Terzaghi (1936a) as

$$\sigma' = \sigma - u \qquad (12.3)$$

where σ is the total stress and u is the pore pressure. The shear strength can be expressed consistently in terms of effective stress:

$$s = c' + \sigma' \tan\phi' = c' + (\sigma - u) \tan\phi' \qquad (12.4)$$

where c' and ϕ' are the strength parameters for effective stress. For a partially saturated soil, the shear strength can be expressed as (Fredlund et al. 1978)

$$s = c' + (\sigma - u_w) \tan\phi' + (u_w - u_a) \tan\phi'' \qquad (12.5)$$

where

u_a = pore-air pressure,
u_w = pore-water pressure, and
ϕ'' = soil property that reflects influence of suction $(u_w - u_a)$ on strength.

When the soil is saturated, $u_a = 0$ and $u = u_w$. For saturated soils, pore pressure consists of the hydrostatic pore pressure related to groundwater level and the excess pore pressure due to applied loads.

When soils are loaded under undrained or partially drained conditions, the tendency to change volume results in an excess pore pressure, which may be positive or negative depending on the type of soil and the stresses involved. General relations between pore pressure and applied stresses have been suggested. For example, Henkel (1960) proposed that

$$\Delta u = B(\Delta \sigma_{oct} + \alpha \Delta \tau_{oct}) \qquad (12.6)$$

where

α = empirical coefficient,
$\tau_{oct} = \frac{1}{3}[(\sigma_1 - \sigma_2)^2 + (\sigma_2 - \sigma_3)^2 + (\sigma_3 - \sigma_1)^2]^{1/2}$,
$\sigma_{oct} = \frac{1}{3}(\sigma_1 + \sigma_2 + \sigma_3)$, and
$\sigma_1, \sigma_2, \sigma_3$ = major, intermediate, and minor principal stresses.

For soils tested in the triaxial apparatus or loaded so that $\Delta\sigma_2 = \Delta\sigma_3$, Skempton (1954) proposed that the excess pore pressure be given by

$$\Delta u = B[\Delta\sigma_3 + A\Delta(\sigma_1 - \sigma_3)] \qquad (12.7)$$

where A is an empirical coefficient related to the excess pore pressure developed during shear and B is an empirical coefficient related to the soil's com-

pressibility and degree of saturation. For saturated soils, $B = 1$. For an elastic material, $A = \frac{1}{3}$. For soils that compress under shear, $A > \frac{1}{3}$, and for soils that dilate under shear, $A < \frac{1}{3}$.

Under the fully drained condition, the excess pore pressure is zero, and pore pressure in saturated soils caused by groundwater flow can usually be evaluated without serious difficulty. Hence, analysis with the effective stress description of shear strength (Equation 12.4) is most useful. For partially drained and undrained conditions, the evaluation of excess pore pressure is often difficult. In some cases, a total-stress description of shear strength may be used. One important case is the undrained loading of saturated soils, for which the undrained shear strength $(s = s_u)$ can be used. This is the common $\phi = \phi_u = 0$ analysis (Skempton and Golder 1948). The shear strength usually changes as the void ratio changes with drainage. If the change results in a higher strength, the short-term, undrained stability is critical and the stability can be expected to improve with time. On the other hand, if drainage produces a decrease in strength, the long-term, drained stability is critical; the undrained shear strength can be used only for short-term or temporary stability. For partially saturated soils, the prediction of pore-air and pore-water pressures is more difficult. Currently, the only reliable method is in situ measurement.

2.3 Common States of Stress and Stress Change

The Mohr-Coulomb criterion does not indicate any effect of the intermediate principal stress (σ_2') on the shear strength. In practical problems, σ_2' may range from σ_3' to σ_1', depending on the geometry of the problem. The direction of the major principal stress also changes during loading. Experimental studies show that the value of σ_2' relative to σ_3' and σ_1' has an influence on the shear strength.

Several common states of stress are shown in Figure 12-1. In the initial state (a), σ_z' is the effective overburden pressure, $\sigma_r' = K_o \sigma_z'$ is the radial or lateral pressure, and K_o is the coefficient of earth pressure at rest. In the stress state beneath the center of a circular loaded area [Figure 12-1(b)], the vertical stress is the major principal stress and the radial stress σ_r' is the minor principal stress. The

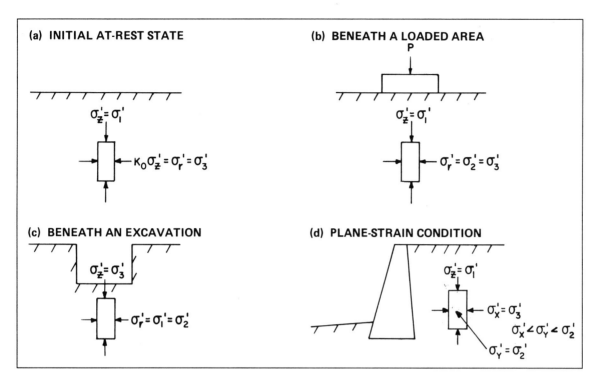

FIGURE 12-1
Common states
of stress.

intermediate principal stress (σ_2') is equal to the minor principal stress (σ_3'). In the stress state below the center of a circular excavation [Figure 12-1(c)], the vertical stress is the minor principal stress and the radial stress σ_r' is the major principal stress. The intermediate principal stress (σ_2') is equal to the major principal stress (σ_1'). Slopes and retaining structures can be approximated by the plane-strain condition in which the intermediate principal strain (ε_2) is zero. Then the intermediate principal stress (σ_2') is σ_y', oriented as shown in Figure 12-1(d), and has a value between σ_1' and σ_3'.

Another important feature in many stability problems is the rotation of the principal axes during loading or excavation and its effect on the shear strength of soft clays (Ladd and Foott 1974). The rotation of principal axes is shown in Figure 12-2. Before the excavation of the cut, the state of stress is represented by that shown in Figure 12-1(a). After excavation, the major principal stress is in the horizontal direction at the toe (Point A, Figure 12-2). Thus, the principal axes are rotated through an angle of 90 degrees; at Point B, a rotation of approximately 45 degrees occurs. At Point C, the original principal stress directions remain unchanged although the values of the stresses change.

2.4 Stress-Strain Characteristics

Two stress-deformation curves are shown in Figure 12-3. A soil sample is sheared under a normal stress σ and a shear stress τ. The shear displacement is Δ. In common practice, the strength of the soil is defined as the peak strength (Points a and b in Figure 12-3) measured in the test. When this is used in a stability analysis, the tacit assumption is that the peak strength is attained simultaneously along the entire rupture surface.

FIGURE 12-2
Directions of
principal stresses
in a slope.

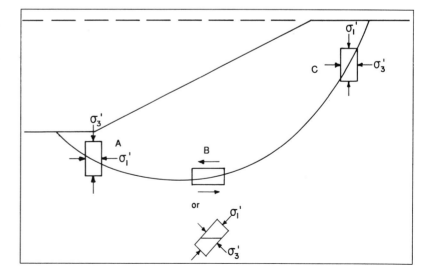

Many soils demonstrate strain-softening behavior, as illustrated by Curve A in Figure 12-3. Any of several mechanisms may be used to explain the strength decrease, but it is important to account for this decrease in design. For strain-softening soils, it is unreasonable to assume that the soil reaches its peak strength simultaneously at all points along a failure surface. In fact, the soil at some points on the rupture surface will suffer displacements greater than Δ_a before the soil at other points reaches this deformation. This phenomenon was called "progressive failure" by Terzaghi and Peck (1948). In the limit of large deformation, the strength at all points will be reduced to that represented by Point c in Figure 12-3. This strength is called the residual strength.

2.5 Effect of Rate of Loading

The difference between the rate of loading applied in a laboratory shear test and that experienced in the slope is usually substantial. Most laboratory and in situ tests bring the soil to failure within several hours or at most a few days. For most slopes the load is permanent, although some dynamic loads may be applied for short durations. The effect of rate of loading on soil strength, excluding effects of consolidation, may be significant. In general, the undrained strength of soils increases as the rate of loading increases; however, this effect depends on the specific material and varies over a wide range (Casagrande and Wilson 1951; Skempton and Hutchinson 1969).

3. LABORATORY MEASUREMENT OF SHEAR STRENGTH

Various methods are available for laboratory measurement of shear strength. The simple methods are designed to determine the shear strength of a sample in a particular state, such as the water content of the soil in situ. These methods are most often used to determine the undrained shear strength (s_u) of saturated cohesive soils. More elaborate laboratory tests allow combinations of normal and shear stresses to be employed and pore pressures to be measured or controlled. Then it is possible to establish the shear strength relation defined by Equation 12.4. The most elaborate tests allow simulation of the field stress or deformation conditions. For example, triaxial compression tests simulate the stress state in Figure 12-1(*b*), and triaxial extension tests simulate the stress state in Figure 12-1(*c*) (see Section 3.2). Plane-strain compression and extension tests and simple shear tests can provide better simulation of the stress states in Figure 12-2 (see Section 3.3). A comprehensive summary of stress states in various laboratory tests and constitutive relations that can be deduced from test results was given by Desai and Siriwardane (1984).

3.1 Simple Shear Tests

Three types of simple shear tests are the unconfined compression, cone, and vane shear tests.

The unconfined compression test is usually performed on a cylindrical sample with a diameter-to-length ratio of 1:2. The sample is compressed axially [Figure 12-4(*a*)] until failure occurs; the shear strength is taken as one-half the compressive strength.

In the cone test, a cone with an angle θ is forced into the soil [Figure 12-4(*b*)] under a force (Q), which may be its own weight. The shear strength is obtained from the relation

$$s_u = \frac{KQ}{h^2} \tag{12.8}$$

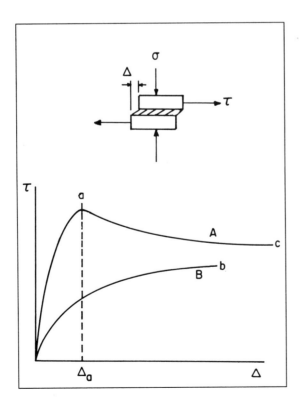

FIGURE 12-3
Typical stress-deformation curves.

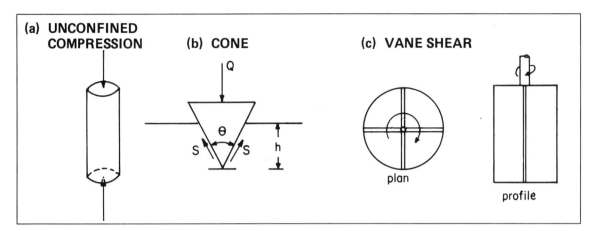

FIGURE 12-4
Simple tests for
measurement of soil
strength.

where h is the penetration and K is the constant that depends on the angle θ and the weight Q. Calibration curves for K were published by Hansbo (1957).

In the laboratory vane test (Cadling and Odenstad 1950), a vane is pushed into the soil specimen, and a torque is applied to the stem to produce shear failure over a cylindrical surface [Figure 12-4(c)]. The shear strength is obtained by equating the torque measured at failure to the moment produced by the shear stresses along the cylindrical surface. According to Cadling and Odenstad, the shear strength for vanes with a diameter-to-height ratio of 1:2 is

$$s = \frac{6}{7}\left(\frac{M}{\pi D^3}\right) \tag{12.9}$$

where M is the torque and D is the diameter of the vane.

Application of the results of these simple tests to the analysis of slope stability should include consideration of the type of soil and loading conditions in situ. The test results are expressed in terms of total stress because pore pressures are not measured. When the soil is brought to failure rapidly under undrained conditions, the shear strength is defined by $s = s_u$. Hence, the application of these test results is commonly limited to saturated cohesive soils under undrained conditions.

It is usually assumed that the measured strength is equal to the in situ strength. However, a major uncertainty is the effect of sampling disturbance on strength. Even "good" samples may suffer strength losses as great as 50 percent (Ladd and Lambe 1964; Clayton et al. 1992; Hight et al. 1992). The effect of sample disturbance is most severe in soft sensitive soils and appears to increase with the depth

from which the sample is taken. The state of stress is another factor that should be considered. The directions of the principal stresses and the orientations of rupture surfaces in these simple tests are not the same. They may also be quite different from directions of the principal stresses and the orientation of the failure surface in a slope. Hence, caution should be exercised when the results of these simple strength tests are applied to slope-stability problems (Bjerrum 1973).

3.2 Triaxial Test

The triaxial test is highly versatile; a variety of stress and drainage conditions can be employed (Bishop and Henkel 1962). The cylindrical soil specimen is enclosed within a thin rubber membrane and placed inside a triaxial cell, as shown in Figure 12-5(a). The cell is then filled with a fluid. As pressure is applied to the fluid in the cell, the specimen is subjected to a hydrostatic stress (σ_3). Drainage from the specimen is provided through the porous stone at the bottom, and the volume change can be measured. Alternatively, if no drainage is allowed, the pore-water pressure can be measured. For partially saturated soils, the pore-water pressure is measured through a fine ceramic stone that prevents air entry. The pore-air pressure is measured through a screen that permits free air entry. Details on testing of partially saturated soils were given by Fredlund and Rohardjo (1993, 260–282).

In triaxial compression, the axial stress (σ_z) is increased by application of a load through the loading ram. From the known stresses at failure ($\sigma_1 = \sigma_z$ and $\sigma_2 = \sigma_3 = \sigma_r$), Mohr circles or other stress plots can be constructed. Results from several triaxial tests, each using a different cell pressure (σ_3), are

FIGURE 12-5
Triaxial test.

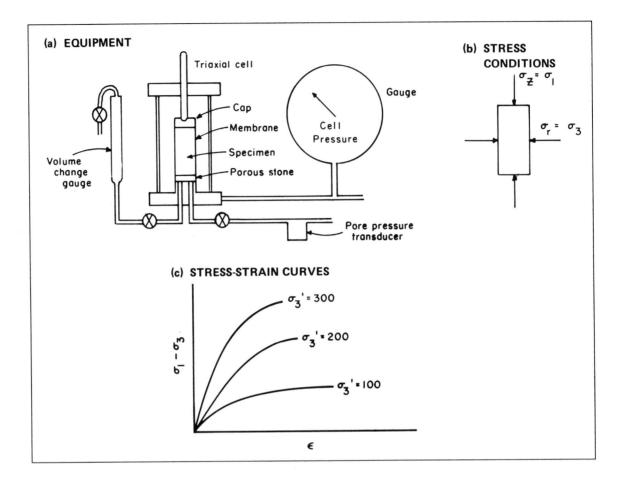

used to obtain the failure envelope. Typical plots of the principal stress difference versus axial strain are shown in Figure 12-5(c). The stress-strain behavior is influenced by the confining pressure, the stress history, and other factors. The sample can also be loaded to failure in triaxial extension by increasing the radial stress while maintaining the axial stress constant; then $\sigma_1 = \sigma_2 = \sigma_r$ and $\sigma_3 = \sigma_z$. Triaxial compression and extension simulate, respectively, the stress states beneath a loaded area, as shown in Figure 12-1(b), and beneath an excavation, as shown in Figure 12-1(c).

Principal stresses and pore pressure at failure are used to construct Mohr circles and to obtain the failure envelope. Figure 12-6 shows a series of failure envelopes. The horizontal axis is $\sigma - u_w$, which is the familar pore pressure for saturated soils. For saturated soils $(u_a - u_w) = 0$. When the soil is partially saturated, $(u_a - u_w)$ is not zero. A series of failure envelopes is obtained for different values of $(u_a - u_w)$, such as those shown in Figure 12-6 for values of $(u_a - u_w)$ equal to a and b.

Tests with the following drainage conditions can be performed with the triaxial apparatus. Tests on saturated soils are considered first.

In the *consolidated-drained test* (also called a drained test or slow test), the soil is allowed to consolidate completely under an effective cell pressure (σ_3'). In the triaxial compression test, the axial stress is increased at a slow rate, and drainage is permitted. The rate should be slow enough that no excess pore pressure is allowed to build up. Since

FIGURE 12-6
Failure envelopes.

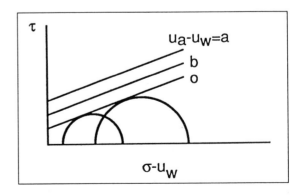

the excess pore pressure is zero in the drained test, effective stresses are known throughout the test and at failure. The results of a series of drained tests can be used to evaluate the effective stress strength parameters in Equation 12.4.

In the *consolidated-undrained test* with pore-pressure measurement (also called the consolidated-quick test), the drainage valves are closed after the initial consolidation of the sample. The axial stress is increased without drainage, and the excess pore pressure is measured. The pore pressure is subtracted from the total axial and radial stresses to give the effective stresses. The effective stresses at failure from a series of tests are used to define the failure criterion, as in the drained test. The consolidated-undrained test is sometimes performed without pore-pressure measurement. Then the effective stresses at failure cannot be determined. The results could be used to estimate the undrained shear strength after consolidation under σ_3'.

In the *unconsolidated-undrained test* (also called the undrained test or quick test), no drainage is allowed during any part of the test. When the cell pressure is applied, a pore-pressure change occurs in the soil. When the axial stress is applied, additional pore-pressure changes occur. If these pore-pressure changes are not measured, the test results can only be interpreted in terms of total stress. At failure, the undrained shear strength (s_u) is taken to represent the strength at the in situ water content. The unconsolidated-undrained test is then similar to the simple shear tests defined in Section 3.1.

The same three types of triaxial tests can be performed on partially saturated soils. However, because the air voids can be compressed even when there is no drainage, the simplifications derived above for undrained loading do not apply to partially saturated soils. For such soils, the pore-air and pore-water pressures must be measured in all tests in order to construct the failure envelope in terms of effective stress.

Triaxial tests are usually begun by increasing the cell pressure (σ_y) to the desired stress level, thus applying an isotropic or hydrostatic stress to the sample. This initial condition may differ from the initial in situ condition [see Figure 12-1(a)], in which the vertical and horizontal principal stresses are different. In situ stresses can be simulated in a triaxial test by using an anisotropic stress state dur-

ing consolidation. Experimental results that show the effect of stress state on strength properties were reviewed by Ladd et al. (1977).

3.3 Plane-Strain Test

The geometry of many geotechnical problems can be approximated by the condition of plane strain, in which the intermediate principal strain (ε_2) is zero. To simulate this condition, plane-strain tests have been developed (Henkel and Wade 1966) in which the sample is consolidated anisotropically with zero lateral strain ($\varepsilon_x = \varepsilon_y = 0$). Then the sample is loaded to failure by increasing either σ_x or σ_z and maintaining $\varepsilon_y = 0$ to simulate, respectively, the stress conditions at Points A or C of Figure 12-2. Plane-strain tests can be conducted under undrained, consolidated-undrained, or drained conditions as described for triaxial tests.

3.4 Direct Shear Test

The direct shear test is shown in Figure 12-7. The soil specimen is enclosed in a box consisting of upper and lower halves; porous stones on the top and the bottom permit drainage of water from the specimen. If the loading is carried out slowly, no excess pore pressure develops and the drained condition is obtained. The potential plane of failure is *a-a*. A normal stress (σ_z') is applied on plane *a-a*

FIGURE 12-7
Direct shear test.

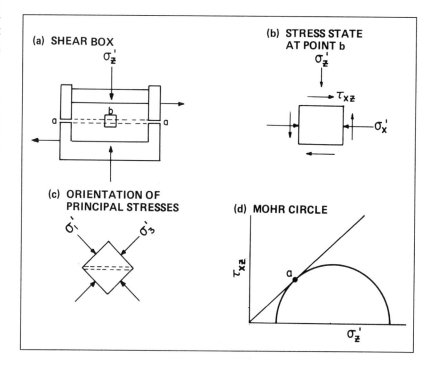

(a) SHEAR BOX

σ_z'

b

a   a

(b) STRESS STATE AT POINT b

σ_z'

τ_{xz}

σ_x'

(c) ORIENTATION OF PRINCIPAL STRESSES

σ_1 σ_3

(d) MOHR CIRCLE

τ_{xz}

a

σ_z'

FIGURE 12-8
Simple shear test.

through a loading head, and the shear stress is increased until the specimen fails along plane *a-a*. A stress-deformation curve is obtained by plotting the shear stress versus the displacement. Because the thickness of shear zone *a-a* is not precisely known, the shear strain cannot be determined. The test gives the value of τ_{xz} at failure. The average vertical stress (σ_z') and shear stress (τ_{xz}) are known, but σ_x' is not known. The directions of the principal stresses for Point *b* [Figure 12-7(*b*)] are approximately as shown in Figure 12-7(*c*). Assuming that Point *a* [Figure 12-7(*d*)] represents the conditions at failure, a Mohr circle can be constructed. The failure envelope is obtained from several tests, each using a different effective normal stress, performed on specimens of the same soil. The values of τ_{xz} at failure are plotted against the values of σ_z'. The foregoing represents the common interpretation of the direct shear test. More elaborate analyses were presented by Hill (1950), Morgenstern and Tchalenko (1967), and Jewell and Wroth (1987).

In saturated clays the direct shear test can be performed at a rapid rate so that the test duration is too short for any appreciable amount of water to flow into or out of the sample. This is an undrained condition, and an excess pore pressure of unknown magnitude is developed in the soil. Consequently, this is essentially a simple test, and the shear stress at failure represents the undrained shear strength (s_u). Test procedures for partially saturated soils were given by Fredlund and Rohardjo (1993, 282–254).

3.5 Simple Shear Test

Several simple shear tests have been developed, but the one described by Bjerrum and Landva (1966) is the most commonly used for undisturbed samples. The cylindrical specimen is enclosed in a rubber membrane reinforced by wire. This enclosure allows the shear deformation to be distributed fairly uniformly through the sample, as shown in Figure 12-8(*a*). The sample is consolidated anisotropically under a vertical stress [Figure 12-8(*a*)] and sheared by application of stress τ_{xz} [Figure 12-8(*b*)]. The simple shear test can be performed under undrained, consolidated-undrained, and drained conditions. Zero volume change during shear in an undrained test can be maintained by continuously adjusting the vertical stress σ_z' during the test. In the simple shear test, the initial principal axes are in the vertical and horizontal direc-

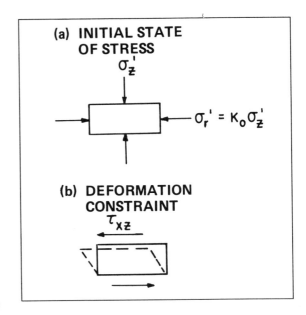

tions. At failure the horizontal plane becomes the plane of maximum shear strain. This condition approximates that at Point *B* of Figure 12-2.

4. IN SITU MEASUREMENT OF SHEAR STRENGTH

A variety of in situ tests are available for measurement of shear strength. Several of these are described in Chapter 10. Although theories have been proposed to interpret the test results, the stress state in most in situ shear tests is not precisely known. Hence empirical correlation is often relied upon for interpretation of the results. Various correlations have been published for different soils. For a comprehensive review of in situ tests, see the discussions by Orchant et al. (1988) and Lunne et al. (1989).

4.1 Standard Penetration Test

The standard penetration test uses a split-spoon sampler, which is driven with a specified energy into the bottom of a borehole. The number of blows (*N*) required to advance the sampling spoon through a distance of 30 cm may be used to estimate the angle of internal friction ϕ' of sands (Peck et al. 1974). More data are available on the correlation of *N* with relative density D_r (Gibbs and Holtz 1957; Marcuson and Bieganousky 1977). Correlation between D_r and ϕ' was given by Schmertmann (1975). Correlation between *N* and shear strength of cohesive soils is not reliable.

4.2 Cone Penetration Test

In the cone penetration test a cone of specified diameter and shape is attached to the end of a drill rod and advanced at a controlled rate. The force required is measured. This test can provide a continuous record of the resistance. For cohesionless soils, bearing capacity equations can be used to compute ϕ' from the penetration resistance. Empirical correlations between cone resistance and ϕ' are also available. Different correlation equations can give very different values of ϕ' (Mitchell and Lunne 1978; Wu et al. 1987a). The undrained shear strength of cohesive soils can be estimated from the cone resistance by the following relation:

$$s_u = \frac{q_c - \sigma_v}{N_k} \qquad (12.10)$$

where

s_u = undrained shear strength,
q_c = cone resistance,
σ_v = total overburden stress, and
N_k = an empirical constant.

The value of N_k ranges between 10 and 20 and depends on the overconsolidation ratio and plasticity index (Aas et al. 1986).

4.3 Field Vane Test

The field vane test is similar to the laboratory vane test described in Section 3.1. It measures the shear strength along a vertical cylindrical surface and the top and bottom surfaces of the cylinder. An equation similar to Equation 12.9 is used to calculate the shear strength. The coefficient depends on the assumption for stress distribution on the cylindri-cal surface and the top and bottom surfaces (Donald et al. 1977). Since the rupture surface in a stability problem may be quite different from that in a vertical cylinder, the shear strength may also be different from that measured in the vane test because of stress and material anisotropy. Correction factors based on an analysis of the stress state were suggested by Wroth (1984), and empirical correction factors were suggested by Bjerrum (1973).

4.4 Pressuremeter Test

In the pressuremeter test (Menard 1965; Baguelin et al. 1972), the cylindrical instrument is expanded against the wall of the borehole until failure in the surrounding soil occurs. Elastic-plastic solutions for expansion of a cylindrical cavity by Gibson and Anderson (1961) and subsequent modifications are used to obtain the shear strength and stress-strain properties of the soil. Consistent results have been obtained for the undrained shear strength of saturated clays and drained shear strength of cohesionless soils (Wroth 1984).

4.5 Borehole Shear Test

The borehole shear test is analogous to the direct shear test. The two halves of a cylinder, which is split lengthwise, are pressed under a controlled pressure against the opposite walls of a borehole. The cylinder is then pulled vertically, shearing the soil (Handy and Fox 1967). The failure surface is not determined; hence the normal and shear stresses are not accurately known. For cohesionless soils with high permeability, the test may be considered a consolidated-drained test. For cohesive soils, the drainage condition is uncertain. The borehole shear test was analyzed by Lutenegger and Hallberg (1981) and Handy et al. (1985).

4.6 Dilatometer Test

The dilatometer is a flat blade that contains a flexible membrane on one side (Marchetti 1975). The blade is pushed into the soil to reach the desired depth, and the membrane is inflated by pressure and pushes back the adjacent soil. The pressures required to "lift off" the membrane and to displace the soil 1 mm are recorded. These pressures are used to estimate the horizontal in situ stress. Empirical correlation between the horizontal in situ stress and s_u / σ_v' is used to estimate the s_u of clays (Marchetti 1980). The force required to advance the dilatometer is measured and used to estimate the bearing capacity and then ϕ' of cohesionless soils (Schmertmann 1982).

5. SHEAR STRENGTH PROPERTIES OF SOME COMMON SOILS

The strength characteristics of natural soils are strongly influenced by the geologic processes of soil formation. Hence, it is possible to identify strength characteristics that are common to soils within broadly defined groups. Groups with well-defined characteristics include saturated cohesionless soils, soft saturated cohesive soils, heavily overconsolidated clays, sensitive soils, and residual soils and colluvium. More detailed discussions of slopes involving specific soil conditions are given in Chapters 19 through 25.

5.1 Saturated Cohesionless Soils

Cohesionless soils, such as gravel, sand, and nonplastic silts, have effective stress failure envelopes that pass through the origin. This means $c' = 0$ in Equation 12.4. The value of ϕ' ranges from about 27 to 45 degrees or more and depends on several factors. For a given soil the value of ϕ' increases as the relative density increases. For different soils at the same relative density, the value of ϕ' is affected by particle-size distribution and particle shape. The value of ϕ' for a well-graded soil may be several degrees greater than that for a uniform soil of the same average particle size and shape. The same is true when a soil composed of angular grains is compared with one made up of rounded grains. The effect of moisture on ϕ' is small and amounts to no more than 1 or 2 degrees (Lambe and Whitman 1969, 147–149; Holtz and Kovacs 1981, 514–519).

The failure envelope, which is a straight line at low pressures, cannot be extended to high confining pressures. Tests with effective normal stresses above 700 kPa indicate that the failure envelope is curved, as shown in Figure 12-9 (Bishop 1966; Vesic and Clough 1968). The high normal stresses apparently cause crushing of grain contacts and result in a lower friction angle. This factor is particularly important in uncemented carbonate sands (Datta et al. 1982). Another important factor

is the difference in the values of ϕ' as measured by different types of tests. The value of ϕ' measured in triaxial tests, which permit change in the radial strain, is 4 to 5 degrees smaller than that measured in plane-strain tests (Ladd et al. 1977).

In ordinary construction sandy and gravelly soils of high permeability can be considered to be loaded in the drained condition. Volume changes occur rapidly, and no excess pore pressures are developed. Without excess pore pressure, effective stresses can be estimated from the groundwater levels. Stability analyses can be performed by using effective stress strength parameters. For silty soils the permeability may be sufficiently low that excess pore pressure will develop during construction. When this is the case, the pore pressure must be measured or estimated if an effective stress analysis is to be performed.

The undrained response of fine sands and silts is important in some situations. Loose saturated sands and silts may develop large excess pore pressures after reaching peak strength. Because of the large pore pressure, the strength drops and the ultimate strength is well below the peak strength. Large displacements occur as a result of this loss of strength. This phenomenon was called *liquefaction* by Casagrande (1936). Large pore pressures may also develop in saturated sands of medium to high density under cyclic loading. In a triaxial test, if an axial stress is repeated for a number of cycles, the pore pressure may become equal to the confining pressure, thus reducing the effective minor principal stress (σ_3') to 0 (Seed and Lee 1967). Large strain increments will occur with each cycle of stress as the pore pressure approaches σ_3. If the soil dilates with strain, the pore pressure soon drops and the soil becomes stable. This phenomenon is called *cyclic mobility* (Castro 1975). The behavior of a sand under cyclic loading may range from liquefaction to cyclic mobility, depending on whether the sand is contractive or dilative.

Liquefaction and cyclic mobility are important considerations in the evaluation of slope stability under earthquake loadings. Lade (1992) and Sladen et al. (1985) formulated conditions for liquefaction in terms of the yield surface and the critical state, respectively. Methods for assessing the potential for liquefaction and cyclic mobility were reviewed by Casagrande (1976), Seed (1979), Poulos et al. (1985), and Seed et al. (1985).

FIGURE 12-9
Failure envelope for cohesionless soils.

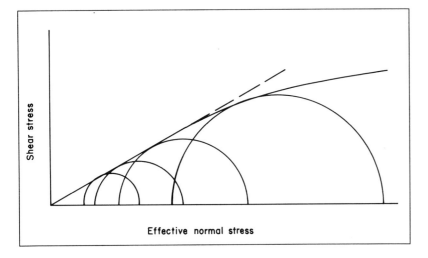

5.2 Normally Consolidated and Lightly Overconsolidated Clays and Clayey Silts

Because of the low permeability of fine-grained soils, undrained or partially drained conditions are common in normal construction. The general characteristics of normally consolidated to lightly overconsolidated clays are described in this section. Very sensitive normally consolidated clays are described in Section 5.4.

A clay soil is considered to be normally consolidated if the consolidation pressure before shear is equal to or greater than the preconsolidation pressure (σ_p'). When a series of drained triaxial tests is conducted on a normally consolidated soil, the failure envelope is a line that passes through the origin [Figure 12-10(a)]; thus, $c' = 0$. If consolidated-undrained tests are performed on a normally consolidated soil, positive excess pore pressures develop. As a result, the undrained shear strength $s_u = \frac{1}{2}(\sigma_1 - \sigma_3)$ will be less than the drained shear strength of a sample initially consolidated under the same stresses. The respective stress paths of the drained and consolidated-undrained tests are shown as ca and cb in Figure 12-10(a). A typical relation between strength and water content is shown in Figure 12-10(b). The results of consolidated-undrained tests can also be used to obtain the ratio s_u/σ_3', which is a constant for normally consolidated soils (Skempton 1948). For lightly overconsolidated soils, s_u/σ_3' is a function of the overconsolidation ratio σ_3'/σ_p' (Ladd and Foott 1974).

If the load or stress change in the field induces positive excess pore pressures, as may be the case for a fill, the undrained strength will be lower than the drained strength. A slope based on an initially stable design can be expected to increase in stability with time as the excess pore pressure dissipates and the strength increases. The increase in strength with consolidation makes possible the construction in stages of embankments over soft soils. The strength gained under each stage of loading permits the addition of the next load increment. The ratio s_u/σ_3' is used to estimate the s_u under a given σ_3'. When a slope is made by excavation, there is a simultaneous increase in shear stress due to the slope and a decrease in mean normal stress due to the unloading of the excavation. In a saturated, normally consolidated soil, the increase in shear stress produces a positive excess pore pressure, and the decrease in normal stress produces a negative excess pore pressure. The net excess pore pressure in various parts of the slope depends on the relative values of these two effects. If the excess pore pressure is negative, the strength will decrease with time and drainage. In this case, the long-term or drained stability will be critical for a normally consolidated clay. Bishop and Bjerrum (1960) described several examples.

Failures of real slopes are good sources of information on the reliability of theoretical models. In a number of careful investigations, the factor of safety of the slope that failed was compared with the measured shear strength. If the theory and soil properties used are correct, the computed safety factor of a slope at failure should be unity. Results from many case histories show that for normally consolidated or lightly overconsolidated homogeneous clays of low sensitivity, total stress analysis with undrained shear strength gives a reasonably accurate estimate of immediate stability. For long-term stability under drained conditions, effective stress analysis with c' and ϕ' also gives satisfactory results. Comprehensive reviews and case histories were given by Bishop and Bjerrum (1960) and Bjerrum (1973), and more recent examples were

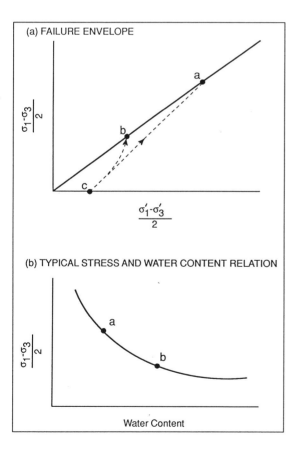

(a) FAILURE ENVELOPE

$\frac{\sigma_1 - \sigma_3}{2}$

$\frac{\sigma_1' - \sigma_3'}{2}$

(b) TYPICAL STRESS AND WATER CONTENT RELATION

$\frac{\sigma_1 - \sigma_3}{2}$

Water Content

FIGURE 12-10 Strength properties of soft saturated clays.

FIGURE 12-11
Strength
properties of stiff
overconsolidated
clays.

presented by Ladd (1991), Pilot et al. (1982), and Tavenas and Leroueil (1980). For most of the reported failures in these case histories, the discrepancy between calculated and observed safety factors is less than 15 percent. Since most of the clays investigated are fairly uniform deposits, an accuracy of ±15 percent may be the best that can be achieved in practice.

It should be noted that earlier studies, such as that by Skempton and Golder (1948), used unconfined compressive strengths obtained from samples of small diameter. These conditions would tend to underestimate the in situ shear strength because of sample disturbance. By comparison, the study by Duncan and Buchignani (1973) used large-diameter samples and unconsolidated-undrained triaxial tests. The computed safety factor was 1.17 for the slope that failed. Various factors that could affect the strength were estimated, and the difference between observed strength and that measured in triaxial tests was attributed to the difference in strain rate. In cases reported by Ladd and Foott (1974), strength anisotropy was accounted for by performing triaxial compression and extension tests and direct simple shear tests. Hence, errors of different nature and magnitude are involved in the estimates of undrained shear strength in the different case histories.

Soft clays may also develop high pore pressures under cyclic loading (Thiers and Seed 1969). Sangrey et al. (1969) showed that a yield surface in effective stress space exists for cyclic stresses. When cyclic stresses reach the yield surface, large strains develop. Measurement and evaluation of strength losses due to complex states of static and cyclic stresses were described by Andersen et al. (1988).

5.3 Heavily Overconsolidated Clays

Most heavily overconsolidated clays show strain softening in their stress-strain curves (Figure 12-3, Curve A). The peak strengths of a series of heavily overconsolidated clay specimens give an effective stress failure envelope shown by Curve A in Figure 12-11(a). The failure envelope is approximately a straight line and if extrapolated to the axis at $\sigma' = 0$ gives a cohesion intercept (c'). Laboratory tests should be performed using normal stresses that are close to the normal stresses in the field because research has shown that the failure envelope for the peak strength of heavily overconsolidated clays is

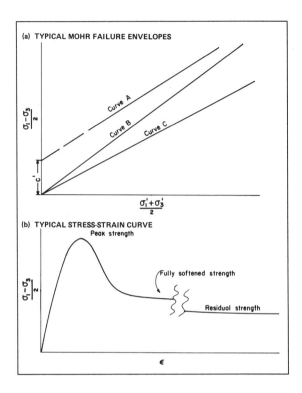

curved in the low-stress region and passes through the origin.

When loaded in the drained condition, a heavily overconsolidated clay will absorb water, and the absorbed water leads to a softening of the clay. The water content will increase significantly as the strength is reduced to the fully softened strength [Figure 12-11(b)]. The failure envelope for fully softened strength is shown as Curve B in Figure 12-11(a). The fully softened strength is close to the strength of the same soil in the normally consolidated condition. The strain necessary to develop fully softened strength in a heavily overconsolidated clay varies from one soil to another, but is at least 10 percent.

When much larger shear displacements take place within a narrow zone, the clay particles become oriented along the direction of shear, and a polished surface, or *slickenside*, forms. In natural slopes, slickensides may be developed along rupture surfaces of old landslides, bedding surfaces, or zones of deformation caused by tectonic forces. Along these surfaces the shear strength may approach the residual strength. The failure envelope, Curve C in Figure 12-11(a), is a line passing through the origin, and the residual angle of internal friction is ϕ_r'. Results of laboratory direct-shear tests indicate that ϕ_r' is dependent on soil mineralogy. As shown in Figure 12-12, "massive minerals," such as quartz, feldspar, and calcite, have high values of ϕ_r' that are close to the values of ϕ' for peak strength. However, clay minerals show significant differences between values of ϕ' and ϕ_r'. The largest difference was found in montmorillonitic clays, for which ϕ_r' was less than 10 degrees (Kenney 1967). The relationship between mineralogical composition and ϕ_r' makes it possible to correlate ϕ_r' with index properties such as plasticity index (Lupini et al. 1981) or liquid limit (Mesri and Cepeda-Diaz 1986). Because of the importance of strain-softening in heavily overconsolidated clays, choice of the strength for a stability analysis of a slope should consider progressive failure as described in Section 2.4. The design strength must be chosen to represent the average of the strengths along the entire rupture surface. For examples of analysis of progressive failure, see studies by Bjerrum (1967) and Duncan and Dunlop (1969).

For most slopes in heavily overconsolidated clays, the excess pore pressure immediately after construction is negative. Thus, the undrained strength will be greater than the drained strength. With drainage, the negative pore pressure dissipates and water is drawn into the sample. As water content increases, strength decreases. The long-term or drained conditions are critical, and there is no assurance that an initially stable slope will remain stable in the long term. Because of the low permeability, the time required to develop the fully

drained condition in the field may be several years (Chandler 1984).

In heavily overconsolidated clays and soft shales, fissures and other discontinuities have an important influence on strength (Terzaghi 1936b). The characteristics of discontinuities in some stiff clays were described by Skempton and Petley (1967). The shear strength of laboratory specimens of such clays is strongly dependent on the number, shape, and inclination of discontinuities in the specimen. The presence of discontinuities is less likely in small specimens, which are often trimmed from intact soil between the discontinuities. Hence, the measured strength of laboratory samples tends to be higher than the in situ strength. When the clay contains discontinuities such as fissures and slickensides, the in situ strength will depend on the frequency and orientation of the discontinuities. Skempton and Petley (1967), Skempton and Hutchinson (1969), Chandler (1984), Wu et al. (1987b), and Burland (1990) provided examples of the in situ strength of clays with discontinuities.

5.4 Very Sensitive Soils

Sensitivity is defined as the ratio of the peak undrained shear strength to the undrained shear strength of the same soil after remolding at constant water content. Causes of clay sensitivity were reviewed by Mitchell and Houston (1969). The most dramatic landslides in sensitive soils occur in

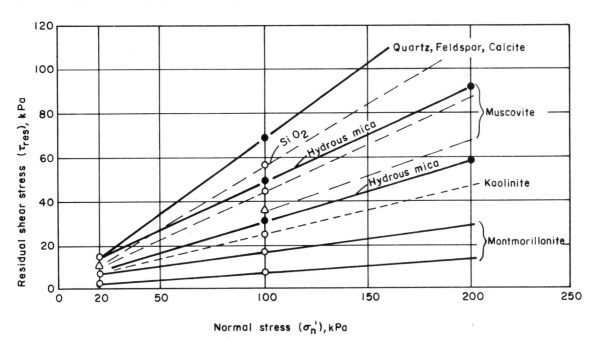

FIGURE 12-12
Residual shear strength of minerals.

Pleistocene marine clays in the Scandinavian countries and in the St. Lawrence River valley of eastern North America. High sensitivity is found in soils that have been subjected to leaching or natural cementation.

Effective stress failure envelopes for sensitive clays differ from those for soft saturated clays, described in Section 5.2. A sensitive clay, consolidated under the K_o condition, has a yield surface Y_o as shown in Figure 12-13. When the stress path for a point reaches the yield curve, such as at Point P, the soil structure is destroyed. Then high excess pore pressures are produced, and the shear strength decreases to Point C on the failure envelope for large strains (Tavenas and Leroueil 1977), which corresponds to the critical state (Schofield and Wroth 1968). In the design of slopes in these soils, it is necessary to keep the stresses within the yield curve. Evaluation of slope stability in very sensitive clays is described in Chapter 24.

5.5 Residual Soil and Colluvium

The weathering of rock produces residual soil. In the initial stages of weathering, coarse rock fragments are produced; the ultimate product of weathering is clay. Between these extremes is a soil composed of a mixture of grain sizes. On flat topography residual soil remains where formed. On slopes the weathered material, composed of soil and rock debris, may move down the face of the slope under the force of gravity. The deposit formed by this process is called *colluvium*. Residual soils have a wide range of properties depending on the parent material and the degree of weathering (Deere and Patton 1971; Mitchell and Sitar 1982).

Residual soils frequently exist in the partially saturated state. Some residual soils have high porosity and high permeability. Large changes in moisture content may occur with dry and wet seasons. The shear strength in terms of total stress has been found to be strongly influenced by the moisture content (Foss 1977; O'Rourke and Crespo 1988). Data on shear strength parameters c', ϕ', and ϕ'' are scarce. Tests by Ho and Fredlund (1982) showed that the component $c' + (u_w - u_a)\tan\phi''$ in Equation 12.5 decreases rapidly with decreasing suction, and c' is generally small at saturation. On the other hand, residual soils that are composed of a mineral skeleton held together by cementation may have an undrained shear strength that is unaffected by the water content. Such soils may show appreciable strength in spite of high water content and high porosity (Wallace 1973).

Colluvium presents some particular problems. The extremely heterogeneous nature of the material makes sampling and testing difficult. Results of tests on the finer matrix are likely to underestimate the shear strength.

Failure of slopes on residual soils and colluvium most frequently occurs during wet periods when there is an increase in moisture content and a decrease in suction. Stability of slopes may be determined by means of effective stress analysis and c', ϕ', and ϕ''. It is necessary to estimate the suction $u_w - u_a$ (Anderson and Pope 1984). In many cases in situ measurement of suction is necessary (Brand 1982; Krahn et al. 1989). If the soil is likely to be saturated, an effective stress analysis for the saturated condition provides a conservative estimate of the stability. If a total stress analysis is used, it is necessary to estimate the strength that corresponds to the in situ moisture content.

REFERENCES

Aas, G., S. Lacasse, T. Lunne, and K. Hoeg. 1986. Use of In Situ Tests for Foundation Design on Clay. In *Use of In Situ Tests in Geotechnical Engineering* (S.P. Clemence, ed.), Geotechnical Special Publication 6, American Society of Civil Engineers, New York, pp. 1–30.

Andersen, K.H., A. Kleven, and D. Heien. 1988. Cyclic Soil Data for Design of Gravity Structures. *Journal of Geotechnical Engineering*, ASCE, Vol. 114, No. 5, pp. 517–539.

Anderson, M.G, and R.G. Pope. 1984. The Incorporation of Soil Physics Models into Geotechnical Studies of Landslide Behavior. In *Proc., Fourth*

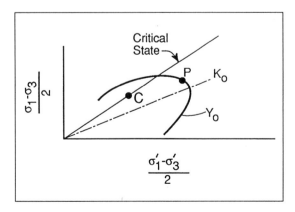

FIGURE 12-13
Strength envelopes of sensitive soils.

International Symposium on Landslides, Canadian Geotechnical Society, Toronto, Vol. 1, pp. 349–353.

Baguelin, F., J.F. Jezequel, E. Le Mee, and A. Le Mehaute. 1972. Expansion of Cylindrical Probes in Cohesive Soils. *Journal of the Soil Mechanics and Foundations Division,* ASCE, Vol. 98, No. SM11, pp. 1129–1142.

Bishop, A.W. 1966. The Strength of Soils as Engineering Materials. *Geotechnique,* Vol. 16, No. 2, pp. 91–128.

Bishop, A.W., and L. Bjerrum. 1960. The Relevance of the Triaxial Test to the Solution of Stability Problems. In *Proc., Research Conference on Shear Strength of Cohesive Soils,* Boulder, Colorado, American Society of Civil Engineers, New York, pp. 437–501.

Bishop, A.W., and D.J. Henkel. 1962. *The Measurement of Soil Properties in the Triaxial Test.* Edward Arnold Publishers, London, 227 pp.

Bjerrum, L. 1967. Progressive Failure in Slopes of Overconsolidated Plastic Clay and Clay Shales. *Journal of the Soil Mechanics and Foundations Division,* ASCE, Vol. 93, No. 5, pp. 1–49.

Bjerrum, L. 1973. Problems of Soil Mechanics and Construction on Soft Clays and Structurally Unstable Soils. In *Proc., Eighth International Conference on Soil Mechanics and Foundation Engineering,* Moscow, Vol. 3, pp. 111–159.

Bjerrum, L., and A. Landva. 1966. Direct Simple-Shear Tests on a Norwegian Quick Clay. *Geotechnique,* Vol. 16, No. 1, pp. 1–20.

Brand, E.W. 1982. Analysis and Design in Residual Soils. In *Engineering and Construction in Tropical and Residual Soils,* American Society of Civil Engineers, New York, pp. 89–143.

Burland, J.B. 1990. On the Compressibility and Shear Strength of Natural Clays. *Geotechnique,* Vol. 40, No. 3, pp. 329–378.

Cadling, L., and S. Odenstad. 1950. The Vane Borer: An Apparatus for Determining the Shear Strength of Clay Soils Directly in the Ground. In *Proc., Swedish Geotechnical Institute,* No. 2, 88 pp.

Casagrande, A. 1936. Characteristics of Cohesionless Soils Affecting the Stability of Earth Fills. *Journal of the Boston Society of Civil Engineers,* Jan., pp. 13–32.

Casagrande, A. 1976. *Liquefaction and Cyclic Deformation of Sands—A Critical Review.* Soil Mechanics Series 88. Harvard University, Cambridge, Mass.

Casagrande, A., and S.D. Wilson. 1951. Effect of Rate of Loading on the Strength of Clays and Shales at Constant Water Content. *Geotechnique,* Vol. 2, No. 3, pp. 251–263.

Castro, G. 1975. Liquefaction and Cyclic Mobility of Saturated Sands. *Journal of the Geotechnical Engineering Division,* ASCE, Vol. 101, No. GT6, pp. 551–569.

Chandler, R.J. 1984. Recent European Experience of Landslides in Heavily Over Consolidated Clays and Soft Rocks. In *Proc., Fourth International Symposium on Landslides,* Canadian Geotechnical Society, Toronto, Vol. 1, pp. 61–81.

Clayton, C.R.I., D.W. Hight, and R.J. Hopper. 1992. Progressive Restructuring of Bothkennar Clay: Implications for Sampling and Reconsolidation Procedures. *Geotechnique,* Vol. 42, No. 2, pp. 219–240.

Datta, M., S.K. Gulhati, and G.V. Rao. 1982. Engineering Behavior of Carbonate Soils of India and Some Observations on Classification of Such Soils. In *Geotechnical Properties, Behavior, and Performance of Calcareous Soils,* Special Technical Publication 777, ASTM, Philadelphia, Pa., pp. 113–140.

Deere, D.U., and F.D. Patton. 1971. Slope Stability in Residual Soils. In *Proc., Fourth Pan-American Conference on Soil Mechanics and Foundation Engineering,* San Juan, Puerto Rico, Vol. 1, pp. 87–170.

Desai, C.S., and H.J. Siriwardane. 1984. *Constitutive Laws for Engineering Materials.* Prentice-Hall, Englewood Cliffs, N.J., 468 pp.

Donald, I.B., D.O. Jordan, R.J. Parker, and C.T. Toh. 1977. The Vane Test—A Critical Appraisal. In *Proc., Ninth International Conference on Soil Mechanics and Foundation Engineering,* Tokyo, Vol. 1, pp. 81–88.

Drucker, D.C., R.E. Gibson, and D.H. Henkel. 1955. Soil Mechanics and Work Hardening Theories of Plasticity. In *Proc., American Society of Civil Engineers,* Vol. 81, No. 798.

Duncan, J.M., and A.L. Buchignani. 1973. Failure of Underwater Slope in San Francisco Bay. *Journal of Soil Mechanics and Foundations Division,* ASCE, Vol. 99, No. SM9, pp. 687–703.

Duncan, J.M., and P. Dunlop. 1969. Slopes in Stiff-Fissured Clays and Shales. *Journal of the Soil Mechanics and Foundations Division,* ASCE, Vol. 95, No. 2, pp. 471–493.

Foss, I. 1977. Red Soil from Kenya as a Foundation Material. In *Proc., Eighth International Conference on Soil Mechanics and Foundation Engineering,* Moscow, Vol. 2, pp. 73–80.

Fredlund, D.G., and H. Rohardjo. 1993. *Soil Mechanics for Unsaturated Soils.* John Wiley & Sons, New York, 517 pp.

Fredlund, D.G., N.R. Morgenstern, and R.A. Widger. 1978. The Shear Strength of Unsaturated Soils. *Canadian Geotechnical Journal*, Vol. 15, No. 3, pp. 313–321.

Gibbs, H.J., and W.H. Holtz. 1957. Research on Determining the Density of Sands by Spoon Penetration Testing. In *Proc., Fourth International Conference on Soil Mechanics and Foundation Engineering*, Butterworths Scientific Publications, London, Vol. 1, pp. 35–39.

Gibson, R.E., and W.F. Anderson. 1961. In Situ Measurement of Soil Properties with the Pressure Meter. *Civil Engineering and Public Works Review*, Vol. 56, pp. 615–618.

Handy, R.L., and N.S. Fox. 1967. A Soil Bore-Hole Direct-Shear Test Device. *Highway Research News*, No. 27, HRB, National Research Council, Washington, D.C., pp. 42–51.

Handy, R.L., J.H. Schmertmann, and A.J. Lutenegger. 1985. Borehole Shear Tests in a Shallow Marine Environment. In *Strength Testing of Marine Sediments*, Special Technical Publication 883, ASTM, Philadelphia, Pa., pp. 140–153.

Hansbo, S. 1957. A New Approach to the Determination of the Shear Strength of Clay by the Fall-Cone Test. In *Proc., Swedish Geotechnical Institute*, No. 14, 47 pp.

Henkel, D.J. 1960. The Shear Strength of Saturated Remolded Clays. In *Proc., Research Conference on Shear Strength of Cohesive Soils*, Boulder, Colorado, American Society of Civil Engineers, New York, pp. 533–554.

Henkel, D.J., and N.H. Wade. 1966. Plane Strain Tests on a Saturated Remolded Clay. *Journal of the Soil Mechanics and Foundations Division*, ASCE, Vol. 92, No. SM6, pp. 67–80.

Hight, D.W., R. Böese, A.P. Butcher, and C.R.I. Clayton. 1992. Disturbance of the Bothkennar Clay Prior to Laboratory Testing. *Geotechnique*, Vol. 42, No. 2, pp. 199–218.

Hill, R. 1950. *The Mathematical Theory of Plasticity*. Oxford University Press, London, 355 pp.

Ho, D.Y.F., and D.G. Fredlund. 1982. Increase in Strength due to Suction for two Hong Kong Soils. In *Engineering and Construction in Tropical and Residual Soils*, American Society of Civil Engineers, New York, pp. 263–295.

Holtz, R.D., and W.D. Kovacs. 1981. *An Introduction to Geotechnical Engineering*. Prentice-Hall, Inc., Englewood Cliffs, N.J., 731 pp.

Jewell, R.A., and C.P. Wroth. 1987. Direct Shear Tests on Reinforced Sand. *Geotechnique*, Vol. 37, No. 1, pp. 53–68.

Kenney, T.C. 1967. The Influence of Mineral Composition on the Residual Strength of Natural Soils. In *Proc., Geotechnical Conference on Shear Strength Properties of Natural Soils and Rocks*, Oslo, Vol. 1, pp. 123–129.

Krahn, J., D.G. Fredlund, and M.J. Klassen. 1989. Effect of Soil Suction on Slope Stability at Notch Hill. *Canadian Geotechnical Journal*, Vol. 26, No. 2, pp. 269–278.

Ladd, C.C. 1991. Stability Evaluation During Staged Construction. *Journal of Geotechnical Engineering*, ASCE, Vol. 117, No. 4, pp. 540–615.

Ladd, C.C., and R. Foott. 1974. New Design Procedure for Stability of Soft Clays. *Journal of the Geotechnical Engineering Division*, ASCE, Vol. 100, No. GT7, pp. 763–786.

Ladd, C.C., R. Foott, K. Ishihara, F. Schlosser, and H.G. Poulos. 1977. Stress-Deformation and Strength Characteristics. In *Proc., Ninth International Conference on Soil Mechanics and Foundation Engineering*, Tokyo, Vol. 2, pp. 421–494.

Ladd, C.C., and T.W. Lambe. 1964. The Strength of "Undisturbed" Clay Determined from Undrained Tests. In *Proc., Symposium on Laboratory Shear Testing of Soils*, Ottawa, Canada, Special Technical Publication 361, ASTM, Philadelphia, Pa., pp. 342–371.

Lade, P.V. 1992. Static Instability and Liquefaction of Loose Fine Sandy Soils. *Journal of Geotechnical Engineering*, ASCE, Vol. 118, No. 1, pp. 51–71.

Lambe, T.W., and R.V. Whitman. 1969. *Soil Mechanics*. John Wiley & Sons, New York, 553 pp.

Lunne, T., S. Lacasse, and N.S. Rad. 1989. SPT, CPT, Pressuremeter Testing and Recent Developments. In *Proc., 12th International Conference on Soil Mechanics and Foundation Engineering*, Rio de Janeiro, A.A. Balkema, Rotterdam, Netherlands, Vol. 4, pp. 2339–2403.

Lupini, J.F., A.E. Skinner, and P.R. Vaughan. 1981. The Drained Residual Strength of Cohesive Soils. *Geotechnique*, Vol. 31, No. 2, pp. 181–214.

Lutenegger, A.J., and G.R. Hallberg. 1981. Borehole Shear Test in Geotechnical Investigations. In *Laboratory Shear Strength of Soil*, Special Technical Publication 740, ASTM, Philadelphia, Pa., pp. 566–578.

Marchetti, S. 1975. A New In Situ Test for the Measurement of Horizontal Soil Deformability. In *Proc., Specialty Conference on In Situ Measurement of Soil Properties*, Raleigh, N.C., American Society of Civil Engineers, New York, Vol. 2, pp. 255–259.

Marchetti, S. 1980. In Situ Tests by Flat Dilatometer. *Journal of the Geotechnical Engineering Division*, ASCE, Vol. 106, No. GT3, pp. 299–321.

Marcuson, W.F. III, and W.A. Bieganousky. 1977. Laboratory Standard Penetration Tests on Fine Sands. *Journal of the Geotechnical Engineering Division*, ASCE, Vol. 103, No. GT6, pp. 565–588.

Menard, L. 1965. Rules for Calculation of Bearing Capacity and Foundation Settlement Based on Pressuremeter Tests. In *Proc., Sixth International Conference on Soil Mechanics and Foundation Engineering*, Montreal, University of Toronto Press, Canada, Vol. 2, pp. 295–299.

Mesri, G., and A.F. Cepeda-Diaz. 1986. Residual Shear Strength of Clays and Shales. *Geotechnique*, Vol. 36, No. 2, pp. 269–274.

Mitchell, J.K., and T.A. Lunne. 1978. Cone Resistance as a Measure of Soil Strength. *Journal of the Geotechnical Engineering Division*, ASCE, Vol. 104, No. GT7, pp. 995–1012.

Mitchell, J.K., and N. Sitar. 1982. Engineering Properties of Tropical Residual Soils. In *Engineering and Construction in Tropical and Residual Soils*, American Society of Civil Engineers, New York, pp. 30–57.

Mitchell, J.K., and W.N. Houston. 1969. Causes of Clay Sensitivity. *Journal of the Soil Mechanics and Foundations Division*, ASCE, Vol. 95, No. SM3, pp. 845–871.

Morgenstern, N.R., and J.S. Tchalenko. 1967. Microscopic Structures in Kaolinite Subjected to Direct Shear. *Geotechnique*, Vol. 17, No. 4, pp. 309–328.

Orchant, C.J., F.H. Kulhawy, and C.H. Trautmann. 1988. *Reliability-Based Foundation Design for Transmission Line Structures*, Vol. 2. Geotechnical Engineering Group, Cornell University, Ithaca, N.Y., 166 pp.

O'Rourke, T.D., and E. Crespo. 1988. Geotechnical Properties of Cemented Volcanic Soil. *Journal of Geotechnical Engineering*, ASCE, Vol. 114, No. 10, pp. 1126–1147.

Peck, R.B., W.E. Hanson, and T.H. Thornburn. 1974. *Foundation Engineering*, 2nd ed. John Wiley & Sons, New York, 514 pp.

Pilot, G., B. Trak, and P. LaRochelle. 1982. Effective Stress Analysis of the Stability of Embankments on Soft Soils. *Canadian Geotechnical Journal*, Vol. 19, No. 4, pp. 433–450.

Poulos, S.J., G. Castro, and J. France. 1985. Liquefaction Evaluation Procedure. *Journal of Geotechnical Engineering*, ASCE, Vol. 111, No. 6, pp. 772–792.

Sangrey, D.A., D.J. Henkel, and M.I. Esrig. 1969. The Effective Stress Response of a Saturated Clay Soil to Repeated Loading. *Canadian Geotechnical Journal*, Vol. 6, No. 3, pp. 241–252.

Schmertmann, J.H. 1975. Measurement of In Situ Shear Strength. In *Proc., Conference on In Situ Measurement of Soil Properties*, Raleigh, N.C., American Society of Civil Engineers, New York, Vol. 2, pp. 57–138.

Schmertmann, J.H. 1982. Method for Determining the Friction Angle in Sands. In *Proc., Second European Symposium on Penetration Testing*, Amsterdam, Vol. 2, pp. 853–862.

Schofield A., and P. Wroth.1968. *Critical State Soil Mechanics*. McGraw-Hill Book Co., New York, 310 pp.

Seed, H.B. 1979. Considerations in the Earthquake-Resistant Design of Earth and Rockfill Dams. *Geotechnique*, Vol. 29, No. 3, pp. 215–263.

Seed, H.B., and K.L. Lee. 1967. Liquefaction of Saturated Sands During Cyclic Loading. *Journal of the Soil Mechanics and Foundations Division*, ASCE, Vol. 92, No. SM6, pp. 105–134.

Seed, H.B., K. Tokimatsu, L.F. Harder, and R.M. Chung. 1985. Influence of STP Procedures in Soil Liquefaction Resistance Evaluation. *Journal of Geotechnical Engineering*, ASCE, Vol. 111, No. 12, pp. 1425–1445.

Skempton, A.W. 1948. A Study of the Geotechnical Properties of Some Post-Glacial Clays. *Geotechnique*, Vol. 1, No. 1, pp. 7–22.

Skempton, A.W. 1954. The Pore-Pressure Coefficients A and B. *Geotechnique*, Vol. 4, No. 4, pp. 143–147.

Skempton, A.W., and H.Q. Golder. 1948. Practical Examples of the $\phi = 0$ Analysis of Stability of Clays. In *Proc., Second International Conference on Soil Mechanics and Foundation Engineering*, Rotterdam, Vol. 2, pp. 63–70.

Skempton, A.W., and D.J. Petley. 1967. The Strength Along Structural Discontinuities in Stiff Clays. In *Proc., Geotechnical Conference on Shear Strength Properties of Natural Soils and Rocks*, Oslo, Vol. 2, pp. 29–46.

Skempton, A.W., and J.N. Hutchinson. 1969. Stability of Natural Slopes and Embankment Foundations. In *Proc., Seventh International Conference on Soil Mechanics and Foundation Engineering*, Mexico City, pp. 291–340.

Sladen, J.A., R.D. Hollander, and J. Krahn. 1985. The Liquefaction of Sands. *Canadian Geotechnical Journal*, Vol. 22, No. 4, pp. 564–578.

Tavenas, F., and S. Leroueil. 1977. Effects of Stress and Time on Yielding of Clays. In *Proc., Ninth International Conference on Soil Mechanics and Foundation Engineering*, Tokyo, Vol. 1, pp. 319–326.

Tavenas, F., and S. Leroueil. 1980. The Behavior of Embankments on Clay Foundations. *Canadian Geotechnical Journal*, Vol. 17, No. 2, pp. 236–260.

Terzaghi, K. 1936a. The Shearing Resistance of Saturated Soils and the Angle Between Planes of Shear. In *Proc., First International Conference on Soil Mechanics and Foundation Engineering*, Cambridge, Harvard University Press, Vol. 1, pp. 54–56.

Terzaghi, K. 1936b. Stability of Slopes of Natural Clay. In *Proc., First International Conference on Soil Mechanics and Foundation Engineering*, Cambridge, Harvard University Press, Vol. 1, pp. 161–165.

Terzaghi, K., and R.B. Peck. 1948. *Soil Mechanics in Engineering Practice*. John Wiley & Sons, New York, 566 pp.

Thiers, G.R., and H.B. Seed. 1969. Strength and Stress-Strain Characteristics of Clays Subjected to Seismic Loads. In *Symposium on Vibration Effects of Earthquakes on Soils and Foundations*, Special Technical Publication 450, ASTM, Philadelphia, Pa., pp. 3–56.

Vesic, A.S., and G.W. Clough. 1968. Behavior of Granular Materials under High Stresses. *Journal of Soil Mechanics and Foundations Division*, ASCE, Vol. 94, No. SM3, pp. 661–668.

Wallace, K.B. 1973. Structural Behaviour of Residual Soils of the Continually Wet Highlands of Papua New Guinea. *Geotechnique*, Vol. 23, No. 2, pp. 203–218.

Wroth, C.P. 1984. The Interpretation of In Situ Soil Tests. *Geotechnique*, Vol. 34, No. 4, pp. 449–489.

Wu, T.H., I-M. Lee, J.C. Potter, and O. Kjekstad. 1987a. Uncertainities in Evaluation of Strength of Marine Sand. *Journal of Geotechnical Engineering*, ASCE, Vol. 113, No. 7, pp. 719–738.

Wu, T.H., R.L. Williams, J.E. Lynch, and P.H.S.W. Kulatilake. 1987b. Stability of Slopes in Red Conemough Shale of Ohio. *Journal of Geotechnical Engineering*, ASCE, Vol. 113, No. 3, pp. 248–264.

Chapter 13

J. MICHAEL DUNCAN

SOIL SLOPE STABILITY ANALYSIS

Analyses of slopes can be divided into two categories: those used to evaluate the stability of slopes and those used to estimate slope movement. Although stability and movement are closely related, two different and distinct types of analyses are almost always used to evaluate them.

1. INTRODUCTION

Stability of slopes is usually analyzed by methods of limit equilibrium. These analyses require information about the strength of the soil, but do not require information about its stress-strain behavior. They provide no information about the magnitude of movements of the slope.

Movements of slopes are usually analyzed by the finite-element method. Understanding the stress-strain behavior of the soils is required for these analyses, and information regarding the strengths of the soils may also be required. Although these methods define movements and stresses throughout the slope, they do not provide a direct measure of stability, such as the factor of safety calculated by limit equilibrium analyses.

The focus in this chapter is on soil slopes and mechanisms involving shear failure within the soil mass. The methods described are applicable to landslides in weak rocks, where the location of the rupture surface is not controlled by existing discontinuities within the mass (see also Chapter 21). Analyses of slopes in strong rock, where instability mechanisms are controlled by existing discontinu-

ities, are described in Chapter 15. Techniques for evaluating rock strength are discussed in Chapter 14, and those for evaluating soil strength are discussed in Chapter 12.

Limit equilibrium and finite-element analyses are described in subsequent sections of this chapter. Methods that are useful for practical purposes are emphasized, and their uses are illustrated by examples.

2. MECHANICS OF LIMIT EQUILIBRIUM ANALYSES

In limit equilibrium techniques, slope stability is analyzed by first computing the factor of safety. This value must be determined for the surface that is most likely to fail by sliding, the so-called *critical slip surface*. Iterative procedures are used, each involving the selection of a potential sliding mass, subdivision of this mass into a series of slices, and consideration of the equilibrium of each of these slices by one of several possible computational methods. These methods have varying degrees of computational accuracy depending on the suitability of the underlying simplifying assumptions for the situation being analyzed.

2.1 Factor of Safety

The factor of safety is defined as the ratio of the shear strength divided by the shear stress required for equilibrium of the slope:

$$F = \frac{\text{shear strength}}{\text{shear stress required for equilibrium}} \quad (13.1)$$

which can be expressed as

$$F = \frac{c + \sigma\tan\phi}{\tau_{eq}} \quad (13.2)$$

in which

F = factor of safety,
c = cohesion intercept on Mohr-Coulomb strength diagram,
ϕ = angle of internal friction of soil,
σ = normal stress on slip surface, and
τ_{eq} = shear stress required for equilibrium.

The cohesive and frictional components of strength (c and ϕ) are defined in Chapter 12.

The factor of safety defined by Equations 13.1 and 13.2 is an overall measure of the amount by which the strength of the soil would have to fall short of the values described by c and ϕ in order for the slope to fail. This strength-related definition of F is well suited for practical purposes because soil strength is usually the parameter that is most difficult to evaluate, and it usually involves considerable uncertainty.

In discussing equilibrium methods of analysis, it is sometimes said that the factor of safety is assumed to have the same value at all points on the slip surface. It is unlikely that this would ever be true for a real slope. Because the assumption is unlikely to be true, describing this assumption as one of the fundamental premises of limit equilibrium methods seems to cast doubt on their practicality and reliability. It is therefore important to understand that Equation 13.1 defines F, and it does not involve the assumption that F is the same at all points along the slip surface for a slope not at failure. Rather, F is the numerical answer to the question, By what factor would the strength have to be reduced to bring the slope to failure by sliding along a particular potential slip surface? The answer to this question can be determined reliably by limit equilibrium methods and is of considerable practical value.

To calculate a value of F as described above, a potential slip surface must be described. The slip surface is a mechanical idealization of the surface of rupture (see Chapter 3, Section 3.1 and Table 3-3). In general, each potential slip surface results

in a different value of F. The smaller the value of F (the smaller the ratio of shear strength to shear stress required for equilibrium), the more likely failure is to occur by sliding along that surface. Of all possible slip surfaces, the one where sliding is most likely to occur is the one with the minimum value of F. This is the *critical slip surface*.

To evaluate the stability of a slope by limit equilibrium methods, it is necessary to perform calculations for a considerable number of possible slip surfaces in order to determine the location of the critical slip surface and the corresponding minimum value of F. This process is described as searching for the critical slip surface. It is an essential part of slope stability analyses.

2.2 Equations and Unknowns

All of the practically useful methods of analysis are methods of slices, so called because they subdivide the potential sliding mass into slices for purposes of analysis. Usually the slice boundaries are vertical, as shown in Figure 13-1. Two useful simplifications are achieved by subdividing the mass into slices:

1. The base of each slice passes through only one type of material, and
2. The slices are narrow enough that the segments of the slip surface at the base of each slice can be accurately represented by a straight line.

The equilibrium conditions can be considered slice by slice. If a condition of equilibrium is sat-

FIGURE 13-1
Division of potential sliding masses into slices (slip surface is idealization of surface of rupture; see Chapter 3, Section 3.1).

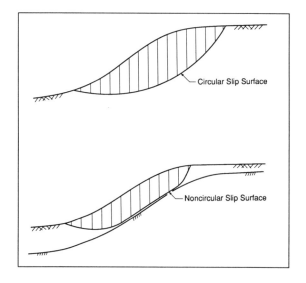

isfied for each and every slice, it is also satisfied for the entire mass. The forces on a slice are shown in Figure 13-2.

The number of equations of equilibrium available depends on the number of slices (N) and the number of equilibrium conditions that are used. As shown in Table 13-1, the number of equations available is $2N$ if only force equilibrium is to be satisfied and $3N$ if both force and moment equilibria are to be satisfied. If only force equilibrium is satisfied, the number of unknowns is $3N - 1$. If both force and moment equilibria are satisfied, the number of unknowns is $5N - 2$. In the special case in which $N = 1$, the problem is statically determinate, and the number of equilibrium equations is equal to the number of unknowns. To represent a slip surface with realistic shapes, it is usually necessary to use 10 to 40 slices, and the number of unknowns therefore exceeds the number of equations. The excess of unknowns over equations is $N - 1$ for force equilibrium analyses and $2N - 2$ for analyses that satisfy all conditions of equilibrium. Thus, such problems are statically indeterminate, and assumptions are required to make up the imbalance between equations and unknowns.

2.3 Characteristics of Methods

The various methods of limit equilibrium analysis differ from each other in two regards:

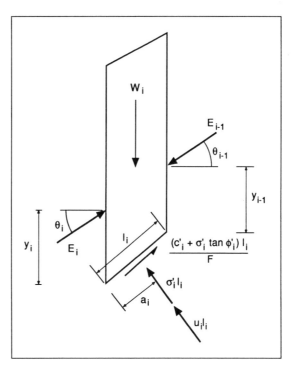

FIGURE 13-2
Forces on slice *i*.

1. Different methods use different assumptions to make up the balance between equations and unknowns, and
2. Some methods, such as the ordinary method of slices (Fellenius 1927) and Bishop's modified method (Bishop 1955), do not satisfy all conditions of equilibrium or even the conditions of force equilibrium.

Table 13-1
Equations and Unknowns in Limit Equilibrium Analysis of Slope Stability

EQUATIONS	UNKNOWNS
Methods That Satisfy Only Force Equilibrium	
N = horizontal equilibrium	N = normal forces on bases of slices
N = vertical equilibrium	$N - 1$ = side forces
	$N - 1$ = side force angles, θ
	1 = factor of safety
$2N$ total equations	$3N - 1$ total unknowns
Methods That Satisfy Both Force and Moment Equilibria	
N = horizontal equilibrium	N = normal forces on bases of slices
N = vertical equilibrium	N = locations of normal forces on bases
N = moment equilibrium	$N - 1$ = side forces
	$N - 1$ = side force angles, θ
	$N - 1$ = locations of side forces on slices
	1 = factor of safety
$3N$ total equations	$5N - 2$ total unknowns

These methods therefore involve fewer equations and unknowns than those shown in Table 13-1. The characteristics of various practically used methods with regard to the conditions of equilibrium that they satisfy, the assumptions they involve, and their computational accuracies are summarized in Table 13-2.

2.4 Computational Accuracy

"Computational accuracy" here refers to the inherent accuracy with which the various methods handle the mechanics of slope stability and the limitations on accuracy that result from the fact that the equations of equilibrium are too few to solve for the factor of safety without using assumptions. The computational accuracy involves only the accuracy with which the shear stress required for equilibrium (τ_{eq}) and the normal stress σ can be evaluated. Computational accuracy is thus distinct from overall accuracy, which involves also the accuracy with which site conditions and shear strengths can be evaluated. By separating issues that determine computational accuracy from the other issues, it is possible to answer the following important questions:

1. Do any of the various limit equilibrium methods provide accurate values of F, and
2. How are the values of F affected by the assumptions involved in the various methods?

Table 13-2
Characteristics of Commonly Used Methods of Limit Equilibrium Analysis for Slope Stability

METHOD	LIMITATIONS, ASSUMPTIONS, AND EQUILIBRIUM CONDITIONS SATISFIED
Ordinary method of slices (Fellenius 1927)	Factors of safety low—very inaccurate for flat slopes with high pore pressures; only for circular slip surfaces; assumes that normal force on the base of each slice is $W \cos \alpha$; one equation (moment equilibrium of entire mass), one unknown (factor of safety)
Bishop's modified method (Bishop 1955)	Accurate method; only for circular slip surfaces; satisfies vertical equilibrium and overall moment equilibrium; assumes side forces on slices are horizontal; $N+1$ equations and unknowns
Force equilibrium methods (Section 6.1.3)	Satisfy force equilibrium; applicable to any shape of slip surface; assume side force inclinations, which may be the same for all slices or may vary from slice to slice; small side force inclinations result in values of F less than calculated using methods that satisfy all conditions of equilibrium; large inclinations result in values of F higher than calculated using methods that satisfy all conditions of equilibrium; $2N$ equations and unknowns
Janbu's simplified method (Janbu 1968)	Force equilibrium method; applicable to any shape of slip surface; assumes side forces are horizontal (same for all slices); factors of safety are usually considerably lower than calculated using methods that satisfy all conditions of equilibrium; $2N$ equations and unknowns
Modified Swedish method (U.S. Army Corps of Engineers 1970)	Force equilibrium method, applicable to any shape of slip surface; assumes side force inclinations are equal to the inclination of the slope (same for all slices); factors of safety are often considerably higher than calculated using methods that satisfy all conditions of equilibrium; $2N$ equations and unknowns
Lowe and Karafiath's method (Lowe and Karafiath 1960)	Generally most accurate of the force equilibrium methods; applicable to any shape of slip surface; assumes side force inclinations are average of slope surface and slip surface (varying from slice to slice); satisfies vertical and horizontal force equilibrium; $2N$ equations and unknowns
Janbu's generalized procedure of slices (Janbu 1968)	Satisfies all conditions of equilibrium; applicable to any shape of slip surface; assumes heights of side forces above base of slice (varying from slice to slice); more frequent numerical convergence problems than some other methods; accurate method; $3N$ equations and unknowns
Spencer's method (Spencer 1967)	Satisfies all conditions of equilibrium; applicable to any shape of slip surface; assumes that inclinations of side forces are the same for every slice; side force inclination is calculated in the process of solution so that all conditions of equilibrium are satisfied; accurate method; $3N$ equations and unknowns
Morgenstern and Price's method (Morgenstern and Price 1965)	Satisfies all conditions of equilibrium; applicable to any shape of slip surface; assumes that inclinations of side forces follow a prescribed pattern, called $f(x)$; side force inclinations can be the same or can vary from slice to slice; side force inclinations are calculated in the process of solution so that all conditions of equilibrium are satisfied; accurate method; $3N$ equations and unknowns
Sarma's method (Sarma 1973)	Satisfies all conditions of equilibrium; applicable to any shape of slip surface; assumes that magnitudes of vertical side forces follow prescribed patterns; calculates horizontal acceleration for barely stable equilibrium; by prefactoring strengths and iterating to find the value of the prefactor that results in zero horizontal acceleration for barely stable equilibrium, the value of the conventional factor of safety can be determined; $3N$ equations, $3N$ unknowns

Studies of computational accuracy have shown the following:

1. If the method of analysis satisfies all conditions of equilibrium, the factor of safety will be accurate within ±6 percent. This conclusion is based on the finding that factors of safety calculated using methods that satisfy all conditions of equilibrium never differ by more than 12 percent from each other or ±6 percent from a central value as long as the methods involve reasonable assumptions. The methods of Morgenstern and Price (1965), Spencer (1967), and Sarma (1973) and the generalized procedure of slices (GPS) (Janbu 1968) satisfy all conditions of equilibrium and involve reasonable assumptions. Studies have shown that values of F calculated using these methods differ by no more than 6 percent from values calculated using the log spiral method and the finite-element method, which satisfy all conditions of equilibrium but are not methods of slices.
2. Bishop's modified method is a special case. Although it does not satisfy all conditions of equilibrium, it is as accurate as methods that do. It is limited to circular slip surfaces.
3. No matter what method of analysis is used, it is essential to perform a thorough search for the critical slip surface to ensure that the minimum factor of safety has been calculated.

2.5 Checking Accuracy of Analyses

When slope stability analyses are performed, it is desirable to have an independent check of the results as a guard against mistakes. Unfortunately, computations using methods that satisfy all conditions of equilibrium are lengthy and complex, too involved for hand calculation. It is therefore more practical to use other, simpler types of analyses that can be performed by hand to check computer analyses.

When analyses are performed using circular slip surfaces, the factor of safety for the critical circle can be checked approximately by using the ordinary method of slices or Bishop's modified method. The ordinary method of slices gives values of F that are lower than values calculated by more accurate methods. For total stress analyses the difference is seldom more than 10 percent, but it may reach 50 percent for effective stress analyses

under conditions involving high pore pressures. Bishop's modified method, only slightly more time-consuming than the ordinary method of slices, is as accurate as the methods that satisfy all conditions of equilibrium.

Charts provide a simple and effective means of checking slope stability analyses. Although approximations and simplifications are always necessary when charts are used, it is usually possible to calculate the factor of safety with sufficient accuracy to afford a useful check on the results of more detailed analyses. It is also useful to perform chart analyses before detailed computer analyses in order to get the best possible understanding of the problem.

When analyses are performed using noncircular slip surfaces, the results can be checked by hand for the critical slip surface using force equilibrium analyses. If Spencer's method is used for the computer analyses, the results for the critical slip surface can be checked by hand using force equilibrium analyses with the same side force inclination. If Morgenstern and Price's method or Janbu's GPS is used for the computer analyses, the force equilibrium hand calculations can be performed using an average side force inclination determined from the computer analyses.

Another means of checking computer analyses of slope stability is to perform independent analyses using another computer program and completely separate input. It is always desirable to have a different person perform the check analyses to make them as independent as possible.

3. DRAINED AND UNDRAINED CONDITIONS

Slope failures may occur under drained or undrained conditions in the soils that make up the slope. If instability is caused by changes in loading, such as removal of material from the bottom of the slope or increase in loads at the top, soils that have low values of permeability may not have time to drain during the length of time in which the loads are changed; thus they may contain unequalized excess pore pressures leading to slope failure. In that case, these low-permeability soils are said to be undrained. Under the same rates of loading, soils with higher permeabilities may have time to drain and will have no significant excess pore pressures.

One of the most important determinations needed for slope stability analyses is that of the drainage conditions in the various soils that form the slope. A rational way of making this determination is to calculate the value of the dimensionless time factor T using the following equation:

$$T = \frac{C_v t}{D^2} \qquad (13.3)$$

where

T = dimensionless time factor,
C_v = coefficient of consolidation (length squared per unit of time),
t = time for drainage to occur (units of time), and
D = length of drainage path (units of length).

D is the distance water would have to flow to drain from the soil zone. Usually D is defined as half the thickness of the soil layer or zone. If the soil zone is drained on only one side, D is equal to the thickness of the zone.

If the value of T is 3.0 or more, the soil zone will drain as rapidly as the loads are applied, and the soil within the zone can be treated as fully drained in the analysis. If the value of T is 0.01 or less, very little drainage will occur during the loading period, and the soil within the zone can be treated as completely undrained in the analysis. If the value of T is between 0.01 and 3.0, partial drainage will occur during the period of time when the loads are changing. In that case both drained and undrained conditions should be analyzed to bound the problem. For normal rates of loading, involving weeks or months, soils with permeabilities greater than 10^{-4} cm/sec are usually drained, and soils with permeabilities less than 10^{-7} cm/sec are usually undrained. Silts, which have permeabilities between 10^{-4} and 10^{-7} cm/sec, are often partially drained.

If instability in a slope is caused by increasing pore-water pressures *within* the slope, failure occurs under drained conditions by definition because adjustment of internal pore pressures in equilibrium with the flow boundary conditions is drainage. Changes in internal water pressures are often the cause of instability in natural slopes, where landslides almost always occur during wet periods. Such problems should be analyzed as drained conditions.

4. TOTAL STRESS AND EFFECTIVE STRESS ANALYSES

Stability of slopes can be analyzed using either effective stress or total stress methods:

- In effective stress analyses the shear strength of the soil is related to the effective normal stress on the potential slip surface by means of effective stress shear strength parameters. Pore pressures within the soil must be known and are part of the information required for analysis.
- In total stress analyses the shear strength of the soil is related to the total normal stress on the potential slip surface by means of total stress shear strength parameters. Pore pressures within the soil mass need not be known and are not required as input for analysis.

In principle, it is always possible to analyze stability by using effective stress methods because the strengths of soils are governed by effective stresses under both undrained and drained conditions. In practice, however, it is virtually impossible to determine accurately what excess pore pressures will result from changes in external loading on a slope (excavation, fill placement, or change in external water level). Because the excess pore pressures for these loading conditions cannot be estimated accurately, it is not possible to perform accurate analyses of stability for these conditions using effective stress procedures.

Total stress analyses for soils that do not drain during the loading period involve a simple principle: if an element of soil in the laboratory (a laboratory test specimen) is subjected to the same changes in stress under undrained conditions as an element of the same soil would experience in the field, the same excess pore pressures will develop. Thus, if the total stresses in the laboratory and the field are the same, the effective stresses will also be the same. Because soil strength is controlled by effective stresses, the strength measured in laboratory tests should be the same as the strength in the field if the pore pressures and total stresses are also the same. Thus, under undrained conditions, strengths can be related to total stresses, obviating the need to specify undrained excess pore pressures.

Although the principle outlined in the preceding paragraph is reasonably simple, experience has shown that many factors influence the pore pressures that develop under undrained loading. These factors include degree of saturation, density, stress history of the soil, rate of loading, and the magnitudes and orientations of the applied stresses. As a result, determining total stress-related undrained strengths by means of laboratory or in situ testing is not a simple matter; considerable attention to detail is required if reliable results are to be achieved. Shear strengths for use in undrained total stress analyses must be measured using test specimens and loading conditions that closely duplicate the conditions in the field. Still, using total stress procedures for analysis of undrained conditions is more straightforward and more reliable than trying to predict undrained excess pore pressures for use in effective stress analyses of undrained conditions.

If a slope consists of soils of widely differing permeabilities, it is possible that the more permeable soils would be drained whereas the less permeable soils would be undrained. In such cases it is logical and permissible in the same analysis to treat the drained soils in terms of effective stresses and the undrained soils in terms of total stresses.

4.1 Shear Strengths

Shear strengths for use in drained effective stress analyses can be measured in two ways:

1. In laboratory or field tests in which loads are applied slowly enough so that the soil is drained (there are no excess pore pressures) at failure, or
2. In laboratory tests such as consolidated-undrained (C-U) laboratory triaxial tests in which pore pressures are measured and the effective stresses at failure can be determined.

Effective stress strength envelopes determined using these two methods have been found to be, for all practical purposes, the same (Bishop and Bjerrum 1960).

Studies by Skempton and his colleagues (Skempton 1970, 1977, 1985) showed that peak drained strengths of stiff, overconsolidated clays determined in laboratory tests are larger than the drained strengths that can be mobilized in the field over a long period of time. Skempton recommended the use of "fully softened" strengths for stiff clays in which there has been no previous sliding. The fully softened strength is measured by remolding the clay in the laboratory at a water content near the liquid limit, reconsolidating it in the laboratory, and measuring its strength in a normally consolidated condition.

Once sliding has occurred in a clay, the clay particles become reoriented parallel to the slip surface, and the strength decreases progressively as sliding

displacement occurs, eventually reaching a low residual value. Residual shear strengths should be used to analyze clay slopes in which slope failure by sliding has already occurred (Skempton 1970, 1977, 1985).

Morgenstern (1992) and others have pointed out that drained shear may cause structural collapse in certain sensitive or structured soils, resulting in development of excess pore pressures at a rate that is so rapid that drainage cannot occur. As a result, it is not possible to mobilize the full effective stress shearing resistance under drained conditions. These difficult soils (loose sands and sensitive and highly structured clays) deserve special attention (see Chapter 24 for sensitive clays).

For soils that are partly saturated, such as compacted clays or naturally occurring clayey soils from above the water table, undrained strengths should be measured using unconsolidated-undrained tests on specimens with the same void ratio and the same degree of saturation as the soil in the field. The undrained strength envelope for such soils is curved, and as a result the values of total stress c and ϕ from such tests are not unique. It is important therefore to use a range of confining pressures in the laboratory tests that corresponds to the range of pressures to which the soil is subjected in the field and to select values of c and ϕ that provide a reasonable representation of the strength of the soil in this pressure range. Alternatively, with some slope stability computer programs, it is possible to use a nonlinear strength envelope, represented by a series of points, without reference to c and ϕ.

For soils that are completely saturated, the undrained friction angle is zero ($\phi_u = 0$). Undrained strengths for saturated soils can be determined from unconsolidated-undrained tests or from C-U tests by the Stress History and Normalized Soil Engineering Properties (SHANSEP) procedure (Ladd and Foott 1974).

Unconsolidated-undrained tests are performed on undisturbed test specimens from the field. These tests define the undrained strength of the soil in the field at the time and location where the samples were obtained, provided that the samples are undisturbed. Although some procedures provide samples that are called "undisturbed," no sample is completely free of disturbance effects. For saturated clays, these effects usually cause the undrained strengths measured in laboratory tests to be smaller than the undrained strength in the field. As a result, unconsolidated-undrained triaxial tests usually give conservative values of undrained strengths.

Ladd and Foott (1974) and others have shown that the effects of disturbance on undrained strength can be compensated for by consolidating clay specimens to higher pressures in the laboratory. Because consolidation to higher pressures produces higher undrained shear strengths, the strengths measured using C-U tests are not directly applicable to field conditions. Ladd and Foott described procedures for determining values of S_u/p (the ratio of undrained strength divided by consolidation pressure) from such tests. The undrained strength in the field is estimated by multiplying the value of S_u/p by the field consolidation pressure. The same procedure can be used to estimate current undrained strengths (using current consolidation pressures) or future undrained strengths (using consolidation pressures estimated for some future condition by means of consolidation analysis).

C-U tests should not be used to determine values of c and ϕ from "total stress" Mohr-Coulomb strength envelopes for use in total stress analyses. This procedure is not valid because the normal stresses used to plot the Mohr's circles of stress at failure are not valid total stresses. These stress circles are plotted using the effective stress at the time of consolidation plus the deviator stress at failure. This procedure is not consistent. The method described by Ladd and Foott avoids the inconsistencies of this method by relating undrained strengths measured in C-U tests to consolidation pressure without the use of c- and ϕ-values. The values of undrained strength (S_u), determined as recommended by Ladd and Foott, are combined with $\phi_u = 0$, the correct undrained friction angle for saturated materials.

4.2 Unit Weights and Water Pressures

The primary requirement in slope stability analyses is to satisfy equilibrium *in terms of total stresses*. This is true whether the analysis is an effective stress analysis (in which shear strengths are related to effective stresses) or a total stress analysis (in which shear strengths are related to total stresses). In either case total stress is the prime variable in terms of the fundamental mechanics of the problem because a correct evaluation of equilibrium

conditions must include both earth and water forces.

For analyses in terms of effective stress, pore pressures are subtracted from total stresses on the base of each slice to determine the values of effective stress, to which the soil strengths are related. For total stress analyses, it is not necessary to subtract pore pressures because the strengths are related to the total stresses.

Slope stability problems can be correctly formulated so that equilibrium is satisfied in terms of total stresses by using total unit weights and external boundary water pressures. Total unit weights are moist unit weights above the water table and saturated unit weights below the water table. When water pressures act on submerged external slope boundaries, these pressures are a component of total stress, and they must be included for correct evaluation of equilibrium in terms of total stresses.

For hydrostatic (no-flow) water conditions, it is possible to perform effective stress analyses by using buoyant unit weights below the water table and ignoring external boundary water pressures and internal pore pressures. Effective stress analyses of nonhydrostatic conditions can be performed in a similar way if seepage forces are also included. These procedures are not as straightforward as using total unit weights and boundary water pressures, however. They involve essentially the same amount of effort in calculation for hydrostatic conditions and much more calculation when seepage forces have to be evaluated. Thus, these procedures are not useful except for one purpose. Using buoyant unit weights and excluding external boundary water pressures and pore pressures for a hydrostatic condition afford a means of checking some of the functions of a computer program.

5. SLOPE STABILITY CHARTS

The accuracy of charts for slope stability analyses is usually at least as good as the accuracy with which shear strengths can be evaluated. Chart analyses can be accomplished in a few minutes, even when shear strengths and unit weights are carefully averaged to achieve the best possible accuracy. Charts thus provide a rapid and potentially very useful means for slope stability analyses. They can be used to good advantage to perform preliminary analyses, to check detailed analyses, and often to make complete analyses.

5.1 Averaging Slope Profile, Shear Strengths, and Unit Weights

For simplicity, charts are developed for simple homogeneous slopes. To apply them to real conditions, it is necessary to approximate the real slope with an equivalent simple and homogeneous slope.

The most effective method of developing a simple slope profile for chart analysis is to begin with a cross section of the slope drawn to scale. On this cross section, using judgment, the engineer draws a geometrically simple slope that approximates the real slope as closely as possible.

To average the shear strengths for chart analysis, it is useful to know, at least approximately, the location of the critical slip surface. The charts contained in the following sections of this chapter provide a means of estimating the position of the critical circle. Average strength values are calculated by drawing the critical circle, determined from the charts, on the slope. Then the central angle of arc subtended within each layer or zone of soil is measured with a protractor. The central angles are used as weighting factors to calculate weighted average strength parameters, c_{avg} and ϕ_{avg}:

$$c_{avg} = \frac{\Sigma \delta_i c_i}{\Sigma \delta_i} \qquad (13.4)$$

$$\phi_{avg} = \frac{\Sigma \delta_i \phi_i}{\Sigma \delta_i} \qquad (13.5)$$

where

c_{avg} = average cohesion (stress units);
ϕ_{avg} = average angle of internal friction (degrees);
δ_i = central angle of arc, measured around the center of the estimated critical circle, within zone i (degrees);
c_i = cohesion in zone i (stress units); and
ϕ_i = angle of internal friction in zone i.

The one condition in which it is preferable not to use these straightforward averaging procedures is the case in which an embankment overlies a weak foundation of saturated clay, with $\phi = 0$. Straightforward averaging in such a case would lead to a small value of ϕ_{avg} (perhaps 2 to 5 degrees), and with $\phi_{avg} > 0$, it would be necessary to use a chart like the one in Figure 13-6, which is based entirely on circles that pass through the toe of the slope, often not critical for embankments on weak satu-

rated clays. With $\phi = 0$ foundation soils, the critical circle usually goes as deep as possible in the foundation, passing below the toe of the slope. In these cases it is better to approximate the embankment as a $\phi = 0$ soil and to use a $\phi = 0$ chart of the type shown in Figure 13-3. The equivalent $\phi = 0$ strength of the embankment soil can be estimated by calculating the average normal stress on the part of the slip surface within the embankment (one-half the average vertical stress is usually a reasonable approximation) and determining the corresponding shear strength at that point on the shear strength envelope for the embankment soil. This value of strength is treated as a value of S_u for the embankment, with $\phi = 0$. The average value of S_u is then calculated for both the embankment and the foundation using the same averaging procedure as described above:

$$(S_u)_{avg} = \frac{\Sigma \delta_i (S_u)_i}{\Sigma \delta_i} \qquad (13.6)$$

where $(S_u)_{avg}$ is the average undrained shear strength (in stress units), $(S_u)_i$ is S_u in layer i (in stress units), and δ_i is as defined previously. This average value of S_u is then used, with $\phi = 0$, for analysis of the slope.

To average unit weights for use in chart analysis, it is usually sufficient to use layer thickness as a weighting factor, as indicated by the following expression:

$$\gamma_{avg} = \frac{\Sigma \gamma_i h_i}{\Sigma h_i} \qquad (13.7)$$

where

γ_{avg} = average unit weight (force per length cubed),
γ_i = unit weight of layer i (force per length cubed), and
h_i = thickness of layer i (in length units).

Unit weights should be averaged only to the depth of the bottom of the critical circle.

If the material below the toe of the slope is a $\phi = 0$ material, the unit weight should be averaged only down to the toe of the slope, since the unit weight of the material below the toe has no effect on stability in this case.

5.2 Soils with $\phi = 0$

Stability charts for slopes in $\phi = 0$ soils are shown in Figure 13-3. These charts, as well as the related ones in Figures 13-4, 13-5, and 13-6, were developed by Janbu (1968).

The coordinates of the center of the critical circle are given by

$$X_0 = x_0 H \qquad (13.8)$$

$$Y_0 = y_0 H \qquad (13.9)$$

where

X_0, Y_0 = coordinates of critical circle center measured from toe of slope,

x_0, y_0 = dimensionless numbers determined from charts at bottom of Figure 13-3, and

H = height of slope.

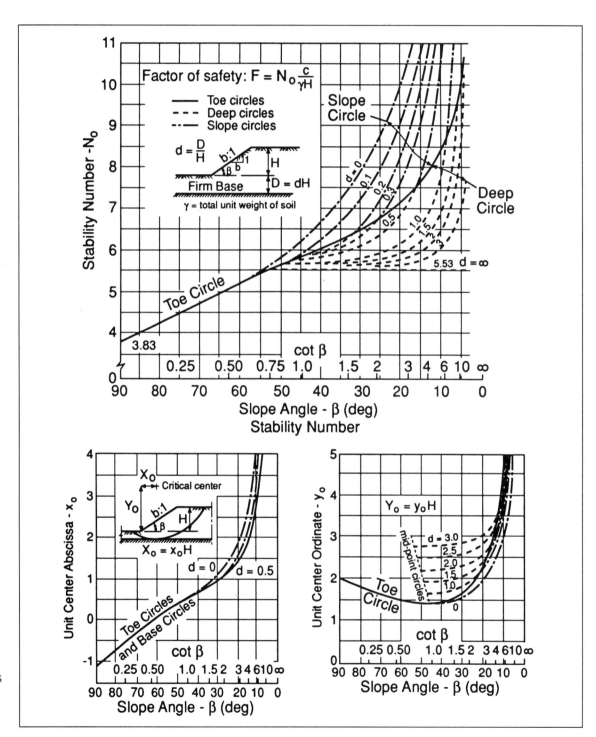

FIGURE 13-3
Slope stability charts for $\phi = 0$ soils (modified from Janbu 1968).

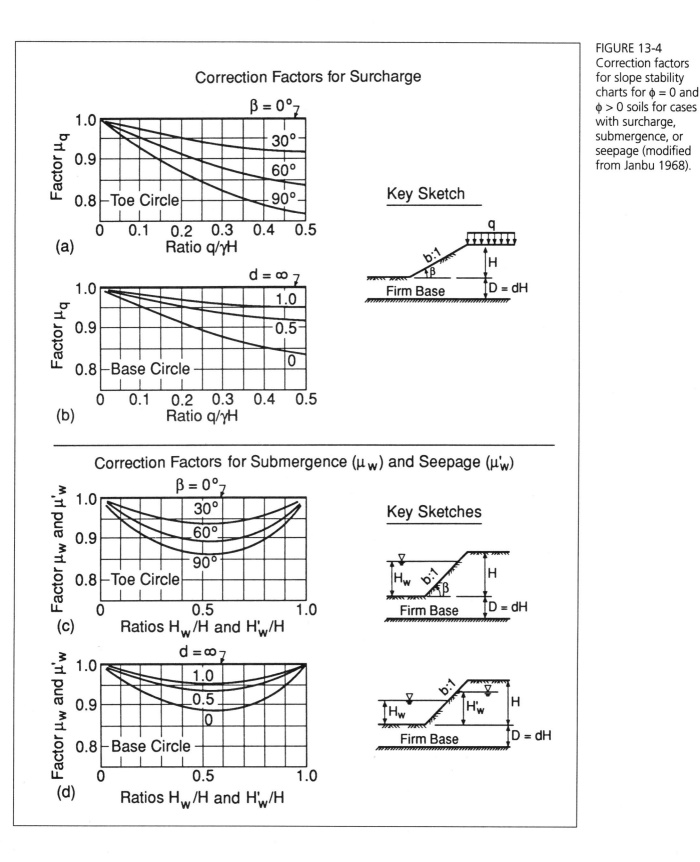

FIGURE 13-4 Correction factors for slope stability charts for $\phi = 0$ and $\phi > 0$ soils for cases with surcharge, submergence, or seepage (modified from Janbu 1968).

FIGURE 13-5
Correction factors
for slope stability
charts for φ = 0 and
φ > 0 soils for cases
with tension cracks
(modified from
Janbu 1968).

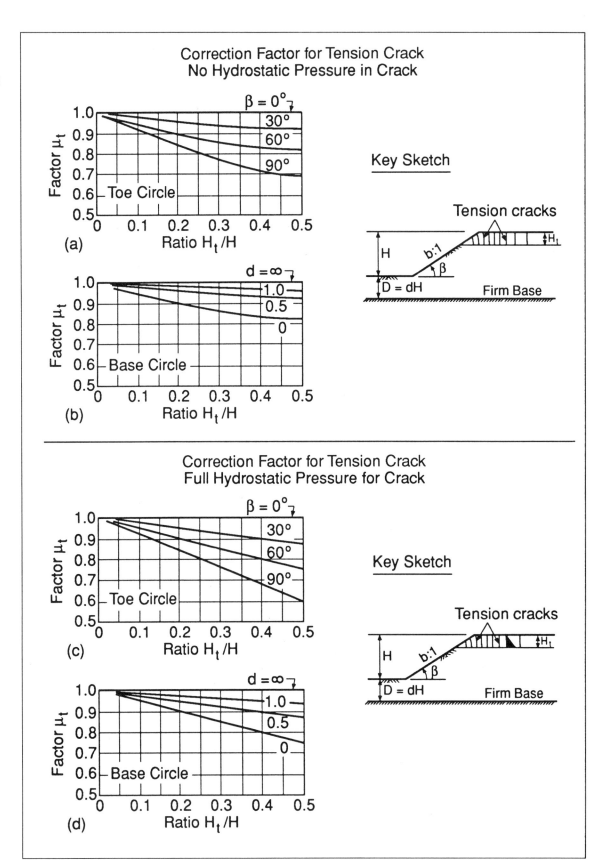

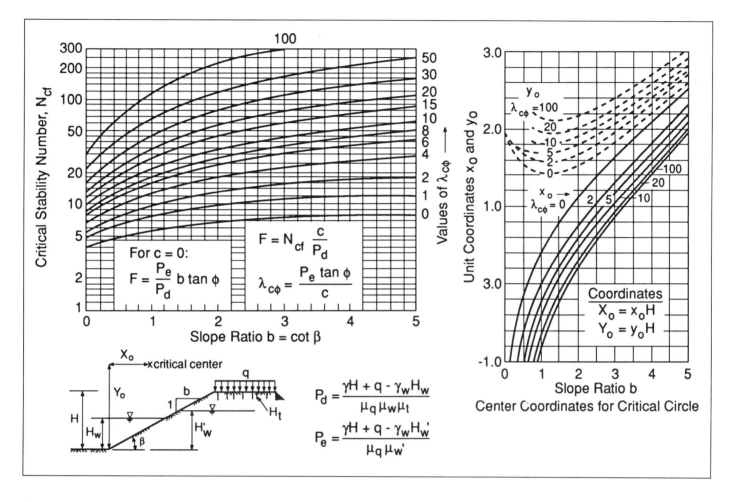

The radius of the critical circle is determined by the position of the center of the circle and the fact that the circle is tangent to a horizontal plane at depth D below the bottom of the slope. Using the chart in Figure 13-3, it is possible to calculate factors of safety for a range of depths (D) to determine which is most critical. If the value of S_u in the foundation does not vary with depth, the critical circle will extend as deep as possible, and only a circle extending through the full depth of the clay layer need be analyzed. If the strength of the foundation clay increases with depth, the critical circle may or may not extend to the base of the layer, and it is necessary to examine various depths to determine which is most critical.

The upper part of Figure 13-3 gives values of the stability number, N_0, which is related to the factor of safety by the expression

$$F = N_0 \frac{c}{P_d} \qquad (13.10)$$

where

F = factor of safety (dimensionless),
N_0 = stability number from Figure 13-3 (dimensionless), and
$c = S_u$ = undrained shear strength (in stress units).

The value of P_d is given by the expression

$$P_d = \frac{\gamma H + q - \gamma_w H_w}{\mu_q \mu_w \mu_t} \qquad (13.11)$$

where

P_d = driving force term (in pressure units),
γ = unit weight (force per length cubed),
H = slope height (length),
q = surcharge pressure on top of slope (in stress units),
γ_w = unit weight of water (force per length cubed),
H_w = depth of water outside slope, measured above toe (length),
μ_q = surcharge correction factor,
μ_w = water pressure correction factor, and
μ_t = tension crack correction factor.

FIGURE 13-6
Slope stability charts for $\phi > 0$ soils (modified from Janbu 1968).

The correction factors μ_q, μ_w, μ_t are given in Figures 13-4 and 13-5.

5.3 Soils with $\phi > 0$

In most cases for slopes in soils with $\phi > 0$, the critical slip circle passes through the toe of the slope. The stability chart shown in Figure 13-6 is based on the assumption that this is true. The coordinates of the center of the critical toe circle are shown in the chart on the right side of Figure 13-6.

Cohesion values, friction angles, and unit weights are averaged using the procedures described at the beginning of this chapter. The factor of safety is calculated using the expression

$$F = N_{cf} \frac{c}{P_d} \qquad (13.12)$$

where N_{cf} is the stability coefficient from the chart on the left-hand side of Figure 13-6 (dimensionless), c is the average cohesion intercept (in stress units), and P_d is defined by Equation 13.11.

The values of N_{cf} shown in Figure 13-6 are determined by the value of b = cotangent of the

slope angle = cot ß and the value of the dimensionless parameter, $\lambda_{c\phi}$. The latter is calculated using the expression

$$\lambda_{c\phi} = \frac{P_e \tan\phi}{c} \qquad (13.13)$$

where c and $\tan\phi$ are the average values of cohesion intercept and friction angle. The value of P_e is defined as

$$P_e = \frac{\gamma H + q - \gamma_w H_w'}{\mu_q \mu_w'} \qquad (13.14)$$

where H_w' is the effective average water level inside the slope measured above the toe of the slope (in length units), μ_w' is the seepage correction factor from the lower part of Figure 13-4, and γ, H, q, and μ_q are as defined previously.

The value of H_w', somewhat difficult to estimate using judgment, is related to H_c, the water level measured below the crest of the slope, as shown in Figure 13-7. When the value of H_c has been determined by field measurements or seepage analyses, the value of H_w' can be determined using the curve given in Figure 13-7.

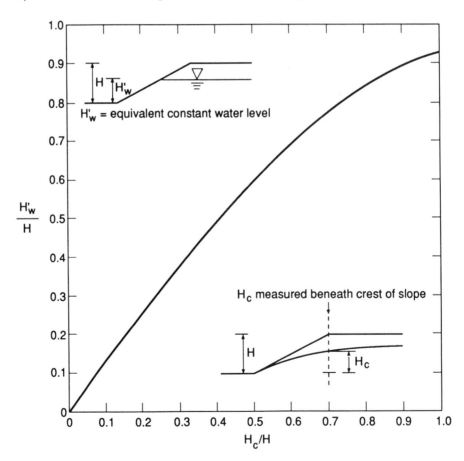

FIGURE 13-7
Chart for determining steady seepage conditions (modified from Duncan et al. 1987).

5.4 Infinite Slope Analyses

Conditions are sometimes encountered in which a layer of firm soil or rock lies parallel to the surface of the slope at shallow depth. In such conditions the slip surface is constrained to parallel the slope, as shown in Figure 13-8. When such slip surfaces are long compared with their depth, they can be approximated accurately by infinite slope analyses. Such analyses ignore the driving force at the upper end of the slide mass and the resisting force at the lower end. The resisting force is ordinarily greater, and infinite slope analyses are therefore somewhat conservative.

The factor of safety for infinite slope analyses can be expressed as

$$F = A \frac{\tan\phi'}{\tan\beta} + B \frac{c'}{\gamma H} \qquad (13.15)$$

where

A, B = dimensionless stability coefficients given in Figure 13-8,
ϕ', c' = effective stress strength parameters for slip surface,
β = slope angle,
γ = unit weight of sliding mass (force per length cubed), and

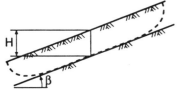

γ = total unit weight of soil
γ_w = unit weight of water
c' = cohesion intercept ⎱ Effective
ϕ' = friction angle ⎰ Stress
r_u = pore pressure ratio = $\frac{u}{\gamma H}$
u = pore pressure at depth H

Steps:

① Determine r_u from measured pore pressures or formulas at right

② Determine A and B from charts below

③ Calculate $F = A \frac{\tan\phi'}{\tan\beta} B \frac{c'}{\gamma H}$

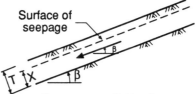

Surface of seepage

Seepage parallel to slope
$$r_u = \frac{X}{T} \frac{\gamma_w}{\gamma} \cos^2\beta$$

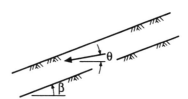

Seepage emerging from slope
$$r_u = \frac{\gamma_w}{\gamma} \frac{1}{1+\tan\beta \tan\phi}$$

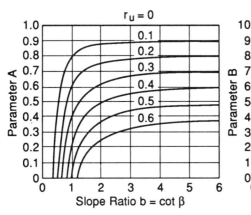

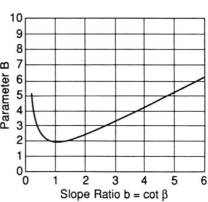

FIGURE 13-8
Stability charts for infinite slopes (modified from Duncan et al. 1987).

H = depth measured vertically from slope surface to slip surface (in length units).

The charts shown in Figure 13-8 can be used for effective stress or total stress analyses. For effective stress analyses, the pore pressures along the slip surface are characterized by the dimensionless pore-pressure ratio, r_u, as indicated in Figure 13-8. For total stress analyses, c and ϕ (total stress shear strength parameters) are used instead of c' and ϕ', and the value of r_u is equal to zero.

5.5 Soils with ϕ = 0 and Strength Increasing with Depth

The undrained shear strengths of normally consolidated clays usually increase with depth. The strength profile for this condition can often be approximated by a straight line, as shown in Figure 13-9. The factor of safety for slopes in such deposits can be expressed as

$$F = N \frac{c_b}{\gamma (H + H_0)} \tag{13.16}$$

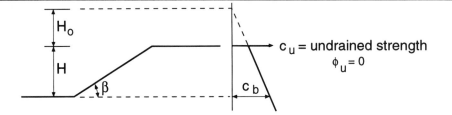

Steps:

① Extrapolate strength profile upward to determine value of H_0, where strength profile intersects

② Calculate $M = H_0/H$

③ Determine stability number N from chart below

④ Determine c_b = strength of bottom of slope

⑤ Calculate $F = N \dfrac{c_b}{\gamma (H + H_0)}$

Use $\gamma = \gamma_{buoyant}$ for submerged slope

Use $\gamma = \gamma_m$ for no water outside slope

Use reduced value of γ for partly submerged slope:

$$\gamma_r = \frac{\gamma H - \gamma_w H_w}{H}$$

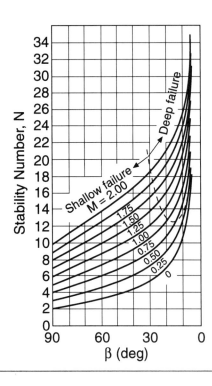

FIGURE 13-9
Slope stability charts for ϕ = 0 and strength increasing with depth (modified from Hunter and Schuster 1968).

where

N = dimensionless stability coefficient from Figure 13-9,

c_b = undrained shear strength (S_u) at elevation of toe of slope (in stress units),

γ = unit weight of soil (force per length cubed),

H = slope height (length), and

H_0 = height above top of slope where strength profile, projected upward, intersects zero strength (in length units).

Figure 13-9 can be used to analyze subaerial slopes or slopes that are submerged in water. For subaerial slopes the total (moist or saturated) unit weight is used. For submerged slopes the buoyant unit weight ($\gamma_b = \gamma_{sat} - \gamma_w$) is used. A reduced value of γ, calculated using the following expression, can be used for partly submerged slopes:

$$\gamma_r = \frac{\gamma H - \gamma_w H_w}{H} \qquad (13.17)$$

where γ_r is the value of unit weight reduced for partial submergence, and the other terms are as defined previously.

6. DETAILED ANALYSES

Detailed analyses of slope stability are needed whenever it is necessary to include details of slope configuration, soil property zonation, or shape of the slip surface that cannot be represented in chart analyses. If a computer and slope stability computer program are available, detailed analyses can be performed efficiently, including a thorough search for the critical slip surface. If a computer is not available, detailed analyses can be performed manually using the procedures described in the following section. Even when analyses are performed using a computer, it is good practice to check the factor of safety for the critical slip surface using manual calculations. The U.S. Army Corps of Engineers (1970) requires such calculations for the critical slip surface for all important slope stability analyses.

6.1 Manual Calculations

It is convenient to use a tabular computation form to perform manual calculations of slope stability. Calculations that are well organized are easy to check. In the following sections, tabular computation forms are given for the ordinary method of

slices, Bishop's modified method, and a force equilibrium method with the same side force inclination on all slices.

6.1.1 Ordinary Method of Slices

A tabular form for the ordinary method of slices is presented in Figure 13-10, and example calculations made using this form are given in Figure 13-11. The slope to which the calculations apply is shown in Figure 13-12.

The slice weights in the second column were calculated on the basis of the dimensions of the slices and the unit weights of the soils within them. A simple method of calculating slice weights, shown in Figure 13-12, uses

$$W = b\Sigma(\gamma_i h_i) \qquad (13.18)$$

where

W = slice weight (force per unit length),

b = width of slice (length),

γ_i = unit weight of soil i (force per length cubed), and

h_i = height of soil layer i where it is subtended by slice (length).

Values of h_i are measured at the center of the slice. Equation 13.18 can be applied to triangular as well as quadrilateral slices.

The base length l and the base angle α for each slice are measured on a scale drawing of the slope and slip surface. The values of c and ϕ for each slice correspond to the type of soil at the bottom of the slice, and slices are drawn so that the base of each slice is in only one type of soil. The value of u for each slice is the average value at the middle of the base. Slices need not be of equal width.

The quantities in the last two columns (N_1 and N_2) can be calculated using the tabulated values for each slice. These two columns are summed, and the factor of safety is calculated as shown in the bottom right corner of the computation form (Figure 13-11). For the example shown, the ordinary method of slices factor of safety is 1.43.

6.1.2 Bishop's Modified Method

A tabular computation form for Bishop's modified method is shown in Figure 13-13, and calculations

Slice No.	W	l	α	c	ϕ	u	N_1	N_2
1								
2								
3								
4								
5								
6								
7								
8								
9								
10								
11								
12								
13								
14								
15								
						$\Sigma =$		

W = weight of slice - kN/m

c = cohesion intercept - kN/m^2

ϕ = friction angle - degrees

u = pore pressure - kN/m^2

α = angle between base of slice and horizontal - degrees

l = length of slip surface segments measured along base of slice - m

$N_1 = W \sin \alpha$

$N_2 = \left[W \cos \alpha - ul \right] \tan \phi + cl$

$F = \Sigma (N_2) / \Sigma (N_1) = \underline{\hspace{1cm}}$

FIGURE 13-10
Tabular computation form for ordinary method of slices.

for the example slope shown in Figure 13-12 are given in Figure 13-14. The first eight columns in the form are the same as those for the ordinary method of slices.

The value of the quantity N_2 for Bishop's modified method depends on the factor of safety, as shown by the expression for N_2 at the bottom of the form.

Because N_2 depends on the factor of safety and the factor of safety also depends on N_2, it is necessary to use repeated trials to calculate the factor of safety by Bishop's modified method. For this reason, four columns are shown for N_2, each corresponding to a different value of the assumed factor of safety, F_a. The first value assumed for F_a was 1.43, the value calculated using the ordinary method of slices; this value is shown at the top of the ninth

column (Figure 13-14). Using the resulting values of N_2 (the values in the ninth column), the value of F_c (the calculated factor of safety) was found to equal 1.51. When this value was used as F_a, F_c was 1.52. When 1.52 was used as the value of F_a, F_c was also 1.52, indicating that the assumed value was correct and no further iteration was needed.

It can be seen that the value of F calculated using Bishop's modified method (1.52) is larger than the value calculated using the ordinary method of slices (1.43), as is usually the case. For effective stress analyses with high pore pressures, the difference may be much larger than the 6 percent difference in this case, indicating significant inaccuracy in the ordinary method of slices for effective stress analyses with high pore pressures.

Slice No.	W	l	α	c	φ	u	N_1	N_2
1	112	5.3	-32.0	35.9	0.0	0	-60	191
2	297	4.9	-22.0	35.9	0.0	0	-111	177
3	499	4.7	-13.0	35.9	0.0	0	-112	169
4	726	4.6	-4.0	35.9	0.0	0	-51	165
5	903	4.6	4.0	35.9	0.0	0	63	165
6	1,028	4.7	13.0	35.9	0.0	0	231	169
7	1,003	4.9	22.0	35.9	0.0	0	376	177
8	818	5.3	32.0	35.9	0.0	0	433	191
9	587	6.7	43.0	4.8	35.0	0	400	333
10	128	5.6	55.0	4.8	35.0	0	105	79
11							0	0
12							0	0
13							0	0
14							0	0
15							0	0
						$\Sigma =$	1274	1816

W = weight of slice - kN/m

c = cohesion intercept - kN/m^2

φ = friction angle - degrees

u = pore pressure - kN/m^2

α = angle between base of slice and horizontal - degrees

l = length of slip surface segments measured along base of slice - m

$N_1 = W \sin \alpha$

$N_2 = [W \cos \alpha - ul] \tan \phi + cl$

$F = \Sigma (N_2) / \Sigma (N_1) = \underline{1.43}$

Numerical problems sometimes arise with Bishop's modified method. These difficulties arise at the top or bottom ends of the slip circle, where the bases of the slices are steep. When these numerical problems arise, the value of the factor of safety calculated by Bishop's modified method may be smaller than the value calculated by the ordinary method of slices, rather than larger. When this happens, it is reasonable to use the F-value determined by the ordinary method of slices rather than that determined by Bishop's modified method. To check for these numerical problems, it is useful to calculate F by the ordinary method of slices to determine whether it is smaller or larger than the value from Bishop's modified method.

6.1.3 Force Equilibrium with Constant Side Force Inclination

The ordinary method of slices and Bishop's modified method can only be used to analyze circular slip surfaces. It is useful to have a method that can be used for manual calculations of noncircular slip surfaces in order to check computer analyses or to perform analyses when a computer is not available. A tabular form for force equilibrium analyses with the same side force inclination on every slice is shown in Figure 13-15, and an example of the use of this form is shown in Figure 13-16. Figure 13-17 shows the slope analyzed in Figure 13-16, a sand embankment on a clay foundation.

FIGURE 13-11
Example computation for ordinary method of slices.

FIGURE 13-12
Example slope: sand
embankment over
clay foundation.

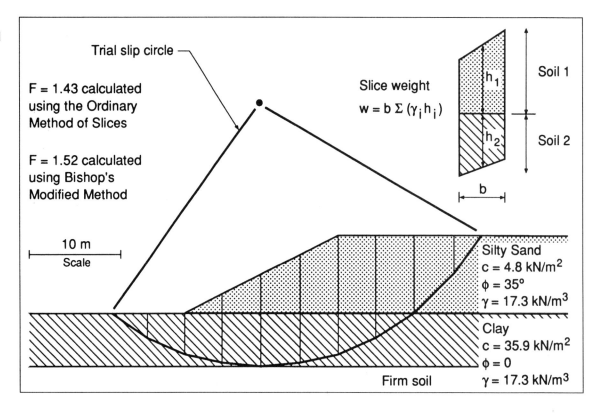

FIGURE 13-12
Example slope: sand embankment over clay foundation.

The seven columns in the upper section of the form contain the same information regarding the slices as shown in the first seven columns of the form for the ordinary method of slices and Bishop's modified method. Note that the assumed value of θ is listed at the top of the upper section of the form. Values of θ must be assumed for all force equilibrium methods.

The calculations for ΔE, the unbalanced force on each slice, involve the five quantities N_0, N_1, N_2, N_3, and N_4. Expressions for these terms are given in the lower part of the form. The values of all of these terms depend on the value of the factor of safety, and iteration is therefore needed to evaluate F. This involves assuming a value for F and checking to see if the forces balance. The forces balance when $\Sigma\Delta E = 0$.

The middle section of the form is labeled Trial 1 and the lower section is labeled Trial 2. Each trial corresponds to a new value of F_a (the assumed value of the factor of safety). The solution is achieved by assuming a value of F_a and using the tabular form to calculate $\Sigma\Delta E$. If $\Sigma\Delta E$ is positive, the assumed factor of safety is too high, and a lower value is assumed for the next trial. If $\Sigma\Delta E$ is negative, a larger value of F_a is assumed for the next

trial. The assumed value of F is the correct value when $\Sigma\Delta E = 0$. If more than two trials are needed to find the correct value of F, as is commonly the case, additional copies of the form are used for those calculations.

For the example shown in Figure 13-17, the factor of safety is 2.44 for $\theta = 10$ degrees. This is shown by the calculations in Figure 13-16. Other calculations, not shown, resulted in values of $F = 1.96$ for $\theta = 0$, and $F = 3.38$ for $\theta = 20$ degrees. The variation of F with θ for this example is shown in Figure 13-18.

It is clear from the results shown in Figure 13-18 that the factor of safety from force equilibrium solutions varies significantly with the assumed value of θ. In this particular example, the computed factor of safety increases by 72 percent (from 1.96 to 3.38) as the side force angle θ is varied from zero to 20 degrees. This illustrates the importance of having a reasonable estimate of the value of θ when the force equilibrium method is used.

The best way to determine the value of θ is by use of the condition of moment equilibrium. Spencer's method, like the method of analysis on which the computation form in Figure 13-15 is based, assumes a constant value of θ. However,

Slice No.	W	l	α	c	ϕ	u	$F_a =$				
							N_1	N_2	N_2	N_2	N_2
1											
2											
3											
4											
5											
6											
7											
8											
9											
10											
11											
12											
13											
14											
15											
						$\Sigma=$					
						$F_c=$					

W = weight of slice - kN/m   c = cohesion intercept - kN/m^2 $N_1 = W \sin \alpha$

ϕ = friction angle - degrees   u = pore pressure - kN/m^2

α = angle between base of slice and horizontal - degrees

$$N_2 = \frac{\left\{ \left[\dfrac{W}{\cos \alpha} - ul \right] \tan \phi + cl \right\}}{\left[1 + \dfrac{\tan \alpha \tan \phi}{F_a} \right]}$$

l = length of slip surface segments measured along base of slice - m $F_c = \Sigma (N_2) / \Sigma (N_1) = \underline{\quad\quad}$

F_a = assumed F F_c = calculated F

Spencer's method satisfies moment equilibrium as well as force equilibrium. Only one value of θ satisfies both force and moment equilibrium. For the example shown in Figure 13-17, the value that satisfies moment equilibrium is θ = 13.5 degrees, and the corresponding factor of safety is F = 2.69. This value of F is the best estimate of the factor of safety for this slope and slip surface. It would differ by no more than about 12 percent from the value of F calculated using any other method that satisfied all conditions of equilibrium and that involved reasonable assumptions.

The trend of F with assumed value of θ shown in Figure 13-18 is typical. As the assumed value of θ for a force equilibrium solution increases, the calculated value of F also increases. It is almost always conservative to assume that θ = 0 degrees, and it is almost always unconservative to assume that θ

equals the slope angle. Thus, although the method outlined by the form given in Figure 13-15 can be used to check Spencer's method calculations precisely, it has limited value as an independent method of calculating F because of the strong dependence of the calculated value of F on the assumed value of θ. The same is true of all force equilibrium methods.

6.2 Computer Analyses

As shown in the preceding sections, slope stability analyses involve lengthy calculations. Use of a computer for these calculations has great potential in two respects:

1. Computer analyses make it possible to perform calculations by advanced methods that satisfy all

FIGURE 13-13
Tabular computation form for Bishop's modified method.

								$F_a =$	1.43	1.51	1.52	
Slice No.	W	l	α	c	ϕ	u	N_1	N_2	N_2	N_2		N_2
1	112	5.3	-32.0	35.9	0.0	0	-60	192	192	192	
2	297	4.9	-22.0	35.9	0.0	0	-111	177	177	177	
3	499	4.7	-13.0	35.9	0.0	0	-112	170	170	170	
4	726	4.6	-4.0	35.9	0.0	0	-51	165	165	165	
5	903	4.6	4.0	35.9	0.0	0	63	165	165	165	
6	1,028	4.7	13.0	35.9	0.0	0	231	170	170	170	
7	1,003	4.9	22.0	35.9	0.0	0	376	177	177	177	
8	818	5.3	32.0	35.9	0.0	0	433	192	192	192	
9	587	6.7	43.0	4.8	35.0	0	400	408	415	416	
10	128	5.6	55.0	4.8	35.0	0	105	108	111	111	
11							0	0	0	0	
12							0	0	0	0	
13							0	0	0	0	
14							0	0	0	0	
15							0	0	0	0	
						$\Sigma=$	1,275	1,925	1,934	1,935	
						$F_c=$		1.51	1.52	1.52	

W = weight of slice - kN/m   c = cohesion intercept - kN/m^2 $N_1 = W \sin \alpha$

ϕ = friction angle - degrees   u = pore pressure - kN/m^2

α = angle between base of slice and horizontal - degrees

$$N_2 = \frac{\left\{ \left[\dfrac{W}{\cos \alpha} - ul \right] \tan \phi + cl \right\}}{\left[1 + \dfrac{\tan \alpha \tan \phi}{F_a} \right]}$$

l = length of slip surface segments measured along base of slice - m

F_a = assumed F F_c = calculated F

$$F_c = \Sigma (N_2) / \Sigma (N_1) = \underline{\quad\quad}$$

FIGURE 13-14
Example computation for Bishop's modified method.

conditions of equilibrium. This advantage reduces the uncertainty resulting from the method of calculation to ±6 percent. Note that this is only the uncertainty due to the method of analysis. The overall uncertainty, including the uncertainty involved in the estimated values of shear strength, is almost always considerably larger.

2. Computer analyses make it possible to conduct a thorough search for the critical circle or critical noncircular slip surface. Because the calculations for a single slip surface take 1 to 3 hr by hand, it is not feasible to search for these values as thoroughly as it is using computer analyses.

6.2.1 Computer and Computer Program Selection

Practically any personal computer can be used for slope stability analyses. A thorough search for the critical slip surface can be accomplished in 10 min or so using a 286-class personal computer and in 1 min or so using a 486-class personal computer.

The most important aspect of computer analysis is the computer program. Programs are being developed continuously to provide a wider range of capabilities and to offer greater ease of use. The optimum computer program for slope stability analysis should have these features:

- It should be based on an advanced method of analysis that satisfies all conditions of equilibrium. It should also be capable of performing calculations using other methods, such as the ordinary method of slices, Bishop's modified method, and force equilibrium methods if the user needs them.
- It should be efficient in terms of the time required for input and output. The cost of com-

$\theta =$

Slice	Weight	l	α	c	ϕ	u
1						
2						
3						
4						
5						
6						

Trial #1 $F_a =$

Slice	N_0	N_1	N_2	N_3	N_4	ΔE
1						
2						
3						
4						
5						
6						

$\Sigma \Delta E =$

Trial #2 $F_a =$

Slice	N_0	N_1	N_2	N_3	N_4	ΔE
1						
2						
3						
4						
5						
6						

$\Sigma \Delta E =$

$$N_0 = \frac{\sin \alpha - \cos \alpha \ \tan \phi \ / \ F}{\cos \alpha + \sin \alpha \ \tan \phi \ / \ F}$$

$$N_1 = W \ N_0$$

$$N_2 = \frac{c \ l}{F} (\cos \alpha + N_0 \sin \alpha)$$

$$N_3 = u \ l \ (\sin \alpha - N_0 \cos \alpha)$$

$$N_4 = \cos \theta + N_0 \sin \theta$$

$$\Delta E = (N_1 - N_2 + N_3) \ / \ N_4$$

θ = side force angle - degrees

F_a = assumed F

W = weight of slice - kN/m

c = cohesion intercept - kN/m^2

ϕ = friction angle - degrees

u = pore pressure - kN/m^2

α = slice base angle - degrees

l = length of slice base - m

FIGURE 13-15
Tabular computation
form for force
equilibrium method
with constant θ.

puter resources for slope stability analyses is usually very small when the cost of equipment and programs is spread over a large number of analyses. The cost of engineering time for input and output is far greater. Therefore it is important that the computer program be designed to minimize the personnel time required for analyses.

• It should be capable of analyzing all the conditions of interest to the user. These will always include undrained conditions, drained conditions, ponded water outside the slope, and internal

FIGURE 13-16
Example
computation for
force equilibrium.

Theta =   10

Slice	Weight	l	α	c	φ	u
1	50.5	3.52	60.00	0.0	30.00	0.00
2	77.1	1.73	45.00	24.0	0.00	0.00
3	189.2	3.66	0.00	24.0	0.00	0.00
4	21.0	1.73	-45.00	24.0	0.00	0.00
5						
6						

Trial #1 F_a =   2.30

Slice	N_0	N_1	N_2	N_3	N_4	ΔE
1	1.03	52.13	0.00	0.00	1.16	44.78
2	1.00	77.10	25.42	0.00	1.16	44.61
3	0.00	0.00	38.11	0.00	0.98	-38.70
4	-1.00	-21.00	25.42	0.00	0.81	-57.22
5						
6						

ΣΔE =   -6.53

Trial #2 F_a =   2.44

Slice	N_0	N_1	N_2	N_3	N_4	ΔE
1	1.06	53.61	0.00	0.00	1.17	45.86
2	1.00	77.10	23.91	0.00	1.16	45.91
3	0.00	0.00	35.85	0.00	0.98	-36.41
4	-1.00	-21.00	23.91	0.00	0.81	-55.36
5						
6						

ΣΔE =   0.00

$$N_0 = \frac{\sin \alpha - \cos \alpha \, \tan \phi \, / F}{\cos \alpha + \sin \alpha \, \tan \phi \, / F}$$

$$N_1 = W \, N_0$$

$$N_2 = \frac{c \, l}{F} (\cos \alpha + N_0 \sin \alpha)$$

$$N_3 = u \, l \, (\sin \alpha - N_0 \cos \alpha)$$

$$N_4 = \cos \theta + N_0 \sin \theta$$

$$\Delta E = (N_1 - N_2 + N_3) / N_4$$

θ = side force angle - degrees

F_a = assumed F

W = weight of slice - kN/m

c = cohesion intercept - kN/m^2

φ = friction angle - degrees

u = pore pressure - kN/m^2

α = slice base angle - degrees

l = length of slice base - m

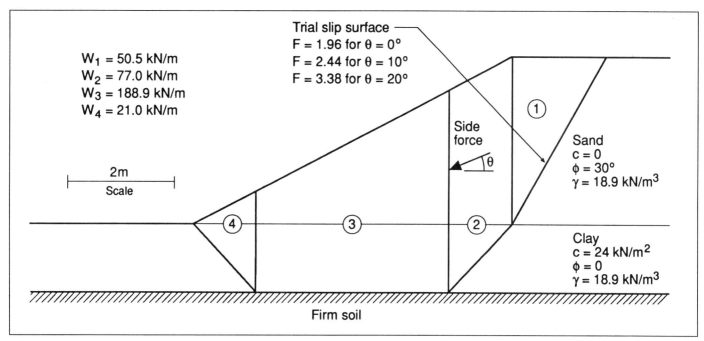

FIGURE 13-17
Noncircular slip surface, force equilibrium analysis.

pore-water pressures. Other conditions that may be of interest include rapid drawdown, slopes with reinforcement, pseudostatic seismic loading, and concentrated or distributed loads on slopes.

Selection of a suitable computer program is an important matter, worthy of careful research and evaluation. It is important to bear in mind that the initial outlay for the computer program is less important than its capabilities, its reliability, and its efficiency as measured in terms of personnel time required to perform analyses. Program distributors and users can provide the information required to make a selection. Because computer programs for slope stability are being improved constantly, acquiring a new program from time to time is a logical method of improving efficiency and effectiveness. Suitable programs may be identified by recommendations of professional colleagues or by review of professional papers and advertisements in technical publications.

6.2.2 Example Computer Analyses

A cross section through Waco Dam, in Texas, is shown in Figure 13-19. When the embankment

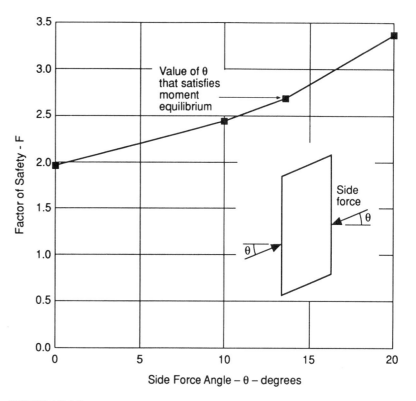

FIGURE 13-18
Variation of F with side force inclination assumed in force equilibrium analyses of sand embankment on clay.

reached the height shown, a slide took place on the downstream slope (Wright and Duncan 1972). The key to the occurrence of the slide was the low undrained shear strength of the Pepper Shale (member of the Woodbine Formation) in the foundation. As shown in Figure 13-19(*a*), the Pepper Shale is highly anisotropic with respect to shear strength. The undrained strength on a horizontal plane (which was measured in laboratory tests using specimens trimmed so that their axes were 30 to 45 degrees from horizontal) was only about 40 percent of the strength measured on conventional vertical test specimens.

FIGURE 13-19
Analysis of slide in Waco Dam, Texas.

The rupture surface of the slide was located in the field by observation of the slide scarp and toe bulge and by inclinometers installed through the slide mass. As a result of the highly anisotropic strength of the Pepper Shale, the slip surface extended horizontally for a considerable distance downstream.

Postfailure analyses of this slide were complicated by two factors: (*a*) the anisotropic strength of the foundation soil and (*b*) the noncircular slip surface. As a result, it was not practical to perform postfailure analyses by hand. With a computer program using Spencer's method, a wide variety of analyses were performed to examine the correspondence among the observed slide, the shear strengths measured in the laboratory, and the most critical slip surface determined by analysis. As shown in Figure 13-19(*b*), the calculated slip surface for $F = 1.00$ was quite similar to the observed slip surface.

A cross section through the Cucaracha landslide at the Panama Canal is shown in Figure 13-20. This slope, which had suffered landslides during excavation of the canal, suffered another slide in 1986 (Berman 1988). The slip surface shown in Figure 13-20 was estimated on the basis of surface observations and inclinometer measurements.

The slide was caused by changes in the groundwater level, with perhaps some influence of loading at the top of the slope, at a location away from the section shown in Figure 13-20. Because the slide was caused by changes in the seepage conditions that resulted in increased pore pressures, it was a drained failure. Therefore the postfailure analyses were performed using drained strengths and the piezometric level that was estimated for the time of the failure.

The slide movement appeared to coincide with bedding surfaces in the Cucaracha shale, which contains numerous slickensides. It therefore seemed appropriate to use residual shear strengths on these bedding surfaces. The residual drained shear strength along the bedding is shown by the lower of the two strength envelopes in Figure 13-20(*a*). The shear strengths on surfaces cutting across bedding were found to be considerably higher, as shown by the upper curve in Figure 13-20(*a*).

The strength envelopes for shear along and across the bedding are nonlinear, and they cannot be represented accurately by single values of *c* and ϕ. Therefore, each strength envelope was represented by a series of points without reference to values of *c* and ϕ.

Postfailure analyses of the 1986 Cucaracha landslide involved a number of complications: noncir-

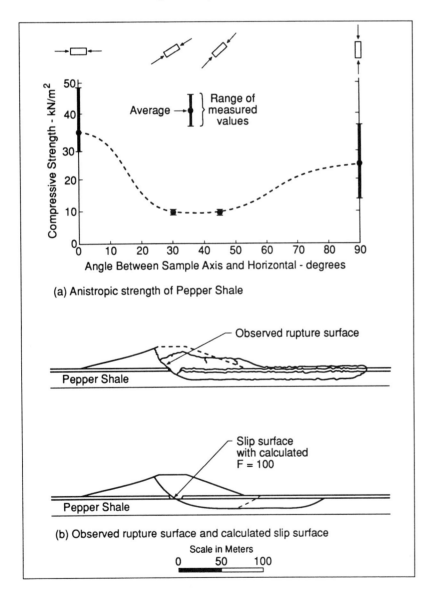

(a) Anistropic strength of Pepper Shale

(b) Observed rupture surface and calculated slip surface

Scale in Meters
0   50   100

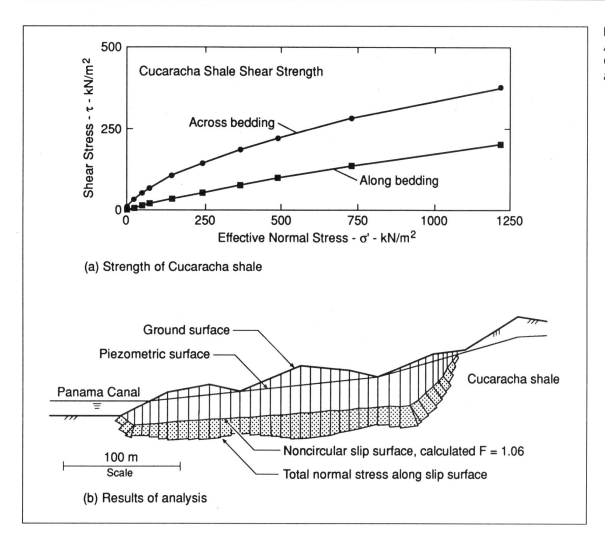

(a) Strength of Cucaracha shale

(b) Results of analysis

FIGURE 13-20
Analysis of 1986
Cucaracha landslide
at Panama Canal.

cular slip surfaces, nonlinear shear strength envelopes, and shear strengths that varied with the orientation of the slip surface. Thorough computer analyses of the slide were performed, confirming the applicability of the shear strengths, and other analyses were made to evaluate the effectiveness of repair schemes (Berman 1988). The results of one of these analyses are shown in Figure 13-20(*b*).

6.2.3 Benefits and Risks of Computer Analyses

Computers afford the most effective means of analysis of slope stability in soil. Computer analyses can be accomplished quickly when a suitable computer program is available; they can accommodate complex conditions of site geometry, seepage, and shear strength; they can be performed using advanced methods that satisfy all conditions

of equilibrium; and they can be used to perform thorough analyses to locate the critical slip surface. However, it should be borne in mind that a computer can also propagate mistakes much faster than is possible by hand. Computer analysis must be checked and verified before the results can be relied upon. Analyses can be checked using charts, manual calculations for the critical slip surface, and independent analysis with another computer program using independent input. Common sense and judgment must always be used to ensure that analysis conditions and results are reasonable.

7. BACK ANALYSIS TO DETERMINE SOIL STRENGTHS

The strength properties of many types of natural soil are difficult to determine reliably by means of laboratory tests. In cases where a landslide has occurred and repair measures are being evaluated, it is often most effective to determine soil strength by back analysis. This process involves three steps:

1. The best estimates possible must be made of the soil strengths and unit weights using the information at hand. Laboratory tests and strength correlations provide an effective basis for these estimates. The slope geometry and phreatic conditions at the time of failure must also be established.
2. The slide should be analyzed using the estimated properties. If the calculated factor of safety is equal to 1.00, the properties and conditions represent a reasonable model of the slide. If the calculated factor of safety is not equal to 1.00, the strengths are adjusted until $F = 1.00$. The adjustment ratio need not be the same for all soils

involved in the slide. Logically, larger adjustments should be made for strengths that are considered to involve greater degrees of uncertainty.
3. When values of soil strength have been determined that give $F = 1.00$ for the conditions at the time of failure, these strengths are used to evaluate repair measures.

This process was used to develop mitigative schemes for the Waco Dam slide discussed previously (Figure 13-19) and the 1986 Cucaracha landslide (Figure 13-20). A further example of back analysis is shown by the Olmsted landslide on the Illinois shore of the Ohio River (Filz et al. 1992).

A cross section through the Olmsted landslide is shown in Figure 13-21. The slide occurred on June 4, 1988, when the river dropped to a low level. The location of the rupture surface was known from

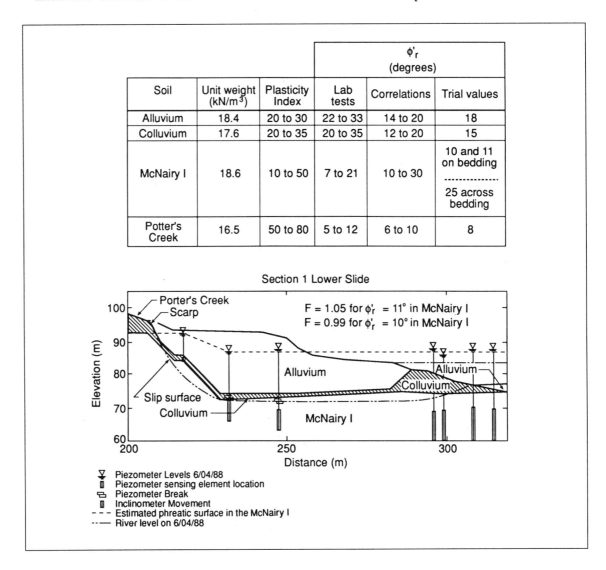

FIGURE 13-21
Back analysis of Olmsted landslide, Illinois.

the position of the scarp at the top of the slide, from movements in inclinometers installed after the slide, and from locations of breaks in piezometer riser tubes. Over most of its length, the rupture surface passed through the McNairy I formation, a sedimentary deposit of flat-lying interbedded sands and clays. The thickness of the individual layers ranged from a few millimeters to nearly a meter.

Laboratory tests were performed on all of the soils involved in the slide, with the results shown in the table at the top of Figure 13-21. It can be seen that there is considerable scatter in the results for all of the soils owing to the variability of the materials, the difficulties involved in obtaining representative and undisturbed specimens for testing, and the layering in McNairy I. On the basis of the results of the tests and correlations with plasticity index, trial values of residual friction angle were established for all of the soils. These are shown in the last column of the table.

The first analysis was performed using Spencer's method with $\phi'_r = 11$ degrees in the McNairy I formation. The calculated factor of safety was 1.05. A second analysis, using $\phi'_r = 10$ degrees, gave $F = 0.99$. Only the value of ϕ'_r for McNairy I was changed for the second analysis because it was considered to have the greatest effect on the results and was therefore the most important parameter in the analyses. The same strength parameter values were also used to analyze four other sections through the slide, where the geometry and piezometric levels were different. The calculated values of F, with $\phi'_r = 10$ degrees in McNairy I, varied from 0.96 to 1.00 for the five sections. The parameters calculated by back analysis were considered to represent the behavior of the soils quite well and were utilized to analyze the effects of drainage and slope flattening to improve the stability of the slide area.

8. RAPID DRAWDOWN ANALYSIS

When the water level in a lake or stream adjacent to a slope drops, the stabilizing influence of the water pressure on the slope is lost. Many failures of dam slopes and submerged natural slopes have occurred during rapid drawdown. Analysis of rapid drawdown is therefore an important consideration for design of dams and other slopes that will be submerged and subject to such changes in water level.

Methods of analysis of stability during rapid drawdown have been evaluated by Duncan et al.

(1990). On the basis of those studies, a method of analysis for rapid drawdown was developed that incorporates these features:

- Undrained strengths are used for soil zones that have values of C_v so low that they are unable to drain during the period of drawdown. Criteria to determine whether a soil zone is drained or undrained are described in Section 3.
- Analyses are performed in three steps:
  - Anisotropic consolidation pressures are determined for each slice using stability and seepage analyses for the predrawdown steady seepage conditions.
  - Values of undrained strength are determined for each slice with its base in a zone of soil that will be undrained, drained strengths are determined for zones of soil that will be drained, and an analysis of stability after drawdown using these strengths is performed.
  - The after-drawdown analyses are repeated using drained strengths for slices where the drained strengths after drawdown are found to be smaller than the undrained strengths.

Thus, the method uses drained or undrained strengths, whichever are lower, for slices where drainage is not expected during drawdown. This technique ensures that the method will not yield unconservative results if the soil drains faster than anticipated.

This method of analysis is logical in its treatment of strengths and stability. Used with Spencer's method for stability computations, it was found to give satisfactory results for Walter Bouldin Dam in Alabama and Pilarcitos Dam in California, both of which suffered embankment slides during rapid drawdown (Duncan et al. 1990).

9. ANALYSIS OF REINFORCED SLOPES

Reinforcing materials, such as geotextiles, geogrids, and steel grids, are widely used to strengthen slopes. Considerable research has been done to evaluate the long-term behavior of these materials and to investigate the mechanisms of interaction by which they improve the stability of slopes. The principal mechanism of reinforcement is limitation of tensile strains at the locations of the reinforcement. Reinforcing materials are commonly placed horizontally within the soil.

Studies of soil reinforcement have shown that the stability of reinforced slopes can be analyzed using the following simple procedure:

1. The long-term ultimate strength of the reinforcing material is determined. This determination involves consideration of short-term strength, creep under load, and the allowable tensile strains considering the type of slope or embankment and the sensitivity of the underlying foundation soil. This long-term limit load is expressed in units of force per unit length of the reinforcement.
2. The long-term limit load is divided by a reinforcement factor of safety, F_R, to produce a factored limit load. The value of F_R logically should depend on the reliability with which the limit load can be evaluated and the possible consequences of failure. Values of F_R near 1.5 are suitable for normal conditions.
3. The factored limit loads are used as known values in conventional limit equilibrium analyses of

slope stability. The forces are applied to the bases of slices in which a potential slip surface cuts across the reinforcement. Studies have shown that very flexible types of reinforcement may be reoriented if slip occurs and dragged down so that they parallel the slip surface where they cross it. If a method of analysis is used that satisfies all conditions of equilibrium, the orientation of the reinforcement force (horizontal or parallel to the slip surface) has essentially no effect on the factor of safety computed (Wright and Duncan 1991). The factor of safety computed in this type of analysis is the same as that discussed earlier in this chapter; it is the factor by which the soil strength would have to be divided to bring the soil into a state of barely stable equilibrium.

An example analysis of a reinforced slope is shown in Figure 13-22, in which cross sections through the St. Alban test embankment are used; this embankment was constructed on the sensitive Champlain Clay (Schaefer and Duncan 1988). To avoid uncertainties resulting from the sensitivity of the clay, the analyses described here were performed using the postpeak residual undrained shear strength of the crust and the normally consolidated clay beneath the crust. As defined by La Rochelle et al. (1974) and Trak et al. (1980), this "residual" undrained strength is measured at 15 percent axial strain in triaxial tests and is equal to about 90 percent of the undrained strength measured in field vane-shear tests.

The sand embankment was reinforced with two horizontal layers of Tensar SR-2 geogrid. Failure took place as the embankment was being raised about 20 days after construction began. For this load duration, the limit load corresponding to 10 percent strain is about 32 kN/m. Analyses were performed at this reinforcement magnitude using Bishop's modified method and Spencer's method. The results of the analyses are shown in Figure 13-22.

In both analyses the reinforcement force was assumed to act horizontally. It can be seen that both analyses resulted in minimum factors of safety very close to unity. For practical purposes, the values of F calculated by the two analyses were the same. Considering the arbitrary choice of 10 percent strain as the failure strain in the reinforcing, the fact that the calculated factors of safety were so close to unity must be considered somewhat fortuitous.

FIGURE 13-22
Analysis of reinforced St. Alban test embankment.

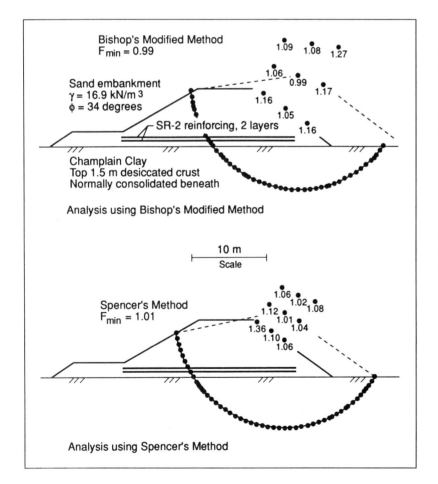

10. THREE-DIMENSIONAL ANALYSES OF SLOPE STABILITY

Although the methods of slope stability analyses discussed in the previous sections are formulated in two dimensions, actual slope failures are three-dimensional. The question therefore arises as to the accuracy and reliability of two-dimensional (2D) analyses applied to three-dimensional (3D) problems. Research studies [for example, the work by Cavounidis (1987)] have shown clearly that factors of safety calculated using 3D analyses are larger than those calculated using 2D analyses, all other things being equal. Implicit in this conclusion is the notion that the 2D section analyzed is the most critical section through the 3D potential sliding mass.

An example is shown in Figure 13-23, which summarizes 2D and 3D analyses of an ellipsoidal slip surface (Hungr et al. 1989). As shown in Figure 13-23(a), the factors of safety for three 2D sections through the sliding mass are $F_2 = 1.10$, 1.00, and 1.19. The central section (Section 2) is the most critical, and the minimum 2D factor of safety is thus the value calculated for this section, $F_2 = 1.00$. Figure 13-23(b) shows the results of a 3D analysis performed by Hungr et al. (1989) using Bishop's modified method extended to three dimensions (Hungr 1987). The shape of the critical ellipsoidal slip surface is shown in Figure 13-23(c). The minimum 3D factor of safety for this case is $F_3 = 1.01$, which is only 1.0 percent higher than the minimum 2D factor of safety.

It is more difficult to perform 3D analyses than 2D analyses. Because 2D analyses give somewhat conservative results ($F_2 \leq F_3$, all other things being equal), they provide a reasonable and sufficiently accurate approach to most practical problems of stability of soil slopes. Use of 2D analyses requires that the section or sections analyzed be selected using judgment regarding which section will be most critical. In many cases, as in that shown in

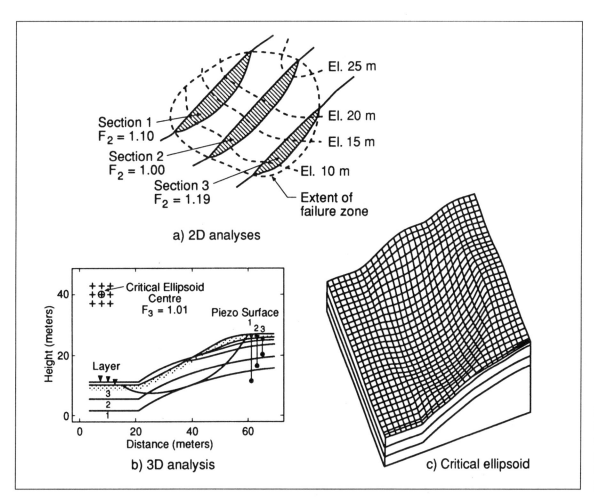

FIGURE 13-23
Comparison of two-dimensional and three-dimensional analyses.

Figure 13-23, the critical 2D section is the one where the slip surface can cut most deeply. In some cases other sections may be more critical. If there is doubt concerning which 2D section is most critical, several should be analyzed.

11. DEFORMATION ANALYSIS

Finite-element analyses have been used since the mid-1960s for analysis of stresses and movements in slopes. A recent review (Duncan 1992) described more than 100 examples of their application. Even though finite-element analyses have been used fairly often in recent years for evaluation of slope deformations, the procedure is not routine. Finite-element analyses are more difficult and time-consuming than slope stability analyses, and they require special expertise if they are to be done successfully and productively.

Experience with finite-element analyses has shown that they are most useful when performed in conjunction with field instrumentation studies. They can be valuable in planning instrumentation programs by showing where the largest movements would be expected to occur and how large they may be. They can also be used for interpreting the results of instrumentation studies. If calculated and measured movements are in agreement at the locations of the instruments, this suggests that the analytical results also may provide reasonable indications of behavior at other locations. Often the more complete finite-element results provide insight into the causes and significance of the measured movements.

An example of a finite-element analysis, summarized in Figure 13-24, concerns Otter Brook Dam in New Hampshire, which deformed considerably during construction (Kulhawy et al. 1969). The embankment, about 40 m high, is essentially homogeneous. It was constructed of compacted silt on a firm foundation of glacial till over rock.

As shown in Figure 13-24, a bridge pier was constructed at about mid-height on the upstream slope of the dam. Because the embankment was constructed above the level of the bridge pier, the embankment deformed under the weight of the added material and the lower part bulged outward. Measurements were made of the movements of the bridge pier as the embankment was constructed. As shown in Figure 13-24(c), these measurements reached about 0.15 m vertically and

about 0.9 m horizontally by the end of construction. Measurements at other elevations on the upstream slope, also shown in Figure 13-24(c), showed that the maximum horizontal movement occurred near mid-height of the embankment.

A finite-element analysis of Otter Brook Dam was performed by Kulhawy et al. (1969) using hyperbolic stress-strain properties determined from conventional laboratory tests. The tests had been performed in support of stability analyses of the dam. The calculated displacements of the upstream slope and the bridge pier are compared with the measured values in Figure 13-24(c). It can be seen that agreement is quite good, indicating that the finite-element analyses afford a reasonable representation of the actual behavior of the embankment.

Because engineered fills are water conditioned and compacted with a reasonable degree of control of density and uniformity of the fill, stress-strain properties can be determined for these materials more precisely than for natural soil deposits, which may be less uniform and may have more complex behavior. As a result, the reliability of calculated movements is not as great for natural soil deposits as for embankments. Very often, even in the case of embankments, the agreement between calculated and measured movements is not as close as that shown in Figure 13-24.

12. USE OF CENTRIFUGE EXPERIMENTS

Centrifuge experiments have been used to study problems of slope stability and deformation behavior since the early 1970s (Schofield 1980). Techniques have been developed for constructing model slopes, varying water levels, measuring shear strengths, and measuring displacements in the models while they are rotating. The techniques were applied to studies of levee failures (Kusakabe et al. 1988), highway embankment settlement and stability (Feng and Hu 1988), settlement and cracking in dams (Shcherbina and Olympiev 1991; Zhang and Hu 1991; Zhu et al. 1991), stability of tailings dams (Liu et al. 1988), mechanisms of flow slides (Schofield 1980), and stabilization of slopes by drainage (Resnick and Znidarcic 1990).

The method has the greatest value for qualitative studies of mechanisms of deformation and failure. Because centrifuge models can be constructed so that they simulate the geometry and

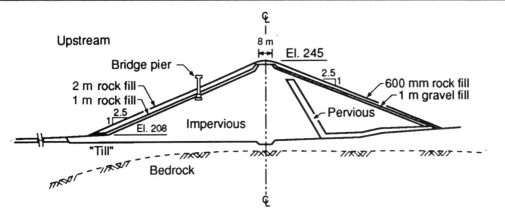

Upstream

Bridge pier

2 m rock fill
1 m rock fill

El. 245

2.5
1

600 mm rock fill
1 m gravel fill

1
2.5
El. 208

Impervious

Pervious

"Till"

Bedrock

8 m

(a) Cross section through Otter Brook Dam

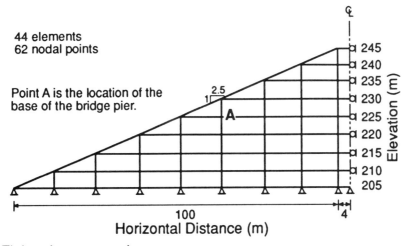

44 elements
62 nodal points

Point A is the location of the
base of the bridge pier.

2.5
1

A

Elevation (m)

245
240
235
230
225
220
215
210
205

Horizontal Distance (m)

100

4

(b) Finite element mesh

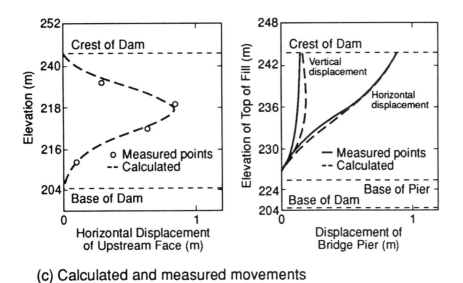

252

Crest of Dam

240

218

216

Elevation (m)

204

Base of Dam

o Measured points
-- Calculated

0   1
Horizontal Displacement
of Upstream Face (m)

248

Crest of Dam

242

Vertical
displacement

Horizontal
displacement

236

230

Elevation of Top of Fill (m)

— Measured points
-- Calculated

224

Base of Pier

Base of Dam

204

0   1
Displacement of
Bridge Pier (m)

(c) Calculated and measured movements

FIGURE 13-24
Finite-element
analysis of Otter
Brook Dam,
New Hampshire.

stresses in prototype slopes, they are capable, in principle, of simulating the modes of deformation and failure to be expected in the field. Thus, they are useful where mechanisms of deformation and failure are unknown.

Limitations on the use of centrifuge experiments to derive quantitative results stem from the difficulties inherent in constructing centrifuge models that accurately mimic the important details of the full-scale prototype and the loading conditions it will experience. These difficulties, plus the fact that the technique is costly and requires highly specialized equipment, appear to be the principal limitations on its use for practical studies of stability and deformation in soil slopes.

REFERENCES

Berman, G. 1988. *Landslide Control on the Panama Canal: Status Report—1988.* Geotechnical Branch, Engineering Division, Panama Canal Commission, Panama City, 26 pp.

Bishop, A. W. 1955. The Use of the Slip Circle in the Stability Analysis of Slopes. *Geotechnique,* Vol. 5, No. 1, pp. 7–17.

Bishop, A. W., and L. Bjerrum. 1960. The Relevance of the Triaxial Test to the Solution of Stability Problems. In *Proc., ASCE Research Conference on the Shear Strength of Cohesive Soils,* Boulder, Colorado, American Society of Civil Engineers, New York, pp. 437–501.

Cavounidis, S. 1987. On the Ratio of Factors of Safety in Slope Stability Analyses. *Geotechnique,* Vol. 37, No. 2, pp. 207–210.

Duncan, J. M. 1992. State-of-the-Art: Static Stability and Deformation Analysis. In *Proc., ASCE Specialty Conference on Stability and Performance of Slopes and Embankments II,* Berkeley, California, Geotechnical Special Publication 31, American Society of Civil Engineers, New York.

Duncan, J. M., A. L. Buchignani, and M. De Wet. 1987. *An Engineering Manual for Slope Stability Studies.* Virginia Polytechnic Institute and State University, Blacksburg, 80 pp.

Duncan, J. M., S. G. Wright, and K. S. Wong. 1990. Slope Stability During Rapid Drawdown. In *Proc., H. Bolton Seed Memorial Symposium,* Berkeley, California, Bitech Publishers, Vancouver, British Columbia, Canada.

Fellenius, W. 1927. *Erdstatische Berechnungen mit Reibung und Kohasion.* Ernst, Berlin.

Feng, G., and Y. Hu. 1988. Centrifugal Model Test of a Highway Embankment on Soft Clay. In *Proc., First International Conference on Geotechnical Centrifuge Modeling,* Paris, pp. 153–162.

Filz, G. M., T. L. Brandon, and J. M. Duncan. 1992. Back Analysis of the Olmsted Landslide Using Anisotropic Strengths. In *Transportation Research Record 1343,* TRB, National Research Council, Washington, D.C., pp. 72–78.

Hungr, O. 1987. An Extension of Bishop's Simplified Method of Slope Stability Analysis to Three Dimensions. *Geotechnique,* Vol. 37, No. 1, pp. 113–117.

Hungr, O., F. M. Salgado, and P. M. Byrne. 1989. Evaluation of a Three-Dimensional Method of Slope Stability Analysis. *Canadian Geotechnical Journal,* Vol. 26, pp. 679–686.

Hunter, J. H., and R. L. Schuster. 1968. Stability of Simple Cuttings in Normally Consolidated Clays. *Geotechnique,* Vol. 18, No. 3, pp. 372–378.

Janbu, N. 1968. *Slope Stability Computations.* Soil Mechanics and Foundation Engineering Report. Technical University of Norway, Trondheim.

Kulhawy, F. H., J. M. Duncan, and H. B. Seed. 1969. *Finite Element Analyses of Stresses and Movements in Embankments During Construction.* Report TE 69-4. Office of Research Services, University of California, 169 pp.

Kusakabe, O., A. Kawashima, and T. Hagiwara. 1988. Levee Failures Modeled in Drum Centrifuge. In *Proc., First International Conference on Geotechnical Centrifuge Modeling,* Paris, pp. 147–152.

Ladd, C. C., and R. Foott. 1974. New Design Procedure for Stability of Soft Clays. *Journal of the Geotechnical Engineering Division,* ASCE, Vol. 100, No. 7, pp. 763–786.

LaRochelle, P., B. Trak, F. Tavenas, and M. Roy. 1974. Failure of a Test Embankment on a Sensitive Champlain Clay Deposit. *Canadian Geotechnical Journal,* Vol. 11, No. 1, pp. 142–164.

Liu, L., W. Jiang, and W. Han. 1988. Centrifugal Model Tests for Stability of Ash Soil Composite Dams. In *Proc., First International Conference on Geotechnical Centrifuge Modeling,* Paris, pp. 141–146.

Lowe, J., and L. Karafiath. 1960. Stability of Earth Dams upon Drawdown. In *Proc., First Pan-American Conference on Soil Mechanics and Foundation Engineering,* Mexico City, Vol. 2, pp. 537–552.

Morgenstern, N. R. 1992. The Evaluation of Slope Stability—A 25 Year Perspective. In *Proc., ASCE Specialty Conference on Stability and Performance of Slopes and Embankments II,* Berkeley, California,

Geotechnical Special Publication 31, American Society of Civil Engineers, New York, pp. 1–26.

Morgenstern, N. R., and V. E. Price. 1965. The Analysis of the Stability of General Slip Surfaces. *Geotechnique*, Vol. 15, No. 1, pp. 79–93.

Resnick, G. S., and D. Znidarcic. 1990. Centrifugal Modeling of Drains for Slope Stabilization. *Journal of Geotechnical Engineering*, ASCE, Vol. 116, No. 11, pp. 1607–1624.

Sarma, S. K. 1973. Stability Analysis of Embankments and Slopes. *Geotechnique*, Vol. 23, pp. 423–433.

Schaefer, V. R., and J. M. Duncan. 1988. Finite Element Analyses of the St. Alban Test Embankments. In *Proc., Symposium on Geosynthetics for Soil Improvement*, National Convention, Nashville, American Society of Civil Engineers, pp. 158–177.

Schofield, A. N. 1980. Cambridge Geotechnical Centrifuge Operations. *Geotechnique*, Vol. 30, No. 3, pp. 227–268.

Shcherbina, V. I., and D. N. Olympiev. 1991. Numerical and Centrifugal Modeling of Cracking in an Earth-and-Rockfill Dam Core. In *Proc., Second International Conference on Geotechnical Centrifuge Modeling*, Boulder, Colorado, pp. 67–74.

Skempton, A. W. 1970. First-Time Slides in Overconsolidated Clays. *Geotechnique*, Vol. 20, No. 4, pp. 320–324.

Skempton, A. W. 1977. Slope Stability of Cuttings in London Clay. In *Proc., Ninth International Conference on Soil Mechanics and Foundation Engineering*, Tokyo, pp. 261–270.

Skempton, A. W. 1985. Residual Strength of Clays in Landslides, Folded Strata, and the Laboratory. *Geotechnique*, Vol. 35, No. 1, pp. 3–18.

Spencer, E. 1967. A Method of Analysis of the Stability of Embankments Assuming Parallel Interslice Forces. *Geotechnique*, Vol. 17, No. 1, pp. 11–26.

Trak, B., P. LaRochelle, F. Tavenas, S. Leroueil, and M. Roy. 1980. A New Approach to the Stability Analysis of Embankments on Sensitive Clays. *Canadian Geotechnical Journal*, Vol. 17, No. 4, pp. 526–584.

U.S. Army Corps of Engineers. 1970. *Engineering and Design—Stability of Earth and Rockfill Dams*. Engineer Manual EM 1110-2-1902. Department of the Army, Corps of Engineers, Office of the Chief of Engineers, Washington, D.C.

Wright, S. G., and J. M. Duncan. 1972. Analysis of Waco Dam Slide. *Journal of the Soil Mechanics and Foundations Division*, ASCE, Vol. 98, No. SM7, pp. 653–665.

Wright, S. G., and J. M. Duncan. 1991. Limit Equilibrium Stability Analyses for Reinforced Slopes. In *Transportation Research Record 1330*, TRB, National Research Council, Washington, D.C., 1991, pp. 40–46.

Zhang, L., and T. Hu. 1991. Centrifugal Modeling of the Pubugan High-Rise Rockfill Dam. In *Proc., Second International Conference on Geotechnical Centrifuge Modeling*, Boulder, Colorado, pp. 45–50.

Zhu, W., J. Yi, D. Zhu, and S. Liu. 1991. Application of Centrifuge Modeling to Predictions of Earth-Rockfill Dam Core Cracking. In *Proc., Second International Conference on Geotechnical Centrifuge Modeling*, Boulder, Colorado, pp. 59–66.

Chapter 14

DUNCAN C. WYLLIE AND
NORMAN I. NORRISH

ROCK STRENGTH PROPERTIES AND THEIR MEASUREMENT

1. INTRODUCTION

The shear strength developed along potential rupture surfaces within a slope has an important influence on the stability of rock slopes. In carrying out stability analyses, it is usually assumed that the rock behaves as a Mohr-Coulomb material in which the shear strength is expressed in terms of the cohesion and friction angle along the rupture surface. The actual values of these two strength parameters are closely related to the geological conditions at each site, and any program to determine rock strength should start with a thorough examination of the geology. In this chapter the relationship between geology and rock strength is discussed, and methods are described that are used to determine shear strength values.

The relationship between the strength of samples that can be tested in the laboratory and the strength of the rock mass is particularly important in determining design strength values for stability analysis. *Rock mass* is the term used to describe the material along the rupture surface. The rock mass may consist of a continuous fracture, a combination of intact rock and fractures, or solely intact rock. It is rarely possible to test the rock mass in the laboratory because of difficulty in obtaining a large, undisturbed sample, and there are few machines available with the required load capacity. Therefore it is necessary to use a combination of laboratory testing of small samples, empirical analysis, and field observations to determine a strength value, or range of values, that is representative of the rock mass on the rupture surface.

1.1 Examples of Rock Masses

Figures 14-1 through 14-4 show four different geological conditions that are commonly encountered in the design and analysis of rock slopes. These are typical examples of rock masses in which the strength of laboratory-size samples may differ significantly from the shear strength along the overall rupture surface. In all four cases, instability occurs as a result of shear movement along a rupture surface that either lies along an existing fracture or passes partially or entirely through intact rock. Figures 14-1 through 14-4 also show that in fractured rock the shape of the rupture surface is influenced by the orientation and length of the discontinuities.

Figure 14-1 shows a strong, massive sandstone containing a set of continuous bedding surfaces that dip out of the slope face. Because the near-vertical cut face is steeper than the dip of the bedding, the bedding surfaces are exposed, or *daylight*, on the face and sliding has occurred, with a tension crack opening along the subvertical, orthogonal joint set. Under these circumstances the shear strength used in stability analysis is that of the bedding surfaces. The strength properties of fracture surfaces are discussed in Section 2 and methods of measuring the friction angle in the laboratory are discussed in Section 3.

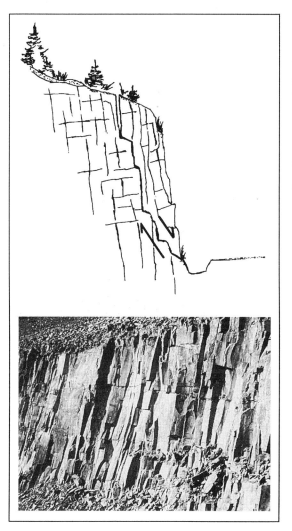

FIGURE 14-1
(far left)
Planar failure on
continuous bedding
surface dipping out
of slope face.

FIGURE 14-2
(left)
Shallow circular
failure in closely
jointed rock mass.

Figure 14-2 shows a slope cut in a strong but closely jointed columnar basalt containing a sub-horizontal joint set and numerous vertical joints. Because the joint sets are discontinuous, no single joint controls the stability of the slope. However, if a rupture surface were to develop in this slope, it would follow a stepped path that would partly lie along fracture surfaces and partly pass through intact rock. The shear strength of this complex rupture surface cannot be determined analytically, so a set of empirical equations has been developed from which the cohesion and friction angle can be calculated with respect to the degree of fracturing and the rock strength. This procedure is described in Section 4.

Figure 14-3 shows a slope cut in a weathered rock in which the degree of weathering varies from residual soil in the upper part of the slope to slightly weathered rock near the toe. For these conditions, the rupture surface lies predominantly in the weaker materials in the upper part of the slope, and in the stability analysis it is necessary to use different strength parameters for the upper and lower portions of the rupture surface. Because the degree of degradation of weathered rock tends to be highly variable, the strength of the rock mass is also variable and can be difficult to determine. Consequently, a means of determining the strength of weathered rock is to carry out a back analysis of slopes in similar material; this approach is described in Section 4.

A fourth geological condition that may be encountered is that of a very weak but intact rock containing essentially no fractures. Figure 14-4 shows a cut face in tuff, a rock formed by the consolidation of volcanic ash. The photograph shows that a geological hammer can be embedded in the face easily, indicating the low strength of this rock. However, because the rock contains no fractures, it has a significant cohesive strength in addition to a moderate friction angle. Therefore, it was possible to cut a stable, vertical face to a height of

FIGURE 14-3
(right)
Circular failure in
residual soil and
weathered rock.

FIGURE 14-4
(far right)
Near-vertical cut
in very weak but
massive tuff
containing no
fractures.

at least 10 m in this material, provided water pressures and erosion were controlled. Slope failure would tend to be restricted to shallow sloughing.

1.2 Mohr-Coulomb Materials

For most shear failures the rock can be assumed to be a Mohr-Coulomb material in which the shear strength of the sliding surface is expressed in terms of the cohesion (c) and the friction angle (ϕ). When an effective normal stress σ' acts on the rupture surface, the shear stress (τ) developed is

$$\tau = c + \sigma' \tan\phi \qquad (14.1)$$

Equation 14.1 is expressed in Figure 14-5 as a straight line that shows the influence of both the effective normal stress and the characteristics of the fracture surface on the shear strength. The main component of the effective normal stress is the weight of the rock. However, if water pressures

act on the potential rupture surface, the effective normal stress is diminished. Tensioned rock bolts can be installed across the fracture to increase the normal stress. These concepts are used in the analysis of rock slope stability and the design of rock slope stabilization measures, as discussed in Chapters 15 and 18, respectively.

Figure 14-5 illustrates typical strength parameters for five geological conditions that range from a low-strength infilled fracture to relatively high-strength fractured rock containing discontinuous fractures to weak intact rock. These five conditions may be described as follows:

1. **Infilled fracture:** If the infilling is a weak clay or fault gouge, the friction angle (ϕ_{inf}) is likely to be low, but there may be some cohesion if the infilling is undisturbed. Alternatively, if the infilling is a strong calcite, for example, which produces a healed surface, the cohesive strength may be significant (see Section 2.3).
2. **Smooth fracture:** A smooth, clean fracture has zero cohesion, and the friction angle is that of the rock surfaces (ϕ). The friction angle of rock

is related to the grain size and is generally low in fine-grained rocks and higher in coarse-grained rocks (see Section 2.1).

3. **Rough fracture:** Clean, rough fracture surfaces have zero cohesion, and the friction angle consists of the rock-material friction angle (ϕ), a component (i) related to the roughness (asperities) of the surface, and the ratio between the rock strength and the normal stress. As the normal stress increases, the asperities are progressively sheared off and the total friction angle diminishes (see Section 2.2).

4. **Fractured rock mass:** The shear strength of a fractured rock mass, in which the rupture surface partially lies on fracture surfaces and partially passes through intact rock, can be expressed as a curved envelope (Figure 14-2). At low normal stresses where there is little confinement of the fractured rock and the individual fragments may move and rotate, the cohesion is low but the friction angle is high because the sliding surface is effectively rough. At higher normal stresses, crushing of the rock fragments begins to take place, with the result that the friction angle diminishes. The shape of the strength envelope is related to the degree of fracturing and the strength of the intact rock (see Section 4.2).

5. **Weak intact rock:** Rock such as the tuff shown in Figure 14-4 is composed of fine-grained material that has a low friction angle. However, because it contains no fractures, the cohesion can be higher than that of a strong but closely fractured rock.

The range of shear strength conditions that may be encountered in rock slopes, as shown in Figures 14-1 through 14-5, clearly demonstrates the importance of examining both the characteristics of the fractures and the rock strength during the site investigation. Surface mapping of outcrops or existing cuts usually provides the most reliable data on rock-mass characteristics because the large exposures show the range of conditions that may be encountered. In contrast, diamond drilling samples only a very small area of the fractures, which may not be representative of the overall fracture characteristics. However, diamond drilling may be required when the potential rupture surface lies deep within the slope or there are limited surface outcrops. Investigation methods for rock are discussed in Chapter 10.

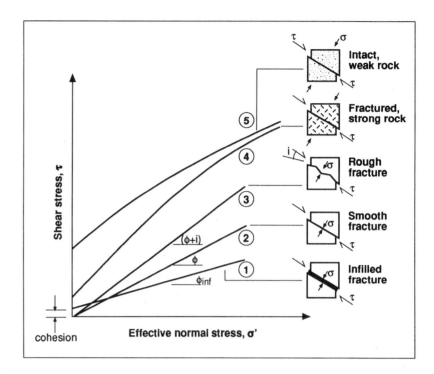

2. SHEAR STRENGTH OF FRACTURES

If geological mapping, diamond drilling, or both identify fractures in a slope on which shear or toppling failures may take place, it will be necessary to determine the friction angle and cohesion of the fracture surface in order to carry out stability analyses and design any necessary remedial work. The investigation program should also obtain information on characteristics of the fracture surface that may modify the shear strength parameters. Important fracture characteristics include continuous length, surface roughness, thickness and characteristics of any infilling material, as well as the effect of water on the properties of the infilling.

The shear strength properties of the various fracture types are described in the following sections.

2.1 Friction Angle of Rock Surfaces

For a planar, clean (no infilling) fracture, the cohesion is zero and the shear strength is defined solely by the friction angle. The friction angle of the rock material is related to the size and shape of the grains exposed on the fracture surface. Thus, a fine-grained rock and rock with a high mica content tend to have a low friction angle, whereas coarse-grained rock has a high friction angle. Table 14-1 shows typical ranges of friction

FIGURE 14-5 Relationships between shear stress and normal stress on rupture surface for five different geological conditions.

Table 14-1
Typical Ranges of Friction Angles for Variety of Rock Types

ROCK CLASS	FRICTION ANGLE RANGE (DEGREES)	TYPICAL ROCK TYPES
Low friction	20 to 27	Schists (high mica content), shale, marl
Medium friction	27 to 34	Sandstone, siltstone, chalk, gneiss, slate
High friction	34 to 40	Basalt, granite, limestone, conglomerate

angles for a variety of rock types (Barton 1973; Jaeger and Cook 1976).

The friction angles in Table 14-1 should be used only as guidelines because actual values vary widely with site conditions. Laboratory testing procedures to determine the friction angle are described in Section 3.

2.2 Surface Roughness

The roughness of natural rock surfaces has a significant effect on their friction angle (Patton 1966). These surface irregularities, which are given the general term *asperities*, produce interlock between fracture surfaces that increases the resistance to sliding. Asperities can be considered in their simplest form as a series of sawteeth. When normal and shear forces are applied to a rock surface containing a clean, sawtooth fracture, the shear strength of the fracture is

$$\tau = \sigma' \tan(\phi + i) \qquad (14.2)$$

where i is the inclination of the sawteeth as shown in Figure 14-6. This relationship shows that the effective friction angle of a rough surface is equal to the sum of the basic friction angle of the rock and the roughness angle.

Figure 14-6 also shows that the asperities can be sheared off, with a consequent reduction in the friction angle. With increasing stress levels, there is a transition from dilation to shearing, and the degree to which the asperities are sheared off depends on both the magnitude of the normal force in relation to the compressive strength of the rock on the fracture surface and the displacement distance. A rough fracture that is initially undisturbed and interlocked will have a friction angle of $(\phi + i)$, known as the *peak shear strength*. With increasing normal stress and displacement, the asperities will be sheared off and the friction angle will progressively diminish to a minimum value of the basic, or residual, friction angle of the rock.

When the value of the roughness angle of a rock surface is measured, it is necessary to decide on the wavelength of the asperities that are important to design. Patton (1966) proposed that asperities be divided into two classes depending on their wavelength. Asperities with wavelengths of less than about 50 to 100 mm are termed *second-order asperities* and may have roughness angles (i_2) as high as 20 to 30 degrees (Figure 14-7). With increasing wavelength of the asperities, the roughness angle diminishes; asperities with a wavelength of about 500 mm or greater are termed *first-order asperities* and will have roughness angles (i_1) of not more than about 10 to 15 degrees.

The importance of asperity wavelength is shown in Figure 14-8. In rock slopes, stress relief and possibly blast damage can cause opening of fractures and dilation of the rock at the face. Also, the stress levels on the rupture surface may be high enough to cause some shearing of the second-order asperities [Figure 14-8(a)]. Under these conditions, the roughness angle corresponding to first-order asperities would be used in design. If, for example,

FIGURE 14-6
Effect of surface roughness and normal stress on friction angle of fracture surface (Wyllie 1992).

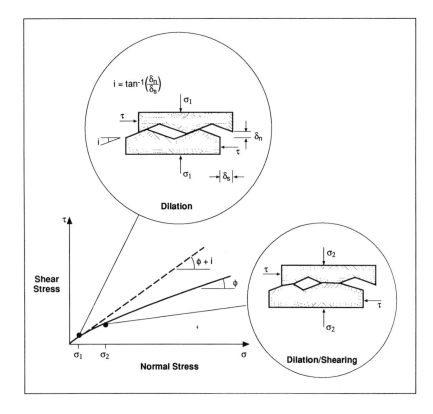

the rupture surface consisted of undulating bedding planes in fine-grained sandstone, the total friction angle may be about 30 degrees (ϕ = 25, i = 5 degrees).

Second-order asperities can have an effect on the stability of reinforced slopes [Figure 14-8(b)]. If the rock mass can be prevented from movement and dilation by the installation of tensioned rock anchors or passive support such as dowels and buttresses (see Chapter 18), interlock along the rupture surface is maintained. Under these conditions the second-order asperities contribute to the shear strength of the potential sliding surface, and the total friction angle of the rough rock surface may be as high as 55 degrees (ϕ = 25, i = 30 degrees).

In order to quantify the relationship among the total friction angle ($\phi + i$), the rock strength, and the normal stress acting on the rupture surface, Barton (1976) studied the shear strength behavior of artificially produced rough, clean "joints." The study showed that the shear strength of a rough rock surface can be defined by the following empirical equation:

$$\tau = \sigma' \tan\left[\phi + JRC \log_{10}\left(\frac{\sigma_j}{\sigma'}\right)\right] \qquad (14.3)$$

where

    JRC = joint roughness coefficient,
      σ_j = compressive strength of the rock at the fracture surface, and
      σ' = effective normal stress.

The roughness of the fracture surface is defined by the joint roughness coefficient, JRC. Barton carried out direct shear tests on a large number of natural discontinuities and calculated JRC values corresponding to the surface roughness of the different shear-test specimens. From these tests a set of typical roughness profiles with specified JRC values was prepared (Figure 14-9). By comparing a fracture surface with these standard profiles, the JRC value can be determined (Barton 1976).

The term [$JRC \log_{10}(\sigma_j/\sigma')$] is equivalent to the roughness angle i in Equation 14.2. At high stress levels when σ_j/σ' = 1 and the asperities are sheared off, the term [$JRC \log_{10}(\sigma_j/\sigma')$] equals zero. At low stress levels the ratio (σ_j/σ') tends to infinity and the roughness component of the strength becomes very large. In order for realistic values of the roughness component to be used in design, the term ($\phi + i$) should not exceed about

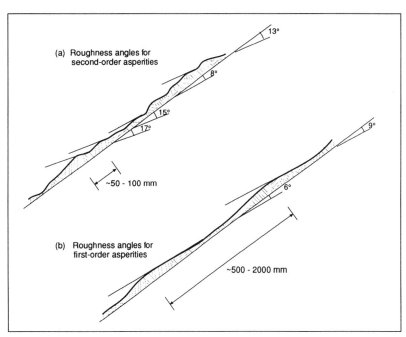

FIGURE 14-7
Definition of first- and second-order asperities on natural rough rock surfaces.

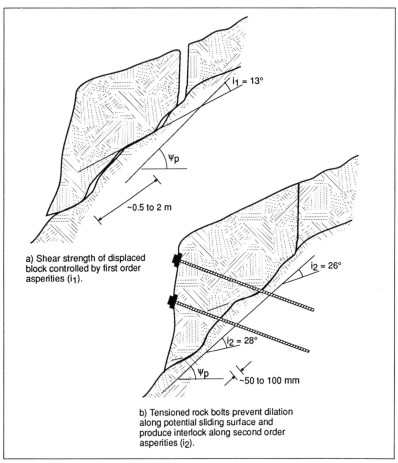

FIGURE 14-8
Roughness angles of first- and second-order asperities on natural rock surface.

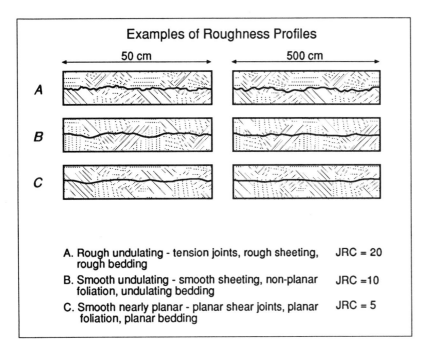

Examples of Roughness Profiles

A. Rough undulating - tension joints, rough sheeting, rough bedding   JRC = 20

B. Smooth undulating - smooth sheeting, non-planar foliation, undulating bedding   JRC =10

C. Smooth nearly planar - planar shear joints, planar foliation, planar bedding   JRC = 5

FIGURE 14-9
Definition of joint
roughness
coefficient, *JRC*
(Barton 1973).

70 degrees and the useful range for the ratio (σ_j/σ') is between about 3 and 100.

2.3 Measurement of Surface Roughness

If there are, at a project site, exposures of the fracture surfaces on which sliding may occur, measurements can be made of the surface roughness. If the exposure is extensive, the actual roughness angle (i) can be measured and used in Equation 14.2 in determining the shear strength of the surface. If the exposures are more limited, measurements of the *JRC* value can be made for use in Equation 14.3. Methods of making these measurements are described below.

A direct measurement of the roughness angle can be made in the field by measuring the variation in the orientation of the fracture surface (Feckers and Rengers 1971; Wyllie 1992). This procedure consists of measuring, in the direction of sliding, the orientation of the surface with varying base lengths. For short base lengths there may be considerable variation in the dip representing a rough surface, whereas for longer base lengths the variation diminishes until the measurements equal the average orientation of the surface. As discussed in Section 2.2, the asperities that would usually be used in design are the second-order asperities with wavelengths of about 500 mm or greater. A plot of base length versus dip angle would show the relationship between these two parameters and provide a guideline for appropriate design values.

When the surface roughness is determined using Equation 14.3, the *JRC* value can be estimated by visual inspection of the fracture and comparison with standard profiles of known *JRC* values, as shown in Figure 14-9. However, because visual comparison with the standard profiles is somewhat subjective and prone to error, two quantitative methods of profile measurement have been established. Tse and Cruden (1979) described a method using a mechanical profilometer, and Mearz et al. (1990) developed a shadow profilometer that records the shape of the surface by means of a video camera and image analyzer.

The method developed by Tse and Cruden is shown in Figure 14-10. The profile is defined by measuring the distance (y_i) of the surface from a fixed reference line at specified equal intervals (Δx) over a length of M intervals. From these measurements the coefficient Z2 is defined as

$$Z2 = \left[\frac{1}{M(\Delta x)^2} \sum_{i=1}^{M}(Y_{i+1} - Y_i)^2\right]^{1/2} \qquad (14.4)$$

The coefficient Z2 is then used to calculate the *JRC* of the fracture surface using the following regression equation:

$$JRC = 32.2 + 32.47 \log(Z2) \qquad (14.5)$$

A study has also been carried out to assess the effect of the size of the sampling interval along the profile on the calculated value of *JRC* (Yu and Vayssade 1991). This study showed that the sampling interval should not exceed 1 mm, and greater accuracy can be achieved by sampling at an interval of 0.25 mm.

2.4 Fracture Infilling

In the preceding section, the surfaces discussed are rough, clean fracture surfaces with rock-to-rock contact and no infilling, in which the shear strength is derived solely from the friction angle of the rock material. However, if the fracture contains an infilling, the shear strength properties of the fracture are often modified, with both the cohesion and friction angle of the surface being influenced by the thickness and properties of the infilling. For example, for a clay-filled fault zone in granite, it would be assumed that the shear strength of the fracture would be that of the clay and not the granite. In the case of a healed, calcite-filled fracture, a high cohesion would be used in design, but only if it was certain that the fracture would remain healed after any disturbance caused by blasting during excavation of the slope.

The presence of infillings along fracture surfaces can have a significant effect on stability. It is most important that infillings be identified in the investigation program and that appropriate strength parameters be used in design. For example, one of the contributing factors to the massive landslide into the Vaiont Reservoir in Italy that resulted in the deaths of about 2,000 people was the presence of low-shear-strength clay along the bedding surfaces of the shale (Trollope 1980).

The effect of the infilling on shear strength depends on both the thickness and strength properties of the infilling material. With respect to the thickness, if it is more than about 25 to 50 percent

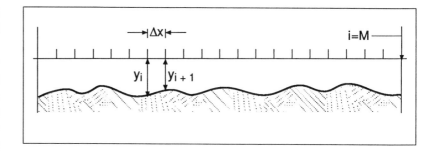

of the amplitude of the asperities, there will be little or no rock-to-rock contact, and the shear-strength properties of the fracture will be the properties of the infilling (Goodman 1970).

Figure 14-11 is a plot of the results of direct shear tests carried out to determine the peak fric-

FIGURE 14-10
Principle of measuring *JRC* of rock surface with mechanical profilometer (modified from Tse and Cruden 1979).
REPRINTED WITH KIND PERMISSION FROM ELSEVIER SCIENCE LTD, THE BOULEVARD, LANGFORD LANE, KIDLINGTON, UNITED KINGDOM

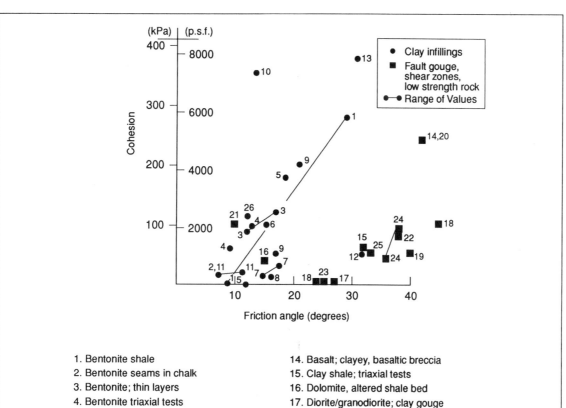

1. Bentonite shale
2. Bentonite seams in chalk
3. Bentonite; thin layers
4. Bentonite triaxial tests
5. Clay, over-consolidated
6. Limestone, 10-20 mm clay infillings
7. Lignite and underlying clay contact
8. Coal measures; clay mylonite seams
9. Limestone; <1 mm clay infillings
10. Montmorillonite clay
11. Montmorillonite; 80 mm clay seam in chalk
12. Schists/quartzites; stratification, thick clay
13. Schists/quartzites; stratification, thick clay
14. Basalt; clayey, basaltic breccia
15. Clay shale; triaxial tests
16. Dolomite, altered shale bed
17. Diorite/granodiorite; clay gouge
18. Granite; clay-filled faults
19. Granite; sandy-loam fault fillings
20. Granite; shear zone, rock and gouge
21. Lignite/marl contact
22. Limestone/marl/lignites; lignite layers
23. Limestone; marlaceous joints
24. Quartz/kaolin/pyrolusite; remolded triaxial
25. Slates; finely laminated and altered
26. Limestone; 10 to 20 mm clay infillings

FIGURE 14-11
Shear strength of filled discontinuities (Barton 1974; Wyllie 1992).

tion angle and cohesion of filled discontinuities (Barton 1974; Wyllie 1992). Examination of the test results shows that the infillings can be divided approximately into two groups, as follows:

- **Clays:** Montmorillonite and bentonitic clays and clays associated with coal measures have friction angles ranging from about 8 to 20 degrees and cohesion values ranging from 0 to about 200 kPa. Some cohesion values were measured as high as 380 kPa.
- **Faults, shears, and breccias:** The material formed in fault zones and shears in rocks such as granite, diorite, basalt, and limestone contains clay as well as granular fragments. These materials have friction angles ranging from about 25 to 45 degrees and cohesion values ranging from 0 to about 100 kPa. Fault gouge derived from coarse-grained rocks such as granites tends to have higher friction angles than those from fine-grained rocks such as limestones.

Some of the tests shown in Figure 14-11 also determined residual shear-strength values. It was found that the residual friction angle was only about 2 to 4 degrees less than the peak friction angle, whereas the residual cohesion was essentially zero.

Shear-strength and displacement behavior is an additional factor to consider regarding the shear strength of filled fractures. In the analysis of the stability of slopes, this behavior indicates whether there is likely to be a reduction in shear strength with displacement. In conditions where there is a significant decrease in shear strength with displacement, slope failure can occur suddenly following a small amount of movement.

Filled fractures can be divided into two general categories, depending on whether there has been previous displacement of the fracture (Barton 1974). These categories are further subdivided into either normally consolidated (NC) or over-consolidated (OC) materials (Figure 14-12).

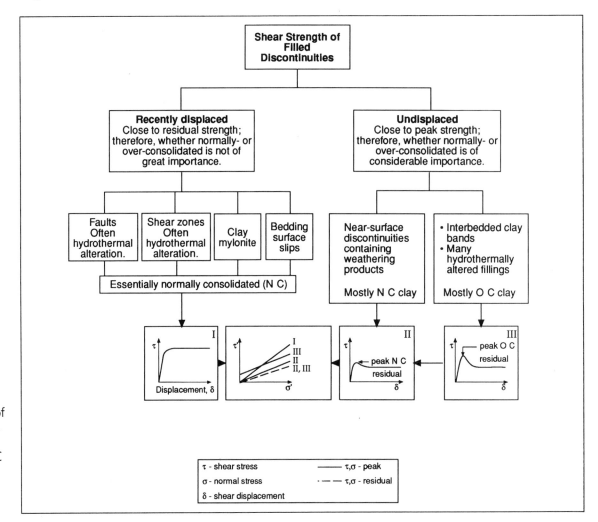

FIGURE 14-12
Simplified division of filled discontinuities into displaced and undisplaced and NC and OC categories (modified from Barton 1974).

2.4.1 Recently Displaced Fractures

Recently displaced fractures include faults, shear zones, clay mylonites, and bedding-surface slips. In faults and shear zones, the infilling is formed by the shearing process, which may have occurred many times and produced considerable displacement. The gouge formed in this process may include both clay-size particles and breccia with the particle orientation and striations of the breccia aligned parallel to the direction of shearing. In contrast, the mylonites and bedding-surface slips are fractures that were originally clay bearing and along which slip occurred during folding or sliding.

For these types of fractures the shear strength is at or close to the residual strength (Figure 14-12, Graph I). Any cohesive bonds that existed in the clay because of previous overconsolidation will have been destroyed by shearing, and the infilling will be equivalent to the normally consolidated state. In addition, strain softening may occur, with increased water content resulting in a further reduction in strength.

2.4.2 Undisplaced Fractures

Infilled fractures that have undergone no previous displacement include igneous and metamorphic rocks that have weathered along fracture surfaces to form clay layers. For example, diabase weathers to amphibolite and eventually to clay. Other undisplaced fractures include thin beds of clay and weak shales that are found with sandstone in interbedded sedimentary formations. Hydrothermal alteration is another process that forms infillings that can include low-strength materials such as montmorillonite and high-strength materials such as quartz and calcite.

The infillings of undisplaced fractures can be divided into NC and OC materials that have significant differences in peak-strength values. This strength difference is illustrated in Figure 14-12 by Graphs II and III. Although the peak shear strength of OC clay infillings may be high, there can be a significant loss of strength due to softening, swelling, and pore-pressure changes on unloading. Unloading occurs when rock is excavated for a slope or foundation, for example. Strength loss also occurs on displacement in brittle materials such as calcite.

2.5 Effect of Water on Shear Strength

The most significant effect of groundwater on slope stability is the reduction in normal stress produced by water pressures acting within the slope. As shown in Equation 14.1, a reduction in the effective normal stress results in a corresponding reduction in the shear strength of the rupture surface. The effect of water pressures on slope design is described in Chapter 15. Generally, the strength properties of rock are the same whether the rock is wet or dry. However, some rocks, such as shales containing swelling clay and evaporites such as gypsum, exhibit a loss of strength in the presence of water.

3. LABORATORY TESTING OF SHEAR STRENGTH

The friction angle of a fracture surface can be determined in the laboratory using a direct shear box of the type shown in Figure 14-13. This is portable equipment that can be used in the field if required and is suited to testing samples with dimensions up to about 75 mm, such as NQ and HQ drill core. The most reliable values are obtained if a sample with a smooth, planar surface is used because it is found that with an irregular surface, the effect of surface roughness can make the test results difficult to interpret.

The test procedure consists of using plaster of paris or sulfur to set the two halves of the sample in a pair of steel boxes (International Society for Rock Mechanics 1981b). Particular care is taken to ensure that the two pieces of core are in their original, matched position and the fracture surface is exactly parallel to the direction of the shear force. A normal load is then applied using the cantilever, and the shear load is gradually increased until a sliding failure occurs. Measurement of the relative vertical and horizontal displacement of the upper block can be made most simply with dial gauges, whereas more precise, continuous displacement measurements can be made with linear variable differential transformers (Hencher and Richards 1989).

Each sample is usually tested three or four times with progressively higher normal loads. When the residual shear stress has been established for each normal load, the sample is reset, the normal load increased, and another shear test conducted. The

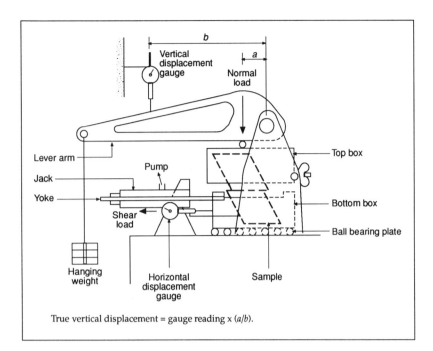

b

Vertical
displacement
gauge

a

Normal
load

Top box

Lever arm

Pump

Jack

Yoke

Bottom box

Shear
load

Ball bearing plate

Hanging
weight

Horizontal
displacement
gauge

Sample

True vertical displacement = gauge reading × (a/b).

FIGURE 14-13
Simple equipment
for performing
direct shear tests
on rock samples
up to about 75 mm
in diameter (Wyllie
1992).

test results are expressed as plots of shear displacement against shear stress from which the peak and residual shear stress values are determined. Each test produces a pair of shear-stress–normal-stress values, which are plotted to determine the peak and residual friction angles of the surface.

Figure 14-14 shows a typical result of a direct shear test on a fracture with a 4-mm-thick sandy-silt infilling. The curves on the upper right are shear-stress–shear-displacement plots showing an approximate peak shear stress as well as a slightly lower, residual shear stress. The sample was undisplaced, so it exhibited a difference between peak and residual strengths (see Figure 14-12). The normal stresses at the peak and residual shear stress values are calculated from the applied normal load and the contact area. When the contact area is calculated, an allowance is made for the decrease in area as shear displacement takes place. For diamond drill core in an inclined hole, the fracture surface is in the shape of an ellipse and the formula for calculating the contact area is as follows (Hencher and Richards 1989):

$$A = \pi ab - \left[\frac{ub(4a^2 - u^2)}{2a} \right]^{1/2} - 2ab\sin^{-1}\left(\frac{u}{2a} \right) \quad (14.6)$$

where

A = gross area of contact,
2a = major axis of ellipse,

2b = minor axis of ellipse, and
u = relative shear displacement.

As a result of the decrease in contact area with shearing, there will be a corresponding increase in the normal stress. This is shown in the upper left diagram of Figure 14-14, where the normal stress for the residual shear strength is greater than that for the peak shear strength.

The measured friction angle is the sum of the friction angle of the rock (ϕ) and the roughness of the surface (i) at each normal stress level. The roughness of the surface is calculated from the plots of shear and normal displacement (δ_s and δ_n, respectively, on the lower right side of Figure 14-14) as follows:

$$i = \tan^{-1}\left(\frac{\delta_n}{\delta_s} \right) \quad (14.7)$$

This value of i is then subtracted from the friction angle calculated from the plot of shear and normal stresses at failure to obtain the friction angle of the rock. Although the shear test can be conducted on a sawed sample so that there is no roughness component, there may be some polishing of the surface resulting in a low value of the friction angle compared with a natural surface.

As shown in Figure 14-14, it is usual, at a minimum, to test each sample at three normal stress levels, with the sample being reset to its original position between tests. When the tests are run at progressively higher normal stress levels, the total friction angle of the surface will diminish with each test if the asperities are progressively sheared. This produces a concave downward normal shear plot as shown in the upper left of Figure 14-14. The degree to which the asperities are sheared off will depend on the level of the normal stress in comparison to the rock strength, that is, the ratio (σ_j/σ') in Equation 14.3. The maximum normal stress that is used in the test is usually the maximum stress level that is likely to develop in the slope.

It is also possible to measure the normal stiffness of the fracture infilling during the direct shear test, as shown in the lower left of Figure 14-14. Normal stiffness is the ratio of the normal stress to normal displacement, or

$$k_n = \frac{\sigma}{\delta_n} \quad (14.8)$$

The plot of σ against δ_n is highly nonlinear and the value of δ_n is the slope of the initial portion of

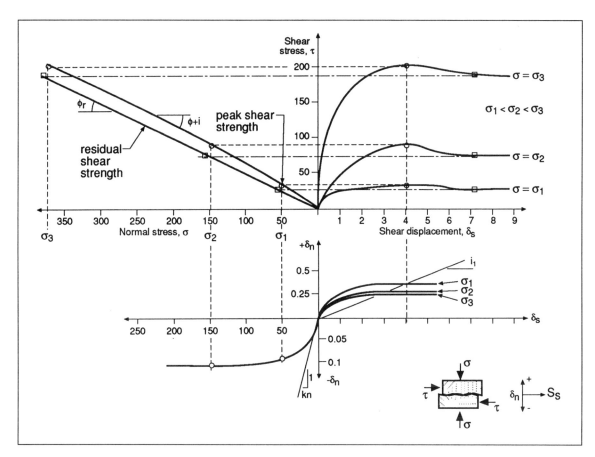

FIGURE 14-14
Results of direct
shear test of filled
discontinuity
showing
measurements
of shear strength,
roughness (*i*),
and normal
stiffness (k_n)
(modified from
Erban and Gill
1988).

the curve. The normal stiffness of a fracture is not usually an issue in rock slope design and is more often used in the estimation of the deformation modulus of a rock mass (Wyllie 1992).

It can be difficult to measure the cohesion of a surface with the direct shear test because if the cohesion is very low, it may not be possible to obtain an undisturbed sample. If the cohesion is high and the sample is intact, the material holding the sample will have to be stronger than the infilling if the sample is to shear. When it is important that the cohesion of a weak infilling be measured, an in situ test of the undisturbed material may be required.

4. SHEAR STRENGTH OF FRACTURED ROCK MASSES

For the geological conditions shown in Figure 14-2, where a cut has been made in fractured rock, there is no distinct fracture surface on which sliding can take place. A rupture surface in this rock mass will comprise natural fractures aligned on the rupture surface together with some

shear failure through intact rock. It is generally too difficult and expensive to sample and test large samples of fractured rock. Consequently, two empirical methods of determining the friction angle and cohesion of rock masses have been developed and are described in this section. In both methods it is necessary to categorize the rock mass in terms of both the intact rock strength and the characteristics of the fractures. Because this requires considerable judgment, it is advisable to compare the strength values obtained by both methods to improve the reliability of values used in design.

4.1 Strength Determination by Back Analysis of Failures

Probably the most reliable method of determining the strength of a rock mass is to back analyze a failed, or failing, slope. This procedure involves carrying out a stability analysis with the factor of safety set at 1.0 and using available information on the position of the rupture surface, the groundwater conditions at the time of failure, and any

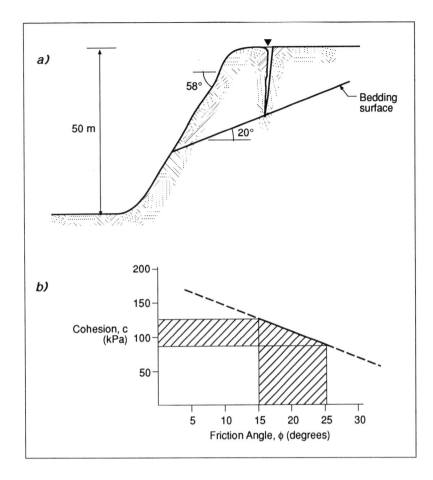

FIGURE 14-15
Example of back
analysis of failed
slope to determine
shear strength of
bedding surface
(modified from
Hoek and Bray
1981): (*a*) section
showing failed
slope; (*b*) plot of
cohesion versus
friction angle.

external forces such as foundation loads and earth-quake motion, if applicable. Methods of stability analysis of rock slopes are discussed in Chapter 15.

Figure 14-15(*a*) shows a back analysis of a slope failure in limestone in which sliding occurred on bedding planes dipping out of the face at an angle of 20 degrees (Hoek and Bray 1981). At the time of failure, the upper bench of the quarry was flooded and the tension crack was full of water, so the forces acting on the slope were known with some confidence.

In back analysis both the friction angle and the cohesion of the rupture surface are unknown, and their values can be estimated by the following method. By carrying out a number of stability analyses with a range of cohesion values, it is possible to calculate a corresponding value for the friction angle (at *F* = 1). From a plot of cohesion against friction angle [Figure 14-15(*b*)], it is possible to select values for the cohesion and friction angle that can be used in design. When the friction angle of the rock can be determined in a direct shear test and the roughness of the surface can be

measured in the field, the cohesion of the rupture surface can be determined with some confidence.

In many cases it may not be feasible to carry out a back analysis of a slope in geological conditions similar to those in which the new slope is to be excavated. In these circumstances, published results of rock mass shear strength can be used in design. Figure 14-16 shows the results of back analyses of slope failures in a variety of geological conditions (as described in Table 14-2) and the shear-strength parameters (ϕ/*c* values) calculated at failure. By adding points to Figure 14-16 for local geological conditions, it is possible to draw up a readily applicable rock mass strength chart for shear failures.

4.2 Hoek-Brown Strength Criteria for Fractured Rock Masses

As an alternative to back analysis to determine the strength of fractured rock masses, an empirical method was developed by Hoek (1983) and Hoek and Brown (1988) in which the shear strength is represented as a curved envelope (Figure 14-17). This strength criterion was derived from the Griffith crack theory of rock fracture, as well as from observations of the behavior of rock masses in the laboratory and in the field (Marsal 1973; Jaeger 1970).

The three parameters defining the curved strength envelope of the rock mass are the uni-axial strength of the intact rock, σ_c, and two dimensionless constants *m* and *s*. The values of *m* and *s* are defined in Table 14-3, in which rock types are shown in the columns, and the strength and degree of fracturing are shown in the rows. The rock types are grouped into five classes depending on the grain size and crystal structure. The quality of the rock mass is defined in Table 14-3 either by the brief descriptions of the geology or by rock-mass rating systems (Bieniawski 1974, 1976) that assign point scores to a number of characteristics of the rock mass.

The equation for the curved shear strength envelope is

$$\tau = (\cot\phi_i' - \cos\phi_i') \frac{m\sigma_c}{8} \qquad (14.9)$$

where τ is the shear stress at failure and ϕ_i' is the instantaneous friction angle at given values of τ and σ'. The value of ϕ_i' is the inclination of the tangent to the Mohr failure envelope at the point (σ', τ) as shown in Figure 14-17 and is given by

$$\phi_i' = \arctan\{4h\cos^2[30 + \tfrac{1}{3}\arcsin(h^{-3/2})] - 1\}^{-1/2}$$

$$\text{(14.10)}$$

where

$$h = 1 + \frac{16(m\sigma' + s\sigma_c)}{3m^2\sigma_c} \qquad \text{(14.11)}$$

The dimensionless constants m and s depend on the rock type and the degrees of fracturing of the rock mass and are defined in Table 14-3. The instantaneous cohesion (c_i') is the intercept of the line defining the friction angle on the shear stress axis and is given by

$$c_i' = \tau - \sigma'\tan\phi_i' \qquad \text{(14.12)}$$

The features of the curved shear-strength envelope are that at low normal-stress levels, the blocks of rock are interlocked and the instantaneous friction angle is high, whereas at higher normal-stress levels, shearing of the rock is initiated with the result that the friction angle diminishes. The instantaneous cohesion progressively increases with the normal stress as a result of the greater confinement and interlock of the rock mass.

The procedure for using the curved strength envelopes in stability analysis is, first, to determine the range of effective normal stresses acting along a potential rupture surface in the slope and, second, to calculate the instantaneous cohesion values and friction angles in this stress range. The stability analysis is carried out in the normal manner except that a number of values of c_i and ϕ_i

Table 14-2
Sources of Shear-Strength Data Plotted in Figure 14-16

POINT NUMBER	MATERIAL	LOCATION	SLOPE HEIGHT (M)	REFERENCE
1	Disturbed slates and quartzites	Knob Lake, Canada	—	Coates et al. (1965)
2	Soil	Any location	—	Whitman and Bailey (1967)
3	Jointed porphyry	Rio Tinto, Spain	50-110	Hoek (1970)
4	Ore body hanging wall in granite rocks	Grangesberg, Sweden	60-240	Hoek (1974)
5	Rock slopes with slope angles of 50 to 60 degrees	Any location	300	Ross-Brown (1973)
6	Bedding planes in limestone	Somerset, England	60	Roberts and Hoek (1972)
7	London clay, stiff	England	—	Skempton and Hutchinson (1969)
8	Gravelly alluvium	Pima, Arizona	—	Hamel (1970)
9	Faulted rhyolite	Ruth, Nevada	—	Hamel (1971a)
10	Sedimentary series	Pittsburgh, Pennsylvania	—	Hamel (1971b)
11	Kaolinized granite	Cornwall, England	75	Ley (1972)
12	Clay shale	Fort Peck Dam, Montana	—	Middlebrook (1942)
13	Clay shale	Gardiner Dam, Canada	—	Fleming et al. (1970)
14	Chalk	Chalk Cliffs, England	15	Hutchinson (1972)
15	Bentonite/clay	Oahe Dam, South Dakota	—	Fleming et al. (1970)
16	Clay	Garrison Dam, North Dakota	—	Fleming et al. (1970)
17	Weathered granites	Hong Kong	13-30	Hoek and Richards (1974)
18	Weathered volcanics	Hong Kong	30-100	Hoek and Richards (1974)
19	Sandstone, siltstone	Alberta, Canada	240	Wyllie and Munn (1979)
20	Argillite	Yukon, Canada	100	Wyllie (unpublished data)

FIGURE 14-16 *(right)* Relationship between friction angles and cohesive strength mobilized at failure for slopes analyzed in Table 14-2 (Hoek and Bray 1981).

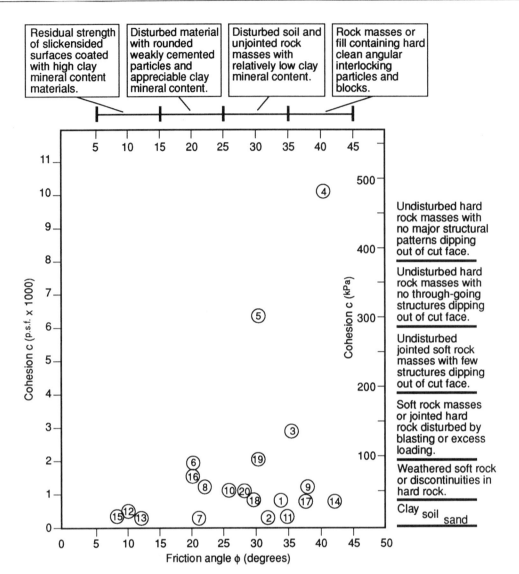

FIGURE 14-17 *(below)* Typical curved shear strength envelope defined by Hoek-Brown theory for rock mass strength (Hoek 1983).

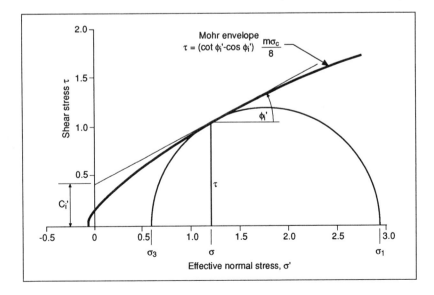

are used corresponding to the variation in normal stress along the rupture surface.

The following is an example of the use of the Hoek-Brown strength criterion to develop shear-strength values for a fractured rock mass. Consider the slope shown in Figure 14-3, where the rupture surface lies within the highly weathered rock. This material could be described as a "poor quality rock mass," and if the rock type is a limestone, then from Table 14-3, the values of the material constants are $m = 0.029$ and $s = 0.000003$. The compressive strength of the rock, as determined by inspection and simple field tests shown in Chapter 9 (Table 9-4), is estimated to be 35 MPa for "moderately weak rock." This strength, together with values for the parameters m and s, is then substituted in Equations 14.10, 14.11, and 14.12 to cal-

TABLE 14-3
Approximate Relationship Between Rock-Mass Quality and Material Constants Used in Defining Nonlinear Strength (Hoek and Brown 1988).

Empirical failure criterion $\sigma'_1 = \sigma'_3 + \sqrt{m\sigma_{u(r)}\sigma'_3 + s\sigma^2_{u(r)}}$ σ'_1 = major principal effective stress σ'_3 = minor principal effective stress $\sigma_{u(r)}$ = uniaxial compressive strength of intact rock, and m and s are empirical constants	MATERIAL CONSTANTS: m, s	CARBONATE ROCKS WITH WELL DEVELOPED CRYSTAL CLEAVAGE *dolomite, limestone and marble*	LITHIFIED ARGILLACEOUS ROCKS *mustone, siltstone, shale and slate (normal to cleavage)*	ARENACEOUS ROCKS WITH STRONG CRYSTALS AND POORLY DEVELOPED CRYSTAL CLEAVAGE *sandstone and quartzite*	FINE GRAINED POLYMINERALLIC IGNEOUS CRYSTALLINE ROCKS *andesite, dolerite, diabase and rhyolite*	COARSE GRAINED POLYMINERALLIC IGNEOUS & METAMORPHIC CRYSTAL-LINE ROCKS - *amphibolite, gabbro gneiss, granite, norite, quartz-diorite*
INTACT ROCK SAMPLES *Laboratory size speciments free from discontinuities* *CSIR rating: RMR = 100* *NGI rating: Q = 500*	m s	7.00 1.00	10.00 1.00	15.00 1.00	17.00 1.00	25.00 1.00
VERY GOOD QUALITY ROCK MASS *Tightly interlocking undisturbed rock with unweathered joints at 1 to 3 m* CSIR rating: RMR = 85 NGI rating: Q = 100	m s	2.40 0.082	3.43 0.082	5.14 0.082	5.82 0.082	8.56 0.082
GOOD QUALITY ROCK MASS *Fresh to slightly weathered rock, slightly disturbed with joints at 1 to 3 m* CSIR rating: RMR = 65 NGI rating: Q = 10	m s	0.575 0.00293	0.821 0.00293	1.231 0.00293	1.395 0.00293	2.052 0.00293
FAIR QUALITY ROCK MASS *Several sets of moderately weathered joints spaced at 0.3 to 1m.* CSIR rating: RMR = 44 NGI rating: Q = 1	m s	0.128 0.00009	0.183 0.00009	0.275 0.00009	0.311 0.00009	0.458 0.00009
POOR QUALITY ROCK MASS *Numerous weathered joints at 30-500 mm some gouge. Clean compacted waste rock* CSIR rating: RMR = 23 NGI raing: Q = 0.1	m s	0.029 0.000003	0.041 0.000003	0.061 0.000003	0.069 0.000003	0.102 0.000003
VERY POOR QUALITY ROCK MASS *Numerous heavily weathered joints spaced <50 mm with gouge. Waste rock with fines* CSIR rating: RMR = 3 NGI rating: Q = 0.01	m s	0.007 0.0000001	0.010 0.0000001	0.015 0.0000001	0.017 0.0000001	0.025 0.0000001

*CSIR Commonwealth Scientific and Industrial Research Organization (Bieniawski 1974)

†NGI Norway Geotechnical Institute (Barton 1974)

culate the cohesion and friction angle at each selected normal-stress value. Table 14-4 shows an example of these calculations using the approximate dimensions of the slope in Figure 14-3.

As part of the ongoing development of the Hoek-Brown strength criterion, a revised rock-mass rating system called the Geological Strength Index (GSI) and revised equations for the strength envelope have been proposed (Hoek 1995). For this report the more familiar Norwegian Geotechnical Institute (NGI) and Commonwealth Scientific and Industrial Research Organization (CSIR) ratings have been retained because of their present wide use in practical applications.

These results demonstrate the variation in shear strength along the rupture surface. When a circular stability analysis of this slope is carried out, the effective normal stress at the base of each slice and the corresponding instantaneous shear strength parameters can be calculated as part of the input for the analysis.

5. ROCK DURABILITY AND COMPRESSIVE STRENGTH

The strength parameters that are usually of greatest significance to the analysis of slope stability are the cohesion and friction angle on the rupture surface, as discussed previously in this chapter. However, both the durability and compressive strength of the rock may be of importance, depending on the geological and stress conditions at the site, as discussed below. Both the procedures discussed next are index tests best used to classify and compare one rock with another; if necessary, the index measurements can be calibrated by more precise tests.

5.1 Slake Durability

A widely occurring class of rock materials, such as shales and mudstone, many of which have a high clay content, are prone to degradation when exposed to weathering processes such as wetting and drying cycles. The degradation can take the form of swelling, weakening, and disintegration and can occur within a period of minutes to years after exposure. The effect of degradation on slope stability can range from surficial sloughing and gradual retreat of the face to slope failures resulting from the loss of strength with time (Wu et al. 1981). In sedimentary formations comprising alternating beds of resistant sandstone and degradable shale, the weathering process can develop overhangs in the sandstone and produce a rock-fall hazard.

A simple index test of the tendency of rock to weather and degrade is the slake durability test (International Society for Rock Mechanics 1981a). It is important that undisturbed samples be used that have not been excessively broken in the sampling procedure or allowed to freeze. The test procedure consists of placing the sample in a wire mesh drum and drying it in an oven at 105°C for 2 to 6 hr. The drum is then partially submerged in water and rotated at 20 revolutions per minute for a period of 10 min. The drum is dried a second time and the loss of weight is recorded. The test cycle is repeated and the slake durability index is calculated as the percentage ratio of final to initial dry sample masses. A low slake durability index indicates that the rock is susceptible to degradation when exposed. For highly degradable rocks it is useful to carry out soil classification tests such as Atterberg limits and X-ray diffraction tests to identify clay mineral types and determine if swelling clays such as bentonites and montmorillonites are present.

Table 14-4
Calculated Values of Instantaneous Cohesion and Friction Angle Using Hoek-Brown Strength Criterion

Parameter	Dimension and Units	Center of Rupture Surface	Upper Rupture Surface
Depth of rupture surface	d (m)	12	11
Dip of rupture surface	ψ_p (degrees)	30	50
Rock unit weight	γ (MN/m³)	24	24
Normal stress	$[d \cdot \gamma \cdot \cos\psi_p]$ (MPa)	250	170
Instantaneous cohesion	c_i' (MPa)	3.86	3.16
Instantaneous friction angle	ϕ_i' (degrees)	42	48

5.2 Compressive Strength

In many slopes in moderately strong to strong rock, the stress level due to gravity loads is significantly less than the strength of the rock. Therefore, there is little tendency for intact rock within the slope to fracture, and the compressive strength is a less important design parameter than the shear strength. The compressive strength of the rock on the rupture surface is only used indirectly in stability analysis when determining the roughness of a fracture (Equation 14.3) and in the application of the Hoek-Brown strength criteria (Equations 14.9 to 14.12). For both of these applications it is satisfactory to use an estimate of the compressive strength because the results are not particularly sensitive to the value of this parameter.

The point-load test is an appropriate method to estimate the compressive strength in which both core and lump samples can be tested (International Society for Rock Mechanics 1981c). The equipment is portable, and tests can be carried out quickly and inexpensively in the field. Because the point-load test provides an index value for the strength, the usual practice is to calibrate the results with a limited number of uniaxial compressive tests on prepared core samples.

If no equipment is available to measure the compressive strength, simple field observations can be used to estimate the strength with sufficient accuracy for most purposes. In Chapter 9 (Table 9-4) a series of field index tests and observations for estimating strength is described, and the corresponding range of compressive strengths is given.

REFERENCES

Barton, N. R. 1973. Review of a New Shear Strength Criterion for Rock Joints. *Engineering Geology*, Vol. 7, pp. 287–332.

Barton, N. R. 1974. *A Review of the Shear Strength of Filled Discontinuities in Rock*. Publication 105. Norwegian Geotechnical Institute, Oslo, 38 pp.

Barton, N. R. 1976. The Shear Strength of Rock and Rock Joints. *International Journal of Rock Mechanics and Mining Sciences & Geomechanics Abstracts*, Vol. 13, pp. 255–279.

Bieniawski, Z. T. 1974. Geomechanics Classification of Rock Masses and Its Application in Tunnelling. In *Proc., Third International Congress on Rock Mechanics*, Denver, National Academy of Sciences, Washington, D.C., Vol. 2, Part 2, pp. 27–32.

Bieniawski, Z. T. 1976. Rock Mass Classifications in Rock Engineering. In *Proc., Symposium on Exploration for Rock Engineering* (Z. T. Bieniawski, ed.), A.A. Balkema, Rotterdam, Netherlands, Vol. 1, pp. 97–106.

Coates, D. F., M. Gyenge, and J. B. Stubbins. 1965. Slope Stability Studies at Knob Lake. In *Proc., Rock Mechanics Symposium*, Department of Mines and Technical Surveys, Ottawa, Ontario, Canada, pp. 35–46.

Erban, P.-J., and K. Gill. 1988. Consideration of the Interaction Between Dam and Bedrock in a Coupled Mechanic-Hydraulic FE-Program. *Rock Mechanics and Rock Engineering*, Vol. 21, No. 2, pp. 99–118.

Feckers, E., and N. Rengers. 1971. Measurement of Large Scale Roughness of Rock Planes by Means of Profilometer and Geological Compass. In *Proc., International Symposium on Rock Fracture*, Nancy, France, International Society for Rock Mechanics, Lisbon, Portugal, Paper 1-18.

Fleming, R. W., G. S. Spencer, and D. C. Banks. 1970. *Empirical Study of the Behavior of Clay Shale Slopes*. Nuclear Cratering Group Technical Report 15. U.S. Department of the Army.

Goodman, R. E. 1970. The Deformability of Joints. In *Determination of the In-Situ Modulus of Deformation of Rock*, Special Technical Publication 477, ASTM, Philadelphia, Pa., pp. 174–196.

Hamel, J. V. 1970. The Pima Mine Slide, Pima County, Arizona. *Abstracts with Programs, Geological Society of America*, Vol. 2, No. 5, p. 335.

Hamel, J. V. 1971a. Kimberley Pit Slope Failure. In *Proc., Fourth Pan-American Conference on Soil Mechanics and Foundation Engineering*, Puerto Rico, Pan-American Conference on Soil Mechanics and Foundation Engineering, New York, Vol. 2, pp.117–127.

Hamel, J. V. 1971b. The Slide at Brilliant Cut. In *Proc., 13th Symposium on Rock Mechanics*, Urbana, Illinois, American Society of Civil Engineers, New York, pp. 487–510.

Hencher, S. R., and L.R. Richards. 1989. Laboratory Direct Shear Testing of Rock Discontinuities. *Ground Engineering*, March, pp. 24–31.

Hoek, E. 1970. Estimating the Stability of Excavated Slopes in Opencast Mines. *Transactions of the Institution of Mining and Metallurgy*, London, Vol. 79, pp. A109–A132.

Hoek. E. 1974. Progressive Caving Caused by Mining an Inclined Ore Body. *Transactions of the Institution of Mining and Metallurgy*, London, Vol. 83, pp. A133–A139.

Hoek, E. 1983. Strength of Jointed Rock Masses. *Geotechnique*, Vol. 33, No. 3, pp. 187–223.

Hoek, E. 1995. Strength of Rock and Rock Masses. *News Journal, International Society of Rock Mechanics*, Vol. 2, No. 2, pp. 4–16.

Hoek, E., and J. Bray. 1981. *Rock Slope Engineering*, 3rd ed. Institution of Mining and Metallurgy, London, 402 pp.

Hoek, E., and E. T. Brown. 1988. The Hoek-Brown Failure Criterion—A 1988 Update. In *Proc., 15th Canadian Rock Mechanics Symposium*, University of Toronto, Canada.

Hoek, E., and L. R. Richards. 1974. *Rock Slope Design Review*. Report to the Principal Government Highway Engineer, Hong Kong. Golder Associates, 150 pp.

Hutchinson, J. N. 1972. Field and Laboratory Studies of a Fall in Upper Chalk Cliffs at Joss Bay, Isle of Thanet. In *Stress-Strain Behaviour of Soil: Roscoe Memorial Symposium*, Cambridge University, G.T. Foulis, United Kingdom.

International Society for Rock Mechanics. 1981a. Suggested Method for Determination of Slake Durability Index. In *Rock Characterization, Testing and Monitoring*, Pergamon Press, London, pp. 92–94.

International Society for Rock Mechanics. 1981b. Suggested Method for Determination of Direct Shear Strength. In *Rock Characterization, Testing and Monitoring*, Pergamon Press, London, pp. 135–137.

International Society for Rock Mechanics. 1981c. Suggested Method for Determination of Uniaxial Compressive Strength of Rock Materials. In *Rock Characterization, Testing and Monitoring*, Pergamon Press, London, pp. 113–120.

Jaeger, J. C. 1970. The Behavior of Closely Jointed Rock. In *Proc., 11th Symposium on Rock Mechanics*, Berkeley, California, American Society of Civil Engineers, New York, pp. 57–68.

Jaeger, J. C., and N. G. W. Cook. 1976. *Fundamentals of Rock Mechanics*. Chapman and Hall, London, 585 pp.

Ley, G. M. M. 1972. *The Properties of Hydrothermally Altered Granite and Their Application to Slope Stability in Open Cast Mining*. M.Sc. thesis. London University.

Marsal, R. J. 1973. Mechanical Properties of Rock Fill. In *Embankment Dam Engineering—Casagrande Volume* (R. C. Hirschfield and S. J. Paulos, eds.), J. Wiley and Sons, New York, pp. 109–200.

Mearz, N. H., J. A. Franklin, and C. P. Bennett. 1990. Joint Roughness Measurement Using Shadow Profilometry. *International Journal of Rock Mechanics and Mining Sciences & Geomechanics Abstracts*, Vol. 27, No. 5, pp. 329–343.

Middlebrook, T. A. 1942. Fort Peck Slide. *Proc., American Society of Civil Engineers*, Vol. 107, Paper 2144, pp. 723.

Patton, F. D. 1966. Multiple Modes of Shear Failure in Rock. In *Proc., First Congress of the International Society for Rock Mechanics*, Lisbon, National Laboratory of Civil Engineering, Lisbon, Portugal, Vol. 1, pp. 509–513.

Roberts, D., and E. Hoek. 1972. A Study of the Stability of a Disused Limestone Quarry Face in the Mendip Hills, England. In *Proc., First International Conference on Stability in Open Pit Mining*, Vancouver, Canada, American Institute of Mining Engineers, New York, pp. 239–256.

Ross-Brown, D. R. 1973. *Slope Design in Open Cast Mines*. Ph.D. thesis. London University, 250 pp.

Skempton, A. W., and J. N. Hutchinson. 1969. Stability of Natural Slopes and Embankment Foundations: State of the Art Report. In *Proc., Seventh International Conference on Soil Mechanics and Foundation Engineering*, Mexico, Mexican Society for Soil Mechanics, Mexico City, Vol. 1, pp. 291–340.

Trollope, D. H. 1980. The Vaiont Slope Failure. *Rock Mechanics*, Vol. 13, No. 2, pp. 71–88.

Tse, R., and D. M. Cruden. 1979. Estimating Joint Roughness Coefficients. *International Journal of Rock Mechanics and Mining Sciences & Geomechanics Abstracts*, Vol. 16, pp. 303–307.

Whitman, R. V., and W. A. Bailey. 1967. Use of Computers for Slope Stability Analysis. *Journal of the Soil Mechanics and Foundations Division*, ASCE, Vol. 93, pp. 475–498.

Wu, T. H., E. M. Ali, and H. S.U. Pinnaduwa. 1981. *Stability of Slopes in Shale and Colluvium*. Project No. EES 576. Ohio Department of Transportation and Federal Highway Administration, 229 pp.

Wyllie, D. C. 1992. *Foundations on Rock*. Chapman and Hall, London, 331 pp.

Wyllie, D. C., and F. J. Munn. 1979. Use of Movement Monitoring to Minimize Production Losses Due to Pit Slope Failures. In *Proc., First International Symposium on Stability in Coal Mining*, Vancouver, Canada, Miller Freeman Publications, pp. 75–94.

Yu, X., and B. Vayssade. 1991. Joint Profiles and Their Roughness Parameters. *International Journal of Rock Mechanics and Mining Sciences & Geomechanics Abstracts*, Vol. 16, No. 4, pp. 333–336.

Chapter 15

NORMAN I. NORRISH AND
DUNCAN C. WYLLIE

ROCK SLOPE STABILITY ANALYSIS

1. INTRODUCTION

Except for the rare case of a completely un-fractured rock unit, the majority of rock masses can be considered as assemblages of intact rock blocks delineated in three dimensions by a system of discontinuities. These discontinuities can occur as unique randomly oriented features or as repeating members of a discontinuity set. This system of structural discontinuities is usually referred to as the structural fabric of the rock mass and can consist of bedding surfaces, joints, foliation, or any other natural break in the rock. In most cases, engineering properties of fractured rock masses, such as strength, permeability, and deformability, are more dependent on the nature of the structural fabric than on the properties of the intact rock. For this reason, practitioners in the field of rock mechanics have developed the following parameters to characterize the nature of the discontinuities that make up the structural fabric:

- **Orientation:** The orientation of a discontinuity is best defined by two angular parameters: dip and dip direction.
- **Persistence:** Persistence refers to the continuity or areal extent of a discontinuity and is particularly important because it defines the potential volume of the failure mass. Persistence is difficult to quantify; the only reliable means is mapping of bedrock exposures.
- **Spacing:** The distance between two discontinu-

ities of the same set measured normal to the discontinuity surfaces is called spacing. Together, persistence and spacing of discontinuities define the size of blocks. They also influence the dilatancy of the rock mass during shear displacement and determine the extent to which the mechanical properties of the intact rock will govern the behavior of the rock mass.
- **Surface properties:** The shape and roughness of the discontinuity constitute its surface properties, which have a direct effect on shear strength.
- **Infillings:** Minerals or other materials that occur between the intact rock walls of discontinuities are termed infillings. Their presence can affect the permeability and shear strength of a discontinuity. Secondary minerals, such as calcite or quartz, may provide significant cohesion along discontinuities. However, these thin infillings are susceptible to damage by blasting, resulting in total loss of cohesion.

For all but very weak rock materials, the analysis of rock slope stability is fundamentally a two-part process. The first step is to analyze the structural fabric of the site to determine if the orientation of the discontinuities could result in instability of the slope under consideration. This determination is usually accomplished by means of stereographic analysis of the structural fabric and is often referred to as *kinematic analysis* (Piteau and Peckover 1978). Once it has been determined that a kinematically possible failure mode is present, the second step requires a limit-equilibrium stability analysis to compare the forces resisting failure with the forces causing failure. The ratio between these two sets of forces is termed the factor of safety, *FS*.

For very weak rock where the intact material strength is of the same magnitude as the induced stresses, the structural geology may not control stability, and classical soil mechanics principles for slope stability analysis apply. These procedures are discussed in Chapter 13.

2. TYPES OF ROCK SLOPE FAILURES

As shown in Figure 15-1, most rock slope failures can be classified into one of four categories depending on the type and degree of structural control:

- *Planar failures* are governed by a single discontinuity surface dipping out of a slope face [Figure 15-1(a)],

- *Wedge failures* involve a failure mass defined by two discontinuities with a line of intersection that is inclined out of the slope face [Figure 15-1(b)],
- *Toppling failures* involve slabs or columns of rock defined by discontinuities that dip steeply into the slope face [Figure 15-1(c)], and
- *Circular failures* occur in rock masses that are either highly fractured or composed of material with low intact strength [Figure 15-1(d)].

Recognition of these four categories of failures is essential to the application of appropriate analytical methods.

3. STEREOGRAPHIC ANALYSIS OF STRUCTURAL FABRIC

From a rock slope design perspective, the most important characteristic of a discontinuity is its

FIGURE 15-1
(facing pages)
Types of rock
slope failures:
(*a*) planar failure,
(*b*) wedge failure,
(*c*) toppling failure,
(*d*) circular failure
(diagrams modified
from Hoek and
Bray 1981).

N.I. NORRISH AND
D.C. WYLLIE

(a)

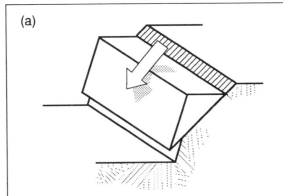

Planar failure in rock in which a discontinuity "daylights" the slope face

(b)

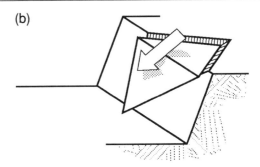

Wedge failure on two intersecting discontinuities with a line of intersection which "daylights" the slope

orientation, which is best defined by two parameters: dip and dip direction [Figure 15-2(a)]. The dip angle refers to the inclination of the plane below the horizontal and thus ranges from 0 to 90 degrees. The dip direction of the plane is the azimuth at which the maximum dip is measured and ranges from 0 to 360 degrees. The dip direction differs from the strike direction by 90 degrees and is the preferred parameter to avoid ambiguity as to the direction of dip. These values are determined by compass measurements on rock outcrops (Chapter 9), oriented drilling techniques, or interpretation of geologic structural trends (Chapter 8).

Interpretation of these geologic structural data requires the use of stereographic projections that allow the three-dimensional orientation data to be represented and analyzed in two dimensions. The most commonly used projections are the equal-area net and the polar net, replications of which are shown in Appendix A.

The rock slope practitioner is referred to the work by Hoek and Bray (1981), Hoek and Brown (1980), and Goodman (1976) for in-depth treatment of the principles of stereographic analysis. A detailed presentation of the procedures for plotting, analyzing, and interpreting data on stereographic projections is beyond the scope of this report; however, these techniques are essential to rock slope design and to the following discussion.

Stereographic presentations remove one dimension from consideration so that planes can be represented by lines and lines represented by points. Stereographic analyses consider only angular relationships between lines, planes, and lines and planes. These analyses do not in any way represent the position or size of the feature.

The fundamental concept of stereographic projections consists of a reference sphere that has a fixed orientation of its axis relative to the north and of its equatorial plane to the horizontal [Figure

(c)

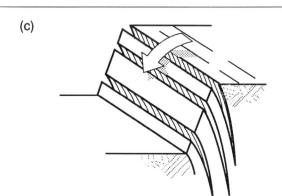

Toppling failure in hard rock with slabs or columns defined by discontinuities that dip steeply into the slope

(d)

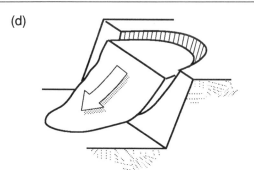

Circular failure in overburden soil, waste rock or heavily fractured rock with no identifiable structural pattern

FIGURE 15-2
Concepts for
stereographic
representation of
linear and planar
features (modified
from Hoek and
Bray 1981).

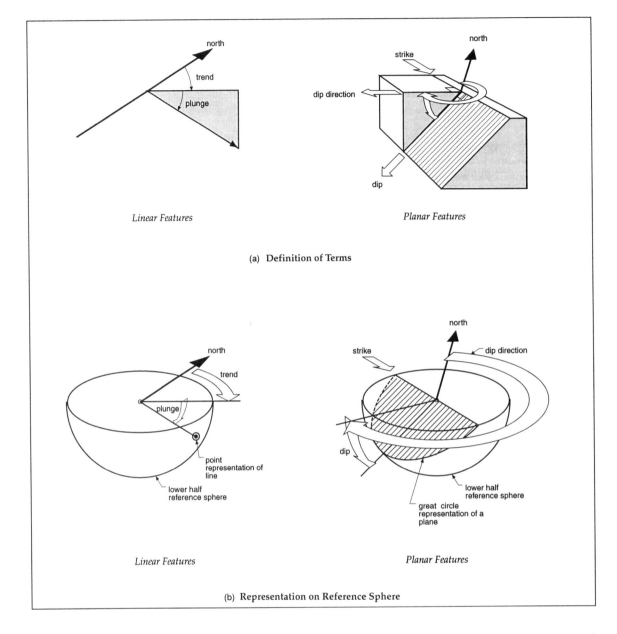

Linear Features

Planar Features

(a) **Definition of Terms**

Linear Features

Planar Features

(b) **Representation on Reference Sphere**

15-2(*b*)]. Linear features with a specific plunge and trend are positioned in an imaginary sense so that the axis of the feature passes through the center of the reference sphere. The intersection of the linear feature with the lower half of the reference sphere defines a unique point [Figure 15-3(*a*)]. Depending on the type of stereographic projection, this point is rotated down to a unique point on the stereonet. For the purposes of this discussion, only equal-area nets will be considered, although the reader should be aware that equal-angle projections can also be used. Linear features with shallow plunges plot near the circumference of the stereographic projection, whereas those with steep plunges plot near

the center. Similarly, planar features are positioned so that the feature passes through the center of the reference sphere and produces a unique intersection line with the lower half of the reference sphere [Figures 15-2(*b*) and 15-3(*b*)]. The projection of this intersection line onto the stereographic plot results in a unique representation of that plane referred to as a *great circle*. Planes with shallow dips have great circles that plot near the circumference of the net, and those with steep dips plot near the center. Planes are used to represent both discontinuities and slope faces in stereographic analyses.

A useful alternative method of representing planes is to use the normal to the plane. This normal in a stereographic projection will be a unique point that is referred to as the *pole to the plane* [Figure 15-3(b)]. Structural mapping data of discontinuities are often plotted in the pole format rather than the great-circle format in order to detect the presence of preferred orientations, thus defining discontinuity sets, and to determine mean and extreme values for the orientations of these sets. As shown in Figure 15-4, this process can be facilitated by contouring to accentuate and distinguish the repetitive features from the random or unique features. Computerized graphical methods greatly facilitate the analysis of large amounts of structural data. However, when fewer than 50 data points are involved, manual plotting and analysis are probably more efficient.

The intersection of two planes defines a line in space that is characterized by a trend (0 to 360 degrees) and plunge (0 to 90 degrees). In stereographic projection, this line of intersection is defined at the point where the two great circles cross, and the trend and plunge of this point are determined by conventional stereographic principles. It is interesting to note that the line of intersection, represented by a point on a stereonet, is the pole to a great circle containing the poles of the two wedge-forming discontinuities.

4. PLANAR FAILURE

Planar failures are those in which movement occurs by sliding on a single discrete surface that approximates a plane [Figure 15-1(a)]. Planar failures are analyzed as two-dimensional problems. Additional discontinuities may define the lateral extent of planar failures, but these surfaces are considered to be release surfaces, which do not contribute to the stability of the failure mass. Alternatively, the planar failure may be located on a nose of rock so that the slope forms the lateral termination of the failure mass.

The size of planar failures can range from a few cubic meters to large-scale landslides that involve entire mountainsides. The K M Mountain landslide in the state of Washington (Lowell 1990), which involved an estimated 1.2 to 1.5 million m³, was controlled by the bedding orientation within a siltstone sequence. The 1965 Hope landslide in southern British Columbia occurred along planar felsite dikes dipping subparallel to the slope and involved approximately 48 million m³ of displaced material (Mathews and McTaggert 1969).

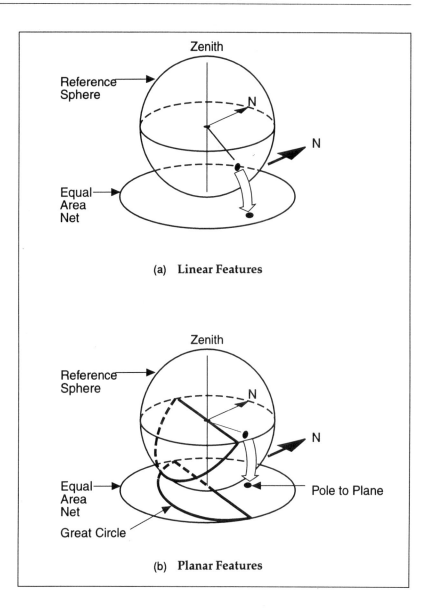

(a) Linear Features

(b) Planar Features

4.1 Kinematic Analysis

The four necessary structural conditions for planar failures can be summarized as follows:

1. The dip direction of the planar discontinuity must be within 20 degrees of the dip direction of the slope face, or, stated in a different way, the strike of the planar discontinuity must be within 20 degrees of the strike of the slope face.
2. The dip of the planar discontinuity must be less than the dip of the slope face and thereby must "daylight" in the slope face.
3. The dip of the planar discontinuity must be greater than the angle of friction of the surface.
4. The lateral extent of the potential failure mass

FIGURE 15-3
Equal-area projections for linear and planar features (modified from Hoek and Bray 1981).

FIGURE 15-4
Examples of symbol
and contour plots of
pole projections in
equal-area
stereographic
projections.

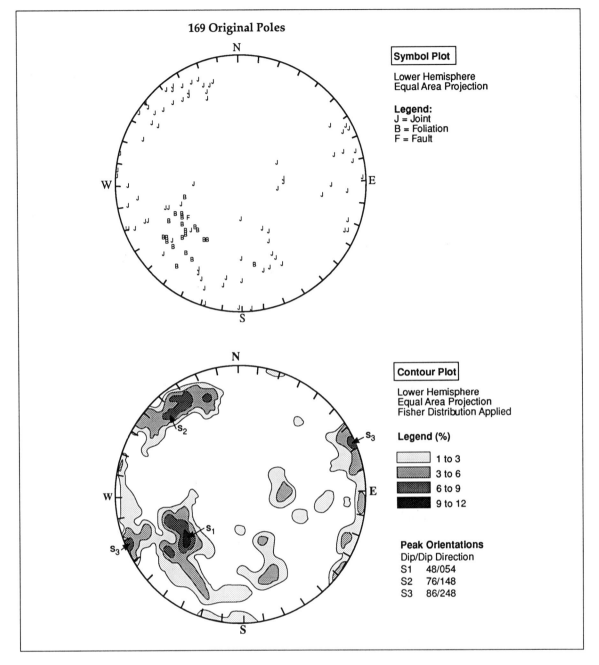

must be defined either by lateral release surfaces that do not contribute to the stability of the mass or by the presence of a convex slope shape that is intersected by the planar discontinuity.

Figure 15-5 illustrates the first three of these conditions; these are the only conditions that can be evaluated by stereographic analysis.

The presence of significant pore-water pressures along the failure surface can in some cases alter the kinematic possibility of planar failure. For example, the introduction of water pressure may cause a fail-

ure even though the dip of the failure plane is less than the frictional strength of the plane. Hodge and Freeze (1977) have presented an in-depth perspective of the influence of groundwater flow systems on the stability of interbedded sedimentary rock slopes.

4.2 Stability Analysis

If the kinematic analysis indicates that the requisite geologic structural conditions are present, stability must be evaluated by a limit-equilibrium

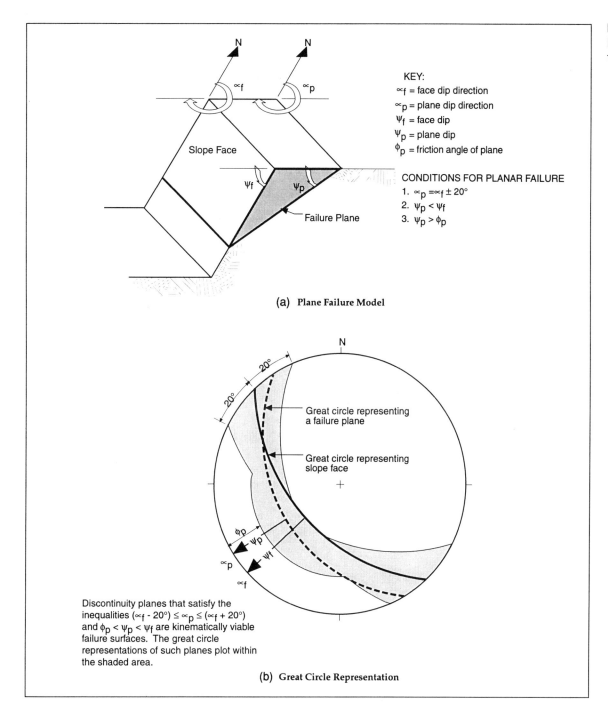

FIGURE 15-5
Kinematic analysis
for planar failure.

KEY:
$\propto_f$ = face dip direction
$\propto_p$ = plane dip direction
ψ_f = face dip
ψ_p = plane dip
ϕ_p = friction angle of plane

CONDITIONS FOR PLANAR FAILURE
1. $\propto_p = \propto_f \pm 20°$
2. $\psi_p < \psi_f$
3. $\psi_p > \phi_p$

Slope Face

Failure Plane

(a) Plane Failure Model

Great circle representing
a failure plane

Great circle representing
slope face

Discontinuity planes that satisfy the
inequalities $(\propto_f - 20°) \leq \propto_p \leq (\propto_f + 20°)$
and $\phi_p < \psi_p < \psi_f$ are kinematically viable
failure surfaces. The great circle
representations of such planes plot within
the shaded area.

(b) Great Circle Representation

analysis, which considers the shear strength along the failure surface, the effects of pore-water pressures, and the influence of external forces such as reinforcing elements or seismic accelerations.

Stability analysis for planar failure requires the resolution of forces perpendicular to and parallel to the potential failure surface. This resolution can be carried out in either two or three dimensions, but the most common case is the former, in which the

stability formulation considers a unit thickness of the slope. Two geometric cases are considered (Figure 15-6), depending on the location of the tension crack relative to the crest of the slope:

- Tension crack present in slope face [Figure 15-6(a)], and
- Tension crack present in upper slope surface [Figure 15-6(b)].

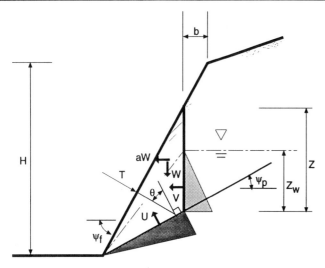

(a) Tension Crack in Slope Face

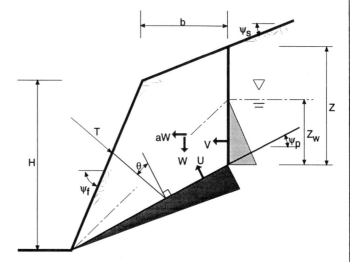

(b) Tension Crack in Upper Slope Surface

The stability equations are as follows:

For Case a:

Depth of tension crack:

$$Z = (H \cot\Psi_f - b)(\tan\Psi_f - \tan\Psi_p)$$

Weight of block:

$$W = (1/2)\gamma_r H^2 [(1 - Z/H)^2 \cot\Psi_p (\cot\Psi_p \tan\Psi_f - 1)]$$

Area of sliding plane:

$$A = (H \cot\Psi_f - b) \sec\Psi_p$$

For Case b:

Depth of tension crack:

$$Z = H + b \tan\Psi_s - (b + H\cot\Psi_f) \tan\Psi_p$$

Weight of block:

$$W = (1/2)\gamma_r (H^2 \cot\Psi_f X + bHX + bZ)$$

$$X = (1 - \tan\Psi_p \cot\Psi_f)$$

Area of sliding plane:

$$A = (H \cot\Psi_f + b) \sec\Psi_p$$

For either Case a or b:

Uplift water force:

$$U = (1/2)\gamma_w Z_w A$$

Driving water force:

$$V = (1/2)\gamma_w Z_w^2$$

Factor of safety:

$$FS = \frac{\{cA + [W(\cos\Psi_p - a\sin\Psi_p) - U - V\sin\Psi_p + T\cos\theta]\tan\phi\}}{[W(\sin\Psi_p + a\cos\Psi_p) + V\cos\Psi_p - T\sin\theta]}$$

where

H = height of slope face;
Ψ_f = inclination of slope face;
Ψ_s = inclination of upper slope face;
Ψ_p = inclination of failure plane;
b = distance of tension crack from slope crest;
a = horizontal acceleration, blast or earthquake loading;
T = tension in bolts or cables;
θ = inclination of bolt or cable to normal to failure plane;
c = cohesive strength of failure surface;
ϕ = friction angle of failure surface;
γ_r = density of rock;
γ_w = density of water;
Z_w = height of water in tension crack;
Z = depth of tension crack;
U = uplift water force;
V = driving water force;
W = weight of sliding block; and
A = area of failure surface.

A simplified groundwater model consists of a measured depth of water in the tension crack defining a phreatic surface that is assumed to decrease linearly toward and exit at the toe of the slope. Other configurations of the phreatic surface can be assumed with consequent modification to the forces U and V (Figure 15-6). The stability equations also incorporate external stabilizing forces, for example, those due to bolts or cables, and destabilizing forces such as those due to seismic ground accelerations.

Of particular note in Figure 15-6 is the location of the tension crack, expressed by the dimension b. In most cases, small-magnitude antecedent movements will define the location of the tension crack so that the geometry for slope analysis can be ascertained. However, in some cases these movements may not have occurred or the tension crack may be covered with surficial soils. In such circumstances, an approximation incorporating the most probable or "critical" tension crack location is as follows (Hoek and Bray 1981):

$$b/H = \sqrt{(\cot\Psi_f \cot\Psi_p)} - \cot\Psi_f \qquad (15.1)$$

The derivation of this equation assumes a dry slope and a horizontal upper slope surface, but as a first approximation, it is probably adequate for most cases. Sensitivity analyses should be carried out when the geometric or other parameters are not well defined.

For either of the two cases shown in Figure 15-6, the factor of safety, FS, is expressed as shown in the box below:

$$FS = \frac{\{cA + [W(\cos\Psi_p - a\sin\Psi_p) - U - V\sin\Psi_p + T\cos\theta]\tan\phi\}}{[W(\sin\Psi_p + a\cos\Psi_p) + V\cos\Psi_p - T\sin\theta]} \qquad (15.2)$$

This expression can be simplified in a number of successive cases as follows:

External forces not present ($a = T = 0$):

$$FS = \frac{[cA + (W\cos\Psi_p - U - V\sin\Psi_p)\tan\phi]}{(W\sin\Psi_p + V\cos\Psi_p)} \qquad (15.3)$$

Dry slope ($U = V = 0$) and external forces not present ($a = T = 0$):

$$FS = \frac{(cA + W\cos\Psi_p\tan\phi)}{(W\sin\Psi_p)} \qquad (15.4)$$

Cohesionless surface ($c = 0$), dry slope ($U = V = 0$), and external forces not present ($a = T = 0$):

$$FS = \frac{(\tan\phi)}{(\tan\Psi_p)} \qquad (15.5)$$

In this simplest case, the factor of safety for the slope is unity when the inclination of the failure surface equals the angle of friction for the surface. This relationship forms the basis of tilt tests, which are used to estimate base friction angles using blocks of rock or intact pieces of core (Franklin and Dusseault 1989).

Alternative methods of stability analysis for planar failures include limit-equilibrium analyses, which incorporate the method of slices to approximate the slope and failure-surface geometry. As an example, Sarma (1979) presented a method that can be applied to both circular and noncircular sliding surfaces. Hoek and Bray (1981) developed graphical solutions for planar failures for the case in which the external forces are zero and the upper slope surface is horizontal.

4.3 Case Example of Planar Failure

Methods of stability analysis for planar failures can be best explained by the use of a case example.

4.3.1 Problem Description

Structural mapping of an existing highway cut reveals the existence of well-developed foliation with a mean orientation of 30 degrees/145 degrees (dip/dip direction). The slope face is 30 m high with an orientation of 70 degrees/135 degrees. Further examination of the 11-degree upper slope indicates the presence of a tension crack 15 m behind the slope crest with water in the tension crack 9 m deep, as shown in Figure 15-7(a). Because of infilling and roughness along the foliation planes, the surfaces are characterized by shear strength parameters that include a friction angle of 25 degrees and a cohesion of 96 kPa. The rock mass is estimated to have a unit weight of 25 kN/m³. The slope designer is required to perform analyses to demonstrate the existing factor of safety. If this evaluation indicates a factor of safety less than 1.5, a slope-stabilization program must be designed to provide this value. The sensitivity of the slope stability to groundwater fluctuations and to seismic accelerations must also be determined.

4.3.2 Solution

A stereographic analysis of the slope and discontinuity orientations indicates that the three conditions necessary for a kinematically possible failure surface are present [Figure 15-7(b)]:

- The dip direction of the foliation and that of the slope face are within 20 degrees,

FIGURE 15-6 (*opposite page*) Two geometric cases of planar failure (modified from Hoek and Bray 1981).

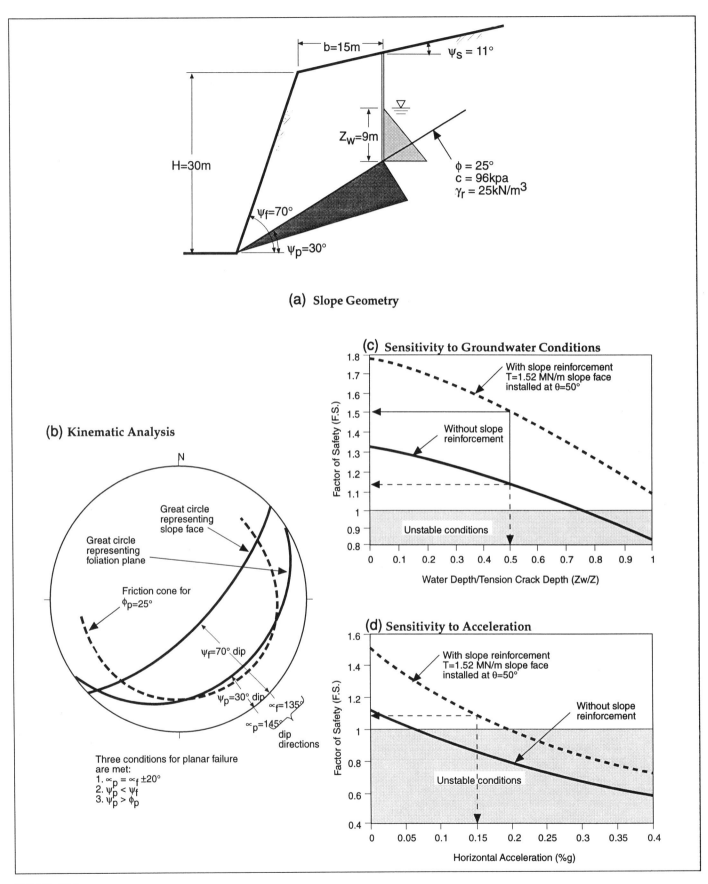

$b=15\text{m}$
$\psi_s = 11°$

$Z_w=9\text{m}$

$H=30\text{m}$

$\phi = 25°$
$c = 96\text{kpa}$
$\gamma_r = 25\text{kN/m}^3$

$\psi_f=70°$

$\psi_p=30°$

(a) Slope Geometry

(b) Kinematic Analysis

N

Great circle representing slope face

Great circle representing foliation plane

Friction cone for $\phi_p=25°$

$\psi_f=70°$ dip

$\psi_p=30°$ dip
$\propto_f=135°$
$\propto_p=145°$
dip directions

Three conditions for planar failure are met:
1. $\propto_p = \propto_f \pm20°$
2. $\psi_p < \psi_f$
3. $\psi_p > \phi_p$

(c) Sensitivity to Groundwater Conditions

With slope reinforcement
T=1.52 MN/m slope face
installed at $\theta=50°$

Without slope reinforcement

Unstable conditions

Water Depth/Tension Crack Depth (Zw/Z)

Factor of Safety (F.S.)

(d) Sensitivity to Acceleration

With slope reinforcement
T=1.52 MN/m slope face
installed at $\theta=50°$

Without slope reinforcement

Unstable conditions

Horizontal Acceleration (%g)

Factor of Safety (F.S.)

FIGURE 15-7
Case example of planar failure.

- The dip of the foliation is less than the dip of the slope face, and
- The dip of the foliation exceeds its angle of friction.

Having determined that a planar failure is possible, the next step is to determine the degree of stability (or instability) by calculating the factor of safety. Evaluation of the stability of the slope requires the solution of the equations represented by Figure 15-6(b). The computations are shown in the box below.

tions given in Section 4.2 into a commercially available spreadsheet program that enables the development of stability graphs such as those in Figure 15-7(c) and (d). As an example, to examine the feasibility of stabilization through slope drainage, the accompanying sensitivity graph [Figure 15-7(c)] demonstrates that the factor of safety ranges from a high of 1.32 for a dry slope to a low of 0.82 for a fully saturated, unreinforced slope. Even if measures to promote permanent drainage of the slope were implemented, a factor of safety of 1.32 would probably be marginal (in

---

Depth of tension crack:

$Z = 30 + 15(\tan 11°) - [15 + 30(\cot 70°)] \tan 30°$
$= 18$ m

Weight of block:

$W = \frac{25}{2}[(30)^2 \cot 70° (X) + (15)(30)(X) + (15)(18)]$

where

$X = (1 - \tan 30° \cot 70°) = 0.79$

then

$W = 11.05$ MN/m

Area of sliding plane:

$A = (30 \cot 70° + 15)(\sec 30°) = 29.9$ m²/m

Uplift water force:

$U = (\frac{1}{2})(10)(9)(29.9) = 1.35$ MN/m

Driving water force:

$V = (\frac{1}{2})(10)(9)^2 = 0.41$ MN/m

Substituting these values into Equation 15.3 for the case where external forces are not present yields

$$FS = \frac{\{(0.096)(29.9) + [(11.05)(\cos 30°) - 1.35 - (0.41)(\sin 30°)](\tan 25°)\}}{[(11.05)(\sin 30° + (0.41)(\cos 30°)]}$$

$FS = 6.61/5.88 = 1.12$

The calculated factor of safety for the existing condition is 1.12, which corresponds to a tension crack half-filled with water ($Z_w/Z = 0.5$). This factor of safety would be considered unacceptable for the long-term performance of the slope, and therefore stabilization should be undertaken.

To analyze stabilization alternatives, it is useful to perform sensitivity analyses that illustrate the dependence of slope stability on various parameters of interest. These analyses are most easily accomplished by programming the stability equa-

most design circumstances), and other stabilization methods should be considered.

Equation 15.2 can be used to calculate the bolting force, T, necessary to provide the required factor of safety of 1.5. For illustrative purposes, the inclination of the bolting force will be set at $\theta = 50$ degrees. Although this is not the optimal orientation, practical considerations would dictate that the anchor holes be drilled below horizontal to facilitate grouting. Substituting these values into Equation 15.2 produces the computations below:

$$1.5 = \frac{\{(0.096)(29.9) + [(11.05)(\cos 30°) - 1.35 - (0.41)(\sin 30°) + T(\cos 50°)](\tan 25°)\}}{(11.05)[\sin 30° + (0.41)(\cos 30°) - T(\sin 50°)]}$$

$$1.5 = \frac{[2.87 + (8.01 + 0.64T)(\tan 25°)]}{(5.88 - 0.77T)}$$

$T = 1.52$ MN/m

It is important to note that T is computed in terms of the force per unit of slope length. This value combined with the selected bolt or anchor capacity defines the required bolt spacing. The slope designer should verify that the reinforced slope will be stable under the saturated slope extreme ($Z_w/Z = 1$). For the case example, the sensitivity analysis in Figure 15-7(c) indicates an acceptable factor of safety of 1.08 under this condition.

The stability of the reinforced slope for a design horizontal acceleration, $a = 0.15g$, is verified using Equation 15.2 according to the computation shown in the box below.

The variation of stability of the reinforced slope with seismic acceleration is shown in the sensitivity graph of Figure 15-7(d). For the case example, a factor of safety of 1.5 under static conditions with a corresponding factor of safety of 1.09 for a design acceleration of 0.15g would generally be considered prudent engineering practice. Figure 15-7(d) also indicates a limiting acceleration equivalent to 0.19g for the reinforced slope.

$$FS = \frac{\{(0.096)(29.9) + [(11.05)(\cos30°) - (0.15)(\sin30°) - 1.35 - (0.41)(\sin30°) + (1.52)(\cos50°)](\tan25°)\}}{(11.05)[\sin30° + (0.15)(\cos30°) + (0.41)(\cos30°) - (1.52)(\sin50°)]}$$

$$FS = 6.68/6.15 = 1.09$$

FIGURE 15-8
Massive wedge failure along I-40 in mountainous sections of eastern Tennessee that closed two lanes for more than two weeks (Moore 1986).

5. WEDGE FAILURE

Wedge failures result when rock masses slide along two intersecting discontinuities both of which dip out of the cut slope at an oblique angle to the cut face, forming a wedge-shaped block [Figures 15-1(b) and 15-8]. Commonly, these rock wedges are exposed by excavations that daylight the line of intersection that forms the axis of sliding, precipitating movement of the rock mass either along both planes simultaneously or along the steeper of the two planes in the direction of maximum dip.

Depending upon the ratio between peak and residual shear strengths, wedge failures can occur rapidly, within seconds or minutes, or over a much longer time frame, on the order of several months. The size of a wedge failure can range from a few cubic meters to very large slides from which the potential for destruction can be enormous (Figure 15-8). Figure 15-9 illustrates schematically the development of a classic wedge failure. Figure 15-9(a) shows a typical rock cut along a highway in which the geologic structure is conducive to wedge failure, resulting in an unstable slope condition. Incipient development of the wedge, defined by intersecting planes of fracture, cleavage, bedding, or all three, results in the formation of a V-shaped wedge of unstable rock with fully developed failure scarps, as shown in Figure 15-9(b). Figure 15-9(c) shows the rapid movement of the rock wedge failure, and its aftermath is shown in Figure 15-9(d).

The formation and occurrence of wedge failures are dependent primarily on lithology and structure of the rock mass (Piteau 1972). Rock masses with well-defined orthogonal joint sets or cleavages in

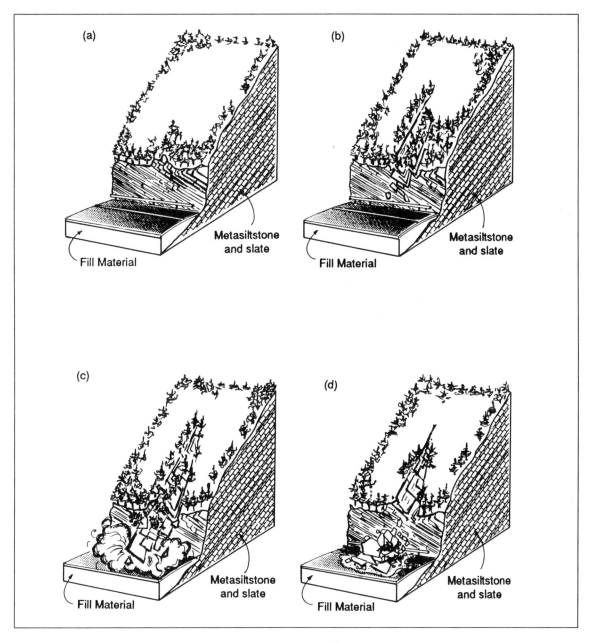

FIGURE 15-9
Stages in
development of
wedge-failure
landslide:
(a) rock slope as
constructed;
(b) incipient wedge
failure; (c) rock
wedge during
failure; (d) rock
slope after wedge
failure (modified
from Moore 1986).

addition to inclined bedding or foliation generally are favorable situations for wedge failure (Figure 15-10). Shale, thin-bedded siltstones, claystones, limestones, and slaty lithologies tend to be more prone to wedge failure development than other rock types. However, lithology alone does not control development of wedge failures.

5.1 Kinematic Analysis

A kinematic analysis for wedge failure is governed by the orientation of the line of intersection of the wedge-forming planes. Kinematic analyses deter-

mine whether sliding can occur and, if so, whether it will occur on only one of the planes or simultaneously on both planes, with movement in the direction of the line of intersection.

The necessary structural conditions for wedge failure are illustrated in Figure 15-11(a) and can be summarized as follows:

1. The trend of the line of intersection must approximate the dip direction of the slope face.
2. The plunge of the line of intersection must be less than the dip of the slope face. Under this

FIGURE 15-10
Exposed rock
face exhibiting
well-developed
intersecting
discontinuities
with potential for
wedge failure
(Moore 1986.)

condition, the line of intersection is said to day-light on the slope.

3. The plunge of the line of intersection must be greater than the angle of friction of the surface. If the angles of friction for the two planes are markedly different, an average angle of friction is applicable. This condition is also shown in Figure 15-11(a).

Because the model represents a three-dimensional shape, no assumptions of the lateral extent of the wedge are required. Stereographic analysis can also determine whether sliding will occur on both the wedge-forming planes or on only one of the two. This procedure is referred to as Markland's test (Hoek and Bray 1981), which is described in Figure 15-11(a).

The presence of significant pore-water pressures along the failure planes can in some cases alter the possibility of kinematic wedge failures. For example, the introduction of water pressure may cause a failure even though the plunge of the intersection line is less than the average frictional strength of the planes.

5.2 Stability Analysis

Once a kinematic analysis of wedge stability using stereographic methods [Figure 15-11(b)] has been performed indicating the possibility of a wedge failure, more detailed stability analyses may be required for the design of stabilization measures. The common analytical technique is a rigid-block analysis in which failure is assumed to be by linear sliding along the line of intersection formed by the discontinuities or by sliding along one of the discontinuities. Toppling or rotational sliding is not considered in the analysis.

The analysis of wedge stability requires that the geometry of the wedge be defined by the location and orientation of as many as five bounding surfaces. These include the two intersecting discontinuities, the slope face, the upper slope surface, and, if present, the plane representing a tension crack [Figure 15-12(a)]. The size of the wedge is defined by the vertical distance from the crest of the slope to the line of intersection of the wedge-forming discontinuities. If a tension crack is present, the location of this bounding plane relative to the slope crest must be specified to analyze the wedge size.

The stability of the wedge can be evaluated using the factor-of-safety concept by resolving the forces acting normal to the discontinuities and in the direction parallel to the line of intersection. These forces include the weight of the wedge, external forces such as foundation loads, seismic accelerations, tensioned reinforcing elements, forces generated by water pressures acting on the surfaces, and the shear strength developed along the sliding plane or planes.

The completely general formulation of wedge stability calculation requires adherence to a strict system of notation. Hoek and Bray (1981) presented the equations for the general analysis as well as a methodology to undertake the calculation in a systematic manner. Because the calculation is extended, this general analysis is best adapted to computer solution. However, in most cases, assumptions can be made that significantly simplify the controlling stability equations so that simple calculator or graphical methods can provide a good indication of the sensitivity of the wedge stability to alternative strength and load combinations. Figure 15-12 (a-e) defines the calculation of wedge stability under various simplifying assumptions.

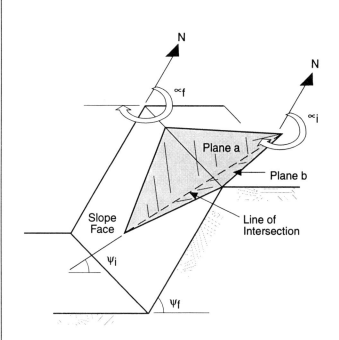

KEY:
$\propto_f$ = dip direction of face
$\propto_a$ = dip direction of plane a
$\propto_b$ = dip direction of plane b
ψ_f = dip of face
ψ_a = dip of plane a
ψ_b = dip of plane b
ϕ = friction angle
ψ_i = plunge of intersection
$\propto_i$ = azimuth of intersection

CONDITIONS FOR WEDGE FAILURE
1. $\propto_i = \propto_f \pm$ (to daylight slope face)
2. $\psi_i < \psi_f$
3. $\phi < \psi_i$

MARKLAND'S TEST
If $\propto_a$ or $\propto_b$ is between $\propto_i$ and $\propto_f$ then sliding will occur on the steeper of planes a and b in the direction of maximum dip; otherwise sliding occurs along line of intersection.

(a) Wedge Failure Model

FIGURE 15-11
Kinematic analysis for wedge failure (modified from Hoek and Bray 1981).

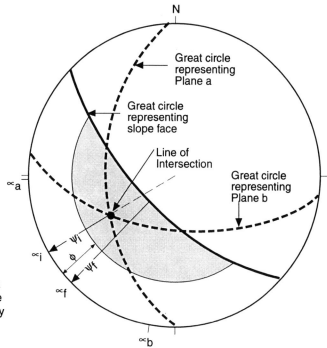

Combinations of discontinuity planes with a line of intersection that "daylights" the slope face, $\propto_i = \propto_f\pm$, and that satisfy the inequality $\phi < \psi_i < \psi_f$ represent kinematically viable wedge failures. The lines of intersection of such planes plot within the shaded area.

(b) Great Circle Representation

FIGURE 15-12
(facing pages)
Stability equations
for wedge failure
(modified from
Hoek and Bray
1981).

(a) General Case

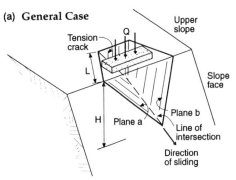

Upper slope
Tension crack
Q
L
Slope face
H
Plane a
Plane b
Line of intersection
Direction of sliding

Use comprehensive solution by Hoek and Bray (1981)
Note: This solution required if external loads to be included.
Typically solved using computer program.

(b) Tension Crack Not Present

Plane b
Plane a
Wedge
Slope face
Note: Saturated slope assumed.
H
Line of intersection

$$FS = \frac{3}{\gamma_r H}(c_a \cdot X + c_b \cdot Y) + (A - \frac{\gamma_w}{2\gamma_r}X)\tan\phi_a + (B - \frac{\gamma_w}{2\gamma_r}Y)\tan\phi_b$$

PARAMETERS:

c_a and c_b are the cohesive strengths of planes a and b
ϕ_a and ϕ_b are the angles of friction on planes a and b
γ_r is the unit weight of the rock
γ_w is the unit weight of water
H is the total height of the wedge
X, Y, A, and B are dimensionless factors which depend upon the geometry of the wedge
ψ_a and ψ_b are the dips of planes a and b
ψ_i is the plunge of the line of intersection

$$X = \frac{\sin\theta_{24}}{\sin\theta_{45} \cdot \cos\theta_{na\cdot 2}}$$

$$A = \frac{\cos\psi_a - \cos\psi_b \cdot \cos\theta_{na} \cdot nb}{\sin\psi_i \cdot \sin^2\theta_{na} \cdot nb}$$

$$Y = \frac{\sin\theta_{13}}{\sin\theta_{35} \cdot \cos\theta_{nb\cdot 1}}$$

$$B = \frac{\cos\psi_b - \cos\psi_a \cdot \cos\theta_{na} \cdot nb}{\sin\psi_i \cdot \sin^2\theta_{na} \cdot nb}$$

(See Figure 15-14 for definition of angles)

(c) Fully Drained Slope

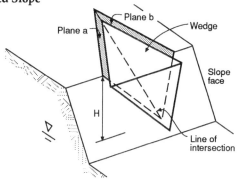

Plane b
Plane a
Wedge
Slope face
H
Line of intersection

(See Figure 15-14 for definition of angles)

$$FS = \frac{3}{\gamma_r H}(c_a X + c_b Y) + A \tan \phi_a + B \tan \phi_b$$

PARAMETERS:

c_a and c_b are the cohesive strengths of planes a and b

ϕ_a and ϕ_b are the angles of friction on planes a and b

γ_r is the unit weight of the rock

H is the total height of the wedge

X, Y, A, and B are dimensionless factors which depend upon the geometry of the wedge

ψ_a and ψ_b are the dips of planes a and b

ψ_i is the plunge of the line of intersection

$$X = \frac{\sin\theta_{24}}{\sin\theta_{45} \cdot \cos\theta_{na\cdot2}}$$

$$Y = \frac{\sin\theta_{13}}{\sin\theta_{35} \cdot \cos\theta_{nb\cdot1}}$$

$$A = \frac{\cos\psi_a - \cos\psi_b \cdot \cos\theta_{na \cdot nb}}{\sin\psi_i \cdot \sin^2\theta_{na \cdot nb}}$$

$$B = \frac{\cos\psi_b - \cos\psi_a \cdot \cos\theta_{na \cdot nb}}{\sin\psi_i \cdot \sin^2\theta_{na \cdot nb}}$$

(d) Friction Only Shear Strengths

$$FS = A \tan \phi_a + B \tan \phi_b$$

(Parameters as above)

Note: Hoek and Bray (1981) have presented graphs to determine factors A and B.

(e) Friction Angle Same for Both Planes

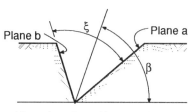

Plane b
Plane a
ξ
β

View along line of intersection

$$FS = \frac{\sin \beta}{\sin (\xi/2)} \cdot \frac{\tan \phi}{\tan \psi_i}$$

PARAMETERS:

ϕ = friction angle

ψ_i = plunge of line of intersection

β = see sketch

ξ = angle between wedge - forming planes

5.3 Case Example of Wedge Failure

Structural mapping of outcrops in the vicinity of a proposed highway cut indicates the presence of five discontinuity sets as follows:

Feature	Dip/Dip Direction
Bedding	48 degrees/168 degrees
Joint set 1	53 degrees/331 degrees
Joint set 2	64 degrees/073 degrees
Joint set 3	42 degrees/045 degrees
Joint set 4	45 degrees/265 degrees

The proposed cut is to be excavated at 0.25H:1V (dip of 76 degrees) with a dip direction of 196 degrees. It is estimated that the average angle of friction of the discontinuities is 30 degrees subject to confirmation by subsequent direct-shear testing.

A kinematic analysis of these data is shown in Figure 15-13. The shaded area is formed by the great circle representing the cut slope face and the friction circle for ϕ = 30 degrees. As previously shown in Figure 15-11, the lines of intersection of kinematically possible wedges plot as points within this shaded area. For the case example, Figure 15-13 shows that the intersection of the great circles representing bedding and joint set 2 (B-J2) and the intersection of the great circles representing bedding and joint set 4 (B-J4) plot

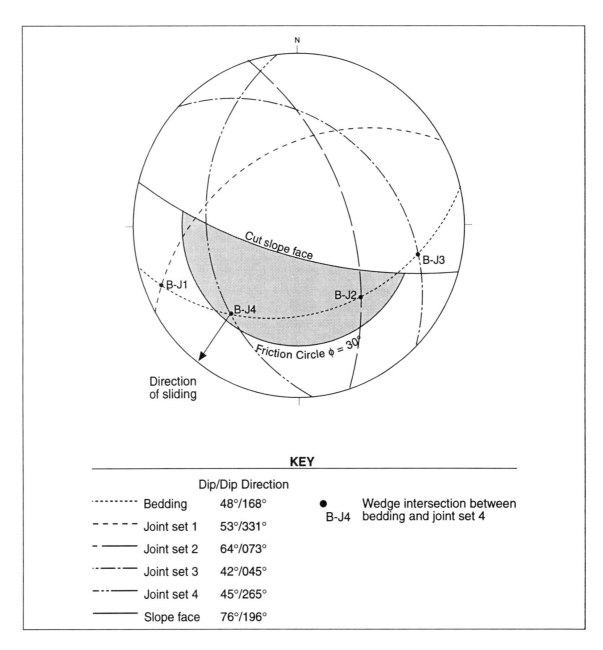

	Dip/Dip Direction		
········· Bedding	48°/168°	●	Wedge intersection between
‐ ‐ ‐ ‐ Joint set 1	53°/331°	B-J4	bedding and joint set 4
‐ —— Joint set 2	64°/073°		
‐·‐·‐ Joint set 3	42°/045°		
‐··‐·· Joint set 4	45°/265°		
—— Slope face	76°/196°		

FIGURE 15-13
Great circle representation for case example of wedge failure.

within the shaded area and therefore form kinematically possible wedge failures.

To determine the factor of safety of the wedge formed by the intersection of bedding and joint set 4 (B-J4), the following additional information is obtained:

- Orientation of upper slope surface (based on mapping): 10 degrees/196 degrees dip/dip direction;
- Total wedge height (based on cross-section analysis): $H = 30$ m;
- Strength of plane a (Joint set 4) (based on laboratory testing): friction angle, $\phi_a = 35$ degrees; cohesion, $c_a = 20$ kPa;
- Strength of plane b (bedding) (based on laboratory testing): friction angle, $\phi_b = 25$ degrees; cohesion, $c_b = 10$ kPa;

- Unit weight of rock (based on laboratory testing): $\gamma_r = 25$ kN/m³; and
- Fully drained slope (based on piezometric measurements).

As shown in Figure 15-12(c), the applicable equation for a wedge without a tension crack is as follows:

$$FS = \frac{3}{(\gamma_r H)}(c_a X + c_b Y) + A \tan\phi_a + B \tan\phi_b \quad (15.6)$$

The values for the factors X, Y, A, and B are determined from a number of angular relationships measured on a stereographic projection as illustrated in Figure 15-14. The computations are shown in the box on the next page.

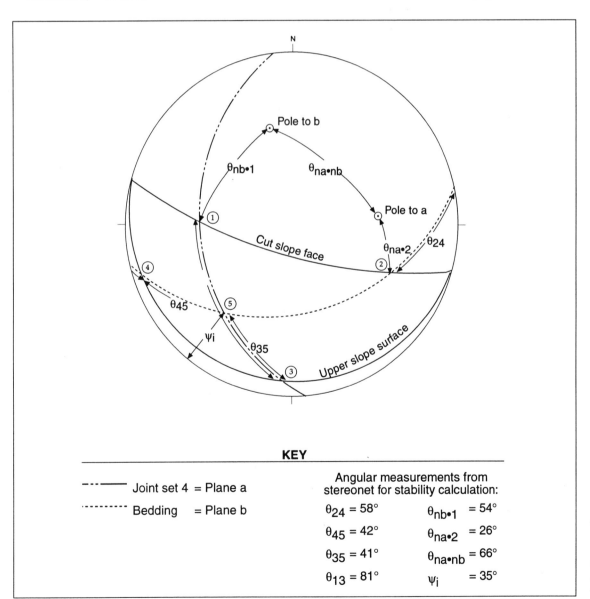

KEY

| Joint set 4 | = Plane a |
| Bedding | = Plane b |

Angular measurements from stereonet for stability calculation:

$\theta_{24} = 58°$ $\theta_{nb\cdot1} = 54°$

$\theta_{45} = 42°$ $\theta_{na\cdot2} = 26°$

$\theta_{35} = 41°$ $\theta_{na\cdot nb} = 66°$

$\theta_{13} = 81°$ $\psi_i = 35°$

FIGURE 15-14 Angular measurements required for stability calculation for case example of wedge failure.

FIGURE 15-15
Primary toppling
modes (Hoek and
Bray 1981).

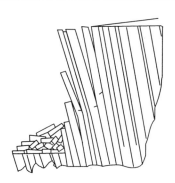

Flexural toppling in hard rock
slopes with well developed
steeply dipping discontinuities.

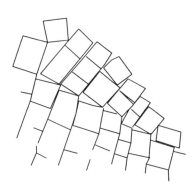

Block toppling in a hard rock
mass with widely spaced
orthogonal joints

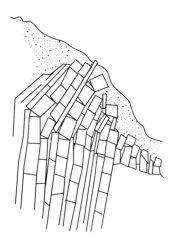

Block-flexure toppling characterized by
pseudo-continuous flexure of long
columns by accumulated motions along
numerous cross joints.

$$X = \frac{(\sin\theta_{24})}{(\sin\theta_{45}\cos\theta_{(na2)})} = \frac{(\sin 58°)}{(\sin 42° \cos 26°)} = 1.41$$

$$Y = \frac{(\sin\theta_{13})}{(\sin\theta_{35}\cos\theta_{(nb1)})} = \frac{(\sin 81°)}{(\sin 41° \cos 54°)} = 2.56$$

$$A = \frac{[\cos\Psi_a - \cos\Psi_b \cos\theta_{(na\cdot nb)}]}{[\sin\Psi_5 \sin^2\theta_{(na\cdot nb)}]}$$

$$= \frac{(\cos 45° - \cos 48° \cos 66°)}{(\sin 35° \sin^2 66°)} = 0.91$$

$$B = \frac{[\cos\Psi_b - \cos\Psi_a \cos\theta_{(na\cdot nb)}]}{[\sin\psi_5 \sin^2\theta_{(na\cdot nb)}]}$$

$$= \frac{(\cos 48° - \cos 45° \cos 66°)}{(\sin 35° \sin^2 66°)} = 0.80$$

Substituting these values into Equation 15.6 gives

$$FS = \frac{\{3[(20)(1.41) + (10)(2.56)]\}}{[(25)(30)]}$$
$$+ (0.91)\tan 35° + (0.80)\tan 25°$$

$$FS = 0.22 + 0.64 + 0.37 = 1.23$$

This computed factor of safety is at the lower bound for most geotechnical engineering practice, and consideration should be given to reinforcement of the slope with tensioned rock anchors (see Chapter 18).

If poor blasting practices were utilized in the excavation process, it is possible that the cohesion along the discontinuities could have been destroyed. In this case, the first term of Equation 15.6 would be zero and the resultant factor of safety would be equal to 1.01. The wedge would probably fail during excavation; this analysis illustrates the importance of carefully controlled blasting in the creation of stable permanent slopes (see Chapter 18).

6. TOPPLING FAILURE

Toppling failures most commonly occur in rock masses that are subdivided into a series of slabs or columns formed by a set of fractures that strike approximately parallel to the slope face and dip steeply into the face [Figure 15-1(c)]. In a toppling failure the rock column or slab rotates about an essentially fixed point at or near the base of the slope at the same time that slippage occurs between the layers. A rarer case of toppling is that of a single column defined by a unique discontinuity such as a fault.

Rock types most susceptible to this mode of failure are columnar basalts and sedimentary and metamorphic rocks with well-developed bedding or foliation planes [Figure 15-1(*c*)]. As described by Hoek and Bray (1981), there are several types of toppling failures, including flexural, block, or a combination of block and flexural toppling (Figure 15-15). Toppling can also occur as a secondary failure mode associated with other failure mechanisms such as block sliding. Examples of these various types of secondary toppling failure are shown in Figure 15-16.

In order for toppling to occur, the center of gravity of the column or slab must fall outside the dimension of its base. Toppling failures are characterized by significant horizontal movements at

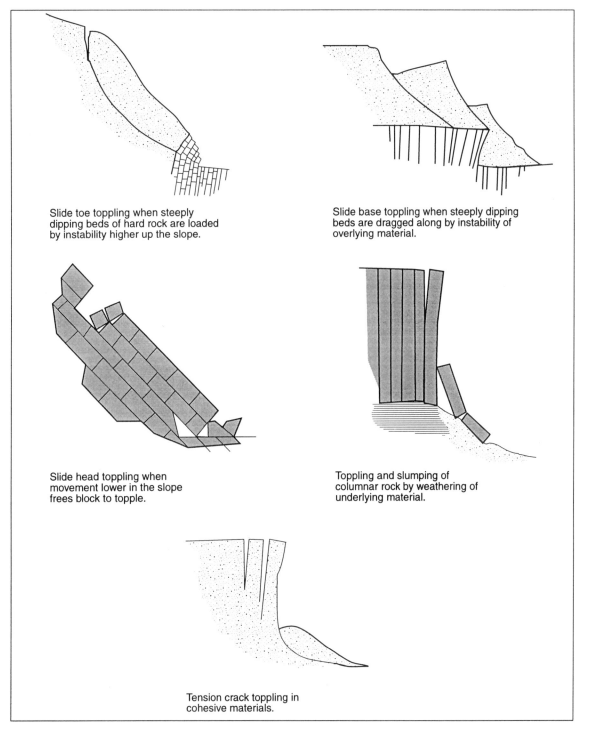

Slide toe toppling when steeply dipping beds of hard rock are loaded by instability higher up the slope.

Slide base toppling when steeply dipping beds are dragged along by instability of overlying material.

Slide head toppling when movement lower in the slope frees block to topple.

Toppling and slumping of columnar rock by weathering of underlying material.

Tension crack toppling in cohesive materials.

FIGURE 15-16
Secondary toppling modes (Hoek and Bray 1981).

the crest and very little movement at the toe. To accommodate this differential movement between the toe and crest, interlayer movement must occur. Thus, shear strength between layers is crucial to the stability of a slope that is structurally susceptible to toppling. Another characteristic of toppling movements is the antecedent development of major tension cracks behind the crest and parallel to the strike of layers. Failure does not occur until there is shear failure of the slabs at the base of the slope. Slopes with rock structure that is susceptible to toppling can be induced to fail by increased pore-water pressures or by erosion or excavation at the toe of the slope.

6.1 Kinematic Analysis

Figure 15-17 indicates the slope parameters that define an analytical model for toppling analysis and the kinematic analysis of toppling using stereonet projection. The slope parameters necessary for analysis of Goodman and Bray's (1976) model of toppling failures are defined in Figure 15-17(a). Of particular note is the presence of a stepped failure base assumed to develop along cross fractures between the columns. The necessary conditions for toppling failure can be summarized as follows:

1. The strike of the layers must be approximately parallel to the slope face. Differences in these orientations of between 15 and 30 degrees have been quoted by various workers, but for consistency with other modes of failure, a value of 20 degrees seems appropriate.
2. The dip of the layers must be into the slope face. Using the dip direction convention, conditions 1 and 2 can be stated as follows: the dip direction of the layers must be between 160 and 200 degrees to the dip direction of the slope face.
3. As stated by Goodman (1980), in order for interlayer slip to occur, the normal to the toppling plane must have a plunge less than the inclination of the slope face less the friction angle of the surface. This condition can be formulated as follows:

$$(90° - \Psi_p) \leq (\Psi_f - \phi_p) \tag{15.7}$$

where

Ψ_p = dip of geologic layers (planes),
Ψ_f = dip of slope face, and
ϕ_p = friction angle along planes.

On the basis of extensive development of nomograms for analysis of toppling, Choquet and Tanon (1985) proposed the following modification to this kinematic condition:

$$(90° - \Psi_p) \leq (\Psi_f - \phi_p + k) \tag{15.8}$$

where $k = 0$ for $\phi_p < 20$ degrees and $k = 3/5(\phi_p - 20$ degrees) for $\phi_p \geq 20$ degrees.

Analogous to planar failures, some limitation to the lateral extent of the toppling failure is a fourth condition for a kinematically possible failure. Because the analysis is two-dimensional, it is usually assumed that zero-strength lateral release surfaces are present or that the potential failure mass is defined by a convex slope in plan.

6.2 Stability Analysis

The analysis of toppling failures has been investigated by several researchers, including Goodman and Bray (1976), Hittinger (1978), and Choquet and Tanon (1985). The analytical procedures are not as clearcut as for other methods of rock slope failure, particularly the concept of factor of safety. In general terms, the techniques check that the center of gravity for a specific column of rock lies within the base area of that column. Columns in which the center of gravity lies outside the base are susceptible to toppling.

The method developed by Goodman and Bray (1976) considers each column in turn proceeding from the crest of the slope to the toe and determines one of three stability conditions: stable, plane sliding, or toppling. The stability condition depends on the geometry of the block, the shear strength parameters along the base and on the sides of the column, and any external forces. Those columns that are susceptible to either sliding or toppling exert a force on the adjacent column in the downslope direction. The analysis is carried out for each column in the slope section so that all the intercolumnar forces are determined. The stability of the slope generally cannot be explicitly stated in terms of a factor of safety. However, the ratio between the friction value required for limiting equilibrium and that available along the base of the columns is often used as a factor of safety for toppling analyses. Figure 15-18 demonstrates the method of analysis for a toppling failure. The reader is referred to Hoek and Bray (1981) for a more comprehensive summary

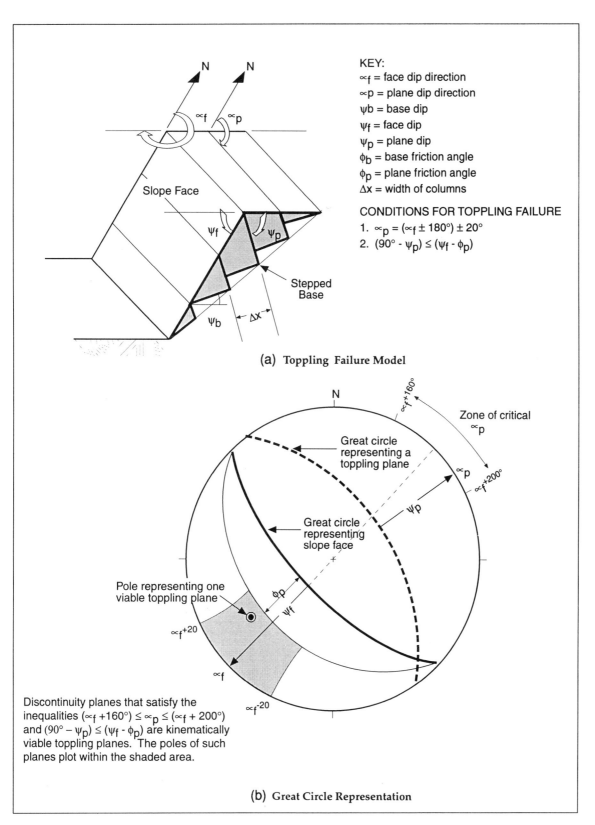

KEY:
$\propto_f$ = face dip direction
$\propto_p$ = plane dip direction
ψ_b = base dip
ψ_f = face dip
ψ_p = plane dip
ϕ_b = base friction angle
ϕ_p = plane friction angle
Δx = width of columns

CONDITIONS FOR TOPPLING FAILURE
1. $\propto_p = (\propto_f \pm 180°) \pm 20°$
2. $(90° - \psi_p) \le (\psi_f - \phi_p)$

Slope Face

ψ_f

ψ_p

Stepped Base

ψ_b

Δx

(a) Toppling Failure Model

N

Zone of critical $\propto_p$

$\propto_f$+160°

Great circle representing a toppling plane

$\propto_p$

$\propto_f$+200°

ψ_p

Great circle representing slope face

ϕ_p

ψ_f

Pole representing one viable toppling plane

$\propto_f$+20

$\propto_f$

$\propto_f$-20

Discontinuity planes that satisfy the inequalities $(\propto_f +160°) \le \propto_p \le (\propto_f + 200°)$ and $(90° - \psi_p) \le (\psi_f - \phi_p)$ are kinematically viable toppling planes. The poles of such planes plot within the shaded area.

(b) Great Circle Representation

FIGURE 15-17
Kinematic analysis for toppling failure.

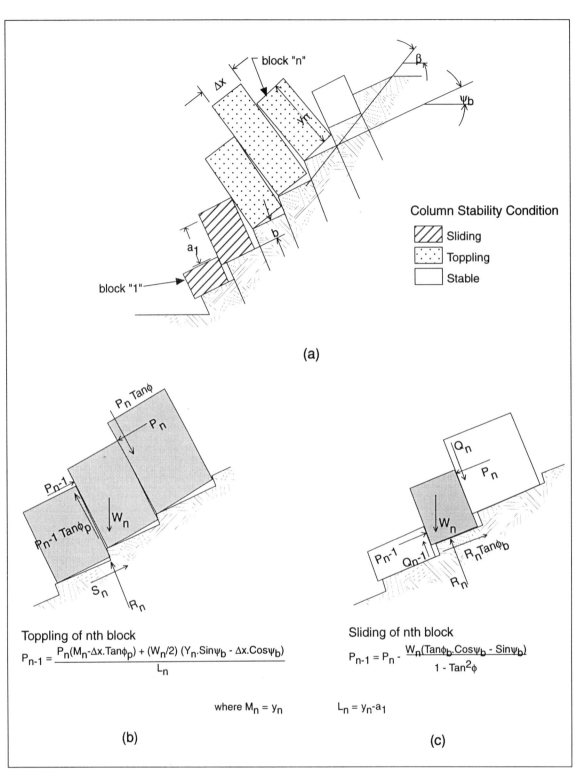

FIGURE 15-18
Stability equations for toppling failure (Hoek and Bray 1981).

of this technique or to Hittinger (1978) for a computer solution of toppling analyses. Wyllie (1992) has expanded this analysis to include external loads and water pressures.

Choquet and Tanon (1985) utilized the computer solution developed by Hittinger (1978) to derive a series of nomograms for the assessment of toppling failures. Unique nomograms were developed based upon the interlayer angle of shearing resistance, ϕ_p. A nomogram for $\phi_p = 30$ degrees is given in Figure 15-19(a). Inherent in these nomograms are the following assumptions:

1. The columns in the model have a constant width, defined as Δx [Figure 15-17(a)].
2. The base of each column forms a stepped failure base with an inclination assumed at +15 degrees [$\Psi_b = 15$ degrees in Figure 15-17(a)].
3. No pore-water pressures are present within the slope.

An example of the use of this method of analysis is illustrated in Figure 15-19. Note that the formulation yields the limiting column width, Δxe, at which toppling failure occurs. The input variables for the example are

Friction angle for plane = ϕ_p = 30 degrees,
Dip of face = Ψ_f = 64 degrees,
Dip of plane = Ψ_p = 60 degrees.

From Figure 15-19(a), the ratio of $H/\Delta xe$ is 10 at the onset of toppling or limiting equilibrium (see the point indicated by the star). Thus, for a slope height of $H = 20$ m, $\Delta xe = 2$ m. If the column width is less than 2 m, the slope will be unstable with respect to toppling. This result is illustrated in Figure 15-19(b). Similarly, for a slope height of $H = 10$ m, the limiting column width would be 1 m.

Choquet and Tanon (1985) suggested that the factor of safety be evaluated by the ratio of the actual column width, Δx, to the theoretical limiting column width, Δxe, according to

$$FS = \frac{(\Delta x)}{(\Delta xe)} \tag{15.9}$$

where for FS > 1, the slope is stable against toppling and for FS < 1, the slope is unstable against toppling.

Figure 15-19(c) shows a sensitivity curve for the dip of the slope face, Ψ_f, as a function of the limit-

ing column width, Δxe, for a constant dip of plane, $\Psi_p = 60$ degrees, and slope height $H = 15$ m. This curve is developed from the nomogram in Figure 15-19(a) by determining the corresponding $H/\Delta xe$ values for each of the Ψ_f curves. The sensitivity curve demonstrates that a slope with a face angle of 67 degrees has a limiting column width of 2 m. If the actual width is greater than this value, the slope will be stable against toppling, and conversely if it is less, the slope will be unstable. By comparison, if the actual column width was 4 m, the slope would be stable at a face angle of 82 degrees.

Stabilization of toppling failures can be accomplished by reducing the aspect ratio of the columns in one of two ways: by reducing the slope height so that column height is reduced or by bolting layers together to increase the base width of the columns. Both of these methods change the column geometry so that the centers of gravity of the columns are within their bases.

7. CIRCULAR FAILURE

For the purposes of this discussion, a circular failure is defined as a failure in rock for which the failure surface is not predominantly controlled by structural discontinuities and that often approximates the arc of a circle [Figure 15-1(d)]. It should be recognized that there is a complete spectrum of structural control of rock failures ranging from completely structurally controlled, to those that are in part structurally controlled and in part material failures, to those in which structure has little or no influence. Rock types that are susceptible to circular failures include those that are partially to highly weathered and those that are closely and randomly fractured. In either case the rock mass is so highly fragmented that the failure surface can avoid passing through the stronger intact rock material. In this sense, a circular failure in rock can be considered much like that through a soil mass with a very large grain size.

The characteristics of circular failure in rock are similar to those for a classical rotational failure in soil except that the failure surface in rock tends to form a shallow, large-radius circle. Figure 15-1(d) illustrates an example of such a failure in a highly weathered basalt. Slope movement at the crest of the slope tends to be very steep, whereas movement at the toe tends to be subhorizontal. In general, signs of slope distress precede a rotational

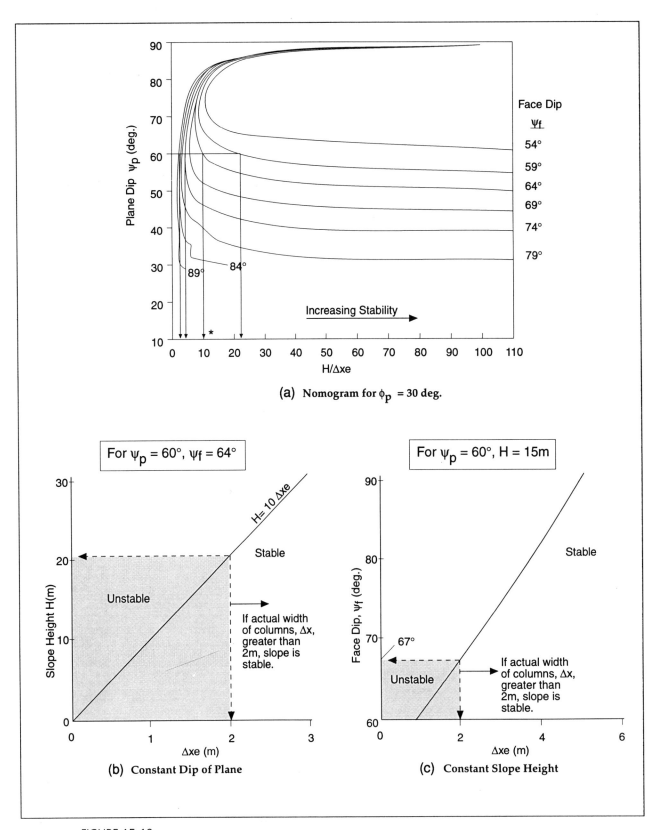

(a) Nomogram for ϕ_p = 30 deg.

For $\psi_p = 60°$, $\psi_f = 64°$

For $\psi_p = 60°$, H = 15m

(b) Constant Dip of Plane

(c) Constant Slope Height

FIGURE 15-19
Nomogram analysis of toppling stability (Choquet and Tanon 1985).

REPRINTED FROM ASHWORTH, E. (ED.), RESEARCH AND ENGINEERING APPLICATIONS IN ROCK MASSES, 26TH U.S. SYMPOSIUM ON ROCK MECHANICS, 26–28 JUNE 1985, SOUTH DAKOTA SCHOOL OF MINES AND TECHNOLOGY, RAPID CITY, 1985, 1340 PP., 2 VOLUMES, A.A. BALKEMA, OLD POST ROAD, BROOKFIELD, VERMONT 05036 failure in rock. These signs of distress include arcuate tension cracks near the crest of the slope, bulging in the toe area of the slide, and longitudinal cracks parallel to the inclination of the slope face. Before failure there usually is a period of accelerating movement, which generally can be monitored in order to predict failure.

7.1 Kinematic Analysis

The kinematic analysis for the circular mode of failure is in reality an analysis to exclude other modes of structurally controlled failure that have been discussed in the foregoing sections. In general, the structural discontinuities do not form distinct patterns and the dominant structures are not oriented with respect to the proposed slope configuration to develop a kinematically possible failure mode. An example of the latter case is the instance of a sedimentary sequence in which the bedding planes dip into the slope. Assuming that no other persistent structures are present, a circular failure model with appropriate allocation of rock-mass strength parameters would be suitable for slope design.

7.2 Stability Analysis

Appropriate methods of analysis for a circular failure in rock depend upon the shear-strength criterion used to characterize the rock-mass material. If the rock mass is assumed to obey the Mohr-Coulomb criterion, any of the analytical techniques developed for soil can be applied. These include stability charts as presented by Duncan (see Chapter 13) and Hoek and Bray (1981) and analytical techniques such as those presented by Janbu (1954), Bishop (1955), Morgenstern and Price (1965), and Sarma (1979). Details of the methodology for applying these various analytical techniques are considered in Chapter 13.

Because of the strong intact strength of the individual blocks and the high degree of interlock inherent in rock masses, it is more often the case that the material is very dilatant as it is sheared. For this reason, many investigators have found it appropriate to characterize the rock mass with a nonlinear shear-strength criterion. Several criteria have been proposed, including those by Landanyi and Archambault (1970) and by Hoek and Brown (1980). If criteria such as these are utilized, an analytical method similar to that pre-

sented in Chapter 14 must incorporate the ability to determine shear strength as a function of normal stress for each slice along the failure surface.

Several workers have presented useful failure charts for the rapid solution of simplified stability problems involving the circular failure mechanism. Each specific chart is applicable to a particular location of the phreatic surface ranging from fully drained to fully saturated. The solutions vary in technique but generally involve the calculation of one or more dimensionless ratios. These values in combination with the slope angle graphically determine corresponding dimensionless ratios from which the factor of safety can be calculated. The reader should refer to Chapter 13 for examples of stability charts for circular failures.

Such stability charts are very useful in rapidly carrying out sensitivity analyses of any of the parameters affecting stability and are also used to back analyze circular slope failures. Typical limitations of such charts include the following:

- A single material type is present throughout the slope,
- The slope inclination is uniform or can be approximated by a single value,
- The upper slope surface is horizontal,
- Pore-water pressures can be treated as a simple phreatic surface (i.e., there is no capability for modeling piezometric surfaces),
- The critical circle is assumed to exit through the toe of the slope or to be bounded by a firm base layer, and
- External loads such as earthquakes or reinforcement cannot be incorporated.

To analyze stability problems where conditions may preclude the use of stability charts generally requires the utilization of a computer program such as XSTABL (Sharma 1994), PC-STABL5 (Carpenter 1986), and LISA (Hammond et al. 1990), and the method developed by Sarma (1979) and adapted by Hoek (1986). The Sarma program is particularly useful because it allows the slope to be subdivided into slices, the boundaries of which can correspond to known geologic structures (i.e., slice boundaries are not constrained to be vertical). The Sarma program also allows distinct strength parameters to be applied to each segment of the sliding surface and to each slice boundary. This program is amenable to analyses with nonlinear shear strength criteria because it allows the user to determine the interslice stresses and to assign appropriate apparent friction and apparent cohesion values from the failure envelope. For examples of this type of analysis, the reader is referred to McCreath (1991) and Wyllie (1992, Chapter 6).

8. OTHER FAILURE MODES AND ANALYTICAL TECHNIQUES

The preceding sections have dealt with the four most common methods of stability analysis for rock slopes. Two additional failure modes that require specific structural fabric are column buckling and the bilinear wedge. For completeness, these failure modes are described next. The balance of this section deals with the issue of slope design under dynamic loading and with specialized analytical techniques based on probabilistic and distinct-element methods.

8.1 Column Buckling Failure

Slopes that contain steeply dipping repetitive discontinuities, such as bedding surfaces that parallel the slope face, may be susceptible to column buckling. Such a slope face is often referred to as a *dip slope*. The failure mode involves planar movement along the discontinuity and a material failure near the toe of the slope by buckling (Figure 15-20).

The necessary structural conditions that cause this failure mode to be kinematically feasible are similar to those for planar failure with the exception that the discontinuity forming the failure surface does not daylight the slope face but is parallel to it. As with the toppling failure mode, buckling failure is most probable in rocks with thin slabs.

This mode of failure is typically analyzed using the classical methods for column stability in which the stress parallel to the axis of a column is equal to Euler's critical stress for buckling (Goodman 1980; Cavers 1981). Parameters in the analysis include the length, l, and thickness, t, of the column; the length of the buckling zone, L; the inclination of the columns, Ψ_p; the shear strength along the columns; and the compressive rock strength (Figure 15-20). Buckling failures have occurred in high-footwall mine slopes in coal-measure sequences. Slopes that have experienced this type of failure are usually more than 200 m high. Only under unusual circumstances involving high-dip slopes in thinly bedded weak rock does buckling failure represent a design consideration for slopes along transportation corridors.

8.2 Bilinear Wedge Failure

In some situations a combination of discontinuities may define a failure mass composed of two blocks as shown in Figure 15-21. The upper active block is separated from the lower passive block by steeply dipping discontinuities that allow interblock movements to occur. The failure surface is usually defined by major planar structures such as faults. The upper block is unstable and exerts a force on the passive block.

The kinematic conditions for the two structures that form the bilinear failure surface are similar to those previously discussed for planar failure. The dip of the upper plane must exceed its angle of friction, and the dip of the lower plane must be less than the angle of friction for that surface. Although the most common configuration of this failure mode is for the lower failure surface to have a dip toward the slope face, a case has been reported by Calder and Blackwell (1980) in which the lower plane dipped into the slope. Survey monitoring demonstrated that at this site, movement of the passive block actually included an upward component.

The unique kinematic condition for this failure mode is that a geologic structure must be present that will allow differential movement between the blocks. At the intersection of the active and pas-

FIGURE 15-20
Column buckling failure mode (modified from Goodman 1980).

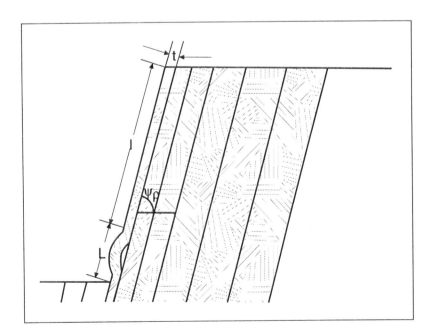

sive blocks on the failure surface, movements probably result in local rock crushing so that a failure arc develops between the two planar segments of the failure surface.

With reference to Figure 15-21, the stability of a bilinear wedge is evaluated by calculating the resultant force, $R3$, which is transferred from the active block to the passive block across Plane 3. The force $R3$ is the minimum to just resist sliding along Plane 1. The factor of safety is calculated by resolving the following forces:

- Resultant force, $R3$;
- Weight of the passive block, Wp;
- Shear force, $R2$; and
- Any uplift or driving forces due to water pressures along Plane 2.

The factor of safety of the bilinear wedge is the ratio of the total resisting force to the total driving force along this plane (CANMET 1977).

This type of failure is relatively uncommon; the reader is referred to the *Pit Slope Manual* (CANMET 1977) and *Introduction to Rock Mechanics* (Goodman 1980) for more detailed discussions.

8.3 Analysis of Earthquake Effects on Rock Slope Stability

Dynamic forces introduced in rock slopes should be considered in slope design analyses for areas susceptible to earthquakes. Design accelerations for a particular site are estimated from consideration of the Maximum Credible Earthquake (MCE) along each fault in the vicinity of the site. For each MCE and fault combination, distance-attenuation relationships are used to develop the resultant acceleration, expressed as a fraction of gravitational acceleration, for the site. The critical MCE-fault combination produces the highest or design acceleration for the site. In critical situations design accelerations can be developed using such seismological techniques. In most cases, seismic zoning maps provide adequate baseline data for design accelerations.

Pseudostatic analysis and Newmark analysis (Newmark 1965) are two commonly used engineering design methods that incorporate the forces due to earthquakes. In theory these methods can be applied to all the failure mechanisms for rock slopes previously discussed.

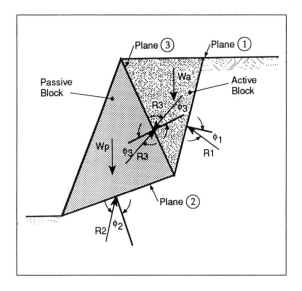

FIGURE 15-21 Bilinear failure mode (modified from CANMET 1977).

8.3.1 Pseudostatic Analysis

In the pseudostatic method, a force is applied to the unstable block equal to the product of the design acceleration and the weight of the block. This force is usually applied horizontally so as to decrease stability (Case i, Figure 15-22). Other investigators (e.g., Mayes et al. 1981) have suggested that the earthquake force be applied in two components, a horizontal component due to the design acceleration and a vertical component acting in an upward direction and equal to two-thirds of the horizontal component. The resultant force is 1.2 times the horizontal force and acts in a direction 34 degrees above the horizontal (Case ii, Figure 15-22). The latter formulation is obviously more conservative than the former.

Although the pseudostatic method is the most common way to include forces due to earthquakes in stability calculations, the method is inherently conservative because the cyclic loading due to the earthquake is replaced by a constant force equal to the maximum transient force. This conservatism is accounted for in the analyses by designing for a factor of safety that is lower for the pseudostatic analysis than for the static analysis. For example, whereas an acceptable static factor of safety for a rock slope may be 1.25 to 1.5, an acceptable factor of safety under pseudostatic conditions may be 1.0 to 1.1.

8.3.2 Newmark Analysis

Newmark's method (Newmark 1965) predicts the displacement of the slope under an acceleration-

FIGURE 15-22
Pseudostatic
analysis.

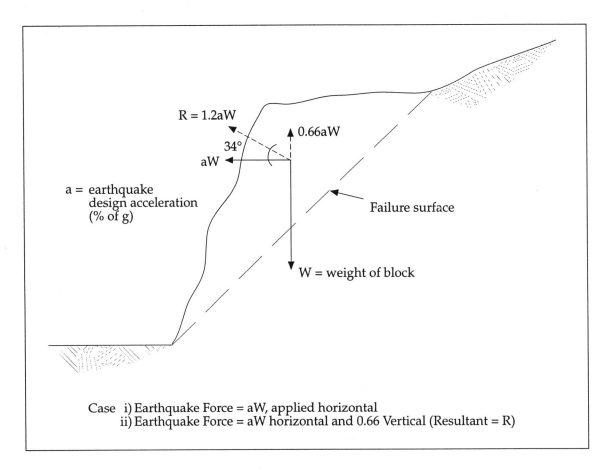

R = 1.2aW

34°

aW

0.66aW

a = earthquake
design acceleration
(% of g)

Failure surface

W = weight of block

Case i) Earthquake Force = aW, applied horizontal
      ii) Earthquake Force = aW horizontal and 0.66 Vertical (Resultant = R)

time record appropriate for the site. In the first step of the analysis, permanent displacement of the slope is assumed to occur only when the earthquake acceleration exceeds the critical acceleration for the geometry and preexisting stability condition of the slope (Jibson 1993). The cumulative permanent displacement of the slope is the sum of those portions of the acceleration-time history that exceed the critical acceleration value (Figure 15-23).

$$a_c = (FS - 1)g \sin\Psi_p \qquad (15.10)$$

where

a_c = critical acceleration,

g = acceleration due to earth's gravity,

FS = static factor of safety, and

Ψ_p = angle from the horizontal that the center of mass of the slide first moves (equal to inclination of failure plane).

Thus a slope with a static FS close to unity has a critical acceleration close to zero and experiences movement under any earthquake motion.

The second step in the Newmark analysis is to calculate the displacement for the ground motion

record selected for the site by double integration of those portions of the record that lie above the critical acceleration, as shown in Figure 15-23. Jibson (1993) has presented both rigorous and simplified methods to calculate the Newmark displacement.

Interpretation of the results of a Newmark analysis requires knowledge of the relationship between shear strength and shear strain for the material along the failure surface. Materials that show ductile behavior may accommodate the calculated displacement with little effect, whereas in brittle materials the displacement may lead to rapid strength loss and failure. For most rock slope design situations, and particularly for reinforced rock slopes, the strength-strain relationship is unknown and the Newmark method is difficult to apply.

8.4 Probabilistic Analysis

General engineering design practice has evolved the concept of factor of safety to determine the margin of conservatism that should be explicitly incorporated into a design. Typical factor-of-safety values used in geotechnical engineering practice for static design in transportation projects are as follows:

Slopes with potential impact on permanent structures: $FS = 1.5$ to 2.0;
Slopes with no potential impact on permanent structures: $FS = 1.25$ to 1.5.

The factor-of-safety concept is appropriate where material properties can be specified and stresses accurately computed. In geotechnical engineering these requisites are not generally present. It is often necessary to perform a testing or field measurement program of limited scope and to use engineering judgment to assign so-called "design values." Because the design analysis incorporates many such variables, each with its own degree of uncertainty, the cumulative impact upon the calculated factor of safety is difficult to assess.

An alternative approach that is beginning to gain broad acceptance is an index to the factor of safety referred to as the *coefficient of reliability* or alternatively as the *probability of failure*. These indexes are expressed as a decimal fraction of unity, and their sum is 1.0. Although either can be used in probabilistic analysis, the adoption of one over the other is an issue of perception. Not surprisingly, the former is more readily accepted by property owners than is the latter.

In this analysis process, the uncertainty in each variable is incorporated into the analytical process by assigning a probability distribution function (PDF) to each variable instead of a single design value. For example, multiple measurements of the angle of friction of a joint surface may yield a mean value of 32 degrees with a standard deviation of 5 degrees. This mean value can be incorporated into the probabilistic analysis as a normal distribution with the stated parameters. The range of all the variables in the analysis can be similarly defined by PDFs, although the type of distribution can be triangular, normal, lognormal, beta, or other (Wyllie 1992). For example, PDFs can be assigned to water levels, discontinuity inclination, unit weight, and cohesion.

The analysis requires selection of a value from each of the PDFs on a random basis and then calculation of a factor of safety. Typically, a Monte Carlo simulation is used to generate random numbers from which each of the variable values is assigned. By repeating this process a large number of times (100 or more), a cumulative distribution

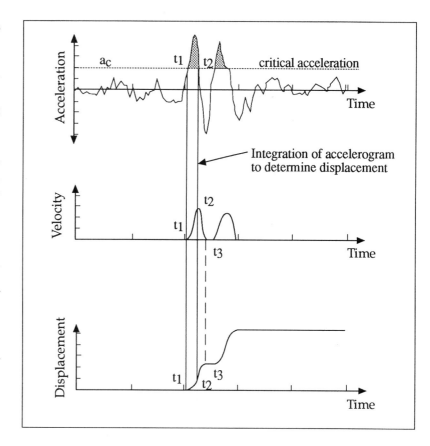

function (CDF) for the factor of safety can be developed. An example of this process is shown in Figure 15-24. As a result of such an analysis, the slope practitioner is in a position to state that the coefficient of reliability is x percent (or that the probability of failure is $100 - x$ percent). This probability of failure can be evaluated in conjunction with the consequences of failure to determine an optimum course of action. In cases where all of the consequences of failure can be expressed in monetary terms, the optimum course of action will be the one that is most cost-effective. Applications of this methodology to transportation corridors have been described by Wyllie et al. (1979) and Roberds (1990, 1991) and an illustrative application to an open pit mine slope by Piteau and Martin (1977).

8.5 Three-Dimensional Analyses

Two-dimensional analyses evaluate geometric, material, and loading conditions along a vertical section aligned normal to the slope. Thus, lateral forces are not considered. In most cases the analysis of a unit thickness of the slope provides an acceptable factor of safety. Three-dimensional

FIGURE 15-23
Newmark analysis
(Jibson 1993).

FIGURE 15-24
Probabilistic slope
stability analysis.

INPUT

Consider Equation 15-4 for planar failure:

$$FS = \frac{(cA + W\cos \Psi_p \tan \phi)}{(W \sin \Psi_p)}$$

Stochastic Variables: (cumulative probability distributions for each variable defined by testing or measurement and fitted to standard distribution type such as normal, log-normal etc.)

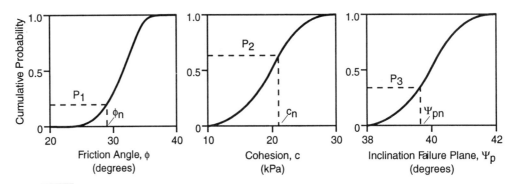

METHOD

Monte Carlo Analysis

1. *For the "nth" FS calculation:*

Generate random numbers P_1, P_2, P_3 between 0 and 1.

From distributions calculate ϕ_n, c_n, ψ_{pn}.

Calculate FS_n

2. Repeat procedure until n = 100, 200, etc.

RESULT

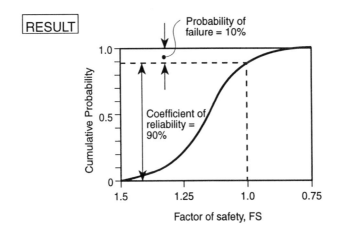

analyses may be warranted where the assumption of uniformity in the transverse direction cannot be made, for example, on slopes with sharp convex or concave curvatures in plan. Two-dimensional analyses of such configurations can either underestimate or overestimate the factor of safety.

Hungr et al. (1989) developed an application of Bishop's simplified method for three-dimensional limit-equilibrium analysis referred to as CLARA. The method subdivides the failure mass into a series of columns rather than slices as is the case in the two-dimensional formulation. The vertical intercolumnar shear forces are neglected in the analysis, which can lead to conservative results when applied to some asymmetrical failure masses in which the energy required to overcome internal deformation of the failure mass contributes to stability. However, for rotational or symmetrical sliding surfaces for which the internal strength can be neglected, the method is useful (Hungr et al. 1989). Hungr et al. (1989) also presented applications of the model to rotational, wedge, and bilinear sliding surfaces.

8.6 Distinct-Element Method

Methods of analysis other than limit equilibrium have been utilized to analyze slopes in which kinematically feasible failure modes are not present, yet observation and monitoring indicate that displacements are occurring. One such method is the distinct-element method described by Cundall (1987) and Lorig et al. (1991). This method does not require a prescribed failure surface in order to reach a solution. The iterative calculation procedure models the progressive failure of slopes, a process that generates a surface of demarcation between a group or groups of stable blocks and a group or groups of unstable blocks.

As stated by Lorig et al. (1991), the three attributes of the distinct-element method that distinguish it from other numerical methods are as follows:

1. The rock mass is composed of individual blocks that can undergo large rotation and large displacements relative to one another,
2. Interaction forces between blocks arise from changes in their relative geometrical configuration, and

3. The solution scheme is explicit in time.

The distinct-element method utilizes a calculation procedure that solves the equations of motion and contact force for each modeled rigid block at each time step [Figure 15-25(a)]. Block interaction at contact points is modeled by strength and deformation parameters. As shown in Figure 15-25(b), the analysis provides the sequential positions and velocities of each block and thus is useful when predictions of slope displacement are required. The method can also be used to incorporate reinforcing elements, to analyze foundation loading on rock, and to carry out dynamic analyses using earthquake velocity records. Computer programs have been formulated for both two-dimensional and three-dimensional distinct-element analyses.

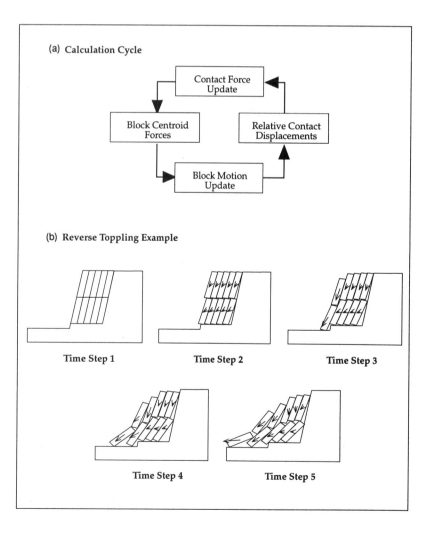

FIGURE 15-25 Distinct-element method (Lorig et al. 1991).

REFERENCES

Bishop, A.W. 1955. The Use of the Slip Circle in the Stability Analysis of Slopes. *Geotechnique,* Vol. 5, No. 1, pp. 7–17.

Calder, P.N., and G.H. Blackwell. 1980. Investigation of a Complex Rock Slope Displacement at Brenda Mines. *Canadian Institute of Mining Bulletin,* Vol. 73, No. 820, pp. 73–82.

CANMET. 1977. *Pit Slope Manual,* Supplement 5-1: *Plane Shear Analysis.* Canada Centre for Mineral and Energy Technology, Ottawa, Ontario, 307 pp.

Carpenter, J. R. 1986. *STABL5/PC STABL5 User Manual.* Joint Highway Research Project Report to Indiana Department of Highways. School of Civil Engineering, Purdue University, West Lafayette, Ind., 46 pp.

Cavers, D. S. 1981. Simple Methods to Analyze Buckling of Rock Slopes. *Rock Mechanics,* Vol. 14, No. 2, pp. 87–104.

Choquet, P., and D. D. B. Tanon. 1985. Nomograms for the Assessment of Toppling Failure in Rock Slopes. In *Proc., 26th U.S. Symposium on Rock Mechanics,* South Dakota School of Mines and Technology, June 26–28, 1985, Rapid City, S.D., A.A. Balkema, Rotterdam, Netherlands, Vol. 1, pp. 19–30.

Cundall, P. A. 1987. Distinct Element Models of Rock and Soil Structure. In *Analytical and Computational Methods in Engineering Rock Mechanics* (E. T. Brown, ed.), George Allen and Unwin, London, pp. 129–163.

Franklin, J. A., and M. B. Dusseault. 1989. *Rock Engineering.* McGraw-Hill, Inc., New York, 600 pp.

Goodman, R. E. 1976. *Methods of Geological Engineering in Discontinuous Rocks.* West Publishing Co., St. Paul, Minn., 472 pp.

Goodman, R. E. 1980. *Introduction to Rock Mechanics.* John Wiley & Sons, New York, 478 pp.

Goodman, R. E., and J. W. Bray. 1976. Toppling of Rock Slopes. In *Proc., Specialty Conference on Rock Engineering for Foundations and Slopes,* Boulder, Colo., American Society of Civil Engineers, New York, Vol. 2, pp. 201–234.

Hammond, C. J., S. Miller, D. Hall, and P. Swetik. 1990. *Level I Stability Analysis (LISA) User's Manual to Version 2.0.* Forest Service, U.S. Department of Agriculture, Ogden, Utah, 256 pp.

Hittinger, M. 1978. *Numerical Analysis of Toppling Failures in Jointed Rock Masses.* Ph.D. thesis. University of California, Berkeley, 297 pp.

Hodge, R. A. L., and R. A. Freeze. 1977. Groundwater Flow Systems and Slope Stability. *Canadian Geotechnical Journal,* Vol. 14, pp. 466–476.

Hoek, E. 1986. General Two-Dimensional Slope Stability Analysis. In *Analytical and Numerical Methods in Rock Engineering* (E. T. Brown, ed.), George Allen and Unwin, London, pp. 95–128.

Hoek, E., and J. W. Bray. 1981. *Rock Slope Engineering,* 3rd ed. Institution of Mining and Metallurgy, London, 402 pp.

Hoek, E., and E. T. Brown. 1980. *Underground Excavations in Rock.* Institution of Mining and Metallurgy, London, 527 pp.

Hungr, O., F. M. Salgado, and P. M. Byrne. 1989. Evaluation of a Three-Dimensional Method of Slope Stability Analysis. *Canadian Geotechnical Journal,* Vol. 26, No. 4, pp. 679–686.

Janbu, N. 1954. Application of Composite Slip Circles for Stability Analysis. In *Proc., European Conference on Stability of Earth Slopes,* Stockholm, Statens Reproduktionsanstalt, Vol. 3, pp. 43–49.

Jibson, R. W. 1993. Predicting Earthquake-Induced Landslide Displacements Using Newmark's Sliding Block Analysis. In *Transportation Research Record 1411,* TRB, National Research Council, Washington, D.C., pp. 9–17.

Landanyi, B., and G. Archambault. 1970. Simulation of Shear Behaviour of a Jointed Rock Mass. In *Proc., 11th Symposium on Rock Mechanics,* American Institute of Mining Engineers, New York, pp. 105–125.

Lorig, L. J., R. D. Hart, and P. A. Cundall. 1991. Slope Stability Analysis of Jointed Rock Using the Distinct Element Method. In *Transportation Research Record 1330,* TRB, National Research Council, Washington, D.C., pp. 1–9.

Lowell, S. M. 1990. The K M Mountain Landslide near Skamokawa, Washington. *Washington Geologic Newsletter,* Vol. 18, No. 4, pp. 3–7.

Mathews, W. H., and K. C. McTaggert. 1969. The Hope Landslide, British Columbia. *Proc., Geology Association of Canada,* Vol. 20, No. 65-B, pp. 65–76.

Mayes, R. L., R. L. Sharpe, and J. D. Cooper. 1981. Seismic Design Guidelines for Highway Bridges. In *Proc., 50th Annual Convention,* Structural Engineers Association of California, Coronado, pp. 93–112.

McCreath, D. 1991. Use of Nonlinear Strength Criteria in Stability Analyses of Bridge Foundation on Jointed Rock. In *Transportation Research Record 1330,* TRB, National Research Council, Washington, D.C., pp. 10–21.

Moore, H. L. 1986. Wedge Failures Along Tennessee Highways in the Appalachian Region: Their Occurrence and Correction. *Bulletin of the Association of Engineering Geologists,* Vol. 23, No. 4, pp. 441–460.

Morgenstern, N. R., and V. E. Price. 1965. The Analysis of the Stability of General Slip Surfaces. *Geotechnique,* Vol. 15, No. 1, pp. 79–93.

Newmark, N. M. 1965. Effects of Earthquakes on Dams and Embankments. *Geotechnique,* Vol. 15, No. 2, pp. 139–159.

Piteau, D. R. 1972. Engineering Geology Considerations and Approach in Assessing the Stability of Rock Slopes. *Bulletin of the Association of Engineering Geologists,* Vol. 9, No. 3, pp. 301–320.

Piteau, D. R., and D. C. Martin. 1977. Slope Stability Analysis and Design Based on Probability Techniques at Cassiar Mine. *Canadian Institute of Mining Bulletin,* Vol. 70, No. 779, pp. 139–150.

Piteau, D. R., and F. L. Peckover. 1978. Engineering of Rock Slopes. In *Special Report 176: Landslides: Analysis and Control* (R. L. Schuster and R. J. Krizek, eds.), TRB, National Research Council, Washington, D.C., pp. 192–234.

Roberds, W. J. 1990. Methods for Developing Defensible Subjective Probability Assessments. In *Transportation Research Record 1288,* TRB, National Research Council, Washington, D.C., pp. 183–190.

Roberds, W. J. 1991. Methodology for Optimizing Rock Slope Preventative Maintenance Programs. In *Proc., Geotechnical Engineering Congress,* Boulder, Colo., American Society of Civil Engineers, New York, pp. 634–645.

Sarma, S. K. 1979. Stability Analysis of Embankments and Slopes. *Journal of the Geotechnical Engineering Division,* ASCE, Vol. 105, No. GT12, pp. 1511–1524.

Sharma, S. 1994. *XSTABL, An Integrated Slope Stability Analysis Program for Personal Computers: Reference Manual Version 5.* Interactive Software Designs, Inc., Moscow, Idaho, 214 pp.

Wyllie, D. C. 1992. *Foundations on Rock.* Chapman & Hall, London, 333 pp.

Wyllie, D. C., N. R. McCammon, and W. L. Brumund. 1979. Planning of Slope Stabilization Programs Using Decision Analysis. In *Transportation Research Record 749,* TRB, National Research Council, Washington, D.C., pp. 34–39.

PART 4

MITIGATION

Chapter 16

JEFFREY R. KEATON AND
GEORGE H. BECKWITH

IMPORTANT CONSIDERATIONS IN SLOPE DESIGN

1. INTRODUCTION

Many slopes are designed to perform specific functions, chiefly to provide safe access with acceptable maintenance, economy, and appearance. The complete design process is likely to involve five stages (Norris and Wilbur 1960):

1. Preparation of a general layout to meet functional requirements;
2. Consideration of possible solutions to satisfy the functional requirements;
3. Formulation of preliminary design incorporating possible solutions;
4. Selection of the optimum solution by comparing the economy, functionality, and appearance of the possible solutions; and
5. Provision of detailed design of the optimum solution.

The chief responsibility of the engineer is to design the optimum solution. Analyses of the slope geometry and response to loading are the means by which the alternative solutions are evaluated. Thus, the details of analysis are important considerations in the slope design process.

Slopes generally are evaluated in terms of the factor of safety against sliding. The design team must define adequate factors of safety for both constructed and natural slopes. Defining an adequate factor of safety poses a philosophical problem of considerable magnitude.

In many branches of engineering, the engineer can specify materials whose performance under particular stress-strain conditions is known with considerable precision. Furthermore, construction procedures can be specified to prevent overstressing components during assembly. Usually the chief uncertainty concerns the assessment of loads to which the completed structure will be subjected while in service. This difficulty is overcome by the use of tabulated loadings, which are usually much greater than any probable loadings but have been approved by engineers as representing extreme conditions. Thus, economical designs are produced with suitably conservative features.

In contrast, the geotechnical engineer and engineering geologist responsible for assessing slope conditions are faced with much greater uncertainties. The materials forming the slope may be highly variable and exhibit a range of material properties. The inherent variability and uncertainty concerning material distribution and properties make adequate characterization of natural slopes especially troublesome and focus attention on the need for an accurate geologic model. However, even in constructed embankments, these uncertainties can lead to slope instabilities. Slope performance may be adversely affected by undetected variations in material properties and environmental conditions or by changes in material properties over time caused by weathering or erosion of the materials or changes in climate conditions.

The geotechnical engineering approach to slope stability evaluation involves the assessment of shear strength (i.e., capacity) rather than of loads (i.e., demand). Except when subjected to external loading caused by earthquake shaking, increased potentiometric conditions, or human activities, slopes are exposed to relatively constant loads due to the weight of the soil and rock materials of which the slope is composed, and changes in loading are not a significant factor controlling slope stability.

The shear strength of the soil and rock materials composing slopes may have a considerable effect on slope stability. Thus, a factor of safety based on the ratio of available shear strength to that which is required to initiate undesired slope movement forms a suitable measure of slope stability. The shear-strength values forming the numerator and denominator of this factor-of-safety ratio may change over time in response to changing environmental conditions. The shear strength of the materials changes with changing moisture conditions or because of changing physical-chemical properties of the materials because of weathering, strain-softening phenomena, fabric changes, or even initial shearing. Even more critical in many cases is the reduction in the effective shear strength caused by increased pore-water pressure. Changes in slope geometry due to natural erosion or human modification may adversely increase the shear stresses favoring slope movement.

In contrast to the factor-of-safety approach, the application of probability theory and statistical methods to determine the probability of slope failure or unacceptable performance is emerging as a potentially useful alternative (Sharp et al. 1981; Bowles and Ko 1984). Quantifying the variability in material properties of geologic formations and discontinuities is a major challenge for the engineering geologist. The properties of constructed embankments, of course, can be controlled and tested to provide a basis for evaluating slope performance.

Bromhead (1992) stated that the critical slope-destabilizing factors, in order of importance for most cases, are

- Increase in pore-water pressures,
- Removal of support at the toe either by erosion or by excavation,
- Changes in soil or rock shear-strength properties,

- Seismic loading, and
- Loading at the head of the slope, most often by other mass movements but occasionally by placement of fill or other "live loading."

Very conservative slope configurations would result if slopes were required to have factors of safety of 1.0 when conditions included maximum conceivable pore-water pressures, conservative (low or residual) shear-strength material properties, excavation at the toe of the slope, additional loading at the top of the slope, and seismic loading. Such a design would not be justifiable on economic grounds. Each of the many factors listed above involves conservative assumptions; thus, the resulting factor of safety is genuinely overconservative.

Furthermore, the individual assumptions must become increasingly conservative in cases where the conditions are poorly understood. This conservatism can result in the extreme situation in which a small, insignificant slope subjected to minimum field investigation can receive a higher factor of safety than a larger, more important slope subjected to extensive and expensive investigation. Finally, large increases in the factor of safety may be unachievable, either economically or physically, in the cases of large slopes. In such situations, acceptance of lower factor-of-safety values may be the only course of action.

These considerations emphasize the importance of careful selection of slope design parameters, which must reflect

1. Site-specific environmental conditions potentially affecting slope stability,
2. Strength properties of the materials forming the slope,
3. Economic considerations, and
4. Other facility design constraints.

2. ENVIRONMENTAL CONDITIONS AFFECTING SLOPE STABILITY

Environmental conditions affecting slope stability include external loads, probable earthquake accelerations, climate conditions, and volcanic eruptions.

2.1 External Loads

The effect of external loads, which may be natural or human-induced, must be incorporated into engineering analyses on which slope design is based.

2.1.1 Natural External Loads

Natural external loads on slopes may change because of erosion and deposition that cause the toe of a slope to become oversteep or material to accumulate on the crest of the slope. An engineering geologic assessment of the slope should reveal the potential for natural erosion and deposition processes. If these processes can be expressed in quantitative terms, their effects may be included in the analysis. If these processes are not quantifiable, their potential may be recognized but no basis exists for including them in the analysis.

2.1.2 Human-Induced Modifications to External Loads

Human activity in the vicinity of a slope, particularly a marginally stable slope, may increase the forces tending to cause failure. In some circumstances, these activities may improve the stability of some slopes. For example, slope improvement measures, such as rock bolts and buttress fills, result in positive forces on slopes. Many common slope improvements are viewed as providing external forces to a slope.

Human-induced modifications that may adversely affect external loads are (a) grading of the existing slope or of adjacent slopes, (b) adjacent construction, (c) construction blast damage, and (d) vibrations of passing vehicles.

Slope regrading may result in a condition similar to that caused by natural erosion and deposition in which the toe becomes oversteep or material is accumulated on the crest. Construction adjacent to the crest of a slope, particularly of heavy facilities such as railroads or stockpiles, can surcharge a slope. Grading adjacent to a slope can consist of excavation or filling; excavation would create oversteepened slopes, whereas filling would create a surcharge. These conditions must be anticipated or prevented so that an appropriate stability analysis can be made.

Excavation blasting can cause excessive fracturing of otherwise reasonably intact rock material. Improper blasting loosens the rock, decreases the density and strength, and increases the permeability. These conditions must be recognized in the field so that slope remediation can be carried out during or shortly after construction. Alternatively,

they must be anticipated or prevented so that an appropriate stability analysis can be made.

Most vibrations of passing vehicles impose small loads on adjacent slopes, but some vibrations can influence slope stability. For example, a freight train moving on tracks a short distance from a steep slope with shallow groundwater may induce vibrations at a frequency and amplitude that could cause permanent deformation or failure of the slope. Such dynamic loads should be assessed and incorporated in the design.

2.2 Earthquake Accelerations

In seismically active regions, design of slope remediation may be controlled by expected earthquake accelerations. These accelerations may be mapped as a probabilistic approach to determining earthquake hazard (Krinitzsky et al. 1993). Maps have been prepared for the United States showing earthquake hazard in terms of the peak horizontal acceleration that has a 10 percent probability of being equalled or exceeded in periods of 50 and 250 years (Algermissen et al. 1990). Generalized maps showing levels of earthquake shaking that have a 10 percent probability of occurring in a period of 50 years are also included in building codes, such as the *Uniform Building Code* (ICBO 1994) and the *Standard Building Code* (SBCCI 1991).

In the deterministic approach, an earthquake scenario, such as the maximum earthquake conceivable in the region or a specific-magnitude earthquake on a specific fault, is presented. Earthquake hazard should be considered in the context of the acceptable risk of damage due to other types of hazards, such as flooding. It may not make sense for a slope repair to be designed for a 5,000-year earthquake acceleration, whereas an adjacent section of highway would be washed out in a 500-year flood. Alternatively, the earthquake-induced damage may be judged to be much more costly to repair than flood-induced damage. In such a case, it would be appropriate to select a larger, less likely design earthquake scenario.

Design earthquake values normally are used in limit-equilibrium stress analyses in which the earthquake accelerations are commonly applied to the slope as continuous (pseudostatic) forces acting in a downslope direction, either horizontally or parallel to the slope. This approach is conservative because earthquake shaking is vibratory in nature. About half of the time the forces act into the slope, tending to make it more stable, as, for example, in earth materials that have strain-independent strength (materials that do not lose strength with strain, such as liquefiable sand). The conservatism is increased when a maximum earthquake acceleration is applied to a slope with worst-case piezometric conditions. Although the maximum earthquake might occur under such conditions, the probability that both conditions would occur simultaneously may be very small. This issue may become even more important if conservative values are also used for the material strength parameters.

The resulting computations may provide design solutions that are unreasonable for the economy of the project. Thus, the design earthquake must be selected with knowledge of the acceptable risk for the project (see Section 4.2).

2.3 Climate Conditions

The importance of climate in slope design has three aspects. First, the influence of precipitation on groundwater is critical for the stability of most slopes. In some instances, precipitation and runoff estimates can be important in predicting the potential and frequency of slope failures, for example, in evaluating the potential for debris flows in some canyons susceptible to these processes. Second, precipitation and runoff estimates are also important in assessing volumes of discharge from irrigation systems in rural areas and from storm-water drains in urban areas. Third, estimates of the cyclic variation of moisture and temperature conditions may also be important in assessing slope stability designs in some situations.

2.3.1 Precipitation and Runoff

Precipitation and runoff estimates often are used to design the optimum size of drainage collection systems on slopes. The most reasonable approach to providing suitable precipitation and runoff parameters for slope stability analysis is to assume a shallow depth to groundwater and a storm recurrence interval that corresponds to the economic importance of preventing a slope failure. Generally, precipitation values should not be used to estimate the probable occurrence of high piezometric levels in the slope.

Regions with steep slopes and high-intensity rainfall may be susceptible to shallow soil slides that have the potential to mobilize into debris flows. Dietrich et al. (1986) identified geologic and geomorphic conditions favorable for shallow soil slides leading to debris flows. Keefer et al. (1987) developed a warning system for debris flows based on relationships between rainfall intensity and duration. Kramer and Seed (1988) discussed static loading conditions leading to soil liquefaction. The potential for damaging debris flows originating on slopes adjacent to a project must be recognized and incorporated into design provisions.

2.3.2 Irrigation and Urban Storm-Water Discharge

Agricultural or landscape irrigation can be a significant source of water that may cause slope failure some distance away, depending on the subsurface geologic conditions (Schuster et al. 1987). Fractured and layered materials with lateral permeability exceeding vertical permeability (except at fractures) can allow groundwater to migrate long distances, build up hydrostatic pressures, and emerge as springs. Material adjacent to the irrigated site that would otherwise remain unsaturated can become saturated also. In those instances in which the designer of a slope has no control over the application of irrigation water, it may be prudent to assume that fully saturated conditions will occur.

Urban storm water and other types of urban water sources (leaking water mains and sewer lines) pose conditions similar to that discussed above for irrigation water. If storm water is discharged on a slope rather than into a stream channel at the base of a slope, erosion is likely to occur. This condition is common on some highways and railroads where storm water is collected on the upslope side of the right-of-way, conveyed under the right-of-way in a buried conduit, and discharged on the downhill slope. After a few storm events, particularly if the events result in flash floods, substantial erosion gullies form on the slope. The oversteepened erosion gullies lead to localized slope failure, ultimately threatening the highway or railroad. In some cases, initial movement of a slope distresses buried water mains or sewer lines, causing them to leak; the leakage in turn promotes instability, which further distresses the buried pipelines.

Pipeline trenches backfilled with free-draining material can act as conduits for subsurface water, directing it to places that otherwise might remain relatively dry (Keaton 1992). Such introduction of water can be an important factor in the design of stable slopes. Backfill with hydraulic properties similar to those of the surrounding soil can alleviate the potential for such trenches to act as conduits. Storm-water culverts can become plugged with sediment, causing backup of water that can overtop and erode embankments (Anderson et al. 1984).

2.3.3 Cyclic Moisture and Temperature Loads

Cyclic loads involving moisture and temperature can cause a gradual change in the strength of earth materials that may become important in stability analyses.

In cold climates freezing and icing are important processes in slope stability evaluations because ice is effectively impermeable and can cause the buildup of hydrostatic pressures that otherwise would not develop. The stability of slopes in permafrost regions is addressed in Chapter 25 of this report. In environments where significant freezing can occur, it is important either to assume full hydrostatic pressure or to provide for a drainage system that will not become blocked by the formation of ice at the discharge points. Rapid melting of a frozen slope can create an equivalent rapid drawdown condition causing slope movement.

A dramatic example of thermal cycles that caused increased slope instability was the failure of a steep cut slope in the South American Andes in which material overlying an undulating bedding plane surface expanded during the heat of the day (Sowers and Carter 1979). The expansion separated the bedding plane so that the undulations at the base of the slide block were no longer seated in the undulations on the underlying rock mass. Failure occurred through intact rock at the "peaks" of the undulations, resulting in loss of life and damage to property.

2.4 Volcanic Eruptions

In areas with active or potentially active volcanoes, eruptions can influence slope stability by ejecting large volumes of volcanic material or by causing regional or local deformation. Volcanic eruptions can generate large volumes of volcanic ash and other ejecta that can bury landscapes or flow long distances. Volcanic ash and ash-derived debris flows can clog drainage devices or create natural dams that can change groundwater flow patterns. Such changes can make significant differences in the parameters used in stability analyses. Volcanic eruptions also can cause regional and local tilting of the ground surface, resulting in steeper slopes than before the volcanic activity. Stresses associated with regional deformation can fracture weakly cemented materials, changing their physical properties.

3. MATERIAL PROPERTIES AND SITE CHARACTERISTICS

Samples collected in the field and submitted for testing in a laboratory often are assumed to be representative of the geologic unit in the field. However, such samples clearly can represent only the intrinsic properties of the material itself, not the properties of the rock or soil mass, including joints, fractures, and other defects. The problem of representativeness becomes even more difficult with materials that cannot be sampled or do not produce samples suitable for laboratory testing. Field tests may be substantially more expensive than laboratory tests. Nevertheless, the results of field tests often are more representative of the geologic conditions. An additional challenge with field tests is interpreting the results to reflect individual geologic units rather than contributions from several geologic units.

3.1 Identification of Properties of Soil or Rock Masses

Complete characterization of a site must include the mass properties of the earth materials. The mass properties include both intrinsic material properties and defects. Material properties consist of soil or rock texture (grain size), mineralogy (including organic material), degree of weathering, porosity, water content, specific gravity, unit weight, and strength of intact material (cohesion intercept and friction angle). Mass properties consist of permeability, slaking, drill-core rock-quality designation, defect condition (orientation, spacing, roughness, openness, and infilling), mass strength, and groundwater conditions. Not all parameters may be necessary for specific analyses. Investigations to identify geologic conditions are discussed in Chapters 8, 9, and 10. Strength properties of soils are discussed in Chapter 12, and those of rocks are discussed in Chapter 14.

3.2 Engineering Analysis of Site Conditions

Engineering analysis of slope systems usually requires information concerning three basic elements: topography, subsurface conditions, and external loads. The identification and assessment of external loads are discussed in Section 2.1.

3.2.1 Topography

Most engineering analyses evaluate simplified two-dimensional models of slope systems. Analyses can be performed on existing topography or on proposed construction geometry. In many computer-aided analyses, the topography is simplified and lateral boundary conditions for the landslide are ignored. The longitudinal profile of an existing landslide commonly is taken to be representative of the slope.

3.2.2 Subsurface Conditions

Accurate assessment of subsurface conditions is fundamental to engineering analysis. The geometry of existing or potential slide surfaces, boundaries of subsurface material units, and piezometric levels, along with the mass properties of the geologic units, including slide surfaces, constitute the most important design parameters. The geometry of slide surfaces and boundaries of geologic units are based on surface mapping and subsurface investigation, discussed in Chapters 9 and 10. The importance of an accurate geologic model on which to base engineering analyses and slope designs cannot be overstated. Design piezometric levels should represent the most extreme condition possible under which the slope must remain stable. The design mass properties of the geologic units should be reasonable but not overly conservative. If worst-case parameters are selected at each opportunity, excessive conservatism in the estimated factor of safety, and hence the anticipated risk of failure, is likely to result.

4. ECONOMIC CONSTRAINTS

Design parameters must be selected to reflect the characteristics of the site under so-called "design conditions." For most sites where slope stability is a primary issue, the design conditions may include cut slopes, fill slopes, earth reinforcement, full saturation of the slope, and earthquake shaking. Mass properties of the earth materials at a site must be based on geologic descriptions of the site and results of field and laboratory tests. Engineering analyses must reflect the inherent geologic uncertainties, acceptable levels of risk, and appropriate conservatism in design.

4.1 Geologic Uncertainties

A characteristic of natural materials is variability. The goal of the geologic evaluation of a site is to describe the nature and distribution of the materials present. Geologic information pertains to surficial and bedrock materials, geologic structures, surface-water features, and correlation of surface and subsurface observations (see Chapter 9). The geologic uncertainties include variabilities in identification of geologic units, material properties, locations of boundaries separating geologic units, and correlation of geologic units across a site or slope area.

Geologists use the principle of multiple working hypotheses to continuously synthesize the geologic information as it is obtained. Even after extensive data collection, several equally valid hypotheses may exist to explain the available data. This aspect of geologic investigation often is frustrating to engineers responsible for designing a specific solution to a complex problem. However, if sufficient information is available, a statistical analysis can be made to develop a mean value and standard deviation of the material properties.

To address the problem that all details concerning subsurface conditions will never be fully known, geologists adopt one of two mapping strategies. Some geologists subdivide the geologic materials into a large number of mappable units, whereas others group the geologic conditions into a few mappable units. Large numbers of subdivided units may exhibit differences so minor that their behavior will be very similar and the mean values of their material properties will be less than one standard deviation apart. Grouped units may show a larger standard deviation, but their mean values may adequately represent the material properties. The engineer may find more utility in the geologic description of fewer units grouped by material properties and may choose to deal with geologic uncertainties in the factor of safety.

4.2 Acceptable Risk

National or local risk standards can be prescribed in the form of design life, average annual number of occurrences, or average return period. The design life of most structures ranges from about 20 to about 50 years, occasionally to as much as 100 years. The design life has major economic implications for two reasons. First, it is the time period over which capital expenditures can be amortized or loans repaid. Second, the reliability of the structure over the design life must be reflected in the conservatism incorporated into the design. In simple terms, structures with short design lives may be designed with some appropriate economical shortcuts, whereas long-design-life structures must have more substantial components throughout.

Acceptable risk for a structure may be defined as the ability to withstand an event that is reasonably likely to occur within its design life but to sustain some damage due to a larger event that is unlikely to occur. This is the concept of *dual-level design*, which provides a rationale for decisions related to economic issues. If the consequence of some level of damage is not tolerable, acceptable risk will be associated with the event or force at the threshold of "tolerable" damage. Acceptable risk can be expressed in terms directly related to the length of the design life. For example, a facility can be designed for an event with an average return period equal to some multiple of the facility's design life. Thus, a facility with a design life of 50 years can be planned for acceptable damage from an event with a return period of 500 years (10 times the design life).

4.3 Design Conservatism

The assessment of landslides and slope stability provides numerous opportunities for conservatism to be introduced. This is particularly true because geology and engineering often are independent activities, so conservative assumptions may be made without being recognized as such by the design engineer. Geologists and engineers tend to describe uncertainty and apply conservatism in different ways. Geologists consider a variety of possible explanations for a given set of facts and select the most reasonable interpretation. Too often, however, the set of facts may be interpreted equally well in two or more ways. Unless the geologist provides the engineer with a succinct discussion of the reasonable interpretations and then makes a best guess as to which is correct, the engineer will be forced to make the guess.

Engineers, on the other hand, use the factor-of-safety method to deal with uncertainty and conservatism. Uncertainty is encountered in many factors of a stability analysis, and the engineer may be unable to accurately quantify the uncertainty. Rather than adjusting each factor to account for the uncertainty, it may be prudent to use average parameters for each factor and apply a single factor to the final result to express the uncertainty. In this way acceptable stability often is described as a factor of safety equal to or greater than 1.5. This factor of safety expresses the ratio between the forces tending to maintain stability and those tending to create instability. An alternative might be to use a value equal to the mean minus one standard deviation for individual parameters, such as the coefficient of friction, the cohesion intercept, or joint roughness. The cumulative effect of conservatism at several points during an analysis might be excessive conservatism.

5. OTHER DESIGN CONSTRAINTS

Selection of design parameters for repair of a slope failure can be influenced by several other considerations, such as location, size, costs, and design standards. An additional consideration that can influence selection of design parameters is the urgency for repair of the slope. A slope failure on a busy section of Interstate highway may have considerably more urgency for repair than a similar slope failure on a less important highway.

5.1 Location

Repair of slope failures in urban settings where public safety is the primary concern may warrant stricter design parameters than those for otherwise similar slope failures in rural or remote settings. However, in many cases, public safety issues may be addressed adequately in rural areas without some of the provisions that may be deemed necessary in urban areas. Slope appearance also may be more important in urban settings than in remote areas. Remote areas, particularly along low-volume roads, provide challenges to produce efficient designs that can be constructed economically. A number of issues are important in controlling or guiding selection of design parameters for a specific location; these issues include the design life, desirability of scheduled maintenance, tolerability of interruption of service during construction, and the importance of the facility.

5.2 Dimensions

Road geometrics such as grades, curve radii, and lane and shoulder widths are factors that control the design of repair schemes for slope failures. Design parameters often are influenced by the scale of the facility and of the existing or potential slope movement. A simple retaining wall and control of surface drainage may be adequate to repair and stabilize a thin, slab-type landslide occurring above a highway, whereas a deep-seated, rotational slide mass with a slide surface that passes under a highway requires a much more elaborate scheme.

5.3 Design Standards

A variety of guidelines are in common use for the design of transportation and other facilities (e.g., AASHTO 1990). Common highway standards include those for minimum lane width, maximum grade, and minimum curve radius. Additional standards exist for designing drainage culverts and pavement sections and for more unusual aspects such as vibrations due to construction blasting. Design manuals exist for general aspects of soil mechanics, including the design of slopes in soils (NAVFAC 1986; USBR 1980), and for geologic investigations (USBR undated; McCauley and Gamble 1981).

5.4 Costs

Costs of facilities include planning, design, construction, and maintenance. If costs are paid from a general pool (usually tax revenue), users may have free access to the facilities. Alternatively, users may be required to pay a fee for the privilege or convenience of using the facilities. The source of

revenue should not be a major factor in design. However, if funds are limited at the time of design, a plan may be formulated in which some simple components are treated with the intention of passing on to maintenance more of the burden of correcting a slope failure. Challenges of this type require accurate understanding of the causes of slope movement and careful attention to design so that each design element provides a foundation for the next element until repair has been completed.

Maintenance can be a significant cost issue. It must be understood at the design phase what constitutes tolerable maintenance. This definition may be difficult because the design team and the maintenance team often are totally independent of each other. Rock-fall catchment ditches, for example, must be cleaned periodically or their effectiveness will be lost. However, access to the ditches must be included as part of the design, or the maintenance team will be unable to fulfill their part of the rock-fall remediation without costly additional effort.

5.5 Landscape Aesthetics

The sensitivity to scars created by access road cuts or excavations for a highway or facility is greater in some areas than in others. The degree to which scars, the size of cuts and fills, and other aesthetic issues need to be considered is a fundamental component of the design concept that must be dealt with at an early phase. If blasting is needed on a steep slope, an early decision must be made regarding how much blasted rock can be allowed to fall down the slope without being retrieved. Similarly, if excavated slopes must be treated to make them look like natural slopes, such as the cut slopes along Interstate 70 in Glenwood Canyon, Colorado, the designer must know early enough to evaluate the geologic feasibility and plan for the aesthetic work.

5.6 Balancing Cut and Fill

In many stability evaluations, the resulting recommendations include the placement in fill embankments of material excavated from cut slopes. The volume of a mass of in-place material, particularly rock, is significantly less than that of the same mass of excavated and recompacted material. Often this increase in volume is underestimated. A broader area of detailed topographic information may be needed to evaluate fill volumes than that for a specific slope stability assessment.

5.7 Disposal of Waste Rock or Soil

Providing suitable locations for disposal of waste rock or soil constitutes a significant project design requirement. The disposal may be an environmental issue as well as an economic one, particularly if the waste rock or soil contains some material considered hazardous, such as lead or sulfide minerals that could generate acid drainage or asbestos minerals that are subject to air-quality and health regulations (Dusek and Yetman 1993).

5.8 Roadside Hazards

One of the project requirements may be that the highway remain open to traffic during investigation and construction for repair or stabilization of slope failures. Traffic patterns and tolerable delays must be incorporated into the design scheme. In addition, the risk to motorists following completion of the repair or stabilization of the slope must be specified so that the appropriate level of construction-slope reliability and the need for signs can be planned.

REFERENCES

ABBREVIATIONS

AASHTO	American Association of State Highway and Transportation Officials
ICBO	International Conference of Building Officials
NAVFAC	Naval Facilities Engineering Command
SBCCI	Southern Building Code Congress International, Inc.
TRB	Transportation Research Board
USBR	Bureau of Reclamation, U.S. Department of the Interior

AASHTO. 1990. *A Policy on Geometric Design of Highways and Streets.* Washington, D.C., 1044 pp.

Algermissen, S.T., D.M. Perkins, P.L. Thenhaus, S.L. Hanson, and B.L. Bender. 1990. *Probabilistic Earthquake Acceleration and Velocity Maps for the United States and Puerto Rico.* U.S. Geological Survey Map MF-2120 (2 sheets, 1:10,000,000).

Anderson, L.R., J.R. Keaton, T.F. Saarinen, and W.G. Wells II. 1984. *The Utah Landslides, Debris Flows and Floods of May and June, 1983.* National Academy Press, National Research Council, Washington, D.C., 96 pp.

Bowles, D.S., and H.Y. Ko. 1984. *Probabilistic Characterization of Soil Properties.* American Society of Civil Engineers, New York, 182 pp.

Bromhead, E.N. 1992. *The Stability of Slopes,* 2nd ed. Blackie Academic and Professional Publishers, Glasgow (imprint of Chapman & Hall, London and New York), 411 pp.

Dietrich, W., C. Wilson, and S. Reneau. 1986. Hollows, Colluvium, and Landslides in Soil Mantled Landscapes. In *Hillslope Processes* (A.D. Abrahams, ed.), Allen & Unwin, Boston, pp. 362–388.

Dusek, C.J., and J.M. Yetman. 1993. Control and Prevention of Asbestos Exposure from Construction in Naturally Occurring Asbestos. In *Transportation Research Record 1424,* TRB, National Research Council, Washington, D.C., pp. 34–41.

ICBO. 1994. *Uniform Building Code.* International Conference of Building Officials, Whittier, Calif.

Keaton, J.R. 1992. Engineering Geology for Design of the Kern River Pipeline in Southwestern Utah. In *Engineering and Environmental Geology of Southwestern Utah* (K.M. Harty, ed.), Publication 21, Utah Geological Association, Salt Lake City, pp. 53–91.

Keefer, D.K., R.C. Wilson, R.K. Mark, E.E. Brabb, W.M. Brown III, S.D. Ellen, E.L. Harp, G.F. Wieczorek, C.S. Alger, and R.S. Zatkin. 1987. Real-Time Landslide Warning During Heavy Rainfall. *Science,* Vol. 238, pp. 921–925.

Kramer, S.L., and H.B. Seed. 1988. Initiation of Soil Liquefaction Under Static Loading Conditions. *Journal of the Geotechnical Engineering Division,* ASCE, Vol. 114, No. GT4, pp. 412–430.

Krinitzsky, E.L., J.P. Gould, and P.H. Edinger. 1993. *Fundamentals of Earthquake Resistant Construction.* John Wiley & Sons, Inc., New York, 299 pp.

McCauley, M.L., and J. Gamble. 1981. *Geologic Hazards and Their Relation to Planning, Location, and Design of Transportation Facilities.* California Department of Transportation, Sacramento (various pagings).

NAVFAC. 1986. *Soil Mechanics.* Naval Facilities Engineering Command Design Manual 7.01 (re-validated by change September 1986), U.S. Navy, Alexandria, Va. (various pagings).

Norris, C.H., and J.B. Wilbur. 1960. *Elementary Structural Analysis.* McGraw-Hill Book Company, New York, 651 pp.

SBCCI. 1991. *Standard Building Code*. Southern Building Code Congress International, Inc., Birmingham, Ala.

Schuster, R.L., A.F. Chleborad, and W.H. Hays. 1987. Irrigation-Induced Landslides in Fluvial-Lacustrine Sediments, South-Central Washington State. In *Proc., Fifth International Conference and Field Workshop on Landslides,* Christchurch, New Zealand, August 1–12, pp. 147–156.

Sharp, K., L.R. Anderson, D.S. Bowles, and R.V. Canfield. 1981. A Model for Assessing Slope Reliability. In *Transportation Research Record 809,* TRB, National Research Council, Washington, D.C., pp. 70–78.

Sowers, G.F., and B.R. Carter. 1979. Paracti Rockslide, Bolivia. In *Rockslides and Avalanches* (B. Voight, ed.), Developments in Geotechnical Engineering, Vol. 14B, Elsevier Scientific Publishing Company, Amsterdam, Vol. 2, pp. 401–417.

USBR. 1980. *Earth Manual*. Bureau of Reclamation, U.S. Department of the Interior, 810 pp.

USBR. Undated. *Engineering Geology Field Manual*. Bureau of Reclamation, Denver, Colo., 598 pp.

Chapter 17

ROBERT D. HOLTZ AND
ROBERT L. SCHUSTER

STABILIZATION OF SOIL SLOPES

1. INTRODUCTION

The basic principles for design and construction of stable slopes in soils are quite well known. The engineering properties of soils as they relate to slope stability are generally understood. Analysis capabilities for slope stability have improved markedly in recent years because of the digital computer. In this report Parts 2, Investigation, and 3, Strength and Stability Analysis, provide important background information for this chapter. Specifically, Chapter 12 (Soil Strength Properties and Their Measurement) gives the procedures for determination of the appropriate soil parameters utilized in the stability analyses that are discussed in detail in Chapter 13. In this chapter the basic principles established in Parts 2 and 3 are applied to the design of stable slopes for new construction of both excavated and embankment slopes. The procedures are also appropriate for the analysis of preconstructed slopes, as well as for design of remedial works and correction of existing landslides.

This chapter is an update of Chapter 8 in Special Report 176 (Gedney and Weber 1978), which in turn built upon earlier reports (Baker and Marshall 1958; Root 1958). Because much of the basic technical information given by Baker and Marshall, Root, and Gedney and Weber is still valid, emphasis in this chapter will be on recent case histories and innovations in slope stabilization techniques since 1978.

2. DESIGN CONSIDERATIONS

Several factors are basic and must be considered in the design of stable slopes. First, because of the nature of soils and the geologic environments in which they are found, virtually every slope design problem is unique (Peck and Ireland 1953; Hutchinson 1977). Second, the procedures used to estimate the stability of an excavated slope are the same as those used to estimate the stability of an embankment slope. These first two factors are true for the analysis of newly constructed slopes as well as for existing slopes and for the design of remedial measures. Third, designing a stable slope includes field investigations, laboratory tests, stability analyses, and proper construction control. Because most of the details involved in this work cannot be standardized, good engineering judgment, experience, and intuition must be coupled with the best possible data gathering and analytical techniques to achieve a safe and economical solution to slope stabilization.

3. FACTOR OF SAFETY

In conventional practice the stability of a slope is expressed in terms of its factor of safety, although in recent years there has been increasing interest in developing a probabilistic assessment of slope reliability (see Chapter 6). In the conventional approach, factors of safety less than 1 obviously indicate failure, or at least the potential for failure, whereas stability is represented by safety factors greater than 1. The choice of the appropriate safety factor for a given slope depends on a number of considerations, such as the quality of the data used in the analysis, which in turn depends on the quality of the subsurface investigations; laboratory and field testing; interpretation of field and laboratory data; quality of construction control; and, in some cases, degree of completeness of information about the design problem. The engineer must also consider the probable consequences of failure. In most transportation situations, slope designs generally require safety factors in the range of 1.25 to 1.50. Higher factors may be required if slope movements have the potential for causing loss of human life or great economic loss or if there is considerable uncertainty regarding the pertinent design parameters, construction quality control, potential for seismic activity, and so forth. Likewise, lower safety factors may be used if the engineer is confident of the accuracy of the input data and if good construction control may be relied upon.

4. DESIGN PROCEDURES AND APPROACHES

Details of slope stability analysis procedures are given in Chapter 13. Analytical techniques allow a comparison of various design alternatives, including the effects of those alternatives on the stability of the slope and on the economy of the possible solutions. In addition, all potential failure modes and surfaces should be considered. As discussed in Chapter 13, preliminary analyses may utilize stability charts with simplified assumptions; such simple stability determinations may be adequate in many cases to decide whether a standard slope angle can be used. More involved analysis and stability calculations may be necessary for more complex problems. In all cases, consideration must be appropriately taken of the environmental conditions to which the slope is likely to be subjected during its entire design life, including changes in soil strength and groundwater conditions, possible seismic activity, or other environmental factors. As a minimum, the analysis should include conditions expected immediately after construction and at some later time after construction.

Approaches to the design of stable slopes can be categorized as follows:

- Avoid the problem,
- Reduce the forces tending to cause movement, and
- Increase the forces resisting movement.

A summary of these three approaches is given in Table 17-1.

5. AVOIDANCE OF THE PROBLEM

A geological reconnaissance is an important part of preliminary project development for many transportation design studies. This reconnaissance should note any evidence of potential stability problems due to poor surface drainage, seepage on existing natural slopes, hillslope creep, and ancient landslides. As noted in Table 17-1, avoiding the landslide problem is an excellent approach if it is considered during the planning phase. However, a large cost may be involved if a landslide problem is detected after the location has been selected and the design completed.

5.1 Ancient Landslides

Ancient landslides can be one of the most difficult landforms to identify and often are the most costly to deal with in terms of construction. Natural geomorphic and weathering processes, vegetation, or human activities may all but obscure these landforms, and careful field investigation is necessary to detect them.

As with the case of talus slopes, which are discussed below and in Chapter 20, old landslides are often barely stable, and they may not have significant resistance to new loadings or other changed conditions that tend to reduce their stability. Such slopes may continue to move, for example, during periods of heavy rainfall, and yet be relatively stable during other parts of the year. Changing natural drainage patterns on the surfaces of old landslides may significantly influence their stability and cause unwanted movements. Thus, the decision to construct transportation facilities over ancient landslides must be carefully investigated and appropriate consideration given to remedial measures and long-term stability.

5.2 Removal of Materials

If relocation or realignment of a proposed facility is not practical, complete or partial removal of the

Table 17-1
Summary of Approaches to Potential Slope Stability Problems (modified from Gedney and Weber 1978)

Category	Procedure	Best Application	Limitations	Remarks
Avoid problem	Relocate facility	As an alternative anywhere	Has none if studied during planning phase; has large cost if location is selected and design is complete; also has large cost if reconstruction is required	Detailed studies of proposed relocation should ensure improved conditions
	Completely or partially remove unstable materials	Where small volumes of excavation are involved and where poor soils are encountered at shallow depths	May be costly to control excavation; may not be best alternative for large landslides; may not be feasible because of right-of-way requirements	Analytical studies must be performed; depth of excavation must be sufficient to ensure firm support
	Install bridge	At sidehill locations with shallow soil movements	May be costly and not provide adequate support capacity for lateral forces to restrain landslide mass	Analysis must be performed for anticipated loadings as well as structural capability
Reduce driving forces	Change line or grade	During preliminary design phase of project	Will affect sections of roadway adjacent to landslide area	—
	Drain surface	In any design scheme; must also be part of any remedial design	Will only correct surface infiltration or seepage due to surface infiltration	Slope vegetation should be considered in all cases
	Drain subsurface	On any slope where lowering of groundwater table will increase slope stability	Cannot be used effectively when sliding mass is impervious	Stability analysis should include consideration of seepage forces
	Reduce weight	At any existing or potential slide	Requires lightweight materials that may be costly or unavailable; excavation waste may create problems; requires right-of-way	Stability analysis must be performed to ensure proper placement of lightweight materials
Increase resisting forces				
Apply external force	Use buttress and counterweight fills; toe berms	At an existing landslide; in combination with other methods	May not be effective on deep-seated landslides; must be founded on a firm foundation; requires right-of-way	Consider reinforced steep slopes for limited right-of-way
	Use structural systems	To prevent movement before excavation; where right-of-way is limited	Will not stand large deformations; must penetrate well below sliding surface	Stability and soil-structure analyses are required
	Install anchors	Where right-of-way is limited	Requires ability of foundation soils to resist shear forces by anchor tension	Study must be made of in situ soil shear strength; economics of method depends on anchor capacity, depth, and frequency
Increase internal strength	Drain subsurface	At any landslide where water table is above shear surface	Requires experienced personnel to install and ensure effective operation	—
	Use reinforced backfill	On embankments and steep fill slopes; landslide reconstruction	Requires long-term durability of reinforcement	Must consider stresses imposed on reinforcement during construction
	Install in situ reinforcement	As temporary structures in stiff soils	Requires long-term durability of nails, anchors, and micropiles	Design methods not well established; requires thorough soils investigation and properties testing

(continued on following page)

Table 17-1 *(continued)*

CATEGORY	PROCEDURE	BEST APPLICATION	LIMITATIONS	REMARKS
Increase internal strength *(continued)*	Use biotechnical stabilization	On soil slopes of modest heights	Climate; may require irrigation in dry seasons; longevity of selected plants	Design is by trial and error plus local experience
	Treat chemically	Where sliding surface is well defined and soil reacts positively to treatment	May be reversible; long-term effectiveness has not been evaluated; environmental stability unknown	Laboratory study of soil-chemical treatment must precede field installations; must consider environmental effects
	Use electroosmosis	To relieve excess pore pressures and increase shear strength at a desirable construction rate	Requires constant direct current power supply and maintenance	Used when nothing else works; emergency stabilization of landslides
	Treat thermally	To reduce sensitivity of clay soils to action of water	Requires expensive and carefully designed system to artificially dry or freeze subsoils	Methods are experimental and costly

unstable materials should be considered. Figure 17-1 shows an example of one such study. Stability analyses indicated that removal of Volume B was more effective than removal of Volume A. Economics, as well as the potential increase in slope stability, will decide the final course of action.

Removal of potentially unstable materials can range from simple stripping of near-surface materials a few meters thick, as shown in Figure 17-2, to more complicated and costly operations such as those encountered in a sidehill cut along the Willamette River in West Linn, Oregon, where a section of I-205 required extensive excavation to depths as great as 70 m (Gedney and Weber 1978).

5.3 Bridging

In some instances, removal of especially steep and long, narrow unstable slopes is simply too costly or too dangerous. One alternative design is to span the unstable area with a structure (bridge) supported on driven piles or drilled shafts placed well below the unstable foundation materials (Baker and Marshall 1958). Site investigations and stability analyses must ascertain that the bridge is indeed founded at sufficient depth below the unstable materials. If the bridge foundation must penetrate through the moving soil, as shown in Figure 17-3, the foundation piling must be designed to withstand the predicted lateral forces, but predicting these forces is not a simple task. Bridging may also

include limited excavation of near-surface unstable materials as well as the use of other stabilization techniques such as subsurface drainage.

Bridging may also be a suitable alternative in very steep mountainous country where construction may cause unsightly scarring because of unstable excavated slopes or excavations that "daylight" high on steep mountain slopes. Several spectacular examples of this bridging approach can be found on the autobahns in the Austrian Alps and on the autostrade in Italy.

6. REDUCTION OF DRIVING FORCES

Since the forces tending to cause movements downslope are essentially gravitational, a simple approach to increasing stability is to reduce the mass of soil involved in the slope. Techniques for this include flattened slopes, benched slopes, reduced excavation depths, surface and subsurface drainage, and lightweight fill (Table 17-1). All of these possibilities reduce driving forces and all have been successfully used at one time or another.

As explained in Chapters 12 and 13, the stability of embankment slopes cannot necessarily be approached in the same manner as that of natural or excavated slopes. For example, the stability of embankment slopes tends to increase with time because of consolidation and the resulting strength increase of the fill and foundation. A notable exception to this would be embankments com-

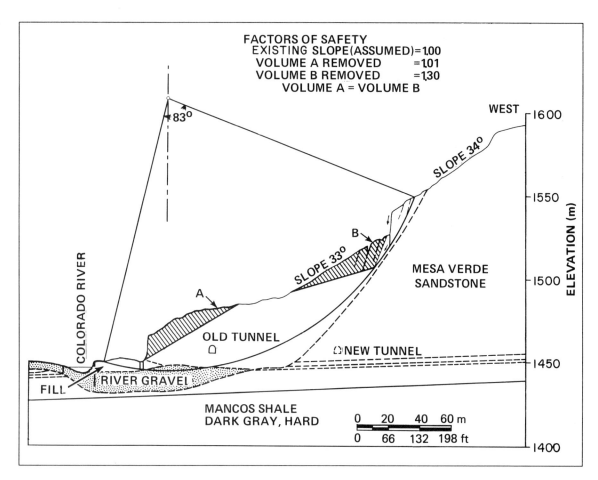

FIGURE 17-1
Stabilization of Cameo slide above railroad in Colorado River Valley, Colorado, by partial removal of materials (Peck and Ireland 1953; Baker and Marshall 1958).

posed of degradable shales and other soft rocks, which may deteriorate with time and result in settlement or even failure (see Chapter 21). In natural or excavated slopes, however, the long-term stability may be significantly less than that available at the end of construction. Design conditions that are appropriate for these cases are discussed in Chapters 12 and 13.

Talus slopes often have marginal stability (Ritchie 1963) and deserve special consideration (see Chapter 20). Runoff from normal rainfall or snowmelt may cause sufficient increase in seepage pressures to initiate movement. Recognition of talus slopes is important in preliminary location designs because of the potential for dangerous movements. Such slopes may also be disturbed by

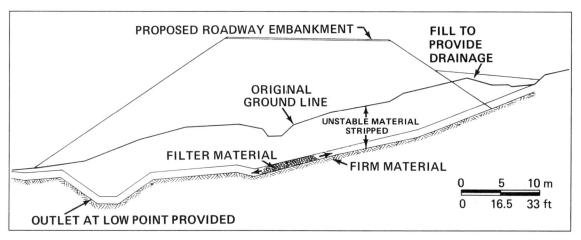

FIGURE 17-2
Stripping of unstable surface material to reduce landslide potential for sidehill embankment (Root 1958).

FIGURE 17-3
Landslide avoidance by bridging near Santa Cruz, California (Root 1958).

FIGURE 17-4
Grade change made during construction to prevent possible failure of excavated slope (Gedney and Weber 1978).

construction activities, and if such activities cannot be avoided, special construction procedures and stabilization methods are necessary. R. Barrett (personal communication, 1992, Colorado Department of Transportation) noted that mechanical stability of cuts in talus slopes is not a major problem; instead, the faces of these cuts tend to ravel. To prevent surface raveling of cuts along I-70 in Glenwood Canyon, Colorado, backfill was placed between the cut surface and the structure for which the cut was made. In addition, the Colorado Department of Transportation is experimenting with improving the resistance of talus to raveling by spraying on fiber-reinforced shotcrete or urethane liquid. Tests on Colorado Highway 82 near Aspen have shown that liquid urethane will pene-

trate talus to a depth of 10 to 30 cm and will improve both stability and resistance to raveling.

6.1 Change of Line or Grade or Both

Early in the design stage, cut-and-fill slopes should be evaluated for potential instability. If conditions warrant, adjustments to the line and grade can be made to minimize or possibly completely eliminate potential stability problems. This approach can also be applied to landslides discovered during and after construction. Feasibility will be controlled by the economics of various alternative solutions. A general example of a grade revision to prevent movement of an excavated slope is shown in Figure 17-4.

A specific example occurred in design and construction of I-70 across Vail Pass, Colorado, where active bedrock landslides were encountered on opposing slopes adjacent to the right-of-way. Stabilization measures included filling the valley and then transferring the thrust of one landslide against the other and placing the stream and highway on the fill (Robinson 1979). Edil (1992) described a case in which regrading was successfully used to stabilize an unstable bluff along Lake Superior in northern Wisconsin.

Line or grade changes often result in reduction of driving forces. For example, a roadway may be relocated away from the toe of the potential or existing landslide to avoid removal of toe support. Changes to cause reduction in driving forces during construction operations are both difficult and expensive. To flatten embankment slopes often

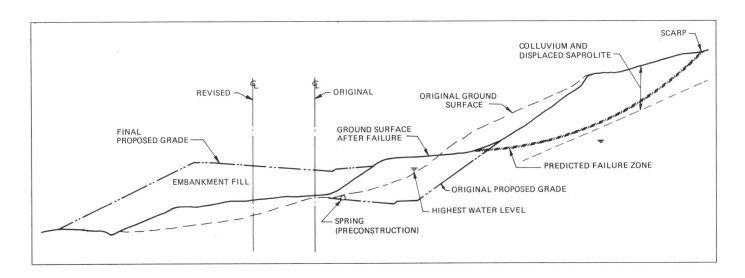

requires additional right-of-way and can involve alignment shifts that affect the design of facilities on either side of the problem area. Thus, changes in line and grade after construction may be so costly as to prevent their use. These comments emphasize the fact that the cost-effectiveness of geotechnical investigations is greatest when they are carried out during the preliminary design phases of the project.

Another technique for reducing the driving forces, especially for known unstable areas, is the partial removal or excavation of a sufficient quantity of slope material at the top of the landslide to ensure stability of the potential sliding mass (Figure 17-1). If an infinite slope condition exists or certain types of flow or debris slides are involved, this method may be ineffective. The quantity of material required to be removed is predicted by trial and error using ordinary stability analysis (Chapters 12 and 13). As usual, economic considerations and potential use of the excavated materials may dictate whether unloading procedures are feasible for a particular project. In some instances—for example, when the project needs additional borrow materials—removal of the entire sliding mass may be feasible provided the material volumes are reasonable. The design of landslide removal should always consider the stability of the slope behind or above the area to be removed. Of course, this technique is most appropriate during the design stage. These procedures usually involve slope flattening; in some cases, benching has been used to reduce the driving forces on a potential or existing landslide. Specific geometric constraints of the site will determine if benching is appropriate. As discussed in the following section, benches also serve to help control surface runoff and provide work areas for placement of horizontal drains.

6.2 Drainage

Because of its high stabilization efficiency in relation to design and construction costs, drainage of surface water and groundwater is the most widely used and generally the most successful slope stabilization method (Committee on Ground Failure Hazards 1985). Of all possible schemes to be considered for the correction of existing or potential landslides, proper drainage is probably the single most important. Drainage will both reduce the weight of the mass tending to cause the landslide

and increase the strength of the materials in the slope. Increasing the strength of soil materials using drainage is discussed in Section 7.2.1.

6.2.1 Surface Drainage

Surface drainage measures require minimal engineering design and offer positive protection to slopes. Thus they are among the first approaches that should be considered in preventing potential stability problems.

Adequate surface drainage is necessary in new excavations as well as in maintenance of constructed slopes where movement has already occurred. The design of cut slopes should always take into consideration the natural drainage patterns of the area and the effect that the constructed slope will have on these drainage patterns. Two conditions that should be evaluated are

1. Surface water flowing across the face of the slope, and
2. Surface water seeping into or infiltrating into the head of the cut.

Both of these conditions cause erosion of the face and increase the tendency for localized failures on the slope face. Diversion ditches and interceptor drains are widely used as erosion control measures, especially in situations where large volumes of runoff are anticipated (Figure 17-5).

Good surface drainage is strongly recommended as part of the treatment of any landslide or potential landslide (Cedergren 1989). Every effort should be made to ensure that surface waters are carried away from a slope. Such considerations become

FIGURE 17-5
Surface drainage of slope using diversion ditch and interceptor drain (Gedney and Weber 1978).

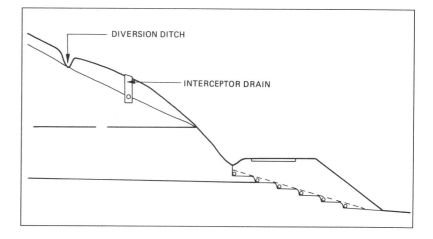

DIVERSION DITCH

INTERCEPTOR DRAIN

especially important when a failure has already occurred. Unless they are sealed, cracks and fissures behind the scarp face of a landslide can carry surface waters into the failure zone and reactivate the landslide. Consequently, reshaping the surface of the landslide mass can be very beneficial because unnoticed cracks and fissures may be sealed and water-collecting surface depressions eliminated.

There are a number of possibilities to treat the surface of the slope itself in order to promote rapid runoff and improve slope stability. Some of these measures are

- Seeding, sodding, or mulching, and
- Using shotcrete, riprap, thin masonry, concrete paving, asphalt paving, and rock fills.

All of these techniques have been used successfully to protect slopes made of degradable shales or claystones and to prevent the infiltration of surface runoff. Techniques for controlling surface runoff are especially effective when used in conjunction with various subsurface drainage techniques.

The Geotechnical Control Office of Hong Kong (Geotechnical Control Office 1984) has presented useful guidelines for the maintenance of surface drainage systems. These guidelines particularly recommend the use of surface channels as opposed to pipes placed on the surface.

6.2.2 Subsurface Drainage

Because seepage forces act to increase the driving force on a landslide, control of subsurface water is of major importance. If the preliminary site investigation reveals the presence of groundwater, if design studies predict potential downslope movements, and if positive subsurface drainage can reduce failure potential, it is worth preparing a suitable design for cost comparison with other alternatives.

Subsurface drainage as a method of lowering the groundwater table within an unstable slope has traditionally consisted of one or more of the following procedures:

1. Drainage blankets and trenches;
2. Drainage wells;
3. Drainage galleries, adits, or tunnels;
4. Subhorizontal (commonly called "horizontal") drains drilled either from the slope surface or from drainage wells or galleries; and
5. Subvertical drains drilled upward from drainage galleries.

Most often these systems drain by means of gravity flow; however, pumps are occasionally used to remove water from low-level collector galleries or wells. Figure 17-6 is a schematic drawing showing some of these techniques.

The effectiveness and frequency of use of the various types of drainage treatments vary according to the hydrogeologic and climatic conditions. It is generally agreed, however, that groundwater constitutes the single most important cause of the majority of landslides. Thus, in many areas of the world the most generally used and successful methods for prevention and correction of landslides consist entirely or partially of groundwater control (Cedergren 1989).

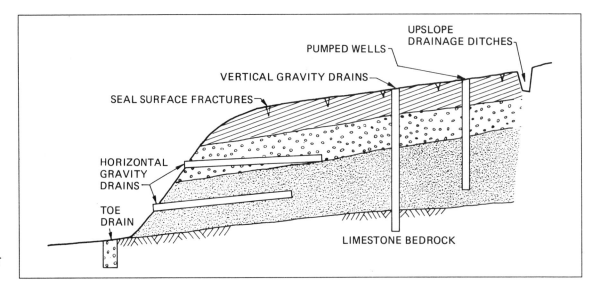

FIGURE 17-6
Schematic diagram of subhorizontal and vertical drains used to lower groundwater in natural slopes (Gedney and Weber 1978).

Subsurface drainage is equally important in cuts and embankments, and most types of subsurface drainage treatments are applicable to the prevention and correction of landslides in both situations. When embankments are constructed on potentially unstable slopes, landslides may occur because the increase in shear stresses imposed by the embankment exceeds the shearing strength of the foundation soils. Instability may also result if the embankment interferes with the natural movement of groundwater. Therefore, in an investigation of possible instability, two factors must be considered:

1. Weak zones in the foundation that may be overstressed by the proposed embankment load, and
2. Increases in pore-water pressure sufficient to cause a significant reduction in the shear strength of the soil.

Because there often is no surface indication of unstable slope conditions, careful subsurface exploration is required if slope instability is to be predicted before construction.

Drainage is sometimes installed to intercept subsurface water outside the limits of excavations, but it frequently is impossible to determine whether such interceptor trenches will effectively cut off all groundwater that might contribute to slope failure. Other methods that may be appropriate in sidehill embankment areas often are excessively expensive in excavation areas. Each case must be examined on an individual basis. The cost of drainage systems is generally lower when these measures are incorporated into the preliminary design process. When they are included as remedial measures during or following construction or landslide movement, the drainage systems may be expensive.

In this section recent advances in subsurface drainage systems are discussed. In addition, a few less common means of drainage such as electroosmosis, vacuum drains, siphons, and the use of geotextiles and geocomposites will be briefly mentioned.

6.2.2.1 Drainage Blankets and Trenches
When an embankment is to be constructed over a surface layer of relatively shallow, weak soil underlain by stable rock or soil, usually the most economical treatment is to strip the unsuitable material, as shown in Figure 17-2. After stripping,

if seepage is evident or if there is a possibility that it may develop during wet periods, a blanket drain of pervious material (clean free-draining sands or gravels) should be placed before the embankment is constructed. If springs or concentrated flows are encountered, drainpipes may also be required.

When conditions are such that drainage blankets are uneconomical, for example, when the depth of unsuitable material to be stripped is large, trench drains may be appropriate. Trenches filled with free-draining material have been used effectively for shallow subsurface drainage for several decades. The trenches are excavated by backhoe-type excavators to depths of 5 to 6 m and by clamshells to greater depths; some type of temporary support system may be required if the excavation is more than 5 to 6 m deep. Cancelli et al. (1987) provided a brief review of the theory for the design of trench drains. Sometimes single drainage trenches perpendicular to the center line of the facility are sufficient; in other situations, larger, more extensive networks of interconnected drains may be necessary. In addition to facilitating drainage, trench drains provide increased resistance to possible sliding because the compacted backfill of the trench section acts as a key into firmer material beneath the trench.

In some cases drainage wells (discussed below) are used under or at the bottom of drainage trenches (Figure 17-7) (Lew and Graham 1988).

6.2.2.2 Drainage Wells
Deep wells are increasingly being used to drain unstable slopes, particularly where the required drainage depths are too deep for economical construction of drainage trenches. Sometimes the wells are drilled fairly close together, essentially to form a drainage gallery, as in the case of I-80 in

FIGURE 17-7
Trench and augered sand drain slope stabilization system for gas pipeline crossing of the Assiniboine River, Canada (modified from Lew and Graham 1988).

REPRINTED FROM CH. BONNARD (ED.), LANDSLIDES/GLISSEMENTS DE TERRAIN, *PROCEEDINGS OF THE FIFTH INTERNATIONAL SYMPOSIUM, LAUSANNE, 10–15 JULY 1988, VOLUME 2, 1988–1989, 1604 PP., 3 VOLUMES, A.A. BALKEMA, OLD POST ROAD, BROOKFIELD, VERMONT 05036

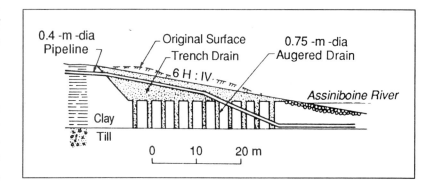

California described by Smith et al. (1970) and Gedney and Weber (1978). A similar system was recently utilized in Kentucky, as described by Greer and Mathis (1992), to correct a landslide. Sometimes large-diameter vertical wells up to 2 m in diameter are used, as in the cases described by Collotta et al. (1988) (Figure 17-8) and Bruce (1992b); these wells can be as much as 50 m deep. Each well is connected to its neighbor either by intersecting "belled" bases, which may overlap, or by horizontal drill holes and drains. Occasionally some shafts are kept open for inspection, monitoring, and maintenance purposes. Schuster (1995) provided a number of additional examples of landslide stabilization using drainage wells.

Vertical drainage wells have also been installed under embankments to accelerate the consolidation of weak and unstable foundation soils. This use is similar to classical sand drains or modern prefabricated vertical (wick) drains (Holtz et al. 1991).

6.2.2.3 Drainage Tunnels, Adits, or Galleries

When the depth to subsurface water is so great that drainage trenches or wells are prohibitively expensive, drainage tunnels are sometimes used. Although originally used as a corrective treatment, drainage tunnels are sometimes constructed as a preventive measure. They are expensive, and thus are less commonly used in highway construction than in large slope-stabilization measures, which are sometimes required for hydroelectric projects. One example is the case described by Millet et al. (1992) in which drainage tunnels, among other techniques, were used to stabilize a large landslide threatening Tablachaca Dam in Peru (Figure 17-9).

Drainage galleries are constructed using conventional tunneling and mining techniques, and in addition to acting as drainage facilities, they serve as exploratory adits for geologists to study landslides at depth. They are also useful for installing monitoring equipment and instrumentation. As with the case of deep drainage wells, gravity may be insufficient for positive drainage, and pumping may be required. Gillon et al. (1992) described a case in which drainage galleries were used for landslide stabilization adjacent to a reservoir in New Zealand (Figure 17-10).

6.2.2.4 Subhorizontal Drains

In recent years small-diameter subhorizontal drains have become very common for landslide stabilization. As was the case with drainage tunnels, they were first installed as corrective measures, and although they are still used for this purpose, subhorizontal drains are now primarily installed as preventive measures against slope instability. Although these drains are frequently called horizontal drains, they really are subhorizontal in that they are typically installed on slopes approximately 2 to 5 degrees above the horizontal. The drill hole is then fitted with a perforated pipe; in the past, steel pipe was commonly used, but in the last few years polyvinylchloride (PVC) plastic pipe has been preferred.

Subhorizontal drains are probably the most commonly used subsurface drainage technique for landslide stabilization. In large excavations that are potentially unstable, the drains are installed as the cut is excavated, often from one or more benches in the slope. Drains may be installed from the ground surface or from drainage galleries,

FIGURE 17-8
Vertical drainage shafts connected to horizontal PVC outlet drains, Florence-Bologna Motorway, Italy (modified from Collotta et al. 1988).

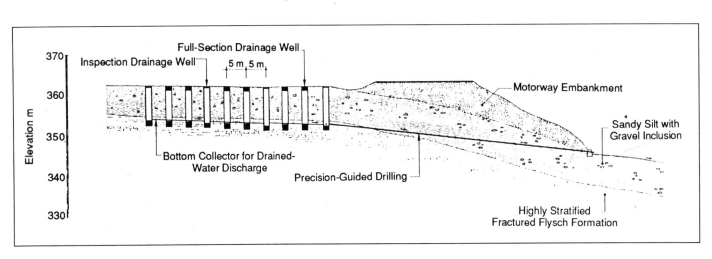

large-diameter wells, and drilled shafts. The typical drain hole is 120 to 150 mm in diameter and is lined with a slotted PVC plastic casing 60 to 100 mm in diameter. Drilling is usually by conventional rotary techniques, although precision drilling may be necessary for very long drains (Sembenelli 1988). Royster (1980) described traditional drilling and installation techniques. Because a filter is ordinarily needed, the annular space between the casing and the soil or rock of the drill hole is a problem. Filling this annulus with sand is difficult if not impossible. Sometimes pipes with very fine slots are used; alternatively, coarse slotting can be protected on the outside by geotextile "stockings."

Several interesting case histories of successful applications of subhorizontal drainage to stabilize landslides have been published, for example, in *Transportation Research Record 783* (Transportation Research Board 1980), which includes examples from California, Colorado, Mississippi, and Tennessee. Reagan and Jutkowfsky (1985) described horizontal-drain practice in New York State. Craig and Gray (1985) reported on horizontal drainage in Hong Kong, an area noted for many unstable natural slopes. Isenhower (1987) described the stabilization of landslides in San Antonio, Texas, using horizontal drains, and Sharma and Buu (1992) used horizontal drains and lightweight fill to stabilize a landslide in Idaho. Roth et al. (1992) described the use of horizontal drains ("hydraugers") to stabilize the famous Pacific Palisades landslide in California.

Smith (1980a, 1980b) presented a comprehensive study of the long-term effectiveness of horizontal-drain installations in California. His primary conclusions were that metal pipe casings will only last approximately 30 to 40 years; slotted PVC pipes will provide longer service life, are less susceptible to corrosion, and appear to allow considerably less sediment to enter the drains. On the basis of California experience, periodic maintenance is required; most drains need to be cleaned once every 5 to 8 years unless exceptionally fine-grained sediments or heavy root growth are present.

6.2.3 Other Less Common Techniques

Schuster (1995) summarized other less common techniques such as electroosmosis, vacuum dewatering, and blasting of rock slopes for improving drainage.

(a)

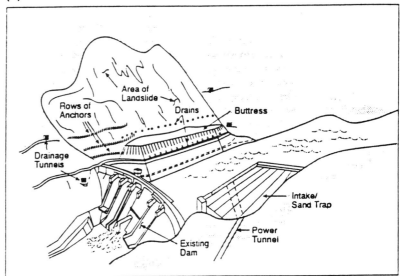

(b)

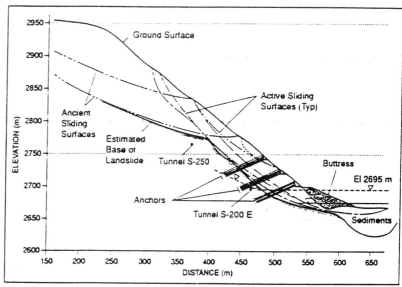

6.2.3.1 Electroosmotic Dewatering

In the mid-1930s the concept of using electroosmosis to promote drainage and consolidation of unstable soils was introduced (Casagrande 1948). Despite some success for slope stabilization (see, for example, Section 7.2.4.2), this process has not received widespread usage, especially for drainage, probably because of installation and operation costs and some remaining technical uncertainties about the process. However, Lo et al. (1991a, 1991b) used specially designed copper electrodes to prevent gas accumula-

FIGURE 17-9
Stabilization of Tablachaca Dam landslide, Peru: (a) overview and (b) cross section (Millet et al. 1992).

FIGURE 17-10
Brewery Creek
landslide-
stabilization
scheme, Clyde
Power Project,
New Zealand
(modified from
Gillon et al. 1992).
REPRINTED FROM
D.H. BELL (ED.),
LANDSLIDES/GLISSEMENTS
DE TERRAIN, *PROCEEDINGS
OF THE SIXTH
INTERNATIONAL
SYMPOSIUM,*
CHRISTCHURCH 10–14
FEBRUARY, 1992, VOLUME 1,
1992–1994, 1800 PP., 3
VOLUMES, A.A. BALKEMA,
OLD POST ROAD,
BROOKFIELD, VERMONT
05036

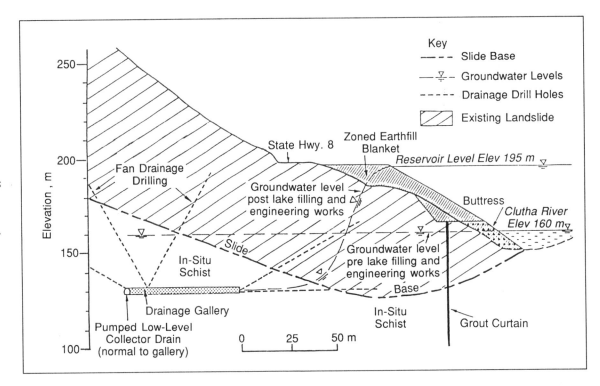

tion around the anode and to allow free water to flow from the cathode without pumping.

6.2.3.2 Vacuum Dewatering

Arutjunyan (1988) reported on the use of vacuum in drill holes to dewater fine-grained soil slopes in the former Soviet Union. The technique has been applied to landslides as a temporary expedient until long-term stabilization could be carried out. The vacuum treatment increases soil suction and accelerates the process of soil consolidation. It was successful to depths of 30 to 35 m when the vacuum was applied for a period of 2 to 4 weeks.

6.2.3.3 Drainage by Siphoning

Siphon drains for slope stabilization were installed at 40 sites in France during the 5 years before 1992 (Gress 1992). These drains have the advantage of being able to raise water to the surface without pumping. Siphoning of water from unstable strata is accomplished by sealed PVC pipe systems. One example of successful use of siphoning to lower the groundwater table under a highway embankment on an unstable slope occurred at Venarey-Les-Laumes near Dijon, France, where five vertical siphon drains spaced at intervals of 10 m lowered the piezometric level from an original depth of 2 m to a depth of 8 m beneath the highway (Gress 1992).

6.3 Geotextiles and Geocomposites

Geosynthetic products can be used for drainage and slope stabilization in many of the situations described above. The use of a geotextile filter is often cheaper and just as effective in situations where graded granular filters are required. Just as with graded granular filters, the geotextile filter should be designed for soil retention, system permeability, and long-term filtration characteristics. Christopher and Holtz (1985, 1989) and Koerner (1990), among others, have presented methods for geotextile filter design.

Geocomposites are products consisting of a geotextile filter to protect the drain and keep it free-flowing throughout its service life and a plastic net or core that provides in-plane drainage. Geocomposites can be installed in trenches on slopes, especially in areas where access is difficult, behind retaining structures, and in other places where interception of seepage is desired. Geocomposites are manufactured in sheets or strips, and thus are more easily and cheaply installed than conventional granular filters and drains. When properly designed, they work as well as, or often better than, conventional aggregate filters and drains.

Figure 17-11 shows drainage applications in which geotextile filters or geocomposite drainage products may be used.

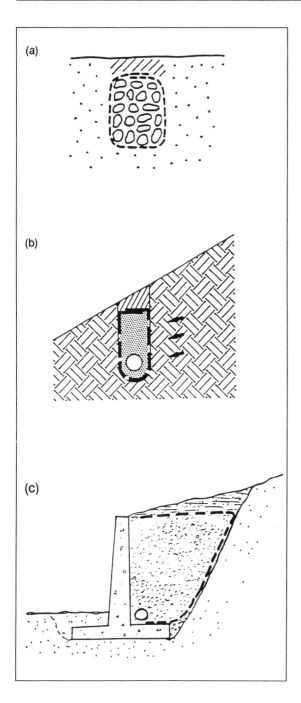

(a)

(b)

(c)

FIGURE 17-11
Drainage
applications using
geotextiles or
geocomposites:
(a) trench drain,
(b) interceptor drain,
and (c) drainage
behind structure
(modified from
Christopher and
Holtz 1989).

panded polystyrene, shredded and chipped tires, and oyster shells and clamshells. The advantages and disadvantages of the use of many of these materials were described by Holtz (1989).

Several examples of the use of lightweight fill materials to stabilize landslides have been given in the literature. Nelson and Allen (1974) reported on a successful landslide correction using chipped bark and sawdust to produce a sidehill embankment weighing about 5 kN/m^3. They used a 0.3-m gravel base under the pavement section; in addition, an asphalt seal was placed on the exposed slope to retard deterioration and pollution. Since then there have been many other similar projects in the Pacific Northwest. Yeh and Gilmore (1992) reported the use of expanded polystyrene (EPS) to correct a large landslide in southern Colorado. The landslide was successfully stabilized by using a counterweight berm at the toe and replacing the landslide material in the embankment with EPS with a unit weight of about 0.2 kN/m^3. Sharma and Buu (1992) used pumice weighing about 12 kN/m^3 in the correction of a landslide on I-15 in Idaho. Shredded waste tires weighing about 6.4 kN/m^3 were used as lightweight fill in the correction of a landslide under a highway embankment on Highway 42 in Oregon (Read et al. 1991). Humphrey and Manion (1992) conducted tests on tire chips to determine their basic properties.

7. INCREASE IN RESISTING FORCES

The third general method for stabilizing earth slopes is to increase the resisting forces on a potential or existing landslide. Although techniques vary widely, they generally function by either

1. Applying a resisting force at the toe of a landslide, or
2. Increasing the internal strength of the soils in the failure zone so that the slope remains stable without external assistance (Table 17-1).

6.4 Reduction of Weight

Especially in embankment construction, the use of lightweight backfill materials can be very effective in reducing the gravitational driving forces tending to cause instability. Lightweight materials that have been successfully used in highway embankments include chipped bark, sawdust, dried peat, fly ash, slag, cinders, cellular concrete, expanded clay or shale, lightweight geologic materials, ex-

Both approaches should be considered during preliminary design investigations to ensure the best technical and economical solution. The techniques described in this section to correct landslides are appropriate for those that occur both during and after construction.

A number of procedures have been developed that increase the resisting forces at the toe of a potentially sliding soil mass. These include various methods of adding mass to the toe areas, various structural retention systems to deflect or redirect the driving forces, and a variety of earth material reinforcement systems. Although reinforced soil structures may be used as buttresses or structural retention systems, an important aspect of their stability is internal, and they will be discussed for convenience under the internal-strength category below.

Methods used to increase the internal strength of the potentially sliding soil mass include subsurface drainage, a wide variety of reinforcing systems, vegetative and biotechnical stabilization, and miscellaneous methods such as chemical treatment, electroosmosis, and thermal stabilization. Sometimes a combination of external and internal treatments is employed. In the following description of these techniques, emphasis is on developments since 1978.

7.1 Application of External Force

Application, or in some cases redirection, of external forces so as to increase the resistance of the slope to potential movements will increase the stability of a slope. Such resisting forces are most often applied to the toe of the potentially moving mass by a variety of methods, including

1. Buttresses, counterweight fills, and toe berms;
2. Structural retention systems such as cantilever and gravity retaining walls, externally braced walls or walls supported by anchors or tiebacks, soil nailing, root piles, conventional piles, and drilled shafts; and
3. A variety of reinforced-soil systems.

7.1.1 Buttresses, Counterweight Fills, and Toe Berms

The principle behind the use of buttresses, counterweight fills, or toe berms is to provide sufficient dead weight or restraint near the toe of the unstable mass to prevent slope movement. In other words, the buttress must be heavy enough to provide the additional component of resistance near the toe of the slope required for stability. Figure 17-12 shows a rock buttress used to stabilize an unstable slope.

The basic design of a buttress is similar to the design for external stability of conventional gravity-retaining structures. The buttress must be stable against

- Overturning,
- Sliding at or below its base, and
- Bearing failure of the foundation.

Conventional soil mechanics analyses for these three possible failure modes should be carried out with the usual factors of safety applied. Although not strictly a factor in stability, a settlement analysis should also be performed if the foundation is compressible to ensure that the final grade of the buttress is consistent with the geometric design requirements of the facility. Depending on the geometry and the internal strength of the buttress, possible internal failure modes should also be checked to ensure that the buttress does not fail by shear within itself.

Buttresses are commonly constructed of blasted quarry rock, boulders and cobbles, and coarse gravel fill. These materials are also commonly used to repair small slope failures, or "pop-outs," in highway cuts. If these occur during construction, the highway contractor usually makes repairs as part of the normal construction contract. If the landslide occurs after construction and is not too large, similar reconstruction is normally carried out by highway maintenance crews.

FIGURE 17-12
Rock buttress used to stabilize unstable slope (Gedney and Weber 1978).

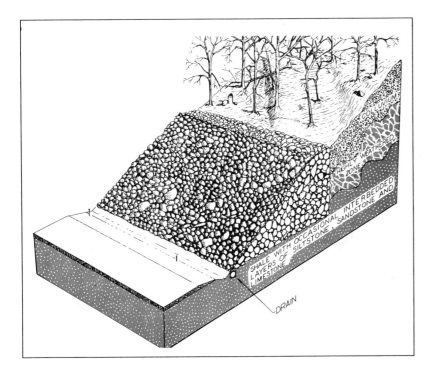

For larger landslides, a more involved geotechnical analysis and design are required. Gedney and Weber (1978) described the reconstruction of a slide in a shale embankment in Indiana on I-74 by the use of a large earth-and-rock counterweight buttress.

Millet et al. (1992) described an interesting use of a large buttress (nearly 500 000 m³) constructed of granular materials to stabilize a large, potentially unstable landslide threatening Tablachaca Dam, a major hydroelectric power dam in Peru (Figure 17-9). The project required treatment of the sediments in the reservoir bottom under the buttress with the use of stone columns. On a more modest scale, Edil (1992) reported on the use of a terraced berm constructed of concrete demolition debris to stabilize an unstable slope along the western shores of Lake Michigan. Kropp and Thomas (1992) used buttress fills to successfully stabilize an existing landslide area in California.

7.1.2 Structural Retention Systems

In situations where a buttress fill is not feasible because of geometry or cost or because it encroaches on adjacent property, conventional retaining structures, piles, and reinforced-soil slopes and walls may provide workable alternative solutions. Because the design principles for retaining walls and deep foundations used for this purpose are fairly well understood, only a few pertinent references will be given.

One major drawback to the use of relatively stiff structural elements for slope stabilization is that they are not able to tolerate much movement. Consequently, elemental systems such as crib walls, gabion walls, and the various soil reinforcement systems are becoming increasingly more common.

Figure 17-13 summarizes the current methods of earth retention; the methods are divided into two groups depending on whether they provide external or internal stabilization. Examples of both types of systems are shown in Figure 17-14. Externally stabilized systems rely on some type of external structural walls or fill against which the stabilizing forces are mobilized. Before the late 1960s, the predominant types of permanent retaining structures were concrete and masonry gravity walls and reinforced-concrete cantilevered walls. More recently, Reinforced Earth and a number of similar reinforcing systems using, for exam-

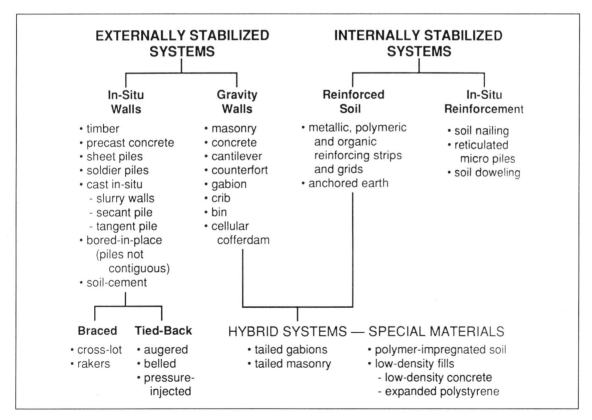

FIGURE 17-13
Classification scheme for earth retention systems (modified from O'Rourke and Jones 1990).

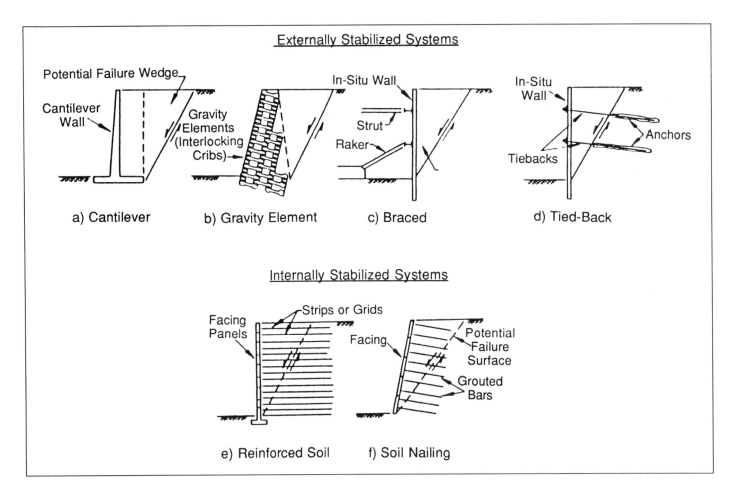

a) Cantilever   b) Gravity Element   c) Braced   d) Tied-Back

e) Reinforced Soil   f) Soil Nailing

FIGURE 17-14
Examples of
externally and
internally stabilized
earth retention
systems (modified
from O'Rourke and
Jones 1990).
REPRINTED WITH
PERMISSION OF
AMERICAN SOCIETY OF
CIVIL ENGINEERS

ple, geosynthetics have become common for earth-retaining structures. Earth-reinforcing systems are discussed in Section 7.2.2.

There were two sessions on the use of walls to control landslides at the National Convention of the American Society of Civil Engineers in Las Vegas in 1982. The papers by Morgenstern (1982) and Schuster and Fleming (1982) are particularly recommended in this connection.

The use of closely spaced vertical driven piles or large-diameter drilled shafts is quite common. They may be placed as a preexcavation retaining system, or they may be used following slope movement to stabilize landslides. If the piles are placed adjacent to each other, the wall is known as a *tangent pile wall*; if there is some spacing that may later be filled in with timber lagging, reinforced-concrete panels, or shotcrete, the wall is known as a *secant pile wall*. These walls may be designed as cantilevers or, more commonly, they are supported by tied-back anchor systems, which are discussed in the following section. In the case of an excava-

tion, the pile wall must be designed to resist the full earth pressure imposed by the soil in the slope as the excavation progresses.

One of the more successful applications of drilled shaft walls to stabilize a slope was described by Palladino and Peck (1972) and Wilson (1970). During the construction of I-5 in Seattle, heavily overconsolidated lacustrine clays were supported by very large diameter (up to 4 m) drilled concrete shafts reinforced with heavy steel H-pile sections [see also the discussion by Gedney and Weber (1978)]. Nethero (1982) described the use of cantilever drilled-shaft walls in the Ohio River Valley for the correction and prevention of slope stability problems. Several interesting case histories of landslide stabilization using drilled-shaft walls were described by Rollins and Rollins (1992). Although the shafts were not very deep, they were designed to function as cantilevers and they performed excellently in all cases.

The use of the finite-element method to analyze drilled piers used for slope stabilization was reported by Oakland and Chameau (1984), and Reese et al. (1992) presented a method for the analysis of drilled shafts used for increasing the stability of slopes. A design procedure for walls embedded in failed slopes was proposed by Isenhower et al. (1989), who reported that such walls supported by 0.5- and 0.6-m-diameter shafts are much more economical than conventional earth-retaining structures used for landslide correction.

According to Gedney and Weber (1978), attempts to stabilize landslides using driven steel or timber piles have seldom been successful. Depending on the geologic conditions at the site, piles may be difficult to drive to the desired depth, and in the case of steel H-piles, for example, it is difficult to drive them vertically or in tangent. In Scandinavia there has been some success in the use of both timber and steel piles to stabilize moving slopes and embankments (Bjerrum 1972); at these sites the subsurface conditions were relatively soft glacial and postglacial marine clays.

7.1.3 Anchor Systems

One of the more effective measures for stabilizing landslides is the use of anchor systems to increase the resisting forces by applying external restraint to a moving soil mass [Figure 17-13 and Figure 17-14(d)]. These developments followed from the use of tiebacks to support walls for temporary excavations for the construction of buildings and other structures.

According to Gould (1990), tieback anchors were developed as early as the 1930s from European experience with permanent rock anchors in dam stabilization. In the early 1960s their use in Milwaukee and New York for building excavations clearly indicated the advantages of an uncluttered working space achieved by eliminating external supports [compare Figure 17-14(c) and 17-14(d)], and this successful application led to rapid developments in the use of high-capacity prestressed anchors to support deep excavations. Tiebacks were also used to stabilize landslides in Europe in the mid-1960s, and, soon after, the technology was being employed in North America for similar purposes. An excellent description of the use of tiebacks for landslide stabilization was provided by Weatherby and Nicholson (1982). They described the design, installation, and corrosion protection that are necessary for permanent ground anchors

in landslide stabilization. A spectacular use of ground anchors was described by Millet et al. (1992) in which high-capacity rock anchors were used together with a granular buttress to stabilize a large landslide at Tablachaca Dam in Peru (Figure 17-9). Interesting case histories describing the use of tiebacks together with drilled shafts and driven H-pile walls were presented by Hovland and Willoughby (1982) and Tysinger (1982).

Other related systems such as soil nailing [Figure 17-14(f)] and micropiles increase the internal strength of the sliding mass rather than applying external force. Thus, these systems are discussed in the following section.

7.2 Increase in Internal Strength

Techniques that are used to increase the internal strength of the potentially unstable soil mass for both backfills and in situ stabilization are discussed next. These include subsurface drainage, soil-reinforcing systems, vegetative and biotechnical stabilization techniques, and miscellaneous techniques such as chemical, electrical, and thermal stabilization.

7.2.1 Subsurface Drainage

As noted by Gedney and Weber (1978), one of the most effective treatments for landslides and other unstable slopes is to increase the shear strength of the soil by means of subsurface drainage. Drainage reduces the excess pore pressures in the soil mass, which increases the effective stress on the potential failure surface. Subsurface drainage techniques are discussed in detail in Section 6.2.2.

7.2.2 Soil Reinforcement

Soil reinforcement is the inclusion of tensile resistant elements in a soil mass to improve its overall shearing resistance. One of the most exciting developments in geotechnical engineering practice in the past 25 years, soil reinforcement is a technically attractive and very cost-effective technique for increasing the stability of natural and embankment soil slopes and for reducing the earth pressures against retaining walls and abutments.

As noted in Figure 17-13, internally stabilized soil systems can be conveniently divided into reinforced soil [Figure 17-14(e)] and in situ reinforcement. In situ reinforcement systems include soil nailing [Figure 17-14(*f*)], micropiles, pin piles, and root piles. Reinforced soil is applicable to situations in which the reinforcement and backfill are placed as the slope or wall is constructed. Common reinforcing elements include steel strips (Reinforced Earth), welded wire sheets, bar mats and meshes, geotextiles, geogrids, and fibers.

Excellent general discussions of internally stabilized soil reinforcement systems include those by Lee et al. (1973), Jones (1985), Mitchell and Villet (1987), Christopher et al. (1990), Mitchell and Christopher (1990), and O'Rourke and Jones (1990). This discussion will deal with recent advances in the use of internally stabilized soil reinforcement systems.

7.2.2.1 Backfill Systems

The concept of reinforcing the backfill behind retaining walls was developed by H. Vidal in France in the mid-1960s. Since then, thousands of walls reinforced with all types of inclusions have been successfully built throughout the world. Reinforced-soil structures have the advantage over more traditional retaining walls (Mitchell and Villet 1987) in that they

1. Are coherent and flexible and thus can tolerate relatively large settlements,
2. Can use a wide range of backfill materials,
3. Are easy to construct,
4. Are relatively resistant to seismic loadings,
5. Can form aesthetically attractive retaining walls and slopes because of a variety of available facing types, and
6. Are very often less costly than conventional retaining structures, especially for high steep slopes and high walls.

Steep slopes of reinforced soil or vertical retaining walls take up less of the right-of-way, and thus they are especially attractive for landslide repairs adjacent to transportation facilities.

Common types of backfill reinforcement inclusions include steel or polymeric strips, steel or polymeric grids, geotextile and geogrid sheets, and steel cables or bars attached to different anchor systems. The two main mechanisms of stress transfer between the reinforcement and the soil are

1. Friction between the surface of the reinforcement and the soil, and

2. Passive soil bearing resistance on reinforcement surfaces that are oriented normal to the direction of relative movement between the reinforcement and the soil.

Because of their inherent flexibility, reinforced-soil structures are considered to be relatively resistant to seismic loadings. However, in some seismically active regions, the use of reinforced slopes is still somewhat restricted because of the lack of definitive research in this area. For reinforced-backfill retaining structures, the connections between the reinforcing and the facing are critical during seismic events (Allen and Holtz 1991).

7.2.2.1.1 Strip Reinforcement   In systems using strip reinforcement, a coherent material is formed by placing the strips horizontally between successive layers of backfill (Figure 17-15). When the French architect and engineer Vidal introduced the concept in the 1960s, he named his development *Terre Armée* (Reinforced Earth). Reinforced Earth has since become almost a generic term, often being used to describe all forms of soil reinforcement. However, in some countries, including the United States and Canada, Reinforced Earth is a trademark (Jones 1985). The first Reinforced Earth wall was built near Nice, France, in 1965. As of 1991, 16,000 Reinforced Earth walls with a

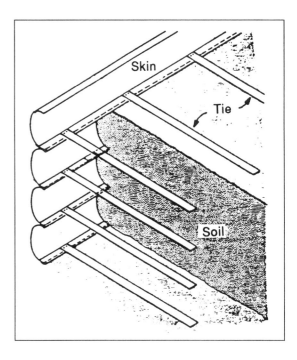

total face area of 9 600 000 m² had been constructed worldwide (D. McKittrick, personal communication, 1991, Reinforced Earth Company, McLean, Virginia). Schlosser (1990) observed that about one-third of the Reinforced Earth wall area in the world has been built in Europe and about one-third in North America.

After the introduction of Reinforced Earth, knowledge of soil reinforcement increased rapidly, primarily because of the research sponsored by government agencies, notably the Laboratoire Central des Ponts et Chaussées (LCPC) in France, the Federal Highway Administration in the United States, and the Department of Transport in the United Kingdom.

Early experiments with fiberglass-reinforced polymers (Schlosser 1990), stainless steel, and aluminum strips for Reinforced Earth walls were not successful, so all Reinforced Earth walls currently are constructed using galvanized steel strips. However, corrosion rates of metals in soils are very difficult to predict, and even galvanized steel is subject to corrosion; thus free-draining sand-and-gravel backfills are specified to reduce corrosion potential. Epoxy-coated steel strips have been developed and may offer higher resistance to corrosion. Elias (1990) discussed the corrosion of epoxy strips as well as traditional steel-strip reinforcing.

In 1973 nonmetallic reinforcing strips were introduced in construction of a highway retaining wall in Yorkshire, England (Holtz 1978; Jones 1978). The reinforcing strips were made of continuous glass fibers embedded lengthwise in a protective coating of epoxy resin. The Paraweb strip, in which the fibers are made of high-tenacity polyester or polyamarid, is another example of a geosynthetic strip reinforcement.

7.2.2.1.2 Sheet Reinforcement In sheet reinforcement, geotextiles are placed horizontally between layers of embankment soils to form a composite reinforced-soil system. Figure 17-16 shows a number of possible uses of geotextiles in reinforced walls and slopes. The mechanism of stress transfer in sheet reinforcement is predominantly friction (Mitchell and Villet 1987; Christopher and Holtz 1989; Christopher et al. 1990). A large variety of geotextiles with a wide range of mechanical properties is available; they include nonwoven needle-punched or heat-bonded polyester and polypropylene and woven polypropylene

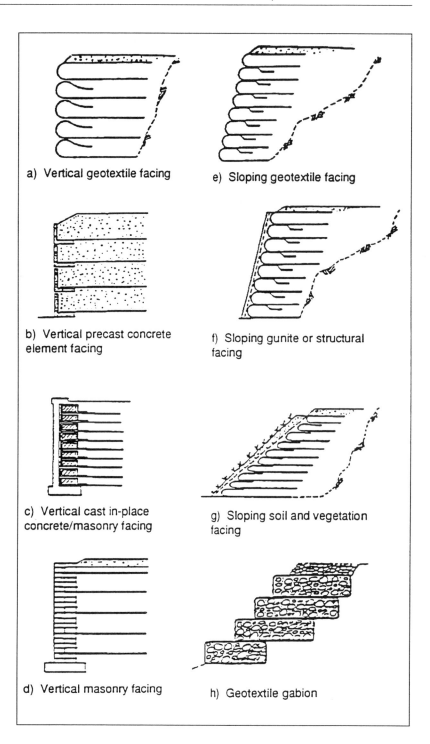

a) Vertical geotextile facing

b) Vertical precast concrete element facing

c) Vertical cast in-place concrete/masonry facing

d) Vertical masonry facing

e) Sloping geotextile facing

f) Sloping gunite or structural facing

g) Sloping soil and vegetation facing

h) Geotextile gabion

and polyester geotextiles (Christopher and Holtz 1989; Koerner 1990). Granular soils ranging from silty sands to gravels commonly are used as backfill, and the most common facings are formed by wrapping the geotextile around the exposed soil [Figure 17-16(a) and 17-16(e)]. Because geotextiles are subject to vandalism and deterioration from ultraviolet light, for long-term protection the

FIGURE 17-16 Possible reinforced walls and slopes using geosynthetic reinforcement (modified from Mitchell and Villet 1987).

exposed material must be covered with shotcrete or asphalt emulsion or with soil and vegetation. Typical applications of geotextile-reinforced walls include landslide stabilization on remote mountain roads, highway retaining walls on steep slopes, embankment walls for temporary or permanent road widening or diversion, and highway embankment walls on soft foundation soils (Mitchell and Villet 1987).

The use of geotextiles in reinforced-soil walls followed shortly after the introduction of Reinforced Earth. The first geotextile-reinforced wall was built in France in 1971. In the United States, the first such wall was one 3.3 m high built by the U.S. Forest Service to reconstruct a failed road fill above the Illinois River in Siskiyou National Forest, Oregon (Bell and Steward 1977). The geotextile was a 440-g/m² needle-punched nonwoven polypropylene that was permeable and resistant to rotting. However, it was subject to deterioration when exposed to ultraviolet light; therefore, the wall was finished with a shotcrete facing to protect it from sunlight. An example of a very successful short-term use of geosynthetic reinforcement was reported by Allen et al. (1992). A 12.6-m-high wall supporting a 6-m-high surcharge was used to preload a compressible foundation for a bridge abutment (Figure 17-17).

During the 1980s, the use of geotextile soil reinforcement for retaining walls increased significantly; more than 80 projects were completed in North America during this period (Yako and Christopher 1988). Not only are geosynthetics thought to be more resistant to corrosion than steel, but geosynthetic walls have been found to be 30 to 50 percent less expensive than most other reinforcement systems. Still, there is some doubt as to the durability and longevity of geosynthetic reinforcing materials because of chemical and biological attack. One of the most comprehensive studies of the effects of outdoor exposure on geotextiles was conducted in Hong Kong (Brand and Pang 1991). The study confirmed that geotextiles should be protected and that their use in critical reinforcement situations should be limited to short-term projects. Elias (1990) provided some information on the durability of geosynthetics, as did Allen (1991). A U.S. Federal Highway Administration study is currently under way on the durability of geosynthetics for soil reinforcement.

7.2.2.1.3 Grid, Bar, and Mesh Reinforcement

In grid reinforcement, polymeric or metallic elements are arranged in a rectangular grid shape. The two-dimensional grid-soil interaction involves both friction along the longitudinal grid elements and passive bearing resistance against the transverse elements. Because of this passive resistance, grids are more resistant to pullout than flat steel strips; however, full passive resistance develops only after relatively large displacements (5 to 10 cm) (Schlosser 1990).

Probably the greatest development in the use of grids for soil reinforcement has been polymeric geogrids. Geogrids are relatively stiff, netlike materials, with open spaces called apertures that usually measure 1 to 10 cm between the elements. They are made of polypropylene, polyethylene, or PVC-coated polyester.

Applications of grids to reinforced walls and slopes are shown in Figure 17-16. The first use of polymeric geogrids was by Japanese engineers in the 1960s to reinforce subsoils for railway embankments (Jones 1985). Because these original grids were made of nonoriented polymers, they were relatively fragile and lacked the necessary tensile strength to serve as reinforcement. However, they were quite effective as compaction aids (Iwasaki and Watanabe 1978). In the 1970s, advances in the formulation of polymers led to significant improvements in their strength and stiffness, and geogrids were developed with polymers oriented in orthogonal directions, which provided increased directional strength. These high-strength geogrids were first used in 1979 to construct a reinforced-soil wall

FIGURE 17-17
Geosynthetic-reinforced wall on I-90 in Seattle, Washington.
COURTESY OF T. M. ALLEN, WASHINGTON STATE DEPARTMENT OF TRANSPORTATION

at a railroad station in Yorkshire (O'Rourke and Jones 1990).

In 1981 geogrids were used to repair slope failures in cuts on the M1 and M4 motorways in England (Murray and Irwin 1981; Murray 1982; Jones 1985). Considerable cost savings resulted because the landslide debris was reused together with the geogrid reinforcement (Figure 17-18). No new backfill material was required. In 1983 the first geogrid wall in the United States was built to stabilize a landslide near Newport on the Oregon coast (Szymoniak et al. 1984). This 9-m-high geogrid wall with a face slope of 80 degrees was selected over other alternatives because

- It had the lowest estimated cost, and
- The open face of the grid wall allowed establishment of vegetation that provided a natural appearance compatible with the surroundings of the adjacent state park.

At about the same time, Forsyth and Bieber (1984) reported on the construction of a geogrid-reinforced slope in California to repair a slope failure. It was 9.5 m high and had a slope of 48 degrees. Figure 17-19(a) shows the slope under construction, and Figure 17-19(b) shows the completed structure. Since the construction of these early walls, more than 300 polymeric geogrid walls and slopes have been constructed in the United States (Mitchell and Christopher 1990). Figure 17-20 shows another, nearly completed landslide repair on a state highway in California.

Several bar-and-mesh reinforcement systems have been developed that rely on both frictional and passive resistance to pullout. In 1974 the first "bar mat" system of soil reinforcement was developed by the California Department of Transportation (Caltrans) to construct a 6-m-high wall along I-5 near Dunsmuir, California; these crude grids were formed by cross-linking steel reinforcing bars to form a coarse bar mat (Forsyth 1978). Laboratory tests by Chang et al. (1977) showed that this bar-and-mesh reinforcement could produce more than five times the pullout resistance of longitudinal bars. In an agreement with the Reinforced Earth Company, the Caltrans bar-and-mesh reinforcement technique was designated Mechanically Stabilized Embankment (MSE). One of the difficulties with MSE in the field has been corrosion of the bar-and-mesh reinforcement. Evolving from the

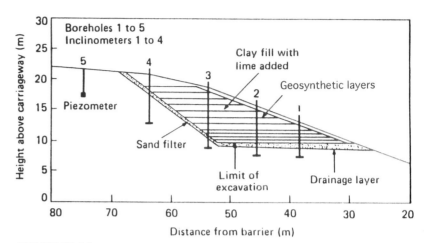

FIGURE 17-18
Location of boreholes and inclinometers after remedial measures (Murray 1982).

FIGURE 17-19
La Honda, California, landslide repair using geogrids: (a) under construction and (b) completed slope face.
COURTESY OF TENSAR EARTH TECHNOLOGIES, INC.

FIGURE 17-20
Nearly completed
landslide repair
using geogrids,
Highway 9 near
Santa Cruz,
California.
COURTESY OF TENSAR
EARTH TECHNOLOGIES, INC.

Caltrans project and other independent developments were the Hilfiker Welded Wire Wall (WWW), Hilfiker Reinforced Soil Embankment (RSE), VSL Retained Earth, and the Georgia Stabilized Embankment systems.

The Hilfiker WWW uses the same type of welded-wire reinforcing mesh that is commonly placed in concrete slabs; the facing is formed of the same mesh that serves as the horizontal reinforcement. The material is fabricated in 2.4-m-wide mats with grid spacings of 15 by 61 cm. To the casual observer, WWW may appear to be a type of gabion wall [Figure 17-16(h)]. Gabion walls, however, are gravity walls made by encasing coarse-grained fill in wire or geogrid baskets; they are based on the principle of confinement and gravity retainment rather than internal tensile reinforcement (Hausmann 1990). The first commercial WWW was built for the Southern California Edison Power Company in 1977 for road repair along a power line in southern California. During the 1980s, the use of WWW for retaining structures expanded rapidly. By 1990 about 1,600 such walls had been completed in the United States (Mitchell and Christopher 1990). The tallest WWW constructed so far is nearly 17 m high (Anderson et al. 1987).

The Hilfiker RSE, which resembles the Caltrans MSE, is a continuous welded-wire reinforcement system with precast concrete facing panels. It was introduced commercially in 1983 on New Mexico State Highway 475 northeast of Santa Fe, where four reinforced-soil structures were built with a total of 1600 m² of wall facing. By 1990 more than 50 additional RSE walls had been constructed (Mitchell and Christopher 1990).

The VSL Retained Earth system utilizes 0.5- to 0.75-m-wide strips of steel grid (bar-mat) reinforcement that is bolted to hexagonal precast concrete panels. The first VSL Retained Earth wall in the United States was constructed in 1983 in Hayward, California. By 1990 more than 600 VSL Retained Earth walls with some 465 000 m² of wall facing had been built in the United States (Mitchell and Christopher 1990). The system is licensed in the United States under a Reinforced Earth patent, but it uses its own patented system to connect the bar-mat reinforcement and the concrete facing panels.

The Georgia Stabilized Embankment was developed by the Georgia Department of Transportation and uses a steel-grid reinforcement and precast-concrete face elements. It is licensed in the United States under a proprietary agreement with the Reinforced Earth Company.

As with other reinforcement systems, geogrids and steel bar mats are susceptible to environmental deterioration and corrosion, and the prudent designer makes certain of the chemical compatibility of the backfill and reinforcement materials.

Full-scale experiments with the sidewalls of used tires hooked together with a bent No. 3 rebar clip to reinforce embankment fills have been carried out by Caltrans (Forsyth and Egan 1976; Forsyth 1978). Sometimes the entire tire casing is used. The tires are placed in layers similar to geotextiles and other geogrids. Results to date have been very promising, although there is some difficulty in properly compacting backfill in and around the tires. The French Ministry of Transport also has used similar systems of used tire casings for reinforcing slopes, embankments, and walls. Research began at the LCPC on *pneusol* (tiresoil) in 1976, and so far more than 80 structures have been built with *pneusol* in France and Algeria (Long 1990).

7.2.2.1.4 Earth-Anchor Reinforcement Embankment soil slopes can be reinforced during construction by slender steel-rod reinforcements bent at one end to form anchors. This type of retaining wall is still in an experimental stage. Stress transfer is mainly by passive resistance on the anchor, which implies that the system provides stability in the same manner as tied-back retaining structures and thus may not be a true reinforced-soil system (Mitchell and Villet 1987). However, the system is discussed here because it is analogous in placement technique to other methods of soil reinforcement in embankments and fills.

The concept of earth-anchor reinforcement was developed and patented by the Transport Research Laboratory (TRL) of the United Kingdom as Anchored Earth (Figure 17-21). The reinforcement consists of 16- to 20-mm-diameter mild-steel bars. The outer end of each bar is threaded to fit into concrete facing panels; the other end is formed into an anchor in the form of a Z or triangle. Anchored Earth is designed to rely only on passive resistance developed against the deformed ends (anchors) of the reinforcing bars. Because Anchored Earth is still in the research and developmental stages, none of its applications can be considered to be routine. However, it does appear to be a promising approach to soil reinforcement, especially on projects where clean granular backfill is not available.

A similar concept was used successfully by Fukuoka and Imamura (1982) in Japan in the late 1970s to construct a 5-m-high fabric-faced retaining wall with multiple anchors. Each of the 20-mm-diameter steel tie bars was attached to a 40- by 40-cm concrete plate embedded in the backfill soil.

7.2.2.1.5 Fiber Reinforcement The engineering use of fiber reinforcement of backfill soil, which is analogous to fiber reinforcement of concrete, is still in the developmental stage. Materials being investigated for possible use include synthetic fibers, continuous synthetic filaments, metallic fibers (metal threads) and powders, and natural fibers (reeds and other plants) (Mitchell and Villett 1987). A recent innovation is a three-dimensional reinforcement technique that was developed in France at the LCPC in 1980 (Schlosser 1990) and known as Texsol. Texsol is made by mixing the backfill soil, usually a clean sand, with a continuous polyester filament with a diameter of 0.1 mm (Leflaive 1982). Approximately 0.1 to 0.2 percent by weight of the composite material consists of the filament and a total length of reinforcement of about 100 km/m^3 of reinforced soil. In the field the sand is deposited using a shotcrete system (without cement) and the filaments are spun from bobbins and carried along with the high-velocity sand.

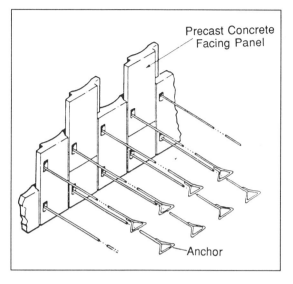

FIGURE 17-21
Schematic diagram of Anchored Earth retaining wall (modified from Murray and Irwin 1981).

The first Texsol wall, with a face angle of 60 degrees to the horizontal, was built in France in 1983 (Leflaive 1988). By the end of 1988, 85 Texsol projects had been built in France, using 100 000 m^3 of Texsol-reinforced soil (Schlosser 1990). Most projects involved landslide repairs and failed-slope reconstruction [Figure 17-22(a)]. These walls and slopes have a high bearing capacity and relatively rapid growth of vegetation on the slope face [Figure 17-22(b)] and are self-healing when subjected to erosion.

Another material that has been suggested for use as fiber reinforcement is bamboo, which is one of the fastest-growing and most replenishable biological materials in existence. Bamboo can also be used as continuous elements in other types of reinforced soil. Fang (1991) has published data on the strength and durability of bamboo as soil reinforcement.

7.2.2.2 In Situ Systems

In situ soil reinforcement methods allow for the reinforcement of existing soil masses. These methods include techniques described as soil nailing, the application of soil anchors, or the use of root piles, micropiles, or pin piles.

7.2.2.2.1 Soil Nailing Soil "nails" are steel bars, rods, cables, or tubes that are driven into natural soil or soft rock slopes or are grouted into predrilled boreholes. Soil nailing can be used to restrain either

• Potentially unstable slopes where little or no movement is occurring but where safety factors

(a)

(b)

FIGURE 17-22 *(above)* Texsol: (a) wall under construction in French Alps; (b) 10-m-high Texsol slope with vegetation 1 year after construction.
COURTESY OF J. L. WALKINSHAW, FEDERAL HIGHWAY ADMINISTRATION

FIGURE 17-23 Soil nails extruding from shotcrete surface of temporary excavation for tunnel portal, I-70, Glenwood Canyon, Colorado.

Soil nailing has been used for in situ stabilization of natural and excavated slopes for nearly 20 years. In North America the system was first used in Vancouver, British Columbia, in the early 1970s for temporary excavation support (Shen et al. 1978). In Europe the earliest reported soil-nailing projects were for retaining wall construction in Spain (1972), France (1973), and Germany (1976) in connection with highway or railway cut-slope construction or temporary support for building excavations (Elias and Juran 1991). Today the technique of soil nailing is widespread in Canada, Germany, France, Great Britain, Japan, and the United States.

The stability of soil-nailed reinforcement relies upon

- Development of friction or adhesion mobilized at the soil-nail interface and
- Passive resistance developed along the surface perpendicular to the direction of the soil-nail relative movement.

Soil nailing is most effective in dense granular and low-plasticity stiff silty clay soils because a top-down sequential construction procedure is commonly used. The soil must have sufficient strength are low enough to indicate a strong possibility for future movement, or

- Creeping slopes, in which movement is actually occurring.

Together with the in situ soil, the nails form coherent reinforced-soil structures capable of stopping the movement of unstable slopes [Figure 17-14(f)] or of supporting temporary excavations (Figure 17-23). Nailing differs from tieback support systems because the nails are passive elements that are not posttensioned; also, the nails are usually spaced more closely than tiebacks. Commonly, one nail is used for each 1 to 6 m² of soil surface area. Stability of the surface between the nails is provided by a thin layer (10 to 15 cm) of shotcrete reinforced with wire mesh (Figure 17-23) and by intermittent large steel washers or panels (which later may be covered with shotcrete).

to be able to stand in a vertical cut of about 2 m without failure or excessive deformation. Thus, soil nailing is generally not cost-effective or practical in loose granular soils and very soft clays (Mitchell and Christopher 1990). A high groundwater table may also present construction difficulties.

Suggestions for design of soil nailing include those by Elias and Juran (1991) and Byrne (1992). Soil-nailing systems are relatively flexible and thus should be resistant to seismic loading. However, knowledge of the dynamic behavior of soil-nailed structures is limited (Felio et al. 1990), and research is needed to develop procedures for earthquake-resistant design. There are no proprietary restrictions on the use of soil nailing. However, some specific installation systems, nails, and facings are patented.

The technique of soil nailing has been used mostly for stabilization of temporary excavations. There is some concern about the corrosion rate of the steel nails used in the process. However, new types of nails and coatings with high resistance to corrosion are being developed. To further increase confidence in the potential use of this method for permanent slope stabilization, additional research is being done on the field performance of soil-nailed structures. For example, in 1986 a 4-year, $4 million national research program named CLOUTERRE was initiated by the French Ministry of Transport to improve the status of knowledge and to develop design and construction guidelines for soil-nailed retention systems (Schlosser and Unterreiner 1991; Schlosser et al. 1992). Plumelle and Schlosser (1991) presented the results of three full-scale experiments from this project.

A number of interesting case histories concerning the use of soil nailing for support of temporary excavations are available in both the French and U.S. literature [for example, Lambe and Hanson (1990)]. Denby et al. (1992) described the design, construction, and performance of a 23-m-deep excavation in Seattle, Washington, that was successfully supported by a temporary soil-nailed wall. Collin et al. (1992) used a timber crib wall as the facing and helical anchors as the soil nails for a permanent earth retention system for a slope. There appear to be no published cases in which soil nailing has been used to stabilize active landslides or moving slopes.

A related system is anchored geosynthetics (spider netting) for slope stabilization (Figure 17-24).

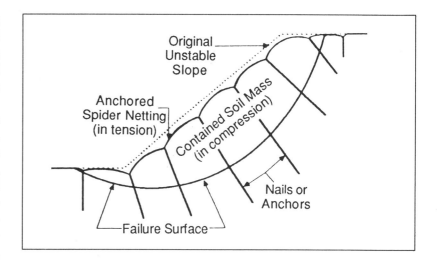

Instead of shotcrete facing, a geotextile or geonet is attached to the nails, and as the nails are installed, they pull the surface netting or geotextile into the soil, putting it into tension and at the same time constraining the near-surface soils in the slope (Koerner and Robins 1986). Design procedures for anchored geosynthetic systems were described by Hryciw and Haji-Ahmad (1992).

7.2.2.2.2 Soil Anchors

Stabilization of soil slopes with long prestressed anchors is increasing. Hutchinson (1984) described the stabilization of a landslide in glacial and glaciofluvial deposits in southern Wales (Figure 17-25). Before treatment in 1980, movements of as much as 15 mm/year were occurring at the head of the landslide and 2 to 5 mm/year was occurring at the toe. Because of severe spatial constraints, anchoring into the underlying bedrock proved to be the most effective stabilization measure.

FIGURE 17-24 *(above)* Schematic cross section of anchored geosynthetic "spider netting" used with soil nails or anchors to stabilize slope (modified from Koerner and Robins 1986).

FIGURE 17-25 Cross section of landslide in Quaternary deposits at Nantgarw, South Wales, United Kingdom (modified from Hutchinson 1977); slide has been stabilized by deep anchors in underlying bedrock.

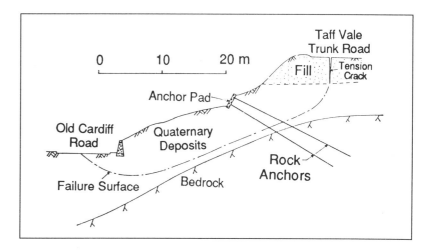

7.2.2.2.3 Root Piles, Micropiles, and Pin Piles

Another approach to reinforcement of in situ soils and soft rocks is the use of root piles, also known as micropiles or pin piles. Root piles are cast-in-place reinforced concrete piles with diameters ranging from 7.5 to 30 cm. In the smaller-diameter range, the insertions are provided with a central reinforcing rod or steel pipe, whereas those with larger diameters may be provided with a reinforcing-bar cage bound with spiral reinforcement. A root-pile system forms a monolithic block of reinforced soil that extends below the critical failure surface (Figure 17-26). In contrast to soil nailing, the reinforcement provided by root piles is strongly influenced by their three-dimensional, rootlike geometric arrangement.

Reticulated Root Piles were originally developed in the 1950s by Lizzi (1977) and were patented by the Italian firm Fondedile of Naples, which introduced and installed the system worldwide (mainly for underpinning). The original patents have now expired. Root piles have been used for slope stabilization only for the last 20 years, and most root-pile slope stabilization works have been constructed within the past 10 years. Aurilio (1987) described the use of root piles for stabilization of a landslide in California.

According to Bruce (1992a), during the past 20 years, U.S. practice using root piles has developed quite differently from the original European development. He provided details from some 20 pro-jects that illustrate recent developments in U.S. pin-pile technology and discussed applications more specific to slope stabilization (Bruce 1992b). An interesting case history of the use of pin piles to control slope movements was described by Pearlman et al. (1992).

7.2.3 Vegetative and Biotechnical Stabilization

Slope stabilization provided directly by vegetation and by biotechnical slope protection (the use of vegetation combined with structural slope-stabilization elements) is reviewed briefly. The basic concepts of vegetative stabilization are not new, but recent research and development now enable more effective use of this technique than in the past. Additional information about bio-stabilization was provided by Gray (1970), Gray and Leiser (1982), Greenway (1987), and Wu (1994a, 1994b).

According to Wu (1994b), vegetation contributes to stability of slopes through (*a*) root reinforcement and (*b*) rainfall interception and evapotranspiration, which reduce pore pressures. Case studies have shown that slope failures can be attributed to the loss of reinforcement provided by tree roots (Wu et al. 1979; Riestenberg and Sovonick-Dunford 1983; Riestenberg 1987). In spite of the fact that Greenway (1987), in his extensive summary of the effects of vegetation on slope stability, included reports that vegetation tends to reduce slope stability, most researchers believe that vegetation is by far a positive aspect in the protection of slopes. Wu (1994b) quantified this protection in terms of root reinforcement and reduction of soil moisture and pore pressures.

Research into the engineering role of vegetation for slope stabilization by the Geotechnical Control Office of Hong Kong may be the most comprehensive such program in the world (Barker 1991). Especially notable are the root reinforcement studies conducted on vegetated slopes by Greenway et al. (1984), Greenway (1987), and Yin et al. (1988). In addition, the *Geotechnical Manual for Slopes* (Geotechnical Control Office 1984) includes information on the mechanical and hydrological effects of vegetation.

In recent years trees have been planted on many slopes worldwide to increase slope stability. One example is an element of a program to cor-

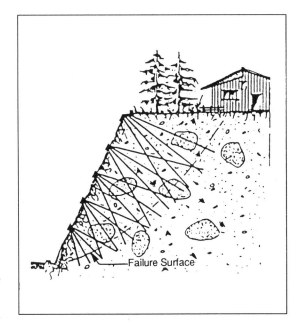

FIGURE 17-26
Schematic cross section illustrating use of root piles for stabilization of slope (modified from Lizzi 1977).

rect an embankment failure on I-77 in Caldwell, Ohio. The slope was planted with black locust seedlings at a spacing of 1.2 m. The long-term objective was to help reduce the soil moisture and to develop root stabilization. As of 1994 the project appeared to be successful.

Another recent well-documented case of the planting of tree seedlings occurred as part of the stabilization program for the Cucaracha landslide in the Gaillard Cut at the Panama Canal. The historic Cucaracha landslide was reactivated in 1986, almost blocking the canal (Berman 1991). As part of a comprehensive stabilization program, beginning in 1987 portions of the Cucaracha landslide and other landslide areas in the Canal Zone were planted with 60,000 fast-growing acacia and gmelina seedlings (Rivera 1991).

Stabilization of slopes by the combined use of vegetation and manufactured structural elements working together in an integrated manner is known as biotechnical slope stabilization. This relatively new concept is generally cost-effective as compared with the use of structures alone; it has increased the environmental compatibility of such treatments and allows the use of indigenous natural materials. Although vegetative treatments alone are usually much less expensive than earth-retaining structures or other constructed slope protection systems, their effectiveness in arresting slope movement or preventing soil loss under extreme conditions may be much lower than that of the structures (Gray and Leiser 1982). Figure 17-27 presents different biotechnical slope stabilization and protection measures and examples.

Grasses and woody plants are used most often in biotechnical stabilization. They have a true reinforcing function and should not be considered merely cosmetic adjuncts to the structure. They may be planted on a slope above a low retaining wall, or the interstices of the structure may be planted with vegetation whose roots bind together the soil within and behind the structure. The stability of all types of retaining structures with open gridwork or tiered facings benefits from such vegetation (Figure 17-27).

Gray and Sotir (1992) described the use of a crushed-rock-blanket toe berm combined with a brush-layer fill to stabilize a road cut along a scenic highway in Massachusetts. The crushed rock was placed at the toe of the cut; the brush-layer fill included stems and branches of plant

CATEGORY		EXAMPLES
LIVE CONSTRUCTION Conventional Plantings		* Grass seeding * Sodding * Transplants
MIXED CONSTRUCTION	Woody plants used as reinforcements and barriers to soil movement	* Live staking * Contour wattling * Brush layering * Soft gabions * Brush mattress
	Plant/ structure associations	* Breast walls with slope face plantings * Revetments with slope face plantings * Tiered structures with bench plantings
	Woody plants grown in the frontal openings or interstices of retaining structures	* Live cribwalls * Vegetated rock gabions * Vegetated geogrid walls * Vegetated breast walls
	Woody plants grown in the frontal openings or interstices of porous revetments	* Joint plantings * Staked gabion matresses * Vegetated concrete block revetments * Vegetated cellular grids * "Reinforced" grass
INERT CONSTRUCTION Conventional Structures		* Concrete gravity walls * Cylinder pile walls * Tie back walls

species, such as willow and dogwood, that rooted readily from cuttings (Figure 17-28). The branches acted as reinforcement and as horizontal drains, and rooting of the embedded stems provided secondary stabilization.

Thomas and Kropp (1992) described a very interesting scheme using vegetation to stabilize a large debris-flow scar in California. At the time the paper was presented in 1992, the vegetation was quite effective and the project was considered by the authors to be a success.

7.2.4 Miscellaneous Stabilization Methods

The following miscellaneous stabilization methods are discussed:

FIGURE 17-27 Classification of different biotechnical slope protection and erosion control measures (Gray and Sotir 1992).

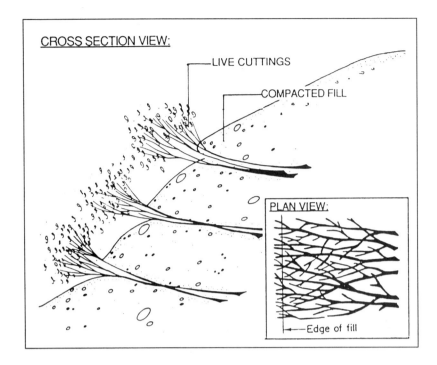

FIGURE 17-28 *(above)*
Schematic of
brush-layer fill
construction (Gray
and Sotir 1992).
REPRINTED WITH
PERMISSION OF
AMERICAN SOCIETY OF
CIVIL ENGINEERS

FIGURE 17-29
Schematic cross
section illustrating
process of lime-slurry
pressure injection
for stabilization of
embankment slope
(modified from
Boynton and
Blacklock 1985).

• Chemical treatment of the soils on the potential sliding surface, including grouting, ion exchange, and injection of lime, portland cement, and fly ash;

• Electrical treatment, primarily electroosmosis; and

• Thermal treatment, including freezing.

7.2.4.1 Chemical Treatment

The use of chemicals to treat unstable soil slopes is an attractive possibility. Because of their success in stabilizing compacted soils in highway and airfield subgrades, portland cement, lime, fly ash, and calcium chloride have been suggested as suitable chemicals for treatment of unstable slopes. However, a major problem in using any of these chem-

icals in slope stabilization has been obtaining adequate injection of the chemical into the soil in the zone of rupture.

Gedney and Weber (1978) presented four case histories on the use of chemicals to stabilize landslides and other unstable slopes. Since that time, however, in spite of significant advances in grouting technology (Baker 1982; Borden et al. 1992), only a few case histories could be located in which the use of grouting techniques for landslide stabilization was described.

In recent years grouting techniques have been used to inject lime slurry into the soil (Figure 17-29) (Boynton and Blacklock 1985). The slurry, which follows natural fracture zones, bedding surfaces, and other surfaces of weaknesses, is injected through 40-mm-diameter pipes fitted with perforated nozzles (Rogers 1991). The pipes are hydraulically pushed into the ground, and the slurry is injected to refusal at depth intervals of 300 to 450 mm. Typical injection pressures range from 350 to 1300 kPa. In this way depths of more than 40 m can be treated.

Blacklock and Wright (1986) discussed restoration of failed soil embankments on Interstate highways in Alabama, Arkansas, and Missouri using the lime- and lime-fly ash-slurry injection method of in situ soil stabilization. Baez et al. (1992) evaluated the use of lime/fly ash (L/FA) slurry injection in the slope rehabilitation of a levee on the lower Chariton River in Missouri. Their studies showed that double injection of the L/FA increased the strength of the levee soil by 15 to 30 percent. Furthermore, in the 4 years since the levee was treated with L/FA injections, no slope failures have occurred in the rehabilitated stretch of the levee, whereas there have been failures in adjacent unstabilized areas.

Lime stabilization was very effective in improving the strength of slope debris that was recompacted in the reconstruction of a landslide in England (Murray 1982). (This is an interesting case history from another viewpoint in that it is the first use of geogrids to increase stability of a failed slope.) Considerable economy was achieved by use of landslide debris as backfill rather than the traditional excavation and removal of the debris and its replacement with higher-quality granular materials at considerable additional cost. A geotextile-wrapped granular drainage blanket was provided behind the grid- and lime-stabilized slope. Probably all three

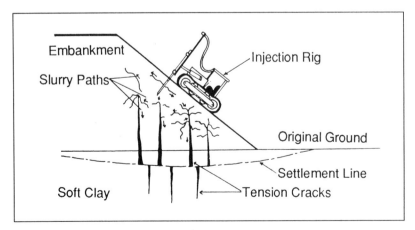

treatment techniques (drainage, lime stabilization, and geosynthetic reinforcement) contributed to the success of the project.

7.2.4.2 Electroosmosis

Another special technique that increases the shearing strength of soils in situ is electroosmosis. The classic papers on electroosmosis were published by Casagrande (1948, 1952, 1953). There are a number of case histories in the literature in which electroosmosis was used to stabilize embankment foundations; these were summarized by Holtz (1989). However, the classic application of electroosmosis has been to stabilize landslides and slopes. An early case history was reported by Casagrande et al. (1961) in which electroosmosis was used to stabilize a 30-m-high slope in organic silt that had developed serious instability during construction of a pier for the Little Pic River bridge on the Trans-Canada Highway in Ontario. Casagrande et al. (1981) also described the use of electroosmosis to stabilize a slope for an 80-m-deep excavation for the core trench of a dam in British Columbia.

As summarized by Veder (1981), Griffin described the use of electroosmosis to stabilize a 30-m-high slope of loose, sensitive silt above the Kootenay Channel power plant in British Columbia, Canada. As a result of electroosmosis, the slope, which originally was barely stable at 1V:5H, was able to be safely steepened to 1V:2H.

In previous discussion concerning electroosmotic dewatering (Section 6.2.3.1), it was noted that Lo et al. (1991a, 1991b) developed special electrodes that allow free water flow without pumping. In field tests conducted on the soft, sensitive Leda clay of eastern Canada, undrained shear strength increased uniformly by about 50 percent in a period of 32 days throughout the depth of the electrodes. Because no pumping was required, both installation and electricity costs were significantly lower than those in previous electroosmosis drainage installations.

7.2.4.3 Thermal Treatment

As reported by Gedney and Weber (1978), experiments were carried out in the former Soviet Union for a number of years using thermal treatment of clays and loessial soils. The technique was used by Romanian engineers for stabilization of clay slopes (Beles 1957). High temperatures dry out the soil and tend to fuse fine-grained particles, which leads to permanent increases in shear strength and consequent increases in slope stability. However, the high cost of thermal treatment precludes its use on all but the most experimental slope remediation problems.

Ground freezing has developed in recent years to be a very effective technique for temporary stabilization of large excavations and tunnels. One of the most complete treatises on ground freezing is by Jessberger (1979).

REFERENCES

Allen, T. A., B. R. Christopher, and R. D. Holtz. 1992. Performance of a 12.6 m High Geotextile Wall in Seattle, Washington. In *Geosynthetic Reinforced Soil Retaining Walls,* Proceedings of the International Symposium, Denver (J.T.H. Wu, ed.), A.A. Balkema, Rotterdam, Netherlands, pp. 81–100.

Allen, T. A., and R. D. Holtz. 1991. Design of Retaining Walls Reinforced with Geosynthetics. In *Proc., Geotechnical Engineering Congress 1991,* Boulder, Colorado (F. G. McLean, D. A. Campbell, and D. W. Harris, eds.), Geotechnical Special Publication 27, American Society of Civil Engineers, New York, pp. 970–987.

Allen, T.M. 1991. Determination of Long-Term Tensile Strength of Geosynthetics: A State-of-the-Art Review. In *Proc., Geosynthetics '91 Conference,* Atlanta, Industrial Fabrics Association International, St. Paul, Minn., Vol. 1, pp. 351–379.

Anderson, R. L., K. D. Sharp, and O. T. Harding. 1987. Performance of a 50-Foot High Welded Wire Wall. In *Soil Improvement—A Ten Year Update* (J. P. Welsh, ed.), Geotechnical Special Publication 12, American Society of Civil Engineers, New York, pp. 280–308.

Arutjunyan, R. N. 1988. Prevention of Landslide Slope Process by Vacuuming Treatment of Disconsolidated Soils. In *Proc., Fifth International Symposium on Landslides,* Lausanne, July 10–15, A.A. Balkema, Rotterdam, Netherlands, Vol. 2, pp. 835–837.

Aurilio, G. 1987. In-Situ Ground Reinforcement with Root Piles. In *Soil Improvement—A Ten Year Update* (J. P. Welsh, ed.), Geotechnical Special Publication 12, American Society of Civil Engineers, New York, pp. 270–279.

Baez, J. I., R. H. Bordeau, and J. H. Henry. 1992. Rehabilitation of Lower Chariton River Levee by Lime/Flyash Slurry Injection. In *Transportation Research Record 1310,* TRB, National Research Council, Washington, D.C., pp. 117–125.

Baker, W. H. (ed.). 1982. Grouting in Geotechnical Engineering. In *Proc., American Society of Civil Engineers Specialty Conference*, New Orleans, American Society of Civil Engineers, New York, 1018 pp.

Baker, R. F., and H. C. Marshall. 1958. Control and Correction. In *Special Report 29: Landslides and Engineering Practice* (E. B. Eckel, ed.), HRB, National Research Council, Washington, D.C., pp. 150–188.

Barker, R. F. 1991. Developments in Biotechnical Stabilization in Britain and the Commonwealth. In *Proc., Workshop on Biotechnical Stabilization*, University of Michigan, Ann Arbor, pp. A-83 to A-123.

Beles, A. A. 1957. Le traitement thermique du sol. In *Proc., Fourth International Conference on Soil Mechanics and Foundation Engineering*, A.A. Balkema, Rotterdam, Netherlands, Vol. 3, pp. 266–267.

Bell, J. R., and J. E. Steward. 1977. Construction and Observations of Fabric Retained Soil Walls. In *Proc., International Conference on the Use of Fabrics in Geotechnics*, Association Amicale des Ingénieurs Anciens Elèves de l'Ecole Nationale des Ponts et Chaussées, Paris, April, Vol. 1, pp. 123–128.

Berman, G. 1991. Landslides on the Panama Canal. *Landslide News* (Japan Landslide Society), No. 5, pp. 10–14.

Bjerrum, L. 1972. Embankments on Soft Ground. In *Proc., Specialty Conference on Performance of Earth and Earth Supported Structures*, West Lafayette, Indiana, American Society of Civil Engineers, New York, Vol. 2, pp. 1–54.

Blacklock, J. R., and P. J. Wright. 1986. Injection Stabilization of Failed Highway Embankments. Presented at 65th Annual Meeting of the Transportation Research Board.

Borden, R. H., R. D. Holtz, and I. Juran (eds.). 1992. *Grouting, Soil Improvement and Geosynthetics: Proceedings of a Conference*, New Orleans, Vols. 1 and 2, Geotechnical Special Publication 30, American Society of Civil Engineers, New York, 1453 pp.

Boynton, R. S., and J. R. Blacklock. 1985. *Lime Slurry Pressure Injection Bulletin*. Bulletin 331. National Lime Association, Arlington, Va., 43 pp.

Brand, E. W., and P. L. R. Pang. 1991. Durability of Geotextiles to Outdoor Exposure in Hong Kong. *Journal of Geotechnical Engineering*, ASCE, Vol. 117, No. 7, pp. 979–1000.

Bruce, D. A. 1992a. Recent Progress in American Pinpile Technology. In *Grouting, Soil Improvement,*

and Geosynthetics: Proceedings of a Conference, New Orleans (R. H. Borden, R. D Holtz, and I. Juran, eds.), Geotechnical Special Publication 30, American Society of Civil Engineers, New York, Vol. 2, pp. 765–777.

Bruce, D. A. 1992b. Two New Specialty Geotechnical Processes for Slope Stabilization. In *Stability and Performance of Slopes and Embankments: Proceedings of a Specialty Conference*, Berkeley, Calif. (R. B. Seed and R. W. Boulanger, eds.), Geotechnical Special Publication 31, American Society of Civil Engineers, New York, pp. 1505–1519.

Byrne, R. J. 1992. Soil Nailing: A Simplified Kinematic Analysis. In *Grouting, Soil Improvement, and Geosynthetics: Proceedings of a Conference*, New Orleans (R. H. Borden, R. D. Holtz, and I. Juran, eds.), Geotechnical Special Publication 30, American Society of Civil Engineers, New York, Vol. 2, pp. 751–764.

Cancelli, A., B. Delia, and P. Sembenelli. 1987. Groundwater Problems in Embankments, Dams, and Natural Slopes: General Report, Session 4. In *Proc., Ninth European Conference on Soil Mechanics and Foundation Engineering*, Dublin, A.A. Balkema, Rotterdam, Netherlands, Vol. 3, pp. 1043–1072.

Casagrande, L. 1948. Electro-osmosis. In *Proc., Second International Conference on Soil Mechanics and Foundation Engineering*, Rotterdam, Gebr. Keesmaat, Haarlem, Netherlands, Vol. 1, pp. 218–223.

Casagrande, L. 1952. Electro-osmotic Stabilization of Soils. *Journal of the Boston Society of Civil Engineers*, Vol. 39, No. 1 (reprinted in *Contributions to Soil Mechanics 1941–1953*, Boston Society of Civil Engineers, 1953, pp. 285–317).

Casagrande, L. 1953. *Review of Past and Current Work on Electro-osmotic Stabilization of Soils*. Soil Mechanics Series 45. Harvard University, Cambridge, Mass., 133 pp.

Casagrande, L., R. W. Loughney, and M. A. J. Matich. 1961. Electro-osmotic Stabilization of a High Slope in Loose Saturated Silt. In *Proc., Fifth International Conference on Soil Mechanics and Foundation Engineering*, Paris, Dunod, Paris, Vol. 2, pp. 555–561.

Casagrande, L., N. Wade, M. Wakely, and R. Loughney. 1981. Electro-osmosis Projects, British Columbia, Canada. In *Proc., 10th International Conference on Soil Mechanics and Foundation Engineering*, Stockholm, A.A. Balkema, Rotterdam, Netherlands, Vol. 3, pp. 607–610.

Cedergren, H. R. 1989. *Seepage, Drainage, and Flow Nets*, 3rd ed. John Wiley and Sons, New York, 465 pp.

Chang, J. C., J. B. Hannon, and R. A. Forsyth. 1977. Pull Resistance and Interaction of Earthwork Reinforcement and Soil. In *Transportation Research Record 640*, TRB, National Research Council, Washington, D.C., pp. 1–7.

Christopher, B. R., S. A. Gill, J. P. Giroud, I. Juran, J. K. Mitchell, F. Schlosser, and J. Dunnicliff. 1990. *Reinforced Soil Structures*, Vol. 1: Design and Construction Guidelines. Report FHWA-RD-89-043. FHWA, U.S. Department of Transportation, 287 pp.

Christopher, B. R., and R. D. Holtz. 1985. *Geotextile Engineering Manual*. Report FHWA-TS-86-203. FHWA, U.S. Department of Transportation, 1024 pp.

Christopher, B. R., and R. D. Holtz. 1989. *Geotextile Design and Construction Guidelines*. Report FHWA-HI-90-001. FHWA, U.S. Department of Transportation, 334 pp.

Collin, J. G., M. A. Gabr, and A. G. MacKinnon. 1992. Timber Crib-Faced Soil-Nailed Retaining Wall. In *Stability and Performance of Slopes and Embankments: Proceedings of a Specialty Conference*, Berkeley, Calif. (R. B. Seed and R. W. Boulanger, eds.), Geotechnical Special Publication 31, American Society of Civil Engineers, New York, pp. 1457–1463.

Collotta, T., V. Manassero, and P. C. Moretti. 1988. An Advanced Technology in Deep Drainage of Slopes. In *Proc., Fifth International Symposium on Landslides*, Lausanne, A.A. Balkema, Rotterdam, Netherlands, Vol. 2, pp. 887–892.

Committee on Ground Failure Hazards. 1985. *Reducing Losses from Landsliding in the United States*. Commission on Engineering and Technical Systems, National Research Council, Washington, D.C., 41 pp.

Craig, D. J., and I. Gray. 1985. *Groundwater Lowering by Horizontal Drains*. GCO Publication 2-85. Geotechnical Control Office, Engineering Development Department, Hong Kong, 123 pp.

Denby, G., D. Argo, and D. Campbell. 1992. Soil Nail Design and Construction, Swedish Hospital Parking Garage, Seattle, Washington. Presented at Seminar on Soil Nailing and Reinforced Soil Walls, Seattle, American Society of Civil Engineers, 11 pp.

Edil, T. B. 1992. Landslide Cases in the Great Lakes: Issues and Approaches. In *Transportation Research Record 1343*, TRB, National Research Council, Washington, D.C., pp. 87–94.

Elias, V. 1990. *Durability/Corrosion of Soil Reinforced Structures*. Report FHWA-RD-89-186. FHWA, U.S. Department of Transportation, 163 pp.

Elias, V., and I. Juran. 1991. *Soil Nailing for Stabilization of Highway Slopes and Excavations*. Report FHWA-RD-89-198. FHWA, U.S. Department of Transportation, 210 pp.

Fang, H.Y. 1991. The Use of Bamboo Inclusions for Earth Reinforcement. In *Proc., Workshop on Biotechnical Stabilization*, University of Michigan, Ann Arbor, pp. 33–62.

Felio, G., M. Vucetic, M. Hudson, P. Barar, and R. Chapman. 1990. Performance of Soil Nailed Walls During the October 17 1989 Loma Prieta Earthquake. In *Proc., 43rd Canadian Geotechnical Conference*, Quebec City, Canadian Geotechnical Society, Rexdale, Ontario, Vol. 1, pp. 165–173.

Forsyth, R. A. 1978. Alternative Earth Reinforcements. In *Proc., Symposium on Earth Reinforcement*, Pittsburgh, American Society of Civil Engineers, New York, pp. 358–370.

Forsyth, R. A., and D.A. Bieber. 1984. La Honda Slope Repair with Grid Reinforcement. In *Proc., International Symposium on Polymer Grid Reinforcement in Civil Engineering*, London, pp. 54–58.

Forsyth, R. A., and J. P. Egan. 1976. Use of Waste Materials in Embankment Construction. In *Transportation Research Record 593*, TRB, National Research Council, Washington, D.C., pp. 3–8.

Fukuoka, M., and Y. Imamura. 1982. Fabric Retaining Walls. In *Proc., Second International Conference on Geotextiles*, Las Vegas, Industrial Fabrics Association International, St. Paul, Minn., Vol. 2, pp. 575–580.

Gedney, D. S., and W. G. Weber. 1978. Design and Construction of Soil Slopes. In *Special Report 176: Landslides: Analysis and Control* (R. L. Schuster and R. J. Krizek, eds.), TRB, National Research Council, Washington, D.C., Chapter 8, pp. 172–191.

Geotechnical Control Office. 1984. *Geotechnical Manual for Slopes*, 2nd ed. Public Works Department, Hong Kong, 295 pp.

Gillon, M. D., C. J. Graham, and G. G. Grocott. 1992. Low Level Drainage Works at the Brewery Creek Slide. In *Proc., Sixth International Symposium on Landslides*, Christchurch, New Zealand, A.A. Balkema, Rotterdam, Netherlands, Vol. 1, pp. 715–727.

Gould, J. P. 1990. Earth Retaining Structures— Developments through 1970. In *Design and Performance of Earth Retaining Structures: Proceedings of a Conference*, Ithaca, N.Y., Geotechnical Special Publication 25, American Society of Civil Engineers, New York, pp. 8–21.

Gray, D. H. 1970. Effect of Forest Clear-Cutting on the Stability of Natural Slopes. *Bulletin of the*

Association of Engineering Geologists, Vol. 7, No. 1, pp. 45–66.

Gray, D. H., and A. T. Leiser. 1982. *Biotechnical Slope Protection and Erosional Control*. Van Nostrand Reinhold, New York, 271 pp.

Gray, D. H., and R. B. Sotir. 1992. Biotechnical Stabilization of Cut and Fill Slopes. In *Stability and Performance of Slopes and Embankments: Proceedings of a Specialty Conference*, Berkeley, Calif. (R. B. Seed and R. W. Boulanger, eds.), Geotechnical Special Publication 31, American Society of Civil Engineers, New York, pp. 1395–1410.

Greenway, D. R. 1987. Vegetation and Slope Stability. In *Slope Stability* (M. G. Anderson and K. S. Richards, eds.), John Wiley, New York, pp. 187–230.

Greenway, D. R., M. G. Anderson, and K. C. Brian-Boys. 1984. Influence of Vegetation on Slope Stability in Hong Kong. In *Proc., Fourth International Symposium on Landslides*, Toronto, A.A. Balkema, Rotterdam, Netherlands, Vol. 1, pp. 399–404.

Greer, D. J., and H. A. Mathis. 1992. Proposed Correction of a Landslide for Relocation of Kentucky Highway 9 Using a Deep Drainage Gallery. In *Transportation Research Record 1343*, TRB, National Research Council, Washington, D.C., pp. 95–100.

Gress, J.C. 1992. Siphon Drain: A Technic for Slope Stabilization. In *Proc., Sixth International Symposium on Landslides*, Christchurch, New Zealand, A.A. Balkema, Rotterdam, Netherlands, Vol. 1, pp. 729–734.

Hausmann, M. R. 1990. *Engineering Principles of Ground Modification*. McGraw-Hill, New York, 632 pp.

Holtz, R. D. 1978. Special Applications. In *Proc., Symposium on Earth Reinforcement*, Pittsburgh, American Society of Civil Engineers, New York, pp. 77–97.

Holtz, R. D. 1989. *NCHRP Synthesis of Highway Practice 147: Treatment of Problem Foundations for Highway Embankments*. TRB, National Research Council, Washington, D.C., pp. 14–16.

Holtz, R. D., M. B. Jamiolkowski, R. Lancellotta, and R. Pedroni. 1991. *Prefabricated Vertical Drains: Design and Performance*. CIRIA Series. Butterworth-Heineman, London, 131 pp.

Hovland, H. J., and D. F. Willoughby. 1982. Slide Stabilization at the Geyser's Power Plant. In *Application of Walls to Landslide Control Problems*, Proceedings of two sessions, American Society of Civil Engineers National Convention, Las Vegas (R. B. Reeves, ed.), American Society of Civil Engineers, New York, N.Y., pp. 77–92.

Hryciw, R. D., and K. Haji-Ahmad. 1992. Design of Anchored Geosynthetic Systems for Slope Stabilization. In *Stability and Performance of Slopes and Embankments, Proceedings of a Specialty Conference*, Berkeley, Calif. (R. B. Seed and R. W. Boulanger, eds.), Geotechnical Special Publication 31, American Society of Civil Engineers, New York, pp. 1464–1480.

Humphrey, D. N., and W. P. Manion. 1992. Properties of Tire Chips for Lightweight Fill. In *Grouting, Soil Improvement, and Geosynthetics: Proceedings of a Conference*, New Orleans (R. H. Borden, R. D. Holtz, and I. Juran, eds.), Geotechnical Special Publication 30, American Society of Civil Engineers, New York, Vol. 2, pp. 1346–1355.

Hutchinson, J. N. 1977. Assessment of the Effectiveness of Corrective Measures in Relation to Geologic Conditions and Types of Slope Movement. *Bulletin of the Association of Engineering Geologists*, Vol. 16, pp. 131–155.

Hutchinson, J. N. 1984. Landslides in Britain and Their Countermeasures. *Journal of Japan Landslide Society*, Vol. 21, No. 1, pp. 1–24.

Isenhower, W. M. 1987. Slope Failures in San Antonio, Texas (preprint). Presented at 12th Southwest Geotechnical Engineers Conference, Scotsdale, Arizona, pp. 1–26.

Isenhower, W. M., S. G. Wright, and M. K. Kayyal. 1989. Design Procedures for Slide Suppressor Walls. In *Transportation Research Record 1242*, TRB, National Research Council, Washington, D.C., pp. 15–21.

Iwasaki, K., and S. Watanabe. 1978. Reinforcement of Railway Embankments in Japan. In *Proc., Symposium on Earth Reinforcement*, Pittsburgh, American Society of Civil Engineers, New York, pp. 473–500.

Jessberger, H. L. (ed.). 1979. *Ground Freezing*. Developments in Geotechnical Engineering, Vol. 26. Elsevier, Amsterdam, 552 pp.

Jones, C. J. F. P. 1978. The York Method of Reinforced Earth Construction. In *Proc., Symposium on Earth Reinforcement*, Pittsburgh, American Society of Civil Engineers, New York, pp. 501–527.

Jones, C. J. F. P. 1985. *Earth Reinforcement and Soil Structures*. Butterworths, London, 183 pp.

Koerner, R. M. 1990. *Designing with Geosynthetics*, 2nd ed. Prentice-Hall, Englewood Cliffs, N.J., 625 pp.

Koerner, R. M., and J. C. Robins. 1986. In Situ Stabilization of Soil Slopes using Nailed Geosynthetics. In *Proc., Third International Conference on Geotextiles*, Öesterreichischer Ingenieur-und

Architektenverein, Vienna, Austria, Vol. 2, pp. 395–400.

Kropp, A., and M. Thomas. 1992. Partial Landslide Repair by Buttress Fill. In *Transportation Research Record 1343*, TRB, National Research Council, Washington, D.C., pp. 108–113.

Lambe, P. C., and L. A. Hanson (eds.). 1990. *Design and Performance of Earth Retaining Structures: Proceedings of a Conference*, Ithaca, N.Y., Geotechnical Special Publication 25, American Society of Civil Engineers, New York, 904 pp.

Lee, K. L., B. D. Adams, and J. M. J. Vagneron. 1973. Reinforced Earth Retaining Walls. *Journal of the Soil Mechanics and Foundations Division*, ASCE, Vol. 99, No. SM10, pp. 745–764.

Leflaive, E. 1982. The Reinforcement of Granular Materials with Continuous Fibers (in French). In *Proc., Second International Conference on Geotextiles*, Las Vegas, Industrial Fabrics Association International, St. Paul, Minn., Vol. 3, pp. 721–726.

Leflaive, E. 1988. Texsol: Already More Than 50 Successful Applications. In *Proc., International Symposium on Theory and Practice of Earth Reinforcement*, Kyushu, Japan, A.A. Balkema, Rotterdam, Netherlands, pp. 541–545.

Lew, K. V., and J. Graham. 1988. River Bank Stabilization by Drains in Plastic Clay. In *Proc., Fifth International Symposium on Landslides*, Lausanne, A.A. Balkema, Rotterdam, Netherlands, Vol. 2, pp. 939–944.

Lizzi, F. 1977. Practical Engineering in Structurally Complex Formations (the In Situ Reinforced Earth). In *Proc., International Symposium on the Geotechnics of Structurally Complex Formations*, Capri, Italy, Associazione Geotecnica Italiana.

Lo, K. Y., H. S. Ho, and I. I. Inculet. 1991a. Field Test of Electroosmotic Strengthening of Soft Sensitive Clays. *Canadian Geotechnical Journal*, Vol. 28, pp. 74–83.

Lo, K. Y., I. I. Inculet, and H. S. Ho. 1991b. Electroosmotic Strengthening of Soft Sensitive Clays. *Canadian Geotechnical Journal*, Vol. 28, pp. 62–73.

Long, N. T. 1990. *Le Pneusol*. Etudes et Recherches des Laboratoires des Ponts et Chaussées, Série Géotechnique GT44. Laboratoire Central des Ponts et Chaussées, Paris, 76 pp.

Millet, R. A., G. M. Lawton, P. C. Repetto, and V. K. Garga. 1992. Stabilization of Tablachaca Dam. In *Stability and Performance of Slopes and Embankments: Proceedings of a Specialty Conference* (R. B. Seed and R. W. Boulanger, eds.), Geotechnical Special Publication 31, American Society of Civil Engineers, New York, pp. 1365–1381.

Mitchell, J. K., and B. R. Christopher. 1990. North American Practice in Reinforced Soil Systems. In *Design and Performance of Earth Retaining Structures: Proceedings of a Conference*, Ithaca, N.Y. (P. C. Lambe and L. A. Hanson, eds.), Geotechnical Special Publication 25, American Society of Civil Engineers, New York, pp. 322–346.

Mitchell, J. K., and W. C. B. Villet. 1987. *NCHRP Report 290: Reinforcement of Earth Slopes and Embankments*. TRB, National Research Council, Washington, D.C., 323 pp.

Morgenstern, N. R. 1982. The Analysis of Wall Supports to Stabilize Slopes. In *Application of Walls to Landslide Control Problems*, Proceedings of two sessions, American Society of Civil Engineers National Convention, Las Vegas (R. B. Reeves, ed.), American Society of Civil Engineers, New York, N.Y., pp. 19–29.

Murray, R. T. 1982. Fabric Reinforcement of Embankments and Cuttings. In *Proc., Second International Conference on Geotextiles*, Las Vegas, Industrial Fabrics Association International, St. Paul, Minn., Vol. 3, pp. 701–713.

Murray, R. T., and M. J. Irwin. 1981. *A Preliminary Study of TRRL Anchored Earth*. Supplementary Report 674. U.K. Transport and Road Research Laboratory, Crowthorne, Berkshire, England.

Nelson, D. S., and W. L. Allen. 1974. Sawdust and Lightweight Fill Material. Report FHWA-RD-74-502. FHWA, U.S. Department of Transportation, 24 pp. (reprinted in *Highway Focus*, Vol. 6, No. 3, pp. 53–66).

Nethero, M. F. 1982. Slide Controlled by Drilled Pier Walls. In *Application of Walls to Landslide Control Problems*, Proceedings of two sessions, American Society of Civil Engineers National Convention, Las Vegas (R. B. Reeves, ed.), pp. 61–76.

Oakland, M. W., and J. L. A. Chameau. 1984. Finite Element Analysis of Drilled Piers Used for Slope Stabilization. In *Laterally Loaded Deep Foundations: Analysis and Performance*, Special Technical Publication 835, ASTM, Philadelphia, Pa., pp. 182–193.

O'Rourke, T. D, and C. J. F. P. Jones. 1990. Overview of Earth Retention Systems: 1970–1990. In *Design and Performance of Earth Retaining Structures: Proceedings of a Specialty Conference*, Ithaca, N.Y. (P. C. Lambe and L. A. Hanson, eds.), Geotechnical Special Publication 25, American Society of Civil Engineers, New York, pp. 22–51.

Palladino, D. J., and R. B. Peck. 1972. Slope Failures in an Overconsolidated Clay, Seattle, Washington. *Geotechnique*, Vol. 22, No. 4, pp. 563–595.

Pearlman, S. L., B. D. Campbell, and J. L. Withiam. 1992. Slope Stabilization Using In-Situ Earth Reinforcements. In *Stability and Performance of Slopes and Embankments: Proceedings of a Conference*, Berkeley, Calif. (R. B. Seed and R. W. Boulanger, eds.), Geotechnical Special Publication 31, American Society of Civil Engineers, New York, pp. 1333–1348.

Peck, R. B., and H. O. Ireland. 1953. Investigation of Stability Problems. *Proc., American Railway Engineering Association*, Vol. 54, pp. 1116–1128.

Plumelle, C., and F. Schlosser. 1991. Three Full-Scale Experiments of French Project on Soil Nailing: CLOUTERRE. In *Transportation Research Record 1330*, TRB, National Research Council, Washington, D.C., pp. 80–86.

Read, J., T. Dodson, and J. Thomas. 1991. *Experimental Project Use of Shredded Tires for Lightweight Fill: Post-Construction Report*. FHWA Experimental Project No. 1. Oregon Department of Transportation, Salem, 31 pp.

Reagan, J., and W. Jutkowfsky. 1985. *Applications and Performance of Horizontal Drain Projects in New York State*. Soil Mechanics Bureau, New York Department of Transportation, Albany, 26 pp.

Reese, L. C., S. T. Wang, and J. L. Fouse. 1992. Use of Drilled Shafts in Stabilizing a Slope. In *Stability and Performance of Slopes and Embankments: Proceedings of a Conference*, Berkeley, Calif. (R. B. Seed and R. W. Boulanger, eds.), Geotechnical Special Publication 31, American Society of Civil Engineers, New York, pp. 1318–1332.

Riestenberg, M. M. 1987. *Anchoring of Thin Colluvium on Hillslopes by Roots of Sugar Maple and White Ash*. Ph.D. dissertation. University of Cincinnati, Cincinnati, Ohio.

Riestenberg, M. M., and S. Sovonick-Dunford. 1983. The Role of Woody Vegetation in Stabilizing Slopes in the Cincinnati Area, Ohio. *Geological Society of America Bulletin*, Vol. 94, pp. 506–518.

Ritchie, A. M. 1963. Evaluation of Rockfall and Its Control. In *Highway Research Record 17*, HRB, National Research Council, Washington, D.C., pp. 13–28.

Rivera, R. 1991. *Reforestation Program—Gaillard Cut*. Panama Canal Commission Report No. 1, 12 pp. plus appendix.

Robinson, C. S. 1979. Geologic Constraints on the Vail Pass Project. In *Highway Research Record 717*, HRB, National Research Council, Washington, D.C., pp. 6–14.

Rogers, C. D. F. 1991. Slope Stabilization Using Lime. In *Proc., International Conference on Slope Stability Engineering: Developments and Applications*, Institution of Civil Engineers, Isle of Wight, Thomas Telford, Ltd., pp. 335–492.

Rollins, K. M., and R. L. Rollins. 1992. Case Histories of Landslide Stabilization Using Drilled Shaft Walls. In *Transportation Research Record 1343*, TRB, National Research Council, Washington, D.C., pp. 114–122.

Root, A. W. 1958. Prevention of Landslides. In *Special Report 29: Landslides and Engineering Practice* (E. B. Eckel, ed.), HRB, National Research Council, Washington, D.C., pp. 113–149.

Roth, W. H., R. H. Rice, and D. T. Liu. 1992. Hydraugers at the Via de las Olas Landslide. In *Stability and Performance of Slopes and Embankments: Proceedings of a Conference*, Berkeley, Calif. (R. B. Seed and R. W. Boulanger, eds.), Geotechnical Special Publication 31, American Society of Civil Engineers, New York, pp. 1349–1364.

Royster, D. L. 1980. Horizontal Drains and Horizontal Drilling: An Overview. In *Transportation Research Record 783*, TRB, National Research Council, Washington, D.C., pp. 16–20.

Schlosser, F. 1990. Mechanically Stabilized Earth Retaining Structures in Europe. In *Design and Performance of Earth Retaining Structures* (P. C. Lambe and L. A. Hanson, eds.), Geotechnical Special Publication 25, American Society of Civil Engineers, New York, pp. 347–378.

Schlosser, F., and P. Unterreiner. 1991. Soil Nailing in France: Research and Practice. In *Transportation Research Record 1330*, TRB, National Research Council, Washington, D.C., pp. 72–79.

Schlosser, F., P. Unterreiner, and C. Plumelle. 1992. French Research Program CLOUTERRE on Soil Nailing. In *Grouting, Soil Improvement, and Geosynthetics: Proceedings of a Conference*, New Orleans (R. H. Borden, R. D. Holtz, and I. Juran, eds.), Geotechnical Special Publication 30, American Society of Civil Engineers, New York, Vol. 2, pp. 739–750.

Schuster, R. L. 1995. Recent Advances in Slope Stabilization. Keynote paper, Session G.3. *Proc., Sixth International Symposium on Landslides*, Christchurch, New Zealand, A.A. Balkema, Rotterdam, Netherlands, Vol. 3, pp. 1715–1745.

Schuster, R. L., and R. L. Fleming. 1982. Geologic Aspects of Landslide Control Using Walls. In *Application of Walls to Landslide Control Problems*, Proceedings of two sessions, American Society of Civil Engineers National Convention, Las Vegas (R. B. Reeves, ed.), American Society of Civil Engineers, New York, N.Y., pp. 1–18.

Sembenelli, P. 1988. General Report: Stabilization and Drainage. In *Proc., Fifth International Symposium on Landslides*, Lausanne, A.A. Balkema, Rotterdam, Netherlands, Vol. 2, pp. 813–819.

Sharma, S., and T. Buu. 1992. The Bud Peck Slide, I-15, Near Malad, Idaho. In *Transportation Research Record 1343*, TRB, National Research Council, Washington, D.C., pp. 123–129.

Shen, C. K., S. Bang, L. R. Herrmann, and K. L. Ronstad. 1978. A Reinforced Lateral Earth Support System. In *Proc., Symposium on Earth Reinforcement*, Pittsburgh, American Society of Civil Engineers, New York, pp. 764–793.

Smith, D. D. 1980a. Long-Term Performance of Horizontal Drains. In *Transportation Research Record 783*, TRB, National Research Council, Washington, D.C., pp. 39–45.

Smith, D. D. 1980b. *The Effectiveness of Horizontal Drains*. Report FHWA-CA-TL-80-16. FHWA, U.S. Department of Transportation, 79 pp.

Smith, T. W., R. H. Prysock, and J. Campbell. 1970. Pinole Slide, I-80, California. *Highway Focus*, Vol. 2, No. 5, pp. 51–61.

Szymoniak, T., J. R. Bell, G. R. Thommen, and E. L. Johnsen. 1984. A Geogrid-Reinforced Soil Wall for Landslide Correction on the Oregon Coast. In *Transportation Research Record 965*, TRB, National Research Council, Washington, D.C., pp. 47–55.

Thomas, M. R., and A. Kropp. 1992. Stabilization of a Debris Flow Scar Using Soil Bioengineering. In *Transportation Research Record 1343*, TRB, National Research Council, Washington, D.C., pp. 101–107.

Transportation Research Board. 1980. *Transportation Research Record 783: Rock Classification and Horizontal Drilling and Drainage*. National Research Council, Washington, D.C., 45 pp.

Tysinger, G. L. 1982. Slide Stabilization 4th Rocky Fill, Clinchfield Railroad. In *Application of Walls to Landslide Control Problems*, Proceedings of two sessions, American Society of Civil Engineers National Convention, Las Vegas (R. B. Reeves, ed.), American Society of Civil Engineers, New York, pp. 93–107.

Veder, C. 1981. *Landslides and Their Stabilization*. Springer-Verlag, New York, 247 pp.

Weatherby, D.E., and P. J. Nicholson. 1982. Tiebacks Used for Landslide Stabilization. In *Application of Walls to Landslide Control Problems*, Proceedings of two sessions, American Society of Civil Engineers National Convention, Las Vegas (R. B. Reeves, ed.), American Society of Civil Engineers, New York, pp. 44–60.

Wilson, S. D. 1970. Observational Data on Ground Movements Related to Slope Instability: Sixth Terzaghi Lecture. *Journal of the Soil Mechanics and Foundations Division*, ASCE, Vol. 96, No. SM5, pp. 1521–1544.

Wu, T.H. 1994a. Slope Stabilization Using Vegetation. In *Geotechnical Engineering: Emerging Trends in Design and Practice* (K.R. Saxena, ed.), Oxford & IBH Publishing Co. Pvt. Ltd., New Delhi, Chap. 15, pp. 377–402.

Wu, T.H. 1994b. Slope Stabilization. In *Slope Stabilization and Erosion Control: A Bioengineering Approach* (R.P.C. Morgan and R.J. Rickson, eds.), E.&F.N. Spon, London, Chap. 7, pp. 221–263.

Wu, T. H., W. P. McKinnell, and D. N. Swanston. 1979. Strength of Tree Roots and Landslides on Prince of Wales Island, Alaska. *Canadian Geotechnical Journal*, Vol. 16, No. 1, pp. 19–33.

Yako, M. A., and B. R. Christopher. 1988. Polymerically Reinforced Retaining Walls in North America. In *Proceedings of the Workshop on Polymeric Reinforcement in Soil Retaining Structures*, Kingston, Ontario, Canada (P. M. Jarrett and A. McGown, eds.), Kluwer Academic Publishers, pp. 239–284.

Yeh, S., and J. Gilmore. 1992. Application of EPS for Slide Correction. In *Stability and Performance of Slopes and Embankments: Proceedings of a Conference*, Berkeley, Calif. (R. B. Seed and R. W. Boulanger, eds.), Geotechnical Special Publication 31, American Society of Civil Engineers, New York, pp. 1444–1456.

Yin, K. P., L. K. Heung, and D. R. Greenway. 1988. Effect of Root Reinforcement on the Stability of Three Fill Slopes in Hong Kong. In *Proc., Second International Conference on Geomechanics in Tropical Soils*, Singapore, pp. 293–302.

Chapter 18

DUNCAN C. WYLLIE AND
NORMAN I. NORRISH

STABILIZATION OF ROCK SLOPES

1. INTRODUCTION

A successful rock slope stabilization program requires the integration of a number of interrelated activities, including geotechnical engineering as well as environmental and safety issues, construction methods and costs, and contracting procedures. Modern methods for design and stabilization of rock slopes were developed in the 1970s (Brawner and Wyllie 1975; Fookes and Sweeney 1976; Groupe d'Etudes des Falaises 1978; Piteau and Peckover 1978; Hoek and Bray 1981) and continue to be refined and developed (Federal Highway Administration 1989; Wyllie 1991; Schuster 1992). These procedures can therefore be used with confidence for a wide range of geological conditions. However, as described in this chapter, it is essential that the methods used be appropriate for the particular conditions at each site.

1.1 Effect of Rock Falls on Transportation Systems

The safe operation of transportation routes in mountainous terrain often requires that measures be taken to control the incidence of rock falls and rock slope failures. Along highways, the types of events and accidents that can be caused by rock falls range from minor falls that damage tires and bodywork, to larger falls that affect vehicles or cause vehicles to swerve off the road, to substantial slope failures that block the facility. The effects of these events can be damage to vehicles, injury to or death of drivers and passengers, economic loss due to road closures, and possibly discharge of toxic substances when transporting vehicles are damaged. Figure 18-1 shows a rock slide that occurred from a height of about 300 m above the road and closed both the road and the railway. The removal of the rock on the road followed by stabilization of the slide area, which brought down further unstable rock, resulted in a total closure time of several weeks with significant economic losses for users of both the road and the railway.

Although the cost of a major slide such as that shown in Figure 18-1 is substantial, the cost of even a single-car accident can be significant. For example, costs may be incurred for hospitalization of the driver and passengers, for repair to the vehicle, and in some cases for legal costs and compensation. Often there are additional costs for stabilization of the slope that will involve both engineering and contracting charges, usually carried out at premium rates because of the emergency nature of the work. In cases where the rock slide results in closure of a road, indirect costs may be incurred, such as loss of revenue for businesses located along the highway. In the case shown in Figure 18-1, the costs included leasing a car ferry to move vehicles around the landslide. For railroads and toll highways, closures result in a direct loss of revenue that may amount to thousands of dollars per hour.

FIGURE 18-1
Railway and highway at Howe Sound, British Columbia, blocked by October 1990 rock fall that originated approximately 300 m above highway.
DENISE HOWARD, VANCOUVER SUN

1.2 Causes of Rock Falls

The California Department of Transportation (Caltrans) made a comprehensive study of rock falls that have occurred on the state highway system to assess both their causes and the effectiveness of the various remedial measures that have been implemented (McCauley et al. 1985). Because of the diverse topography and climate within California, Caltrans records provide a useful guideline on the stability conditions of rock slopes and the causes of falls. Table 18-1 shows the results of a study of 308 rock falls on California highways in which 14 different causes of instability were identified.

Of the 14 causes of California rock falls identified, 6 are directly related to water, namely, rain, freeze-thaw, snowmelt, channeled runoff, differential erosion, and springs or seeps. One cause is indirectly related to rainfall—the growth of tree roots in cracks, which can open fractures and loosen blocks of rock on the slope face. These seven causes of rock falls together account for 68 percent of the total falls.

These statistics are confirmed by the authors' experience in the analysis of rock-fall records over a 19-year period on a major railroad in western Canada, in which approximately 70 percent of the events occurred during the winter. Weather conditions during the winter included heavy rainfall, prolonged periods of freezing temperatures, and daily freeze-thaw cycles in the fall and spring. The results of a similar study carried out by Peckover (1975) are shown in Figure 18-2. They clearly show that the majority of rock falls occurred between October and March, the wettest and coldest time of the year in western Canada.

The other major group of factors affecting stability in the California study were the particular

Table 18-1
Causes of Rock Falls on Highways in California

Cause	Percentage of Total
Rain	30
Freeze-thaw	21
Fractured rock	12
Wind	12
Snowmelt	8
Channeled runoff	7
Adverse planar fracture	5
Burrowing animals	2
Differential erosion	1
Tree roots	0.6
Springs or seeps	0.6
Wild animals	0.3
Truck vibrations	0.3
Soil decomposition	0.3

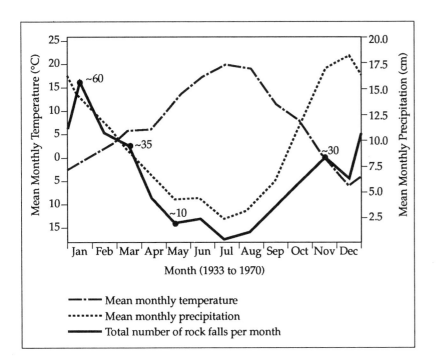

FIGURE 18-2
Correlation of
number of rock falls
with temperature
and precipitation
on railway lines in
Fraser Canyon,
British Columbia
(Peckover 1975).

geologic conditions at each site, namely, fractured rock, adverse planar fracture (a fracture dipping out of the slope face), and soil decomposition. These three factors represented 17 percent of the falls, and the total rock falls caused by water and geologic factors accounted for 85 percent of the falls. These statistics demonstrate that water and geology are the most important influences in rock slope stability.

It appears that the study in California was carried out during a time when there were no significant earthquakes, which frequently trigger rock falls and cause displacement and failure of slopes (Van Velsor and Walkinshaw 1991). The mechanism by which earthquakes trigger landslides is described in Chapter 4, and methods of incorporating earthquake forces into the design of rock slopes are demonstrated in Chapter 15.

2. PLANNING SLOPE STABILIZATION PROGRAMS

In mountainous areas where there are numerous rock-fall hazards that may result in a significant cost to the operator of the transportation system, a stabilization program is often justified. The situation may be particularly severe where the slopes have been open for 20 to 30 years and are in an unstable condition either because of poor initial construction practices or because of weathering.

Under these circumstances, stabilization work can take many years to complete. In order to make the best use of available funds, it often is beneficial to set up a systematic program that identifies the most hazardous slopes. Then annual stabilization work can be scheduled, with the most hazardous sites being worked on early in the program. The components of such a program make up six steps, as shown in Figure 18-3.

The objective of the program shown in Figure 18-3 is to be proactive in identifying and stabilizing slopes before rock falls and accidents occur. This requires a careful examination of the site to identify the potential hazard and estimate the likely benefit of the stabilization work. In contrast, a reactive program places the emphasis on areas in which rock falls and accidents have already occurred and the hazard may be significantly diminished. The benefits of a reactive program are likely to be less than those of a proactive program.

An effective proactive approach to stabilization requires a consistent, long-term program under the direction of a team experienced in both the engineering and construction aspects of this work. Another important component of this work is to keep accurate records, with photographs, of slope conditions, rock falls, and stabilization work. This information will document the location of hazardous areas and determine the long-term effectiveness of the program in reducing the incidence of rock falls. These records can be most conveniently handled using data-base programs that readily allow updating and retrieval (see Section 2.2).

2.1 Rock Slope Inventory Systems

The relative risk of rock falls at a site as compared with other sites can be used in selecting priorities. Early work on this topic by Brawner and Wyllie (1975) and Wyllie (1987) was developed by Pierson et al. (1990) into a process for the rational management of the rock slopes along transportation systems, which has been named the Rockfall Hazard Rating System (RHRS). The first step in this process is to make an inventory of the stability conditions of each slope so that they can be ranked according to their rock-fall hazard (Figure 18-3).

The rock-fall areas identified in the inventory are ranked by scoring the categories as shown in Table 18-2. These categories represent the significant elements of a rock slope that contribute to the overall hazard. The four columns correspond to logical breaks in the hazard represented by each category. The rating criteria scores increase exponentially from 3 to 81 points and represent a continuum of points from 1 to 100. An exponential system allows for a rapid increase in score, which quickly distinguishes more hazardous sites. Using a continuum of points allows flexibility in evaluating the relative impact of conditions that are variable by nature. Some categories require a subjective evaluation, whereas others can be directly measured and then scored. A brief description of each of the categories of the RHRS follows. It should be noted that the system was developed on the basis of customary units; however, conversion to metric specifications is generally straightforward.

2.1.1 Slope Height

Category 1 represents the vertical height of the slope. Measurement is to the highest point from which rock fall is expected. If rocks are coming from the natural slope above the cut, the cut height plus the additional slope height (vertical distance) are used.

2.1.2 Ditch Effectiveness

The effectiveness of a ditch is measured by its ability to prevent falling rock from reaching the traveled way. In estimating the ditch effectiveness, factors to consider are (a) slope height and angle; (b) ditch width, depth, and shape; (c) anticipated block size and quantity of rock fall; and (d) effect of slope irregularities (launching features) on falling rocks. A launching feature can negate the benefits expected from a fallout area. Valuable information on ditch performance can be obtained from maintenance personnel and reference to Figure 18-15 (see Section 5.1).

2.1.3 Average Vehicle Risk

Average vehicle risk (AVR) represents the percentage of time that a vehicle will be present in the rock-fall section, which is obtained from the following relationship:

INVENTORY OF STABILITY CONDITIONS

Assign point scores to ten parameters describing slope and hazard conditions (rock fall hazard rating).

Section 2.1

ANALYZE ROCK FALL HAZARDS

Analyze inventory database to determine locations and causes of rock falls.

Section 2.2

PLAN STABILIZATION WORK

Plan work according to hazard rating, site location and type of stabilization work.

Sections 2.3, 2.4, 2.5

DECISION ANALYSIS TO DETERMINE OPTIMUM STABILIZATION METHODS

Compare effectiveness of stabilization measures in reducing rock fall hazards.

Section 2.3

DESIGN STABILIZATION MEASURES AND PREPARATION OF CONTRACT DOCUMENT

Categorize stabilization measures according to reinforcement, removal, and protection methods.

Sections 3, 4, 5

CONSTRUCTION SERVICES

Perform construction on cost-plus contracts where modifications to stabilization measures to suit actual conditions are revealed as work progresses. Use unit price contracts where specific tasks such as bolting, shotcrete required.

Section 6

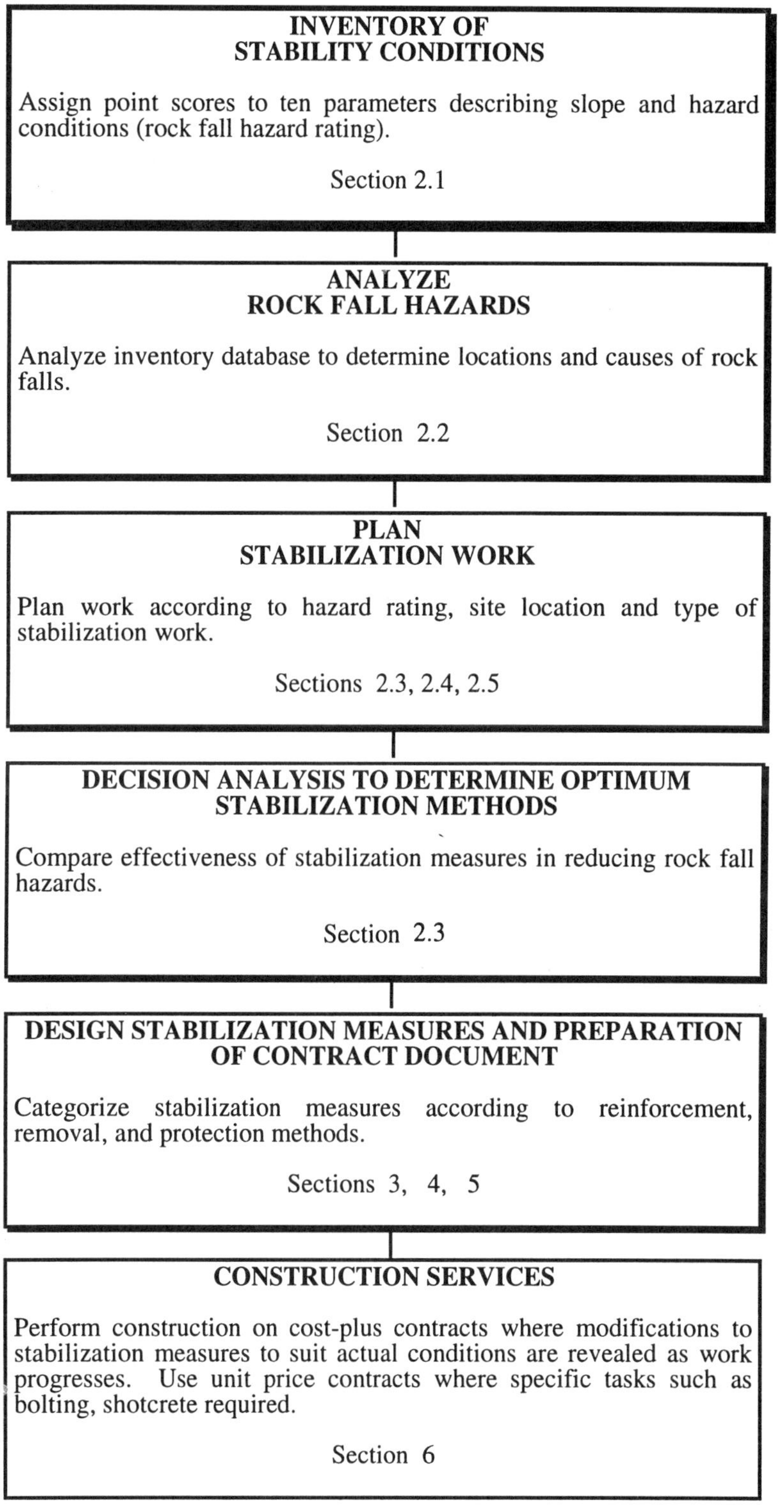

FIGURE 18-3 Rock slope stabilization program for transportation systems.

$$\text{AVR (\%)} = \frac{\text{ADT (cars/day)} \times \text{slope length (mi)}}{\text{posted speed limit (mph)} \times 24 \text{ (hr/day)}} \times 100\% \qquad (18.1)$$

Table 18-2
Summary Sheet of Rockfall Hazard Rating System (Pierson et al. 1990)

	RATING CRITERIA BY SCORE			
CATEGORY	POINTS 3	POINTS 9	POINTS 27	POINTS 81
1. Slope height (m)	7.5	15	23	>30
2. Ditch effectiveness	Good catchment	Moderate catchment	Limited catchment	No catchment
3. Average vehicle risk (% of time)	25	50	75	100
4. Decision sight distance (% of design value)	Adequate	Moderate	Limited	Very limited
5. Roadway width including paved shoulders (m)	13.5	11	8.5	6
6. Geologic case characteristics Case 1				
Structural condition	Discontinuous joints, favorable orientation	Discontinuous joints, random orientation	Discontinuous joints, adverse orientation	Continuous joints, adverse orientation
Rock friction	Rough, irregular	Undulating	Planar	Clay infilling or slickensided
Case 2				
Structural condition	Few differential erosion features	Occasional erosion features	Many erosion features	Major erosion features
Difference in erosion rates	Small	Moderate	Large	Extreme
7. a. Block size (m)	0.3	0.6	1	1.2
b. Volume of rock fall or event (m³)	3	6	9	12
8. Climate and presence of water on slope	Low to moderate precipitation; no freezing periods; no water on slope	Moderate precipitation or short freezing periods or intermittent water on slope	High precipitation or long freezing periods or continual water on slope	High precipitation and long freezing periods or continual water on slope and long freezing periods
9. Rock-fall history	Few falls	Occasional falls	Many falls	Constant falls

A rating of 100 percent means that on average a vehicle can be expected to be within the section 100 percent of the time.

2.1.4 Decision Sight Distance

Sight distance is the shortest distance along a roadway that an object is continuously visible to the driver. The sight distance can change appreciably throughout a rock-fall section. Horizontal and vertical curves, together with obstructions such as rock outcrops and roadside vegetation, can severely limit the available sight distance. Decision sight distance (DSD) is used to determine the length of roadway (in meters) needed to make a complex or instantaneous decision. DSD

is critical when obstacles on the road are difficult to perceive or when unexpected or unusual maneuvers are required.

The relationship between DSD and the posted speed limit used in the inventory system has been modified from the *Policy on Geometric Design of Highways and Streets* of the American Association of State Highway and Transportation Officials (AASHTO 1984).

2.1.5 Roadway Width

The available maneuvering room to avoid a rock fall is measured perpendicular to the highway centerline from one edge of the pavement to the other and includes the shoulders if they are paved.

When the roadway width is not constant, the minimum width is used.

2.1.6 Geologic Case Characteristics

Generally, there are two cases of conditions that cause rock fall. Case 1 includes slopes in which joints, bedding planes, or other discontinuities are the dominant structural features. In Case 2 differential erosion or oversteepened slopes are the dominant conditions that control rock fall. The case that best fits the slope should be used in the evaluation. If both types are present, both are scored but only the worst case (highest score) is used in the rating.

Case 1

Structural condition is characterized by adverse joints, those with orientations and lengths that promote planar, circular, block, wedge, or toppling failures. Rock friction on a discontinuity is governed by the characteristics of the rock material as well as by the surface roughness and properties of any infilling (see Section 2 of Chapter 14).

Case 2

Structural condition is characterized by differential erosion or oversteepening, which is the dominant condition that leads to rock fall. Erosion features include oversteepened slopes, unsupported rock units, or exposed resistant rocks. The different rates of erosion within a slope directly relate to the potential for a future rock-fall event. The score should reflect how quickly erosion is occurring; the size of rocks, blocks, or units being exposed; the frequency of rock-fall events; and the amount of material released during a rock fall.

2.1.7 Block Size and Volume of Rock Fall per Event

The measurements in Category 7 should be representative of whichever type of event is most likely to occur. The scores should also take into account any tendency of the blocks of rock to break up as they fall down the slope.

2.1.8 Climate and Presence of Water on Slope

Water and freeze-thaw cycles both contribute to the weathering and movement of rock materials. If water is known to flow continually or intermittently on the slope, the slope is rated accordingly. Areas receiving less than 50 cm per year are low-precipitation areas. Areas receiving more than 125 cm per year are considered high-precipitation areas.

2.1.9 Rock-Fall History

Historical information is an important check on the potential for future rock falls. As a better data base of rock-fall occurrences is developed, more accurate conclusions for the rock-fall potential can be made.

2.2 Data-Base Analysis of Rock Slope Inventory

It is common practice to enter the results of the rock slope inventory into a computer data base. The data base can be used both to analyze the data contained in the inventory and to facilitate updating of the inventory with new information on rock falls and construction work. The following are some examples of data-base analyses:

- The slopes can be ranked in order of decreasing point score to identify the most hazardous slopes;
- Correlations can be found, for example, between rock-fall frequency and such factors as weather conditions, rock type, and slope location;
- The severity of rock falls can be assessed from analysis of delay hours caused by falls;
- The effectiveness of stabilization work can be assessed by determining how soon rock falls reoccur after stabilization work has been carried out.

The results of such analyses of the rock slope inventory data base can be used to plan future stabilization work. For example, mobilization costs for construction equipment can be minimized by selecting sites within one contract that are close together and that require similar types of equipment.

2.3 Selection of Stabilization Measures

Methods of slope stabilization fall into three categories: reinforcement, rock removal, and protection. Figure 18-4 includes 16 of the more common stabilization measures divided into these categories.

FIGURE 18-4
Categories of rock
slope stabilization
measures.

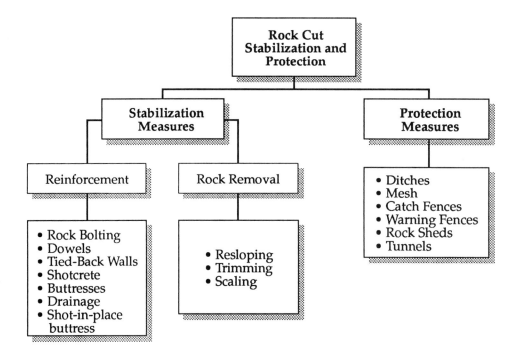

It is important that the appropriate stabilization method be used for the particular conditions at each site. For example, where the slope is steep and the toe is close to the highway or railway, there will be no space to excavate a catch ditch or construct a fence. Alternative stabilization measures may be to remove loose rock, secure it in place with bolts, or cover the slope with mesh. It is generally preferable to remove loose rock and eliminate the hazard, but only if this will form a stable face and not undermine other potentially loose rock on the face.

When stabilization measures that are appropriate for a site are selected and designed, geotechnical, construction, and environmental issues must be considered. The geotechnical issues—geology, rock strength, groundwater, and stability analysis—are discussed in Chapters 14 and 15. Construction and environmental issues, which can affect the costs and schedule of the work, must be addressed during the design phase of the project in order for the work to be carried out as efficiently as possible. Issues that are frequently important are equipment access, available work time during traffic closures, and disposal of waste rock and soil.

Another factor to consider in the selection of stabilization measures is the optimum level of work. For example, a minor scaling project will remove the loosest rock on the slope face, but if the rock is susceptible to weathering, this work may have to be repeated every 3 to 5 years. Alterna-

tively, a more comprehensive program can be carried out using shotcrete and bolting in addition to scaling. Although the initial costs of this second program will be higher, its effects will last longer, perhaps for 20 years. Alternative stabilization programs such as these, as well as the alternative of doing no work, can be compared by the use of decision analysis, which is a systematic procedure for evaluating alternative courses of action taking into account both the cost and design life of the stabilization work, as well as the probability that rock falls will occur, causing accidents and their costs (Wyllie et al. 1980; Roberds 1991).

2.4 Construction Issues

The following is a brief discussion of some construction issues that may have a significant influence on stabilization work, depending on actual site conditions.

2.4.1 Blasting

Damage to rock faces by excessively heavy blasting is a frequent cause of instability in the years following excavation of a slope. Methods of controlled blasting, such as preshear and trim blasting, as described in Section 4.2, can be used to excavate a slope to a specified line with minimal damage to the rock behind the face.

2.4.2 Topography

The topography at a site can significantly affect both the slope design and the type and extent of stabilization work. For example, if there is a continuous slope above the crest of a cut, stabilization work that involves laying back the cut will have the effect of increasing the cut height. This increase in cut height will require a larger catch ditch and may result in additional stability problems, especially if there is a substantial layer of soil or weathered rock at the surface.

2.4.3 Construction Access

In the design of excavations and stabilization work, it is necessary to determine the type of equipment that is likely to be required to carry out the work and how this equipment will be used at the site. For example, if it is planned to excavate a substantial volume of rock in order to lay back a slope, it is likely that airtrac drills and excavators will have to work on the slope. In steep terrain it may be found that the construction of an access road for this equipment would be costly and cause additional instability. Alternatively, if stabilization work is planned using large-diameter rock bolts, it is essential that suitable drilling equipment have access to the site. For example, on steep faces, holes with a diameter larger than about 100 mm will have to be drilled with heavy equipment supported by a crane or anchored cables. If it is not possible to use heavy drilling equipment, it will be necessary to drill smaller-diameter holes with lightweight equipment, such as downhole hammers or hand-held percussion drills.

2.4.4 Construction Costs

Cost estimates for stabilization work must take into account both the costs of the work on the slope and indirect costs such as mobilization and traffic closures. For example, when minor stabilization work is carried out on a steep face, traffic closures can be minimized by having the construction crews work from ropes secured behind the crest of the slope rather than from a crane located on the road, which may block two to three lanes of traffic.

2.5 Environmental Issues

A number of procedures can be implemented during both the design and construction phases of a rock slope project to minimize the effects of this work on the environment. These procedures may add to the direct cost of the work, but can have benefits of producing a more visually attractive highway and reducing environmental impact and public opposition.

2.5.1 Waste Disposal

The least expensive method of disposing of waste rock produced by excavation and scaling operations in mountainous terrain is by dumping it down the slope below the highway. However, disposing of waste rock in this manner has a number of drawbacks. First, a steeply sloping pile of loose rock may be a visual scar on the hillside that can be difficult to vegetate. Second, the waste rock may become unstable if not adequately drained or keyed into the existing slope; the material may move a considerable distance if it fails, which could endanger facilities downslope. Third, when the road is located in a river valley, the dumped rock may fall into the river and have a deleterious effect on fish populations. In order to minimize these impacts, it is sometimes required that the rock excavated from the slope be hauled to designated, stable waste sites.

Another problem that may need to be addressed in the disposal of waste rock is acid-water drainage. In some areas of North Carolina and Tennessee, for example, some argillite and schist formations contain iron disulfides; percolation of water through fills constructed with this rock produces low-pH, acidic runoff. One method that has been tested to control this condition is to mix the rock with lime to neutralize the acid potential and then to place the blended material in the center of the fill (Byerly and Middleton 1981). Sometimes it is necessary to encase the rock-lime mixture in an impervious plastic membrane.

2.5.2 Aesthetics

A series of steep, high rock cuts above a highway may have a significant visual impact when viewed both by the road user and the local population. In scenic areas it may be desirable to incorporate appropriate landscaping measures in the design of the rock cuts in order to minimize their visual impact (Norrish and Lowell 1988).

3. ROCK REINFORCEMENT

Figure 18-5 shows a number of reinforcement techniques that may be implemented to secure potentially loose rock on the face of a rock cut. The common feature of all these techniques is that they minimize the relaxation and loosening of the rock mass that may take place as a result of excavation and unloading (Hoek 1983). Once relaxation has been allowed to take place, there is a loss of interlock between the blocks of rock and a significant decrease in the shear strength. Figure 14-8(*b*) shows the effect of installing rock bolts to maintain the interlock on high-angle, second-order asperities. Once relaxation has taken place, it is not possible to reverse the process. For

this reason, reinforcement of rock slopes is most effective if it is installed before excavation—a process known as prereinforcement.

The different applications of untensioned (prereinforcement) bolts and tensioned bolts are shown in Figure 18-6. Prereinforcement of a benched excavation can be achieved by installing bolts as each bench is excavated. Installation of fully grouted but untensioned bolts at the crest of the cut before excavation [Figure 18-6(*b*)] prevents loss of interlock of the rock mass because the bolts are sufficiently stiff to prevent movement on the natural fractures (Moore and Imrie 1982; Spang and Egger 1990). However, when blocks have moved and relaxed, it is necessary to install tensioned bolts in order to prevent further displacement and loss of interlock [Figure 18-6(*a*)]. The advantages of using untensioned bolts are the lower costs and quicker installation compared with tensioned bolts.

3.1 Rock Bolts

Tensioned rock bolts are installed across potential failure surfaces and anchored in sound rock beyond the surface. The application of a tensile force in the bolt, which is transmitted into the rock by a reaction plate at the rock surface, produces compression in the rock mass and modifies the normal and shear stresses across the failure surface. The effect of the forces produced by the rock bolt when installed at a dip angle flatter than the normal to the potential failure surface is to increase the resisting force and decrease the driving force. It is found that the required bolt force to produce a specified factor of safety is minimized when the sum of the dip angle of the bolt (Ψ_T) and the dip angle of the failure surface (Ψ_P) is equal to the friction angle (see inset, Figure 18-5). Savings in bolting costs usually can be realized by installing bolts at the optimum angle (Ψ_{opt}) rather than at an angle normal to the failure surface.

The three main requirements of a permanent, tensioned rock-bolt installation are that

1. There be a method of anchoring the distal end of the anchor in the drill hole,
2. A known tension be applied to the bolt without creep and loss of load over time, and
3. The complete anchor assembly be protected from corrosion for the design life of the project.

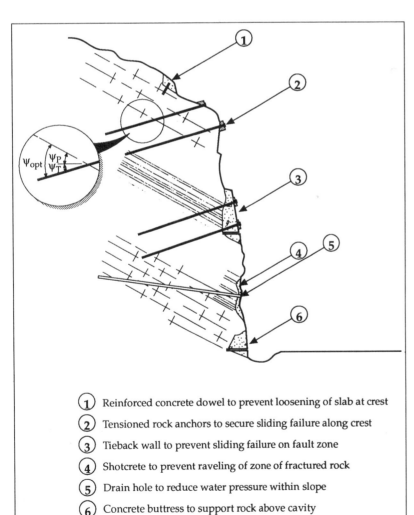

FIGURE 18-5
Rock slope reinforcement methods.

(1) Reinforced concrete dowel to prevent loosening of slab at crest

(2) Tensioned rock anchors to secure sliding failure along crest

(3) Tieback wall to prevent sliding failure on fault zone

(4) Shotcrete to prevent raveling of zone of fractured rock

(5) Drain hole to reduce water pressure within slope

(6) Concrete buttress to support rock above cavity

These three aspects of rock-anchor design and installation are discussed below.

3.1.1 Anchorage

Methods of securing the distal end of a bolt in the drill hole include resin, mechanical, and cement-grout anchors. The selection of the appropriate anchor depends on such factors as the required capacity of the anchor, speed of installation, strength of the rock in the anchor zone, access to the site for drilling and tensioning equipment, and level of corrosion protection required.

3.1.1.1 Resin Anchors

The resin anchor system includes a plastic cartridge about 25 mm in diameter and 200 mm long that contains a liquid resin and a hardener that set when mixed together (Figure 18-7). Setting times vary from about 1 to 5 min to as much as 90 min, depending on the reagents used. The setting time is also dependent on the temperature, with fast-setting resin hardening in about 4 min at a temperature of -5°C and in about 25 sec at 35°C.

The installation method consists of inserting a sufficient number of cartridges into the drill hole to fill the annular space around the distal end of the bolt (the bar). It is important that the hole diameter in relation to the bar size be within specified tolerances so that complete mixing of the resin is achieved when the bar is spun. This fit usually precludes the use of coupled anchors because the hole diameter to accommodate the coupling would be too large for complete resin mixing. The bar is spun as it is driven through the cartridges to mix the resin and form a rigid solid that anchors the bar in the hole. The required speed of rotation is about 60 rpm, and spinning is continued for about 30 sec after the bar has reached the end of the hole. The maximum bolt length is about 12 m because most drills cannot rotate longer bars at sufficient speed to mix the resin. It is possible to install a tensioned, resin-grouted bolt by using a fast-setting (about 2 min) resin for the anchor and a slower-setting (30 min) resin for the remainder of the bar. The bolt is tensioned between the setting of the fast and slow resins.

The primary advantages of resin anchorage are the simplicity and speed of installation, with support of the slope being provided within minutes of spinning the bolt. The disadvantages are the lim-

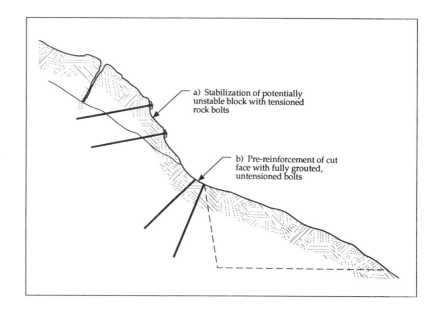

ited length and tension capacity of the bolt (400 kN) and the fact that only rigid bars can be used. Furthermore, the resin is not as effective as cement grout for corrosion protection of steel. Unlike cement grout, resin does not provide the high pH protective layer against corrosion, and it cannot be verified that the cartridges completely encapsulate the steel.

3.1.1.2 Mechanical Anchors

Mechanical anchors consist of a pair of steel wedges that are pressed against the walls of the drill hole. For the Williams anchor shown in Figure 18-8(*a*), the installation procedure is to first drill to the specified diameter so that when the bar is installed, the cone threaded on the bar is in con-

FIGURE 18-6
Reinforcement of a rock slope with (*a*) tensioned rock bolts and (*b*) fully grouted, untensioned rock bolts.

FIGURE 18-7
Resin cartridges for anchoring rock bolts.

FIGURE 18-8
Examples of rock
bolts with (a)
Williams hollow
core mechanical
wedge anchor and
(b) Dywidag
grouted anchor.
WILLIAMS FORM
HARDWARE AND ROCK
BOLT COMPANY AND
DYWIDAG SYSTEMS
INTERNATIONAL

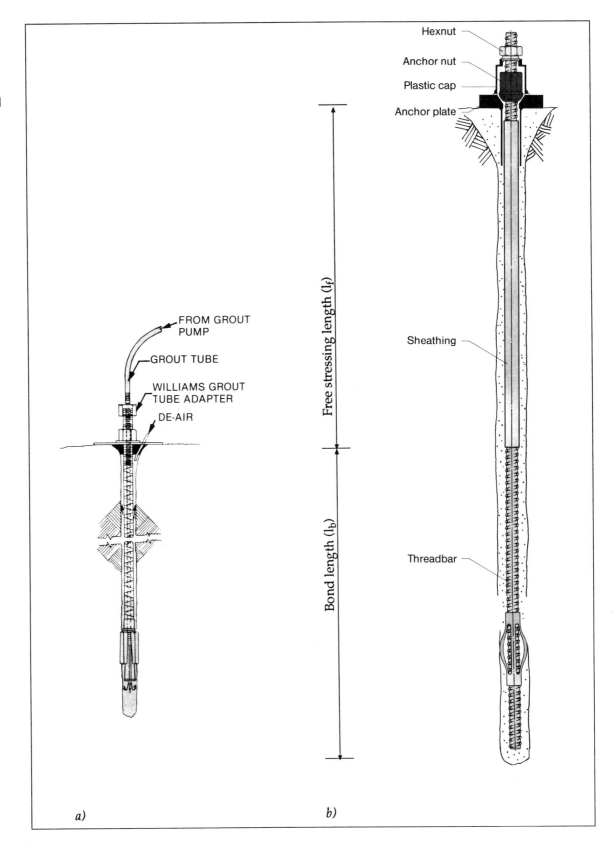

tact with the walls of the hole. When the bar is torqued, the cone moves along the bar and expands the wedges against the walls of the hole to anchor the bar.

The advantages of mechanical anchors are that installation is rapid and tensioning can be carried out as soon as the anchor has been set. Grouting can then be done using a grout tube attached to the bar or through the center hole in the case of the Williams anchor. The disadvantages of mechanical anchors are that they can be used only in medium-to-strong rock in which the anchor will grip, and the maximum working tensile load is about 200 kN. Mechanical anchors for permanent installations must always be fully grouted because the wedge will creep and corrode, resulting in loss of support.

3.1.1.3 Cement-Grout Anchors

Cement grout is the most common method of anchoring long-service-life rock bolts because the materials are inexpensive and installation is simple. Cement-grout anchorage can be used in a wide range of rock and soil conditions, and the cement protects the steel from corrosion. Figure 18-8(b) shows a typical rock-bolt installation with cement-grout anchorage and centering sleeves to ensure complete encapsulation of the steel. The grout mix usually consists of nonshrink cement and water at a water:cement ratio in the range of 0.4 to 0.45. This ratio will produce a grout that can be readily pumped down a small-diameter grout tube, yet produce a high-strength, continuous grout column with minimal bleed of water from the mix. Admixtures are sometimes added to the grout to reduce shrinkage and bleeding and to increase the viscosity.

Tensioned rock bolts include a free-stressing length (l_f) and a bond length (l_b) [Figure 18-8(b)], with the full bond length being beneath the potential failure surface. Figure 15-6 shows the method of calculating the factor of safety of a planar failure in which tensioned bolts are installed. When the bolt is at a flatter angle than the normal to the failure surface, the bolt tension increases the normal force and decreases the shear-displacing force on the surface. The design of rock anchors to sustain the necessary tension load requires the selection of an appropriate hole and bar diameter and free-stressing and anchor lengths with respect to the geological conditions.

For cement-grout-anchored bolts, the stress distribution along the bond length is highly nonuniform; the highest stresses are concentrated in the proximal end of the anchor and ideally the distal end of the anchor is unstressed (Farmer 1975; Aydan 1989). However, it is found that the required length of the bond zone can be calculated with the simplifying assumption that the shear stress at the rock-grout interface is uniformly distributed along the anchor and is given as follows:

$$\tau_a = \frac{T}{\pi d_h l_b} \quad \text{or} \quad l_b = \frac{T}{\pi d_h l_b} \qquad (18.2)$$

where

T = design tension force,
d_h = hole diameter,
τ_a = allowable bond stress, and
l_b = bond length.

Values of τ_a can be estimated from the uniaxial compressive strength (σ_u) of the rock in the anchor zone according to the following relationship (Littlejohn and Bruce 1975):

$$\tau_a = \frac{\sigma_u}{30} \qquad (18.3)$$

Approximate ranges of allowable bond stress (τ_a) related to rock strength and rock type are presented in Table 18-3.

The diameter of the drill hole is partially determined by the available drilling equipment but must also meet certain design requirements. The hole diameter should be large enough to allow the anchor to be inserted without driving or hammering and be fully embedded in a continuous column of grout. A hole diameter significantly larger than the anchor will not materially improve the design and will result in unnecessary drilling costs and possibly excessive grout shrinkage. A guideline for a suitable ratio between the diameter of the hole (d_h) and the diameter of the anchor (d_a) is

$$0.4 \leq \frac{d_a}{d_h} \leq 0.6 \qquad (18.4)$$

The working shear strength at the steel-grout interface of a grouted deformed bar is usually greater than the working strength at the rock-grout interface. For this reason, the required anchor length is typically determined from the stress level developed at the rock-grout interface.

Table 18-3
Allowable Bond Stresses in Cement-Grout Anchors (Wyllie 1991)

ROCK STRENGTH AND TYPE	ALLOWABLE BOND STRESS (MPA)	COMPRESSIVE STRENGTH RANGE (MPA)
Strong	1.05–1.40	>100
Medium	0.7–1.05	≈50–100
Weak	0.35–0.7	≈20–50
Granite, basalt	0.55–1.0	
Dolomitic limestone	0.45–0.70	
Soft limestone	0.35–0.50	
Slates, strong shales	0.30–0.45	
Weak shales	0.05–0.30	
Sandstone	0.30–0.60	
Concrete	0.45–0.90	

3.1.2 Tensioning

When tensioned rock anchors are to be installed, it is important that a procedure be carried out to check that the full design load is applied at the required depth and that there will be no loss of load with time. A suitable testing procedure has been drawn up by the Post Tensioning Institute (1985) that includes the following four types of test: performance, proof, creep, and lift-off.

The performance and proof tests consist of a cyclic testing sequence to a maximum load of 150 percent of the design load, in which the deflection of the head of the anchor is measured as the anchor is tensioned. The purpose of these tests is to check that the anchor can sustain a load greater than the design load and that the load in the anchor is transmitted into the rock at the location of the potential failure surface. The creep test checks that there will be no significant loss of load with time, and the lift-off test checks that the tension applied during the testing sequence has been permanently transferred to the anchor. The Post Tensioning Institute (1985) provides acceptance criteria for each of the four tests, and it is necessary that each anchor meet all the acceptance criteria.

The usual method of tensioning rock bolts is to use a hollow-core hydraulic jack that allows the applied load to be precisely measured as well as cycles the load and holds it constant for the creep test. It is important that the hydraulic jack be calibrated before each project to check that the indicated load is accurate. The deflection of the anchor head is usually measured with a dial gauge to an accuracy of about 0.05 mm; the dial gauge is mounted on a reference point that is independent of the anchor. Figure 18-9 shows a typical test arrangement for a tensioned cable anchor including a hydraulic jack and a dial gauge set up on a tripod.

3.1.3 Corrosion Protection

Corrosion protection is provided for almost all permanent anchors to ensure their longevity. Even if anchors are not subject to corrosion at the time of installation, conditions may change, which must be accounted for in design. The following is a list of conditions that will usually create a corrosive environment for steel anchors (Hanna 1982):

- Soils and rocks that contain chlorides,
- Seasonal changes in the groundwater table,
- Marine environments where sea water contains chlorides and sulfates,
- Fully saturated clays with high sulfate contents,
- Passage through ground types that possess different chemical characteristics, and
- Stray direct electrical current that develops galvanic action between the steel and the surrounding rock.

The corrosive environments described above can be quantified in terms of the pH value and the resistivity of the site. In highly acidic ground (pH < 4), corrosion by pitting is likely, whereas in slightly alkaline ground (pH just greater than 7), sulfate-reducing bacteria flourish, producing a corrosive environment. Corrosion potential is also related to the soil resistivity by the magnitude of current that can flow between the steel and the soil. In general, the degree of corrosiveness decreases as follows (King 1977):

FIGURE 18-9
Typical arrangement for measuring applied tension and deflection of head of multistrand anchor during performance and proof tests of grouted anchor.

Organic soil > clays > silts > sands > gravels

A number of rock-anchor manufacturers have proprietary corrosion protection systems, all of which meet the following requirements for long-term reliability:

- There will be no breakdown, cracking, or dissolution of the protection system during the service life of the anchor;
- The protection system can be fabricated either in the plant or on the site in such a manner that the quality of the system can be verified;
- The anchor can be installed and stressed without damage to the protection system; and
- The materials used in the protection system will be inert with respect to both the steel anchor and the surrounding environment.

Methods of protecting steel against corrosion include galvanizing, applying an epoxy coating, and encapsulating the steel in cement grout. Cement grout is commonly used for corrosion protection, primarily because it creates a high pH environment that protects the steel by forming a surface layer of hydrous ferrous oxide. In addition, cement grout is inexpensive and simple to install and has sufficient strength for most applications and a long service life. Because of the brittle nature of grout and its tendency to crack, particularly when loaded in tension or bending, the protection system is usually composed of a combination of grout and a plastic [high-density polyethylene (HDPE)] sleeve. In this way the grout produces the high pH environment around the steel, and the plastic sleeve provides protection against cracking. In order to minimize the formation of shrinkage cracks that reduce corrosion resistance of the grout, nonshrink grouts are usually used for all components of the installation. Figure 18-10 shows an example of a three-layer corrosion protection system for a rock anchor. The strands are encapsulated in a grout-filled HDPE sheath, and the outer annular space between the sheath and the rock is filled with a second grout layer.

3.2 Dowels

Loosening and failure of small blocks of rock on the slope face can be prevented by the installation of passive dowels at the toe of the block. Dowels are composed of lengths of reinforcing steel

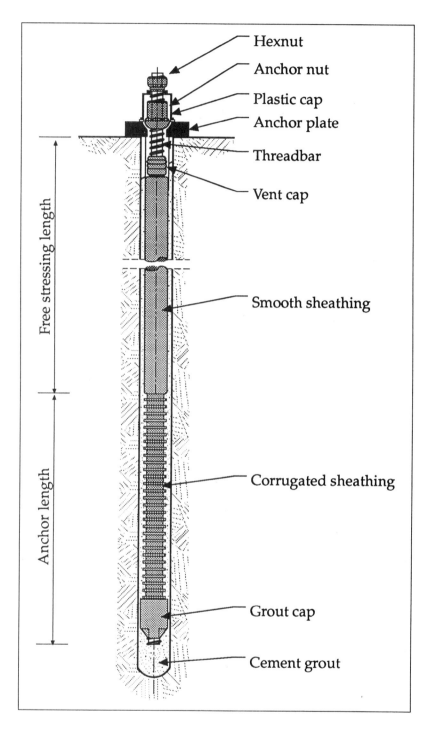

Hexnut
Anchor nut
Plastic cap
Anchor plate
Threadbar
Vent cap

Free stressing length

Smooth sheathing

Anchor length

Corrugated sheathing

Grout cap

Cement grout

grouted into holes drilled in the underlying, stable rock, with a cap of reinforced concrete encasing the exposed steel (Figure 18-5, Method 1). It is important that the concrete be in intimate contact with the rock that it is supporting so that movement and loss of interlock on the potential rupture surface are minimized. The reinforcing steel dowels used to anchor the concrete to the rock are usually about 25 mm in diameter, embed-

FIGURE 18-10
Corrosion protection system for rock anchor.
DYWIDAG SYSTEMS INTERNATIONAL

ded about 0.5 m into sound rock, and spaced about 0.5 to 0.8 m apart.

Because dowels provide only passive shear resistance to sliding, they are used to support slabs of rock with thicknesses up to 1 to 2 m. Dowels are most effective when there has been no prior movement of the rock so that there is interlock on the potential sliding surface. For slabs thicker than 1 to 2 m or where there has been movement, the required support may be provided more reliably by tensioned anchors.

3.3 Tieback Walls

Method 3 in Figure 18-5 is used where there is potential for a sliding failure in closely fractured rock. Tensioned rock bolts are required to support this portion of the slope, but the fractured rock may degrade and ravel from under the reaction plates of the anchors; thus, eventually the tension in the bolts will be lost. In these circumstances, a reinforced concrete wall can be constructed to cover the area of fractured rock, and then the holes for the rock anchors can be drilled through sleeves in the wall. Finally, the anchors are installed and tensioned against the face of the wall. The wall acts as both a protection against raveling of the rock and a large reaction plate for the rock anchors.

Since the purpose of the concrete wall is to distribute the anchor loads into the rock, the reinforcement for the tieback wall should be designed so that there is no cracking of the concrete from the concentrated loads of the anchor heads or from bending between anchors. It is also important that there be drain holes through the concrete to prevent buildup of water pressure behind the wall.

3.4 Shotcrete

Zones or beds of closely fractured or degradable rock can be protected by applying a layer of shotcrete to the rock face (Figure 18-5, Method 4). The shotcrete controls both the fall of small blocks of rock and progressive raveling that will produce large, unstable overhangs on the face. However, shotcrete provides little support against sliding of the overall slope; its primary function is surface protection. Shotcrete is a pneumatically applied, fine-aggregate mortar (less than 13 mm aggregate size) that is usually placed in a 75- to 100-mm layer (American Concrete Institute (1983).

The effectiveness of shotcrete depends to a large degree on the condition of the rock surface to which it is applied. The surface should be free of loose and broken rock, soil, vegetation, and ice and should be damp to improve the adhesion between the rock and the shotcrete. It is important that drain holes be drilled through the shotcrete to prevent buildup of water pressure behind the face; the drain holes are usually about 0.5 m deep and are located on 1- to 2-m centers. In massive rock it is important that the drain holes be drilled before the shotcrete is applied and that they be located so as to intersect fractures that carry water. The holes are temporarily plugged with wooden pegs or rags while the shotcrete is applied.

For all permanent applications, shotcrete should be reinforced to reduce the risk of cracking and spalling. The two common methods of reinforcement are welded-wire mesh and steel fibers. Welded-wire mesh is fabricated from light-gauge (3.5-mm-diameter) wire on 100-mm centers and is attached to the rock face on about 1- to 2-m centers with steel pins, complete with washers and nuts, grouted into the rock face. The mesh must be closely attached to the rock surface and fully encased in shotcrete, with care being taken to eliminate voids within the shotcrete. On irregular rock faces it can be difficult to attach the mesh closely to the rock surface. In these circumstances the mesh can be installed between two layers of shotcrete, with the first layer creating a smoother surface to which the mesh can be more readily attached.

An alternative to mesh reinforcement is to use steel fibers that are a component of the shotcrete mix and form a reinforcement mat throughout the shotcrete layer (Wood 1989). The fibers are manufactured from high-strength carbon steel with a length of 30 to 38 mm and diameter of 0.5 mm. To resist pullout, the fibers have deformed ends or are crimped. The principal function of steel fibers is to significantly increase the shear, tensile, and postcrack strength of the shotcrete compared with unreinforced shotcrete. These loading conditions tend to develop when the fractured rock loosens behind the face.

3.5 Buttresses

When a rock fall has occurred that forms a cavity in the slope face, it may be necessary to construct a concrete buttress in the cavity to prevent further falls (Figure 18-5, Method 6). The buttress serves two functions: to retain and protect areas of weak rock and to support the overhang. Buttresses should be designed so that the thrust from the rock to be supported loads the buttress in compression. In this way bending moments and overturning forces are eliminated, and there is no need for heavy reinforcement of the concrete or for tiebacks anchored in the rock.

For the buttress to prevent relaxation of the rock, it should be founded on a clean, sound rock surface. If this surface is not at right angles to the direction of thrust, it should be anchored to the base using steel pins to prevent sliding. Also, the top should be poured so that it is in contact with the underside of the overhang. In order to meet this second requirement, it may be necessary to place the last pour through a hole drilled downward into the cavity from the rock face and to use a nonshrink agent in the mix.

3.6 Drainage

As shown in Table 18-1, groundwater in rock slopes is often a contributory cause of instability. The usual method of reducing water pressures is to drill drain holes at the toe of the slope to create a series of outlets for the water.

The most important factor in the design of drain holes for rock slopes is to locate them so that they intersect the fractures that are carrying the water; little water is contained in the intact rock. For the conditions shown in Figure 18-5, the drain holes are drilled at a shallow angle to intersect the more continuous fractures that dip out of the face. If the holes were drilled at a steeper angle parallel to these fractures, the drainage would be less effective because the holes would only intersect the steeply dipping, less continuous fractures.

There are no formulas from which to calculate the required spacing of drill holes because it is more important that the holes be located to suit the geological conditions at the site. As a guideline, holes are usually drilled on a spacing of about 3 to 10 m to a depth of at least one-third of the slope height. The holes are often lined with a perforated casing, with the perforations sized to minimize infiltration of fines that are washed from fracture infillings. Another important aspect of the design of drain holes is the disposal of the seepage water. If this water is allowed to infiltrate the toe of the slope, it

may result in degradation of low-strength materials or produce additional stability problems downstream of the drains. Depending on site conditions, it may be necessary to collect all the seepage water in a manifold and dispose of it some distance from the slope.

Other methods of reducing water pressures in rock slopes may be appropriate, depending on site conditions. For example, if surface runoff infiltrates open tension cracks, it is often worthwhile to construct diversion ditches behind the crest of the slope and to seal the cracks with clay or plastic sheeting. For large slides it may not be possible to significantly reduce the water pressure in the slope with small drain holes. In these circumstances a drainage tunnel can be driven into the base of the slide from which a series of drain holes is drilled up into the saturated rock. For example, the Downie Slide in British Columbia has an area of about 7 km^2 and a thickness of about 250 m. There was concern regarding the stability of the slope when the toe was flooded by the construction of a dam. A series of drainage tunnels with a total length of 2.5 km was driven at an elevation just above the high-water level of the reservoir. From these tunnels, 13 500 m of drain holes was drilled to reduce groundwater pressures within the slope. These drainage measures have been effective in reducing the water level in the slide by as much as 120 m and in reducing the rate of movement from 10 mm/year to about 2 mm/year (Forster 1986).

An important aspect of slope drainage is to install piezometers to monitor the effect of drainage measures on the water pressure in the slope. For example, one drain hole with a high flow may only be draining a small, permeable zone in the slope and monitoring would show that more holes would be required to lower the water table throughout the slope. Conversely, monitoring may show that in low-permeability rock, a small seepage quantity that evaporates as it reaches the surface may be sufficient to reduce the water pressure and significantly improve stability conditions.

3.7 "Shot-in-Place" Buttresses

On landslides where the surface of rupture is a well-defined geological feature such as a continuous bedding surface, stabilization may be achieved by blasting this surface to produce a "shot-in-place" buttress (Aycock 1981). The friction angle of the broken rock is greater than that of the smooth and planar bedding surface because of the greater effective roughness of the broken rock. If the total friction angle is greater than the dip of the surface of rupture, sliding is prevented. Fracturing of the rock also enhances drainage.

The method of blasting involves drilling a pattern of holes through the sliding surface and placing an explosive charge at this level that is just sufficient to break the rock. This technique requires that the drilling begin while it is still safe for the drills to reach the slope and before the rock becomes too broken for the drills to operate.

FIGURE 18-11
Rock-removal methods for slope stabilization.

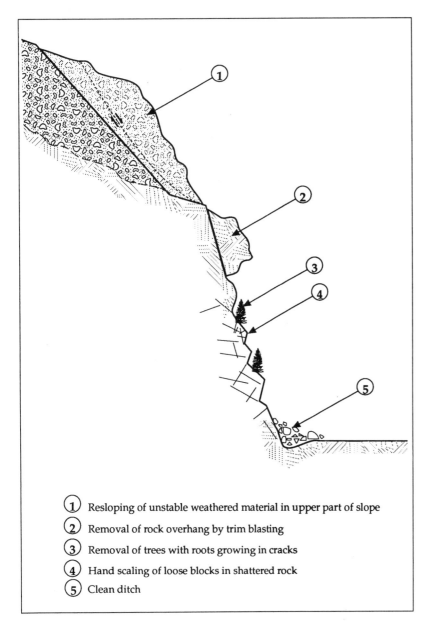

① Resloping of unstable weathered material in the upper part of slope
② Removal of rock overhang by trim blasting
③ Removal of trees with roots growing in cracks
④ Hand scaling of loose blocks in shattered rock
⑤ Clean ditch

4. ROCK REMOVAL

Stabilization of rock slopes can be accomplished by the removal of potentially unstable rock; Figure 18-11 shows typical removal methods including

- Resloping zones of unstable rock,
- Trim blasting overhangs, and
- Scaling individual blocks of rock.

Methods of rock removal and circumstances in which removal should and should not be used are described in this section. In general, rock removal is a preferred method of stabilization because the work eliminates the hazard and no future maintenance is required. However, removal should only be used where it is certain that the new face will be stable and where construction on active transportation routes can be carried out efficiently and safely.

4.1 Resloping and Unloading

When there is overburden or weathered rock in the upper portion of a cut, it is often necessary to cut this material at an angle flatter than that of the more competent rock below (Figure 18-11, Method 1). Careful attention should be given to investigating this condition because the thickness and properties of the overburden or weathered rock can vary considerably over short distances. Contract disputes may arise if there are major design changes during construction because of a variation in excavation quantities.

Another condition that should be considered during design is the rock weathering that takes place after the cut has been excavated. An initially stable, steep slope may become unstable some years after construction, at which time resloping may be difficult to carry out. Where a slide has developed, it may be necessary to unload the crest of the cut to reduce its height and diminish the driving force.

Resloping and unloading are usually carried out by equipment such as excavators and bulldozers. Consequently, the cut width must be designed to accommodate suitable equipment on the slope with no danger of collapse of the weak material; this width would usually be at least 6 m. Safety requirements for equipment access usually preclude the excavation of "sliver" cuts in which the toe of the new cut coincides with that of the old cut.

The design procedure for resloping and unloading starts with back analysis of the unstable slope. By setting the factor of safety of the unstable slope to 1.0, it is possible to calculate the rock-mass strength parameters (see Chapter 14). This information can then be used to calculate the required reduced slope angle or height, or both, that will produce the required factor of safety. For most slopes on highway and railway projects, an acceptable factor of safety for long-term stability is about 1.5.

4.2 Trimming

Failure of a portion of a rock slope may form an overhang on the face (Figure 18-11, Method 2), which may be a hazard to traffic if it also were to fail. The following is a brief discussion of methods of controlled blasting that can be used to excavate rock to narrow tolerances without damaging the rock behind the face.

Figure 18-12 shows the difference in stability conditions between heavily blasted rock (upper face) and controlled blasting in which there are

no blast-induced fractures behind the face (lower face). Controlled blasting involves drilling a series of closely spaced, parallel holes on the required break line to evenly distribute the explosive on the face. The spacing between holes on the final line is generally in the range of 10 to 12 times the drill-hole diameter. Hole diameters on most stabilization projects range from about 40 mm for hand-held pneumatic drills to 80 mm for light, track-mounted drills. It is also important to limit the hole length to about 3 m for hand-held drills and to about 9 m for track drills, so that hole deviation is not excessive. However, rock with a pronounced structure can also cause excessive hole deviation even in short holes.

In controlled blasting a low-velocity explosive with a diameter about one-half that of the hole is used. The air gap around the explosive provides a cushion that minimizes shattering of the rock around the hole but generates sufficient energy to fracture the rock between the final line holes. The ratio between the hole diameter and the explosive diameter is termed the *decoupling ratio*. At a decoupling ratio of 2, the pressure generated in the sides of the borehole is about an order of magnitude less than that when the explosive is packed in the hole. As a guideline the explosive load in the final line holes should be in the range of 0.1 to 0.5 kg/m^2 of face area (Langefors and Kihlstrom 1967; Hemphill 1981; Federal Highway Administration 1985). It is normal practice to stem the holes to minimize venting of the explosive gases.

It is usual in controlled blasting to detonate each hole on a single delay, with the detonation sequence starting at a free face and progressing back to the final line. With a delay interval between holes of about 25 msec, there will be little risk of sympathetic detonation between holes and sufficient time for the broken rock to displace before the subsequent detonation. This methodology helps to limit the shock energy that is transmitted to the rock behind the face.

An alternative detonation sequence is to detonate the final line first in the sequence, sometimes on a single delay. This procedure is termed *preshear blasting*. Preshearing should only be used when the total burden is at least two to three times the hole depth to ensure that there is adequate weight of rock to confine the explosive energy. Instances of displacement of the entire burden by preshear blasting have been recorded. Also, preshearing is

FIGURE 18-12 Comparison of stability conditions of blasted rock face: *upper face,* heavy fracturing and overbreak; *lower face,* stable, presheared face.

generally not used in closely fractured rock because there is little relief for the explosive gases, which can cause damage to the rock behind the final line. Furthermore, ground vibrations produced by the instantaneous detonation of the preshear blasting can be greater than those produced by controlled blasting with a single hole per delay.

4.3 Scaling

Scaling describes the removal of loose rock, soil, and vegetation on the face of a slope using hand tools such as scaling bars, shovels, and chain saws. On steep slopes workers are usually supported by ropes anchored at the crest of the slope and tied to belts and carabiners around their waists (Figure 18-13). The most appropriate type of rope for these conditions is made of steel-core hemp that is highly resistant to cuts and abrasion. The scalers work their way down the face to ensure that there is no loose rock above them.

A staging suspended from a crane is an alternative to using ropes for the scalers to reach the face.

FIGURE 18-13
High scaler
suspended on rope
and belt while
removing loose rock
on steep rock slope.

The crane is located on the road at the toe of the slope if there is no access to the crest of the slope. The disadvantage of using a crane rather than ropes is the expense and the fact that with the outriggers extended, it can occupy several lanes of the road with consequent disruption to traffic. Also, scaling from a staging suspended from a crane can be less safe than using ropes because the scalers are not able to direct the crane operator to move quickly in the event of a rock fall from the face above them.

An important component of a scaling operation in wet climates is the removal of all trees and vegetation growing on the face and to a distance of several meters behind the crest of the slope. Tree roots growing in fractures on the rock face can force open the fractures and eventually cause rock falls. Also, movement of the trees by the wind produces leverage by the roots on loose blocks. The general loosening of the rock on the face by tree roots also permits increased infiltration of water, which in temperate climates will freeze and expand and cause further opening of the cracks. As shown in Table 18-1, approximately 0.6 percent of the rock falls on the California highway system can be attributed to root growth.

4.4 Rock-Removal Operations

When rock-removal operations are carried out above active highways or railroads or in urban areas, particular care must be taken to prevent injury or damage from falling rock. Usually all traffic should be stopped while rock removal is in progress and until the slope has been made safe and the road cleared of debris. When pipelines or cables are buried at the base of the slope, it may be necessary to protect them, as well as pavement surfaces or rail track, from the impact of falling rock. Adequate protection can usually be provided by placing a cover of sand and gravel to a depth of about 1.5 to 2 m. For particularly sensitive structures, additional protection can be provided by a mat of tires or timber buried in sand.

During blasting it often is necessary to take precautions to control rock fragments thrown into the air by the explosions. These fragments, called *flyrock*, may damage buildings and utilities such as overhead transmission lines. Flyrock can be controlled by means of appropriate stemming lengths, burden distances, and detonation sequences and by the use of blasting mats. Blasting mats are fabricated from rubber tires or conveyor belts chained or wired together.

The advantage of removing unstable rock in comparison with stabilization by the installation of tensioned rock anchors, for example, is that removal can be a permanent stabilization measure. However, removal of loose rock on the face of a slope will be an effective stabilization measure only if there is no risk of undermining the upper part of the slope and if the rock forming the new face is sound. Method 4 in Figure 18-11 is an example of a case in which rock removal should be carried out with care. It would be safe to remove the outermost loose rock provided that the fracturing was caused by blasting and extended only to shallow depth. However, if the rock mass is deeply fractured, continued scaling will soon develop a cavity that will undermine the upper part of the slope.

Removal of loose rock on the face of a slope is not effective when the rock is highly degradable, such as shale. In these circumstances exposure of a new face will start a new cycle of weathering and instability. For this condition, a more appropriate stabilization method would be protection of the face with shotcrete and rock bolts or with a tieback wall. Design of the protection measure should consider its longevity compared with the design life of the facility and the stability of the slope, because shotcrete alone will not provide support against sliding.

5. PROTECTION MEASURES AGAINST ROCK FALLS

An effective method of minimizing the hazard of rock falls is to let the falls occur and to control their distance and direction of travel. Methods of rock-fall control and protection of facilities at the toe of the slope include catchment ditches and barriers, wire mesh fences, mesh hung on the face of the slope, and rock sheds. A common feature of all these protection structures is their energy-absorbing characteristics, which either stop the rock fall over some distance or deflect it away from the facility that is being protected. As described in this section, it is possible, by the use of appropriate techniques, to control rocks with diameters of as much as 2 to 3 m falling from heights of several hundred meters and striking with energies as high as 1 MJ. Rigid structures, such as reinforced concrete walls or fences with stiff attachments to fixed supports, are rarely appropriate for stopping rock falls.

5.1 Rock-Fall Modeling

Selection and design of effective protection measures require the ability to predict rock-fall behavior. An early study of rock falls was made by Ritchie (1963), who drew up empirical ditch design charts related to the slope dimensions, as described in Section 5.3. In the 1980s, the prediction of rock-fall behavior was enhanced by the development of a number of computer programs that simulate the behavior of rock falls as they roll and bounce down slope faces (Piteau 1980; Wu 1984; Descoeudres and Zimmerman 1987; Spang 1987; Hungr and Evans 1988; Pfeiffer and Bowen 1989; Pfeiffer et al. 1990).

Figure 18-14 shows an example of the output from the Colorado Rockfall Simulation Program (CRSP). The larger plot shows the path of a single falling rock, which lands and comes to rest on the outer slope of the ditch. The inset shows the distribution for 100 rock falls with bounce heights at each 1.7-m horizontal interval measured from the top of the slope. Similar histograms can be obtained for the bounce height and rock velocity at the analysis point (in the example, this point is located at a 21.6-m horizontal distance from the top of the slope). The input for the program includes definitions of the slope and ditch geometry, irregularity (roughness) of the face, attenuation characteristics of the slope materials, and the size and shape of the block. The degree of variation in the shape of the ground surface is modeled by randomly varying the surface roughness in relation to the block size between fixed limits for each of a large number of runs.

The results of analyses such as those shown in Figure 18-15, together with geological data on block sizes and shapes, can be used to estimate the dimensions of a ditch or its optimum position and the required height and strength of a fence or barrier. In some cases it may also be necessary to verify the design by constructing a test structure. In the following sections, types of ditches, fences, and barriers and the conditions in which they can be used are described.

5.2 Benched Slopes

Excavation of intermediate benches on rock cuts usually increases the rock-fall hazard. Benches can

FIGURE 18-14
Example of analysis
of rock-fall behavior
using Colorado
Rockfall Simulation
Program (CRSP) with
analysis point at
horizontal distance
of 21.6 m from crest
(Pfeiffer and Bowen
1989).

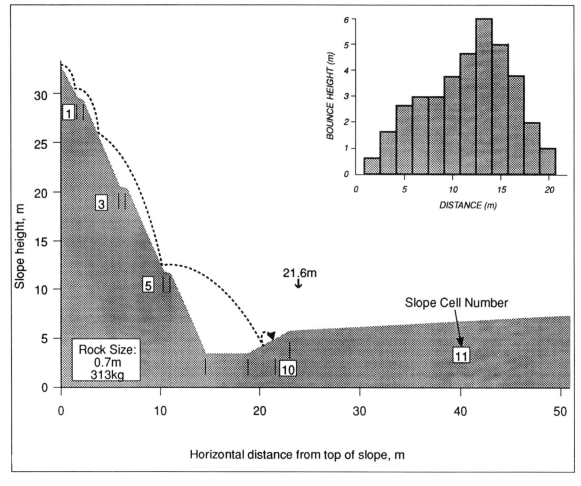

be a hazard when the crests of the benches fail because of blast damage and the failed benches leave irregular protrusions on the face. Rock falls striking protrusions tend to bounce away from the face and land a considerable distance from the toe. When the narrow benches fill with debris, they will not be effective in catching rock falls. It is rarely possible to remove this debris because of the hazardous conditions of working on narrow, discontinuous benches.

5.3 Ditches

A catch ditch at the toe of the slope is often a cost-effective means of stopping rock falls when there is adequate space for it. The required dimensions of the ditch, as defined by the width of the base and the depth, are related to the height and face angle of the slope (Richie 1963). A ditch design chart developed from field tests shows the effect of the slope angle on the path that rock falls tend to follow and how this influences ditch de-

sign (Figure 18-15). For slope angles steeper than about 75 degrees, the rocks tend to stay close to the face and land near the toe of of the slope. For slope angles between about 55 and 75 degrees, rocks tend to bounce and spin, with the result that they can land a considerable distance from the toe and a wide ditch is required. For slope angles between about 40 and 55 degrees, rocks will tend to roll down the face and into the ditch, and a steep outer face is required to prevent them from rolling out. Where the slope is irregular witih protrusions on the face, the ditch dimensions should be increased from those shown on the chart to account for less predictable bounces of falling rock, or the slope can be modeled as shown in Figure 18-14 to determine required ditch dimensions.

Recent updates of Ritchie's ditch design chart incorporate highway geometric requirements as well as cut dimensions (Washington State Department of Transportation 1986), and the Oregon Department of Transportation has carried out an extensive series of tests of ditch require-

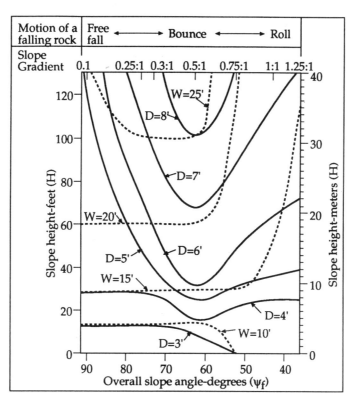

FIGURE 18-15
Design chart to
determine required
width and depth of
rock catch ditches
in relation to height
and face angle of
slope (Ritchie 1963).

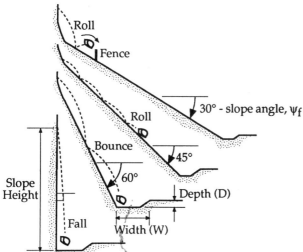

ments for rock falls on 1/4:1 cut faces (Pierson et al. 1994).

5.4 Barriers

A variety of barriers can be constructed to either enhance the performance of excavated ditches or form catchment zones at the toes of slopes. The required type of barrier and its dimensions depend on the energy of the falling blocks, the slope di-mensions, and the availability of construction materials. Commonly used examples are gabions and concrete blocks or geofabric-and-soil barriers.

The function of a barrier is to form a ditch with a vertical outer face that traps rolling rock. Barriers are particularly useful at the toes of flatter slopes where falling rock rolls and spins down the face but does not bounce significantly. Such rocks may land in a ditch at the toe of the slope but can roll up the sloping outer side onto the road; a vertical barrier helps to trap such falls.

5.4.1 Gabions and Concrete Blocks

Gabions are rock-filled baskets, typically measuring 1 m by 1 m in cross section, that are often constructed on site with local waste rock (Threadgold and McNichol 1985). Advantages of gabions are the ease of construction on steep hillsides and irregular foundations and their capacity to sustain considerable impact from falling rock. However, gabions are not immune to damage by impacts of rock and maintenance equipment, and repair costs can become significant. Concrete barriers are also used extensively on transportation systems for rock-fall containment because they are often readily available and can be placed quickly. However, concrete blocks are somewhat less resilient than gabions.

Gabions and concrete blocks or concrete Jersey barriers form effective protection from falling rock with diameters up to about 0.75 m. Figure 18-16 shows an example of a ditch with two layers of gabions along the outer edge forming a 1.5-m-high barrier (Wyllie and Wood 1981).

5.4.2 Geofabric-and-Soil Barriers

Various barriers have been constructed using geofabric and soil layers, each about 0.6 m thick, built up to form a barrier that may be as high as 4 m. By wrapping the fabric around each layer, it is possible to construct a barrier with vertical front and back faces; the impacted face can be strengthened with such materials as railway ties, gabions, and

rubber tires (Figure 18-17). The capacity of a barrier of this type to stop rock falls depends on its mass in relation to the impact energy and on its capacity to deform without failing. The deformation may be both elastic deformation of the barrier components and shear displacement at the fabric layers or on the base.

Extensive testing of prototype barriers by the Colorado Department of Transportation has shown that limited shear displacement occurs at the fabric layers and that they can withstand high-energy impacts without significant damage (Barrett and White 1991). For example, a 1.8-m-wide geofabric barrier stopped rock impacts delivering 950 kJ of energy.

5.5 Rock Catch Fences and Attenuators

During the 1980s various fences and nets suitable for installation on steep rock faces, in ditches, and on talus runout zones were developed and thoroughly tested (Smith and Duffy 1990; Barrett and White 1991; Hearn 1991; Kane and Duffy 1993). A design suitable for a particular site depends on the topography, anticipated impact loads, bounce height, and local availability of materials. A common feature of these designs is that with the absence of any rigid components, they exhibit energy-absorbing characteristics. When a rock collides with a net, there is deformation of the mesh, which then engages energy-absorbing components over an extended time of collision. This significantly increases the capacity of these components to stop rolling rock and allows the use of lighter, lower-cost elements in construction. The total impact energy is the sum of the kinetic and rotational energies; the rotation of the block is also significant in determining the amount of damage at the point of impact.

5.5.1 Woven Wire-Rope Nets

Nets with high-energy-absorption capacity constructed with woven wire-rope mesh or ring mesh (submarine netting) have been developed. They can be located either in the ditch beside the road or railroad or on the steep slopes above it. The components of these nets are a series of steel I-beam posts on about 6-m centers anchored to the foundation with grouted bolts. Additional support for the posts can be provided by up-slope

FIGURE 18-16
Rock catch ditch with 1.5-m-high gabion wall along outer edge (Wyllie and Wood 1981).

anchor ropes incorporating friction brakes that are activated during high-energy events to help dissipate the impact forces (Figure 18-18).

The mesh is a two-layer system composed of a 50-mm chain-link mesh and a woven wire-rope net constructed typically with 8-mm-diameter wire rope in a diagonal pattern on 100- to 200-mm centers. The wire rope and net dimensions will vary with expected impact energies and block sizes. An important feature of the wire-rope net is the method of fixing the intersection points of the wire rope with high-strength crimped fasteners. The mesh is attached to the posts by lacing it on a continuous-perimeter wire rope that is attached to brackets at the top and bottom of each post.

An extensive series of full-scale tests conducted on mesh nets has shown that they have the capacity to stop rocks with impact energies as high as 800 kJ without sustaining significant damage (Smith and Duffy 1990; Duffy and Haller 1993). This energy is equivalent to rock weighing 3000 kg traveling at a velocity of 23 m/sec. The tests also showed that rock falls could readily be cleared out from behind the net by unthreading the perimeter rope and lifting the mesh.

5.5.2 Flex-Post Rock-Fall Fence

The Flex-post fence (Hearn 1991) is composed of a series of posts constructed of bundles of seven-wire prestressing strands grouted into lengths of steel pipe (Figure 18-19). A portion of the strand bundle is left ungrouted, forming a flexible spring element that can withstand large rotations of the post without damage. The posts are spaced at about 5 m and are installed by being grouted into drill holes or pits to depths of about 1.0 m. Pairs of diagonal wire-rope stays installed between the tops and bottoms of adjacent posts help to dissipate impact energies over an extended length of the fence. The mesh is standard 50 mm square attached to the posts with intertwined steel wires. When small rocks strike the fence, the energy is absorbed by the mesh and wire stays; for larger rocks, the mesh is stretched taut and the posts bend. The fence then rebounds to its initial vertical position, sometimes throwing the rock back into the ditch.

The Flex-post fence is a low-cost, low-energy rock-fall barrier that should be designed preferably so that the highest impact energy does not result in

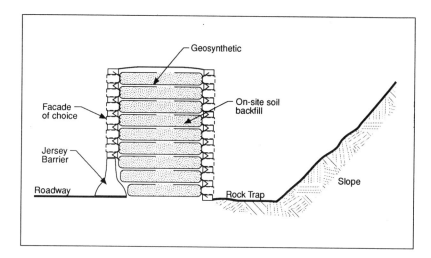

a deflection of the posts of more than about 45 degrees. This deflection limit provides protection from rock-fall events in which a number of boulders strike the fence at approximately the same time.

5.5.3 Fences on Talus Slopes

Rock falls down talus slopes tend to roll and stay close to the slope surface. In these conditions a lightweight chain-link fence placed either on the slope or along the outside edge of the ditch will decelerate or catch the rock. The position of the fence depends on the size of the blocks and the height of

FIGURE 18-17
Rock-fall barrier constructed with soil wrapped in geofabric reinforcing strips and timber protective facade (Barrett and White 1991)

FIGURE 18-18
Side view of woven wire-rope rock catch fence (Smith and Duffy 1990).

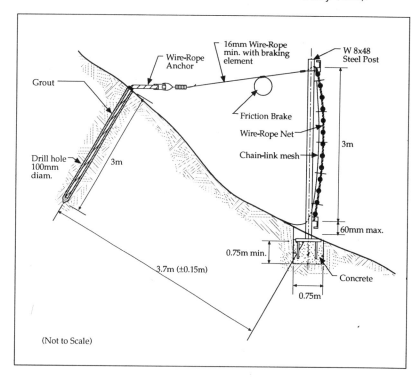

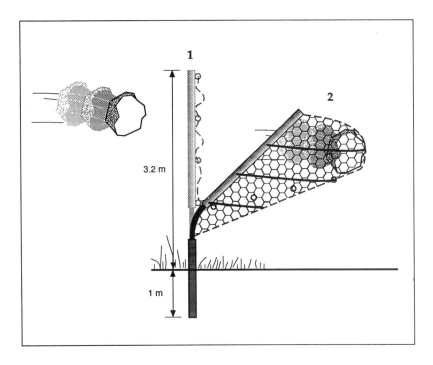

FIGURE 18-19
Colorado Flex-post
rock-fall fence
(Hearn 1991).

the fall. Such fences are used extensively in the states of Washington and Oregon (Lowell 1987).

5.5.4 Rock-Fall Attenuators

When rocks fall down a narrow gully or chute bounded by stable rock walls, it is possible to install a variety of relatively lightweight fences that slow down and absorb the rock-fall energy. The general method of construction is to install a pair of anchors to support a wire rope from which the fence, spanning the gully, is suspended. For rock falls with fragment dimensions up to about 200 mm, it is possible to use chain-link mesh draped down the chute from the anchor rope; wire-rope mesh can be used for larger blocks. Falling rocks are gradually brought to a halt as they bounce and roll under the mesh.

For blocks as large as 1 m, an attenuator fence utilizing waste automobile tires has proved to be successful in a number of installations in Colorado (Andrew 1992). The fence consists of a wire rope from which a number of steel rods are suspended each holding a stack of tires mounted on rims (Figure 18-20). The stacks of tires are arranged so that they form a continuous barrier across the chute. When the attenuator is struck by a falling rock, the kinetic energy is dissipated by a combination of the compression of the rubber tires and the movement of the stacks of tires, which swing on the suspension cable rope. In

order to improve the aesthetics of the fence, a second fence constructed from hanging timbers can be located just downslope from the tire attenuator (Figure 18-20).

Maintenance requirements and worker safety should be considered in the design of fence systems. A properly designed system should not need frequent repairs if the actual impacts are within the tolerances of design energies. However, clearing out of accumulated rock is necessary for any system. Typically, fixed barriers such as geofabric walls require room behind them for clearing operations. In contrast, woven wire-rope fences do not have this requirement because of their modular design, which allows the nets to be cleared from the front by the removal or lifting of each panel in turn.

5.6 Draped Mesh

Wire mesh hung on the face of a rock slope can be an effective method of containing rock falls close to the face and preventing them from bouncing onto the road (Ciarla 1986). Figure 18-21 shows an example of mesh installed on a near-vertical face. Because the mesh absorbs some of the energy of the falling rock, the required dimensions of the ditch at the toe of the slope are considerably reduced from those shown in Figure 18-15. When there is no ditch, the lower end of the mesh should be no more than about 0.6 m above the toe to prevent rocks from rolling onto the road.

Chain-link mesh is suitable on steep faces for controlling rock falls with dimensions less than about 0.6 m, and woven wire rope may be suitable for rock with dimensions up to about 1 m. For larger blocks, ring nets can be used. For very high installations, chain-link mesh can be reinforced with lengths of wire rope. In all cases the upper edge of the mesh or net should be placed close to the source of the rock fall so that the blocks will have little momentum when they strike the mesh.

The features of a mesh installation are as follows. Anchors are installed along the slope above the anticipated rock-fall source. The anchors can consist of steel loop-eye rock bolts grouted into holes drilled in the rock or concreted into larger-diameter holes excavated in soil. Support cables consisting of an upper-perimeter cable and vertical reinforcement cables are attached to the anchors. The mesh is secured to the cables with lacing wire, hog rings, or other types of specialty fasteners. The mesh is not anchored at the bottom of the slope or at intermediate points. This permits rocks to work their way down to the ditch rather than accumulating behind the mesh; the weight of such accumulations can cause the mesh to fail.

5.7 Warning Fences

Fences and warning signals that are triggered by falling rock are sometimes used to protect railroads and highways (Figure 18-22). The warning fence consists of a series of timber posts and cantilever crossarms that support rows of wires spaced about 0.5 m apart. At least one wire will be broken by rocks rolling or bouncing down the face. The wires are connected to a signal system that displays a red light when a wire is broken. The signal light is located far enough from the rock slope that drivers have time to stop and then to proceed with caution before they reach the rock-fall location.

Warning fences are most applicable on transportation systems where traffic is light enough to accommodate occasional closures of the line. However, the use of warning fences as a protection measure has a number of disadvantages. The signal lights must be located a considerable distance from the slope, and falls may occur after the traffic has passed the light. False alarms can be caused by minor falls of rock or ice, and maintenance costs can be significant.

5.8 Rock Sheds and Tunnels

In areas of extreme rock-fall hazard where stabilization work would be very costly, construction of a rock shed or even relocation of the highway or railroad into tunnels may be justified. Figure 18-23 shows an example of a number of rock sheds constructed to protect a railroad. The sheds were built with timber roofs that are inclined at a slightly shallower angle than the chutes down which the rocks are falling. The function of the roof is to direct the rocks over the track rather than to withstand a direct impact.

Figure 18-24 shows a possible construction procedure for a rock shed where there are steep slopes above and below the highway. In this case the roof will sustain the direct impact of falling rock and must be designed to withstand impact loads. The critical feature of the design is the relationship between the energy of the falling rock and the

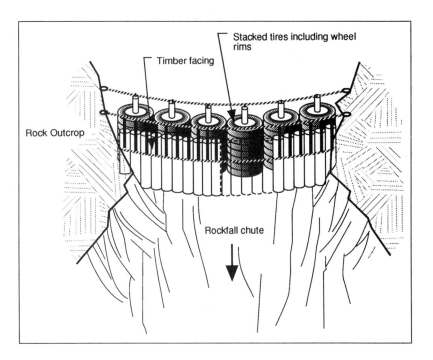

energy-absorbing characteristics of the components of the shed—anchorage of the roof to the cliff face, the roof, and the footings in the slope below the road. A stable anchorage requires that the anchors themselves, and the rock face to which the roof is anchored, withstand both the

FIGURE 18-20
Rock-fall attenuator constructed in chute (Andrew 1992).

FIGURE 18-21
Wire mesh hung on vertical face to direct falling rock into ditch at toe of slope.

FIGURE 18-22
Typical warning
fences at toe of
steep rock slope to
detect rock falls on
railway.

FIGURE 18-23
Rock sheds
constructed with
steeply sloping roofs
that deflect rock
falls over
railway tracks.
CANADIAN NATIONAL
RAILWAY

Alternatively, a lighter-weight crushable element, such as a cellular metal panel, may be used. The columns supporting the lower end of the roof are founded on rock-socketed piers that extend through the fill into sound rock. This type of foundation provides the required support with minimal settlement.

In locations at which it is impractical to construct a rock shed, it may be necessary to drive a tunnel to bypass the hazard zone. For example, a railway in British Columbia drove a 1200-m-long tunnel to avoid a section of track located on a narrow bench between a steep, unstable rock cliff above a 400-m-deep lake. Major rock falls were a hazard to train operations and had caused track closures lasting as long as two weeks (Leighton 1990).

6. CONTRACTING PROCEDURES

Specifications and contract administration for projects involving rock slope stabilization have particular requirements that differ from those for most excavation projects. Some contractual requirements of these projects and how they may be addressed in the specifications are described in the following sections.

6.1 Flexibility

Contracts on most stabilization projects must be flexible with regard to both the scope of the work and the procedures required because these may only be determined as the work progresses. For example, scaling may be specified for an area of loose, broken rock, but the actual scaling time will depend on the extent of the loose rock that becomes evident once scaling begins. Furthermore, the number and length of bolts will be assessed from the dimensions and shear-strength characteristics of the potentially loose blocks remaining on the face after scaling is complete. It is particularly difficult to estimate the scope of the stabilization work when the slope is partially covered with soil and vegetation and when it is too dangerous to reach the slope during the investigation phase of the project.

One form of stabilization in which the scope of the work can be defined with some precision is resloping, in which an existing slope is cut back at a flatter angle. If a series of cross sections is prepared, the excavation volume can be specified.

dead load of the structure and the live load produced by the rock-fall impact. The impact load will include both a normal load and some tensile component. The gravel on the roof acts as an energy absorbent and protection for the concrete, but its thickness must be balanced against the weight that must be supported by the structure.

6.2 Payment Methods

In order to prepare a flexible contract, it is necessary to have a method of payment that allows for possible changes in the quantities and methods of work. This requirement can usually be met by requesting bids on unit prices for items the quantities of which are uncertain. An estimate is given for the expected quantity of an item, and the product of this quantity and the unit bid price forms the basis for comparing bids and making payments. A formula should be included for increasing unit prices if actual quantities are less than those estimated or for decreasing unit prices if quantities are greater than those estimated. This adjustment takes into account the contractor's fixed overhead, which is independent of the quantity of work.

Work items often bid on unit prices include scaling hours for a crew of a specified size, length of rock bolts, volume of shotcrete, and standby time on transportation routes when work has to stop for traffic openings. Work items often bid as lump sums include mobilization and demobilization and concrete volume for buttresses of specified dimensions.

Cost-plus-fixed-fee is an appropriate type of contract for emergency work. In this type of contract the contractor is paid specified costs for labor, equipment, and materials related to the direct costs plus an additional fee. This fee is a profit-plus management fee as reimbursement to the contractor for overhead costs incurred from the performance of the work. Items making up the fee include, but are not limited to, salaries, rent, taxes, and interest on money borrowed to finance the project. For work on emergencies the fee may be negotiated on a percentage of the cost. However, in order to control costs, the scope of the work should be defined as soon as possible and either a fixed fee should be negotiated or the contract should be converted into a lump-sum contract.

6.3 Preselection

When the stabilization contract requires work on steep slopes and specialized skills such as trim blasting and installation of tensioned rock bolts, it is usually advantageous to have the work performed by experienced contractors. Therefore, if it is permitted by procurement regulations, only suitably qualified contractors should be invited to submit bids.

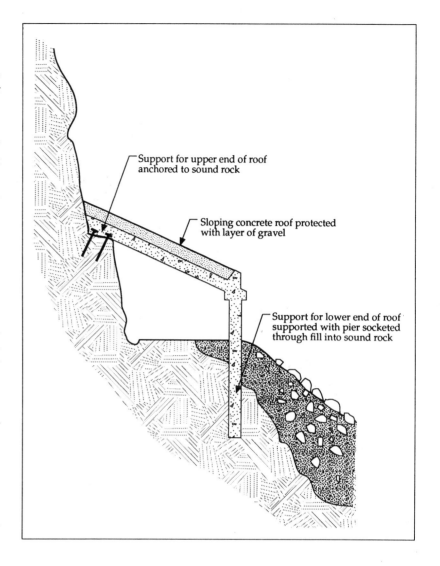

FIGURE 18-24
Rock shed showing possible anchoring and foundation construction methods.

ACKNOWLEDGMENTS

Important contributions to this chapter were made by R. K. Barrett, Colorado Department of Transportation (fences and barriers); J. D. Duffy, California Department of Transportation (fences); S. M. Lowell, Washington Department of Transportation (fences and ditches); and L. A. Pierson, Oregon Department of Transportation (slope inventory systems).

REFERENCES

American Concrete Institute. 1983. *Specifications for Materials, Proportioning and Application of Shotcrete*. ACI Report 506.2-77, revised. Detroit, Mich.

American Society of State Highway and Transportation Officials. 1984. *A Policy on Geometric Design of Highways and Streets*. Washington, D.C., 1087 pp.

Andrew, R. D. 1992. Selection of Rockfall Mitigation Techniques Based on Colorado Rockfall Simulation Program. In *Tranportation Research Record 1343*, TRB, National Research Council, Washington, D.C., pp. 20–22.

Aycock, J. H. 1981. Construction Problems Involving Shale in a Geologically Complex Environment—State Route 32, Appalachian Corridor "S," Grainger County, Tennessee. In *Proc., 32nd Annual Highway Geology Symposium*, Gatlinburg, Tenn., pp. 6–58.

Aydan, O. 1989. *The Stabilization of Rock Engineering Structures by Rock Bolts*. Ph.D. thesis. Department of Geotechnical Engineering, Nagoya University, Japan, 204 pp.

Barrett, R. K., and J. L. White. 1991. Rock Fall Prediction and Control. In *Proc., National Symposium on Highway and Railway Slope Maintenance*, Association of Engineering Geologists, Chicago, Ill., pp. 23–40.

Brawner, C. O., and D. C. Wyllie. 1975. Rock Slope Stability on Railway Projects. In *Proc., American Railway Engineering Association Regional Meeting*, Vancouver, B.C., American Railway Engineering Association, Washington, D.C., 8 pp.

Byerly, D. W., and L. M. Middleton. 1981. Evaluation of the Acid Drainage Potential of Certain Precambrian Rocks in the Blue Ridge Province. In *Proc., 32nd Annual Highway Geology Symposium*, Gatlinburg, Tenn., pp. 174–185.

Ciarla, M. 1986. Wire Netting for Rock Fall Protection. In *Proc., 37th Annual Highway Geology Symposium*, Helena, Mont., Montana Department of Highways, pp. 100–118.

Descoeudres, F., and T. Zimmerman. 1987. Three-Dimensional Calculation of Rock Falls. In *Proc., Sixth International Congress on Rock Mechanics*, Montreal, Canada, International Society for Rock Mechanics, Lisbon, Portugal, pp. 337–342.

Duffy, J. D., and B. Haller. 1993. Field Tests of Flexible Rockfall Barriers. In *Proc., International Conference on Transportation Facilities through Difficult Terrain*, Aspen, Colo., A.A. Balkema, Rotterdam, Netherlands, pp. 465–473.

Farmer, I. W. 1975. Stress Distribution Along a Resin Grouted Anchor. *International Journal of Rock Mechanics & Geomechanics Abstracts*, Vol. 12, pp. 347–351.

Federal Highway Administration. 1985. *Rock Blasting*. U.S. Department of Transportation, 355 pp.

Federal Highway Administration. 1989. *Rock Slopes: Design, Excavation and Stabilization*. Report FHWA-TS-89-045. U.S. Department of Transportation.

Fookes, P. G., and M. Sweeney. 1976. Stabilization and Control of Local Rock Falls and Degrading Rock Slopes. *Quarterly Journal of Engineering Geology*, Vol. 9, pp. 37–55.

Forster, J. W. 1986. Geological Problems Overcome at Revelstoke, Part Two. *Water Power and Dam Construction*, Aug., pp. 42–45.

Groupe d'Etudes des Falaises. 1978. *Eboulements et Chutes de Pierres sur les Chutes: Recensement des Parades*. Rapport de Recherche LPC No. 81. Ministère de l'Environment et du Cadre de Vie Ministère des Transports, Laboratoire Central des Ponts et Chaussées, Paris.

Hanna, T. H. 1982. *Foundations in Tension—Ground Anchors*. Trans Tech Publications/McGraw-Hill Book Company, Clausthal-Zellerfeld, West Germany, 573 pp.

Hearn, G. 1991. *CDOT Flex-post Rockfall Fence*. Report CDOH-R-UCB-91-6. Colorado Department of Transportation, Denver, 105 pp.

Hemphill, G. B. 1981. *Blasting Operations*. McGraw-Hill, New York, 258 pp.

Hoek, E. 1983. Strength of Jointed Rock Masses. *Geotechnique*, Vol. 33, No. 3, pp. 187–223.

Hoek, E., and J. Bray. 1981. *Rock Slope Engineering*, 2nd ed. Institute of Mining and Metallurgy, London, U.K., 402 pp.

Hungr, O., and S. G. Evans. 1988. Engineering Evaluation of Fragmental Rock Fall Hazards. In *Proc., Fifth International Symposium on Landslides*, Lausanne, A.A. Balkema, Rotterdam, Netherlands, pp. 685–690.

Kane, W. F., and J. D. Duffy. 1993. *Brugg Low Energy Flexible Rock Fall Barrier Tests*. Technical Research Report 93-01. University of the Pacific, Stockton, Calif., 20 pp.

King, R. A. 1977. *A Review of Soil Corrosiveness with Particular Reference to Reinforced Earth*. Supplementary Report 316. U.K. Transport and Road Research Laboratory, Crowthorne, Berkshire, England.

Langefors, U., and B. Kihlstrom. 1967. *The Modern Technique of Rock Blasting*. John Wiley and Sons, New York, 405 pp.

Leighton, J. C. 1990. New Tunnel at Shalath Bluff on BC Rail's Squamish Subdivision. In *Proc., Eighth Annual Meeting of the Tunnelling Association of Canada*, Vancouver, B.C., Oct., Bitech Publishers, Ltd., pp. 255–266.

Littlejohn, G. S., and D. A. Bruce. 1975. Rock Anchors—State of the Art. Part 1, Design. *Ground Engineering*, Vol. 8, No. 4, pp. 41–48.

Lowell, S. M. 1987. Development and Application of Ritchie's Rock Fall Catch Ditch Design. In *Proc., Federal Highway Administration Rockfall Mitigation Seminar*, FHWA Region 10, Portland, Oreg., 15 pp.

McCauley, M. L., B. W. Works, and S. A. Naramore. 1985. *Rockfall Mitigation*. Report FHWA/CA/TL-85/12. FHWA, U.S. Department of Transportation, 147 pp.

Moore, D. P., and A. S. Imrie. 1982. Rock Slope Stabilization at Revelstoke Damsite. In *Transactions, 14th International Congress on Large Dams*, ICOLD, Paris, Vol. 2, pp. 365–385.

Norrish, N. I., and S. M. Lowell. 1988. Aesthetic and Safety Issues for Highway Rock Slope Design. In *Proc., 39th Annual Highway Geology Symposium*, Park City, Utah, 31 pp.

Peckover, F. L. 1975. *Treatment of Rock Falls on Railway Lines*. Bulletin 653. American Railway Engineering Association, Washington, D.C., pp. 471–503.

Pfeiffer, T. J., and T. D. Bowen. 1989. Computer Simulation of Rockfalls. *Bulletin of the Association of Engineering Geology*, Vol. 26, No. 1, pp. 135–146.

Pfeiffer, T. J., J. D. Higgins, and A. K. Turner. 1990. Computer Aided Rock Fall Hazard Analysis. In *Proc., Sixth International Congress of the International Association of Engineering Geology*, Amsterdam, A.A. Balkema, Rotterdam, Netherlands, pp. 93–103.

Pierson, L. A., S. A. Davis, and R. Van Vickle. 1990. *The Rockfall Hazard Rating System: Implementation Manual*. Technical Report FHWA-OR-EG-90-01. FHWA, U.S. Department of Transportation.

Pierson, L. A., S. A. Davis, and T. J. Pfeiffer. 1994. *The Nature of Rock Falls as the Basis for New Fallout Area Design Criteria for 0.25:1 Slopes*. Report FHWA-OR-GT-95-05. FHWA, U.S. Department of Transportation, 31 pp.

Piteau, D. R. 1980. Slope Stability Analysis for Rock Fall Problems: The Computer Rock Fall Model for Simulating Rock Fall Distributions. In *Rock Slope Engineering*, Part D, Reference Manual FHWA-TS-79-208, FHWA, U.S. Department of Transportation, pp. 62–68.

Piteau, D. R., and F. L. Peckover. 1978. Engineering of Rock Slopes. In *Special Report 176: Landslides: Analysis and Control* (R.L. Schuster and R.J. Krizek, eds.), TRB, National Research Council, Washington, D.C., Chap. 9, pp. 192–225.

Post Tensioning Institute. 1985. *Recommendations for Prestressed Rock and Soil Anchors*, 2nd ed. Phoenix, Ariz., 57 pp.

Ritchie, A. M. 1963. Evaluation of Rock Fall and Its Control. In *Highway Research Record 17*, HRB, National Research Council, Washington, D.C., pp. 13–28.

Roberds, W. J. 1991. *Methodology for Optimizing Rock Slope Preventive Maintenance Programs*. Geotechnical Special Publication 27. American Society of Civil Engineers, New York, pp. 634–45.

Schuster, R. L. 1992. Recent Advances in Slope Stabilization. Keynote paper, Session G.3. In *Proc., Sixth International Symposium on Landslides*, Christchurch, New Zealand, A.A. Balkema, Rotterdam, Netherlands.

Smith, D. D., and J. D. Duffy. 1990. *Field Tests and Evaluation of Rockfall Restraining Nets*. Division of Transportation Materials and Research, Engineering Geology Branch, California Department of Transportation, Sacramento.

Spang, R. M. 1987. Protection Against Rock Fall—Stepchild in the Design of Rock Slopes. In *Proc., Sixth International Congress on Rock Mechanics*, Montreal, Canada, International Society for Rock Mechanics, Lisbon, Portugal, pp. 551–557.

Spang, K., and P. Egger. 1990. Action of Fully-Grouted Bolts in Jointed Rock and Factors of Influence. *Rock Mechanics and Rock Engineering*, Vol. 23, pp. 201–229.

Threadgold, L., and D. P. McNichol. 1985. The Design and Construction of Polymer Grid Boulder Barriers to Protect a Large Public Housing Site for the Hong Kong Housing Authority. In *Polymer Grid Reinforcement*, Thomas Telford Press, London, pp. 212–219.

Van Velsor, J. E., and J. L. Walkinshaw. 1991. Accelerated Movement of a Large Coastal Landslide Following the October 17, 1989 Loma Prieta Earthquake in California. In *Transportation Research Record 1343*, TRB, National Research Council, Washington, D.C., pp. 63–71.

Washington State Department of Transportation. 1986. *Roadway Design Manual*, Section 640. Olympia, Wash.

Wood, D. F. 1989. Shotcrete—Steel Fiber Reinforced with Silica Fume Additives. *Tunnelling Association of Canada Newsletter*, April.

Wu, S-S. 1984. Rockfall Evaluation by Computer Simulation. In *Transportation Research Record 1031*, TRB, National Research Council, Washington, D.C., pp. 1–5.

Wyllie, D. C. 1987. Rock Slope Inventory System. In *Proc., Federal Highway Administration Rockfall Mitigation Seminar*, FHWA Region 10, Portland, Oreg., 25 pp.

Wyllie, D. C. 1991. Rock Slope Stabilization and Protection Measures. In *Proc., National Symposium on Highway and Railway Slope Stability*,

Association of Engineering Geologists, Chicago, pp. 41–63.

Wyllie, D. C., and D. F. Wood. 1981. Preventive Rock Blasting Protects Track. *Railway Track and Structures*, Sept., pp. 34–40.

Wyllie, D. C., N. R. McCammon, and W. Brumund. 1980. Planning Slope Stabilization Programs Using Decision Analysis. In *Transportation Research Record 749*, TRB, National Research Council, Washington, D.C., pp. 34–39.

PART 5

SPECIAL CASES AND MATERIALS

Chapter 19

PHILIP C. LAMBE

RESIDUAL SOILS

1. INTRODUCTION

Because residual soils weather from parent bedrock, the soil profile represents a history of the weathering process. Profile classification systems distinguish states of weathering and separate the profile vertically into different zones. The permeability and shear strength gradually change with depth, which controls both the local seepage response to rainfall infiltration and the location of the shear surface. Soil profile thickness and properties depend upon parent bedrock, discontinuities, topography, and climate. Because these factors vary horizontally, the profile can vary significantly over relatively short horizontal distances. In addition, deep profiles form in tropical regions where weathering agents are especially strong and the advanced stages of chemical weathering form cemented soils called laterites. Technical papers often refer to both tropical soils and residual soils. In this chapter general characteristics of residual soils are considered and illustrated with examples drawn from Brazil, Hong Kong, and North Carolina.

Figure 19-1 shows the typical weathering profiles that develop over igneous and metamorphic rocks labeled according to Deere and Patton's (1971) profile classification system, which divides the profile into three zones: *residual soil, weathered rock,* and *unweathered rock.* In addition, Deere and Patton present 12 other weathering-profile classification systems proposed by workers from other regions. For example, Brand (1985) divided Hong Kong soils into six material grades and four profile zones.

In 1990 a working party of the Engineering Group of the Geological Society of London issued an extensive report entitled *Tropical Residual Soils* (Geological Society of London 1990), which proposed the classification of tropical residual soils presented in Figure 19-2. This classification contains a series of grades identified by roman numerals I through VI. Comparison of the Deere and Patton classification (Figure 19-1) with the Geological Society of London working party classification (Figure 19-2) reveals the potential for significant confusion. Both classifications use roman numerals to identify different portions of the weathering profile. However, Deere and Patton use roman numeral I to identify the most intensely weathered material at the top of the profile, whereas the working party uses roman numeral I to define fresh rock at the base of the profile. Obviously the "grades" of the working party and the "zones" of Deere and Patton attempt to define the same phenomena.

In this chapter Deere and Patton's system as defined in Figure 19-1 will be used exclusively, although it is recognized that other systems, such as the classification by the Geological Society of London working party defined in Figure 19-2, may work better in some regions. Papers describing landslides in residual soils often spend more time qualitatively describing the profile and the horizontal variability than in quantitatively reporting mea-

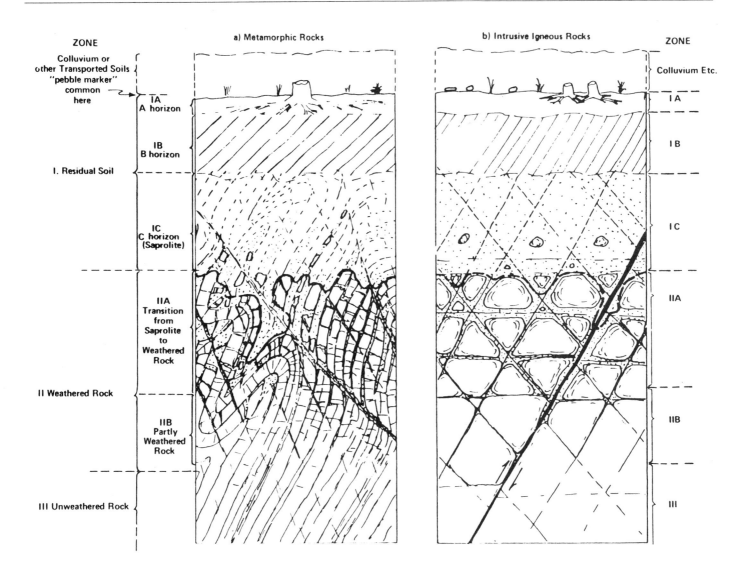

FIGURE 19-1
Typical weathering
profile for
metamorphic and
igneous rocks
(Deere and Patton
1971).

sured pore pressures and shear-strength parameters. Brand (1985) concluded that the ability to predict the performance of slopes in residual soils is only poor to fair.

Table 19-1 shows the relative permeability and shear strength for materials within different weathering profile zones as defined by Deere and Patton (1971). The subdivision of Zone I, residual soil, into A, B, and C horizons follows the convention used by soil scientists who study weathering processes in the formation of soils. Soils in A and B horizons have experienced significant chemical weathering that results in a breakdown of rock into silt- and clay-sized particles. Rainfall infiltration leaches chemicals from the A horizon, which then accumulate in the B horizon. Because weathering has removed most traces of bedrock

structure, the soils in these horizons resemble the silty clays, sandy silts, and silty sands found in transported soils. Soils in horizon C, or saprolite, behave differently than transported soils do. Deere and Patton emphasized that the profile classification has to be added to more traditional soil classifications during exploration. Generally, unambiguous classification requires continuous sampling or inspection of open cuts in test pits or road cuts. Brand and Phillipson (1985) provided a good overview of practices used around the world to sample and test residual soils.

Saprolites retain the structure of parent bedrock, but with only a trace of the original bond strength. De Mello (1972) suggested that standard soil properties tests performed on thoroughly mixed specimens do not effectively represent

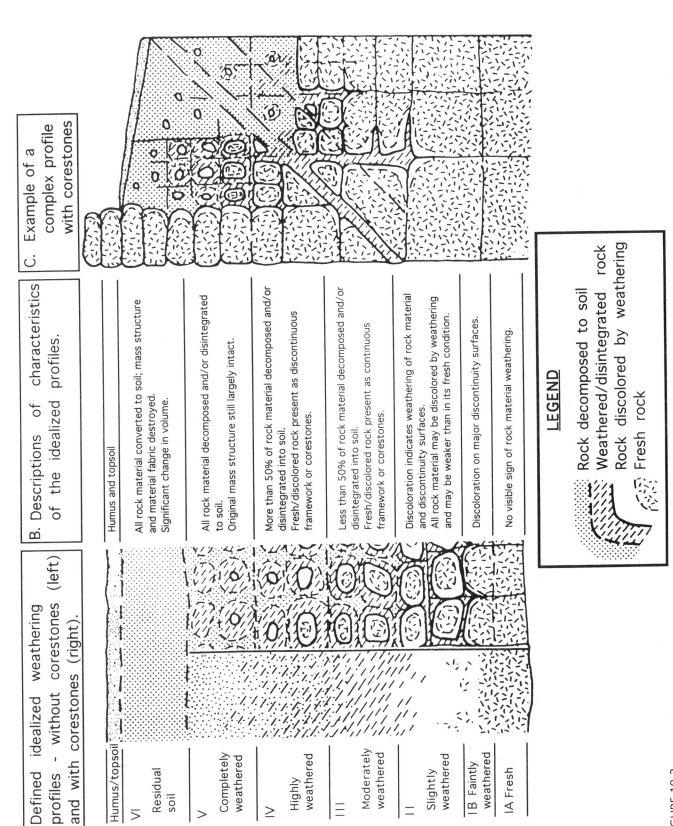

A. Defined idealized weathering profiles - without corestones (left) and with corestones (right).

B. Descriptions of characteristics of the idealized profiles.

C. Example of a complex profile with corestones

Humus/topsoil	Humus and topsoil
VI Residual soil	All rock material converted to soil; mass structure and material fabric destroyed. Significant change in volume.
V Completely weathered	All rock material decomposed and/or disintegrated to soil. Original mass structure still largely intact.
IV Highly weathered	More than 50% of rock material decomposed and/or disintegrated into soil. Fresh/discolored rock present as discontinuous framework or corestones.
III Moderately weathered	Less than 50% of rock material decomposed and/or disintegrated into soil. Fresh/discolored rock present as continuous framework or corestones.
II Slightly weathered	Discoloration indicates weathering of rock material and discontinuity surfaces. All rock material may be discolored by weathering and may be weaker than in its fresh condition.
IB Faintly weathered	Discoloration on major discontinuity surfaces.
IA Fresh	No visible sign of rock material weathering.

LEGEND

Rock decomposed to soil
Weathered/disintegrated rock
Rock discolored by weathering
Fresh rock

FIGURE 19-2
Tropical residual soil classification proposed by Geological Society of London working party (Geological Society of London 1990).

Table 19-1
Description of Weathering Profile for Igneous and Metamorphic Rocks (modified from Deere and Patton 1971)

ZONE	DESCRIPTION	ROCK QUALITY DESIGNATION[a] (NX CORE %)	PERCENT CORE RECOVERY[b] (NX CORE)	RELATIVE PERMEABILITY	RELATIVE STRENGTH
I, Residual soil					
IA, A Horizon	Topsoil, roots, organic material; zone of leaching and eluviation; may be porous	Not applicable	0	Medium to high	Low to medium
IB, B Horizon	Characteristically clay enriched; also accumulation of Fe, Al, and Si; hence may be cemented; no relict structures present	Not applicable	0	Low	Commonly low (high if cemented)
IC, C Horizon (saprolite)	Relict rock structures retained; silty grading to sandy material; less than 10% core stones: often micaceous	0 or not applicable	Generally 0–10	Medium	Low to medium (relict structures very significant)
II, Weathered rock					
IIA, Transition	Highly variable, soil-like to rocklike; fines commonly fine to coarse sand (gruss); 10 to 90% core stones; spheroidal weathering common	Variable, generally 0–50	Variable, generally 10–90	High	Medium to low where weak and relict structures present
IIB, Partly weathered rock (PWR)	Rocklike soft to hard rock; joints stained to altered; some alteration of feldspars and micas	Generally 50–75	Generally 90	Medium to high	Medium to high[b]
III, Unweathered rock	No iron stains to trace along joints; no weathering of feldspars or micas	Over 75 (generally 90)	Generally 100	Low to medium	Very high[b]

[a] The descriptions provide the only reliable means of distinguishing the zones.
[b] Only intact rock masses with no adversely oriented geologic structures.

saprolite soil properties. Both Sowers (1954, 1963) and Vaughan (1985a) have correlated properties to void ratio rather than to Atterberg limits because the void ratio represents the in situ state. Figure 19-1 shows that saprolites also retain relict discontinuities or joints present in parent bedrock. These discontinuities significantly influence the permeability and shear strength of the soil mass. In some cases laboratory-sized specimens provide misleading permeability coefficients and shear strengths. In addition, highly micaceous residual soils derived from gneisses expand elasti-

cally when unloaded during sampling. This expansion breaks weak bonds and decreases the soil's laboratory-measured shear strength and stiffness (Bressani and Vaughan 1989).

Zone IIA, or upper weathered-rock zone, provides a transition from saprolite to partly weathered rock (PWR). In Zone IIA the soil becomes progressively coarser with depth and has more core stones and clearer discontinuities. Although arbitrary boundaries separate different zones, the profile exhibits trends in changing properties. Zone IIA materials are described as soils, but it is difficult to provide undisturbed samples of them for laboratory strength and permeability testing. Stability analyses of residual soil slopes fall in the continuum between soil mechanics and rock mechanics. Saprolites behave like soils but often slide along discontinuities. In contrast, Zone IIA transition materials exhibit both rocklike and soil-like properties. Since Zone IIA materials often have higher permeabilities than the overlying saprolites, they influence seepage in cut slopes.

Slopes composed of materials classified as Zone IIB (PWR) and Zone III (unweathered rock) are best treated as rock slopes. The depth of the boundary between Zones IIB and III significantly influences the cost of excavation in residual soil (White and Richardson 1987).

Figure 19-3 shows four common landslide types found in residual soil slopes (Deere and Patton 1971). Generally, investigations in specific regions identify and classify the common landslide types; Deere and Patton's classification system provides a general framework. Type a landslides

occur primarily in shallow Zone IA and IB soils, where colluvium from previous landslides has not covered the profile. Landslides frequently occur during heavy rainfall, and the saturated soil often moves as debris flows or mud flows. Jones (1973) described how many Type a landslides killed more than 1,700 people during heavy rainfall in January 1967 in the Serra das Araras region of Brazil. Type b landslides fail along relict joints located in Zone IC (saprolite), Zone IIA (transition), or Zone IIB (PWR). The joint pattern controls the shape of the landsliding mass. Although rainfall significantly influences pore pressures, landsliding can occur well after heavy rains when two- or three-dimensional seepage causes slope failures. In contrast, vertical infiltration of rainfall is the most frequent cause of shallow Type a and b landslides.

Type c and d landslides resemble Type a and b landslides but occur in colluvium-covered weathered profiles. In these residual soil profiles, the overlying colluvium usually resembles the texture of the underlying residual materials but lacks the

FIGURE 19-3 Common types of landslides in weathered rock, residual soil, and colluvium (Deere and Patton 1971). REPRINTED WITH PERMISSION OF AMERICAN SOCIETY OF CIVIL ENGINEERS

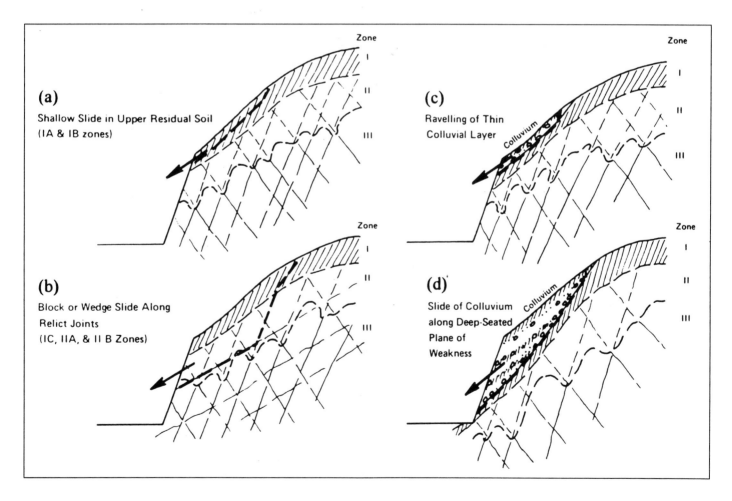

rocklike structure and relict joints. Previous landsliding mixes the colluvium with cobbles and core stones, making sampling difficult. Type *c* landslides differ little from Type *a* landslides except that the material properties are harder to measure. For Type *d* landslides, the colluvial cover significantly influences seepage. At the Boone landslide in North Carolina, described in Section 4.3, the colluvium had a higher permeability than the underlying saprolite. In contrast, at the Balsam Gap landslide, also in North Carolina, the colluvium had a lower permeability and acted as a low-permeability cap. Mapping the colluvium thickness provides a useful description of the horizontal variability controlling both current seepage and historical slide activity. Brand (1985) found that engineers often classify colluvium as residual soil.

Deere and Patton's classification system considers only cut slopes in residual soil profiles. Frequently, highway construction requires fill slopes over residual soils. Lumb (1975) included fill slopes placed upon residual soil slopes as one common and particularly dangerous landslide type in Hong Kong. Generally, residual soils have much higher values for in-place mass permeabilities than the same materials have once they have been transported, mixed, and compacted. Therefore, compacted fills often act as low-permeability caps, disrupting the natural seepage pattern and increasing pore pressures. The resulting landslides resemble Deere and Patton's Type *d* landslides, but in these cases the fill acts as the overlying colluvium.

2. SAPROLITE PROPERTIES

The stability of A and B horizon soils in Zone I may be effectively assessed by traditional soil mechanics analysis methods. The stability of Zone IIB (PWR) and Zone III (unweathered rock) materials may be effectively assessed by rock mechanics analysis methods. However, saprolite and transition materials fall in the continuum between soil mechanics and rock mechanics. Core stones make sampling difficult, and laboratory-sized specimens sometimes provide poor estimates of mass strength and permeability. The soil behavior described in this section refers primarily to saprolites but also applies to the soil matrix in transition soils.

Saprolites are generally unsaturated, weakly bonded, and heterogeneous soils with relict joint systems. Mitchell and Sitar (1982) reviewed the

engineering properties of tropical soils, qualitatively defined six stages of weathering, and described the corresponding changes in mineralogy and soil properties. Vaughan et al. (1988) reported that bond strength and void ratio control behavior of residual soils and that geologic origins of these soils cause the following primary characteristics:

1. The bonding derived from soil evolution and found at equilibrium with the current in situ stress state influences the soil's strength and stiffness,
2. The stress history during soil formation has little effect on the soil's current properties,
3. Soils have widely varying mineralogy and grain strength, and
4. In situ soils have widely varying void ratios.

Vaughan et al. (1988) experimentally examined the behavior of weakly bonded soils by artificially manufacturing test soils from a quartz and kaolin sand mixed with kaolin slurry and then fired in a furnace.

Figure 19-4 shows a conceptual description of soil behavior for a weakly bonded soil. In Figure 19-4(*a*), void ratio, *e*, versus mean effective stress, *p'*, the solid curve represents an oedometer test on a weakly bonded specimen. The dashed curve labeled *d* represents the compression curve from a loosest initial condition of the destructured soil, whereas the dashed curve *e* represents the compression curve for a destructured soil with an initial void ratio equal to that of the weakly bonded specimen. The weakly bonded specimen experiences no compression until it reaches the first yield point. Figure 19-4(*b*) provides a hypothesized comparison of bond strength with interparticle stress. At the first yield point the bond strength starts to drop because small interparticle strains weaken bonds, whereas at the second yield point the bond stress equals the bond strength, and bonds start breaking. The curves of reference void ratio *e* versus effective stress *p'* in Figure 19-4 provide boundaries for classifying soil behavior. When soils have *p'*, *e* states that plot below the loosest destructured curve (labeled *d*), they behave as stable soils. When they plot above that curve, they behave as metastable soils.

Although this classification system does not currently provide quantitative correlations with shear strength, in the future it may provide a more effective framework for empirical correlations than do Atterberg limits, which effectively describe transported soil properties. Vaughan et al. (1988) made several important observations concerning evaluations of shear strength of saprolites. The small strains induced during sampling may weaken bonds and lower the shear-strength values measured in the laboratory. In addition, residual soils often cannot sustain high capillary suctions, and the resulting unloading experienced during sampling causes expansion and breaks bonds. Finally, the laboratory-measured yield stresses are generally lower than those measured by in situ tests. Although these yield-stress values often prove valuable for evaluating deformations, the entire failure envelope must be known in order to estimate safety against instability.

Brand (1985) suggested that bonding causes saprolites at low effective stresses to have higher strengths than those predicted by triaxial tests run at high effective stresses and interpreted using a straight-line Mohr-Coulomb failure envelope. Figure 19-5 shows how the straight-line envelope can underestimate the actual curved failure envelope that applies at low effective stress. Vaughan (1985a) reported that weak bonding probably causes a small effective stress cohesion intercept, c', that can be estimated from drained unconfined tests. Figure 19-6 shows that a horizontal stress path at constant q, $(\sigma_1 - \sigma_3)/2$, and decreasing p', $(\sigma'_1 + \sigma'_3)/2$, corresponds to slope failures caused by rising pore pressures. Typical triaxial tests fail samples along the wrong stress path and result in unrealistically high p' values at failure.

Two additional factors that influence the shear strength of saprolites are the unsaturated state of the soil and relict joints. Lumb (1975) tested both saturated and unsaturated residual soil specimens in drained triaxial tests. He found that the straight-line envelopes determined from these tests gave drained friction-angle values (ϕ_d) that were independent of saturation, S, but that the drained cohesion values, c_d, showed a statistical correlation with saturation for soils derived from volcanic rhyolites and with saturation and void ratio for soils derived from granite. Lumb's test results showed that for all saprolites overlying granites, $\phi_d = 32$ degrees, whereas for unsaturated soils, $c_d = 0$ to 75 kPa and for saturated soils, $c_d = 0$ to 50 kPa. For saprolites overlying volcanic rocks, $\phi_d =$

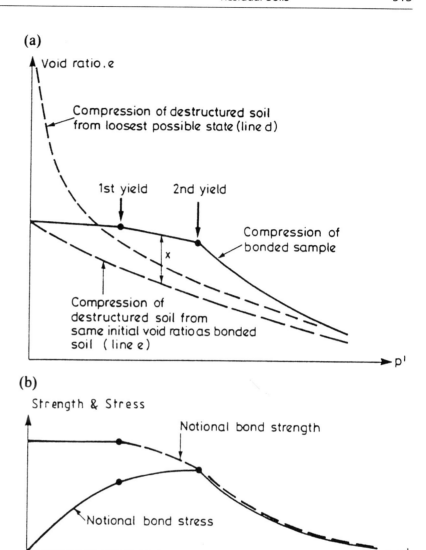

36 degrees, whereas for unsaturated soils, $c_d = 60$ kPa and for saturated soils, $c_d = 25$ kPa. Sowers and Richardson (1983) reported the 67 percent confidence line for results of drained triaxial tests on both saturated and unsaturated specimens of saprolite derived from a range of different gneiss and schist parent rocks sampled during the design and construction of the Metropolitan Atlanta Rapid Transit Authority routes in Atlanta, Georgia. For saturated specimens, $\phi' = 32$ degrees and $c' = 0$ kPa, whereas for unsaturated specimens, $\phi' = 30$ degrees and $c' = 14$ kPa. The saprolites derived from igneous rocks in Hong Kong and those formed on metamorphic rocks in Georgia all have ϕ' values of 30 to 35 degrees but with widely vary-

FIGURE 19-4 Conceptual description of soil behavior for weakly bonded soil: (a) oedometer test on weakly bonded specimen; (b) comparison of bond strength with interparticle stress (Vaughan et al. 1988).

ing cohesion intercepts. The cohesion values represent an empirical fit to a curved envelope.

The cohesion intercept often represents an apparent cohesion resulting from capillary suction in the soils. Fredlund (1987) ran staged triaxial tests on unsaturated undisturbed saprolite specimens by setting constant values of total confining pressure,

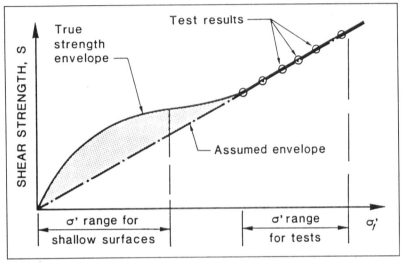

FIGURE 19-5
Underestimation of shear strengths because of incorrect strength envelope deduced from data obtained at too high normal stresses (Brand 1985).

REPRINTED FROM *PROCEEDINGS OF THE 11TH INTERNATIONAL CONFERENCE ON SOIL MECHANICS AND FOUNDATION ENGINEERING*, SAN FRANCISCO, 12–16 AUGUST 1985, VOLUME 5, 1985–1988, 3153 PP., 5 VOLS., A.A. BALKEMA, OLD POST ROAD, BROOKFIELD, VERMONT 05036

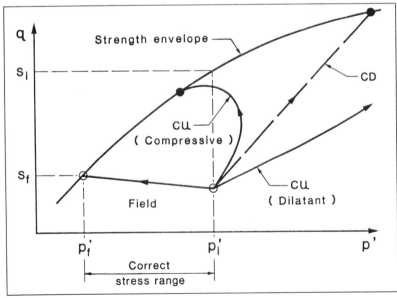

FIGURE 19-6
Comparison between stress paths for rain-induced slope failure and for triaxial tests (Brand 1981).

REPRINTED FROM *PROCEEDINGS OF THE 10TH INTERNATIONAL CONFERENCE ON SOIL MECHANICS AND FOUNDATION ENGINEERING*, STOCKHOLM, 15–19 JUNE 1981, VOLUME 3, 1981–1982, 3542 PP., 4 VOLS., A.A. BALKEMA, OLD POST ROAD, BROOKFIELD, VERMONT 05036

σ, pore-air pressure, u_a, and pore-water pressure, u_w, and then measuring the deviator stress while shearing the specimens at a constant strain rate of 0.001 to 0.004 percent per minute. Fredlund increased the deviator stress until it reached a clear peak, unloaded the sample, adjusted σ and u_a while maintaining u_w constant, and then sheared the specimen again. Figure 19-7 shows the results for a multistage test performed on a decomposed granite specimen taken from a Hong Kong slope. The shear strength can be described by the following equation:

$$\tau = c + (\sigma - u_a)\tan\phi' \qquad (19.1)$$

In the above equation, the cohesion, c, depends upon both the effective stress cohesion intercept, c', and the matrix suction, $u_a - u_w$, according to the following equation:

$$c = c' + (u_a - u_w)\tan\phi^b \qquad (19.2)$$

The ϕ' value equals the slope of envelopes tangent to Mohr circles plotted on τ versus $(\sigma - u_a)$ axes and equals 33.4 degrees for the tests shown in Figure 19-7. These tests were run at a constant $(\sigma_3 - u_a)$ of 140 kPa. The slope of the c versus $(u_a - u_w)$ plot, ϕ^b, equals 16.2 degrees, and the intercept, c', equals 43 kPa.

Abramento and Carvalho (1989) ran multistage drained triaxial tests on unsaturated specimens of weathered migmatites from the Serra do Mar region of Brazil at $(\sigma_3 - u_a)$ of 10 kPa. They used a curved line to determine the cohesion intercept, c, in kilopascals by the following equation:

$$c = 2.5\,(u_a - u_w)^{0.5} \qquad (19.3)$$

These test results suggest that c' equals 0.0. Both of these descriptions show that the shear strength of unsaturated saprolites depends upon three stress variables. Unfortunately, using this description in stability calculations means that the in situ matrix suction, $(u_a - u_w)$, must be measured or predicted.

Relict joints represent the final element influencing the mass shear strength of saprolites. Sandroni (1985) divided discontinuities into three classes according to their size. *Relict structures* represent faults, joints, dikes, and lithological contacts that range in dimension from meters to tens of meters. *Macrostructures* are visible to the naked eye and range in size from centimeters to 1 to 2 m. Isotropic macrostructures can be described as monothonous, stained, veiny, mottled, or porous, and anisotropic macrostructures can be described as

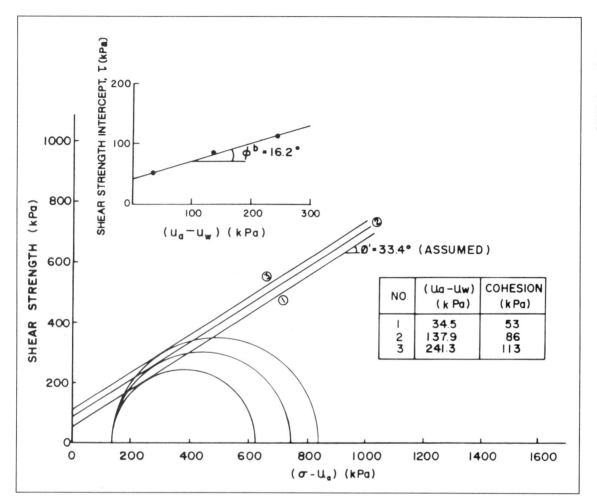

FIGURE 19-7
Mohr circles and
determination of ϕ^b
for decomposed
Hong Kong granite
(Fredlund 1987).
REPRINTED WITH
PERMISSION OF JOHN WILEY
& SONS, LTD.

laminated, banded, schistose, lenticular, or folded. Other macrostructures include remnants, cavities, and concretions. Finally, *microstructures* range from microns to tens of millimeters and can be viewed only by either a hand lens or a microscope.

All three discontinuity types can significantly reduce the mass shear strength below values measured in small-sized laboratory specimens. St. John et al. (1969) described the thin black seams often found in residual soils derived from igneous and metamorphic rocks. The undrained shear strength measured on 10-cm-diameter specimens of saprolite overlying gneiss with black seams oriented at 45 degrees was 50 to 67 percent of the undrained shear strength measured on intact specimens. These black seams had visible slickensides, suggesting previous displacements. Boyce (1985) ran ring-shear tests on tropical residual soils to measure the limiting residual shear strength controlling such slickensided surfaces. When he plotted ϕ_r values versus either clay fraction or plasticity index, he found that his results plotted both above

and below the measured values of ϕ_r for sedimentary soils. Decomposed phyllites had lower ϕ_r values than sedimentary soils of equal clay fraction, and volcanic soils containing allophane and halloysite had higher ϕ_r values.

3. PORE PRESSURES

Rainfall frequently triggers landslides in residual soils. Figure 19-8 shows how rainfall infiltration can increase positive pore pressures by raising either a perched or a regional water table. Townsend (1985) reported that for some naturally occurring residual soils with field permeabilities of 10^{-4} to 10^{-5} cm/sec, compaction decreased the permeability to 10^{-5} to 10^{-7} cm/sec. Brand (1985) added that naturally occurring highly transmissive zones, or pipes, in residual soils can make mass permeabilities very high. Therefore pore pressures in residual soil slopes often react quickly to heavy rainfall. Figure 19-9 shows that heavy rainfall for two days raised the measured piezometric head in a Hong

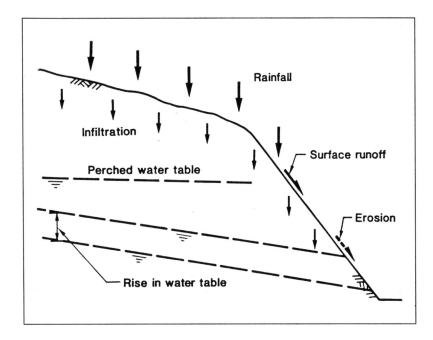

FIGURE 19-8
(above)
Diagrammatic
representation of
effects of rainfall on
slope (Brand 1985).
REPRINTED FROM
PROCEEDINGS OF THE 11TH
INTERNATIONAL
CONFERENCE ON SOIL
MECHANICS AND
FOUNDATION
ENGINEERING, SAN
FRANCISCO, 12–16 AUGUST
1985, VOLUME 5, 1985–1988,
3153 PP., 5 VOLS., A.A.
BALKEMA, OLD POST ROAD,
BROOKFIELD, VERMONT
05036

cohesion. Lumb (1962) proposed that a wetting front advances during heavy rainfall, as shown in Figure 19-10. During a rainfall exceeding the infiltration capacity of the soil for time t, the front will advance a distance h, as given by the following equation:

$$h = \frac{kt}{n(S_f - S_o)} \qquad (19.4)$$

where h and t are as defined previously and
k = coefficient of permeability,
n = porosity,
S_f = final saturation, and
S_o = initial saturation.

Equation 19.4 suggests that the wetting front advances faster when antecedent rainfall has increased S_o. Lumb (1975) reported that decomposed volcanic soils have an effective mass permeability equal to 1.5×10^{-4} cm/sec and that if the rainfall rate exceeds 130 mm/day for a three-day period, providing a total accumulation of about 400 mm, the wetting front will advance 4 m into the soil. The mass permeability in decomposed granite equals 8.0×10^{-4} cm/sec and the same 400 mm of rain needs to accumulate within a 14-hr period to advance the wetting front 4 m. Vaughan (1985b) demonstrated by simple analysis that a profile that has continuously decreasing permeability with depth can lead to instability, whereas increasing permeability with depth provides natural drainage. Although Lumb and Vaughan have analyzed vertical infiltration into slopes, these methods do not automatically provide a good quantitative prediction method for many specific slopes.

Kong slope 5 m in only 18 hr, and the piezometric head dropped quickly when the rains stopped. Such rapid changes make it difficult to measure pore pressures in residual soil slopes at the point of slope failure. Either automatic-reading piezometers or a system that can record the highest rise during and following heavy rainfall is required. Brand (1985) described a series of small Halcrow buckets chained together and suspended in an open-standpipe piezometer. The highest water-level rise is recorded by measuring the depth to the highest filled bucket.

Shallow failures can occur on slopes composed of unsaturated saprolite when rainfall infiltration eliminates the suction and lowers the apparent

Brand (1985) evaluated the current ability to predict the performance of residual soil slopes. He reported that there are good methods of analysis and that it is possible to predict failure modes based upon observed failures in similar slopes and to estimate shear strength fairly well but it is not possible to effectively estimate pore pressures at the point of incipient slope failure. Because there is only a poor to fair ability to predict slope performance, design practices often depend upon empirical and semiempirical methods. For example, four design approaches have been used in Hong Kong:

- Correlation of slope failures with rainfall pattern,
- Terrain evaluation using geomorphological mapping,

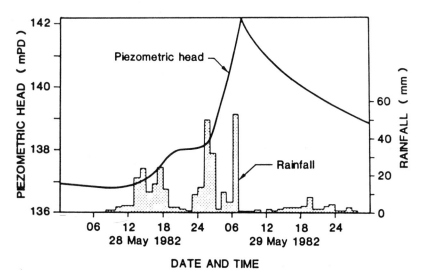

DEGREE OF SATURATION

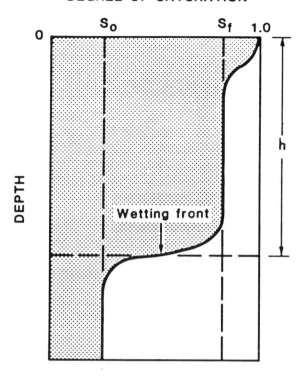

- Semiempirical methods developed from evaluation of past failures, and
- Soil mechanics analysis.

In Hong Kong soil mechanics analyses are used most often to evaluate remedial measures for correcting or stabilizing slope failures and for analyzing the stability of cut slopes that are next to critical structures. The other three methods are used for land use planning and for designing cut slopes along transportation routes.

4. CASE HISTORIES

Case histories of slope stability analyses for slopes in residual soils are widely dispersed throughout the technical literature. Also, as noted in the previous sections of this chapter, residual soils may exhibit a considerable range in properties, and experience gained in one region should be used with caution in others. In the following sections some of the issues discussed earlier in this chapter are illustrated briefly with case histories from Brazil, Hong Kong, and the state of North Carolina in the United States. These examples begin to demonstrate the regional differences in residual

soils and some of the similarities of the factors leading to slope instability in residual soils.

4.1 Brazil

Vargas and Pichler (1957) described residual soil and rock landslides occurring in Santos, Brazil. They divided the landslides occurring in Brazil into three classes:

- Class 1 landslides involve creep of surficial residual soil,
- Class 2 slides involve detritus (colluvium) at lower elevations in old slide areas, and
- Class 3 slides involve sudden rupture of residual soil overlying bedrock.

At Monte Serrate, a 50-m-high Class 3 landslide occurred in February 1928 when 720 mm of rain fell during the month. This landslide was back analyzed using standard soil-stability analysis methods for circular failure surfaces. The results showed that failure occurred when the pore pressure ratio, r_u, reached 0.35 if ϕ = 42 degrees and c = 39 kPa, as measured in direct shear tests. Da Costa Nunes et al. (1979) provided a wider overview of landslides found in Brazil and divided slides into classes according to their geological regions. Vargas (1974) concluded that rotational or planar slides in residual soils fail at the end of the rainy season either during or following rainfall more intense than 100 mm/day. More severe and faster-moving flows that remove the residual soil mantle fail during catastrophic rainstorms with intensities greater than 50 mm/hour.

Jones (1973) reported on some especially severe landslides that occurred over a 100-km² area along the Serra das Araras escarpment, a subdivision of Serra do Mar; these landslides killed an estimated 1,700 people. On January 22 and 23, 1967, 250 mm of rain fell during 3.5 hr and triggered more than 10,000 shallow landslides classified as debris avalanches, debris flows, and mud flows. These landslides removed from less than 1 m to up to several meters of residual soil overlying the well-banded biotitic, feldspathic, and garnetiferous gneisses that were cut by aplitic veins and diabase dikes. The weathering profiles showed an abrupt transition between the Zone IC and Zone IIA materials and the Zone III materials. Specimens taken from the landslide area were classified

FIGURE 19-10 Representation of advance of wetting front as water infiltrates soil (Lumb 1975).

FIGURE 19-9 (opposite page) Rapid changes in groundwater table measured in Hong Kong slope (Brand 1985).
REPRINTED FROM PROCEEDINGS OF THE 11TH INTERNATIONAL CONFERENCE ON SOIL MECHANICS AND FOUNDATION ENGINEERING, SAN FRANCISCO, 12–16 AUGUST 1985, VOLUME 5, 1985–1988, 3153 PP., 5 VOLS., A.A. BALKEMA, OLD POST ROAD, BROOKFIELD, VERMONT 05036

according to the Unified Soil Classification System as low-plasticity silts and sandy silts (ML and SM-ML).

Wolle and Hachich (1989) quantitatively evaluated translational landslides occurring in the same Serra do Mar region. The weathering profile typically had 1 to 1.5 m of colluvial cover overlying 1 to 2 m of saprolite that, in turn, overlaid migmatites. Open fractures in the Zone IIA transition material made the mass permeability high, and the water table occurred at depths of 20 to 30 m in the fractured bedrock. Dense vegetative cover intercepted infiltration and prevented erosion. Observed landslides occurred on the 40- to 45-degree slopes and were 7 to 25 m wide, over 100 m long, and about 1 m deep. The threshold precipitation in this region equaled 180 mm/day when antecedent rain had moistened the soil. The relatively high permeabilities averaged 5×10^{-4} cm/sec, and during intense rainfall the wetting front advanced 1.0 to 1.5 m into the soil profile in 4 to 12 hr. During the rainy season, soils had saturation values of 60 to 80 percent, whereas laboratory tests suggested that no suction occurred when saturations reached 92 to 95 percent. Calculations showed that the slopes failed when rainfall infiltration decreased suctions at a 1-m depth to values ranging from 1 to 2 kPa. Measurements during the rainy season suggested suctions of 1 to 10 kPa. Wolle and Hachich also estimated the influence of root strength on stability and concluded that the sliding soil must reach a width of at least 6 to 15 m before overcoming root reinforcement. In general, it seems that these landslides support Lumb's (1975) proposed hypothesis that landsliding is caused by the advance of a wetting front during intense and sustained rainfall.

4.2 Hong Kong

Lumb (1975) examined slope failures in Hong Kong and correlated their occurrence with rainfall intensity. He divided rainfall events into four categories:

- *Disastrous events* causing more than 50 recorded landslides in one day,
- *Severe events* causing 10 to 50 landslides in one day,
- *Minor events* causing fewer than 10 landslides, and
- *Isolated events* causing only a single slip.

Figure 19-11 shows these rainfall events plotted as the 24-hr rainfall versus the 15-day antecedent rainfall. Using rainfall data obtained from records of the Royal Observatory, Lumb identified bounds between different levels of event severity. Disastrous events occurred when the 24-hr rainfall exceeded 100 mm and the 15-day antecedent rainfall exceeded 350 mm. Severe events were triggered by rainfall of 100 mm/day following 200 mm of antecedent rainfall.

Lumb (1975) divided the Hong Kong landslides into three types. All three types are landslides that are generally less than 3 m deep and have a thickness-to-length ratio of less than 0.15. Type 1 landslides represent fill embankments constructed over residual soil; these move following heavy rainfall. Type 2 landslides occur in natural residual soil or colluvium, extend to or beyond the crest of the slope, and fail following heavy rainfall. The Type 2 failures generally occur in decomposed volcanics but move more slowly than the Type 1 saturated fills. Type 3 landslides occur within a cut face, do not extend beyond the crest of the slope, and often occur in areas of decomposed granite. Type 3 failures occasionally occur during dry periods when some other water source saturates the decomposed granite. Type 3 failures are smaller and much less destructive than either the Type 1 or Type 2 landslides. Lumb (1975) also described two much deeper landslides that differ from his three primary types. These last two moved more slowly and involved seepage patterns much more complex than simple vertical rainfall infiltration.

FIGURE 19-11
Relationship between rainfall and landslides in Hong Kong (Lumb 1975).

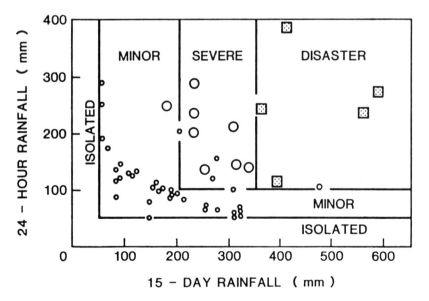

Brand (1985) further examined the landslide risk in Hong Kong using three distinct methods. He first reexamined the earlier correlations by Lumb (1975) using data collected from 46 automatic recording rain gauges distributed throughout Hong Kong and considered 1-hr rainfalls, 24-hr rainfalls, and antecedent rainfall for periods varying up to 30 days. In addition, he identified the precise time of landslide events by using the Fire Services Department reports of calls. Figure 19-12 shows average landslides per day and average casualties per day plotted versus maximum hourly rainfall. This plot shows a clear threshold of 70 mm/hour of rainfall to trigger multiple landslides with subsequent loss of life. Brand (1985) concluded that localized short-duration rainfalls caused the majority of landslides and that antecedent rainfall had little influence. He further concluded that in Hong Kong storm intensities greater than 70 mm/hour trigger landslides and that daily rainfalls of greater than 100 mm/day often indicate that a short intense rainfall has occurred during that day.

The third landslide prediction method investigated by Brand (1985) is based on semiempirical charts and requires detailed data on the height and angle of slopes. Figure 19-13 shows a semiempirical plot developed from a study of 177 slopes, including failed slopes (solid circles) and unfailed slopes (open circles). Solid lines portray the progression of Hong Kong's design guidelines over the years. In 1950 slopes were cut at 75 degrees, or 1(H):4(V), whereas in more recent guidelines the design slope angle depends upon slope height. This plot clearly shows that for cut slopes in residual soils, traditional stability analyses lead to flatter slopes than those recommended by existing semiempirical design practices.

Both Lumb (1975) and Brand (1985) provided detailed geotechnical information necessary to analyze the performance of individual slopes. However, as described earlier, limitations in predicting pore pressures at the point of incipient failure make these stability predictions uncertain. Figure 19-14 shows typical Hong Kong soil profiles overlying both granites and volcanics. The granites can decompose to depths greater than 30 m and the unweathered granite has joint spacings of 2 to 10 m. The water table typically is found near the upper surface of the unweathered rock; thus the full soil profile can be unsaturated. Only the

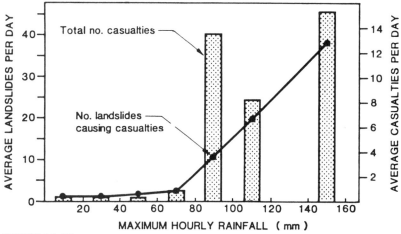

FIGURE 19-12
Number of landslides causing casualties and number of casualties for period 1963–1983 in Hong Kong compared with maximum hourly rainfall (Lumb 1975).

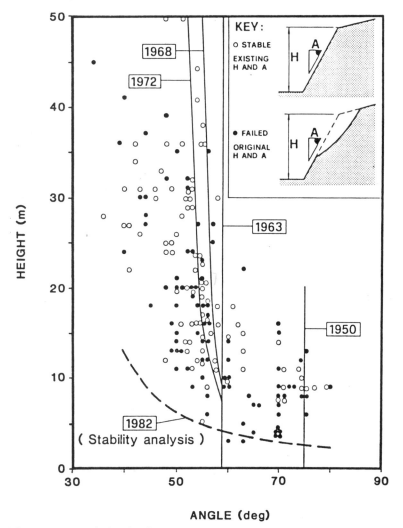

FIGURE 19-13 Relationship between height and slope angle for 177 Hong Kong cut slopes (Brand and Hudson 1982).

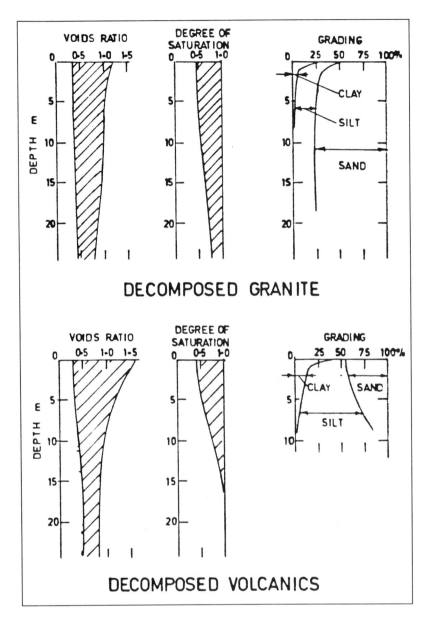

FIGURE 19-14
Bulk properties of
Hong Kong residual
soils (Lumb 1975).

top few meters of soil show any clay fraction, and the soils are classified as silty sands. The volcanics have closer joint spacings of 0.2 to 1 m and typically weather to depths of 10 m, rarely more than 20 m. These soils are classified as sandy silts and have slightly higher void ratios. Colluvium covers the residual soils on the lower slopes. Colluvium overlying decomposed granites has a similar texture to the residual soils, whereas the colluvium overlying the volcanics has more rounded cobbles and small core stones than the underlying residual soils. Lumb (1975) measured an average coefficient of permeability of 8.0×10^{-4} cm/sec in decomposed granites by both field in situ tests

and laboratory tests on undisturbed specimens. Because the volcanics have a closer joint spacing, their average field-measured permeability was 1.5×10^{-4} cm/sec, some two orders of magnitude greater than the average permeability value for these volcanic materials measured in the laboratory, which was 2.0×10^{-6} cm/sec.

Brand (1985) analyzed six specific landslides for which field investigations determined the existing slope geometry and underlying soil profiles and laboratory testing determined the shear strength. Three cases (numbered 1 to 3) involved slopes that had not failed, but their calculated factors of safety ranged from 0.80 to 0.95. Three additional cases (numbered 4 to 6) involved slopes that had failed, but their calculated factors of safety ranged from 1.24 to 1.30. In the first three cases, Brand concluded that either the acting strengths or the suctions were larger than those used in the stability analyses. In Case 4 the actual weathered soil depth was greater than expected from the preliminary borings. In Case 5 sliding occurred along slickensided relict joints, and in Case 6 subvertical relict joints encouraged vertical seepage and influenced the observed failure mode. Therefore, in all these cases, the exploration either did not correctly determine the soil weathering profile or did not accurately identify soil strength properties or groundwater conditions.

4.3 North Carolina

The North Carolina Department of Transportation has experienced recurring landslides at their highway cuts. In 1987 and 1988 two landslides were investigated to better define the failure mode, the acting shear strength, and the pore pressures at the point of slope failure for use in future designs (Lambe and Riad 1991). These two landslides, known as the Balsam Gap landslide and the Boone landslide, were both slow-moving and required no immediate repair. Both landslides moved along shear surfaces just below the colluvium-saprolite interface. Piezometers installed at both sites measured zero or negative pore pressures acting on the shear surfaces.

Figure 19-15 shows a plan view of the Balsam Gap landslide and locates the borings made to take samples and install instrumentation. This 1.75(H):1(V) slope had been excavated north of US 23/74. The slope started moving in 1980 and

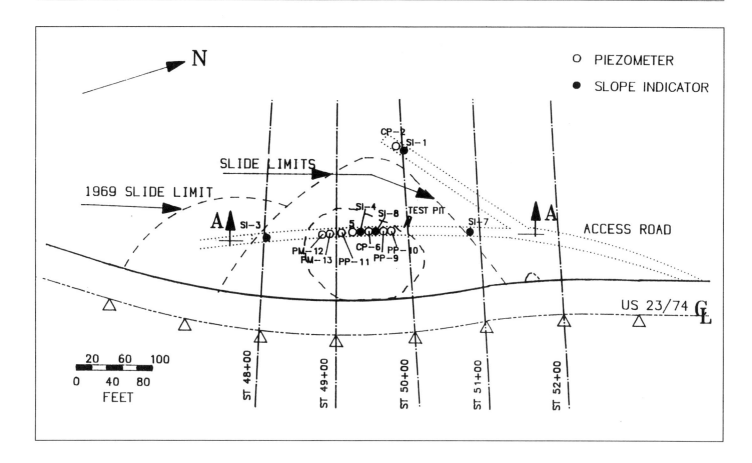

FIGURE 19-15
Plan view of Balsam
Gap landslide
(Lambe and Riad
1991).
REPRINTED WITH
PERMISSION OF
AMERICAN SOCIETY OF
CIVIL ENGINEERS

involved a failure volume of 15 000 m³. Rainfall during the years from 1979 to 1988 averaged 1140 mm/year, but only monthly rainfall data were available from nearby rainfall recording stations. Figure 19-16 shows the cross section and soil profile that were determined from borings and a test pit. The test pit was excavated to examine the 5-mm-thick slickensided shear surface located just below the colluvium cover. The interface between the saprolite's rocklike structure and the colluvium's thoroughly mixed appearance was readily identified.

The residual soil overlaid coarse granitic gneiss bedrock. Figure 19-17 shows the soil profile based upon the Standard Penetration Test profile in Boring 4 and tests on tube samples taken from Borings 5 and 6. The colluvium was classified as a sandy clay (CL-ML), and the saprolite was classified as a silty sand (SM). Falling-head permeability tests performed on trimmed consolidation specimens measured an average permeability of 3×10^{-4} cm/sec for the saprolite and a permeability of 2×10^{-7} cm/sec for the colluvium. Therefore, the colluvium acted as a low-permeability cap over the saprolite. Drained direct-shear tests on the saprolite

measured three envelopes of the type proposed by Skempton (1970). For the fully softened envelope, $\phi' = 27$ degrees and $c' = 19$ kPa. Analyzing noncircular failure surfaces, such as the one shown in Figure 19-16, resulted in the resistance envelopes in Figure 19-18. A resistance envelope portrays the average shear stress and average effective normal stress acting along a family of failure surfaces for an assumed pore-pressure distribution (Casagrande 1950). Lacking any measured pore pressures during heavy rainfall, resistance envelopes were calculated for three different pore-pressure levels:

- $r_u = 0.0$ was used to establish a lower bound,
- $r_u = 0.25$ was used to provide information on "average" conditions, and
- $r_u = 0.5$ was used to establish an upper bound.

These back analyses suggest that the failures may be explained by $r_u = 0.32$, with fully softened shear strength acting along the failure surface shown in Figure 19-16 and with an average effective normal stress along the failure surface of 48 kPa.

The weathering profile of the Boone landslide differed from the profile at Balsam Gap because a

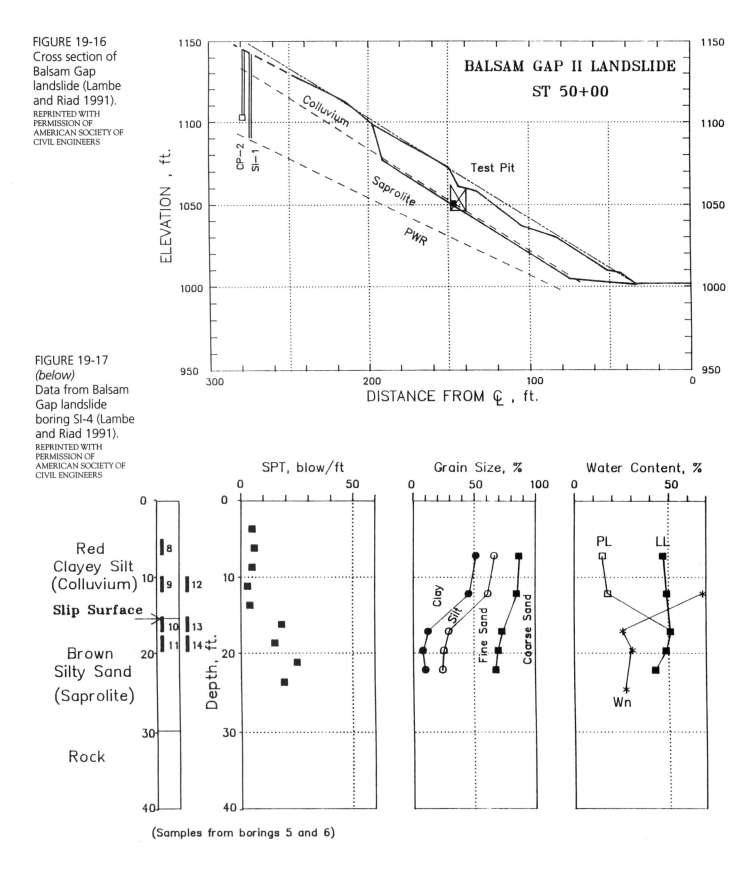

FIGURE 19-16
Cross section of
Balsam Gap
landslide (Lambe
and Riad 1991).
REPRINTED WITH
PERMISSION OF
AMERICAN SOCIETY OF
CIVIL ENGINEERS

FIGURE 19-17
(below)
Data from Balsam
Gap landslide
boring SI-4 (Lambe
and Riad 1991).
REPRINTED WITH
PERMISSION OF
AMERICAN SOCIETY OF
CIVIL ENGINEERS

BALSAM GAP II LANDSLIDE
ST 50+00

(Samples from borings 5 and 6)

more pervious colluvium covered the saprolite and the saprolite had weathered from a biotitic-granitic gneiss. The saprolite was classified as a silt and had an average permeability of 1×10^{-5} cm/sec. Gravel- and cobble-sized particles prevented sampling of the colluvium, which appeared more pervious than the underlying saprolite. Vertical rainfall infiltration through the colluvium could have caused a perched water table on the saprolite. The shear surface was 300 mm thick and appeared as a wet seam at the top of the saprolite, in contrast to the clearly slickensided surface found at Balsam Gap. Back analyses suggested that failure occurred when r_u reached 0.24 and the shear strength fell to the residual envelope with ϕ' = 29 degrees and c' = 2 kPa. At failure, the average effective normal stress, σ'_n, equaled 19 kPa.

For these two slow-moving landslides, the best information came from excavated test pits that intersected the failure surface from which block samples of the failure surface were obtained and trimmed so that the failure surface could be visually inspected and tested in the laboratory. Necessary safety precautions must be taken before entering such test pits, and temporary portable shoring, such as aluminum hydraulic shoring, should be used to hold the trenches open.

REFERENCES

Abramento, M., and C.S. Carvalho. 1989. Geotechnical Parameters for the Study of Natural Slopes Instabilization at "Serra do Mar", Brazil. In *Proc., 12th International Conference on Soil Mechanics and Foundation Engineering*, Rio de Janiero, A.A. Balkema, Rotterdam, Netherlands, pp. 1599–1602.

Boyce, J.R. 1985. Some Observations on the Residual Strength of Tropical Soils. In *Proc., First International Conference on Geomechanics in Tropical Lateritic and Saprolitic Soils*, Brasilia, Brazil, Vol. 1, pp. 229–237.

Brand, E.W. 1981. Some Thoughts on Rain-Induced Slope Failures. In *Proc., 10th International Conference on Soil Mechanics and Foundation Engineering*, Stockholm, Sweden, A.A. Balkema, Rotterdam, Netherlands, Vol. 3, pp. 373–376.

Brand, E.W. 1985. Predicting the Performance of Residual Soil Slopes. In *Proc., 11th International Conference on Soil Mechanics and Foundation Engineering*, San Francisco, A.A. Balkema, Rotterdam, Netherlands, Vol. 5, pp. 2541–2578.

Brand, E.W., and R.R. Hudson. 1982. Chase—An Empirical Approach to the Design of Cut Slopes in Hong Kong Soils. In *Proc., Seventh Southeast Asian Geotechnical Conference*, Hong Kong, Vol. 1, pp. 1–16.

Brand, E.W., and H.B. Phillipson. 1985. Review of International Practice for Sampling and Testing of Residual Soils. In *Sampling and Testing of Residual Soils* (E.W. Brand and H.B. Phillipson, eds.), Scorpion Press, Hong Kong, pp. 7–21.

Bressani, L.A., and P.R. Vaughan. 1989. Damage to Soil Structure during Triaxial Testing. In *Proc., In Proc., 12th International Conference on Soil Mechanics and Foundation Engineering*, Rio de Janiero, Brazil, A.A. Balkema, Rotterdam, Netherlands.

Casagrande, A. 1950. Notes on the Design of Earth Dams. *Journal of the Boston Society of Civil Engineers*, Vol. 37, pp. 405–429.

Da Costa Nunes, J.J., A.M.M. Costa Couto e Fonseca, and R.E. Hunt. 1979. Landslides of Brazil. In *Rockslides and Avalanches* (B. Voight, ed.), Elsevier Scientific Publishing Co., pp. 419–446.

Deere, D.U., and F.D. Patton. 1971. Slope Stability in Residual Soils. In *Proc., Fourth Pan American Conference on Soil Mechanics and Foundation Engi-*

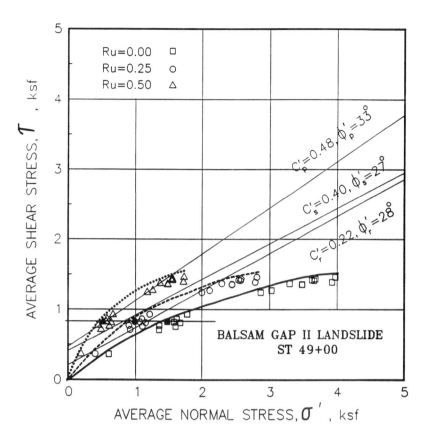

FIGURE 19-18 Resistance envelope for Balsam Gap landslide (Lambe and Riad 1991).

neering, San Juan, Puerto Rico, American Society of Civil Engineers, New York, Vol. 1, pp. 87–170.

De Mello, V.F.B. 1972. Thoughts on Soil Engineering Applicable to Residual Soils. In *Proc., Third Southeast Asian Conference on Soil Engineering*, Hong Kong, Hong Kong Institution of Engineers and Southeast Asian Geotechnical Society, pp. 5–34.

Fredlund, D.G. 1987. Slope Stability Analysis Incorporating the Effect of Soil Suction. In *Slope Stability: Geotechnical Engineering and Geomorphology* (M.G. Anderson and K.S. Richards, eds.), John Wiley & Sons, New York, pp. 113–144.

Geological Society of London. 1990. Tropical Residual Soils: Geological Society Engineering Group Working Party Report. *Quarterly Journal of Engineering Geology*, Geological Society of London, Vol. 23, No. 1, pp. 4–101.

Jones, F.O. 1973. *Landslides of Rio de Janeiro and the Serra das Araras Escarpment, Brazil*. U.S. Geological Survey Professional Paper 697, 42 pp.

Lambe, P.C., and A.H. Riad. 1991. Back Analysis of Two Landslides in Residual Soil. In *Proc., Ninth Pan American Conference on Soil Mechanics and Foundation Engineering*, American Society of Civil Engineers, New York, pp. 425–439.

Lumb, P. 1962. Effect of Rain Storms on Slope Stability. In *Proc., Symposium on Hong Kong Soils*, Hong Kong, Hong Kong Joint Group of the Institutions of Civil, Mechanical, and Electrical Engineers, pp. 73–87.

Lumb, P. 1975. Slope Failures in Hong Kong. *Quarterly Journal of Engineering Geology*, Geological Society of London, Vol. 8, pp. 31–65.

Mitchell, J.K., and N. Sitar. 1982. Engineering Properties of Tropical Residual Soils. In *Engineering and Construction in Tropical and Residual Soils, Honolulu, Hawaii*, American Society of Civil Engineers, New York, pp. 30–57.

Sandroni, S.S. 1985. Sampling and Testing of Residual Soils in Brazil. In *Sampling and Testing of Residual Soils* (E.W. Brand and H.B. Phillipson, eds.), Scorpion Press, Hong Kong, pp. 31–50.

Skempton, A.W. 1970. First-Time Slides in Over-Consolidated Clays. *Geotechnique*, Vol. 20, No. 3, pp. 320–324.

Sowers, G.F. 1954. Soil Problems in the Southern Piedmont Region. *Proc., ASCE*, Vol. 80, Separate No. 416.

Sowers, G.F. 1963. Engineering Properties of Residual Soil Derived from Igneous and Metamorphic Rocks. In *Proc., Second Pan American Conference on Soil Mechanics and Foundation Engineering*, Brazil, American Society of Civil Engineers, New York, Vol. 1, pp. 39–61.

Sowers, G.F., and T.L. Richardson. 1983. Residual Soils of Piedmont and Blue Ridge. In *Transportation Research Record 916*, TRB, National Research Council, Washington, D.C., pp. 10–16.

St. John, B.J., G.F. Sowers, and C.E. Weaver. 1969. Slickensides in Residual Soils and Their Engineering Significance. In *Proc., Seventh International Conference on Soil Mechanics and Foundation Engineering*, Sociedad Mexicana de Mecánica de Suelos, Mexico City, Vol. 2, pp. 591–597.

Townsend, F.C. 1985. Geotechnical Characteristics of Residual Soils. *Journal of the Geotechnical Engineering Division*, ASCE, Vol. 111, No. 1, pp. 77–94.

Vargas, M. 1974. Engineering Properties of Residual Soils from the South-Central Region of Brazil. In *Proc., Second International Congress of the International Association of Engineering Geology*, Sao Paulo, Brazil, Associação Brasileira de Geologia de Engenharia, Vol. 1.

Vargas, M., and E. Pichler. 1957. Residual Soil and Rock Slides in Santos (Brazil). In *Proc., Fourth International Conference on Soil Mechanics and Foundation Engineering*, Butterworths Scientific Publications, London, Vol. 2, pp. 394–398.

Vaughan, P.R. 1985a. Mechanical and Hydraulic Properties of In-Situ Residual Soil General Report. In *Proc., First International Conference on Geomechanics in Tropical Lateritic and Saprolitic Soils*, Brasilia, Brazil, Comisión Federal de Electricidad, Mexico City, Vol. 3, pp. 231–263.

Vaughan, P.R. 1985b. Pore Pressures due to Infiltration into Partially Saturated Slopes. In *Proc., First International Conference on Geomechanics in Tropical Lateritic and Saprolitic Soils*, Brasilia, Brazil, Comisión Federal de Electricidad, Mexico City, Vol. 2, pp. 61–71.

Vaughan, P.R., M. Maccarini, and S.M. Mokhtar. 1988. Indexing the Engineering Properties of Residual Soil. *Quarterly Journal of Engineering Geology*, Geological Society of London, Vol. 21, pp. 69–84.

White, R.M., and T.L. Richardson. 1987. Predicting the Difficulty and Cost of Excavation in the Piedmont. In *Foundations and Excavations in Decomposed Rock of the Piedmont Province* (R.E. Smith, ed.), American Society of Civil Engineers, New York, pp. 15–36.

Wolle, C.M., and W. Hachich. 1989. Rain-Induced Landslides in Southeastern Brazil. In *Proc., 12th International Conference on Soil Mechanics and Foundation Engineering*, Brasilia, Brazil, A.A. Balkema, Rotterdam, Netherlands, pp. 1639–1642.

Chapter 20

A. KEITH TURNER

COLLUVIUM AND TALUS

1. INTRODUCTION

Physical and chemical weathering of bedrock produce disaggregated particles that accumulate on the land surface in the absence of erosional processes adequate to remove them. In locations with sufficient topographic relief, gravitational forces acting on these disaggregated particles cause them to move down the slopes and accumulate as distinctive deposits along the lower portions of slopes, in topographic depressions, and especially at the bases of cliffs. The terms *colluvium* or *colluvial materials* are used to refer to deposits that have been transported by gravitational forces.

The characteristics of colluvial materials vary according to the characteristics of the bedrock sources and the climate under which weathering and transport occur. Generally, colluvium is weakly stratified and consists of a heterogeneous mixture of soil and rock fragments ranging in size from clay particles to rocks a meter or more in diameter. These deposits are often only marginally stable. Because they are found along the lower portions of valley sides, such deposits frequently need to be partially excavated to allow passage of transportation facilities (Figure 20-1). The resulting cut slopes are commonly unstable and require constant monitoring and maintenance.

In humid tropical climates rapid chemical weathering promotes the formation of deep residual soils, or residuum. Creep of such soils on slopes may cause them to gradually take on the characteristics of colluvial materials. In cold climates colluvium is affected by the presence of permanently frozen ground, or *permafrost*. As discussed in Chapter 25, frozen ground conditions impede internal drainage and often lead to saturation of the surficial layers. *Solifluction*, the slow flowage of saturated surficial materials, produces characteristic lobate topographic patterns reflecting slope instability.

Instability of colluvial slopes in tropical conditions is discussed in Chapter 19, and Chapter 25 contains a discussion of such slopes in very cold climates. Information provided in this chapter

FIGURE 20-1
Colluvial slope along Denver and Rio Grande Western Railroad, Glenwood Canyon, Colorado.
J.D. HIGGINS, COLORADO SCHOOL OF MINES, GOLDEN, COLORADO

should be combined with that in Chapters 19 and 25 for a complete evaluation of these materials.

1.1 Definition of Colluvium

Bates and Jackson defined colluvium as follows:

> A general term applied to any loose, heterogeneous, and incoherent mass of soil material and/or rock fragments deposited by rain wash, sheet wash, or slow continuous downslope creep, usually collecting at the base of gentle slopes or hillsides. (Bates and Jackson 1980, 125)

Typically, colluvium is a poorly sorted mixture of angular rock fragments and fine-grained materials. These deposits rarely are more than 8 to 10 m thick, and they usually are thinnest near the crest and thickest near the toe of each slope (Figure 20-2).

Colluvium may be the most ubiquitous surficial deposit. Costa and Baker (1981) reported estimates that colluvium covers more than 95 percent of the ground surface in humid temperate regions and from 85 to 90 percent of the ground surface in semiarid mountainous areas. At the base of slopes, colluvium interfingers with alluvial deposits and may actually constitute the major portion of these deposits.

1.2 Definition of Talus and Scree

FIGURE 20-2
Idealized cross section of colluvial deposit.

Relatively rapid physical fragmentation of bedrock exposed on cliffs frequently produces large numbers of rock fragments ranging in size from small to very large. After falling from the cliff, these frag-

FIGURE 20-3
Talus slope in western Colorado.
R. D. ANDREW, COLORADO DEPARTMENT OF TRANSPORTATION

ments accumulate at the base to form a wedge-shaped deposit (Figure 20-3). A number of terms are used to describe these deposits, *talus* and *scree* being the most common. The term *talus* is derived from the French word for a slope on the outside of a fortification; originally it referred to the landform rather than the material. More recently, *talus* has been used to describe the material itself and *talus slope* has been used to define the landform. Often the cliff is not perfectly straight but contains recesses or ravines that tend to funnel the rock fragments into a chute (Figure 20-4); these fragments form a conical pile called a *talus cone* (Figure 20-5). In many cases multiple chutes form overlapping cones, creating a complex intermixing of talus materials (Figure 20-6).

Some geologists use *talus* and *scree* interchangeably, but there is considerable confusion concerning the definition and use of both terms. Bates and Jackson defined talus as

> rock fragments of any size or shape (usually coarse and angular) derived from and lying at the base of a cliff or very steep rocky slope. Also, the outward sloping and accumulated heap or mass of such loose broken rock, considered as a unit, and formed chiefly by gravitational falling, rolling, or sliding. (Bates and Jackson 1980, 638)

According to Bates and Jackson, scree is defined as follows:

> A term commonly used in Great Britain as a loose equivalent of *talus* in each of its senses: broken rock fragments; a heap of such frag-

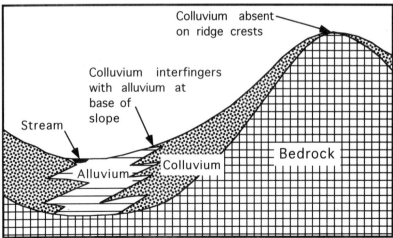

FIGURE 20-4
Rock chute as source of talus, Glenwood Canyon, Colorado.
J.D. HIGGINS, COLORADO SCHOOL OF MINES, GOLDEN, COLORADO

FIGURE 20-5
Talus cone, Glenwood Canyon, Colorado.
J.D. HIGGINS, COLORADO SCHOOL OF MINES, GOLDEN, COLORADO

FIGURE 20-6
Multiple overlapping talus cones forming complex talus deposit, Glenwood Canyon, Colorado. Construction of Interstate-70 is under way at base of this deposit.
R. D. ANDREW, COLORADO DEPARTMENT OF TRANSPORTATION

ments; and the steep slope consisting of such fragments. Some authorities regard scree as the material that makes up a sloping land feature known as talus; others consider scree as a sheet of any loose, fragmented material lying on or mantling a slope and talus as that material accumulating specifically at the base of, and obviously derived from, a cliff or other projecting mass. (Bates and Jackson 1980, 562)

Talus fragments can range in size to include huge boulders tens of meters across; however, a lower size limit has not been well defined. Typically, large fragments are found near the toe of the deposit and smaller particles are found near the top because the greater momentum and larger radius of the bigger particles permit them to travel further (Figures 20-3 and 20-4). The maximum angle of slope that can be held by a loose pile of rock fragments, regardless of their size, is termed the *angle of repose*. Talus slopes composed of fragments of varying sizes typically exhibit angles of repose that are rarely less than 34 degrees or greater than 37 degrees (Strahler 1971, 583). The angles of most talus slopes appear relatively uniform from top to bottom (Figure 20-3). However, several authors suggest that these slopes may be slightly concave, with steeper angles on the highest portions.

1.3 Terms Used in This Chapter

In this chapter the terms *colluvium* or *colluvial deposits* will be considered the most generic for all deposits created by the accumulation of disaggregated particles largely by gravitational forces. For such deposits composed of predominantly coarse rocky fragments, the term *talus* will generally be used to refer to both the landform and the material, and the term *talus slope* will be used only when it is necessary to distinguish between the landform and the material. The term *scree* will not be used.

2. SLOPE INSTABILITY IN COLLUVIUM

Given the ubiquitous nature of colluvium and its distribution within many hilly or mountainous areas, it should not be surprising that the instability of colluvial slopes has resulted in major economic losses and on some occasions the loss of human life (Costa and Baker 1981).

The most common class of landslides in colluvium involves two distinct components of movement: an initial shallow rotational or translational slide followed by flowage of the disturbed mass (Ellen and Fleming 1987; Ellen 1988). This failure mode results in a relatively small circular scar

at the location of the initial slide, a long narrow track leading directly down the hillside formed by the flowage of the liquefied soil and debris, and a zone of deposition in the minor drainage channel at the base of the slope (Figure 20-7). Numerous individual such slides provide large volumes of debris that clog stream channels. Liquefaction of this material causes it to remobilize and produce large and rapidly moving debris flows that destroy lives and property downstream.

Various terms have been used for such landslides in the past, including *debris avalanche* (Sharpe 1938), *soil slip/debris flow* (Campbell 1975), *flow slide* (Hutchinson 1968), *soil avalanche* (Wentworth 1943; Keefer 1984), and *disintegrating soil slip* (Kesseli 1943). In this chapter these landslides will be referred to as *debris flows* to conform with the terminology developed in Chapter 3. However, it should be noted that in these landslides both sliding and flowing are critical to the process. Sliding determines the timing and location of the initiation of the movement, whereas flowing determines movement path and rate. A more complete term for these landslides would be *soil slide–debris flow*; however, the simpler term *debris flow* will be used except in cases where a distinction between slide and flow is useful.

Debris flows involving colluvium may move with velocities exceeding 3 m/sec, and their composition will reflect not only the colluvium materials, which may range from fine-grained soils to coarse-grained debris, but also abundant coarse organic materials (such as trees) and materials from within and adjacent to stream channels (including buildings and automobiles). Accordingly, debris flows may include movements that others have termed *mudflows*, *debris avalanches*, or *debris torrents* (Sharpe 1938; Swanston and Swanson 1976; Varnes 1978). This simplification appears justified because all these processes are similar (Costa 1984; Ellen and Wieczorek 1988, 4).

Much colluvium in midlatitude regions originated during Pleistocene glacial periods when a mantle of rubble was formed by vigorous frost action on bedrock exposed beyond the ice-sheet margins. In areas subjected to glaciation, colluvial deposits overlie, and thus are younger than, glacial deposits. In investigations of slope stability in midlatitude regions, it must be recognized that many of these colluvial deposits formed under different climate regimes. Since their physical properties reflect

FIGURE 20-7
Principal features of a soil slide–debris flow (modified from Smith 1988).

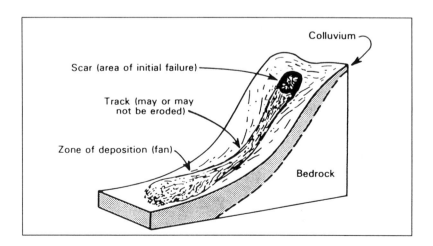

Colluvium

Scar (area of initial failure)

Track (may or may not be eroded)

Zone of deposition (fan)

Bedrock

both their age and their origin, their stability may be marginal under current climate conditions.

Sidle et al. (1985) identified five natural factors that have the greatest influence on the stability of colluvial slopes:

1. Soil properties, especially the hydrologic and mineralogic conditions that affect engineering soil behavior and strength properties of the colluvium;
2. Geomorphology, including the geologic and tectonic setting, slope gradient, and shape;
3. Hydrology, especially soil water recharge and effective evapotranspiration rates that reflect local climate and vegetation conditions;
4. Vegetative cover, including the reinforcing effect of root systems and the loss of such strength when roots deteriorate following timber harvesting or fire; and
5. Seismicity, especially the potential for liquefaction of marginally stable soils on steep slopes.

These natural factors are important on all slopes, but they must be evaluated in special ways when slopes are formed with colluvial materials. Section 2.3 contains a discussion of commonly found characteristics of colluvial deposits that must be considered when field investigations are conducted or strength properties are assigned for stability calculations. Geomorphic factors, especially the importance of topographic hollows in many situations, are evaluated in Section 2.4. Hydrology and the importance of vegetative cover to the stability of colluvial slopes are discussed in Sections 2.5 and 2.6, and the frequently overlooked effects of fire on the strength and erosion resistance of colluvium are discussed in Section 2.7. Seismic shaking frequently results in widespread failures of colluvial slopes, but since this topic is addressed in Chapter 4, only a brief discussion is included in Section 2.8 of this chapter.

2.1 Studies of Landslides in Colluvium

Before the 1960s, the importance of landslides in colluvium was largely ignored except by researchers in southern California (Eaton 1935; Sharpe 1938; Kesseli 1943). Several collections of technical reports now exist that address colluvial landslides from several aspects and at various levels of detail. The Geological Society of America

has published three major collections of studies of colluvial landslides (Coates 1977; Costa and Wieczorek 1987; Schultz and Jibson 1989). Books by Sidle et al. (1985) and Anderson and Richards (1987) also contain considerable information about such landslides as well as examples of landslides that do not involve colluvium. Numerous references will be made to these publications in the subsequent discussion.

Japanese researchers have also conducted detailed evaluations of the landslide danger on colluvial slopes (Sassa 1984; Sassa et al. 1981; Takahashi 1978). Reviews and assessments of these Japanese studies were provided by Chen (1987) and Okunishi and Okimura (1987).

Unusually large rainstorms have triggered catastrophic landslide events in colluvium at several locations around the world. These events led to considerable economic damage and loss of life. They were thus studied in some detail to determine their underlying causes in the hope that this information would prove useful in preventing future disasters. These studies form the basis for much of the current knowledge and theories concerning landslide processes involving colluvium.

Jones (1973) was the first to emphasize the potential severity of colluvial landslides in areas beyond southern California. He reported on the tens of thousands of landslides that occurred along the Serra das Araras escarpment in Brazil following a severe storm on January 22–23, 1967. Debris flows developed on valley sideslopes in colluvial and residual soils several meters thick overlying the crystalline bedrock. The steep vegetation-covered slopes were nearly denuded as soil and rock debris flowed into local valleys. Once deposited in the valleys, the materials became remobilized and formed huge debris flows that caused severe damage to roads and a major power plant and killed more than 1,700 people (Jones 1973; Costa and Baker 1981). Very similar thin landslides and debris flows were observed in saturated residual soils in southwestern Colombia (Figure 20-8). In this case, however, the landslides were triggered by earthquake shaking rather than intense rainfall.

In August 1969, Hurricane Camille caused severe damage in the Blue Ridge Mountains region of Virginia (Williams and Guy 1973). Thick colluvium and weathered bedrock were destabilized by a 24-hr precipitation that exceeded 600 mm, and debris flows surged rapidly down heavily

FIGURE 20-8
Thin landslides and debris flows on steep slopes of saturated residual soils at headwaters of San Vicente River, southwestern Colombia, triggered by 1994 Paez earthquake.
R. L. SCHUSTER, U.S. GEOLOGICAL SURVEY

FIGURE 20-9
Debris flows along East Branch of Hat Creek, Virginia, caused by Hurricane Camille in 1969 (Williams and Guy 1973, Figure 14a).

forested natural slopes (Figure 20-9). Most of the 150 people killed by this storm were overwhelmed by these rapidly moving debris flows (Costa and Baker 1981).

An unusually powerful storm affected the San Francisco Bay region of California in January 1982. It caused numerous colluvial landslides that resulted in considerable property damage and several deaths. These landslides were subjected to exten-

sive investigations (Smith and Hart 1982; Ellen and Wieczorek 1988). Other studies in the eastern United States (Schultz and Jibson 1989) and in Puerto Rico (Jibson 1989) have been reported. More recently, there has been a series of studies on colluvial landslides in the Cincinnati area of Ohio (Baum 1994; Fleming and Johnson 1994; Haneberg and Gökce 1994; Riestenberg 1994).

Southern California has had long experience with colluvial landslides that occurred after fires had removed slope vegetation. Much research has been conducted to evaluate the effects of fire on the rate of landsliding on colluvial slopes in southern and central California (DeBano et al. 1967; Campbell 1975; DeBano 1980; Wells 1987) as well as in Scotland, New South Wales, and Oregon (Fairbairn 1967; Brown 1972; Mersereau and Dyrness 1972; Bennett 1982).

2.2 Description and Mechanics of Colluvial Landslide Processes

Several workers have noted that the initiation and style of movement of landslides on colluvial slopes reflect the thickness of the colluvium (Ellen and Fleming 1987; Fleming and Johnson 1994). Landslides involving thick colluvium are usually relatively slow-moving rotational slides that rarely exhibit subsequent disaggregation and flowage. In contrast, landslides on slopes mantled with thin colluvium exhibit an initial translational sliding failure and a subsequent disaggregation of this sliding mass to form a debris flow. This *soil slide–debris flow* movement is the most commonly observed failure mode involving colluvial slopes. The mechanism causing the mobilization and flowage of the sliding mass has been the subject of considerable investigation and debate (Johnson and Rahn 1970; Morton and Campbell 1974; Campbell 1975; Sassa 1984; Ellen and Fleming 1987; Ellen 1988; Howard et al. 1988).

Howard et al. (1988) defined the transformation of an initially rigid sliding soil mass into a debris flow as consisting of five stages (Figure 20-10). The first stage [Figure 20-10(a)] occurs as the colluvium slides along, or just above, the soil-bedrock contact. The sliding process is controlled by tension cracks in the colluvium and by high porewater pressures caused by a permeability contrast between soil and bedrock. With continued movement, the second stage [Figure 20-10(b)] evolves as internal shearing causes a reduction in strength

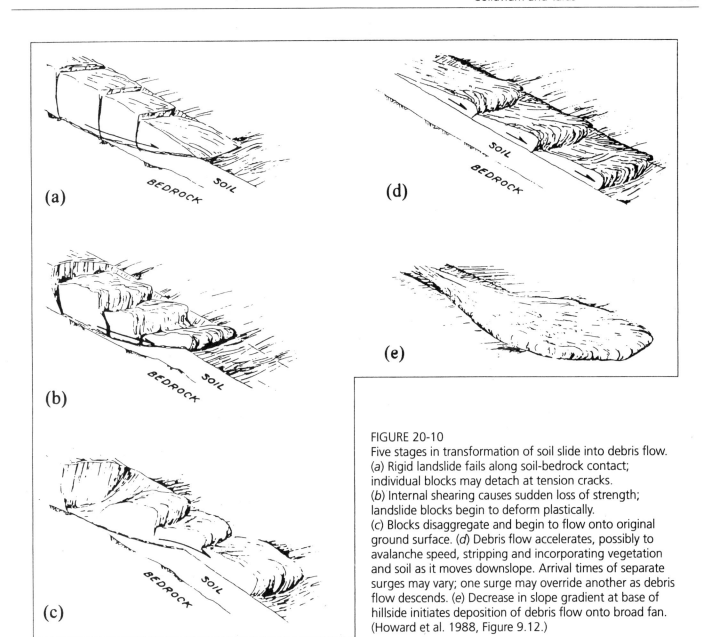

FIGURE 20-10
Five stages in transformation of soil slide into debris flow.
(a) Rigid landslide fails along soil-bedrock contact; individual blocks may detach at tension cracks.
(b) Internal shearing causes sudden loss of strength; landslide blocks begin to deform plastically.
(c) Blocks disaggregate and begin to flow onto original ground surface. (d) Debris flow accelerates, possibly to avalanche speed, stripping and incorporating vegetation and soil as it moves downslope. Arrival times of separate surges may vary; one surge may override another as debris flow descends. (e) Decrease in slope gradient at base of hillside initiates deposition of debris flow onto broad fan. (Howard et al. 1988, Figure 9.12.)

of the colluvium. Increasing levels of plastic deformation result (Johnson and Rahn 1970; Campbell 1975). Shearing causes dilation of the soil mass. Water from tension cracks is drawn into the dilating mass, further weakening and disaggregating the soil material. In stage three, the soil material begins to flow over the ground surface [Figure 20-10(c)]. As the initial mass continues to flow down the slope, it commonly accelerates and incorporates soils and debris, forming, in stage four, a debris flow [Figure 20-10(d)]. At the base of the slope, the decreased gradient causes deposition of the debris in stage five [Figure 20-10(e)].

Ellen and Fleming (1987) and Ellen (1988) reported on investigations of the numerous colluvial landslides that occurred in the San Francisco Bay Area following a severe storm in January 1982. These investigations combined field observations with laboratory testing of soil samples to produce a better understanding of the mechanisms controlling landslides in colluvium.

Field observations confirmed that the most common form of landsliding was a combination *soil slide–debris flow*. The initial failure produced a shallow, relatively small landslide scar caused by translational sliding of the colluvium over the bedrock.

Most of these initial soil slides mobilized completely to form debris flows that drained the scars, leaving them empty of failed material. The debris flows ran rapidly downhill in relatively narrow tracks. However, in some cases the mobilization of the sliding material was less complete. Such partially mobilized flows left dislocated masses of materials in the initial landslide scar and carried intact blocks of material down the slope within the debris flow. Reliable eyewitness reports suggested that rates of mobilization varied from just a few seconds to more than 24 hr (Ellen and Fleming 1987).

Ellen and Fleming wished to identify reasons for such differences in degree and speed of mobilization and flow of the initial soil slide materials. Previous studies (Rodine 1974; Pierson 1981) suggested that some clay in the colluvium is necessary to sustain flow. However, Ellen and Fleming suspected that there must be an upper limit to the amount of clay. They suggested that increased clay content adds cohesive strength to soils, which impedes remolding and also reduces permeability. Ellen and Fleming considered reduced permeability as critical because most soils having normal densities dilate when subjected to shearing and must then take up additional water to continue flowing. Reduced permeability would interfere with the rapid intake of water.

In the Ellen and Fleming study, the clay contents of 50 soil samples ranged from about 3 to 35 percent; however, 98 percent of the samples had clay contents greater than 8 percent. When samples obtained from very slow-moving debris flows in cut slopes or old landslides were removed from consideration, the largest clay content was 25 percent. Ellen and Fleming concluded that the range in clay contents, from 8 to 25 percent, was still so broad that clay content by itself was inadequate to define colluvial slopes that might be susceptible to landsliding. They looked for some additional quantitative measures that might help define this susceptibility.

Rodine (1974) and Johnson (1984) proposed a mobility index (MI) defined as

$$MI = \frac{\text{Saturated water content of in-place soil}}{\text{Water content required for that soil to flow}}$$

(20.1)

Although it is obvious that flow is most apt to occur when $MI \geq 1.0$, they demonstrated that it can take place when MI is as low as 0.85. However,

the determination of MI required soils testing to obtain a value for the denominator, the water content required for the soil to flow. This appeared inconvenient to Ellen and Fleming, and they proposed an approximate mobility index (AMI) defined as

$$AMI = \frac{\text{Saturated water content of in-place soil}}{\text{Liquid limit}}$$

(20.2)

Ellen and Fleming (1987) justified this change on practical grounds. Qualitatively, the liquid limit represents a water content at which soils exhibit marginally fluid behavior at near-surface conditions. Quantitatively, the liquid limit corresponds to a shear strength of about 2 kPa, and if this shear-strength value is used in the equation describing shear flow proposed by Johnson (1970), flow is predicted for colluvium slabs thicker than about 20 cm on slopes of 30 degrees for typical soil materials having saturated unit weights of 20 kN/m^3 (Ellen and Fleming 1987, 34). Since this critical thickness value of 20 cm was considerably less than the 1-m thicknesses typically observed in the San Francisco Bay Area, Ellen and Fleming believed that this validated the AMI concept.

Ellen and Fleming suggested that the observed differences in degree and rates of mobilization could be explained by differences in soil structure or fabric. Soils that have a relatively flocculated or loose fabric will collapse when subjected to even small shear movements. Under wet conditions, the collapse leads to a rapid liquefaction due to increased pore pressure. Collapsible soils having low permeability are especially susceptible, since rapid drainage of the excess water is unlikely. In contrast, relatively dense colluvial soils dilate when subjected to shear. The increased volume requires water to be drawn into the soil mass in order for flowage to continue. This water inflow requires time, especially if the soil has a relatively low permeability.

From the foregoing considerations, Ellen and Fleming (1987) concluded that loose, contractive colluvium with relatively low permeability is the most likely source of rapidly mobilized debris flows. These soils have void ratios greater than those at the liquid limit and hence when saturated have AMI values greater than 1.0. Liquefaction is almost instantaneous, occurring after strains as small as 1 percent, and the debris flows empty the initial landslide scars and flow rapidly downslope forming debris deposits with smooth surfaces. In contrast, dilative, denser colluvium is less likely to mobilize rapidly. Shear deformation along the base of the sliding mass is usually inadequate to remobilize the entire soil mass. Selective mobilization along the flanks and margins is possible, creating blocks within the mass. Cracks may channel water to selected areas, further encouraging selective remobilization. At the toe of the sliding slab, liquefaction may result from the greater volume of available water coupled with greater strains and remolding caused by oversteepening and bending as the slab moves out of the scar (Figure 20-11). Consequently, dilative, denser colluvial deposits produce slower-moving debris flows containing lumps or blocks of undeformed material. In some cases these denser colluvial soils produce more slablike, broader landslide areas (Figure 20-12).

2.3 Characteristics of Colluvium Affecting Slope Stability

Colluvium is formed by the movement and deposition of particles by gravity, although water and wind are sometimes secondary transportation agents. Colluvium may accumulate one small grain at a time or as a result of large catastrophic movement of materials, especially debris flows. In rugged terrain, colluvium can be merged and interlayered with alluvium in debris fans and cones associated with minor drainages. The aprons or wedges of colluvium may be modified over time by deposition, flooding, and dissection of stream channels and by internal composition changes resulting from weathering. Within any given colluvial deposit, particle gradation and density vary with depth and in irregular patterns across the extent of the deposit.

Colluvium frequently contains stratigraphic horizons that represent changes in deposition rates. For example, long periods of inactivity may allow the development of organic soil horizons that are then covered during periods of intense deposition. These horizons can be observed in excavations but are difficult to discern by conventional drilling methods.

Colluvium formed in arid regions often results from a combination of gravity-driven grain-by-grain downslope movements with debris-flow materials rapidly deposited from rare flash floods. These floods are of short duration and lose their moisture

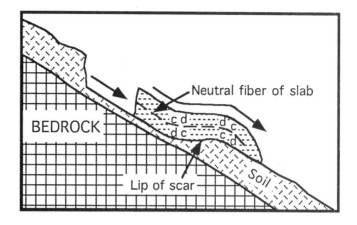

rapidly by evaporation and infiltration. The resulting deposits often exhibit a very flocculated or loose soil structure that is stable only as long as it remains dry. Wetting of these soils, such as may be caused by diversion of water from a paved highway surface, will cause the flocculated structure to collapse to a much smaller volume. This collapse may further lead to temporary liquefaction of the soils and their rapid erosion by flowage. These arid colluvial soils are said to be subject to *hydrocompaction*. Hydrocompaction may lower the ground surface by as much as 1 m (Figure 20-13).

Hydrocompaction is obviously damaging to transportation facilities. Its potential can be readily determined by evaluating in situ soil densities. Stabilization can be achieved by saturating the materials before construction, but a considerable period of time may be required for complete saturation and compaction. Also, stabilization requires large volumes of water, which may not be available in arid regions where hydrocompaction risk is most frequent. Even after compaction is achieved, these soils remain highly erodible, so careful design of culverts and water collection and dispersal systems is necessary.

Shales and other soft rocks generally produce fine-grained colluvium, often containing significant proportions of clay. The slow downslope creep of colluvium results in a progressive alignment of mineral grains and the creation of numerous microscopic shear surfaces. These surfaces greatly reduce the normally expected shear strength of colluvial materials, and yet the surfaces are hard to identify or retain in samples. Fleming and Johnson reported on their examination of colluvial landslide materials from the Cincinnati, Ohio, area:

FIGURE 20-11 Schematic downslope cross section of soil slide–debris flow as it leaves scar, showing zones of dilation *d* and contraction *c* in slab as it passes over lip. Position of neutral fiber depends on soil behavior (modified from Ellen 1988, Figure 6.30).

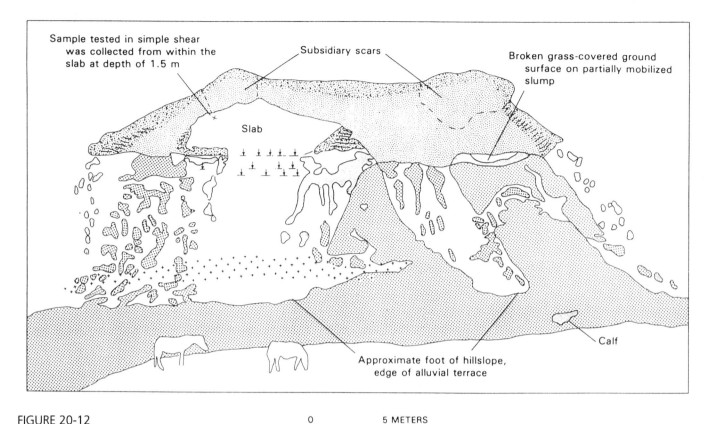

FIGURE 20-12
Sketch based on
photograph of
landslide in shallow
colluvium along
Salmon Creek, Marin
County, California.
Hillside slopes
toward observer at
about 22 degrees.
Both margins of
landslide have
mobilized as
debris flows; slab
remaining within
arcuate scar above
horses has slid
downward about
1 m (Ellen 1988,
Figure 6.35B).

0   5 METERS

APPROXIMATE SCALE AT SCAR

EXPLANATION

Ground surface covered by grass or brush

Deposits largely from light-colored soil

Dark soil exposed in scar

Remnants of slip surface at base of dark soil

Light-colored soil and bedrock exposed in scar, and lesser light-colored deposits on floor of scar

Area of seepage

Deposits from dark soil

Slip surface at toe of slab

Boundary of subsidiary scars

The failure surfaces are paper thin, highly polished, and striated. At all scales of observation, roughness is evident on the surfaces. At a scale of a few micrometers, the surfaces have small steps between shiny surfaces that appear to be analogous to chatter marks on rocks in fault zones. At a larger scale, the surfaces are scratched by fossil fragments and sand-size grains in the colluvium. At a still larger scale, the surfaces bifurcate around rock fragments, and multiple surfaces occur where the shape of the failure surface is changing. (Fleming and Johnson 1994, B-1)

Fleming and Johnson (1994) differentiated between landslides that occurred in thin colluvium (less than about 2 m thick) and those that occurred in thicker colluvium. They noted that the thick-colluvium landslides moved relatively slowly, perhaps only a few centimeters a year, and exhibited multiple overlapping scarps at their heads, multiple toe bulges downslope, and indistinct flanks. In contrast, the landslides in the thin colluvium moved more rapidly, typically exhibited a "stick-slip" behavior in which they accelerated after initial movement, and often moved completely out of their initial failure scars. These landslides had well-defined flanks, multiple head scarps, and a single toe.

Fleming and Johnson conducted laboratory tests on soil samples collected near the failure zones. Strength values varied according to the way the samples were collected and tested. Minimum strength values were obtained from a ring-shear device using samples that had had their coarse fractions removed. These estimates of residual shear strength values matched values obtained by back calculation of stability for the thick-colluvium landslides. The measured, laboratory-derived shear-strength values appeared insufficient to support the colluvium on the thin-colluvium slopes. Fleming and Johnson suggested that additional strength provided by tree roots, roughness of the failure surface, or a small residual cohesion might explain the discrepancy between the apparent strength and the slope stability.

2.4 Geomorphic Factors

A uniform thickness of colluvium is rarely present on any slope, and thicker deposits may effectively mask minor hollows in the underlying bedrock surface (Figure 20-14). The presence of such hollows may materially affect the stability of a colluvium-covered slope (Dietrich et al. 1986). Hack and Goodlett (1960) proposed that slopes be divided geomorphically into hollows, noses, and side slopes. Hollows are "any area on the slope in which the contours are concave outward away from the ridge" and they "generally occur in the valley axis at the stream head as a sort of extension of the channel" (Hack 1965, 20). Noses are areas "on the interfluve in which the slopes are convex outward toward the valley" (Hack 1965, 21), and side slopes are areas with straight contours located between noses and either hollows or channels.

The importance of hollows as locations susceptible to the initiation of colluvial landsliding varies, both locally and between regions, because of complex interactions and small variations in geologic, topographic, and climate factors. Hollows frequently appear to be important sources of debris flows. However, their importance as initiators of shallow colluvial landslides is more variable.

FIGURE 20-13
Results of saturation test on hydrocompactible colluvial soils in western Colorado; land surface has collapsed about 1 m.
A.K. TURNER, COLORADO SCHOOL OF MINES, GOLDEN, COLORADO

Hack and Goodlett (1960) first noted that hollows appeared to be an important factor in initiating debris flows in the Appalachian Mountains of Virginia. Their observations were subsequently confirmed by others (Woodruff 1971; Williams and Guy 1973; Lessing et al. 1976; Bogucki 1977; Pomeroy 1981, 1984). The importance of hollows was also revealed by studies in other regions: in coastal Alaska (Swanston 1967, 1969), British Columbia (O'Loughlin 1972), and the Oregon Coast Range (Pierson 1977; Deitrich and Dunne 1978; Swanson et al. 1981; Swanson and Roach 1985). Studies conducted in the San Francisco Bay Area following the damaging storm of January 1982 focused on hollows as a source of debris (Smith and Hart 1982; Reneau and Dietrich 1987; Ellen and Wieczorek 1988; Smith 1988).

In general, it appears that hollows are most important in areas where hill slopes are underlain by fairly massive bedrock and there is a regular dissection of the topography into minor drainages, and hollows are less important where the bedrock is variable and the topography is less regular. Along the Oregon coast, where the bedrock is composed of volcanic rocks including lava flows, tuffs, and breccias, Marion (1981) reported that side slopes were more important sites of landslide initiation than hollows. Jacobson (1985) reported that shallow landslides tended to initiate on benches on side

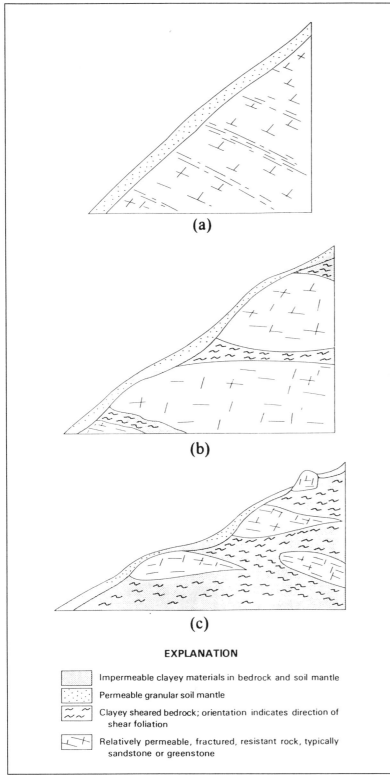

FIGURE 20-14
Schematic downslope cross sections of hillsides in (a) hard,
(b) intermediate, and (c) soft bedrock terrain, showing how hollows may
mask bedrock irregularities (Ellen et al. 1988, Figure 7.5).

slopes formed of stratified sandstone and shale and not along hollow axes. Pomeroy (1981, 1984) reported that in areas of Pennsylvania where sandstone and shale are interbedded, the topography lacks a regular division into noses and hollows, and in these regions hollows are not important factors in initiating landslides. However, in nearby regions where steeper slopes are underlain by more sandstone and less shale, hollows appear to be the most important source of landslides.

In the San Francisco region, about two-thirds of the shallow colluvial landslides are initiated in hollows (Smith and Hart 1982; Reneau and Dietrich 1987; Smith 1988). This finding suggests that the mapping and identification of hollows be given priority in producing landslide hazard maps for this region. Hollows act not only as sources of landslides but also as conduits for debris flows that frequently result from the initial landsliding. Landslide hazard and susceptibility mapping procedures based on analysis of drainage patterns have been implemented (Hollingsworth and Kovacs 1981; Baldwin et al. 1987; Smith 1988).

2.5 Hydrology

Intense rainfall is cited as the most common triggering mechanism for landslides involving colluvium (Jones 1973; Campbell 1975; Fukuoka 1980; Brand et al. 1984; Cannon and Ellen 1985; Church and Miles 1987; Neary and Smith 1987; Wieczorek 1987). Much work has been done to relate debris-flow occurrences to rainfall intensity and duration (Campbell 1975; Mark and Newman 1988; Cannon 1988; Cannon and Ellen 1988; Wieczorek and Sarmiento 1988). Campbell (1975) suggested that at least 267 mm of rainfall is necessary in southern California after a dry season to bring the soils to field capacity, the water content at which water drains from the soil at the same rate as it is added. Campbell further demonstrated that a rainfall intensity of 5 to 6 mm/hr is necessary to trigger debris flows once field capacity has been achieved.

Kesseli (1943) postulated that slope failure of colluvium occurs when rainfall intensity is great enough to cause water to percolate into the colluvium at a rate that exceeds the rate at which the water can percolate into bedrock (Figure 20-15). This condition forms a temporary perched water table in the colluvium, and a downslope seepage force develops within the soil mass. Excess pore

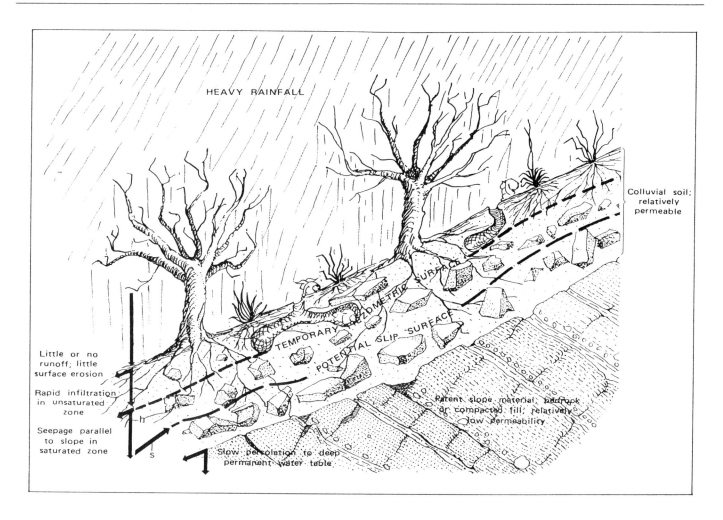

pressures develop, the soil becomes buoyant, and the resisting forces are decreased. This Kesseli model has been adopted by Campbell (1975) and many other workers to explain the initiation of soil sliding in colluvium, the most commonly observed initial failure mechanism.

Water can be concentrated at certain locations within a colluvial slope by various means. Topographic effects, especially the presence of hollows, are discussed in Section 2.4. Ellen (1988) noted that the flow of water within the colluvium may be concentrated by several other factors: the shape of the bedrock surface, breaks in the slope, differences in permeability within both the colluvium and the bedrock, and human-induced modifications to the slope. This last factor includes many of the modifications resulting from the construction of transportation facilities, such as cut-slope geometries and installation of ditches and culverts that modify the patterns of surface flow and concentrate flows at specific locations. Designers of

transportation facilities that cross colluvium must carefully consider the possible effects of the proposed drainage collection and discharge systems on the future stability of these slopes.

The relationship between groundwater conditions within the bedrock and initiation of colluvial landslides has also been noted (Ellen 1988; Ellen et al. 1988). Permeability contrasts within the bedrock may result in artesian conditions. Seepage of water from bedrock into the colluvium is more likely where the bedrock has layers with strong permeability contrasts that dip out of the slope. Seepage rates within bedrock aquifers are usually quite low, but Ellen (1988) stated that at least some of the bedrock seepage that affected initial soil slides in the San Francisco Bay Area appeared to have been fed by storm rainfall.

2.6 Vegetation

Plant root systems may increase slope stability for colluvial materials on steep slopes, and the nearly

FIGURE 20-15
Buildup of perched water table in colluvium during heavy rainfall (Campbell 1975, Figure 16).

total removal of forest cover by clearcutting may cause increased slope instability. Nevertheless, vegetation growing on a slope was traditionally considered to have a negligible effect on slope stability and was ignored in slope stability computations (Greenway 1987). However, studies in the mountainous forested regions of the United States, Canada, Japan, and New Zealand have shown that neglecting the effects of vegetation on steep slopes mantled with relatively thin colluvial soils can yield stability estimates that are considerably in error. Still, vegetation-slope interactions are complex, and this complexity has hampered attempts to quantify such interactions in slope-stability analyses (Greenway 1987).

The importance of roots for the stability of shallow colluvial soils on steep slopes and the effects of forest removal have been documented for western North America (Swanston 1970; O'Loughlin 1974b; Ziemer and Swanston 1977; Burroughs and Thomas 1977; Wu et al. 1979; Gray and Megahan 1981; Sidle and Swanston 1982). Similar studies have been conducted in Japan (Endo and Tsuruta 1969; Kitamura and Namba 1976; Nakano 1971) and in New Zealand (O'Loughlin 1974a; O'Loughlin and Pearce 1976; O'Loughlin et al. 1982; Selby 1982). Recent studies have reported on the role of roots in slope stability of colluvial soils in the Cincinnati, Ohio, area (Riestenberg 1994).

Generally, it is claimed that root systems contribute to soil strength by providing an additional cohesion component, referred to as ΔC. Available data, including the data presented in Table 20-1 from studies in New Zealand, suggest that roots increase the cohesion but do not influence the angle of internal friction (Sidle et al. 1985). Bishop and Stevens (1964) identified a fourfold to fivefold increase in the number of shallow landslides on steep hillsides in southeastern Alaska within 10 years of clearcutting. They suggested that the increase was at least partially attributable to root deterioration.

Some studies have attempted to quantify the strength of soil-root systems. Large metal shear boxes of various sizes and designs have been used in field tests to shear undisturbed blocks of soil containing roots (Endo and Tsuruta 1969; O'Loughlin 1974a, 1974b; Waldron and Dakessian 1981; O'Loughlin et al. 1982; Waldron et al. 1983). These tests demonstrated that the majority of the roots that broke failed in tension rather than by shear. This finding confirms many observations of actual landslides in which failed roots were reported around colluvial landslide scars. Estimates of the magnitude of ΔC provided by roots have been made by several investigators using different methods. These estimates are summarized in Table 20-2.

Sidle et al. (1985) concluded that roots influence soil strength and increase colluvial slope stability in three ways:

1. They bind unstable soil mantles to stable subsoils or substrata. This effect is most pronounced when a potential failure zone exists within the rooting zone.
2. They form a dense interwoven network in the upper 30 to 50 cm of soil, and this network provides lateral membrane strength that tends to reinforce the shallow soil mass and hold it in place.
3. Roots of individual trees may provide deeper anchors, stabilizing soil arches that extend across the slope (Figure 20-16). The trees act as buttresses or pile anchors. It may be possible to apply soil-arching restraint theory to determine a critical spacing of trees that maximizes the stabilization of a colluvial slope.

Vegetation may also lead to slope instability. Keller (1979, 117–118) reported that shallow soil slips are especially common in southern California on steep cut slopes covered with ice plant (*Carpobrotus edulis*). The stems of this

Table 20-1
Shear-Strength Parameters for New Zealand Beech Forest Soils with and Without Competent Roots (Sidle et al. 1985)

Shear-Strength Parameter	Intact Forest with Competent Roots	Clearcut Area Without Competent Roots
Apparent cohesion (kPa)	6.6	3.3
Internal friction angle (degrees)	36	36

Table 20-2
Results of Studies of Root Strength Factor ΔC (Sidle et al. 1985)

INFORMATION SOURCE	SOIL-VEGETATION SITUATION	ΔC (kPa)	METHOD USED
Swanston (1970)	Mountain till soils under conifers in Alaska	3.4-4.4	Back calculations
O'Loughlin (1974a)	Mountain till soils under conifers in British Columbia	1.0-3.0	Back calculations
Endo and Tsuruta (1969)	Cultivated loam soils (nursery) under Alder	2.0-12.0	Direct shear tests
Wu et al. (1979)	Mountain hill soils under conifers in Alaska	5.9	Tensile root-strength equations
Waldron and Dakessian (1981)	Clay loams in small containers growing pine seedlings	5.0	Direct shear tests
O'Loughlin et al. (1982)	Shallow stony loam hill soils under mixed evergreen forests, New Zealand	3.3	Direct shear tests
Gray and Megahan (1981)	Sandy loam soils under conifers in Idaho	10.3	Tensile root-strength equations
Burroughs and Thomas (1977)	Mountain and hill soils under conifers in west Oregon and Idaho	3.10-17.5	Root density information
Sidle and Swanston (1982)	Stony mountain soil under conifers and brush	2.2	Back calculations
Waldron et al. (1983)	Clay loam over gravel in large columns growing 5-year pine seedlings	8.9-10.8	Direct shear tests
Waldron et al. (1983)	Clay loam in large columns growing 5-year pine seedlings	9.9-11.8	Direct shear tests

creeping, low succulent plant contain sacs or beads that are swollen with water (Spellenberg 1979, 318). During wet winter months, the shallow-rooted ice plant absorbs great quantities of water into their vascular systems, adding considerable weight and increasing downslope driving forces. Failures commonly result in a shallow translational slide of several centimeters of soil and roots that extends a considerable distance across the face of the slope.

Other investigations also suggest that the type of vegetation covering a slope may affect slope stability. Studies in southern California by Campbell (1975) reported that soil slips were three to five times more frequent on grass-covered slopes than on chaparral-covered slopes. On grass-covered slopes, the landslides were shorter and wider and the failures began where there were lower slope angles than on chaparral-covered slopes composed of similar soils. However, the complete removal of vegetation, especially the effects of removal of vegetation by fire (discussed in the following section), suggests that vegetation generally has a positive effect on the stability of colluvial slopes.

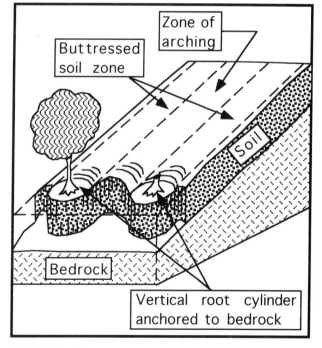

FIGURE 20-16
Schematic diagram showing soil anchoring, buttressing, and arching by trees on colluvial slope (modified from Greenway 1987, Figure 6.23).

2.7 Fire

Increased instability of colluvial slopes has been observed following naturally occurring or human-induced fires. Wildfires are a common occurrence in many ecosystems; for example, Hanes (1977) estimated that the chaparral vegetation of southern California burns every 10 to 40 years. Human-induced fires are a widely used component of land management practices in many parts of the world. Fire is used to clear vegetation before development, to burn residues after cropping or timber harvesting, to aid in the establishment or maintenance of preferred vegetation species, and to reduce wildfire hazard by controlled, prescribed burning. As a consequence, fire is a fairly common occurrence at many landslide sites.

Many soils exhibit water-repellent properties after burning. On a typical burned slope, the surficial soil is loosely compacted and easily wettable. However, a few millimeters below the surface, a layer of soil forms that is water repellent, or *hydrophobic* (Figure 20-17). This water repellency is created when waxy organic molecules formed by the burning of vegetative litter in and on the soil surface are driven down into the soil, where they coat individual soil particles. The waxy organic coating resists penetration by water and forms a barrier to infiltration of the water into the regolith below (DeBano 1980). DeBano et al. (1967) indicated that water repellency in soil develops at a temperature of about 200°C and increases to a maximum at about 370°C. At higher temperatures

FIGURE 20-17 Schematic diagram of postfire colluvial soil slope, showing effects of hydrophobic soil layer.

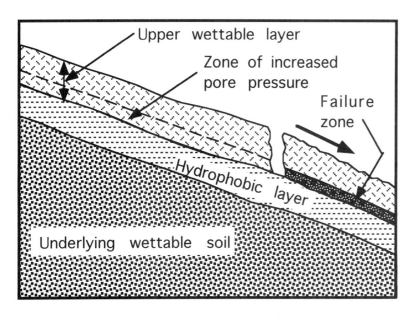

the water repellency is destroyed, so the existence, thickness, and continuity of a water-repellent layer within the colluvium depend on the intensity and duration of the fire.

The most widely documented cases of increased landsliding following fires are based on studies conducted in the Los Angeles region of southern California (Wells 1987). The mediterranean climate there with its warm, dry summers and winter rainy season causes the peak fire season to immediately precede a period of increased precipitation. Early rains fall on barren, freshly burned slopes. These effects were first reported by Eaton (1935), who correctly identified the flooding responses as debris flows. From Eaton's descriptions, it is clear that the phenomenon was recognized and understood by those working in the Los Angeles area but not by those in other regions (Wells 1987).

The problem of slope instability after fires and subsequent flooding by debris flows is not confined to southern California. Brown (1972) reported debris flows and increased sedimentation rates from freshly burned watersheds in Australia. Increased landsliding was observed in Scotland where hill heathland was burned to prevent tree regeneration (Fairbairn 1967). Mersereau and Dyrness (1972) and Bennett (1982) reported the effects on slope instability of slash burning following clearcut timber harvesting in the Cascade Mountains and Coast Range of western Oregon. Cannon et al. (1995) provided a detailed evaluation of postfire debris flows that inundated a 3-mi stretch of Interstate-70 west of Glenwood Springs, Colorado. Following a period of extremely intense rainfall, debris flows originated in areas near the highway that had been subjected to an intense fire in July 1994 that burned more than 800 ha. These debris flows engulfed more than 30 vehicles and their drivers, pushing several into the adjacent Colorado River, and at one location formed an alluvial fan deposit within the river channel (Figure 20-18). Fortunately, no deaths and only a few minor injuries resulted. However, this major highway, which carries about 15,000 vehicles a day, was completely closed for more than 16 hr before emergency cleanup crews succeeded in opening one lane in each direction to traffic.

In colluvium the relationship between fire and landsliding is complex. Decreased infiltration capacity due to the development of a water-repellent zone generally occurs at moderately to severely burned sites. The decreased infiltration capacity has

FIGURE 20-18
Cleanup activities after debris flows blocked Interstate-70 west of Glenwood Springs, Colorado. These debris flows occurred with first heavy rains after large forest fire had affected slopes above roadway.
BOB ELDERKIN, BUREAU OF LAND MANAGEMENT

two opposing effects on slope instability: a possible decrease in landsliding due to reduced moisture within the colluvium and a possible increase in sheet erosion and shallow landsliding due to saturation and erosion of the surface layers. In southern California, Rice (1977) observed an initial decrease in landsliding immediately following a fire and peak landslide activity about 2 to 3 years later. This increased slope instability was explained as a result of decomposition of former root systems and development of root systems by new vegetation that opened passages through the water-repellent layers, allowing increased infiltration into the colluvium.

Greatly accelerated rates of surface erosion by both wet and dry processes have been observed during the days and weeks following a fire (Wells 1987). During and immediately following a fire, significant amounts of hillslope debris are delivered to minor stream channels by a process called *dry ravel.* This process involves downslope movement by gravitational forces of individual soil grains, aggregates of grains, and coarser fragments. It is particularly active on steeper slopes (those more than 30 degrees) and in dry, noncohesive soils such as those derived from granites, pumice, and certain sandy sedimentary rocks (Sidle 1980).

The cause of dry ravel has been related to lower bulk densities of nonwettable materials, leading to decreased angles of internal friction. Thus, materials that were at or near their natural angle of repose before the fire would experience dry ravel during and after the burning (Sidle et al. 1985).

The effects of dry ravel are increased by the rapid development of extensive rill networks on many slopes during the season immediately after a fire. These rill networks form in the surficial soil layers in response to the underlying shallow, water-repellent soil. They form extremely rapidly, sometimes in minutes, and provide an efficient means of moving surficial materials into the minor stream channels (Wells 1987). Once the stream channels become clogged with sediment derived from both dry ravel and rill erosion processes, saturation of these materials and initiation of debris flows down the channels are the obvious consequences.

More material may be added to the stream channels by soil slide—debris flows along the valley sides. Such slope failures are also a consequence of the presence of shallow, water-repellent layers in the colluvium. The material merely adds to the volume of material already clogging the stream channels and further increases the potential for disastrous debris flows downstream. This combination of events helps explain why comparatively large postfire debris flows result from very small storms or after relatively short periods of rainfall. Frequently these debris flows disrupt or even destroy transportation facilities (Figure 20-18).

2.8 Seismicity

Earthquake shaking, a major triggering factor for many types of landslides, is discussed in Chapter 4. Many of the examples used in Chapter 4 are landslides involving materials other than colluvium and so are beyond the scope of this chapter. However, numerous landslides on colluvial slopes are common when moderate to large earthquakes occur in hilly regions. For example, more than 2,000 landslides and avalanches were reported in a 5000-km^2 zone around the epicenter of the 1964 Good Friday earthquake in Alaska (Hackman 1965) and more than 1,000 landslides occurred in a 250-km^2 region during the 1971 San Fernando Valley earthquake in southern California (Morton 1971).

Sidle et al. (1985) suggested that inadequate attention had been paid to documenting the intensity of landslides caused by earthquake shaking or to the geomorphic, ecologic, and socioeconomic significance of such landslides. This lack of attention seems to be no longer the case. Chapter 4 contains several references to recent (post-1980) studies of the socioeconomic effects and geotechnical issues of earthquake-induced landslides. Relevant studies concerning landslides involving colluvium and residual soils include case histories in Guatemala and Ecuador (Harp et al. 1981; Crespo et al. 1991; Schuster 1991) and reports of studies following two recent earthquakes in California—the Loma Prieta earthquake near San Francisco in October 1989 and the Northridge earthquake north of Los Angeles in January 1994 (Plafker and Galloway 1989; Harp and Jibson 1995; Keefer and Manson in press). The Loma Prieta earthquake triggered an estimated 2,000 to 4,000 landslides on colluvial slopes that blocked a major highway and many secondary roads in the region (Plafker and Galloway 1989; Keefer and Manson in press). Interpretation of aerial photographs has identified more than 12,000 landslides with a width of at least 3 m that were triggered by the 1994 Northridge earthquake (Harp and Jibson 1995).

3. SLOPE INSTABILITY IN TALUS

Classic examples of massive talus deposits are common in high mountainous regions, where steep slopes, durable bedrock, lack of vegetation, and harsh climate combine to produce abundant rock fragments (Figure 20-19). In these regions movements of talus fragments are frequent and rock-fall hazards are considerable. Fortunately, transportation facilities rarely cross such regions and thus these unstable and hazardous conditions affect them infrequently. However, in regions of relatively high relief, transportation facilities commonly must pass through valleys and canyons in order to find acceptable routes at lower elevations. Talus deposits are commonly encountered along such valley and canyon routes (Figures 20-20 and 20-21).

FIGURE 20-19
Extensive and active talus deposits in mountainous region.
K.E. KOLM, COLORADO SCHOOL OF MINES, GOLDEN, COLORADO

The recent construction of Interstate-70 (I-70) through Glenwood Canyon in the state of Colorado encountered extensive talus deposits, and the experiences gained there have considerably enhanced the knowledge of geotechnical characteristics of talus and appropriate construction procedures where talus is encountered. Glenwood Canyon was eroded through a series of sedimentary and igneous rocks by the Colorado River. It reaches a depth of more than 1000 m and is characterized by steep slopes and cliffs, with extensive colluvium and talus deposits at their bases. The canyon provides an important low-gradient east-west route through an otherwise rugged plateau region. A railroad and an existing highway shared the narrow canyon bottom with the Colorado River. The scenic nature of the canyon raised environmental concerns that required the new four-lane Interstate highway to be constructed in such a way as to minimize impacts on canyon features. As a consequence, major roadway structures traverse and are founded on extensive talus slopes (Figure 20-22).

Existing knowledge was insufficient to design the required structures. Only a few studies are available that relate talus properties to the design and construction of transportation facilities (Ritchie 1963). Consequently, exploration techniques and foundation treatment methods were developed during the Glenwood Canyon project, several of which are discussed later in this chapter.

Talus is colluvium composed of predominantly large fragments. Talus thus shares many of the characteristics of colluvium discussed in Section 2.3. Hard, massive bedrock formations frequently produce very coarse, open-graded talus deposits. With time, the coarse fragments of these deposits may degrade or finer materials may be added by wind or water transport so that these deposits slowly become infilled with a matrix of fine-grained materials. The degree of infilling of these talus deposits may vary horizontally and vertically. In other cases, the talus may initially be formed from a mixture of coarse and fine particles.

Thus, talus deposits may range from extremely coarse-grained deposits with large voids to much more dense deposits consisting of coarse fragments embedded in a matrix of finer particles. Following the concepts proposed by Clift (1994) for describing rockfills, these talus deposits may be classified as rock-supported, transitional, or matrix-supported (Figure 20-23).

FIGURE 20-20
Typical talus deposit along road in valley.
R.D. ANDREW, COLORADO DEPARTMENT OF TRANSPORTATION

FIGURE 20-21
Talus deposit in canyon setting, western Colorado.
R.D. ANDREW, COLORADO DEPARTMENT OF TRANSPORTATION

Rock-supported talus is often inherently unstable and may be hazardous even to walk across. The weight of the deposit is transmitted as point loads among the fragments. These concentrated loadings may result in crushing or breakage of fragments or even in sudden realignments to a more stable configuration. In contrast, matrix-supported talus transmits most of the load by compaction of the relatively fine-grained matrix materials, and the larger fragments are supported, or "float," within the

FIGURE 20-22
Completed Interstate-70 crossing extensive talus, Glenwood Canyon, Colorado.
A.K. TURNER, COLORADO SCHOOL OF MINES, GOLDEN, COLORADO

(a)

Unfilled space between larger particles

Loose matrix material

(b)

Somewhat-compressed matrix material

(c)

Matrix material compressed to a maximum

FIGURE 20-23
Schematic diagrams of three classes of talus: (a) rock-supported, (b) transitional, and (c) matrix-supported (modified from Clift 1994, Figure 15).

matrix. Transitional talus has intermediate properties; the matrix materials may be loosely compacted since they only partially fill many of the voids, and the weight of the deposit is transmitted mostly by the contact between the large fragments. Each of these three talus categories behaves differently in a kinematic sense. Often a single talus deposit will contain sections from all three categories. Because the surface of most talus deposits is armored with larger-sized particles, the degree of infilling with depth is difficult to predict.

Estimating strength and stability properties of talus is extremely difficult. Stabilizing or densifying open-graded talus may be accomplished by grouting. However, experience with I-70 in Glenwood Canyon demonstrated that determining the best method of injecting grout, estimating the volume of grout required, or defining improved bearing capacity involved much trial and error, in situ field testing, and experience with the characteristics of talus in the area. Gaining suitable access for the necessary exploration equipment is often difficult on talus and may be hazardous in some cases (Figure 20-24).

Ritchie (1963) noted that talus becomes stratified because of the gradual accumulation of rock fragments. Once a rock fragment comes to rest on the sloping surface of a talus deposit, Ritchie (1963, 23) stated, "it never, of itself, begins to roll again (although much of the material may move about by debris sliding). On mature natural slopes, rock roll is seldom a problem for highways located near their base." Ritchie emphasized that excavations into talus slopes are especially troublesome. Talus slopes reflect the natural angle of repose of the talus material, and generally cut slopes are constructed at a steeper angle. Even so, these cut slopes extend large distances away from the transportation facility, and once a rock begins to roll on such a slope, it cannot stop until it reaches the base of the cut. Many cut slopes are constructed slightly steeper than the inclination of the crude stratification of the talus. This constructed slope angle causes individual rock fragments to ravel, or pop out, from the face of the cut, resulting in continuous and progressive failure of the cut. Ritchie recommended that some form of slope protection for the traveling public is most desirable, if not mandatory, where such cut slopes are necessary in talus.

Talus slopes present other hazards. Although most talus is unlikely to initiate debris flows such as those in finer-grained colluvium, debris flows often issue from rock chutes above talus deposits. In severe cases these debris flows may extend completely across the talus slope and cause damage to transportation facilities located on the lower portions of the talus deposit.

4. SUBSURFACE INVESTIGATIONS

Traditional technologies, such as drilling, test pits, trenches, and geophysical methods, are all employed to obtain information on the physical characteristics of colluvium. However, each method has limitations in evaluating colluvial deposits. Internal heterogeneity of colluvial deposits coupled with the tendency for the strength of colluvium to degrade over time increase uncertainty and risk concerning characterizations of engineering properties. Colluvium deposited in water-deficient environments may exhibit a flocculated or honeycomb soil-particle structure that can lead to hydrocompaction.

4.1 Planning Subsurface Investigations

Engineering geologists and geotechnical engineers should plan very carefully any subsurface investigations for site-specific determination of the engineering properties of colluvium in order to take into account and reflect the following characteristics of this material:

- Colluvium at depth may be deeply weathered and is often more dense (infilled);
- The moisture profile can increase to saturation with depth;
- Internal friction and cohesion may vary horizontally and vertically within the deposit;
- Climate and parent materials are major contributors to the engineering parameters of colluvium;
- Typical slope angles, particle size and distribution, strength properties, and moisture content are all related to the origin and location of the colluvium;
- Colluvial deposits are often only marginally stable, and their relatively steep slopes may make equipment access difficult or even impossible; and
- Colluvial slopes frequently are located in areas subject to rock-fall and debris-flow hazards, so precautions to protect field investigation personnel may be necessary.

FIGURE 20-24
Exploratory drilling on talus, Glenwood Canyon, Colorado.
COLORADO DEPARTMENT OF TRANSPORTATION

By recognizing these characteristics and planning investigative procedures to account for them, the engineering geologist or geotechnical engineer is more likely to develop a successful investigation. For example, exploratory drill holes should be extended to greater than typical depths to check for unusual conditions, and areas subjected to subsurface investigations should be extended to add confidence to global stability calculations. Many years of water-level measurements may be required to provide confidence in the estimated maximum water levels assumed in design calculations.

4.2 Drilling Methods

Drilling in colluvium, and especially in talus, is difficult, relatively expensive, and often does not provide the geologist with a complete profile of the deposit. Drilling in fine-grained colluvium can be accomplished with conventional tools, such as solid and hollow flight augers and split-tube samplers. However, the supervising engineering geologist must be prepared for sudden changes in the properties of the deposit and to expect zones with no drilling recovery.

In Glenwood Canyon, Colorado, numerous talus deposits were encountered during construction of I-70. At one point near the center of the canyon, a large ventilation structure for two tunnels was to be located in the bottom of a narrow tributary canyon that had extensive talus deposits. Construction planning required a detailed subsurface evaluation of these talus deposits. These explorations provided much new experience concerning the best methods for investigating the nature of talus deposits.

The Colorado Department of Transportation (CDOT) cooperated with manufacturers to develop wire-line core drilling bits and procedures to successfully drill the test holes in these talus deposits. Severe problems resulted from the use of traditional exploratory drilling methods. In the initial exploration, soil-sampling drill rigs equipped with tri-cone bits and drive casings, or casings with diamond shoes, were used. These methods resulted in very slow production rates and poor-quality samples. When traditional diamond coring bits were used, they were damaged by heat buildup during periodic losses of drilling fluid. At this location, loss of drilling fluid was a common occurrence because of void spaces within the talus. Although void spaces are rarely found deep within a talus deposit, some voids exceeding 3 m were found at depth during this investigation. Uneven or eccentric loadings are detrimental to most rock-coring bits; thus, in spite of the best efforts of the drilling personnel, the diamond bit was frequently damaged because it became placed between a boulder and a void.

4.3 Test Pits and Trenches

Test pits and trenches often provide better observations of the profile of a colluvial deposit than can be obtained by examining samples removed from exploratory holes. Visual inspection of the walls of a trench can lead to more accurate design of sampling and testing programs and ultimately to a more accurate determination of engineering properties. The geologist and the designer are cautioned to consider that any excavations in colluvium may collapse rapidly. These deposits are among the most dangerous in which to perform steep-sided excavations.

4.4 Geophysical Exploration Methods

Geophysical methods often are not satisfactory or especially reliable for determining subsurface interfaces such as water horizons and bedrock contacts. The third dimension of a hillside deposit adds considerable uncertainty to the interpretation. These methods are better used to supplement drilling and test pit data.

5. DESIGN CONSIDERATIONS FOR COLLUVIUM

Although colluvium is unconsolidated and therefore a soil according to engineering classifications, construction in colluvium is typically much more difficult than in other major deposits, including alluvial, glacial, lacustrine, or eolian soils. Colluvium is more likely to exhibit unfavorable properties such as particle size and strength variations, large void spaces, and unexpected excavation failures.

5.1 Creep

Creep is a phenomenon of concern when structures are placed on colluvial deposits. It is known that some long-term translation or movement of materials can occur, especially within the near-surface regime. However, little definitive information exists on this phenomenon. A series of inclinometers were installed by CDOT through a major talus deposit in Glenwood Canyon, Colorado. These inclinometers were monitored for 4 years to support design calculations for foundations of a planned extensive viaduct along the base of the deposit. No movement was observed during the monitoring period. However, this single short-term program certainly cannot be used to conclude that this talus deposit will not exhibit creep in the future. The viaduct foundations were designed to accommodate lateral loadings that would result from some creep movements of the talus.

5.2 Installation of Ground Anchors

Installation of ground anchors in talus can cost an order of magnitude more than a similar installation in a clay soil. Drilling of horizontal or subhorizontal holes into colluvium, especially talus, requires special equipment and experienced crews (Figures 20-25 and 20-26). During construction of

Interstate-70 through Glenwood Canyon, many of the slope stabilization systems required either temporary or permanent retaining walls that relied on tiebacks into talus or colluvial slopes for their stability (Figures 20-26 and 20-27). Where talus deposits were encountered, conventional ground anchor installations employing stage grouting were not feasible because of the numerous void spaces and zones of very loose matrix material. Most of the grout was being lost, resulting in inadequate bond strength and a lack of corrosion protection.

A system that used geotextile "socks" to contain grout, known as grout containment devices (GCDs), was ultimately employed to install these anchors. Experiments were conducted with many different types of geotextiles, including both woven and nonwoven products, to determine which performed best as GCDs. Most of the nonwoven GCDs ruptured against the sharp fragments in the talus deposits under the high pumping pressures required to inject the grout. The most effective method of installation was to grout a woven geotextile material in short sections (approximately 1.5 m long) as the drill casing was removed. In coarse talus, the only other technique that was successful involved a proprietary expander-body anchor that utilized a steel GCD.

5.3 Pile Foundations

Foundations for structures that must be placed on colluvium must account for potential nonuniform bearing capacities beneath them. Nonuniform bearing capacities lead to unexpected eccentricities that may result in distress or failure of the structure. The depth and configuration of most colluvial deposits favor a pile-foundation design. Installation of deep pile foundations can be very difficult and expensive if large boulders, which may be randomly dispersed within many of these deposits, are encountered.

In Glenwood Canyon, the foundations for an extensive viaduct had to be placed on a talus deposit (Figure 20-28). Preblasting techniques were developed by CDOT and successfully used to allow the driving of large-diameter piles through this talus. The method involved predrilling a small-diameter hole at the location of each steel pile using wire-line drilling. The hole was cased with polyvinylchloride (PVC) pipe and backfilled with drill cuttings. Gelatin dynamite was attached

FIGURE 20-25
Large specialized rig drilling horizontal ground anchor holes in talus, Glenwood Canyon, Colorado.
COLORADO DEPARTMENT OF TRANSPORTATION

FIGURE 20-26
Smaller rig drilling ground anchor holes in talus for temporary slope-retention devices (*square concrete pads, upper part of photograph*), Glenwood Canyon, Colorado.
COLORADO DEPARTMENT OF TRANSPORTATION

FIGURE 20-27
Installation of permanent retaining walls that utilize tiebacks into talus, Glenwood Canyon, Colorado.
COLORADO DEPARTMENT OF TRANSPORTATION

FIGURE 20-28
Interstate-70 viaduct founded on talus, Glenwood Canyon, Colorado.
J.D. HIGGINS, COLORADO SCHOOL OF MINES, GOLDEN, COLORADO

to reinforced detonator cords and lowered into the hole, which was then stemmed. Detonation of the charges fractured any large talus fragments so the piling could be driven through the small broken fragments. The process was repeated as necessary until the pile had been driven to the desired tip elevation. This procedure obviously added to the cost for installation of these steel piles. However, when compared with other alternatives, it was a feasible and cost-effective solution.

5.4 Control of Groundwater and Surface Water

Control of groundwater and surface water is a critical design consideration. Every effort must be made to avoid increasing water sources in the colluvial soils. Design intentions must be enforced by construction personnel. One common mistake is to allow the concentration of roadway runoff into pipes that discharge into minor drainages. Improperly designed drainage structures may cause erosion or even landslides, and collection of water around foundations can lead to damaged structures. A properly designed drainage system should take into consideration its relationship to the geologic history of the deposit.

5.5 Dangers of Rock Fall and Debris Flow

When evaluating the stability of colluvial and talus deposits, the engineering geologist must be concerned with rock-fall and debris-flow hazards. A historical perspective that emphasizes the origin of these deposits through downslope movement of materials must be emphasized to designers of transportation facilities. In spite of careful evaluation and design in Glenwood Canyon, rock-fall incidents have caused minor damage to roadway structures (Figure 20-29) and killed three occupants of one vehicle when it was hit by a large rock.

ACKNOWLEDGMENTS

The long-term support and interest of many members of the Colorado Department of Transportation, especially Robert K. Barrett, Richard D. Andrew, and John B. Gilmore, made possible the development of several sections of this chapter. Their willingness to share their experiences during the construction of Interstate-70 through Glen-

FIGURE 20-29
Minor rock-fall
damage at base of
talus slope,
Interstate-70,
Glenwood Canyon,
Colorado.
J.D. HIGGINS, COLORADO
SCHOOL OF MINES, GOLDEN,
COLORADO

wood Canyon is greatly appreciated, and they provided several of the illustrations used in this chapter. The author's colleagues at the Colorado School of Mines, Kenneth E. Kolm and Jerry D. Higgins, generously offered additional photographs.

REFERENCES

Anderson, M.G., and K.S. Richards, eds. 1987. *Slope Stability: Geotechnical Engineering and Geomorphology*. John Wiley and Sons, Inc., New York, 648 pp.

Bates, R.L., and J.A. Jackson, eds. 1980. *Glossary of Geology*, 2nd ed. American Geological Institute, Falls Church, Va., 751 pp.

Baldwin, J.E., II, H.F. Donley, and T.R. Howard. 1987. On Debris Flow/Avalanche Mitigation and Control, San Francisco Bay Area, California. In *Debris Flows/Avalanches: Process, Recognition, and Mitigation* (J.E. Costa and G.F. Wieczorek, eds.), Reviews in Engineering Geology, Vol. 7, Geological Society of America, Boulder, Colo., pp. 223–236.

Baum, R.L. 1994. *Contribution of Artesian Water to Progressive Failure of the Upper Part of the Delhi Pike Landslide Complex, Cincinnati, Ohio*. U.S. Geological Survey Bulletin 2059-D, 14 pp.

Bennett, K.A. 1982. *Effects of Slash Burning on Surface Soil Erosion Rates in the Oregon Coast Range*. M.S. thesis. Oregon State University, Corvallis, 70 pp.

Bishop, D.M., and M.E. Stevens. 1964. *Landslides on Logged Areas on Southeast Alaska*. Research Paper NOR-1. Forest Service, U.S. Department of Agriculture, Juneau, Alaska, 18 pp.

Bogucki, D.J. 1977. Debris Slide Hazards in the Adirondack Province of New York State. *Environmental Geology*, Vol. 1, pp. 317–328.

Brand, E.W., J. Premchitt, and H.B. Phillipson. 1984. Relationship Between Rainfall and Landslides in Hong Kong. In *Proc., Fourth International Symposium on Landslides*, Canadian Geotechnical Society, Toronto, Vol. 1, pp. 377–384.

Brown, J.A.H. 1972. Hydrologic Effects of a Brushfire in a Catchment in Southeastern New South Wales. *Journal of Hydrology*, Vol. 15, pp. 77–96.

Burroughs, E.R., and B.R. Thomas. 1977. *Declining Root Strength in Douglas Fir after Falling as a Factor in Slope Stability*. Research Paper INT-190. Forest Service, U.S. Department of Agriculture, Ogden, Utah, 27 pp.

Campbell, R.H. 1975. *Soil Slips, Debris Flows, and Rainstorms in the Santa Monica Mountains and Vicinity, Southern California*. U.S. Geological Survey Professional Paper 851, 51 pp.

Cannon, S.H. 1988. Regional Rainfall-Threshold Conditions for Abundant Debris-Flow Activity. In *Landslides, Floods, and Marine Effects of the Storm of January 3–5, 1982, in the San Francisco Bay Region, California* (S.D. Ellen and G.F. Wieczorek, eds.), U.S. Geological Survey Professional Paper 1434, pp. 35–42.

Cannon, S.H., and S.D. Ellen. 1985. Rainfall Conditions for Abundant Debris Avalanches in the San Francisco Bay Region, California. *California Geology*, Vol. 38, No. 12, pp. 267–272.

Cannon, S.H., and S.D. Ellen. 1988. Rainfall That Resulted in Abundant Debris-Flow Activity During the Storm. In *Landslides, Floods, and Marine Effects of the Storm of January 3–5, 1982, in the San Francisco Bay Region, California* (S.D. Ellen and G.F. Wieczorek, eds.), U.S. Geological Survey Professional Paper 1434, pp. 27–34.

Cannon, S.H., P.S. Powers, R.A. Pihl, and W.P. Rogers. 1995. *Preliminary Evaluation of the Fire-Related Debris Flows on Storm King Mountain, Glenwood Springs, Colorado.* U.S. Geological Survey Open File Report 95-508, 38 pp.

Church, M., and M.J. Miles. 1987. Meteorological Antecedents to Debris Flow in Southwestern British Columbia: Some Case Studies. In *Debris Flows/Avalanches: Process, Recognition, and Mitigation* (J.E. Costa and G.F. Wieczorek, eds.), Reviews in Engineering Geology, Vol. 7, Geological Society of America, Boulder, Colo., pp. 63–80.

Chen, C. 1987. Comprehensive Review of Debris Flow Modelling Concepts in Japan. In *Debris Flows/ Avalanches: Process, Recognition, and Mitigation* (J.E. Costa and G.F. Wieczorek, eds.), Reviews in Engineering Geology, Vol. 7, Geological Society of America, Boulder, Colo., pp. 13–30.

Clift, A.E. 1994. *In Situ Properties and Settlement Characteristics of Rockfill.* Ph.D. thesis No. T-4375. Colorado School of Mines, Golden, 201 pp. plus appendices.

Coates, D.R., ed. 1977. *Landslides.* Reviews in Engineering Geology, Vol. 3, Geological Society of America, Boulder, Colo., 278 pp.

Costa, J.E. 1984. Physical Geomorphology of Debris Flows. In *Developments and Applications of Geomorphology* (J.E. Costa and P.J. Fleisher, eds.), Springer-Verlag, Berlin, pp. 268–317.

Costa, J.E., and V.R Baker. 1981. *Surficial Geology: Building with the Earth.* John Wiley and Sons, New York, 498 pp.

Costa, J.E., and G.F. Wieczorek, eds. 1987. *Debris Flows/Avalanches: Process, Recognition, and Mitigation.* Reviews in Engineering Geology, Vol. 7, Geological Society of America, Boulder, Colo., 239 pp.

Crespo, E., T.D. O'Rourke, and K.J. Nyman. 1991. Effects on Lifelines. In *The March 5, 1987 Ecuador Earthquakes: Mass Wasting and Socioeconomic Effects*, National Academy Press, Washington, D.C., pp. 83–99.

DeBano, L.F. 1980. *Water Repellent Soils: A State-of-the-Art.* Pacific Southwest Forest and Range Experiment Station General Technical Report PSW-46. Forest Service, U.S. Department of Agriculture, Berkeley, Calif., 21 pp.

DeBano, L.F., J.F. Osborn, J.S. Krammes, and J. Letey. 1967. *Soil Wettability and Wetting Agents...Our Current Knowledge of the Problem.* Pacific Southwest Forest and Range Experiment Station Research Paper PSW-43. Forest Service, U.S. Department of Agriculture, Berkeley, Calif., 13 pp.

Dietrich, W.E., and T. Dunne. 1978. Sediment Budget for a Small Catchment in Mountainous Terrain. *Zeitschrift für Geomorphologie,* Supplement, Vol. 29, pp. 191–206.

Dietrich, W.E., C.J. Wilson, and S.L. Reneau. 1986. Hollows, Colluvium, and Landslides in Soil-Mantled Landscapes. In *Hillslope Processes* (A. Abrahams, ed.), 16th Annual Geomorphology Symposium, Binghampton, N.Y., Allen and Unwin Ltd., Winchester, Mass., pp. 361–388.

Eaton, E.C. 1935. Flood and Erosion Control Problems and Their Solution. *Transactions, American Society of Civil Engineers,* Vol. 101, pp. 1302–1330.

Ellen, S.D. 1988. Description and Mechanics of Soil Slip/Debris Flows in the Storm. In *Landslides, Floods, and Marine Effects of the Storm of January 3–5, 1982, in the San Francisco Bay Region, California* (S.D. Ellen and G.F. Wieczorek, eds.), U.S. Geological Survey Professional Paper 1434, pp. 63–112.

Ellen, S.D., S.H. Cannon, and S.L. Reneau. 1988. Distribution of Debris Flows in Marin County. In *Landslides, Floods, and Marine Effects of the Storm of January 3–5, 1982, in the San Francisco Bay Region, California* (S.D. Ellen and G.F. Wieczorek, eds.), U.S. Geological Survey Professional Paper 1434, pp. 113–132.

Ellen, S.D., and G.F. Wieczorek. 1988. *Landslides, Floods, and Marine Effects of the Storm of January 3–5, 1982, in the San Francisco Bay Region, California.* U.S. Geological Survey Professional Paper 1434, 310 pp.

Ellen, S.D., and R.W. Fleming. 1987. Mobilization of Debris Flows from Soil Slips, San Francisco Bay Region, California. In *Debris Flows/Avalanches: Process, Recognition, and Mitigation* (J.E. Costa and G.F. Wieczorek, eds.), Reviews in Engineering Geology, Vol. 7, Geological Society of America, Boulder, Colo., pp. 31–40.

Endo, T., and T. Tsuruta. 1969. The Effect of the Tree's Roots on the Shear Strength of Soil. In *Annual Report for 1968*, Hokkaido Branch, Forest Experiment Station, Sapporo, Japan, pp. 167–182.

Fairbairn, W.A. 1967. Erosion in the River Findhorn Valley. *Scottish Geographic Magazine*, Vol. 83, No. 1, pp. 46–52.

Fleming, R.W., and A.M. Johnson. 1994. *Landslides in Colluvium*. U.S. Geological Survey Bulletin 2059-B, 24 pp.

Fukuoka, M. 1980. Landslides Associated with Rainfall. *Geotechnical Engineering*, Vol. 11, pp. 1–29.

Gray, D.H., and W.F. Megahan. 1981. *Forest Vegetation Removal and Slope Stability in the Idaho Batholith*. Research Paper INT-271. Forest Service, U.S. Department of Agriculture, Ogden, Utah, 23 pp.

Greenway, D.R. 1987. Vegetation and Slope Stability. In *Slope Stability* (M.G. Anderson and K.S. Richards, eds.), John Wiley and Sons, New York, pp. 187–230.

Hack, J.T. 1965. *Geomorphology of the Shenandoah Valley, Virginia and West Virginia, and Origin of the Residual Ore Deposits*. U.S. Geological Survey Professional Paper 484, 84 pp.

Hack, J.T., and J.C. Goodlett. 1960. *Geomorphology and Forest Ecology of a Mountain Region in the Central Appalachians*. U.S. Geological Survey Professional Paper 347, 66 pp.

Hackman, R.J. 1965. Interpretation of Alaskan Post-Earthquake Photographs. *Photogrammetric Engineering*, Vol. 31, pp. 604–610.

Haneberg, W.C., and A.O. Gökce. 1994. *Rapid Water-Level Fluctuations in a Thin Colluvium Landslide West of Cincinnati, Ohio*. U.S. Geological Survey Bulletin 2059-C, 16 pp.

Hanes, T.L. 1977. Chaparral. In *Terrestrial Vegetation of California* (M.G. Barbour and J. Major, eds.), John Wiley and Sons, New York, pp. 417–470.

Harp, E.L., and R.W. Jibson. 1995. *Inventory of Landslides Triggered by the 1994 Northridge, California, Earthquake*. U.S. Geological Survey Open-File Report 95-213, 16 pp., 2 plates.

Harp, E.L., R.C. Wilson, and G.F. Wieczorek. 1981. *Landslides from the February 4, 1976, Guatemala Earthquake*. U.S. Geological Survey Professional Paper 1204-A, 35 pp.

Hollingsworth, R., and G.S. Kovacs. 1981. Soil Slumps and Debris Flows: Prediction and Protection. *Bulletin of the Association of Engineering Geologists*, Vol. 28, No. 1, pp. 17–28.

Howard, T.R., J.E. Baldwin, II, and H.F. Donley. 1988. Landslides in Pacifica, California, Caused by the Storm. In *Landslides, Floods, and Marine Effects of the Storm of January 3–5, 1982, in the San Francisco Bay Region, California* (S.D. Ellen and G.F. Wieczorek, eds.), U.S. Geological Survey Professional Paper 1434, pp. 163–184.

Hutchinson, J.N. 1968. Mass Movement. In *Encyclopedia of Geomorphology* (R.W. Fairbridge, ed.), Reinhold Press, New York, pp. 688–695.

Jacobson, R.B. 1985. *Spatial and Temporal Distribution of Slope Processes in the Upper Buffalo Creek Drainage Basin, Marion County, West Virginia*. Ph.D. thesis. Johns Hopkins University, Baltimore, Md., 484 pp.

Jibson, R.W. 1989. Debris Flows in Southern Puerto Rico. In *Landslide Processes of the Eastern United States and Puerto Rico* (A.P. Schultz and R.W. Jibson, eds.), Special Paper 236, Geological Society of America, Boulder, Colo., pp. 29–55.

Johnson, A.M. 1970. *Physical Processes in Geology*. Freeman, Cooper and Co., San Francisco, 577 pp.

Johnson, A.M. 1984. Debris Flow. In *Slope Instability* (D. Brunsden and D.B. Prior, eds.), John Wiley and Sons, New York, pp. 257–361.

Johnson, A.M., and P.H. Rahn. 1970. Mobilization of Debris Flows. *Zeitschrift für Geomorphologie*, Supplement, Vol. 9, pp. 168–186.

Jones, F.O. 1973. *Landslides of Rio de Janeiro and the Serra das Araras Escarpment, Brazil*. U.S. Geological Survey Professional Paper 697, 42 pp.

Keefer, D.K. 1984. Landslides Caused by Earthquakes. *Geological Society of America Bulletin*, Vol. 95, No. 4, pp. 406–421.

Keefer, D.K., and M.W. Manson. In press. Landslides Produced by the 1989 Loma Prieta, California, Earthquake—Regional Distribution and Characteristics. In *The October 17, 1989, Loma Prieta, California, Earthquake: Landslides and Stream Channel Change* (D.K. Keefer, ed.) (U.S. Geological Survey Professional Paper 1551-C), Vol. 2 of *The Loma Prieta Earthquake of October 17, 1989*, U.S. Geological Survey Professional Papers.

Keller, E.A. 1979. *Environmental Geology*, 2nd ed. C.E. Merrill Publishing Company, Columbus, Ohio, 548 pp.

Kesseli, J.E. 1943. Disintegrating Soil Slips of the Coast Ranges of Central California. *Journal of Geology*, Vol. 51, No. 5, pp. 342–352.

Kitamura, Y., and S. Namba. 1976. A Field Experiment on the Uprooting Resistance of Tree Roots. In *Proc., 77th Meeting of the Japanese Forest Society* (English translation by J.M. Arata and R.R. Ziemer), Forest Service, U.S. Department of Agriculture, Arcata, Calif., pp. 568–570.

Lessing, P., B.R. Kulander, B.D. Wilson, S.L. Dean, and S.M. Woodring. 1976. *West Virginia*

Landslides and Slide Prone Areas. Environmental Geology Bulletin 15. West Virginia Geological and Economic Survey, Morgantown, 64 pp.

Marion, D.A. 1981. *Landslide Occurrence in the Blue River Drainage, Oregon.* M.S. thesis. Oregon State University, Corvallis, 114 pp.

Mark, R.K., and E.B. Newman. 1988. Rainfall Totals Before and During the Storm: Distribution and Correlations with Damaging Landslides. In *Landslides, Floods, and Marine Effects of the Storm of January 3–5, 1982, in the San Francisco Bay Region, California* (S.D. Ellen and G.F. Wieczorek, eds.), U.S. Geological Survey Professional Paper 1434, pp. 17–26.

Mersereau, R.C., and C.T. Dyrness. 1972. Accelerated Mass Wasting after Logging and Slash Burning in Western Oregon. *Journal of Soil and Water Conservation,* Vol. 27, No. 3, pp. 112–114.

Morton, D.M. 1971. Seismically Triggered Landslides in the Area above the San Fernando Valley. In *The San Fernando, California, Earthquake of February 9, 1971,* U.S. Geological Survey Professional Paper 733, pp. 99–104.

Morton, D.M., and R.H. Campbell. 1974. Spring Mudflows at Wrightwood, Southern California. *Quarterly Journal of Engineering Geology,* Vol. 7, No. 4, pp. 377–384.

Nakano, T. 1971. Natural Hazards: Report from Japan. In *Natural Hazards—Local, National, Global* (G.F. White, ed.), Oxford University Press, New York, pp. 231–243.

Neary, D.G., and L.W. Smith, Jr. 1987. Rainfall Thresholds for Triggering a Debris Avalanching Event in the Southern Appalachian Mountains. In *Debris Flows/Avalanches: Process, Recognition, and Mitigation* (J.E. Costa and G.F. Wieczorek, eds.), Reviews in Engineering Geology, Vol. 7, Geological Society of America, Boulder, Colo., pp. 81–92.

Okunishi, K., and T. Okimura. 1987. Groundwater Models for Mountain Slopes. In *Slope Stability* (M.G. Anderson and K.S. Richards, eds.), John Wiley and Sons, New York, pp. 265–286.

O'Loughlin, C.L. 1972. A Preliminary Study of Landslides in the Coast Mountains of Southwestern British Columbia. In *Mountain Geomorphology: Geomorphological Processes in the Canadian Cordillera* (H.O. Slaymaker and H.J. McPherson, eds.), Tantalus Research, Vancouver, British Columbia, Canada, pp. 101–111.

O'Loughlin, C.L. 1974a. The Effect of Timber Removal on the Stability of Forest Soils. *Journal of Hydrology of New Zealand,* Vol. 13, No. 2, pp. 121–123.

O'Loughlin, C.L. 1974b. A Study of Tree Root Strength Deterioration Following Clearfelling. *Canadian Journal of Forestry Research,* Vol. 4, No. 1, pp. 107–113.

O'Loughlin, C.L., and A.J. Pearce. 1976. Influence of Cenozoic Geology on Mass Movement and Sediment Yield Response to Forest Removal, North Westland, New Zealand. *Bulletin of the International Association of Engineering Geologists,* Vol. 14, pp. 41–46.

O'Loughlin, C.L., L.K. Rowe, and A.J. Pearce. 1982. Exceptional Storm Influences on Slope Erosion and Sediment Yields in Small Forest Catchments, North Westland, New Zealand. In *National Conference Publication 82/6,* Institution of Engineers, Barton, Australian Capital Territory, pp. 84–91.

Pierson, T.C. 1977. *Factors Controlling Debris-Flow Initiation on Forested Hillslopes in the Oregon Coast Range.* Ph.D. thesis. University of Washington, Seattle, 166 pp.

Pierson, T.C. 1981. Dominant Particle Support Mechanisms in Debris Flows at Mt. Thomas, New Zealand, and Implications for Flow Mobility. *Sedimentology,* Vol. 28, No. 1, pp. 49–60.

Plafker, G., and J.P. Galloway. 1989. *Lessons Learned from the Loma Prieta, California, Earthquake of October 17, 1989.* U.S. Geological Survey Circular 1045, 48 pp.

Pomeroy, J.S. 1981. *Storm-Induced Debris Avalanching and Related Phenomena in the Johnstown Area, Pennsylvania, with Reference to Other Studies in the Appalachians.* U.S. Geological Survey Professional Paper 1191, 22 pp.

Pomeroy, J.S. 1984. *Storm-Induced Slope Movements at East Brady, Northern Pennsylvania.* U.S. Geological Survey Bulletin 1618, 16 pp.

Reneau, S.L., and W.E. Dietrich. 1987. The Importance of Hollows in Debris Flow Studies: Examples from Marin County, California. In *Debris Flows/Avalanches: Process, Recognition, and Mitigation* (J.E. Costa and G.F. Wieczorek, eds.), Reviews in Engineering Geology, Vol. 7, Geological Society of America, Boulder, Colo., pp. 165–180.

Rice, R.M. 1977. Forest Management to Minimize Landslide Risk. In *Guidelines for Watershed Management* (S.H. Kunkle and J.L. Thames, eds.), United Nations Food and Agriculture Organization, Rome, Italy, pp. 271–287.

Riestenberg, M.M. 1994. *Anchoring of Thin Colluvium by Roots of Sugar Maple and White Ash on Hillslopes in Cincinnati.* U.S. Geological Survey Bulletin 2059-E, 25 pp.

Ritchie, A.M. 1963. Evaluation of Rockfall and Its Control. In *Highway Research Record 17*, HRB, National Research Council, Washington, D.C., pp. 13–28.

Rodine, J.D. 1974. *Analysis of the Mobilization of Debris Flows*. Ph.D. thesis. Stanford University, Stanford, Calif., 226 pp.

Sassa, K. 1984. The Mechanism Starting Liquefied Landslides and Debris Flows. In *Proc., Fourth International Symposium on Landslides*, Canadian Geotechnical Society, Toronto, pp. 349–354.

Sassa, K., A. Takei, and H. Mauri. 1981. Influences of Underground Erosion on Instability of a Crystalline Schist Slope. In *Proc., International Symposium on Weak Rock; Soft, Fractured, and Weathered Rock*, A.A. Balkema, Rotterdam, Netherlands, pp. 543–548.

Schultz, A.P., and R.W. Jibson, eds. 1989. *Landslide Processes of the Eastern United States and Puerto Rico*. Special Paper 236. Geological Society of America, Boulder, Colo., 102 pp.

Schuster, R.L. 1991. Introduction. In *The March 5, 1987, Ecuador Earthquakes* (R.L. Schuster, ed.), Natural Disaster Studies, National Research Council, Washington, D.C., No. 5, Chap. 1, pp. 11–22.

Selby, M.J. 1982. *Hillslope Materials and Processes*. Oxford University Press, New York, 264 pp.

Sharpe, C.S.F. 1938. *Landslides and Related Phenomena: A Study of the Mass Movement of Soil and Rock*. Columbia University Press, New York, 136 pp.

Sidle, R.C. 1980. *Slope Stability on Forest Land*. Pacific Northwest Extension Publication PNW 209. Oregon State University, Corvallis, 23 pp.

Sidle, R.C., A.J. Pearce, and C.L. O'Loughlin. 1985. *Hillslope Stability and Land Use*. Water Resources Monograph 11. American Geophysical Union, Washington, D.C., 140 pp.

Sidle, R.C., and D.N. Swanston. 1982. Analysis of a Small Debris Slide in Coastal Alaska. *Canadian Geotechnical Journal*, Vol. 19, pp. 167–174.

Smith, T.C. 1988. A Method for Mapping Relative Susceptibility to Debris Flows, with an Example from San Mateo County, California. In *Landslides, Floods, and Marine Effects of the Storm of January 3–5, 1982, in the San Francisco Bay Region, California* (S.D. Ellen and G.F. Wieczorek, eds.), U.S. Geological Survey Professional Paper 1434, pp. 185–194.

Smith, T.C., and E.W. Hart. 1982. Landslides and Related Storm Damage January 1982, San Francisco Bay Region. *California Geology*, Vol. 36, No. 7, pp. 139–152.

Spellenberg, R. 1979. *The Audubon Society Field Guide to North American Wildflowers, Western Region*, 7th ed. Knopf, New York, 862 pp.

Strahler, A.N. 1971. *The Earth Sciences*, 2nd ed. Harper & Row, New York, 824 pp.

Swanson, F.J., and C.J. Roach. 1985. Frequency of Debris Avalanches from Clearcut Hollows, Siuslaw River Basin, Oregon (abstract). *Eos*, Vol. 66, p. 900.

Swanson, F.J., M.M. Swanson, and C. Woods. 1981. Analysis of Debris-Avalanche Erosion in Steep Forest Lands: An Example from Mapleton Oregon, USA. In *Erosion and Sediment Transport in Pacific Rim Steeplands* (T.R.H. Davies and A.J. Pearce, eds.), Publication 132, International Association of Hydrological Sciences, Delft, Netherlands, pp. 67–75.

Swanston, D.N. 1967. *Soil-Water Piezometry in a Southeast Alaska Landslide Area*. Pacific Northwest Forest and Range Experiment Station Research Note PNW-68. Forest Service, U.S. Department of Agriculture, Portland, Oreg., 17 pp.

Swanston, D.N. 1969. *Mass Wasting in Coastal Alaska*. Pacific Northwest Forest and Range Experiment Station Research Paper PNW-83. Forest Service, U.S. Department of Agriculture, Portland, Oreg., 15 pp.

Swanston, D.N. 1970. *Mechanics of Debris Avalanching in Shallow Till Soils of Southeast Alaska*. Pacific Northwest Forest and Range Experiment Station Research Paper PNW-103. Forest Service, U.S. Department of Agriculture, Portland, Oreg., 17 pp.

Swanston, D.N., and F.J. Swanson. 1976. Timber Harvesting, Mass Erosion, and Steepland Forest Management in the Pacific Northwest. In *Geomorphology and Engineering* (D.R. Coates, ed.), Dowden Hutchinson and Ross, Stroudsburg, Pa., pp. 199–221.

Takahashi, T. 1978. Mechanical Characteristics of Debris Flow. *Journal of the Hydraulics Division*, ASCE, Vol. 104, No. HY8, pp. 1153–1169.

Varnes, D.J. 1978. Slope Movement Types and Processes. In *Special Report 176: Landslides: Analysis and Control* (R.L. Schuster and R.J. Krizek, eds.), TRB, National Research Council, Washington, D.C., pp. 11–33.

Waldron, S.J., and S. Dakessian. 1981. Soil Reinforcement by Roots: Calculation of Increased Soil Shear Resistance from Root Properties. *Soil Science*, Vol. 132, No. 6, pp. 427–435.

Waldron, S.J., S. Dakessian, and J.A. Nemson. 1983. Shear Resistance Enhancement of 1.22-meter

Diameter Soil Cross Sections by Pine and Alfalfa Roots. *Journal of the Soil Science Society of America*, Vol. 47, No. 1, pp. 9–14.

Wells, W.G., II. 1987. The Effects of Fire on the Generation of Debris Flows in Southern California. In *Debris Flows/Avalanches: Process, Recognition, and Mitigation* (J.E. Costa and G.F. Wieczorek, eds.), Reviews in Engineering Geology, Vol. 7, Geological Society of America, Boulder, Colo., pp. 105–114.

Wentworth, C.K. 1943. Soil Avalanches on Oahu, Hawaii. *Bulletin of the Geological Society of America*, Vol. 54, No. 1, pp. 53–64.

Wieczorek, G.F. 1987. Effect of Rainfall Intensity and Duration on Debris Flows in Central Santa Cruz Mountains, California. In *Debris Flows/Avalanches: Process, Recognition, and Mitigation* (J.E. Costa and G.F. Wieczorek, eds.), Reviews in Engineering Geology, Vol. 7, Geological Society of America, Boulder, Colo., pp. 93–104.

Wieczorek, G.F., and J. Sarmiento. 1988. Rainfall, Piezometric Levels, and Debris Flows Near La Honda, California, in Storms Between 1975 and 1983. In *Landslides, Floods, and Marine Effects of the Storm of January 3–5, 1982, in the San Francisco Bay Region, California* (S.D. Ellen and G.F. Wieczorek, eds.), U.S. Geological Survey Professional Paper 1434, pp. 43–62.

Williams, G.P., and H.P. Guy. 1973. *Erosional and Depositional Aspects of Hurricane Camille in Virginia, 1969.* U.S. Geological Survey Professional Paper 804, 80 pp.

Woodruff, J.F. 1971. Debris Avalanches as an Erosional Agent in the Appalachian Mountains. *Journal of Geography*, Vol. 70, pp. 399–406.

Wu, T.H., W.P. McKinnel, and D.N. Swanston. 1979. Strength of Tree Roots and Landslides on Prince of Wales Island, Alaska. *Canadian Geotechnical Journal*, Vol. 16, No. 1, pp. 19–33.

Ziemer, R.R., and D.N. Swanston. 1977. *Root Strength Changes after Logging in Southeast Alaska.* Pacific Northwest Forest and Range Experiment Station Research Note PNW-306. Forest Service, U.S. Department of Agriculture, Portland, Oreg., 10 pp.

Chapter 21

JOHN L. WALKINSHAW AND
PAUL M. SANTI

SHALES AND OTHER DEGRADABLE MATERIALS

1. INTRODUCTION

The designers of earthwork must take precautions when the materials at hand cannot be classified as rock or as soils in terms of their behavior in slopes or in civil engineering works in general. In their in situ form, the geologic formations may have names or appearances that imply rocklike behavior. Once disturbed, however, some of these formations retain the character of rock, but others may degrade to soil-size particles in a time frame that is relevant to the long-term performance of slopes built in, on, or with these materials.

The currently available methods of identifying, classifying, and treating these degradable materials so as to reduce the risk of slope failure are discussed in this chapter. Sedimentary rocks, which constitute the bulk of degradable materials worldwide, are discussed first. Other degradable materials, including weathered igneous and metamorphic rocks, are discussed in less detail. Emphasis is placed on the successful use of these materials in embankments and on their treatment in the formation of cut slopes. In geologic terms, all of the soils and rocks in the earth's crust are "degradable" materials, since all materials modify over geologic time. However, on the human time scale, only comparatively rapid degradation of a strong or hard rocklike material into a weaker soil-like material is of concern to the designer of stable slopes. Certain sedimentary rocks can exhibit a loss of strength that can be of several orders of magnitude within a

time frame that may be as short as only a few hours. Such a rapid breakdown is easily identified by straightforward laboratory tests. Other rock materials show no appreciable change in strength over many years. Unfortunately for engineers, the problem materials fall somewhere between these extremes of rock behavior.

Predicting the behavior of degradable materials has been the subject of much research since the early 1960s, and this research thrust continues today. Much of this research can be tied to the construction of large transportation facilities. In the United States, the development of the Interstate highway system required much higher cuts and embankments than had been common in the past. Problem geologic materials that previously could be economically addressed by avoidance, minor mitigation, or maintenance created the need for new engineering solutions. In Central America the construction of the Panama Canal resulted in slope problems that continued decades after construction and are also associated with the properties of degradable materials (Berman 1991).

2. GEOLOGICAL CONSIDERATIONS

Degradable materials were not considered in detail in the 1978 landslide report, but the basic principles that govern them were identified:

Before one can completely comprehend the particular problems of stability, one must under-

555

stand the lithology of the physical properties not only of the rock mass itself but of all the materials in the mass.... A sedimentary rock sequence, for example, is markedly different from an igneous series or a metamorphic complex. Each particular type is characterized by a certain texture, fabric, bonding strength, and macro and micro structures. The most important rock properties are the nature of the mineral assemblage and the strength of the constituent minerals; a rock material cannot be strong if its mineral constituents are weak or if the strength of the bonds between the minerals is weak. (Piteau and Peckover 1978, 194)

The strength of the bonds between the minerals is also related to the geologic history of the formation of interest. The resulting hardness is generally due to long-term consolidation under external pressures and not to cementing minerals (McCarthy 1988). Degradable materials can be grouped according to two broad geologic sources, those derived from sedimentary rocks and those derived from igneous and metamorphic rocks.

2.1 Degradable Materials from Sedimentary Rocks

Shales constitute about one-half of the volume of sedimentary rocks in the earth's crust. They are exposed or are under a thin veneer of soil over a third of the land area (Franklin 1981). Shales are by far the most pervasive and problematic degradable material. As early as 1948, Taylor stated (53): "Shale itself is sometimes considered a rock but, when it is exposed to the air or has the chance to take on water, it may rapidly decompose."

In the American Geological Institute's *Glossary of Geology*, shale is defined as

a fine-grained detrital sedimentary rock, formed by the consolidation (esp. by compression) of clay, silt, or mud. It is characterized by finely laminated structure, which imparts a fissility approximately parallel to the bedding. . . . [It is composed of] an appreciable content of clay minerals and detrital quartz. [Shale includes rocks such as] claystone, siltstone, and mudstone. (Bates and Jackson 1980, 573)

Referring to Huang (1962), Hopkins noted that

typically, shales are composed of about one-third quartz, one-third clay minerals, and one-

third miscellaneous substances. The principal minerals of shales, such as quartz, clay minerals, and hydrated oxides (such as bauxite and limonite), are formed by the weathering of feldspars and mafic igneous rocks. Some associated minerals such as calcite, dolomite, pyrite, illite, and glauconite are formed during and after deposition of the primary minerals. (Hopkins 1988, 8)

Unfortunately, many of the shale particles are less than 1 μm in diameter, and consequently study of their mineralogy is difficult or impossible by simple visual observation. The resulting geologic field classification of shales does not reliably relate to engineering properties.

Terzaghi and Peck described very clearly the geologic processes that lead to the problem properties of shales:

As the thickness of the overburden increases from a few tens of feet to several thousands, the porosity of a clay or silt deposit decreases; an increasing number of cohesive bonds develops between particles as a result of molecular interaction, but the mineralogical composition of the particles probably remains practically unaltered. Finally, at very great depth, all the particles are connected by virtually permanent, rigid bonds that impart to the material the properties of real rock. Yet, all the materials located between the zones of incipient and complete bonding are called shale. Therefore, the engineering properties of any shale with a given mineralogical composition may range between those of a soil and those of a real rock. (Terzaghi and Peck 1967, 425-426)

These two authors further suggested using an immersion test on intact samples to obtain the relative performance of otherwise "identical sedimentary deposits." As will be discussed in Section 3, this was the direction taken by many researchers of that time.

Just as increasing loads over geologic time play an important role in the interparticle bonding of shale formations, the reverse process, unloading, has significant effects. During the removal of load, "the shale expands at practically constant horizontal dimensions" (Terzaghi and Peck 1967, 426).

During expansion, the interparticle bonds are broken, and joints form at fairly regular spacings. At depths on the order of 30 m, the joints are spaced meters apart and are closed. Closer to the surface, intermediate joints form because of differential movements between the blocks. These joints open, allowing moisture to penetrate. The increase in moisture content may reduce the shear strength, and, if so, new fissures are formed. The final result and slope of any exposed face depend on the interparticle bonding remaining in the shale formation.

2.2 Degradable Materials from Igneous and Metamorphic Rocks

Because sedimentary rocks (and shales in particular) are typically formed relatively near the earth's surface and without the extreme heat and pressure that occur at depth, they tend to be mineralogically stable near the surface. Weathering of these materials then involves either a reversal of the consolidation pressure or a dissolution of cement bonds holding the grains or mineral groups together. In contrast, igneous and metamorphic rocks are created under temperature and pressure conditions that are drastically different from conditions at the surface. Macias and Chesworth described the implications of the difference:

> One might therefore expect that they would weather more readily than sedimentary materials.... Generally however, expectations in this regard are not fulfilled. Soils form more readily on sedimentary rocks than on other types and the reason is obviously hydrodynamic. For chemical weathering to take place to any significant degree, water must circulate through the rock, and the open structure of most sedimentary materials is more conducive to this than is the restricted porosity of most igneous and metamorphic rocks...the igneous rocks most susceptible (to weathering) are those with an open structure such as the non-welded pyroclastics. Again, in the metamorphic regime the importance of hydrodynamics is shown in that a vertical disposition of filiation encourages a more facile descent of aqueous solutions and a more rapid weathering, than a horizontal foliation. (Macias and Chesworth 1992, 283)

Weathering of igneous and metamorphic rocks is generally divided into two categories: physical and chemical. Ollier (1969) described in detail several types of physical weathering, including sheeting or spalling (fracturing parallel to a free surface created by erosion, excavation, tunneling, etc.), frost weathering (extension of fractures by expansion of freezing water), salt weathering (extension of fractures by the growth of salt crystals), and isolation (partial disintegration of the rock caused by the volume changes accompanying temperature changes).

Ollier (1969) also described types of chemical weathering, including solution (dissolution of soluble minerals, particularly salts and carbonates), oxidation and reduction (chemical alteration of minerals to form oxides or hydroxides), hydration (incorporation of water to create a new mineral), chelation (leaching of ions such as metals), and hydrolysis (reactions between minerals and the component ions of water).

As noted by Macias and Chesworth (1992), chemical weathering, which brings about mineralogical changes in igneous and metamorphic rocks, is usually more crucial than physical weathering in defining the strength properties of the materials. Physical weathering, however, does provide avenues for water to enter the rock by the creation and extension of fractures and subsequently encourages the more rapid progress of chemical weathering by an increase in surface area exposed to water.

Obviously some minerals, and therefore some rocks, are more susceptible than others to the weathering processes described above. For instance, Bowen's Reaction Series (Goldich 1938), which describes crystallization of magma, may be reversed to model the weathering process: calcium plagioclase weathers more readily than sodium plagioclase, and olivine weathers more readily than biotite, which weathers more readily than muscovite, which weathers more readily than quartz. Thus, rocks containing high percentages of calcium plagioclase or olivine will weather faster than rocks containing high percentages of sodium plagioclase or quartz.

Detailed descriptions of the weathering products of minerals have been provided by Macias and Chesworth (1992), Ollier (1969), and Carroll (1970). In general, most silicates (feldspars and micas in particular) weather into clay minerals. Under more extreme conditions, such as those found in tropical or humid climates, or after long periods of time in a geologic sense, they break down further into oxides and hydroxides of aluminum and iron. The specific types of clay minerals formed depend to a great degree on the parent materials, the pH, and the extent of saturation.

Because the weathering products of rocks are largely a function of the mineralogical composition, certain igneous and metamorphic rock types that share common minerals may share similar weathering characteristics. The following observations may assist in predicting general weathering characteristics of igneous and metamorphic rocks (Ollier 1969; Macias and Chesworth 1992):

1. **Granite and diorite:** Because granites typically exhibit massive structure, they also typically develop unloading fractures when exposed at the surface. Continued physical weathering increases the surface area exposed to chemical weathering. Chemical weathering alters feldspars and micas into clays, whereas quartz persists as a sand. According to Ollier (1969, 81), "weathering often follows the joints, and isolated joint blocks weather spheroidally, leaving 'corestones' of unaltered granite in the center."

2. **Gneiss and amphibolite:** In igneous and metamorphic rocks, feldspars and pyroxenes tend to weather rapidly, amphiboles weather at an intermediate rate, and quartz and accessory minerals are persistent. Ollier (1969, 82) noted that gneiss, in particular, "is rarely as well jointed as granite, so unloading is not common, or at least harder to detect. Minerals are segregated into bands, and bands of the most weatherable mineral affect the total rock strength—a property that often proved troublesome in engineering."

3. **Schist, slate, and phyllite:** Ollier (1969, 82) noted that "these [schists] have marked fissility along the 'schistosity' and this is very important in weathering. They contain some very resistant minerals but weathering is moderately easy. Frost weathering can rapidly break up schist."

4. **Basalt and peridotite:** According to Ollier (1969, 81), "Basalt is attacked first along joint planes, leading eventually to spheroidal weathering. All the minerals are eventually converted to clay and iron oxides, with bases released in solution, and as there is no quartz in the original rock, the ultimate weathering product is often a brown base rich, heavy soil." Peridotite shares mineralogical characteristics with basalt and may be anticipated to weather similarly.

The weathered rock product referred to as *saprolite* is of particular interest in evaluating engineering properties. Saprolites are "rotten rocks," or rocks in which the rock structure is preserved but many of the less durable minerals have altered to clay. Saprolites generally form less stable slopes than their parent rocks because of the increased amount of clay and loss of interlocking structure. They also maintain zones of weakness by preserving the general rock structure, or new zones of weakness may be created by preferential weathering along bands of less stable minerals.

Saprolites that preserve zones of weakness from the original rock structure or contain intact, unweathered blocks may be expected to behave like degradable materials. Saprolites that do not have these characteristics may be expected to behave like deeply weathered soils, whose properties are better described by a system that addresses tropical soils, as discussed in Chapter 19.

This brief geological background clearly establishes that the evaluation of degradable materials is complex and that no single approach to determining the long-term behavior is likely to work for every formation. Thus, many researchers have concentrated on developing identification and classification methods that have regional applications. Local experience and understanding are keys to success when building through, on, or with these materials.

3. IDENTIFICATION AND CLASSIFICATION

The investigative procedures for identifying or recognizing potential slope stability problems in shale formations are similar to those described in other chapters of this report. The focus in this section will be on reviewing laboratory and field tests developed for sedimentary rocks, shales in particular.

3.1 Shales

3.1.1 Identification

From a visual reference, the natural topography of regions underlain by shales displays certain characteristics. Terzaghi and Peck (1967, 426) stated: "On shales of any kind, the decrease of the slope angle to its final equilibrium value takes place primarily by intermittent sliding. The scars of the slides give the slopes the hummocky, warped appearance known as *'landslide topography.'*"

In the 1978 landslide report, Rib and Liang (1978) described the typical landforms of shale landscapes and their interpretation from aerial photography. If available, aerial photographs are excellent tools for identifying potential problem sites provided that the user is trained to recognize certain characteristics associated with the diagnostic features.

For thick, uniform shale beds, Rib and Liang described the associated landforms (1978, 57): "Clay shales are noted for their low rounded hills, well-integrated treelike drainage system, medium tones, and gullies of the gentle swale type." Ray (1960, 16) noted that shales "have relatively dark photographic tones, a fine-textured drainage, and relatively closely and regularly spaced joints."

However, it has also been observed that shales are particularly susceptible to landsliding when interbedded with pervious rocks such as sandstones or limestones. In this case, Rib and Liang noted:

> Interbedded sedimentary rocks show a combination of the characteristics of their component beds. When horizontally bedded, they are recognized by their uniformly dissected topography, contourlike stratification lines, and treelike drainage; when tilted, parallel ridge-and-valley topography, inclined but parallel stratification lines, and trellis drainage are evident. (Rib and Liang 1978, 57)

3.1.2 Laboratory Tests and Classification Systems

Since the late 1960s, there have been numerous attempts to develop tests to assist design engineers in the difficult task of classifying argillaceous shales and predicting their performance in embankment or cut slopes. These issues are of interest to the mining industry as well as to transportation engineers. The main objective has been to find tests that will reliably differentiate between durable shales that may be treated as rock and those with limited durability that are degradable on a human time scale.

Underwood (1967) discussed in detail the limitations of the various geological, chemical, and mineralogical classification methods of that time. He suggested grouping shales according to significant engineering properties (strength, modulus of elasticity, potential swell) and according to common laboratory tests (moisture content, density, void ratio, permeability). He recognized, however, that developing such a scheme would be a significant effort and he recommended (1967, 116) that "a comprehensive study involving the cooperation of government agencies, private engineering firms, and universities, is needed to produce a satisfactory engineering classification for compaction shales, especially the clay shales."

In the early 1970s, major research efforts were under way at Purdue University sponsored by the Indiana Highway Department and at the U.S. Army Engineer Waterways Experiment Station sponsored by the Federal Highway Administration. Numerous reports resulted from this research, including those by Bragg and Zeigler (1975), Shamburger et al. (1975), Strohm et al. (1978), and Strohm (1978). Because of the widespread occurrence of problem shales and their almost infinite variation of behavior, researchers in a number of state transportation departments, other public agencies, private companies, and universities have continued to refine these earlier studies, to revise the proposed tests, and to apply them to their respective areas.

Although it has been known for decades that certain shales deteriorate rapidly upon immersion in water, Franklin and Chandra (1972), Lutton (1977), and Franklin (1981) have made significant contributions to establishing specific tests for slaking of shales. The tests from these studies are described briefly below.

3.1.2.1 Slake Test

The slake test was originally developed to provide an indication of material behavior during the stresses of alternate wetting and drying, which, to some degree, simulate the effects of weathering. The test procedure and applications have been discussed by numerous authors, including Chapman et al. (1976), Withiam and Andrews (1982), and Hopkins and Deen (1984). The procedure is briefly described below:

1. Choose six pieces of shale *each* weighing about 150 g or the largest pieces available to have a total of 150 g in each group.
2. Identify and photograph each piece beside a millimeter scale.
3. Dry shale to a constant weight at 105°C and record dry weight. (Note: drying the sample is

an important step; the following step of the test must not be started with a field-moist sample.)

4. Place each specimen in a separate jar and cover with distilled water. The condition of the specimens should be checked for the first 10 min, then at 1, 2, 4 or 8, and 24 hr.
5. Remove the specimens from the water and check for any change in pH of the water.
6. Dry to a constant weight and record weight of shale specimens retained on 2-mm (No. 10) sieve. (Note: recording weight of intermediate cycles is desirable so that the results may be compared with those of the slake durability test, described next.)
7. Photograph specimens if significant degradation has occurred or at the end of the last test cycle.
8. Repeat procedure four additional times, or until total degradation, to make five cycles.
9. Calculate the slake index for each of the six samples and take the average:

$$S_I = \frac{(\text{original weight} - \text{final weight})}{\text{original weight}} \times 100$$

This simple test will usually identify the poorly performing shales in a matter of hours. If the specimens are quite resistant, however, this test is time consuming and requires qualitative judgment as to its performance.

3.1.2.2 Slake Durability Test

Franklin (1981) suggested a more severe and less time-consuming test known as the slake durability test, which is summarized below. Obviously, shales that fall apart in the slake test need not be subjected to the slake durability test.

For the slake durability test, a wire-mesh drum made with 2-mm (No. 10) mesh is rotated while partially submerged in a trough of water. The axis of the 140-mm-diameter drum is 20 mm above the water surface.

1. Select 10 pieces of shale (40 to 60 g each) with a total weight of approximately 500 g.
2. Identify and photograph the group beside a millimeter scale.
3. Place the shale fragments in the drum. Weigh drum and shale together. Place drum in an oven and dry the shale to a constant weight at 110°C.
4. Compute natural moisture content; then mount drum in trough.

5. Rotate the drum at 20 revolutions per minute for 10 min.
6. Remove the drum from the water, rinse, dry in oven, and weigh drum and remaining shale.
7. Repeat the cycle four more times to produce five cycles, but calculate the slake durability index (I_D) after each cycle. Photograph as necessary.
8. Calculate the durability index as follows:

$$I_D = \frac{\text{weight of shale remaining inside drum}}{\text{original weight of sample}} \times 100$$

Run at least two specimens from each sample of shale and take the average of their durability indexes.

The test proposed by Franklin has been standardized and is described in ASTM D-4644-87 (1992). In this newer test it is recommended that only two cycles be performed before the slake durability index is computed.

From these two tests several agencies have developed classification systems that allow them to determine the method or methods by which they will treat shales in embankment construction. These treatments are discussed in Section 4. In addition, several researchers have proposed classification systems and slope stability evaluations that depend on a number of other laboratory tests, including jar slake, rate of slaking, Atterberg limits, free-swell tests, and point-load strength. The procedures for jar slake, point-load strength, and free-swell tests are described below. Atterberg-limit tests are common in current engineering practice, so no procedural description is given. However, a word of caution should be expressed. Chapman et al. (1976) have noted that Atterberg limits when evaluated for degradable materials are often a function of the energy input and mode of preparation; thus variation in the test can be introduced by the operator.

3.1.2.3 Jar Slake Test

The following procedure for the jar slake test was described by Wood and Deo (1975):

1. A piece of oven-dried shale is immersed in enough water to cover it by 15 mm. [It is important that the shale sample be oven dried. Lutten (1977) reported that damp material is relatively insensitive to degradation in this test when compared with dry material.]
2. After immersion, the piece is observed continuously for the first 10 min and carefully during

the first 30 min. When a reaction occurs, it happens primarily during this time frame. A final observation is made after 24 hr.

3. The condition of the piece is categorized (complete breakdown, partial breakdown, no change), as follows:

Jar Slake Index I_J	Behavior
1	Degrades to a pile of flakes or mud
2	Breaks rapidly, forms many chips, or both
3	Breaks slowly, forms few chips, or both
4	Breaks rapidly, develops several fractures, or both
5	Breaks slowly, develops few fractures, or both
6	No change

The reproducibility of the jar slake test was evaluated by Dusseault et al. (1983).

The U.S. Office of Surface Mining Reclamation and Enforcement (1991) has defined "durable rock" as rock that does not slake in water as in the jar slake test. Welsh et al. (1991) proposed a strength-durability classification system that includes the point-load test to "clearly differentiate between strong-durable and weak or nondurable materials." The principal advantages of the test are that the equipment can be taken to the field, irregular samples can be used, and it is inexpensive to perform.

On the basis of work by Olivier (1979), Welsh et al. (1991) selected a dual-index system to categorize rock into three classes (Class I, nondurable and weak; Class II, conditionally stable; and Class III, durable and strong). In addition to the jar slake test, a graph (Figure 21-1) is used that plots the results of the point-load versus the free-swell test. The two tests are described briefly next.

FIGURE 21-1 Strength-durability classification of jar slaking (Welsh et al. 1991). REPRINTED WITH PERMISSION OF THE AMERICAN SOCIETY OF CIVIL ENGINEERS

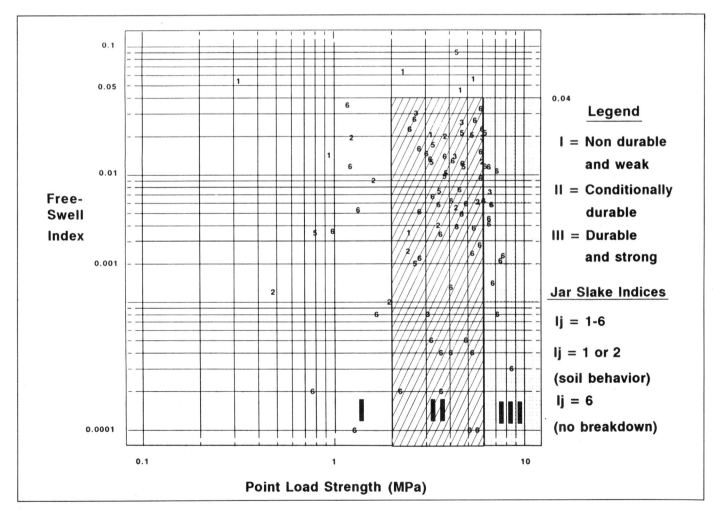

3.1.2.4 Point-Load Test

The history and development of the equipment and the suggested method for determining the point-load index were described more completely by Broch and Franklin (1972). The test was developed principally to be used in the field on rock core or irregular lumps ranging in size from 25 to 100 mm.

The point-load apparatus (Figure 21-2) compresses the rock sampled between the two points of cone-shaped platens. The shape of the cone has been standardized. The radius of curvature of the cone tip (5 mm) is the most critical dimension. The angle of the cone (60 degrees) is of importance only if significant penetration of the cone occurs during testing.

The apparatus must be equipped to measure the distance, D, between the platens at failure to within an accuracy of ± 0.5 mm. The load is applied hydraulically using a small hand pump and high-pressure ram with low-pressure seals to reduce inaccuracies of load measurement. The load, P, is determined from a gauge monitoring the hydraulic pressure in the jack. A maximum-pressure indicator needle on the face of the gauge is necessary to accurately record the maximum pressure or load at failure. The apparatus should have a capacity of 50 kN. After both the distance D and the failure load P have been measured, the point-load index, I_S, is determined:

$$I_S = P/D^2$$

where P is the point load at failure and D is the distance between platens at failure of the sample. It should be noted that for hard rock, the initial tested diameter d and the measured distance D are essentially the same.

To reduce scatter in the results, the samples should respect certain length-to-diameter (l/d) ratios: $l/d \geq 1.4$ for cores and l/d between 1.0 and 1.4 for rock lumps when measured perpendicular to the loaded axis. In addition, when samples indicate to the geologist the potential for significant anisotropic mechanical behavior, samples should be divided and tested in groups to measure the strengths in each direction.

Since strength test results are influenced to some degree by the size of the specimen tested, Broch and Franklin (1972) proposed that the results be adjusted to a reference diameter of 50 mm. In their paper, a number of graphs showed the size effects reported from numerous tests on different rock types. They preferred using graphs to adjust the equation results rather than using factors in the formula because the formula is in stress units and has some theoretical justification. The graphs showed that the point-load strength decreases with increasing diameter. Consequently, when lumps are tested, the authors recommended using or preparing samples so that the initial dimension d be as close as possible to 50 mm.

In order to obtain a statistically valid average value of I_S, at least 20 samples from the same formation should be tested (Oakland and Lovell 1982).

3.1.2.5 Free-Swell Test

The free-swell test is performed on intact core or from a bulk sample sawed into a rectangular prism. The minimum size should be 10 times the maximum grain size or 15 mm, whichever is greater. For direct testing purposes, NX-diameter core (54 mm) is generally acceptable. To measure the maximum swell, an axis perpendicular to the bedding laminations is chosen. The sample is oven-dried, carefully measured, and placed in water, and the volumetric strain is computed from measurements taken after 12 hr of soaking. Olivier (1979) reported that at least 75 percent of the maximum free swelling occurs within the first 2 to 4 hr of the test. Welsh et al. (1991) reported that for Appalachian shales the proportion of swelling ranged from 80 to 99.1 percent, with an average of 90.8 percent by the end of 4 hr. Consequently, they recommended using the shorter-term test for those shales and approximating the 12-hr results by multiplying the 4-hr results by 1.1.

In the proposed classification shown in Figure 21-1, durable and strong rock (Class III) has a

FIGURE 21-2 Point-load apparatus.

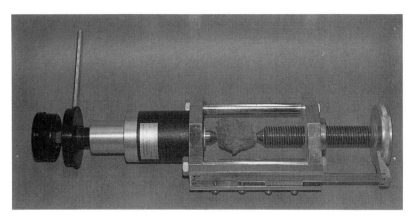

swell of 4 percent or less and a rock strength equal to or greater than 6 MPa and exhibits rocklike behavior during the jar slake test (ranking higher than 2).

In this classification system, any shales with properties less than those of Class III should not be used in drain applications. Therefore, particular care must be taken to avoid placing these materials in or near drainage features used in landslide mitigation works.

Class I material, nondurable and weak rock, has the following characteristics:

- Fails the jar slake test (behaves like a soil),
- Fails during sample preparation for either the free-swell test or point-load test,
- Produces a value less than 2 MPa in the point-load test, or
- Has a free swell greater than 4 percent.

"Hard" shales, as defined by these simple tests, are not all without problems. As discussed by Strohm et al. (1978), the water in the slake durability and jar slake tests should be checked for pH. A pH less than 6.0 indicates an acid condition, and the shale mineralogy should be checked for minerals that can cause chemical deterioration (Shamburger et al. 1975). Chemical deterioration of hard shales in Virginia with $I_d > 90$ percent was studied by Noble (1977), who soaked samples in dilute solutions of concentrated sulfuric acid (18 M) and distilled water as a classification test. He found that a 25 percent solution was more reactive and gave the same ranking in degree of deterioration as the modified sulfate soundness test (ASTM C88).

Noble recommended that hard, dark-colored shales be checked for iron sulfides and chlorite as a clay mineral, since this combination can have great potential for rapid weathering. Upon oxidation and access to water, shales containing iron sulfides (e.g., as pyrite) produce sulfuric acid, which dissolves the chlorite. These chemical soaking tests should be considered in classifying hard shales contemplated for rock fill on important projects where long-term settlement must be kept to a minimum. In contrast to acid reaction, some shales in the western United States have dispersive tendencies (Shamburger et al. 1975) and may react adversely in alkaline (high pH) water.

3.2 Igneous and Metamorphic Rock

Little research has been undertaken to quantify how the degree of weathering of igneous and metamorphic rock affects their engineering properties. Cawsey and Mellon (1983) provided an overview of research in experimental weathering of basic igneous rocks. They noted the merits and weaknesses of various tests to reproduce the effects of weathering. Dearman (1976) discussed the use of a weathering classification in the characterization of rock, and Dearman et al. (1978) provided an evaluation of engineering properties based on a visual classification applied to granites (Table 21-1).

To some degree weathered igneous and metamorphic materials may be characterized by the same tests used to characterize shales (Section 3.1.2). This is particularly true for igneous and metamorphic rocks, which have an abundance of clay minerals and few core stones or unweathered blocks. However, a number of differences should be highlighted:

1. Researchers of degradable rocks have investigated shales and designed their classification and testing systems to apply to shales. Correlations with tests applied to weathered igneous and metamorphic rocks have not been established.
2. Weathering rates, and therefore long-term engineering behavior, depend highly on the original mineralogy, all other factors, including climate, being equal. Consequently, fresh rock cuts may weather dramatically near the surface. Similarly, increasing the exposed surface area of the rock fragments by excavation and crushing before their placement as fill may also accelerate subsequent weathering rates.
3. Control of water not only will improve porepressure characteristics (as in shales) but will also reduce weathering rates, further improving longer-term behavior.

4. ENGINEERING DESIGN CONSIDERATIONS

Using the laboratory tests described above, several researchers have proposed procedures for cut-slope and embankment design and construction using degradable materials. Although these proposals do not have the benefit of a wealth of engineering precedent and experience, they provide general guidelines.

A preliminary evaluation of the characteristics of shales as indicated by a number of standard laboratory tests has been described by Underwood (1965) and is shown in Table 21-2. This evaluation may be applied with care to nonshale degradable rock materials.

4.1 Embankment Design

Numerous workers have sought to correlate laboratory tests with construction design parameters for degradable materials. Much of the early work done at the U.S. Army Engineer Waterways

Table 21-1
Rock Mass Properties of Weathered Granites and Gneisses (modified from Dearman et al. 1978)

ENGINEERING PROPERTY	FRESH, I	SLIGHTLY WEATHERED, II	MODERATELY WEATHERED, III	HIGHLY WEATHERED, IV	COMPLETELY WEATHERED, V	RESIDUAL SOIL, VI
Foundation conditions	Suitable for concrete and earthfill dams	Suitable for concrete and earthfill dams	Suitable for small concrete structures, earthfill dams	Suitable for earthfill dams	Suitable for low earthfill dams	Generally unsuitable
Excavatability	In general, blasting necessary	In general, blasting necessary	Generally blasting needed, but ripping may be possible depending upon the jointing intensity	Generally ripping and/or scraping necessary	Scraping	Scraping
Slope design[a]	1/4:1 H:V	1/2:1 to 1:1 H:V	1:1 H:V	1:1 to 1.5:1 H:V	1.5:1 to 2:1 H:V	1.5:1 to 2:1 H:V
Tunnel support	Not required unless joints are closely spaced or adversely oriented	Not required unless joints are closely spaced or adversely oriented	Light steel sets on 0.6- to 1.2-m centers	Steel sets, partial lagging, 0.6- to 0.9-m centers	Heavy steel sets, complete lagging on 0.6- to 0.9-m centers; if tunneling below water table, possibility of soil flow into tunnel	Heavy steel sets, complete lagging on 0.6- to 0.9-m centers; if tunneling below water table, possibility of soil flow into tunnel
Drilling rock quality designation (RQD), %	75, usually 90	75, usually 90	50–75	0–50	0 or does not apply	0 or does not apply
Core recovery (NX), %	90	90	90	Up to 70 if a high percent of core stones; as low as 15 if none	15 as sand	15 as sand
Drilling rates (m/hr) (diamond NX), 2½-in. percussion	2–4 5–7	2–4 8	8–10 12–15	8–10 12–15	10–13 17	10–13 17
Permeability	Low to medium	Medium to high	Medium to high	High	Medium	Low
Seismic velocity (m/sec)	3050–5500	2500–4000	1500–3000	1000–2000	500–1000	500–1000
Resistivity[b] (ohm-m)	340	240–540	180–240	180–240	180	180

NOTE: Weathering grades (shown here as column headings) are based on Dearman (1976).

[a] Benches and surface protection structures are advisable, particularly for more highly weathered material. The presence of through-going adversely oriented structures is not taken into account.

[b] Tends to be determined by joint openness and water-table depth.

Experiment Station for the Federal Highway Administration was summarized into technical guidelines by Strohm et al. (1978), who divided the design of shale embankments into five areas: foundation benching, drainage provisions, material usage, compaction requirements, and slope inclination. The following discussion utilizes a similar grouping.

4.1.1 Benching and Drainage

Typical recommended benching and drainage provisions are shown in Figures 21-3 and 21-4, where both longitudinal and transverse (cut-to-fill) situations are illustrated for shales interbedded with sandstone. Obviously the benches must be made into stable ground and the drainage rock must be

Table 21-2
Engineering Evaluation of Shales (modified from Underwood1965 and Wood and Deo 1975)

PHYSICAL PROPERTIES			PROBABLE IN SITU BEHAVIOR [a]						
LABORATORY TESTS AND IN SITU OBSERVATIONS	UNFAVORABLE RANGE OF VALUES	FAVORABLE RANGE OF VALUES	HIGH PORE PRESSURE	LOW BEARING CAPACITY	TENDENCY TO REBOUND	SLOPE STABILITY PROBLEMS	RAPID SLAKING	RAPID EROSION	TUNNEL SUPPORT PROBLEMS
Compressive strength (psi)	50 to 300 (0.3 to 2 MPa)	300 to 5,000 (2 to 34 MPa)	X	X					
Modulus of elasticity (psi)	20,000 to 200,000 (140 to 1400 MPa)	200,000 to 2×10^{-6} (1400 to 14,000 MPa)		X					X
Cohesive strength (psi)	5 to 100 (0.03 to 0.7 MPa)	100 to > 1,500 (0.7 to > 10 MPa)			X	X			X
Angle of internal friction (degrees)	10 to 20	20 to 65			X	X			X
Dry density (pcf)	70 to 110 (1.1 to 1.8 g/cm³)	110 to 160 (1.8 to 2.6 g/cm³)	X					X(?)	
Potential swell (%)	3 to 15	1 to 3			X	X		X	X
Natural moisture content (%)	20 to 35	5 to 15	X			X			
Coefficient of permeability (cm/sec)	10^{-5} to 10^{-10} (3×10^{-7} to 3×10^{-12} ft/sec)	$>10^{-5}$ ($>3 \times 10^{-7}$ ft/sec)	X			X	X		
Predominant clay minerals	Montmorillonite or illite	Kaolinite or chlorite	X			X			
Activity ratio = (plasticity index/ clay content)	0.75 to >2.0	0.35 to 0.75				X			
Wetting and drying cycles	Reduces to grain sizes	Reduces to flakes					X	X	
Spacing of rock defects	Closely spaced	Widely spaced		X		X		X(?)	X
Orientation of rock defects	Adversely oriented	Favorably oriented		X		X			X
State of stress	> Existing overburden load	= Overburden load			X	X			X

[a] Expected problems indicated by X.

FIGURE 21-3
Longitudinal bench
drainage tailored to
stratification of
seepage layers
(modified from
Strohm et al. 1978).

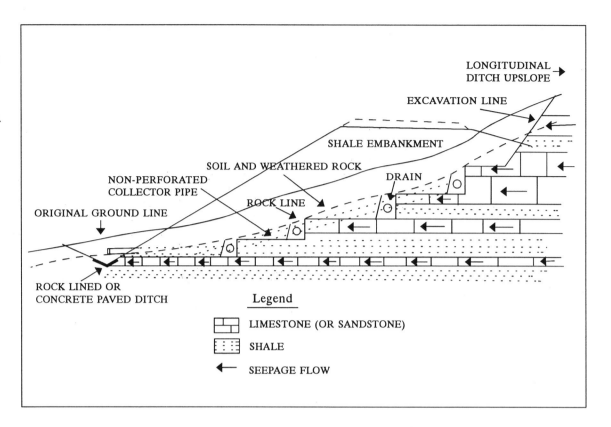

FIGURE 21-4
Transverse bench
drainage tailored to
stratification of
seepage layers,
cut-to-fill (modified
from Strohm et al.
1978).

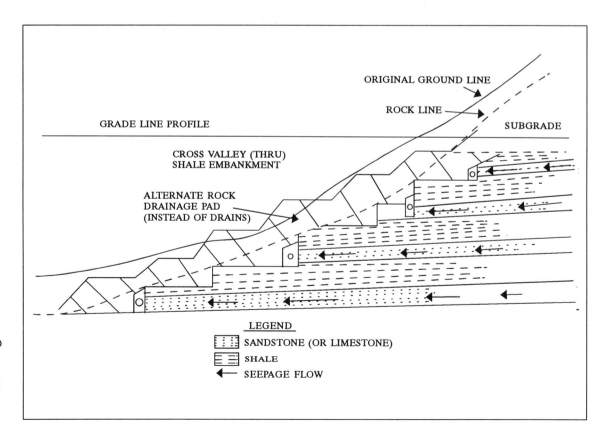

durable and not degrade with time. Requirements for this rock might follow the recommendations of Welsh et al. (1991). A typical design is shown in Figure 21-5, reproduced from a report by the Oregon Department of Transportation (Machan et al. 1989). Here the select durable rock embankment is placed to a level above the flood stage of an adjacent river, essentially eliminating the potential for wet and drying cycles in the shale embankment material placed above.

4.1.2 Material Use

During project development a material use plan should be prepared to cover excavation requirements and ensure that nondurable rocks are placed as soils in thin lifts and that durable rocks are placed as rock fill. The alternative is to place all materials as soils while meeting maximum gradation size and minimum compaction criteria. Wasting of degradable materials, such as shales, can generally be avoided by proper treatment and use. Exceptions might involve extremely wet clay shales, which may not be economically dried by disking or other means.

Strohm et al. (1978) developed design criteria based on the slake durability index, I_D, and the jar slake index, I_J, as follows:

- $I_D > 90$, $I_J = 6$: These materials can be used as rock fill as long as soil- and gravel-size materials do not exceed 20 to 30 percent of total lift. Too much fine material prevents the rock-to-rock contact necessary for stability and causes long-term settlement.
- $I_D = 60–90$, $I_J = 3–5$: These are hard, nondurable intermediate shales that require special treatment, typically including a high degree of compaction by heavy rollers (see Section 4.1.3).
- $I_D < 60$, $I_J \leq 2$: These materials need to be compacted as soil in thin lifts.

A number of other authors have prepared graphs or tables to help designers prepare the material plans. Lutton (1977) provided an estimate of allowable lift thicknesses as a function of slake durability index (Figure 21-6). The shale rating system proposed by Franklin (1981) (Figure 21-7) groups rock materials according to slake durability index and either plasticity index or point-load strength. Various groupings are assigned a shale rating, R, which is used to derive a number of slope design parameters. For instance, Figure 21-8 provides an estimate of allowable lift thicknesses based on Franklin's shale rating.

Santi and Rice (1991) used a modification of the slake index test to provide a preliminary clas-

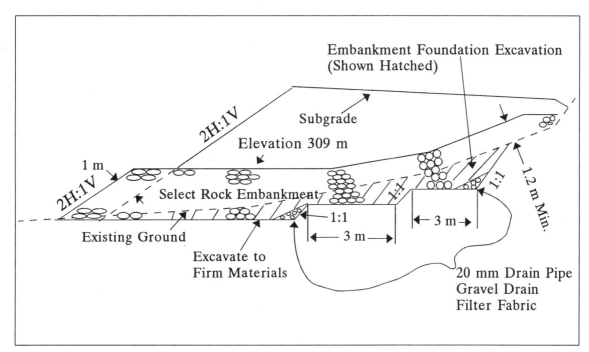

FIGURE 21-5 Oregon benching and drainage detail (modified from Machan et al. 1989).

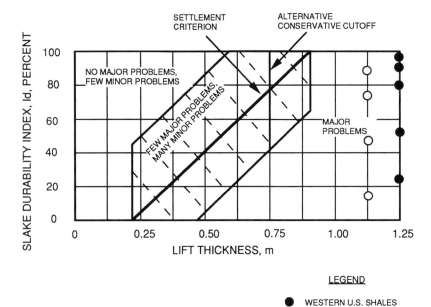

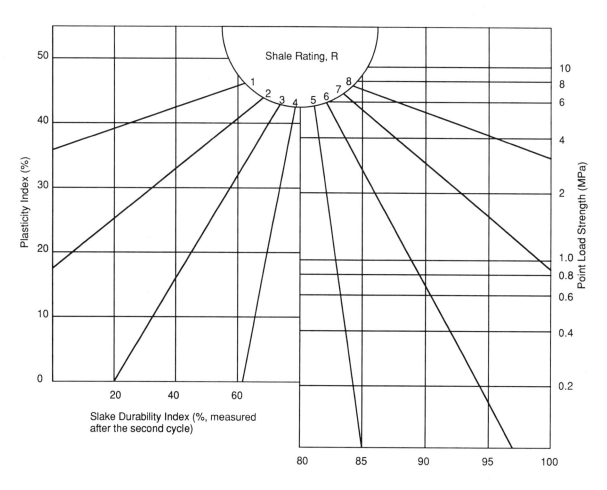

FIGURE 21-6
(above)
Criteria for
evaluating
embankment
construction on
basis of slaking
behavior of
materials (modified
from Lutton 1977).

FIGURE 21-7
Shale rating chart.
Sample of shale is
assigned rating
depending on its
slake durability
and strength if
slake durability
is > 80 percent, or
depending on its
slake durability
and plasticity if
slake durability is
< 80 percent
(modified from
Franklin 1981).

sification of degradable materials. As is shown in Figure 21-9, they suggested plotting the one-cycle slake value (representing the current state of weathering of a material) against the difference between the five-cycle and one-cycle values (representing the susceptibility to weathering of a material). They term this difference the *slake differential*. The resultant graph is then divided into sections denoting expected material behavior. Such a classification emphasizes the continuum between soil and rock behavior. It also indicates subsequent laboratory tests that are likely to further characterize the material.

Hopkins (1984) tested Kentucky shales extensively in order to correlate slake durability with the California bearing ratio (CBR) used by the state of Kentucky for pavement design. Mathematical expressions were developed to define suitable relationships for design. These expressions convert three different indexes for slake durability to predictions of the Kentucky CBR values.

4.1.3 Compaction

In the contract documents for a construction project, it is generally good practice to call for test sections or field-scale tests to evaluate construction materials and methods. The test sections are used to develop the required compaction methods and control procedures for nondurable shales before major earthwork is started.

Most specifications require a minimum density of 95 percent of the maximum dry density as determined by AASHTO T-99 and a moisture content within 2 percent of optimum. Some specifications make a point of indicating a value below optimum. However, by their very nature, these materials degrade excessively during laboratory processing and compaction. This degradation can result in unrealistically high dry densities. The tests should be started with the in situ or natural water content, since that is what will be used on the project. It is also necessary to use fresh sample material for each determination of moisture content because of material degradation.

In the field it has been found that generally it is necessary to use two different types of rollers to obtain the specified density. A static or vibratory sheepsfoot roller weighing around 25 000 kg is

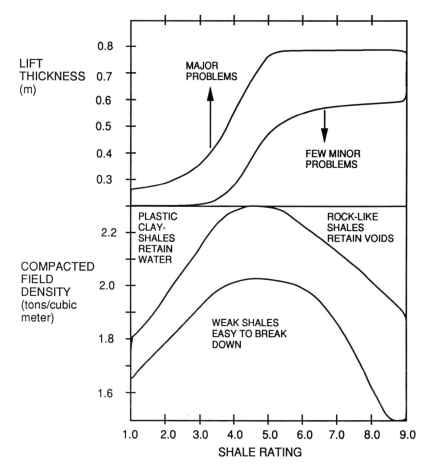

FIGURE 21-8 *(above)* Tentative correlations among shale quality, lift thicknesses, and compacted densities (modified from Franklin 1981).

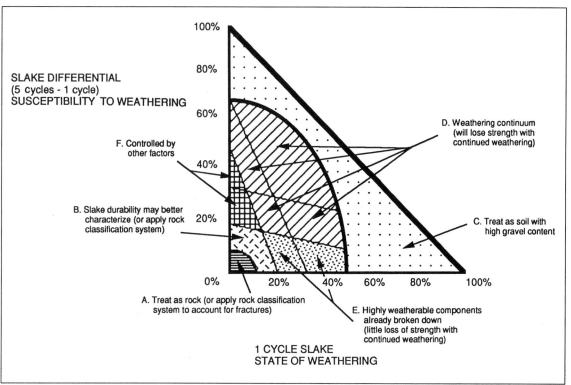

FIGURE 21-9 Implication of slake differential partitioning. Characteristics of each category and appropriate testing may be estimated as shown (modified from Santi and Rice 1991).

needed to break down large rock fragments. Two to four or more passes may be needed, typically followed by a 46 000-kg pneumatic-tired roller for another two to four passes, which compacts the now-soil-like materials. Loose lift thicknesses are normally specified in the 0.2- to 0.3-m range. The quantity of water to be added or dried off by disking must also be evaluated. All of the above considerations support the test-section approach for determining the proper procedures to use in the field with the equipment provided by the contractor.

4.1.4 Slope Design

In this discussion of embankment slope design the conventional procedures outlined in Chapter 13 of this report have been modified. Two figures from Franklin (1981) have been included. Figure 21-10 uses the shale rating, R, obtained from Figure 21-7 to estimate a range of values for cohesion and angle of internal friction. Figure 21-11 provides an estimate of allowable slope angles and embankment heights as a function of R.

Perry and Andrews related the mode of slaking to slope stability problems observed in mine spoils ranging in age from 2 to 10 years:

Little or no stability problems were found where slab or block slaking dominated [degra-

dation to thick, blocky fragments]. Where chip slaking was dominant [degradation to thin, flat segments], the mass appeared to be relatively stable. The chips form an interlocking matrix which is resistant to bulk movement. When slaking to inherent grain size [degradation to fine-grained particles] was found to be the primary mode, stability problems were observed, as evidenced by slips, slides, and similar features. (Perry and Andrews 1982, 27)

In addition to the observations on mass stability, Perry and Andrews (1982) related slaking to erosion problems. Sheet, rill, and gully erosion were observed to occur to varying degrees on all spoils. Where a high proportion of materials that slaked to their inherent grain size was encountered, the most severe erosion was observed, whereas spoils with a high percentage of slake-resistant rocks were least affected. A "pebble pavement" created by an armoring of the surface with resistant small chunks was observed to be effective in controlling sheet erosion.

As a continuation of his earlier study, Hopkins (1988) performed numerous tests on some 40 different shales in an attempt to present predictive equations of engineering parameters from various index tests. For instance, he found that the natural water content of an unweathered shale was a good predictor of important engineering properties.

In addition, Hopkins selected nine types of shales for triaxial testing on remolded specimens compacted to standard-, modified-, and low-energy compaction. The behavior of these complex materials required experience and engineering judgment for interpretation of the results. Hopkins stated:

Since ϕ' and c' values defined by the (σ'_1/σ'_3) failure criterion are generally higher and lower, respectively, than values obtained from the $(\sigma'_1 - \sigma'_3)$ failure criterion, then it is unclear which set of ϕ' and c' to use in a given stability analysis. (Hopkins 1988, 105)

The principal difference lies in the values of c' obtained, which can have a significant influence on the value of the safety factor computed for the slope. Because of this, Hopkins recommended that designers use both sets of parameters in their stability analyses to determine which is the most conservative. In their reports, designers were en-

FIGURE 21-10
Trends in shear-strength parameters of compacted shale fills as function of shale quality (modified from Strohm et al. 1978 and Franklin 1981).

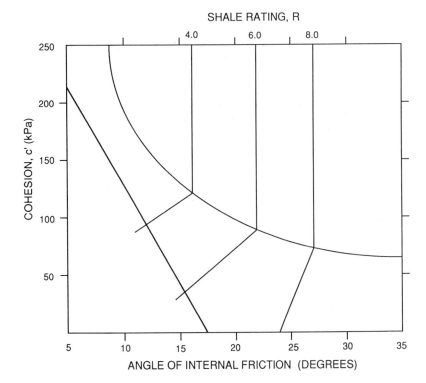

couraged to give the method used in the defining ϕ' and c' obtained from their laboratory tests.

4.2 Cut-Slope Design

The design of cut slopes in degradable materials, particularly in clay shales, is as complex as their geology, which was briefly outlined in Sections 2.1 and 2.2. As described by Terzaghi and Peck (1967, 426), upon unloading, the shales expand horizontally, take on moisture, decrease in strength, and consequently are notorious for delayed and progressive failures.

Duncan and Dunlop (1969) examined the influence of the initial stresses on the stresses near the excavated slope using the finite-element method. In particular, their analyses were conducted to study the behavioral differences of the excavated slope in materials with low and high initial horizontal stresses. Other studies have shown that in heavily overconsolidated clays, the horizontal stresses may be 1.5 to 3 times the current overburden pressures at the site and that in the Bearpaw shale and similar rocks of western North America, horizontal stress has been measured at 1.5 times the overburden pressure at a depth of 20 m.

Duncan and Dunlop concluded:

> Shear stresses around excavated slopes are much larger for conditions representing a heavily overconsolidated clay (high initial horizontal stresses) than for conditions representing a normally consolidated clay (low initial horizontal stresses). The results indicate that shear stresses large enough to cause failure may develop at some points within the slopes, even though the factor of safety according to the usual $\phi_u = 0$ method is considerably greater than unity. (Duncan and Dunlop 1969, 489)

Great care must therefore be used when determining shear strengths of intact shale specimens and when using these values in slope stability calculations. The shear-strength envelopes must be based on residual strength values from tests advanced to large strains.

In view of the difficulties in obtaining and testing samples representative of the materials in a large slope, many engineering firms and transportation departments invoke the local "experience factor" in the design of slopes for minimal maintenance over the long term.

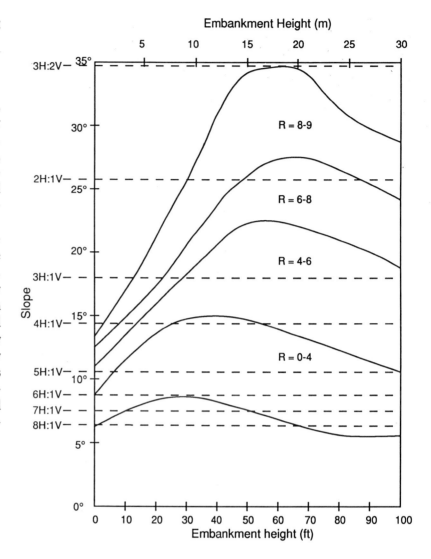

The number of variables is obviously quite large, and only a few examples are illustrated here. Franklin (1981) addressed this topic. Figure 21-12, reproduced from Franklin's paper, provides an estimate of allowable cut-slope angles as a function of the shale rating R and the bedding and joint orientations. State highway departments typically address cut-slope design by using standard plans included in their contracts. These are used to guide the excavation process and locate benches. Typically, the top surface of each bench is located at the top of the degradable materials. Several examples from the Kentucky Department of Highways *Geotechnical Manual* (1993) are shown in Figures 21-13 through 21-19. If rock fall is anticipated, the reader should refer to Chapter 18, Stabilization of Rock Slopes.

FIGURE 21-11 Trends in stable embankment slope angle as a function of embankment height and quality of shale fill (modified from Franklin 1981).

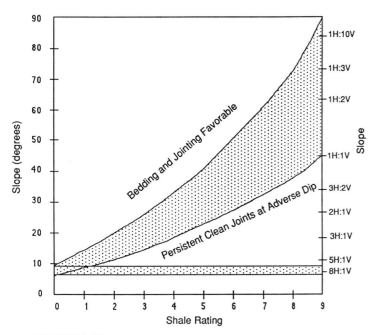

FIGURE 21-12
Trends in stable cut-slope angle as function of character of shale (modified from Franklin 1981).

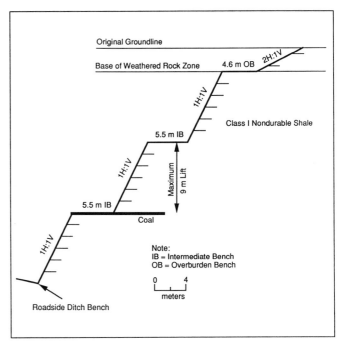

FIGURE 21-14
Typical slope configuration in Class II nondurable shale (modified from Kentucky Department of Highways 1993).

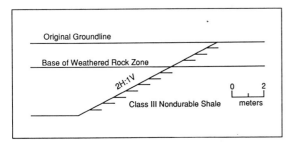

FIGURE 21-13
Typical slope configuration in Class III nondurable shale (modified from Kentucky Department of Highways 1993).

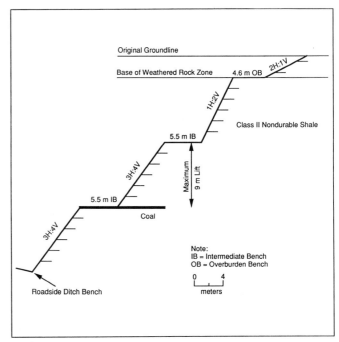

FIGURE 21-15
Typical slope configuration in Class I nondurable shale (modified from Kentucky Department of Highways 1993).

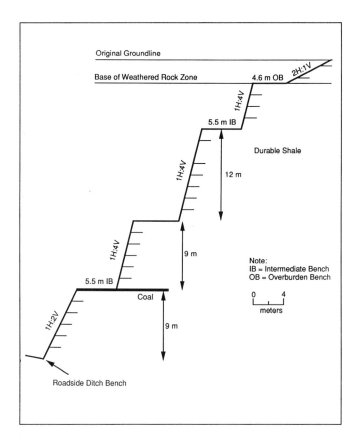

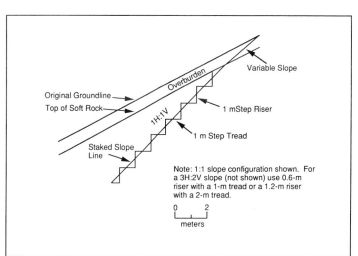

FIGURE 21-16
(left)
Typical slope configuration in durable shale
(modified from Kentucky Department of Highways 1993).

FIGURE 21-17
(below left)
Typical slope configuration in massive limestone or sandstone
(modified from Kentucky Department of Highways 1993).

FIGURE 21-18
(below)
Typical slope configuration for serrated slopes, which are utilized as means of controlling erosion and establishing vegetation on material that can be excavated by bulldozing or ripping (modified from Kentucky Department of Highways 1993).

5. CONCLUSION

It should be clear that the topic of constructing with, in, or through degradable materials is a complex one. There are few absolutes, and one must trust the experience of others. A great deal of reliance is placed on knowing the behavior of slopes in the vicinity and the geologic conditions at the site. It is hoped that the information in this chapter along with that in Chapter 15 on rock slope stability analysis will provide guidance for a successful project.

FIGURE 21-19
Through cut with dipping bedding planes (modified from Kentucky Department of Highways 1993).

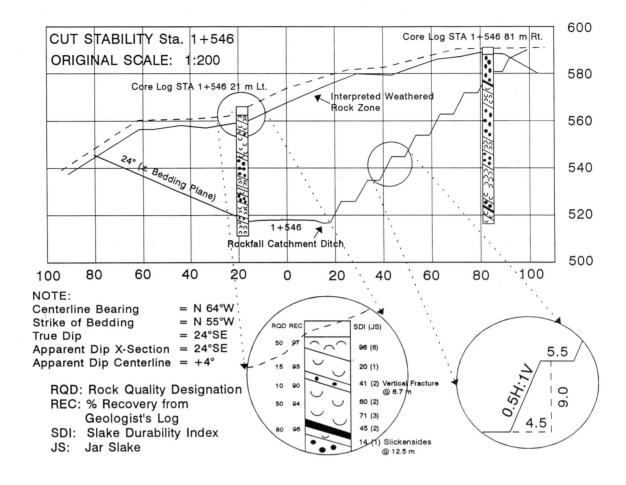

REFERENCES

ASTM. 1992. Standard Test Method for Slake Durability of Shales and Other Similar Weak Rocks. ASTM Designation D-4644-87. In *ASTM Book of Standards*, Volume 4.08, *Soil and Rock; Dimension Stone; Geosynthetics*, ASTM, Philadelphia, Pa., pp. 951–953.

Bates, R.L., and J.A. Jackson, eds. 1980. *Glossary of Geology*, 2nd ed. American Geological Institute, Falls Church, Va., 751 pp.

Berman, G. 1991. Landslides on the Panama Canal. *Landslide News* (International Newsletter of the Japan Landslide Society), No. 5, Aug., pp. 10–14.

Bragg, G.H., Jr., and T.H. Zeigler. 1975. Evaluation and Remedial Treatment of Shale Embankments (Report FHWA-RD-75-62). In *Design and Construction of Compacted Shale Embankments*, U.S. Department of Transportation, Vol. 2, 245 pp.

Broch, E., and J.A. Franklin. 1972. The Point-Load Strength Test. *International Journal of Rock Mechanics and Mining Sciences*, Vol. 9, pp. 669–697.

Carroll, D. 1970. *Rock Weathering*. Plenum Press, New York, 203 pp.

Cawsey, D.C., and P. Mellon. 1983. A Review of Experimental Weathering of Basic Igneous Rocks. In *Residual Deposits: Surface Related Weathering Processes and Materials* (R.C.L. Wilson, ed.), Blackwell Scientific Publications, Oxford, pp. 19–24.

Chapman, D.R., L.E. Wood, C.W. Lovell, and W.J. Sisiliano. 1976. A Comparative Study of Shale Classification Tests and Systems. *Bulletin of the Association of Engineering Geologists*, Vol. 13, No. 4, pp. 247–266.

Dearman, W.R. 1976. Weathering Classification in the Characterization of Rock: A Revision. *Bulletin of the International Association of Engineering Geologists*, Vol. 13, pp. 373–381.

Dearman, W.R., F.J. Baynes, and T.Y. Irfan. 1978. Engineering Grading of Weathered Granite. *Engineering Geology*, Vol. 12, pp. 345–374.

Duncan, J.M., and P. Dunlop. 1969. Slopes in Stiff-Fissured Clays and Shales. *Journal of the Soil Mechanics and Foundations Division*, ASCE, Vol. 95, No. SM2 (March), pp. 467–492.

Dusseault, M.B., P. Cimolini, H. Soderberg, and D.W. Scafe. 1983. Rapid Index Test for Transitional Material. *Geotechnical Testing Journal*, ASCE, Vol. 6, No. 2, pp. 64–72.

Franklin, J.A. 1981. A Shale Rating System and Tentative Applications to Shale Performance. In *Transportation Research Record 790*, TRB, National Research Council, Washington, D.C., pp. 2–12.

Franklin, J.A., and R. Chandra. 1972. The Slake Durability Test. *International Journal of Rock Mechanics and Mining Science*, Vol. 9, pp. 325–341.

Goldich, S.A. 1938. A Study in Rock Weathering. *Journal of Geology*, Vol. 46, pp. 17–23.

Hopkins, T.C. 1984. *Relationship Between Kentucky CBR and Slake Durability*. University of Kentucky Research Report UKTRP-84-24, Lexington.

Hopkins, T.C. 1988. *Shear Strengths of Compacted Shales*. University of Kentucky Research Report UKTRP-88-1, Lexington.

Hopkins, T.C., and R.C. Deen. 1984. Identification of Shales. *Geotechnical Testing Journal*, ASTM, Vol. 7, No. 1, pp. 10–18.

Huang, W.T. 1962. *Petrology*. McGraw-Hill, New York.

Kentucky Department of Highways. 1993. *Geotechnical Manual*. Department of Highways, Transportation Cabinet, Commonweath of Kentucky, Frankfort, 157 pp.

Lutton, R.J. 1977. Slaking Indexes for Design (Report FHWA-RD-77-1). In *Design and Construction of Compacted Shale Embankments*, U.S. Department of Transportation, Vol. 3, 94 pp.

Machan, G., T. Szymoniak, and B. Siel. 1989. *Evaluation of Shale Embankment Construction Criteria*. Experimental Feature Final Report OR 83-02. Oregon Department of Transportation, Salem.

Macias, F., and W. Chesworth. 1992. Weathering in Humid Regions, with Emphasis on Igneous Rocks and Their Metamorphic Equivalents. In *Weathering Soils and Paleosols* (P. Martini and W. Chesworth, eds.), Elsevier, Amsterdam, pp. 283–306.

McCarthy, D.F. 1988. *Essentials of Soil Mechanics and Foundations: Basic Geotechnics*, 3rd ed., Prentice-Hall, Englewood Cliffs, N.J., 614 pp.

Noble, D.F. 1977. *Identifying Shales That Weather Rapidly When Disturbed by Construction-Accelerated Weathering of Tough Shales*. Final Report VHTRC 78-R20. Virginia Highway and Transportation Research Council, Charlottesville.

Oakland, M.W., and C.W. Lovell. 1982. Standardized Tests for Compacted Shale Highway Embankments. In *Transportation Research Record 873*, TRB, National Research Council, Washington, D.C., pp. 15–22.

Olivier, H.J. 1979. A New Engineering-Geological Rock Durability Classification. *Engineering Geology*, Vol. 14, pp. 255–279.

Ollier, C.D. 1969. *Weathering*. American Elsevier, New York, 304 pp.

Perry, E.F., and D.E. Andrews. 1982. Slaking Modes of Geologic Materials and Their Impacts on Embankment Stability. In *Transportation Research Record 873*, TRB, National Research Council, Washington, D.C., pp. 22–28.

Piteau, D.R., and F.L. Peckover. 1978. Engineering of Rock Slopes. In *Special Report 176: Landslides: Analysis and Control* (R.L. Schuster and R.J. Krizek, eds.), TRB, National Research Council, Washington, D.C., pp. 192–225.

Ray, A.G. 1960. *Aerial Photographs in Geologic Interpretation and Mapping*. U.S. Geological Survey Professional Paper 373, 230 pp.

Rib, H.T., and Ta Liang. 1978. Recognition and Identification. In *Special Report 176: Landslides: Analysis and Control* (R.L. Schuster and R.J. Krizek, eds.), TRB, National Research Council, Washington, D.C., pp. 34–80.

Santi, P.M., and R.H. Rice. 1991. Preliminary Classification of Transitional Materials. In *Proc., 27th Symposium on Engineering Geology and Geotechnical Engineering*, Logan, Utah, April 10–12, Idaho State University, Pocatello, pp. 8-1 to 8-13.

Shamburger, J.H., D.M. Patrick, and R.J. Lutten. 1975. Survey of Problem Areas and Current Practices (Report FHWA-RD-75-61). In *Design and Construction of Compacted Shale Embankments*, U.S. Department of Transportation, Vol. 1, 292 pp.

Strohm, W.E., Jr. 1978. Field and Laboratory Investigations, Phase III (Report FHWA-RD-78-140). In *Design and Construction of Compacted Shale Embankments*, U.S. Department of Transportation, Vol. 4, 154 pp.

Strohm, W.E., Jr., G.H. Bragg, Jr., and T.H. Zeigler. 1978. Technical Guidelines (Report FHWA-RD-78-141). In *Design and Construction of Compacted Shale Embankments*, U.S. Department of Transportation, Vol. 5, 216 pp.

Taylor, D.W. 1948. *Fundamentals of Soil Mechanics*. John Wiley & Sons, New York, 700 pp.

Terzaghi, K., and R. Peck. 1967. *Soil Mechanics in Engineering Practice*. John Wiley & Sons, New York, 729 pp.

Underwood, L.B. 1965. Machine Tunneling on Missouri Riverdams. *Journal of the Construction Division*, ASCE, Vol. 91, No. CO1 (May), pp. 1–28.

Underwood, L.B. 1967. Classification and Identification of Shales. *Journal of the Soil Mechanics and Foundations Division*, ASCE, Vol. 23, No. SM6 (Nov.), pp. 97–116.

U.S. Office of Surface Mining Reclaimation and Enforcement. 1991. Permanent Program Performance Standards—Surface Mining Activities. In *Code of Federal Regulations* (7/1/91ed.), Title 30, Chapter VII, Part 816, Section 816.71: Disposal of excess spoil: General requirements, and Section 816.73: Disposal of excess spoil: Durable rock fills.

Welsh, R.A., Jr., L.E. Vallejo, L.W. Lovell, and M.K. Robinson. 1991. The U.S. Office of Surface Mining (OSM) Proposed Strength-Durability Classification System. In *Proc., Symposium on Detection of and Construction at the Soil/Rock Interface* (W.F. Kane and B. Amadei, eds.), ASCE Geotechnical Special Publication No. 28, American Society of Civil Engineers, New York.

Withiam, J.L., and D.E. Andrews. 1982. Relevancy of Durability Testing of Shales to Field Behavior. In *Transportation Research Record 873*, TRB, National Research Council, Washington, D.C., pp. 36–42.

Wood, L.E., and P. Deo. 1975. A Suggested System for Classifying Shale Materials for Embankments. *Bulletin of the Association of Engineering Geologists*, Vol. 12, No. 1, pp. 39–55.

Chapter 22

STEVEN G. VICK

HYDRAULIC TAILINGS

1. INTRODUCTION

Tailings, defined as sand-, silt-, or clay-sized solid wastes from mineral-processing operations that were originally produced, handled, and deposited in slurry form, represent a special case of slope stability in that they are artificially constructed deposits of manufactured soil. Tailings are distinct from other types of coarse mining wastes or smelter slag that are dumped or piled under dry conditions. Tailings deposits share many geotechnical characteristics of such other hydraulic fills as offshore drilling islands, hydraulic-fill water dams, dredged material deposits, and certain other industrial wastes (e.g., fly ash and papermill sludge). These related materials are not discussed here, but pertinent information can be found in two publications from the American Society of Civil Engineers: *Geotechnical Practice for Disposal of Solid Waste Materials* (ASCE 1977) and *Hydraulic Fill Structures* (Van Zyl and Vick 1988).

Tailings are often identified by their association with mining and milling operations. Where encountered along transportation routes, tailings deposits may affect the stability or safety of these facilities by (*a*) causing instability of excavated cuts, (*b*) creating potential landslide hazards from adjacent deposits, (*c*) increasing subgrade compressibility and settlement, and (*d*) possessing adverse environmental characteristics (high erodibility, heavy-metal pollution, and acid drainage).

2. CHARACTERIZATION AND PROPERTIES

For most metalliferous ores, tailings represent the end product of crushing and grinding operations after extraction of the valuable mineral particles. For other ores, such as phosphates, coal, or construction aggregates, tailings are produced by simple washing and screening operations. In either case, the maximum particle size typically varies from 0.1 to 1.0 mm, with the relative proportions of sand- and silt-sized particles governed by the type of processing required for optimum mineral extraction efficiency. Mechanical grinding produces particles that are highly angular; clay is contained in the tailings only if present in the parent ore, and clay content is usually significant only for tailings produced by washing operations. Specific gravity of tailings solids varies from as low as 1.5 for coal to almost 4.0 for tailings derived from sulfide-type orebodies where pyrite is rejected during processing. Vick (1990) summarized gradation and index properties for tailings from a variety of ore types.

2.1 Construction of Tailings Impoundments

During operation of a milling facility, the tailings slurry is usually pumped to a tailings impoundment at a pulp density of 15 to 50 percent solids by weight. Within the impoundment the solids settle from suspension, and the supernatant water collects in a *decant pond* from which it is discharged or recycled to the mill process. Central to geotechnical characterization of any tailings deposit is the manner in which discharge and deposition of the tailings slurry have been conducted, and conventional exploration and sampling techniques alone cannot substitute for an understanding of these operations.

Usually the tailings slurry is discharged from the embankment perimeter, typically as shown in Figure 22-1. To the extent that a sand fraction may be present in the slurry, these particles tend to settle nearest the point of discharge, with finer particles carried farther in the slurry stream. This sedimentation pattern gives rise to an above-water beach of *sand tailings* (defined as those with less than 50 percent passing the No. 200 sieve) and a zone of *slimes* (more than 50 percent passing the No. 200 sieve) extending from the more distant regions or the above-water beach to beneath the decant pond, as shown in an idealized manner in Figure 22-2. The degree of particle-size segregation on the beach, and therefore the extent to which the sand tailings and slimes are differentiated, depends on such factors as gradation of the original slurry and its solids content (Abadjiev 1985). The

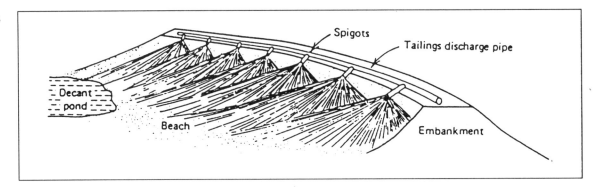

FIGURE 22-1 Tailings slurry discharge and beach deposition (modified from Vick 1990).

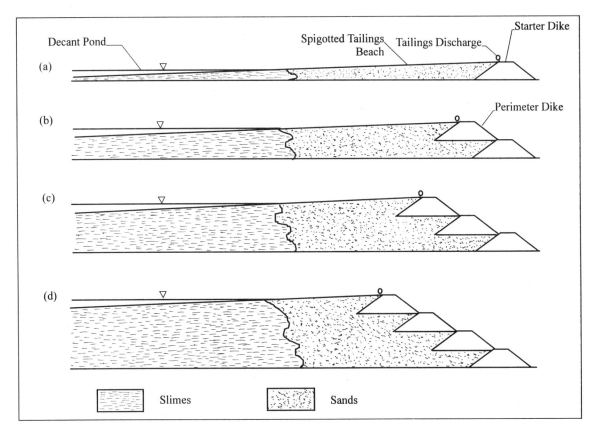

FIGURE 22-2 Sequential tailings dam raising and tailings deposition showing idealized sand and slimes regions (modified from Vick 1990).

transition between zones of sand tailings and slimes in a tailings deposit is gradual, interlayered, and indistinct and is governed by the various techniques and locations of tailings slurry discharge adopted during operation of the impoundment. Nevertheless, characterization of a tailings deposit into regions of predominantly sand and slimes is fundamental to establishing broad ranges of physical, index, and engineering properties and to targeting subsurface exploration in key areas.

2.2 Sand Tailings Characteristics

Sand tailings derived from beach deposition are usually loose, with relative densities ranging from 30 to 50 percent. Upon initial deposition, they typically exhibit corrected standard penetration test (SPT) blowcounts, expressed as $(N_1)_{60}$, of 3 to 5. There is evidence to support some increase in $(N_1)_{60}$ values with time due to "aging" effects (Troncoso 1990).

2.3 Slimes Characteristics

Slimes, on the other hand, qualify as low-plasticity to nonplastic silts for tailings derived from grinding of metalliferous ores. Other ores and processes, most notably Florida phosphatic ores, exhibit less favorable plasticity, sedimentation, and consolidation characteristics in their clay slimes (Bromwell and Raden 1979; Schiffman et al. 1988). A void ratio of 0.8 to 1.3 is typical of many slimes deposits, with $(N_1)_{60}$ values often ranging from 1 to 3. Of particular significance to slope engineering issues is the high specific retention of void water exhibited by slimes. Once sedimentation and consolidation processes are complete, further gravity drainage seldom results in appreciable reduction in water content or gain in strength except in the zone affected by surface desiccation, which seldom extends below depths of 1 to 2 m. Thus, slimes deposits, even in dry climates, have remained soft and virtually fully saturated for up to 60 years after deposition has ceased and surface water has been removed. This feature, along with the associated low undrained shear strength, makes slimes the most troublesome tailings material from the point of view of slope engineering or embankment construction. Slimes deposits also exhibit high compressibility, although rates of consolidation are usually relatively rapid.

2.4 Typical Strength Characteristics

Strength characteristics governing slope stability vary according to sand or slimes characterization. For most types of sand tailings, the drained (effective-stress) friction angles typically range from 33 to 38 degrees. These relatively high values for loose materials are attributable to particle angularity. Sand tailings display virtually no effective cohesion intercept (c') except in special cases involving chemical cementation, such as gypsum tailings or tailings affected by oxidation of sulfide minerals (usually pyrite). Vick (1990) extensively summarized drained strengths for various tailings types.

Provided that the nature, origin, and depositional characteristics of sand tailings deposits are fully understood, estimates of drained strength based on published values for similar tailings types are often sufficient for effective-stress analyses. Slope stability is seldom sufficiently sensitive to ordinary variations in drained strength to justify extensive laboratory testing.

For slimes tailings, either drained or undrained strength may be pertinent to slope stability, depending on such factors as the degree of desiccation-induced overconsolidation that the slimes may have experienced during deposition. For slimes derived from metalliferous ores, drained (effective-stress) angles of internal friction are generally in the range of 28 to 37 degrees (Vick 1990). Slimes exhibit little or no effective cohesion.

The undrained strength (s_u) of slimes is, however, sensitive to details of the original deposition process. For slimes deposited within and beneath the decant pond, normally consolidated conditions prevail after dissipation of deposition-induced excess pore pressures, and the variation of undrained strength with depth can be characterized by stress-normalized ratios (s_u/σ_c') ranging from about 0.20 to 0.27 (Vick 1990; Ladd 1991). In contrast, slimes deposited above water or on intermittently saturated portions of the beach or otherwise influenced by local desiccation during deposition may exhibit far more irregular undrained strength patterns, which may be difficult to assess (MacIver 1961). Although recovery of undisturbed samples of saturated slimes for laboratory testing is possible using fixed-piston samplers, vibration during sample transport or sample extrusion frequently causes gross disturbance and densification. Draining of sample tubes in the field before transport and freezing of recovered samples have both been used successfully to reduce these effects. Alternatively, Wahler and Associates (1974) and MacIver (1961) provided examples of undrained strength assessment for slimes by field vane-shear testing. However, thin sand seams are often present, especially in beach transition zones, and these may make drainage conditions during vane-shear testing difficult to ascertain. In conjunction with conventional sampling, the cone penetrometer test (CPT) is very useful for identifying interlayering and other stratigraphic details of tailings deposits.

3. SLOPE STABILITY AND ENVIRONMENTAL ISSUES

Slope stability problems related to tailings deposits most commonly arise in connection with dams that contain the deposited slurry, and a rich literature on tailings-dam stability exists, including work by Morgenstern (1985), Kealy and Soderberg (1969), Soderberg and Busch (1977), Klohn (1972), Vick (1990), and Corp et al. (1975). Brawner (1979) described several failures and repair case histories. Seismic stability of tailings dams is of considerable importance because of the liquefaction susceptibility of tailings deposits; examples of current practice in dynamic analysis were provided by Finn et al. (1990), Troncoso (1988), Lo et al. (1988), and Vick et al. (1993).

Although tailings-dam design and analysis methods apply in principle to transportation-related slope engineering situations involving tailings, important differences exist. Chief among these are the following:

1. Route alignments on or through tailings deposits most often involve excavated cuts,
2. Tailings may be considered as a borrow source for embankment fill, and
3. Tailings deposits are inactive and abandoned rather than actively operated in conjunction with ongoing mining operations.

3.1 Stability of Excavated Cuts

Stability of excavated cuts in inactive tailings deposits is ordinarily controlled by the cohesionless nature of the sand and the undrained strength behavior of the slimes. Conventional limit-equilibrium procedures of slope stability analysis

apply to tailings deposits, but more important than the computational procedure is the adoption of appropriate strength parameters. Ladd (1991) provided a thorough discussion of effective stress analyses versus undrained strength analyses in actively operating tailings impoundments and demonstrated that pore pressures due to undrained shearing must be addressed in assessing slope stability. These findings apply to excavated slopes in tailings slimes as well and emphasize the previously discussed importance of stress history and undrained strength resulting from the original circumstances of slimes deposition.

Equal in importance to calculating stability of cuts in tailings deposits is evaluating the feasibility of excavating them. The typically dry and even dusty surface of abandoned tailings deposits may produce a misleading impression. Although sand tailings eventually dewater by gravity drainage given sufficient time and foundation permeability, even perched water remaining in sand layers can produce "running" conditions on excavated faces. The same is true for slimes, which may remain fully saturated only a few feet below the desiccated surface crust long after the impoundment has been abandoned. Both saturation and low undrained shear strength of slimes seriously impede the mobility of wheeled construction equipment during excavation, and costly and time-consuming dragline or backhoe excavation techniques must often be used. These problems have been avoided during large-scale remining of tailings by hydraulic excavation (McWaters 1990), but this requires that a new impoundment be constructed to retain the excavated tailings slurry.

3.2 Use of Tailings as Fill

Tailings from abandoned deposits have been considered for use as highway fill. Dry sand tailings have been found to have suitable compaction characteristics and engineering properties. Pettibone and Kealy (1971) presented typical engineering properties for compacted tailings fill and described the use of tailings borrow for construction of a 6-km section of Interstate-90 near Kellogg, Idaho.

However, in the construction of highway fills using tailings, their highly erosive nature must be considered (see Section 3.4). A roadway fill constructed from tailings at Silverton, Colorado, was rapidly eroded when a culvert designed to carry the flow of a small stream became blocked with debris

FIGURE 22-3
Erosion of highway fill constructed from tailings, Colorado State Highway 110, Silverton, Colorado, May 1985. Road embankment was destroyed in a few hours when culvert designed to carry flow of Boulder Creek became blocked with debris during a period of high spring runoff. Fill acted as temporary dam but was quickly eroded once flowing water overtopped it. Reconstruction of roadway with new culvert took several weeks.
COURTESY OF *THE SILVERTON STANDARD AND THE MINER*

FIGURE 22-4
View across eroded section of highway fill constructed from tailings, Colorado State Highway 110, Silverton, Colorado, May 1985. Light-colored slope (*background*) is part of active tailings disposal, which has been constructed with relatively steep slope. Remains of wooden supports for former tailings discharge pipes are visible on slopes. Mill is situated behind tailings, and access to it was disrupted by this failure.
COURTESY OF *THE SILVERTON STANDARD AND THE MINER*

during a spring storm (Figures 22-3 and 22-4). The blockage was not detected during the night and the stream overflowed the fill and rapidly eroded it. Once started, the erosion occurred so rapidly that nothing could be done to prevent it. Reconstruction took several weeks. This road was not a major highway but did provide the only access between an active gold mine and its mill as well as the only feasible access to recreation sites in a mountain valley. The erosion of the road fill thus caused some measurable economic hardship.

3.3 Toxicity of Tailings

Use of tailings as highway fill or disposal of excavated tailings may present environmental issues related to toxic liability. Potential toxicity of tailings depends on the geochemistry of the orebody and the nature of the metallurgical processing and cannot be generalized even for ores of the same basic type. Although probably the majority of tailings deposits are chemically innocuous, some tailings may contain high levels of such constituents as arsenic, lead, cadmium, fluoride, or molybdenum, which can be harmful if ingested by humans, grazing animals, or certain plants. Such exposure can be enhanced when disturbed tailings are dispersed by wind. The solubility, mobility, and potential for transport of many heavy metals within groundwater or surface-water environments are greatly enhanced by the low pH levels generated by sulfide oxidation of some abandoned deposits, especially those rich in pyrite, pyrrhotite, or marcasite minerals. Exposure of such tailings surfaces to the atmosphere accelerates oxidation and acidification through complex chemical and biological reactions. Gadsby et al. (1990) provided an overview of recent research and practice in this area.

3.4 Erosion Susceptibility of Tailings

Environmental issues also derive from the highly erosive nature of tailings slopes, and special measures may be necessary to reduce and control sediment transport. Because of their relatively fine and uniform grain size, tailings are erodible by water and wind. Slopes as flat as 3H:1V are frequently considered necessary for erosion control and establishment of vegetation. Erosion considerations rather than mass stability may ultimately be found to govern slope design for tailings.

Slope erosion notwithstanding, eroded material from adjacent flat tailings surfaces can affect transportation facilities. Tailings eroded by water can quickly plug ditches and culverts, and wind erosion has been known to cause visibility-related safety hazards in dry climates. The rapid erosion of a highway fill constructed from tailings near Silverton, Colorado, was mentioned in Section 3.2. This rapid erosive failure resulted from a blocked culvert, which dammed a small stream and caused it to overtop the highway fill. The adjacent active tailings disposal site is shown in Figure 22-4. This site was constructed with comparatively steep slopes. Wind erosion on these slopes produced large dust clouds under relatively low wind velocities. The steepness of the slopes hampered revegetation efforts. With the recent final closure of the mine and mill, these slopes are being regraded to lower angles and revegetation is beginning to help stabilize these tailings. Such revegetation of exposed tailings surfaces can be complicated by the absence of nutrients, poor textural characteristics, and sometimes by the high residual metal concentrations in the tailings, which may be toxic to plants. Successful revegetation of large areas of tailings often requires specialized studies, including field test plots. Smaller areas can be quickly revegetated by covering tailings surfaces with topsoil, although this may be quite expensive.

4. FAILURE OF TAILINGS IMPOUNDMENTS

Slope failure and breaching of actively operating tailings dams are frequently accompanied by flowing of the saturated tailings deposit. These flows have produced extensive damage and loss of life in the past, but damage to adjacent transportation facilities has historically been limited to temporary closure of railroads and highways. Perhaps the greatest potential hazard may be posed by certain types of actively operating tailings dams where flows triggered by seismic failure may affect lifeline transportation facilities such as highways or pipelines and impede postearthquake disaster response and recovery efforts. Vick (1991) presented methods of estimating flow runout distances in the event of tailings embankment failure. These procedures have been used to address the impacts of potential tailings flows on adjacent highways for existing and proposed tailings dams in Utah and Montana (Vick, unpublished study).

The potential for flow failures from abandoned tailings deposits is less clear but is the topic of current research (Troncoso 1990). Dobry and Alvarez (1967) studied many tailings deposits subjected to strong earthquake ground motion in Chile and documented a number of liquefaction failures and flows from actively operating tailings impoundments. However, with inactive deposits, slope effects were limited to deformation and cracking. No flow failures are known to have occurred for abandoned tailings deposits that did not retain impounded water, under earthquake loadings or otherwise (USCOLD 1994).

Nevertheless, in one recent case involving two abandoned tailings deposits in Missouri that were constructed using unusual tailings deposition procedures and that retained surface water, dynamic analyses and flow-failure runout assessments were performed to evaluate the potential for inundation of a downstream community resulting from tailings-dam failure triggered by a seismic event (Vick et al. 1993). It was found that a proposed highway embankment downstream from the dams could be designed to retain a possible flow and thus substantially reduce the hazard to downstream residents. Thus, if potential failure hazards from tailings dams are evaluated and recognized in advance, judicious planning and layout of transportation facilities can mitigate their effects.

5. MITIGATION MEASURES

Construction of cuts and fills on or through tailings deposits requires addressing stability, subgrade compressibility, construction feasibility, and environmental factors. The characteristics of tailings deposits can be less favorable than those of natural soils in most of these respects. Various soil improvement methods have been considered for tailings deposits, many of which were summarized by Mitchell (1981, 1988). However, the applicability of virtually all such techniques is limited to specific characteristics of soil, saturation, or both. No single technique has been found universally feasible for both sand and slimes tailings under all conditions. Avoidance of tailings deposits during route selection, with due consideration for the risks that adjacent impoundments may pose, therefore remains a primary mitigation strategy.

REFERENCES

ABBREVIATIONS

ASCE   American Society of Civil Engineers
USCOLD U.S. Committee on Large Dams

Abadjiev, C. 1985. Estimation of the Physical Characteristics of Deposited Tailings in the Tailings Dam of Nonferrous Metallurgy. In *Proc., 11th International Conference on Soil Mechanics and Foundation Engineering,* San Francisco, A.A. Balkema, Rotterdam, Netherlands, pp. 1231–1238.

ASCE. 1977. *Geotechnical Practice for Disposal of Solid Waste Materials.* New York, 885 pp.

Brawner, C. 1979. Design, Construction, and Repair of Tailings Dams for Metal Mine Waste Disposal. In *Current Geotechnical Practice for Mine Waste Disposal,* American Society of Civil Engineers, New York, pp. 53–87.

Bromwell, L., and D. Raden. 1979. Disposal of Phosphate Mining Wastes. In *Current Geotechnical Practice for Mine Waste Disposal,* American Society of Civil Engineers, New York, pp. 88–112.

Corp, E.L., R.L. Schuster, and M.M. McDonald. 1975. *Elastic-Plastic Analysis of Mine-Waste Embankments.* Report of Investigations 8069. Bureau of Mines, U.S. Department of the Interior, 98 pp.

Dobry, R., and L. Alvarez. 1967. Seismic Failures of Chilean Tailings Dams. *Journal of the Soil Mechanics and Foundation Division,* ASCE, Vol. 93, No. SM6, pp. 237–260.

Finn, W., M. Yogendrakumar, R. Lo, and R. Ledbetter. 1990. Seismic Response Analysis of Tailings Dams. In *Proc., International Symposium on Safety and Rehabilitation of Tailings Dams,* International Committee on Large Dams (ICOLD), Sydney, Australia, BiTech Publishers, Vancouver, British Columbia, Canada, pp. 7–33.

Gadsby, J., J. Malick, and S. Day (eds.). 1990. *Acid Mine Drainage: Designing for Closure.* BiTech Publishers, Vancouver, British Columbia, Canada, 495 pp.

Kealy, C., and R. Soderberg. 1969. *Design of Dams for Mill Tailings.* IC 8410. Bureau of Mines, U.S. Department of the Interior, 49 pp.

Klohn, E. 1972. Design and Construction of Tailings Dams. *Transactions, Canadian Institute of Mining and Metallurgy,* Vol. 75, pp. 50–66.

Ladd, C. 1991. Stability Evaluation During Staged Construction. *Journal of Geotechnical Engineering,* ASCE, Vol. 117, No. 4, pp. 537–615.

Lo, R., E. Klohn, and W. Finn. 1988. Stability of Hydraulic Sandfill Tailings Dams. In *Hydraulic Fill Structures*, Geotechnical Special Publication 21 (D. Van Zyl and S. Vick, eds.), American Society of Civil Engineers, New York, pp. 549–572.

MacIver, B. 1961. How the Soils Engineer Can Help the Mill Man in the Construction of Proper Tailings Dams. *Engineering and Mining Journal*, May, pp. 85–90.

McWaters, T. 1990. Developing Magma's Tailings Leach Operation. *Mining Engineering*, Sept., pp. 1075–1080.

Mitchell, J. 1981. Soil Improvement: State of the Art Report. In *Proc., 10th International Conference on Soil Mechanics and Foundation Engineering*, Stockholm, Sweden, A.A. Balkema, Rotterdam, Netherlands, Vol. 4, pp. 509–565.

Mitchell, J. 1988. Densification and Improvement of Hydraulic Fills. In *Hydraulic Fill Structures*, Geotechnical Special Publication 21 (D. Van Zyl and S. Vick, eds.), American Society of Civil Engineers, New York, pp. 606–633.

Morgenstern, N. 1985. Geotechnical Aspects of Environmental Control. In *Proc., 11th International Conference on Soil Mechanics and Foundation Engineering*, San Francisco, A.A. Balkema, Rotterdam, Netherlands, pp. 155–185.

Pettibone, H., and C. Kealy. 1971. Engineering Properties of Mine Tailings. *Journal of the Soil Mechanics and Foundation Division*, ASCE, Vol. 97, No. SM8, pp. 1207–1225.

Schiffman, R., S. Vick, and R. Gibson. 1988. Behavior and Properties of Hydraulic Fills. In *Hydraulic Fill Structures*, Geotechnical Special Publication 21 (D. Van Zyl and S. Vick, eds.), American Society of Civil Engineers, New York, pp. 166–202.

Soderberg, R., and R. Busch. 1977. *Design Guide for Metal and Nonmetal Tailings Disposal*. IC 8755.

Bureau of Mines, U.S. Department of the Interior, 136 pp.

Troncoso, J. 1988. Evaluation of Seismic Behavior of Hydraulic Fill Structures. In *Hydraulic Fill Structures*, Geotechnical Special Publication 21 (D. Van Zyl and S. Vick, eds.), American Society of Civil Engineers, New York, pp. 475–491.

Troncoso, J. 1990. Failure Risks of Abandoned Tailings Dams. In *Proc., International Symposium on Safety and Rehabilitation of Tailings Dams*, International Committee on Large Dams (ICOLD), Sydney, Australia, BiTech Publishers, Vancouver, British Columbia, Canada, pp. 34–47.

Van Zyl, D., and S. Vick (eds.). 1988. *Hydraulic Fill Structures*, Geotechnical Special Publication 21, American Society of Civil Engineers, New York, 1068 pp.

Vick, S. 1990. *Planning, Design, and Analysis of Tailings Dams*. BiTech Publishers, Vancouver, British Columbia, Canada, 369 pp.

Vick, S. 1991. Inundation Risk from Tailings Flow Failures. In *Proc., IX Pan-American Conference on Soil Mechanics and Foundation Engineering*, Viña del Mar, Chile, Sociedad Chilena de Geotecnica, Vol. 3, pp. 1137–1158.

Vick, S., R. Dorey, W. Finn, and R. Adams. 1993. Seismic Stabilization of St. Joe State Park Tailings Dams. In *Geotechnical Practice in Dam Rehabilitation* (L. Anderson, ed.), Geotechnical Special Publication 35, American Society of Civil Engineers, New York, pp. 402–415.

USCOLD (Tailings Dam Committee). 1994. *Tailings Dam Incidents*. Denver, Colorado.

Wahler and Associates. 1974. *Evaluation of Mill Tailings Disposal Practices and Potential Dam Stability Problems in Southwestern U.S.—Investigation Report, Kennecott Copper Corp., Magna Tailings Dam, Magna, Utah*. Bureau of Mines, U.S. Department of the Interior, NTIS PB-243 076, 226 pp.

Chapter 23

JERRY D. HIGGINS AND
VICTOR A. MODEER, JR.

LOESS

1. INTRODUCTION

The purpose of this chapter is to review the engineering character of loess and typical cut-slope stability problems that have been observed and to suggest design approaches with special consideration for cut slopes related to transportation facilities. The discussion relies heavily on the experience of state transportation departments in Missouri, Tennessee, Illinois, and Washington.

Loess is an eolian deposit composed primarily of silt-sized particles. Loess deposits constitute approximately 11 percent of the land surface of the earth and 17 percent of the soil deposits of the United States (Turnbull 1965). Extensive areas of Europe, Asia, New Zealand, the Arctic, and the Antarctic are covered by loess deposits with thicknesses commonly between 20 and 30 m. However, the most extensive loess deposits are in China (Hobbs 1943; Schultz and Frye 1965; Eden and Fulkert 1988). The most extensive deposits in the United States cover parts of the Midwest as well as portions of Washington, Oregon, Idaho (Figure 23-1), and Alaska.

Loess is characterized by a loose structure that consists of silt and fine sand particles coated by a clay binder. Because of the clay coating on the silt and sand particles, there is little true intergranular contact, particularly at low confining pressures. Thus, most of the strength is attributable to the clay binder. At low water contents, high negative pore pressures develop in the binder, producing relatively high shear strengths. As a result, as long as it remains unsaturated, loess has the ability to stand in vertical cut slopes and to support relatively large loads as a foundation material. However, upon wetting, the negative pore pressures are eliminated, reducing effective stresses, and thus the strength, as water content within the clay fraction increases to near saturation.

Sudden settlement from wetting (*hydroconsolidation*) is also a result of the loose structure. Upon wetting, the reduction in shear strength allows the granular material to reorient, producing a denser soil with a substantially lower void ratio, which can result in large settlements. It was this tendency to consolidate, particularly with respect to earth-dam

FIGURE 23-1
Loess deposits in the conterminous United States (modified from Turnbull 1965).

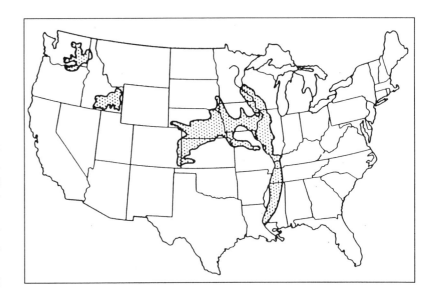

585

foundations, that led to much of the original research on the physical properties of loessial soils performed by the U.S. Bureau of Reclamation in the early 1950s and 1960s (Leighton and Willman 1950; Hansen et al. 1959).

Slope stability problems in loess deposits appear to be common and range in scale from massive landslides to small slides and flows or erosion problems. A number of very large landslides in loessial materials have been documented in the literature, only a few examples of which are noted here.

One of the most infamous landslides in loess deposits occurred in 1920 in Gansu Province, China, as a result of the Richter magnitude (**M**) 8.5 Haiyuan earthquake. Approximately 650 loess collapses (slides and flows) killed about 100,000 people (see Chapter 2) (Close and McCormick 1922; Tungsheng 1988). In the Tadzhik Republic of the former USSR, the 1949 **M** 7.5 Khait earthquake triggered numerous massive debris avalanches and flows (many of which were in loess deposits) that buried 33 villages (Leonov 1960; R. L. Schuster, personal communication, 1993, U.S. Geological Survey, Denver, Colorado). In March 1983, 45 million m³ of loess collapsed from the crest of a hill in Gansu Province. The debris, consisting of a 100-m-thick slab of rain-saturated loess, dropped 270 m and traveled laterally 1.7 km at speeds of 7 to 12 m/sec. The slide buried four villages, dammed a river, and killed 220 people (Adams 1988).

A rare type of landslide in loess occurred in 1989 in a suburb of Dushanbe in the Tadzhik Republic. The terrain was hilly and the highly porous silt (loess) had been wetted by irrigation. An **M** 5.5 earthquake caused liquefaction of the collapsible loess deposits, which triggered a chain of landslides. The landslides turned into a very large mud flow, which buried more than 100 houses 5 m deep (Ishihara et al. 1990).

Although such large and catastrophic loessial landslide problems have not occurred in the United States, a number of slope stability problems related to loess have been of concern to engineering geologists and civil engineers for many years. Many of the problems recorded in the literature are related to transportation issues. The slope problems range from moderate to small slides and flows to erosion. The impact of these problems ranges from concern for safety of the public to environmental damage (soil erosion and stream sedimentation) or, in some cases, to aesthetics.

2. LITERATURE REVIEW

2.1 Formation of Loess

As shown in Figure 23-1, the distribution of loess deposits in the United States centers around major waterways, which during the Late Pleistocene were points of egress for meltwater from continental glaciers. Glacial erosion of the land surface produced soil and rock debris that was transported by the meltwater. Rivers did not have the capacity to transport the large volume of sediment, causing formation of braided streams and outwash plains. Thermal effects of rising temperatures during the glacial retreat created winds that induced eolian transport of materials from the outwash plains. It is generally accepted that these winds eroded and transported silt- and clay-sized materials and deposited them in a loose loessial structure. The eolian source of deposition has been substantiated by the sorting of loess with distance from the source; that is, loess deposits characteristically become more fine grained with distance from the source (Smith 1942; Schultz and Frye 1965).

2.2 Fundamental Properties

Comprehensive studies of the engineering and index properties of loess were made by Holtz and Gibbs (1951), followed by Royster (1963), Royster and Rowan (1968), the Missouri State Highway Department (1978), and Higgins et al. (1985, 1987). Holtz and Gibbs were primarily interested in consolidation. Basic laboratory tests, including gradation, specific gravity, plasticity, and shear strength, were performed on a large number of samples. Grain size characteristics revealed that the majority of samples (71 percent) fell within gradation limits arbitrarily defined as silty loess. Samples found to be finer than the boundaries established for silty loess were termed clayey loess (21 percent), and coarser samples were categorized as sandy loess (8 percent). A grain-size distribution chart delineating these subdivisions is presented in Figure 23-2.

Holtz and Gibbs (1951) also conducted Atterberg-limit tests and plotted the results on a plasticity chart. The plot exhibited a concentration of points representing soils with plasticity indexes ranging from 5 to 12 percent and liquid limits between 28 and 34 percent. Examination of these data in conjunction with the gradation analysis revealed that the concentration was indicative of silty loess. Furthermore, a more poorly defined grouping of higher plasticity index and liquid limits was found to coincide with gradation curves in the clayey loess range. Thus, a useful relationship between gradation and plasticity was established.

Holtz and Gibbs found that shear strength, as determined by triaxial testing, varied considerably with water content and to a lesser extent with dry density. As would be expected, strength increased with decreasing water content and increasing density. The angle of internal friction remained fairly constant, ranging from 30 to 34 degrees, whereas the cohesion intercept increased rapidly as water content decreased.

Although Holtz and Gibbs believed that strength was not affected substantially by the gradational subdivisions (sandy, silty, and clayey loess), it is evident that this is not true at low water contents when a total stress analysis is employed. Undrained cohesion increases with increasing clay content for a given water content below saturation, resulting in higher strength. However, this effect is at least partially offset by the tendency of clayey loess to maintain a higher natural water content than that of silty or sandy loess (Missouri State Highway Department 1978).

Gibbs and Holland (1960) expanded the data base of the original work by Holtz and Gibbs. The results of additional laboratory tests and of plate-load and pile-driving tests were incorporated into their report. Conclusions pertinent to strength properties remained essentially unchanged.

The relationship between water content and shear strength was more thoroughly examined by Kane (1968), who presented what he termed the "critical water content" concept. Unconsolidated-undrained (UU) triaxial shear tests were conducted at various water contents on undisturbed samples obtained from a site near Iowa City, Iowa. In addition, tests to measure negative pore-water pressures and volumetric variation with water content were performed. It was found that above a given water content (the critical water content) the clay binder was volumetrically stable. Below the critical water content, the clay fraction shrinks as a result of desiccation. The critical water content can also be described as the water content at which the clay binder becomes saturated. This mechanism provides a reasonable explanation of the variation in strength with water content. Because the clay

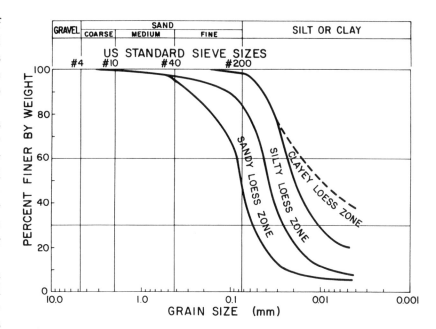

binder is nearly saturated once the critical water content is reached, any additional water fills voids between particles and has little effect on effective stress levels until complete saturation is achieved. As water content decreases below critical, desiccation occurs and negative pore-water pressures are developed in the clay binder. As the water content decreases, the negative pore pressure increases, which increases effective confining pressure and thus increases strength. The applicability of the concept of critical water content to slope stability problems in loess has been noted by several investigators (Missouri State Highway Department 1978; Higgins et al. 1985, 1989; Higgins and Fragaszy 1988). They found the critical water content for Missouri and Washington loess to be in the range of 17 to 20 percent.

FIGURE 23-2
Range in grain-size distribution for loess from Missouri River Valley (modified from Holtz and Gibbs 1951).

2.3 Index Properties and Shear Strength

General trends in index and shear strength properties of loess have been determined for many locations in the central United States, eastern Washington, and Alaska.

2.3.1 Clay Composition

Montmorillonite, or a combination of montmorillonite and illite, is the dominant clay mineral for loess deposits in the central United States (Handy et al. 1955; Gibbs and Holland 1960; Crumpton and Badgley 1965; Krinitzsky and Turnbull 1965; Bandyopodyopadhyay 1983). The clay fraction of Alaskan loess is predominantly chlorite with minor amounts of illite and possibly kaolinite (Davidson and Roy 1959).

2.3.2 Calcium Carbonate

Calcite deposits in loess are generally found as discrete grains, root fillings, or nodules (Holtz and Gibbs 1951; Gibbs and Holland 1960). In some cases a continuous layer or crust of calcite may form at shallow depths below the ground surface. This crust generally forms as the result of ground-surface evaporation and has been observed on cut slopes in Mississippi (Krinitzsky and Turnbull 1965) and eastern Washington (Higgins et al. 1989).

In eastern Washington, it appears that the presence or absence of calcium carbonate is at least partially related to mean annual precipitation or evaporation. In areas averaging greater than 51 cm of precipitation per year, no samples containing calcium carbonate were encountered. Conversely, a large majority of samples collected in areas where mean precipitation was less than 38 cm contained a significant quantity of calcite.

Where present in sufficient quantities, calcite increases the dry strength of loess. However, because of the discontinuous nature of calcite accumulation, care must be taken when strength properties are determined for a specific site. If test samples are obtained from a location containing a higher-than-average percentage of calcite, the tests may indicate a strength higher than the actual strength for the site as a whole.

2.3.3 Gradation

Gradation characteristics of loess from different locations in the United States are presented in Figures 23-3 through 23-9. In general, the range in grain-size distribution falls within the bounds established by Holtz and Gibbs (1951), with the exception of Alaskan loess, which tends to be slightly coarser. The maximum grain size in loess deposits examined within the conterminous United States is 2.00 mm (No. 10 sieve), whereas analyses on Alaskan loess exhibit an upper limit of 9.53 mm (3/8-in. sieve). It is thought that the particles larger than 2.00 mm in Alaskan loess may be carbonate concretions formed after deposition of the loess (Sheller 1968).

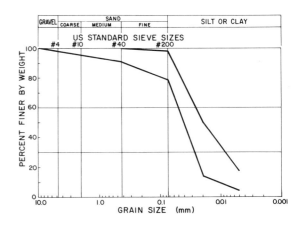

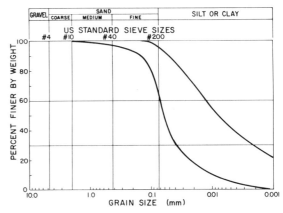

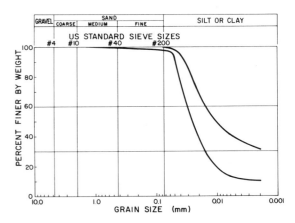

FIGURE 23-3 *(top)*
Range in grain-size distribution for loess from Alaska (modified from Morrison 1968).

FIGURE 23-4 *(center)*
Range in grain-size distribution for loess from Kansas (modified from Bandyopodyopadhyay 1983).

FIGURE 23-5 *(bottom)*
Range in grain-size distribution for loess from southwestern Iowa (modified from Davidson and Handy 1952).

2.3.4 Plasticity

The results of Atterberg-limit tests from various investigations are presented in Figure 23-10. Although substantial variability can be observed in the range of plasticity from location to location, the approximately linear relationship between liquid limit and plasticity index appears to be similar for all locations.

2.3.5 Shear Strength

Shear strength is the most important engineering property of any soil when slope stability is considered. Unfortunately, shear strength is difficult to quantify for loessial soils when saturation is less than 100 percent. The angle of internal friction reported in early studies remains relatively constant, between 28 and 36 degrees, whereas the cohesion intercept varies inversely with water content. Higgins et al. (1987) found shear strength values for UU tests on eastern Washington silty and clayey loess. Cohesion ranged from 21 to 69 kPa and the angle of internal friction ranged between 9 and 21 degrees. Consolidated-undrained tests showed a range for cohesion of 0.5 to 31 kPa and for internal friction of 27 to 29 degrees.

The highest values of cohesion were reported for loess samples with high density, high clay content, and low moisture content. Cohesion values as high as 448 kPa have been observed (Holtz and Gibbs 1951). Conversely, low-density silty loess samples tested at high water contents produced Mohr envelopes with cohesion intercepts equal to zero.

Although the variability of the angle of internal friction appears to be confined to a narrow range (with some exceptions), some controversy exists regarding the strength of low-density silty loess at high water contents. Some authors observed a complete loss of strength below a minimum consolidation pressure of approximately 69 kPa (Holtz and Gibbs 1951; Clevenger 1956; Gibbs and Holland 1960; Royster and Rowan 1968), whereas other authors did not (Olson 1958; Akiyama 1964; Kane 1968). Some investigators explained the lack of shear strength at low consolidation pressures by stating that a minimum degree of consolidation is required to produce grain-to-grain contact and thus shearing resistance (Holtz and Gibbs 1951; Clevenger 1956; Gibbs and Holland 1960).

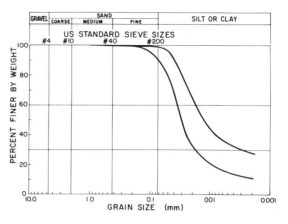

FIGURE 23-6
Range in grain-size distribution for loess from east-central Iowa (modified from Handy et al. 1955).

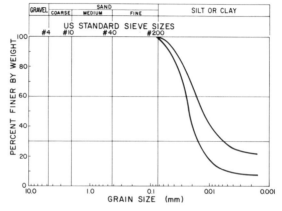

FIGURE 23-7
Range in grain-size distribution for loess from Mississippi (modified from Krinitzsky and Turnbull 1965).

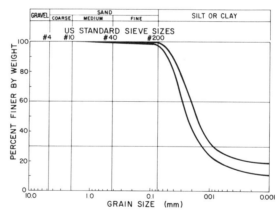

FIGURE 23-8
Range in grain-size distribution for loess from northeast Iowa (modified from Handy et al. 1955).

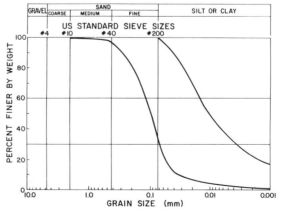

FIGURE 23-9
Range in grain-size distribution for loess from eastern Washington (Higgins et al. 1985).

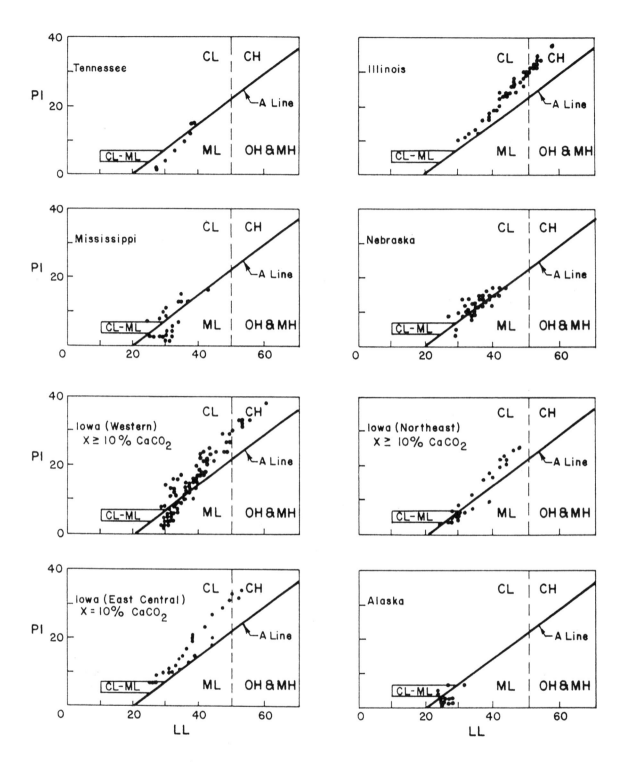

FIGURE 23-10
Plasticity data for loessial deposits throughout the United States
(modified from Sheller 1968).

2.3.6 Variability of Index Properties with Distance from Source

Although index properties within a given deposit of loess tend to vary within narrow bounds, some trends have been noted with respect to the distance from the source of the loess. Investigations in the Midwest, the South, and the state of Washington (Dahl et al. 1952; Krinitzsky and Turnbull 1965; Handy 1976; Higgins et al. 1985, 1989) indicated an increase in clay content and decrease in total thickness with distance from the source. Studies in Iowa (Dahl et al. 1952), Mississippi (Krinitzsky and Turnbull 1965), and Washington (Higgins et al. 1985, 1989) noted that as clay content increases, silt content decreases, with sand content being a uniform, minor constituent.

Variation in clay content tends to affect engineering and other properties. As clay content increases, natural water content and density also increase (Davidson and Handy 1952; Higgins et al. 1985, 1989). Thus, natural water content and density can be expected to increase with distance from the source.

A linear relationship between thickness and the logarithm of distance from the source has been established (Handy 1976) for midwestern loess deposits. Although this relationship is undoubtedly true for some deposits, in many cases thickness is highly variable on a local scale with no clear trends discernible. When a high degree of variability is encountered, it is usually due to hummocky terrain with maximum thicknesses near the crests of hills and thin deposits in intervening depressions (Kolb 1965).

2.3.7 Density

Typical densities for eastern Washington loess were reported by Higgins and others (1987). Dry densities for 12 samples of silty loess ranged from 1.13 to 1.57 Mg/m^3, and for three samples of clayey loess the range was 1.36 to 1.44 Mg/m^3.

2.3.8 Depth Effects

It is not possible to generalize about the variation in textural composition with depth other than to say that changes in the relative percentage of the sand, silt, and clay constituents with depth are usually minor. In some areas sand content was found to increase slightly with depth, whereas clay content remained constant or showed a minor decrease (Lyon et al. 1954; Higgins et al. 1985). Conversely, some vertical sections exhibited a uniform percentage of sand with depth, whereas clay content remained constant but increased slightly at the base of the unit (Hansen et al. 1959). Except for isolated cases, constituent percentages did not vary more than 7 to 8 percent.

In-place density and natural water content demonstrate a more consistent trend with depth than does textural composition. Ignoring fluctuations in the upper 2 m, both density and water content tend to increase with depth.

3. REVIEW OF CUT-SLOPE DESIGN PRACTICE

In the United States, comprehensive design criteria that specifically address the unique engineering properties of loessial soils are applied by a limited number of state transportation departments. A summary of design procedures employed by various states is presented in Table 23-1.

The results of three comprehensive studies (Royster and Rowan 1968; Missouri State Highway Department 1978; Higgins et al. 1985) in three states (Tennessee, Missouri, and Washington) on the design of cut slopes in loess have been published. The findings of all three concerning loess properties and slope behavior showed only minor differences, usually because of different climatic conditions.

In general, the design specifications recommended by these studies are based on four main criteria: (a) gradation boundaries similar to those established by Holtz and Gibbs (1951) for silty and clayey loess, (b) the critical-water-content concept developed by Kane (1968), (c) experience from studying the relationship between slope performance and physical properties of loess soils, and (d) surface-water drainage.

It was concluded that vertical slopes perform well in silty loess when water contents are maintained below about 17 percent (the critical water content). Slightly higher water contents (less than 20 percent) could be tolerated with a favorable exposure (between southeast and southwest). Flattened slopes should be used for silty loess with high water contents and for all cuts in clayey loess. Although 2H:1V slopes are generally flat enough

Table 23-1
Design Procedures Used by Various State Departments of Transportation for Cuts in Loess

STATE	DESIGN PROCEDURE
Arkansas	No set design procedure indicated; vertical cuts often utilized with benches added for cuts exceeding 8 m in height; bench widths approximately 3 to 5 m wide and slope gently toward backslope
	Drainage ditches usually constructed near top of cut and at toe, with ditch paving utilized on benches to channel drainage; collection pipes used to convey runoff from top of cut and benches to road level
Indiana	Experience with cut slopes in loess limited; vertical cuts employed with 3-m-wide benches for cuts over 5 m high; paved interceptor ditches installed above cut to prevent runoff flow over face
Iowa	No special design techniques used for loess; normal backslope design 5 m wide with benches every 8 m of vertical height; benches sloped to back of cut; drains used when loess overlies relatively impermeable layer to drain contact
Kansas	Slopes of 3H:1V or flatter usually employed, although some 2H:1V benched slopes have been constructed; slope performance generally good; however, some erosional problems experienced when runoff flowed over slope face
Missouri	Both vertical cuts and flattened slopes (<2.5H:1V) based on grain-size distribution and water content; slope-stability analysis used for saturated deposits; drainage design specified above and below cut slope
Nebraska	Although vertical cuts constructed for some minor roadways, slopes flattened to 3H:1V typically constructed; flattened slopes perform well once vegetative cover established; use of dikes to ditches preferred when runoff must be restricted on cut face
Ohio	No special considerations; 2H:1V slopes used
Tennessee	Slopes cut to 2H:1V or flatter with 5- to 8-m benches at 6- to 8-m intervals; drainage design specified above and below cut slope
Washington	Both vertical cuts and flattened slopes (<2.5H:1V) constructed on basis of grain-size distribution and water content; slope-stability analysis used for saturated deposits or cuts >15 m high; drainage design specified for slopes above and below cut slope

from a stability point of view, slopes flattened to 2.5H:1V undergo much less degradation from erosion and allow better establishment of a vegetative cover. Where water tables are truncated, much flatter slopes may be required because of seepage forces.

When a vertical cut is employed, it is typically benched when the total height exceeds 10 m. Benches are placed approximately every 6 m vertically with a width of 5 to 8 m. The benches serve the dual purpose of limiting erosion and providing increased stability for the overall slope.

The Missouri and Washington investigations suggested some problems with the drilling and sampling methods most commonly used to obtain "undisturbed" samples, particularly at high water contents. Undisturbed samples were obtained by both Shelby tubes and Denison double-tube samplers, and their densities were compared with those for hand-cut samples. In all cases downhole samples were found to be denser, indicating compression during sampling. The largest discrepan-

cies between hand-cut and downhole samples were found when water contents were in excess of 18 percent (Missouri State Highway Department 1978). The Washington study found that all Shelby-tube samples had a higher density than hand-cut samples; however, no specific correlations were noted (Higgins et al. 1987). A similar investigation on the sampling effect on Bulgarian loess (Milovic 1971) obtained similar results.

The general conclusion of the Missouri and Washington reports regarding near-vertical versus flattened slopes is that the use of near-vertical cuts should be restricted but not eliminated. Provided that the deposit meets the gradation and moisture criteria outlined, a near-vertical cut is the preferable choice, particularly since silty loess tends to erode severely on flattened slopes. However, efficient drainage of surface water away from the slope face is essential.

Although Royster (1963) and Royster and Rowan (1968) generally agreed with these conclusions, the Tennessee Department of Transportation (TDOT) design criteria do not include near-vertical cut slopes. Conventional flattened slopes are commonly constructed, probably for aesthetics and maximum stability. However, TDOT has developed useful guidelines for surface drainage of loess slopes, many of which were adopted or modified by the Washington Department of Transportation.

4. FAILURE MECHANISMS

Studies in Missouri and Washington (Missouri State Highway Department 1978; Higgins and Fragaszy 1985, 1988; Beard et al. 1986; Higgins et al. 1989) on slope design in loess recorded numerous observations of slope-failure modes. Additional examples are included here from Illinois and other portions of the midwestern United States. These observations define the typical types of slope failures and the causative failure conditions that can be expected in loess slopes. An understanding of these failure modes will aid the engineer and engineering geologist in identifying potentially unstable conditions and in the design of stable cut slopes in loess.

4.1 Erosion

Erosion damage (including piping) is a common form of slope degradation, primarily in silty loess, although clayey loess is susceptible also in both dry and moist climates. In eastern Washington many large erosion gullies have been initiated by small piping failures originating 2 to 3 m behind the top of the cut face. As the pipe enlarges, the surface caves, forming a gully. These gullies are sometimes initiated by animal burrows that are intersected by the slope cut. The piping phenomenon often occurs in moist climates when the slope cut truncates a water table.

The loess piping mechanism normally results in relatively small volumes (1 to 5 m³) of displaced material. However, in Illinois, piping in saturated loess has displaced volumes of 200 to 300 m³. The piping manifests itself by an accumulation of redeposited loess at the piping outfall. The displaced mass can result in creation of either an erosion channel or a horizontal pipe coupled with a vertical sink. Erosion channels may be as much as 6 m deep, and vertical sinks may be up to 3 m in diameter. The larger vertical sinks either occur in areas

of ponding water or are created by material removed through the horizontal pipe.

Major erosional features are found to be the result of improper drainage or channel protection. In eastern Washington, serious erosion problems were observed in long cuts transecting small drainage basins that had no provisions for conveying excess runoff away from the cut. The highly erosive nature of silty loess requires the diversion of runoff from what normally would be considered insignificant drainage areas. Figure 23-11 illustrates the result of truncating a small drainage area without providing a means of conveying excess runoff from the slope or providing erosion protection.

Figure 23-12, which shows gully erosion along US-12 near Walla Walla, Washington, demon-

FIGURE 23-11 Gully erosion resulting from truncation of a small drainage basin near Dusty, Washington.

FIGURE 23-12
Gully erosion where small side drainage was truncated by highway cut to right of photograph, US-12 near Walla Walla, Washington.

strates the progressive nature of erosion in loessial soils. Erosion was initiated where a small side drainage was truncated by the highway cut (to the right in the photograph) and has progressed rapidly up gradient. Erosion will continue until the overall gradient is decreased below that causing channel scour.

Figures 23-13 and 23-14 illustrate two important points. In Figure 23-13 the majority of the erosion in the left portion of the photograph is due to piping. To the right of the small pipes, gully erosion is beginning to develop. Vegetation in the

base of the depression, as well as the shape of the developing gully, indicates that the gully originated as a pipe. When the pipe enlarged to the point where the overlying material could no longer be supported, caving occurred and the overlying vegetation was deposited in the depression. This mechanism is very common in the formation of erosion gullies in silty loess.

The second point is that extensive erosion problems may develop caused by runoff from extremely small drainage basins. Figure 23-14 shows the drainage area above the erosion shown in Figure 23-13. When the area was first examined in June, before the harvest of winter wheat in late July and August, there was no indication that drainage was concentrating flow toward the problem area. Figure 23-14, taken following harvest in the early autumn, shows a minor basin draining into the existing erosion zone.

Figure 23-15 demonstrates the extent of damage when proper drainage is not provided. The majority of the damage appears to be due to a single summer storm event; a combination of piping and surface erosion is taking place. Piping inlets originated a substantial distance behind the cut, which exemplifies the need for construction of drainage ditches an adequate distance behind the cut slope.

Figures 23-16 and 23-17 show erosion damage to near-vertical cuts in Mississippi and Louisiana where water ponded near the crest of the slope. If natural drainage is directed away from a loess slope, erosion is usually minimal (Figure 23-18).

The Missouri study (Missouri State Highway Department 1978) reported extensive erosion damage on sodded V-shaped benches in silty loess. The slopes were cut nearly vertical and benched. However, drainage channels on the benches eroded severely, which can lead to failure of the entire bench. Paved ditches have been used in Missouri and Illinois to mitigate this effect.

Another serious erosional problem in loessial soils in any climate, particularly on flattened slopes, is surface erosion during construction and up to the time of establishment of a good vegetative cover. Even with adequate drainage, extensive damage may result from raindrop impact and rill erosion. This problem is most critical in silty loess, but clayey loess is also susceptible.

The ability of a slope to hold a vegetative cover appears to be related to the steepness of the slope. The relationship is complicated by factors such as

FIGURE 23-13
Piping erosion developing into gully erosion (US-12 near Walla Walla, Washington).

FIGURE 23-14
Runoff source causing erosion in Figure 23-13 (small depression at center of photograph).

exposure, type of vegetative cover, soil type, and climatic conditions. There is no quantitative method that accurately predicts the maximum angle at which a slope can be cut and still retain a continuous vegetative cover. However, experience from transportation departments in Washington, Missouri, Illinois, and other states indicates that slopes of 2.5H:1V or flatter are reasonable for establishment and maintenance of vegetation.

4.2 Shallow Slope Movements

Higgins et al. (1985, 1989) found shallow downslope movements to be common in the eastern part of the Washington loess deposit where precipitation is nearly 500 mm annually, but these failures are not common in the midwestern United States. This type of failure is largely due to late-winter and early-spring climatic conditions. Precipitation and snowmelt are common at this time of year, resulting in a thin layer of thawed, saturated soil overlying either frozen or unsaturated soil. The layer of thawed, saturated soil, along with any overlying vegetation, tends to slide or flow downhill.

This form of failure appears to be primarily related to the amount of precipitation, the clay content, and the slope angle. In the western two-thirds of the Washington loessial deposit where precipitation is less than 38 cm annually, downward movements of shallow slopes were rarely observed. Clay content increases from west to east, as does precipitation. Increases in clay content, resulting in lower permeability combined with increased precipitation, raise the likelihood of obtaining the near-surface saturated conditions required for failure.

Two different forms of shallow slope failure have been observed in eastern Washington and appear to be related to soil type. In silty loess it is not uncommon to see sheets of vegetative cover 5 to 15 cm thick with an arcuate upper boundary move downslope (Figure 23-19). In this form, only minor damage was observed. However, over the years these small failures can expose large areas of soil to erosion, and damage increases considerably over time.

In clayey loess, shallow slides and flows are common and result in major slope degradation. Movement appears to be a mud-flow phenomenon with both large- and small-scale failures. Higgins

FIGURE 23-15
Area of large-scale piping near Endicott, Washington; flattened vegetation indicates flow over slope as well as piping.

FIGURE 23-16
Erosion damage to near-vertical cut slope in Mississippi.

FIGURE 23-17
Erosion damage to near-vertical cut in Louisiana.

FIGURE 23-18
Minimal erosion
damage where
natural surface
drainage is away
from slope face
(east of Walla Walla,
Washington).

FIGURE 23-20
(below) Small-scale
mud flow in
highway cut on
US-195 near
Pullman,
Washington; flow
approximately 3 m
long.

et al. (1985, 1989) observed at least minor damage in many of the clayey loess slopes over 3 m high and steeper than 2.5H:1V.

Small-scale slides, flows, or both, such as that in Figure 23-20, were found to be the most common form of failure in eastern Washington. Failures were typically 0.3 to 3 m wide with the depth of failure ranging between several centimeters and about a meter. The initial failure was normally followed by increased erosion due to the loss of vegetative cover, with severe gully erosion a common result. In some cases, failure is due to saturation of the soil under the snowpack as well as to weight of the snowpack on the underlying saturated slope (Figure 23-21).

FIGURE 23-19
Fairly common shallow failures with arcuate scarps in silty loess near Colfax, Washington.

Although not as common, large-scale slides and flows have been observed in Washington. The failure mechanism is thought to be the same as that for small-scale failures with a low depth-to-width ratio for the failure surface. Figure 23-22 shows a relatively large failure. The effect of steepness on stability of slopes cannot be emphasized strongly enough. In all cases where extensive damage was observed in eastern Washington, slopes were approximately 2H:1V or steeper. No instances of failure were observed for slopes flatter than 2.5H:1V. These observations agree with experience in the midwestern United States, where slopes greater than 2.5H:1V have proven to be too steep to maintain good vegetative cover in loessial soils (Royster and Rowan 1968; Missouri State Highway Department 1978).

Exposure has an influence on these shallow slides and flows. It is not uncommon to find that two opposing cuts with similar slopes and drainage, one facing north and the other south, are affected very differently. The north-facing slope will invariably demonstrate a greater degree of degradation because of shallow slides and flows than the south-facing slope, because slopes facing north typically have higher average water contents than do slopes with any other orientation. Figures 23-23 and 23-24 show south- and north-facing slopes, respectively, cut in clayey loess at approximately 2H:1V. Note that in Figure 23-23 erosion has occurred where small patches of vegetation and soil have slipped away. However, the north-facing slope (Figure 23-24) has experienced numerous shallow failures that have stripped the vegetation. Near the slope crest are scarps from larger failures that have occurred over the past several years and have significantly steepened the slope. The result of the shallow slides is exposure of bare soil and rapid formation of erosion gullies, some of which range up to 0.6 m in depth.

The practice of constructing drainage ditches immediately adjacent to the toes of slopes appears to contribute to this type of failure. Periodic highway maintenance requires removal of sediment from ditches, which may result in undercutting of the toe. In addition, drainage ditches directly adjacent to the toe may raise the water content of the toe materials, further encouraging failure.

4.3 Volume Change

Failure of slopes as a result of shrinkage cracks that develop in clayey loess is common in the humid midwestern United States. Blocks and slabs of soil fall from near-vertical faces of loess along shrinkage cracks that have developed from drying of the slope face (Figure 23-25). Shrinkage potential of a soil is a function of soil density, moisture content, and clay content. This type of failure often occurs shortly after slope faces of wet, clayey loess are exposed to drying. In Missouri this phenomenon occurs most often in clayey loess with liquid limits near 35 percent (Missouri State Highway Department 1978).

4.4 Undercutting

Undercutting is common in the midwestern United States in both silty and clayey loess with liquid limits ranging from 31 to 34 percent. The failure is progressive, originating as local sloughing near the

FIGURE 23-21
Failure of slope under melting snowdrifts on US-195 near Pullman, Washington.

FIGURE 23-22
Relatively large-scale mud flow located near Pullman, Washington; undisturbed soil at right indicates depth of failure at approximately 1 m.

base of a near-vertical slope; the shallow failure then oversteepens or undercuts the original slope, which eventually causes material upslope to fail. Undercutting is most frequently found where a vertical rise truncates a water table or a stratum with very low cohesive strength (such as sand) or high water content (Missouri State Highway Department 1978).

4.5 Freezing Damage

Frost damage was found to be a serious problem on near-vertical slopes in silty loess in Missouri

FIGURE 23-23
South-facing slope on US-195 near Pullman, Washington, showing little degradation.

FIGURE 23-24
North-facing slope, opposite cut in Figure 23-23, exhibiting severe gully erosion.

FIGURE 23-25
Slabbing of near-vertical cut along Mississippi State Highway 220.

(Missouri State Highway Department 1978). Failure is initiated by formation of an ice lens at a shallow depth within a slope, thawing of the ice lens raises pore-water pressures, and failure occurs by slump along the resulting surface of weakness. The resultant damage leaves soil exposed to erosion and further damage.

Four conditions are required for this type of failure:

1. The soil must transmit sufficient water during freezing periods to enable ice lenses to form, for example, in silt with high frost-heave susceptibility;

2. Water must be available, for example, at the intersection of a cut slope by the water table;
3. Freezing must occur over relatively long periods, for example, on north-facing slopes; and
4. The slope must be steep enough to allow failure after the ice lens melts.

4.6 Rotational Slides

Rotational slides on loess slopes have occurred in both moist and dry climates; however, they appear to be more common in the moist climate of the midwestern United States than in eastern Washington. Generally, this type of failure is associated with saturated (or nearly saturated) conditions on steep slopes. Occasionally, rotational failures occur in clayey loess in the Washington loess deposit. Water-table depths tend to be great throughout the loess deposit area. However, in the eastern section of the deposit where seasonal precipitation is relatively high and the loess has at least 20 percent clay content, some failures do occur (even on undisturbed slopes). Figure 23-26 shows an example of a rotational failure (a slump and earth flow) on a natural slope. The slope was inclined approximately 2H:1V, the head scarp was approximately 5 m high, and the failed material flowed about 15 m from the base of the scarp. As is common in this area, the failure occurred in a region where runoff and groundwater seepage from the surrounding field could be expected to collect and saturate the soils. Geotechnical investigations for slopes should be designed to detect such groundwater conditions.

4.7 Summary

The foregoing discussion indicates that the majority of experience with cut slopes in loess involves those that are less than 15 m high. This experience shows that the most significant influences on stability are surface drainage, soil water content, and grain-size distribution. Silty loess soils should perform well in near-vertical cuts (¼H:1V) if they are protected by surface-drainage structures and if natural moisture contents remain less than 17 to 20 percent. Cuts in clayey loess should perform well if they do not exceed approximately 2.5H:1V and are protected by a cover of vegetation, and if concentrated runoff flow is not directed across the slope face. The examples presented illustrate that major performance problems are due to (a) poor or inadequate surface drainage systems for cuts in silty loess (and in a few cases for cuts in clayey loess) and (b) oversteepened slopes for cuts in clayey loess. The experience with cut slopes in sandy loess is inadequate to suggest any design criteria other than conventional soil-cut designs.

The designer of slopes in loess should be fully aware of the following soil properties and should design cut slopes to avoid the resulting problems:

1. Loess is highly erodible, and the flow of water, even at low to moderate volumes and gradients (less than 5 percent and 5 to 10 percent, respectively), can cause severe erosion;
2. Disturbance of the natural soil structure by grading, operation of heavy equipment, farming activities, and so forth, makes the soil more susceptible to erosion;
3. Saturation of the soil softens the clay binder and greatly decreases the strength, leading to slope failure or accelerated erosion; and
4. Silty loess soils are highly susceptible to failure by piping.

FIGURE 23-26 Rotational failure and earthflow in loess near Pullman, Washington.

5. SITE CHARACTERIZATION

5.1 Office Study

Before a site visit, published information on the project area should be reviewed. Much generally can be learned about the geotechnical properties of the soils in an area before a site visit, which will then help the slope designer plan the field investigation program.

Typical sources of information on soil conditions can be found in the research reports listed in the reference section to this chapter, U.S. Department of Agriculture county soil surveys, or geologic maps published by the U.S. Geological Survey or the appropriate state geological survey. The county soil conservation agent can also be a good source of information. This information will help the designer determine whether significant thicknesses of loessial soils are present and provide general knowledge of soil properties at the proposed site (grain-size distribution, percentage of clay-sized material, soil stratigraphy, permeability, drainage character, depth to bedrock, agricultural use, and irrigation potential). The general knowledge gained from these data should be used to plan an efficient field investigation to define the specific site conditions.

5.2 Field Reconnaissance

After the office study, the site should be visited to verify information gained in that study and to determine soil type and stratigraphy, natural drainage conditions, land use and potential for its change, and presence of structures or activities near the top of the backslope. Observations of geologic processes at or near the site that could affect the project are important. These may include erosion problems on nearby slopes, seeps, and condition of existing cut slopes in the general area. These observed conditions should be considered in the cut-slope design.

5.3 Field Exploration Program

Sampling should be done in the most efficient manner, which is dependent on the size and type of cut slope that is proposed. Soil samples are required for determination of the natural water content, grain-size distribution, and Atterberg limits for standard cut-slope projects. Cuts in excess of 15 m high or with nonuniform stratigraphy may require stability analysis, and therefore undisturbed samples may be needed for shear-strength tests. Sampling methods in loess are summarized in Table 23-2.

Sampling for shallow cuts can be accomplished by hand augering and for the face of an existing cut by means of horizontal test holes. These holes must extend a minimum of 1 m into the face and should be spaced so that the deposit is characterized both vertically and laterally. Major cuts should be sampled by means of test holes to a minimum of 3 m below grade. Sampling should be continuous in the top 2 m of the hole and every 1.5 m thereafter. There should be a minimum of two holes per cut, normally spaced 60 to 100 m apart (150 m maximum). Samples for shear-strength analysis should be hand cut to avoid densification of the sample by sample-tube collection methods.

6. LABORATORY TESTING

Laboratory tests are used to evaluate soil index and engineering properties that are related to the ultimate engineering behavior of a cut slope and the recommended design. For cut slopes higher than 15 m or for nonuniform stratigraphy, strength tests may be required.

Routine laboratory tests required for cut-slope design include Atterberg limits, sieve analysis, and hydrometer analysis for classification of the soil as silty, clayey, or sandy loess. Determination of natural moisture content is also required. These soil properties determine the appropriate slope design for cuts no higher than 15 m and supply useful information for the stability analysis and design of higher cuts.

Table 23-2
Sampling Methods in Loess

PARAMETER AND SAMPLING METHOD	RELIABILITY	COST
Density		
Type 1: dry hollow-stem auger boring, tube sampling (Higgins et al. 1987)	Very poor	Low
Type 2: dry hollow-stem auger boring, <5-in. pitcher-type sampling with lined tube (Casias 1987)	Fair	Moderate
Type 3: dry hollow-stem auger boring, > 5-in. pitcher-type sampling with lined tube (Casias 1987)	Good	Moderate to high
Type 4: block samples (Higgins et al. 1987; Gibbs and Holland 1960)	Excellent	High (requires excavation)
Gradation, index parameters		
Type 1	Excellent	Low
Types 2, 3, 4	Excellent	Moderate to high
Strength, consolidation		
Type 1	Very poor to unacceptable	Low
Type 2	Poor	Moderate
Type 3	Fair to good	Moderate to high
Type 4	Excellent	High

Loessial soils require slightly different testing procedures than more conventional soils. These differences arise in testing for strength and consolidation properties. Standard procedures are used for gradation determination and index-property testing.

The sensitive, collapsible nature of loess requires extreme care in handling samples. Some disturbance will occur in the preparation of samples for density determinations and triaxial and consolidation testing; such disturbance can be minimized only by care in handling.

The collapse sensitivity of loess adds another parameter that should be evaluated. Collapse sensitivity can be formally evaluated by parallel triaxial tests in which one sample is tested dry and another is saturated at some point in the test. Double oedometer tests (Holtz and Gibbs 1951) can be performed to evaluate the effects of water in causing collapse.

Natural moisture content is an indicator of the degree of disturbance. A moisture content between the shrinkage limit and the plastic limit indicates a high likelihood that the deposit is undisturbed. Moisture contents midway between the plastic and liquid limit (liquidity index of 0.5) indicate an extremely metastable condition. A higher moisture content (liquidity index of 0.8 or higher) generally initiates collapse.

If deep cuts are planned (greater than 15 m) or if water contents are expected to be greater than critical (17 percent), it may be necessary to run strength tests on undisturbed samples. Undisturbed strength testing should ideally model the state of stress and groundwater regime. Triaxial testing should be performed using compression or extension loading as the stress state in the slope dictates. The residual shear strength should be determined by extending the triaxial-test strain beyond the peak value to capture the residual strength value.

Groundwater conditions will indicate whether drained or undrained conditions are required if saturated or high-moisture loess is tested. The relatively dry nature of undisturbed loess requires triaxial testing without saturation and disallows pore-pressure measurements. When groundwater conditions are such that achieving a moisture content above 17 percent is considered unlikely, a total stress analysis is the preferable approach. Conversely, where water contents in excess of

17 percent are anticipated, an effective-stress analysis should be performed on saturated samples.

Triaxial testing with K_o-consolidation has not been reported in the literature, likely because of difficulty in determining the appropriate K_o-value for the porous-structured loess. The negative pore pressure in the undisturbed loess clay binder normally far exceeds the overburden pressure, and the strength is derived from this high negative pore pressure. Practically, it suffices to consolidate dry, undisturbed loess under isotropic conditions for triaxial testing.

The unconfined compression test is used frequently in loessial soil testing. This test provides an indication of the available strength to enable loess to stand vertically. Drawbacks of this test include lack of residual strength data and a failure to indicate strength anisotropy (Matalucci et al. 1970).

7. FIELD TESTING

The Iowa borehole shear-testing device and the cone penetrometer test have been used to determine shear-strength parameters in loess (Higgins et al. 1987; Modeer et al. 1991). The results obtained by these methods have been compared favorably with the results of triaxial testing of block samples from the same soils.

8. LOESS SLOPE DESIGN

The most common procedures for designing slopes to account for the unique properties of loess incorporate soil mechanics with a significant amount of empiricism. This heuristic approach has served many transportation departments and engineering organizations very well, considering the time and expense of an alternative extensive field and laboratory endeavor. Each agency or organization has either modified existing loess slope design procedures or expended significant effort to develop original design procedures. All of these procedures are remarkably similar, and those recommended below are based on a combination of experience and established procedures from the Missouri, Illinois, and Washington transportation departments.

If the soil at the site of the proposed cut is a silty loess with water content below critical (< 17 percent can be used as an approximation on the basis of experience in the Midwest), near-vertical cuts (¼H:1V) may be considered. If utilized, near-vertical cuts should be benched on approximately 6-m vertical intervals (or at the approximate midpoint) when the total height of the cut exceeds 9 m. Benches should be 3 to 5 m wide and gently sloped (10H:1V) toward the back of the cut (Figure 23-27). In cases where benched cuts are required, the benches should be seeded or sodded. Benches should maintain a longitudinal gradient for drainage that should not exceed 3 to 5 percent. If the drainage system above the slope crest is properly constructed (as discussed in the next section), it should not be necessary to employ erosion-control methods in excess of vegetative cover unless extremely long cuts are made. The selection of near-vertical cuts for silty loess is an attempt to avoid potentially severe erosion problems if the slope is flattened. However, it should be noted that maintenance of the near-vertical slope is difficult.

If water content exceeds critical (17 percent) or if the soil is a clayey loess, flattened slopes should

be utilized. Generally, 2.5H:1V slopes would perform adequately, but if a water table is intercepted, flatter slopes may be required because of seepage forces (Figure 23-27). A flattened cut should be seeded immediately following construction. In addition, a protective cover such as a straw mulch or a synthetic material should be placed over the slope. These covers serve the dual purpose of preventing raindrop impact, a major cause of erosion on newly opened cuts, and helping to retain moisture required to initiate a vegetative cover.

The design of cuts (near-vertical or flattened) deeper than about 15 m or cuts that intersect a water table or high-moisture-content zones should be based on a detailed stability analysis. When groundwater conditions are such that achieving a saturated condition is unlikely, a total-stress analysis is the recommended approach. Conversely, when groundwater conditions make saturation possible, an effective-stress analysis should be used.

FIGURE 23-27
Typical sections for cut slopes in silty and clayey loess (modified from Higgins and Fragaszy 1988).

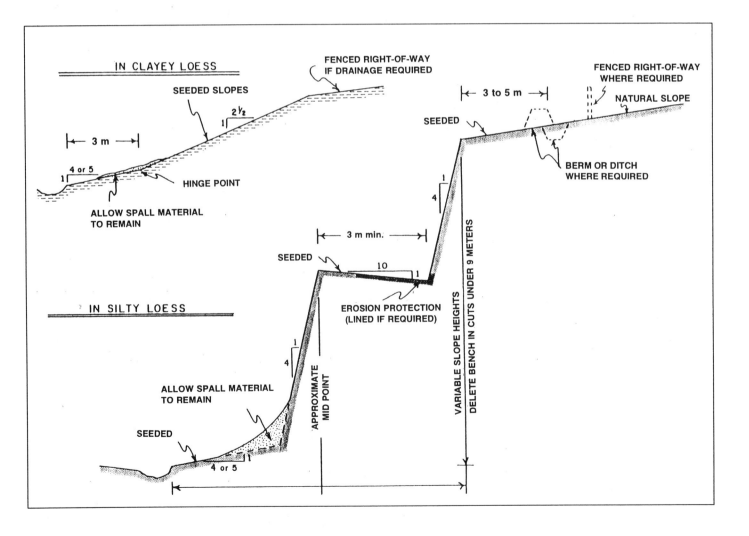

If a slope stability analysis is to be performed on a silty loess that maintains a water content below critical (< 17 percent), a wedge-type failure is perhaps the most likely geometry. When saturation at depth is encountered for either silty or clayey loess, a rotational failure may occur; however, because of possible anisotropy of the soils, the geometry of the failure surface may deviate from circular. Subsurface data from the borehole should aid in evaluating whether a circular or planar surface should be used in the stability analysis.

A near-vertical slope design based on strength-test results on undisturbed loess must take into account the potential for factors that will reduce the available strength. Increasing the design factor of safety or limiting the height of individual cuts (benching) is a common method to account for these intangibles. For example, the designer may use normal global factors of safety for all slopes but limit the heights of vertical faces rather than designing the height on the basis of strength testing. Reduced strength parameters may be used for design in addition to an increased design factor of safety. For example, Illinois uses a limiting factor of safety of 2 for vertical slopes.

These design recommendations are for homogenous materials and do not account for sandy layers or sandy loess. Under these conditions, a conventional soil mechanics approach should be applied.

9. DRAINAGE DESIGN

Designers of surface-water drainage systems for cut slopes in loess should use the following recommendations as a basis for design:

1. Prevent water (sheet wash) from flowing over the face of a cut in silty loess. Prevent concentrated flows from washing over a cut face in silty or clayey loess.
2. Do not allow water to collect in or saturate the soil within 3 to 5 m of the top of the cut face in silty loess. Water has been observed to contribute to piping.
3. Do not allow water to collect against the toe of the cut in silty or clayey loess. Saturation of the soils in the toe can cause sloughing of the slope.
4. Do not direct flow into unprotected channels in silty or clayey loess but line them with vegetation or synthetic or natural material, or deep

gullies may appear within a short period of time (i.e., 1 to 4 years).
5. Avoid disturbance of soil structure and natural vegetation at the crest of the slope cut as much as possible during and after construction to maintain soil strength and erosion resistance.

On the basis of observations of the erosion problems around cut slopes discussed above, the following design criteria are suggested with respect to the drainage required for cut slopes in loess:

1. Damage from sheet wash over the face of flattened slopes (2.5H:1V) in clayey loess is minor if a vegetation cover is maintained. Therefore, surface-water diversion above the cut is not necessary unless gullies, swales, channels, and so on, will concentrate flow onto the slope face.
2. For cuts in silty loess (¼H:1V), placement of drainage ditches or berms 3 to 5 m behind the top of the cut slope is recommended if the drainage area above the cut is inclined toward the cut. The drain should be U-shaped or flat-bottomed and lined to protect it from erosion (Figure 23-28). Also, a gradient must be maintained so that water does not stand and saturate the slope.
3. Drains that convey surface water around the sides of cut slopes often have moderate (5 to 10 percent) to steep (> 10 percent) gradients. In many cases the erosion protection required in these channels will be more substantial than for those at the heads of cuts. These steep drains may require paving with asphalt or concrete or lining with filter fabric, crushed rock, and so forth.
4. If natural drainage channels are truncated by a cut, the drainage system should be adequate to transmit the flow around the cut face (which may require considerable right-of-way) or across the cut face in lined channels or structures. Direct flow across the cut face must be avoided. All drainage structures should be located in a fenced right-of-way for protection, and the area should be seeded or have the natural vegetation preserved to maintain the soil structure and strength and to prevent traffic from farm equipment or other vehicles from damaging or destroying the drainage system and protective cover.
5. Toe drainage can be accomplished with ditches (U-shaped or flat-bottomed) located approximately 3 m from the toe of the slope. The ground slope between the toe of the slope and

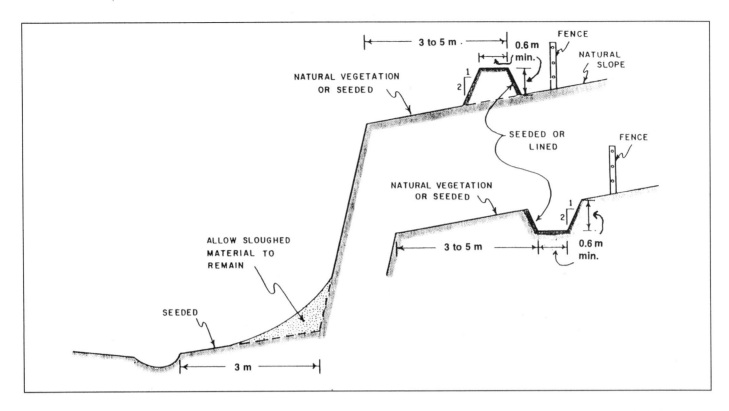

FIGURE 23-28
Drainage design for
cut slopes in loess;
fence is to protect
drainage structures
from traffic
(modified from
Higgins and
Fragaszy 1988).

the ditch should be gently inclined toward the ditch (between 4H:1V and 5H:1V is recommended). Any material that spalls downslope between the toe and the ditch should be left in place to protect the toe.

10. CONSTRUCTION-CONTROL CONSIDERATIONS

Considering the adverse effects of water on the stability of loess, cut slopes are best constructed during the dry season. The drainage structure above the crest of the cut should be constructed before the opening of the cut and with as little disturbance to the surrounding vegetation as possible. Once the cut is made, construction equipment should be kept away from the crest.

A flattened cut should be seeded immediately following construction. In addition, a protective cover such as a straw mulch or a geosynthetic material should be placed over the slope. Any area stripped of vegetation, such as a ditch or berm, should be covered with the appropriate material as soon as possible to avoid excessive erosion.

Slopes should be cut uniformly (there should be no compound slopes) to avoid concentration of erosion and undercutting. Also, if animal holes intersected by the cut daylight above the crest where water may easily enter (such as in the drainage

structure), they should be backfilled with low-permeability fines to avoid development of erosion pipes.

11. MAINTENANCE AND REPAIR

Because of their highly erosive nature, loess slopes deteriorate very rapidly once erosion has been initiated. Thus, it is very important to repair any erosion damage as soon as it is discovered. Maintenance may require repairs to, enlargement of, and removal of deposited silt from existing ditches. Increased erosion protection, such as installation of liners in ditches or in some cases construction of drainage facilities where they were previously believed to be unnecessary, may be required.

Vegetative cover requires periodic attention. In order to maintain a heavy ground cover, fertilizer must be applied every 3 to 5 years (on the basis of experience in the midwestern United States). In addition, some areas will not seed well the first time and may require a second and possibly a third seeding.

Sediment should be removed carefully from toe ditches on existing slopes to avoid undercutting the toe of the slope. Even minor undercutting will cause at least some sloughing, and therefore the grader blade should not contact the slope cut.

Repair of loess piping failures as well as erosion damage on all loess slopes should disturb the soil structure as little as possible and should reduce areas of concentrated surface water flow. Attempts to regrade loess slopes have had only marginal success. It is impossible to regrade damaged near-vertical slopes. These slopes require redirection of surface water from the damaged slope and removal of failed material. The vertical face may be reestablished in the undisturbed portions of the loess.

Large vertical sink areas (developed from piping) can be repaired by placing a geosynthetic liner (with an appropriately designed infiltration rate), backfilling with graded aggregate, placing an impervious cover, and redirecting drainage. Repair measures have failed only when surface drainage was not adequately diverted or the drainage system was not maintained, allowing infiltration.

REFERENCES

Adams, J. 1988. Landslide Hazards in the Loess Country of Gansu, China. *Ground Failure*, No. 5, Committee on Ground Failure Hazards, National Research Council, Washington, D.C., pp. 5–7.

Akiyama, F.M. 1964. *Shear Strength Properties of Western Iowa Loess*. M.S. thesis. Iowa State University, Ames.

Bandyopodyopadhyay, S.S. 1983. Geotechnical Evaluation of Loessial Soils in Kansas. In *Transportation Research Record 945*, TRB, National Research Council, Washington, D.C., pp. 29–36.

Beard, L.D., J.D. Higgins, R.J. Fragaszy, A.P. Killian, and A.J. Peters. 1986. Physical Properties of Southeastern Washington Loess Related to Cut Slope Design. In *Transportation Research Record 1089*, TRB, National Research Council, Washington, D.C., pp. 29–38.

Casias, T.J. 1987. *Results of Research in Sampling Loessial Soil for In-Place Unit Weight Determinations*. Report REC-ERC-87-5. Geotechnical Branch, Engineering and Research Center, Bureau of Reclamation, U.S. Department of the Interior, Denver, Colo., 37 pp.

Clevenger, W.A. 1956. Experiences with Loess as a Foundation Material. *Journal of the Soil Mechanics and Foundations Division*, ASCE, Vol. 82, No. 1025, pp. 1–26.

Close, U., and E. McCormick. 1922. Where the Mountains Walked. *National Geographic Magazine*, Vol. 41, No. 5, pp. 445–464.

Crumpton, C.F., and W.A. Badgley. 1965. *A Study of the Clay Mineralogy of Loess in Kansas in Relation to Its Engineering Properties*. Research Department, State Highway Commission of Kansas, Topeka, 68 pp.

Dahl, A.R., R.L. Handy, and D.T. Davidson. 1952. Variation of Loess Thickness and Clay Content in Southern Iowa. *Proceedings of the Iowa Academy of Science*, Vol. 64, pp. 393–399.

Davidson, D.T., and R.L. Handy. 1952. Property Variations in the Peorian (Wisconsin) Loess of Southwestern Iowa. *Proceedings of the Iowa Academy of Science*, Vol. 59, pp. 248–265.

Davidson, D.T., and C.J. Roy. 1959. *The Geology and Engineering Characteristics of Some Alaskan Soils*. Engineering Experiment Station Bulletin 186. Iowa State University, Ames, 67 pp.

Eden, D.N., and R.J. Fulkert (eds.). 1988. *Loess: Its Distribution, Geology, and Soils*. A.A. Balkema, Rotterdam, Netherlands.

Gibbs, H.J., and W.Y. Holland. 1960. *Petrographic and Engineering Properties of Loess*. Engineering Monograph 28. Bureau of Reclamation, U.S. Department of the Interior, Denver, Colo., 42 pp.

Handy, R.L. 1976. Loess Distribution by Variable Winds. *Bulletin of the Geological Society of America*, Vol. 87, pp. 915–927.

Handy, R.L., C.A. Lyon, and D.T. Davidson. 1955. Comparisons of Petrographic and Engineering Properties of Loess in Southwest, East-Central, and Northeast Iowa. *Proceedings of the Iowa Academy of Science*, Vol. 62, pp. 279–297.

Hansen, J.A, A.R. Dahl, and D.T. Davidson. 1959. Further Studies of Loess in Iowa: Thickness, Clay Content, and Engineering Classification. *Proceedings of the Iowa Academy of Science*, Vol. 65, pp. 317–322.

Higgins, J.D., and R.J. Fragaszy. 1988. *Design Guide for Cut Slopes in Loess of Southeastern Washington*. Report WA-RD 145.2. Washington State Department of Transportation, Olympia, 65 pp.

Higgins, J.D., R.J. Fragaszy, and L.D. Beard. 1985. *Development of Guidelines for Cuts in Loess Soils (Phase 1)*. Report WA-RD69.1. Washington State Department of Transportation, Olympia, 90 pp.

Higgins, J.D., R.J. Fragaszy, and L.D. Beard. 1989. Engineering Geology of Loess in Southeastern Washington. In *Engineering Geology in Washington*, Bulletin 78, Washington Division of Geology and Earth Resources, Olympia, Vol. 2, pp. 887–898.

Higgins, J.D., R.J. Fragaszy, and T. Martin. 1987. *Engineering Design in Loess Soils of Southeastern Washington*. Report WA-RD 145.1. Washington State Department of Transportation, Olympia, 121 pp.

Hobbs, W.H. 1943. The Glacial Anticyclones and the European Continental Glacier. *American Journal of Science*, Vol. 241, No. 5, pp. 333–336.

Holtz, W.G., and H.J. Gibbs. 1951. *Consolidation and Related Properties of Loessial Soils*. Special Technical Publication 126. ASTM, Philadelphia, Pa., pp. 9–26.

Ishihara, K., S. Okusa, N. Oyagi, and A. Ischuk. 1990. Liquefaction-Induced Flow Slide in the Collapsible Loess Deposit in Soviet Tajik. *Soils and Foundations* (Japanese Society of Soil Mechanics and Foundation Engineering), Vol. 30, No. 4, pp. 73–89.

Kane, H. 1968. *A Mechanistic Explanation of the Physical Properties of Undisturbed Loess*. University of Iowa Research Project HR-126. Iowa State Highway Commission, Iowa City, 113 pp.

Kolb, R. 1965. Physical Properties and Engineering Characteristics of Mississippi Loess. In *Field Trip Guidebook: Mississippi Alluvial Valley and Terraces*, Geological Society of America, Boulder, Colo.

Krinitzsky, E.L., and W.J. Turnbull. 1965. Loess Deposits of Mississippi. In *Field Trip Guidebook: Mississippi Alluvial Valley and Terraces*, Geological Society of America, Boulder, Colo.

Leighton, M.M., and H.B. Willman. 1950. Loess Formations of the Mississippi Valley. *Journal of Geology*, Vol. 58, No. 6, pp. 599–623.

Leonov, N.N. 1960. The Khait Earthquake of 1949 and the Geological Conditions of Its Origination. *Bulletin of the Academy of Sciences, USSR, Geophysics Series*, pp. 247–283 [translated and published by American Geophysical Union, Washington, D.C.].

Lyon, C.A., R.L. Handy, and D.T. Davidson. 1954. Property Variations in the Wisconsin Loess of East-Central Iowa. *Proceedings of the Iowa Academy of Science*, Vol. 61, pp. 291–312.

Matalucci, R.V., M. Abdel-Hady, and J.W. Shelton. 1970. Influence of Microstructure of Loess on Triaxial Shear Strength. *Engineering Geology*, Vol. 4, No. 4, pp. 341–351.

Milovic, D.M. 1971. Effect of Sampling on Some Loess Characteristics. In *Proc., Fourth Asian Regional Conference on Soil Mechanics and Foundation Engineering* (Specialty Session: Quality Soil Sampling), Bangkok, pp. 17–20.

Missouri State Highway Department. 1978. *Development of Design Criteria for Cut Slopes in Loess*. Jefferson City, 78 pp.

Modeer, V.A., M.C. Lamie, and L.D. Flowers. 1991. The Electric Cone Penetrometer: Experience in Varied Geotechnical Conditions in Southern Illinois. In *Proc., 34th Annual Meeting*, Association of Engineering Geologists, Chicago, Ill., pp. 100–109.

Morrison, L.L. 1968. Procedures and Problems of Highway Soils Engineering on Loessial Terrain in Alaska. In *Highway Research Record 212*, HRB, National Research Council, Washington, D.C., pp. 33–38.

Olson, G.R. 1958. *Direct Shear and Consolidation Tests of Undisturbed Loess*. M.S. thesis. Iowa State University, Ames.

Royster, D.L. 1963. Engineering Characteristics of Loessial Soils in Western Tennessee. In *Proceedings of the 45th Annual Tennessee Highway Conference*, Bulletin 29, Engineering Experiment Station, University of Tennessee, Knoxville, pp. 12–38.

Royster, D.L., and W.H. Rowan. 1968. Highway Design and Construction Problems Associated with the Loessial Soils of West Tennessee. In *Highway Research Record 212*, HRB, National Research Council, Washington, D.C., pp. 28–32.

Schultz, C.B., and J.C. Frye. 1965. *Loess and Related Eolian Deposits of the World*. University of Nebraska Press, Lincoln.

Sheller, J.B. 1968. Summarization and Comparison of Engineering Properties of Loess in the United States. In *Highway Research Record 212*, HRB, National Research Council, Washington, D.C., pp. 1–9.

Smith, G.D. 1942. *Illinois Loess—Variations in Its Properties and Distribution*. Agricultural Experiment Station Bulletin 490. University of Illinois, Champaign-Urbana.

Tungsheng, L. 1988. *Loess in China*, 2nd ed. China Ocean Press, Beijing; Springer-Verlag, pp. 189–190.

Turnbull, W.J. 1965. Construction Problems Experienced with Loess Soils. In *Highway Research Record 212*, HRB, National Research Council, Washington, D.C., pp. 10–27.

◣ Chapter 24

GUY LEFEBVRE

SOFT SENSITIVE CLAYS

1. INTRODUCTION

Sensitivity is defined as the ratio of the undisturbed to the remolded shear strength of a soil and thus expresses the loss of strength when a soil is remolded. A variety of clay deposits may exhibit sensitivity, but some marine clay deposits exhibit very high sensitivity and very low remolded shear strength. Such deposits are termed *quick clays*, which is a direct translation of the term *kvikkleira* used by Norwegian investigators to describe these deposits, common in their country. The Norwegians define quick clays as those soils exhibiting a sensitivity greater than 30 and having a remolded shear strength less than 0.5 kPa (Norsk Geoteknisk Forening 1974). This definition requires the remolded soil to behave like a fluid. It should be noted that this definition of quick clay considers soil behavior only, ignoring geographic location, depositional environment, or any other factors related to origin or material.

In nonsensitive soils, landslide masses generally come to rest at the toe of the slope and act like a stabilizing mass in the new slope geometry. For sensitive clays, however, the remolding involved in the mass movement results in a drastic reduction in shear resistance, causing the remolded clay to behave like a thick liquid so that the slide mass moves away and leaves the new slope unsupported. A new instability may result and a series of retrogressive failures will be triggered that can extend far beyond the crest of the initial slope. In highly sensitive clays, therefore, it is not just the risk of a slope failure that is of concern, but also the area that can be affected by retrogressive sliding.

Their high sensitivity combined with the fluidity of the remolded materials makes quick clays very susceptible to retrogressive landsliding. Rapid and dramatic geotechnical failures are the result (Figure 24-1). This type of retrogressive failure has been referred to as a flow slide or earth flow. The latter term is preferred, since it is in line with current international landslide terminology (see Cruden and Varnes, Chap. 3 in this report). Several cases have been documented in which such retrogressive failures affected areas larger than 20 ha and extended more than 500 m beyond the crest of the initial slopes (Eden and Jarrett 1971; Eden et al. 1971; Tavenas et al. 1971; Brooks et al. 1994; Evans and Brooks 1994).

At present, the risk and extent of retrogression are difficult to evaluate. However, on the basis of studies conducted in eastern Canada, retrogression may be considered to be a consequence of an initial slope failure. In terms of analysis, the first step is to quantitatively examine the initial slope stability. The second step is to evaluate, in a semiquantitative manner, the risk of retrogression that could develop following initial slope failure. In addition to numerical analysis, a certain comprehension of the formation of slopes in sensitive clay deposits is helpful in order to put the processes into a rational framework.

FIGURE 24-1
Aerial view of Lemieux landslide, Ontario, Canada, late morning, Wednesday, June 23, 1993, involving failure of 2.5 to 3.5 million m³, much of which flowed into South Nation valley causing impoundment of river *(foreground)*. Failure took place within 1 hour on June 20, temporarily damming river with debris. Approximate maximum height of river impoundment is shown during overtopping of dam. Ground surface within failure is about 12 m below original ground surface and slopes gently toward river. Debris flowed about 1.6 km upstream and 1.7 km downstream of landslide (Brooks et al. 1994; Evans and Brooks 1994).

PHOTOGRAPH 1993-254F
COURTESY OF GEOLOGICAL
SURVEY OF CANADA

2. DESCRIPTION OF SENSITIVE CLAY DEPOSITS

The existence of sensitive clay deposits is related to

1. Mineral sources and depositional conditions responsible for an open soil fabric and a relatively high water content and
2. Geochemical factors responsible for a reduction of the liquid limit and the remolded shear strength.

These factors naturally restrict the geographic distribution and geologic characteristics (including composition, stratigraphy, groundwater geochemistry, and groundwater volumes) of soft sensitive clay deposits.

2.1 Geographic Distribution

Although soft clay deposits are found in many parts of the world, the most extensive areas of highly sensitive clays occur in Scandinavia and eastern Canada. They have also been reported in Alaska, Japan, the former Soviet Union, and New Zealand (Torrance 1987). Small areas of similar deposits may still be discovered, but it appears unlikely that any large areas of such deposits remain undiscovered.

2.2 Geologic Origin

Sensitive clay deposits are generally young marine clays deposited in preglacial and postglacial bodies of water that existed during the retreat of the last Wisconsin ice sheet between 18,000 and 6,000 years before the present (BP). Many of the topographic depressions left by the glaciers formed freshwater lakes, but some were connected to the oceans at some stage.

The weight of the continental ice sheets caused depression of the land surface up to several hundred meters. Isostatic rebounding of the land surfaces occurred after the ice retreated, but this is a relatively slow process that is still continuing in some regions. The enormous amounts of water incorporated into the continental ice sheets caused the ocean volumes to be somewhat less than at present, and the ocean levels dropped worldwide by at least 120 m (Kenney 1964). However, the rate of ice-sheet melting and the consequent rise in sea level were much more rapid than the rebounding of the land surface following ice-sheet removal.

Isostatic rebound and eustatic sea-level changes after the last deglaciation have been well documented (Kenney 1964; Andrews 1972; Hillaire-Marcel and Fairbridge 1978; Quigley 1980). The isostatic rebound was much larger than the increase in sea level. The net result of the interaction between changing land elevations and sea levels was a period of inundation and subsequent reemergence of certain portions of the continental land masses in the time between the melting of the continental ice sheets and the present day.

Sensitive clays are considered to have been deposited in marine, or at least brackish, bodies of water. In North America the last glaciation covered most of Canada, parts of the conterminous northern United States, and part of Alaska. The largest deposits of postglacial marine clays were formed in the Champlain Sea, which occupied the St. Lawrence lowlands from the Gulf of St. Lawrence to the region around the city of Ottawa during the period approximately 12,500 to 10,000 years BP (Elson 1969; Gadd 1975; Hillaire-Marcel 1979).

2.3 Groundwater Geochemistry

The emergence of these postglacial marine clay formations above sea level altered their groundwater regimes, and because of hydraulic or concentration gradients, their pore-water salt concentration, which was initially as high as 32 g/L, was reduced. Present-day pore-water salinity in many sensitive clay deposits has been reduced to values below 1 g/L. Reduction of pore-water cations (salinity) is known to reduce the liquid limits (Rosenqvist 1966) and the remolded shear strength of clays.

Because of an open fabric and high water content retained since deposition, some lacustrine clays and unleached marine clays can also be fairly sensitive. The soft Barlow-Ojibway lacustrine clay in northern Québec Province, for example, has a water content of as much as 100 percent, a plasticity index of 40, a liquidity index of 1.5, and a sensitivity of about 20. However, geochemical factors, such as salinity reduction, are considered necessary for the development of extrasensitive or quick clays, which have a liquidity index greater than 2 and flow when remolded. However, it should be noted that geochemical factors other than salinity reduction, such as the introduction of dispersants, can also lead to an increased liquidity index and sensitivity.

2.4 Composition and Mineralogy

Because of their geologic origin, sensitive clays frequently have a characteristic composition and mineralogy. The silt fraction is most often high, typically in the range of 30 to 70 percent. The fraction with grain sizes smaller than 2 μm generally contains more rock flour than clay minerals, and the clay minerals are predominantly illite (Quigley 1980; Locat et al. 1984).

2.5 Stratigraphy

In large depositional basins, marine clay deposits are generally massive and notably uniform. In small basins, or in the proximal areas of large basins, marine clay deposits can be interlayered with sand and gravel and even with glacial till (Gadd 1975). For most slope stability analyses, the stratigraphy of typical postglacial soft clay deposits can generally be simplified (Figure 24-2). The massive clay formation is confined between two relatively pervious boundary layers. The upper boundary consists of either a weathered fissured crust or sand and silt derived from the final depositional stages. The lower boundary is formed either by till deposited on the bedrock by glaciers or by fissured bedrock. The overall coefficient of permeability in the boundary layers is generally at least two orders of magnitude higher than that in the massive clay deposit.

FIGURE 24-2 Simplified down-valley cross section of typical soft marine clay deposits of eastern Canada (modified from Lefebvre 1986).

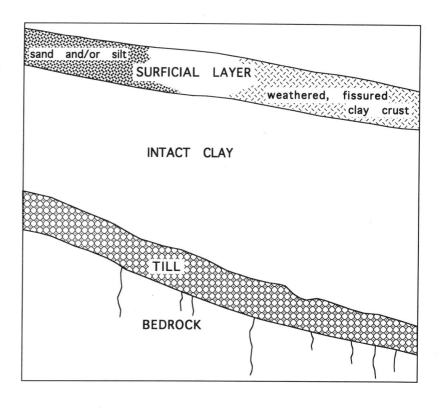

3. VALLEY FORMATION AND GROUNDWATER REGIME

The process of clay deposition in shallow marine or brackish environments usually produced an initial deposit with little or no topographic relief and gentle slopes. Subsequent emergence of these deposits subjected them to subaerial erosion. Today the slopes of geotechnical concern in these sensitive clay deposits are the result of erosion by rivers or streams that cut valleys through these deposits. In the Champlain Sea region of eastern Canada, the preconsolidation pressure is, in general, in agreement with the maximum filling level of the valley (Eden and Crawford 1957; Crawford and Eden 1965). In other areas the apparent preconsolidation reflects either cementation (Locat and Lefebvre 1986; Lefebvre et al. 1988) or secondary consolidation (Bjerrum 1967). Stages of valley formation are illustrated in a simplified manner in Figure 24-3, where the valley bottom is seen to gradually approach the lower pervious boundary layer.

Valleys in soft sensitive clays can be found today in different stages of formation, and the groundwater regime evolves according to the valley development. This evolution of the groundwater regime plays an important role in the general state of slope stability and in the morphology of the landsliding that does occur. The groundwater regime in the vicinity of a slope not only determines the pore-pressure distribution in the clay deposits, but also is a prime factor in the general slope stability along the valley sides. In sensitive clay areas the critical location for the water table is often close to ground surface. The groundwater regime can vary widely from one site to another, ranging from a general underdrainage to high artesian conditions. It is difficult to interpret and to extrapolate piezometric readings without knowing the pattern of the groundwater regime. Simple considerations of typical present-day topographic profiles and of the valley erosion process, as shown in Figure 24-3, are helpful in predicting and understanding the groundwater regime.

Once the boundary conditions have been assumed, it is indeed a simple matter to produce a flow net because for practical purposes the permeability of marine clay deposits can be considered isotropic (Tavenas et al. 1983). Figure 24-4 shows groundwater flow patterns for the geometric conditions corresponding to the three stages of valley development shown in Figure 24-3. These flow nets have been determined numerically assuming a permeability in the lower boundary layer of 10 times the permeability in the clay.

In the early stage of valley formation, the groundwater regime is not significantly influenced by the lower pervious boundary [Figure 24-4(a)] because the bottom of the valley is well above it. However, as the bottom of the valley is eroded closer to the lower boundary [Figure 24-4(b)], high artesian pressures develop. At this stage, because pore-pressure discharge or dissipation is restricted by the relatively impervious clay, the pore pressure in the boundary layer under the bottom of the valley and the lower part of the slope tends to equilibrate with the head existing behind the slope

FIGURE 24-3
Stages of valley formation in marine clay deposits (modified from Lefebvre 1986):
(a) early phase,
(b) intermediate phase,
(c) advanced phase.

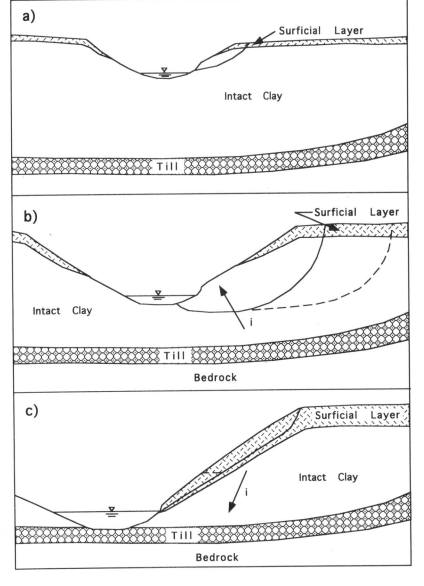

crest. Such a condition is obviously detrimental to stability and favors a deep failure surface. In fact, the artesian pressures evaluated from the flow net of Figure 24-4(b) are sufficient to blow up the bottom of the valley.

Groundwater within the lower boundary layer discharges into the river once the valley formation reaches a sufficiently advanced stage [Figure 24-4(c)]. In the vicinity of the slope, the piezometric head is slightly higher than the river level and creates a downward gradient in the clay formation. The surface infiltration into the upper pervious boundary materials is much larger than the flow that can percolate through the intact clay formation. Thus, the water table is essentially unaffected by the downward drainage and generally remains close to the surface inside the upper boundary layer, fluctuating with the season. In the upper boundary layer, the flow can be assumed to be parallel to the ground surface and under almost hydrostatic conditions.

Groundwater regimes along valleys in their final stages of formation have been documented at sev-

eral sites, especially in the Tyrrell Sea basin of northern Québec Province where valleys appear to have evolved more rapidly because of a clay that is more erodible than that in the St. Lawrence valley (Lefebvre 1986). The groundwater regime deduced from a large number of piezometers at one site in the Broadback River valley is presented in Figure 24-5(a) along with the stratigraphy obtained from several borings. Figure 24-5(b) presents the results of a numerical analysis by the finite-element method. The permeability coefficients used in these analyses were obtained by permeability tests in open-tube piezometers.

Both the numerical analysis and the piezometric data confirm that strong underdrainage developed in the clay formation when the underlying till formation discharged into the river. The groundwater table remained close to the ground surface in the wet season so that essentially hydrostatic conditions prevailed in the surficial crust. Under the crest of the slope, the pore pressure increased with depth in the upper part of the intact clay formation and then decreased toward the

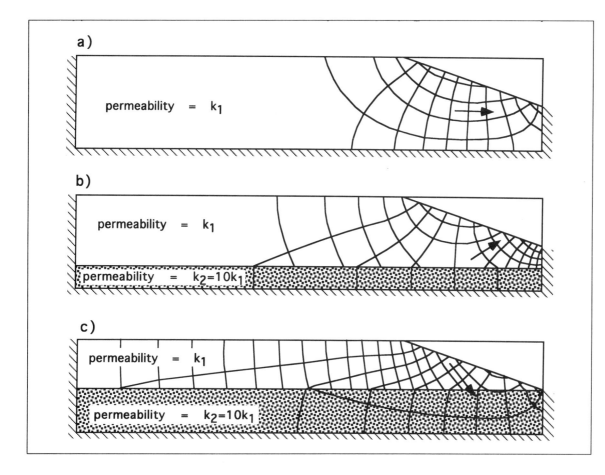

FIGURE 24-4
Flow nets showing groundwater regimes in vicinity of clay slopes corresponding to stages of valley formation shown in Figure 24-3 (modified from Lafleur and Lefebvre 1980):
(a) early phase,
(b) intermediate phase,
(c) advanced phase.

FIGURE 24-5
Equipotential lines
obtained from (*a*)
piezometer readings
November 1980 and
(*b*) finite-element
analyses at
Broadback River site,
northern Québec
Province, Canada
(modified from
Lefebvre 1986).

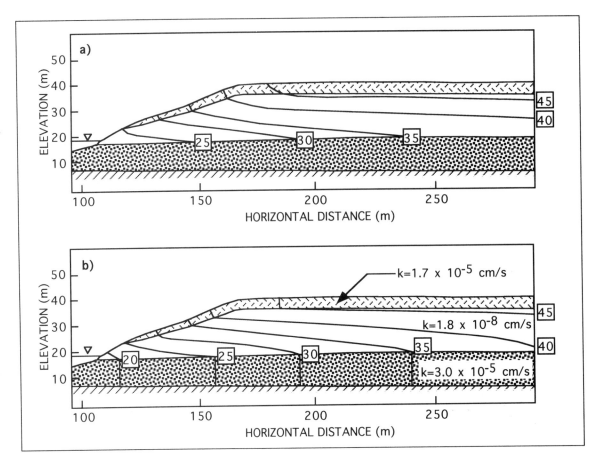

lower boundary value. Such general underdrainage is obviously beneficial for stability.

In contrast, the artesian pore pressure that is characteristic of the intermediate stage of valley formation [Figure 24-4(*b*)] creates conditions that may lead to deep landslides involving large masses of sensitive clays and to subsequent retrogressive failures.

When the lower-boundary material discharges into the river, at the later stages of valley formation [Figure 24-4(*c*)], the valley walls have reached a phase of equilibrium and stability. Because of important underdrainage at this stage, landsliding in response to bank erosion is generally shallow and does not penetrate deeply into the sensitive intact clay. Risk of retrogression is then fairly low. At this stage, slopes as high as 30 m and as steep as 30 degrees have been found to stand in normally consolidated deposits (Lefebvre et al. 1991). Cases of retrogressive failure have been identified in regions reflecting this final stage of valley formation, but only under conditions of exceptionally severe toe erosion and undercutting (Lefebvre et al. 1991).

4. STABILITY ANALYSIS

As noted in Chapter 13, slope stability methods evaluate the state of stress on a potential failure surface and compare the shear stresses with the available shear strength in the form of a safety factor. The stress evaluation for effective stress analysis is more complex because the effective normal stress on the potential failure surface must be calculated in order to evaluate the shear strength. In such analyses, realistic pore-pressure distribution and shear-strength parameters must be used to obtain a correct evaluation of the available shear strength.

4.1 Shear Strength

The shear-strength parameters must correctly model the shear strength that can be mobilized on the failure surface of a slope that has been formed by the erosion processes described in the previous sections. For that purpose, one must distinguish between those conditions causing an initial failure of a slope that has existed for some time and those conditions causing the almost instantaneous retrogressive failures taking place after a first slide.

As shown in Figure 24-6, starting from the maximum level of sediment filling, the slope that exists today is the result of a slow process of erosion that has decreased the confining stress and increased the shear stress. The stress state on a soil element lying on a potential failure surface can thus be considered the end point of a drained stress path (A to B, Figure 24-6). When stress state B approaches the strength envelope closely enough, a slight reduction in effective stress or increase in shear stress caused by any number of reasons is sufficient to make the slope susceptible to incipient failure.

Actual failure may never occur, or it may occur following a slight additional change to the stress state resulting from such causes as a higher water table during a wet season or erosion at the toe of the slope. This failure will obviously be undrained. Because of the overconsolidated state of the soil, if the failure is triggered by an undrained increase of shear stress due to rapid toe erosion, the undrained stress path (BO) will not deviate much from the drained stress path (BN). Thus, an effective stress analysis that implies a drained stress path remains valid until the incipience of failure and even then does not introduce a significant error into the analysis of an initial landslide.

However, the situation is completely different for retrogressive failures. The shear stresses are applied almost instantaneously and create an undrained loading. Stability consideration for the role of retrogression must therefore be based on the undrained shear strength.

4.2 Effective Stress Parameters

The effective stress parameters for analysis of an initial failure must describe the shear strength for the range of stresses existing on a potential failure surface; these parameters are generally determined by drained triaxial compression tests. The reconsolidation, or cell, pressure must be fairly low in order to have a state of stress at failure in the range of those stresses existing in the field on the potential failure surface. For this purpose, triaxial specimens are commonly reconsolidated at cell pressures ranging from 5 to 30 kPa in order to have, at failure, a state of stress that is in the overconsolidated range.

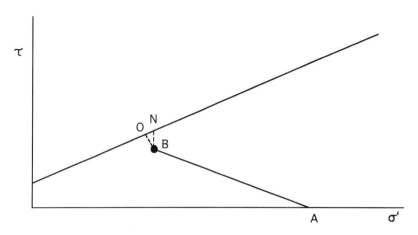

FIGURE 24-6 Approximate stress path of clay element lying on potential failure surface in slope.

When tested under low confining stress in the overconsolidated range, sensitive clays exhibit brittle behavior (Figure 24-7). A sharp peak strength is mobilized at relatively small axial deformation on a triaxial stress-strain curve; the strength rapidly decreases after the peak and stabilizes on a plateau. The shear strength remains approximately constant beyond an axial deformation of 4 to 5 percent. The strength of these eastern Canadian sensitive clays at large deformations is routinely determined at an axial deformation of 8 percent. Study of many case records for locations in eastern Canada has shown that it is the large-deformation, or postpeak, strength that is mobilized during the initial failure of a natural slope or long-term excavation slope (Lefebvre and

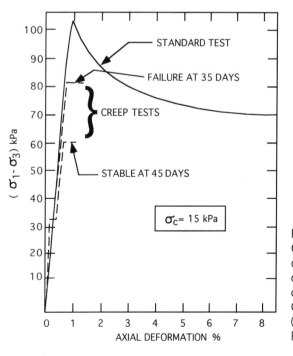

FIGURE 24-7 Creep and standard drained triaxial tests on sensitive Nicolet clay, eastern Canada (modified from Philibert 1976).

LaRochelle 1974; Lefebvre 1981). Skempton (1964) reached a similar conclusion, namely, that for stiff overconsolidated clays, it is the strength at large deformations that is mobilized during a slope failure.

The use of the postpeak strength in performing stability analyses of sensitive clays can be justified by progressive failure along the failure surface or by the fact that the peak strength in sensitive clay cannot be mobilized under sustained load. Triaxial drained creep tests performed on sensitive St. Lawrence clays (Figure 24-7) have shown that creep failure develops for any loading above the strength level corresponding to large deformations (Lefebvre 1981). In any event, the strength at large deformations may be considered to form a stability threshold in terms of either deformation or time. The postpeak strength is generally determined by means of an isotropically consolidated triaxial drained test with a rate of loading of 1 to 2 percent per day to allow drainage; the test is continued to an axial deformation of 10 to 12 percent.

Sampling disturbance is generally a significant factor with sensitive clay in the overconsolidated state. The peak strength is, for example, much affected by sampling disturbance. However, good commercial tube sampling is sufficient to determine the strength at large deformations. When tube sampling methods are used, the scatter in the results may be larger than when block samples are used. In principle, the large-deformation strength envelope can be determined for overconsolidated samples by using undrained tests with pore-pressure measurement. However, difficulties arise in obtaining this determination accurately by such methods because the stress paths all tend to reach the failure envelope at the same stress state.

4.3 Normalized Shear Strength

A large data bank has accumulated over the years on the postpeak shear strengths of the sensitive Champlain Sea clays of eastern Canada. For specimens reconsolidated in the laboratory to the same confining pressure, the strength at large deformations appears to be influenced by the stress history; that is, it increases with preconsolidation pressure. Figure 24-8 presents test results on block sample specimens of Champlain Sea clays from nearly 20 sites (Lefebvre 1981). The data are normalized by the preconsolidation pressure, σ'_p. The data from all these sites, located over a region about 700 km long, plot in a surprisingly narrow band. They represent a range of natural water content from 40 to 90 percent, plasticity index from 5 to 40, and preconsolidation pressure from 100 to 600 kPa.

The normalized postpeak envelope (Figure 24-8) exhibits some curvature, with the shear resistance initially increasing rapidly with effective stress and then flattening toward the normally consolidated envelope. Because this normalized envelope is common to all Champlain Sea clay sites, it can be used as a first approximation to derive the strength parameters from σ'_p. As indicated by the curvature of the envelope, for the same confinement stress range, ϕ' increases with σ'_p. If the parameters are determined for an identical confinement pressure (σ_3') range of 5 to 20 kPa, the normalized envelope of Figure 24-8 provides an estimate of ϕ' that increases from 29 to 43 degrees, whereas σ'_p increases from 100 to 400 kPa and c' remains approximately constant at 7 kPa. These values are summarized in Table 24-1. Test results from tube samples generally correlate with the envelope defined in Figure 24-8 but exhibit a larger scatter.

The postpeak strength defined for the soft sensitive Champlain clays is surprisingly high and appears somewhat paradoxical when compared with the residual strength determined for stiff overconsolidated clays, which is defined by a zero cohesion intercept and a value for ϕ' that sometimes is as low as 10 to 15 degrees. The high postpeak c' and ϕ' values for the Champlain clay are primarily related to its mineralogy. The high silt and rock flour content and the relatively low fraction of illitic clay minerals result in a fairly large friction angle. In addition, as shown by the correlations established with σ'_p, the shear strength at large deformations is still influenced by the stress history and may be different from a true residual strength.

The correlations established for Champlain clay should be checked before they are adopted to describe the characteristics of sensitive clays from other regions of the world. From limited data obtained from analyses of block samples of sensitive clays from four sites in the Tyrrell Sea basin in northern Québec Province, the postpeak parameters for the Tyrrell Sea clays appear to be slightly lower than those for Champlain clays. Figure 24-9 shows a comparison of the normalized postpeak strength envelopes for the Tyrrell Sea and Champlain Sea clays.

Marine clay deposits are generally massive and fairly homogeneous; thus the use of a circular failure surface is generally appropriate. Bishop's modified method is most often used for stability analyses of soft clay slopes at sites in eastern Canada. The relatively high ϕ' that is used for the stability analysis of an initial failure in sensitive clay slopes points to the importance of a good understanding of the pore-pressure distribution.

5. ESTABLISHING RISK OF RETROGRESSION

Although the analysis of an initial landslide failure on a slope composed of sensitive clay appears fairly straightforward, there is much more uncertainty in the evaluation of the risk of retrogression following this initial failure. The potential retrogression distance is indeed an important parameter in assessing the risk of damage.

The risk of retrogression is broadly related to two conditions. First, the clay has to be sensitive enough that the debris of the initial landslide will move away, or flow out, instead of creating a stabilizing mass at the toe of the newly created slope. Second, the newly created slope must have an unstable geometry, or a geometry such that the induced shear stresses exceed the undrained shear strength of the clay.

5.1·Role of Sensitivity and Remolded Shear Strength

Because factors other than clay sensitivity, such as slope geometry and stratigraphy, play a role, it is difficult to establish a sensitivity threshold below which the debris from the initial landslide will not move away and prevent retrogression. However, large retrogressive failures, for example, the typical "bottleneck" earth flow, develop generally in very sensitive or quick clays. In fact, the remolded shear strength controls the susceptibility of the earth mass to flow. Figure 24-10 presents values of sensitivity as a function of the remolded shear strength for a dozen sites of slope failure in Champlain clays (Grondin 1978). Those sites where retrogression has or has not developed are also identified. The remolded shear strengths and sensitivities shown in Figure 24-10 were determined in the laboratory. At sites where retrogressive failures developed, the remolded shear strength was lower than 1 kPa and the sensitivity was generally higher than 40. Figure

Table 24-1
Postpeak Strength Parameters for Sensitive Clays from Champlain Sea Deposits of Eastern Canada (modified from Lefebvre 1981)

RANGE OF CONSOLIDATION STRESS (kPa)	σ_p' (kPa)	c_m' (kPa)	ϕ_m' (degrees)
5–20	100	7.4	28.7
5–20	200	7.7	34.7
5–20	300	7.7	39.8
5–20	400	7.5	43.6

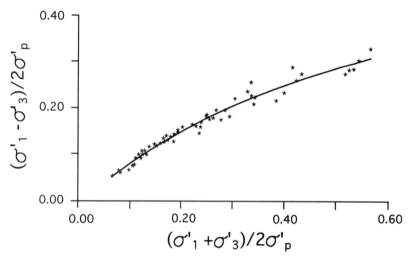

FIGURE 24-8
Normalized postpeak strength envelope from block samples of Champlain Sea sensitive clay (Lefebvre 1981).

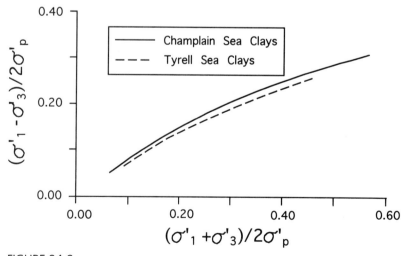

FIGURE 24-9
Comparison of normalized postpeak strength envelopes for Champlain Sea and Tyrell Sea sensitive clays.

FIGURE 24-10
Relation among type
of landslide failure,
sensitivity, and
remolded shear
strength of sensitive
clays (modified from
Grondin 1978).

FIGURE 24-10
Relation among type
of landslide failure,
sensitivity, and
remolded shear
strength of sensitive
clays (modified from
Grondin 1978).

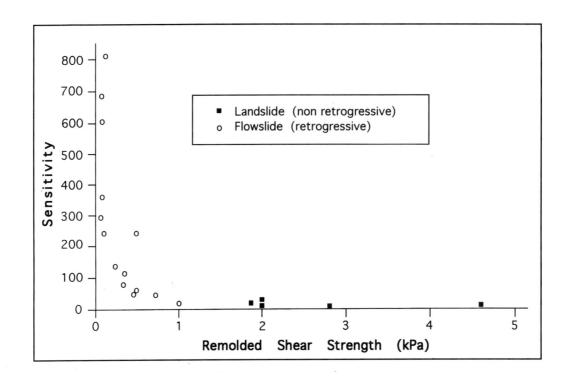

FIGURE 24-11
(below)
Risk of retrogression
in terms of
sensitivity and
liquidity index; data
points are those
shown in
Figure 24-10
(modified from
Grondin 1978).

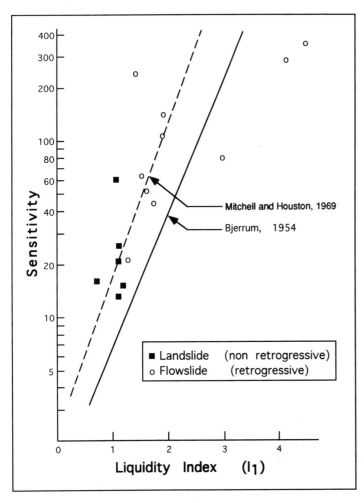

24-10 clearly shows that the quick clays with re-molded shear strengths lower than 0.2 kPa are more prone to flow than those with remolded shear strengths between 0.5 and 1.0 kPa.

At its liquid limit, the remolded shear strength of a typical clay is about 1.7 kPa (Wroth and Wood 1978). The data shown in Figure 24-10 indicate that the liquidity index has to be above unity for a site to have retrogression potential. Relationships between sensitivity and liquidity index have indeed been proposed by many authors (Skempton and Northey 1953; Bjerrum 1954; Mitchell and Houston 1969). Some of these relations are reproduced in Figure 24-11 together with the data for the sites shown in Figure 24-10. The criteria of Mitchell and Houston (1969) suggest that a sensitivity of 40 corresponds to a liquidity index of about 1.5 to 2.0. Clays at sites where retrogression developed (from Figure 24-10) generally have a sensitivity higher than 40 and a liquidity index exceeding 1.5. Therefore, the liquidity index, which is routinely available, is a good indicator of the susceptibility of these clays to flow.

5.2 Stability Number

The susceptibility of the clay to flow is not sufficient to induce retrogressive earth-flow failures. The slope geometry after the initial failure must be such that the shear stress in the new slope exceeds the undrained shear strength of the clay. Therefore, the potential distance of retrogression is difficult to evaluate. Nevertheless, the risk of retrogression after the occurrence of an initial failure can be assessed at least qualitatively. From analytical studies and examination of more than 40 case records, Mitchell and Markell (1974) proposed that retrogression can occur only if the stability number, $N_s = \gamma H/Cu$, is larger than 6 within the depth of potential failure.

This stability number is similar to the coefficient used in the evaluation of bearing capacity, as if the stress γH imposed by the new slope were creating an undrained extrusion of the clay below. The larger the value of $\gamma H/Cu$, the larger will be the retrogression distance before equilibrium. Mitchell and Markell (1974) found a general relationship between $\gamma H/Cu$ and the distance of retrogression, although other factors such as sensitivity, stratigraphy, and topography also appear to play a significant role in limiting retrogression.

The groundwater regime influences the effective-stress profile and consequently the undrained shear strength profile, especially in normally consolidated deposits. In the final stage of valley formation, deposits may be affected by strong underdrainage, as described earlier, which causes the undrained shear strength to increase rapidly with depth in the lower half of the clay deposit. Under these circumstances, the most critical position for a retrogression surface will be much higher on the slope than would otherwise be the case because of the unique relationships between the $\gamma H/Cu$ values and elevation (or depth).

The liquidity index and the ratio $\gamma H/Cu$ are probably the most useful parameters for assessing the risk of occurrence of retrogressive failures. However, because retrogressive landslides can only be triggered after the initial failure has occurred, securing the stability of a slope against an initial failure is the logical approach in preventing large retrogressive earth flows.

6. CONCLUSIONS

In sensitive clay deposits, landslides are an important component in the development of young river valleys. The primary cause of instability is changes of geometry related to natural erosion or sometimes to human intervention. In slopes in which stability has already become marginal, some slight seasonal variations in the groundwater regime may trigger a landslide. For the long term, however, the groundwater regime in the vicinity of slopes evolves with the development of valleys in a way that affects the general stability of the valley walls and the morphology of the landslides. Detailed piezometric readings are required to permit a rational interpretation of the groundwater regime. Piezometric data should not be considered isolated data values but a reflection of a groundwater regime controlled by boundary conditions.

Approaches to stability analyses of sensitive clay slopes are based largely on experience and the study of many case records. However, consideration of the genesis of clay slopes justifies the type of analysis as well as the approach for determination of the shear-strength parameters. Landslides in sensitive clays are notorious for large and very rapidly enlarging retrogressive earth flows that can be triggered as a result of the initial failure. Expertise has been developed in the evaluation of stability against an initial failure using effective stress analysis, proper pore-pressure data, and postpeak strength parameters. However, it remains difficult to assess whether an initial failure will generate a sequence of retrogressive earth flows. Slopes composed of sensitive clays with a stability number $\gamma H/Cu$ larger than 6 and a liquidity index higher than 1.5 are generally considered as presenting a risk of retrogressive failure following an initial failure.

REFERENCES

Andrews, J.T. 1972. Post-Glacial Rebound. In *The National Atlas of Canada*, Canada Department of Energy, Mines, and Resources, Ottawa, Ontario, pp. 35–36.

Bjerrum, L. 1954. Geotechnical Properties of Norwegian Marine Clays. *Geotechnique*, Vol. 4, pp. 49–69.

Bjerrum, L. 1967. Engineering Geology of Norwegian Normally Consolidated Marine Clays as Related to Settlements of Buildings. *Geotechnique*, Vol. 17, pp. 81–118.

Brooks, G.R., J.M. Aylsworth, S.G. Evans, and D.E. Lawrence. 1994. *The Lemieux Landslide of June 20, 1993, South Nation Valley, Southeastern Ontario— A Photographic Record*. Miscellaneous Report 56. Geological Survey of Canada, Ottawa, Ontario, 18 pp.

Crawford, C.G., and W.J. Eden. 1965. A Comparison of Laboratory Results with In Situ Properties of Leda Clay. In *Proc., Sixth International Conference on Soil Mechanics and Foundation Engineering*, Montreal, Canada, University of Toronto Press, Vol. 1, pp. 31–35.

Eden, W.J., and C.B. Crawford. 1957. Geotechnical Properties of Leda Clay in the Ottawa Area. In *Proc., Fourth International Conference on Soil Mechanics and Foundation Engineering*, London, England, Butterworths Scientific Publications, London, Vol. 1, pp. 22–57.

Eden, W.J., and P.M. Jarrett. 1971. *Landslide at Orleans, Ontario*. Technical Paper 321. Division of Building Research, National Research Council of Canada, Ottawa, Ontario.

Eden, W.J., E.D. Fletcher, and R.J. Mitchell. 1971. South Nation River Landslide. *Canadian Geotechnical Journal*, Vol. 8, No. 3, pp. 446–451.

Elson, J.A. 1969. Late Quaternary Marine Submergence of Quebec. *Revue de Géographie de Montréal*, Vol. 23, pp. 247–259.

Evans, S.G., and G.R. Brooks. 1994. An Earthflow in Sensitive Champlain Sea Sediments at Lemieux, Ontario, June 20, 1993, and Its Impact on the South Nation River. *Canadian Geotechnical Journal*, Vol. 31, No. 3, pp. 384–394.

Gadd, N.R. 1975. Geology of Leda Clay. In *Mass Wasting: Proc., Fourth Guelph Symposium on Geomorphology*, Geo Abstracts Ltd., Norwich, England, pp. 137–151.

Grondin, G. 1978. *Etude de la susceptibilité aux coulées d'argile*. M.Sc. thesis. Department of Civil Engineering, Université de Sherbrooke, Québec, Canada.

Hillaire-Marcel, C. 1979. *Les mers post-glaciaires du Québec: quelques aspects*. Thèse de doctorat d'Etats Sciences Naturelles, Université Pierre et Marie Curie, Paris, France (2 vols).

Hillaire-Marcel, C., and R.W. Fairbridge. 1978. Isostacy and Eustacy of Hudson Bay. *Geology* (Geological Society of America), Vol. 6, pp. 117–122.

Kenney, T.C. 1964. Sea Level Movements and the Geologic Histories of the Post-glacial Marine Soils at Boston, Nicolet, Ottawa and Oslo. *Geotechnique*, Vol. 14, pp. 203–230.

Lafleur, J., and G. Lefebvre. 1980. Groundwater Regime Associated with Slope Stability in Champlain Clay Deposits. *Canadian Geotechnical Journal*, Vol. 17, No. 1, pp. 44–53.

Lefebvre, G. 1981. Strength and Slope Stability in Canadian Soft Clay Deposits. *Canadian Geotechnical Journal*, Vol. 18, No. 3, pp. 420–442.

Lefebvre, G. 1986. Slope Instability and Valley Formation in Canadian Soft Clay Deposits. *Canadian Geotechnical Journal*, Vol. 23, No. 3, pp. 261–270.

Lefebvre, G., and P. La Rochelle. 1974. The Analysis of Two Slope Failures in Cemented Champlain Clays. *Canadian Geotechnical Journal*, Vol. 11, No. 1, pp. 89–108.

Lefebvre, G., C.C. Ladd, and J.J. Paré. 1988. Comparison of Field Vane and Laboratory Undrained Shear Strength in Soft Sensitive Clays. In *Laboratory and Field Vane Shear Strength Testing*, Special Technical Publication, ASTM, Philadelphia, Pa., pp. 233–246.

Lefebvre, G., P. Rosenberg, J. Paquette, and J.-G. Lavallée. 1991. The September 5, 1988, Landslide on the La Grande River, James Bay, Québec, Canada. *Canadian Geotechnical Journal*, Vol. 28, No. 2, pp. 263–275.

Locat, J., and G. Lefebvre. 1986. The Origin of Structuration of the Grande-Baleine Marine Sediments Quebec, Canada. *Quarterly Journal of Engineering Geology*, Vol. 19, pp. 365–374.

Locat, J., G. Lefebvre, and G. Ballivy. 1984. Mineralogy, Chemistry and Physical Properties Interrelationships of Some Sensitive Clays from Eastern Canada. *Canadian Geotechnical Journal*, Vol. 21, No. 3, pp. 530–540.

Mitchell, J.K., and W.N. Houston. 1969. The Causes of Clay Sensitivity. *Journal of the Soil Mechanics and Foundations Division*, ASCE, Vol. 95, No. SM3, Paper 6568, pp. 845–871.

Mitchell, R.J., and A.R. Markell. 1974. Flowslides in Sensitive Soils. *Canadian Geotechnical Journal*, Vol. 11, No. 1, pp. 11–31.

Norsk Geoteknisk Forening. 1974. *Retningslinjer for Presentasjon av Geotekniske Undersökelser*. Oslo, 16 pp.

Philibert, A. 1976. *Etude de la résistance au cisaillement d'une argile Champlain*. M.Sc. thesis. Faculté des Sciences Appliquées, Université de Sherbrooke, Québec, Canada.

Quigley, R.M. 1980. Geology, Mineralogy and Geochemistry of Canadian Soft Soils: A Geotechnical Perspective. *Canadian Geotechnical Journal*, Vol. 17, No. 2, pp. 261–285.

Rosenqvist, I. Th. 1966. Norwegian Research into the Properties of Quick Clay—A Review. *Engineering Geology* (Elsevier), Vol. 1, pp. 445–450.

Skempton, A.W. 1964. Long-term Stability of Clay Slopes. *Geotechnique*, Vol. 14, pp. 75–102.

Skempton, A.W., and R.D. Northey. 1953. The Sensitivity of Clays. *Geotechnique*, Vol. 3, pp. 30–53.

Tavenas, F., J.-Y. Chagnon, and P. La Rochelle. 1971. The Saint-Jean-Vianney Landslide: Observations and Eye-Witness Accounts. *Canadian Geotechnical Journal*, Vol. 8, No. 3, pp. 463–478.

Tavenas, F., P. Jean, P. Leblond, and S. Leroueuil. 1983. The Permeability of Natural Soft Clays, Part II: Permeability Characteristics. *Canadian Geotechnical Journal*, Vol. 20, pp. 645–660.

Torrance, J.K. 1987. Quick Clays. In *Slope Stability* (M.G. Anderson and K.S. Richards, eds.), John Wiley and Sons, New York, pp. 447–473.

Wroth, C.P., and D.M. Wood. 1978. The Correlation of Index Properties with Some Basic Engineering Properties of Soils. *Canadian Geotechnical Journal*, Vol. 15, pp. 137–145.

Chapter 25

RUPERT G. TART, JR.

PERMAFROST

1. INTRODUCTION

The term *permafrost* was first proposed by Muller as a conveniently short word to refer to perennially frozen ground:

> Permanently frozen ground or permafrost is defined as a thickness of soil or other superficial deposit, or even of bedrock, at a variable depth beneath the surface of the earth in which a temperature below freezing has existed continuously for a long time (from two to tens of thousands of years). Permanently frozen ground is defined exclusively on the basis of temperature, irrespective of texture, degree of induration, water content, or lithologic character.... The expression "permanently frozen ground" however is too long and cumbersome and for this reason a shorter term "*permafrost*" is proposed as an alternative. (Muller 1947, 3)

The formation and existence of permafrost are largely controlled by climate and terrain but are also greatly affected by human intervention. Permafrost occurs in the northern and southern latitudes that are generally within about 35 degrees of the North and South Poles. Permafrost underlies about one-fifth of the land surface of the world, including about half of Canada (Brown 1970). The distribution of permafrost in the Northern Hemisphere is shown in Figure 25-1.

Permafrost is also found at high elevations in lower latitudes. This high-altitude permafrost is called *alpine permafrost* and is found, for example, in the Rocky Mountains of the United States and in the Alps of Europe. Alpine permafrost is more widespread and thicker on north-facing slopes, but variations in snow cover and vegetation on different slopes complicate the situation. In a totally different environment, permafrost is known to occur offshore in many northern areas of the world, including the eastern Canadian Arctic and under the Beaufort Sea (Samson and Tordon 1969; MacKay 1972; Hunter et al. 1976; Osterkamp and Harrison 1976).

Geocryology is defined as the study of earth materials at temperatures below freezing (Bates and Jackson 1980, 256). The geocryology of landslides is of greatest significance in cold regions where the mean annual air temperature is near or below freezing and the depth of freezing is significant with respect to the landslide structure. In areas where the ground is permanently frozen, natural thawing or thawing induced by human activities can be a significant destabilizer.

Earth material properties can change radically upon freezing (Andersland and Anderson 1978; Johnson 1981; Eyles 1983; Harris 1987). Depending on specific site conditions, freezing can be either a stabilizer or a destabilizer of earth materials (Johnston et al. 1981). Shear strengths of soils increase when soils freeze. Freezing ground can also cap and restrict groundwater flow, increasing pore pressures in soils remaining unfrozen below the freeze front. Conversely, shear strengths of soil

620

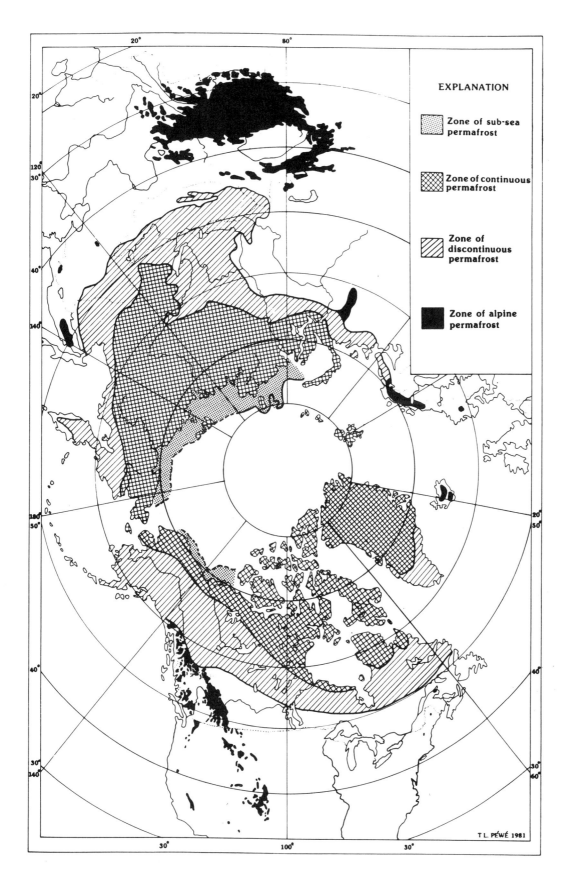

FIGURE 25-1
Distribution of
permafrost in
Northern
Hemisphere
(National Academy
of Sciences 1983).

EXPLANATION

Zone of sub-sea
permafrost

Zone of continuous
permafrost

Zone of
discontinuous
permafrost

Zone of alpine
permafrost

T. L. PÉWÉ 1981

masses reduce as soils thaw, and thawing can relieve pore pressures that develop during periods of freezing. Thawing may promote increased drainage of saturated soils and thus may have the effect of increasing the shear strength of the mass. Therefore, specific sites must be examined closely to determine whether stability or instability may be caused by thermal changes.

2. OCCURRENCE OF PERMAFROST

The temperature within the earth increases gradually with depth. Geothermal gradients ranging from about 1°C per 22 m to 1°C per 160 m, depending on the type of soil or rock and the effect of past climate periods, have been measured in permafrost regions of the Northern Hemisphere (Jessop 1971; Judge 1973). The depth of permafrost can be estimated if the mean annual ground temperature is known and the local geothermal gradient can be estimated.

Figure 25-2 presents an example of ground temperature distribution in a permafrost region. This temperature profile is sometimes called a "trumpet" curve because of its characteristic shape. Annual fluctuations in air temperature produce corresponding fluctuations in ground temperature, but vegetation and snow covering the surface decrease these

FIGURE 25-2
Ground temperature
regime in
permafrost
(after Brown et al.
1981).

fluctuations, which also decrease with increasing depth, reflecting the thermal characteristics of the soil or rock. The limits of summer and winter temperature extremes converge below the zone of annual surface-temperature influence at a point known as the *depth of zero annual amplitude* and follow the geothermal gradient below that depth. As shown in Figure 25-2, the annual temperature fluctuations cause the annual melting and refreezing of the near-surface materials. This zone is called the *active layer*, and the top of the permanently frozen materials is called the *permafrost table*.

The permafrost region of North America (Figure 25-1) is divided into two principal zones: the continuous permafrost zone in the north and the discontinuous permafrost zone in the south (Brown and Péwé 1973).

2.1 Continuous Permafrost

In the continuous permafrost zone, permafrost exists everywhere beneath the land surface except in newly deposited unconsolidated sediments where the climate has just begun to influence the ground thermal regime. In this zone permafrost can extend to great depths. For example, in areas on the North Slope of Alaska, permafrost extends more than 350 m below the ground surface, and in the northern parts of the Canadian Arctic Archipelago it extends to depths as great as 600 m (Judge 1973). At the southern limit of the continuous permafrost zone, permafrost is typically 60 to 100 m thick. The active layer is relatively thin, varying from 0.3 to 1 m. The permafrost temperature at the depth of zero amplitude ranges from −5°C in the south to −15°C in the extreme north.

2.2 Discontinuous Permafrost

In the discontinuous permafrost zone, permafrost exists together with layers of unfrozen ground known as *taliks*. The extent of permafrost terrain in an area depends on the history of the climate for the area and the condition of the ground surface. In the southern fringe areas of this zone, permafrost occurs as scattered islands ranging in size from a few square meters to a few hectares. Regions favorable to permafrost are usually covered with a thick vegetative mat that insulates the permafrost strata from the current above-freezing mean annual surface temperatures. When this mat

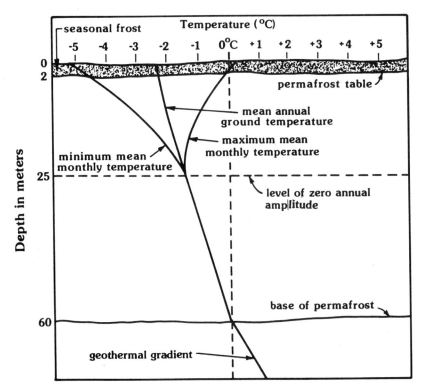

is removed or disturbed, the permafrost degrades rapidly. When the insulating mat remains in place, the permafrost can be preserved for many years. Further north, the permafrost becomes more continuous and underlies a greater variety of terrain types.

Discontinuous permafrost ranges in thickness from a few centimeters in the southernmost areas to 60 to 100 m at its boundary with the continuous permafrost zone. Thicknesses may vary rapidly and irregularly over short distances. The depth of the active layer may exceed 3 m depending on local climate and surface conditions. The active layer may not always extend to the permafrost table; at some locations a zone of unfrozen ground known as the *residual thaw layer* is found between the base of the active layer and the permafrost table (Brown et al. 1981).

2.3 Saline Permafrost

Because permafrost is defined by temperature, it is not necessary for the soil fluids to become frozen solid for the strata to be referred to as permafrost. Salts in pore water are the most frequent cause of permafrost containing unfrozen soil fluids. The salt concentration of seawater is sufficient to cause a depression of the freezing point of about 1.7°C, and as the salt concentration increases, further depression of the freezing point occurs. As saline permafrost pore water freezes, salt is rejected in front of the freeze front, resulting in the formation of pockets of highly saline unfrozen pore water between strata containing fully solidified ice. These conditions are common along the Arctic Ocean coastline in Alaska and adjacent areas in Canada where they have caused some difficulties for engineers designing facilities for existing small communities or associated with petroleum extraction (McPhail et al. 1976; Walker et al. 1983; Nixon and Lem 1984; Sellmann and Hopkins 1983).

3. CHARACTERISTICS OF FROZEN GROUND

From an engineering standpoint the physical properties of permafrost are a prime concern. Frozen rock formations can have strength characteristics similar to those of unfrozen formations when the rock quality is high. Formations with low rock quality can contain significant volumes of ice, which can produce high strengths when frozen but much lower strengths when thawed. In addition,

there can be a volume change in the formation as a result of the thawing and subsequent draining.

Most soils show markedly different properties in their frozen and unfrozen states. Ice-rich strata have properties similar to ice, whereas the properties of dry, clean granular soils can be similar to those of thawed granular soils. Field investigation techniques for frozen ground, including sampling, borehole logging, and field testing of frozen soil properties, have been discussed by Ladanyi and Johnston (1978).

3.1 Classification

A classification of frozen soils that has been widely used in North America was presented by Pihlainen and Johnston (1963) and Linnell and Kaplar (1963). Linnell and Kaplar suggested the following three steps for description and classification of frozen soils:

1. The soil phase is identified independently of the frozen state using the Unified Soil Classification System,
2. Soils characteristics resulting from freezing of the material are added to the description, and
3. Ice found in the frozen materials is described.

The frozen-soil characteristics (step 2) are based on two major groups: soils in which segregated ice is not visible to the naked eye (Group N) and soils in which segregated ice is visible (Group V). Depending on the degree to which the ice in Group N imparts solidification and rigidity to the soil mass, soils may be subdivided into those that are poorly bonded (Group Nf) or well bonded (Group Nb).

Figure 25-3 shows the key elements of the Linnell and Kaplar frozen soil classification system. This system has been adopted in ASTM D4083 and can be used with all frozen soils to estimate their engineering properties when frozen and upon thawing. Those strata with little or no ice will be relatively stable upon thawing. For example, some clayey silts that were overconsolidated when frozen have shown very slight strength changes when they thawed.

In contrast, strata with ice lenses larger than 25 mm will not be thaw stable since these strata will undergo significant volume changes and strength reductions as the ice melts and drains from the formation.

1. DESCRIBE SOIL INDEPENDENT OF FROZEN STATE	Classify Soil by The Unified Soil Classification System			
	Major Group		Subgroup	
	Description [b]	Designation	Description [b]	Designation
2. MODIFY SOIL DESCRIPTION BY DESCRIPTION OF FROZEN SOIL [a]	Segregated ice not visible by eye	N	Poorly bonded or friable	Nf
			Well Bonded — No excess ice	Nbn
			Well Bonded — Excess ice	Nbe
	Segregated ice visible by eye (ice less than 25 mm thick)	V	Individual ice crystals or inclusions	Vx
			Ice coatings on particles	Vc
			Random or irregularly oriented ice formations	Vr
			Stratified or distinctly oriented ice formations	Vu
3. MODIFY SOIL DESCRIPTION BY DESCRIPTION OF SUBSTANTIAL ICE STRATA [a]	Ice greater than 25 mm thick	ICE	Ice with soil inclusions	ICE + Soil Type
			Ice without soil inclusions	ICE

(a) Reference ASTM D4083.

(b) Description of ice within frozen soil is based on visual examination of the sample in the field.

FIGURE 25-3 Summary of classification procedure for frozen soil (Linnell and Kaplar 1963).

3.2 Strength Properties

Behavior of frozen soil is controlled primarily by temperature and ice content and salinity. As the ice content increases, the soil mass acts more like pure ice. As ice content decreases and temperatures rise, the frictional components of the soil particles begin to contribute to the ground behavior. The degree of bonding of fine sands and silts can be evaluated by blowcounts obtained in the standard penetration test. Strata composed of fine sands and silts that are poorly bonded because of depression of the freezing point by saline pore fluids or because of temperatures near 0°C tend to have blowcounts of less than 100 blows per 30 cm (100 blows per foot in cus-

tomary units). Figure 25-4 presents some typical data on blowcount versus salinity from a coastal deposit of silty fine sand with an average temperature of about −2°C (28°F). A temperature-depth profile of the site is presented in Figure 25-5.

3.3 Creep

A major consideration with frozen ground is the creep characteristics of ice and frozen soil. Ice will flow under sustained load; thus ice-rich soils must be investigated to determine their creep characteristics. As the temperature of frozen ground approaches 0°C, the rate of creep increases. Predictions of creep rates for frozen ground were made

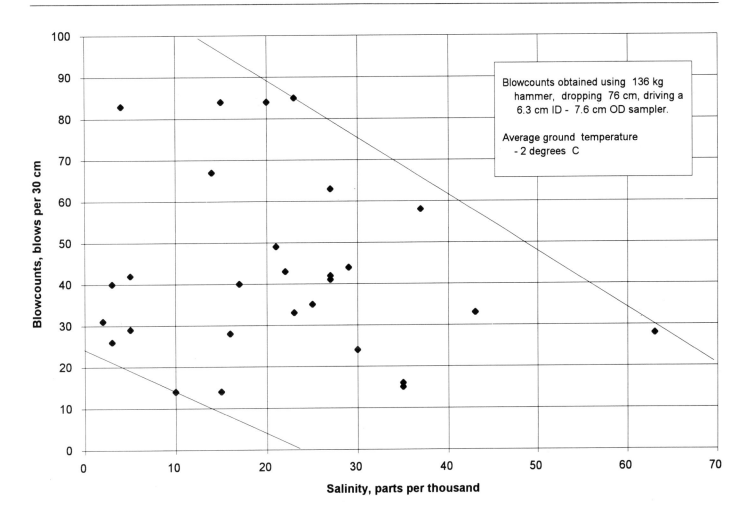

FIGURE 25-4
Blowcount versus
salinity data at
saline permafrost
site in western
Alaska.

by Nixon and Lem (1984) as well as by many others. The soil and ice parameters are not well defined and can be time consuming and difficult to obtain. As shown in Figure 25-6, the data can be scattered, and very slight changes in near-freezing temperatures can cause significant changes in the creep rate.

3.4 Frost Action

Frost action in soils includes the detrimental processes of frost heaving, which result mainly from the accumulation of moisture in the soil in the form of ice lenses at the freeze front during the freezing period, and thaw-weakening, or the decrease in strength when seasonally frozen soil thaws. Frost action is an extremely complex process. Many studies of frost action and related problems have been undertaken (Johnson 1952; Jessberger 1970). Frost heaving and thaw-weakening of soils produce significant damage and frequently require costly main-

tenance. A number of landslides have been caused by these processes.

Three basic conditions must exist for frost action to occur (Johnston et al. 1981):

1. The soil must be frost-susceptible,
2. There must be a water supply, and
3. The soil temperature must be sufficiently low to allow some soil water to freeze.

Frost-susceptible soils are most commonly identified by criteria based on grain size. A grain-size criterion originally proposed by Casagrande (1931) appears to offer the most useful rule of thumb for identifying frost-susceptible soils (Linnell and Kaplar 1959). It has been used by the U.S. Army Corps of Engineers to create a frost design soil classification. This frost classification system is frequently used in the design of pavements and is presented in Figure 25-7.

Casagrande stated:

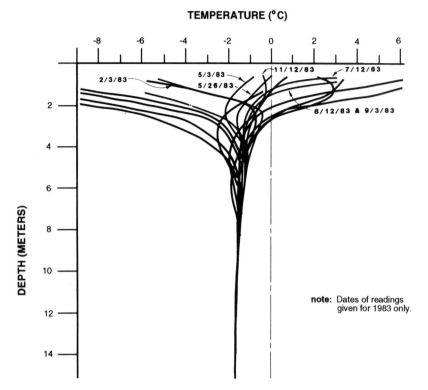

TEMPERATURE (°C)

FIGURE 25-5
Ground temperature
data at saline
permafrost site in
western Alaska
(Krzewinski and
Tart 1985).
REPRINTED WITH
PERMISSION OF
AMERICAN SOCIETY OF
CIVIL ENGINEERS

Under natural freezing conditions and with a sufficient water supply one should expect considerable ice segregation in non-uniform soils containing more than three percent of grains smaller than 0.02 mm and in very uniform soils containing more than ten percent smaller than 0.02 mm. No ice segregation was observed in soils containing less than one percent of grains smaller than 0.02 mm, even if the ground water level was as high as the frost line. (Casagrande 1931, 169)

As noted by Casagrande, the migration of water in both unfrozen and frozen soils is complex and basic to frost action and heaving.

Thaw-weakening is also a complex process and may occur in soils, especially clays, whether or not ice lenses have formed during an earlier freezing period. Ice-rich permafrost has very low permeability. The permeability is controlled by temperature and typically ranges from 10^{-7} cm/sec to less than 10^{-11} cm/sec, as shown in Figure 25-8.

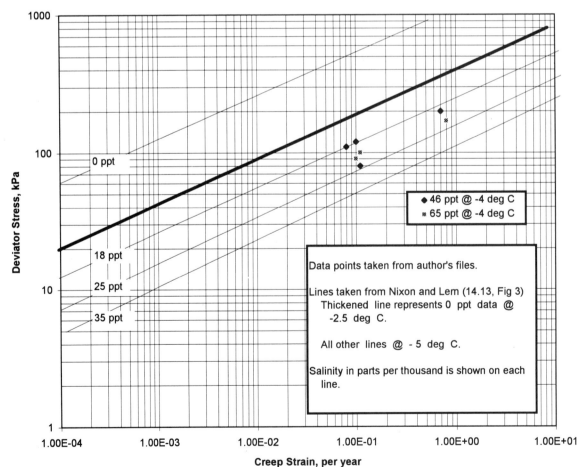

FIGURE 25-6
Creep strain data
from port site in
western Alaska
superimposed on
trends presented by
Nixon and Lem
(1984).

FROST GROUP	GENERAL SOIL TYPE	PERCENTAGE FINER THAN 0.02 mm BY WEIGHT	TYPICAL UNIFIED SOIL CLASSIFICATION
F1	Gravelly soils	3 to 10	GW, GP, GW-GM, GP-MG
F2	(a) Gravelly soils	10 to 20	GM, GW-GM, GP-GM
	(b) Sands	3 to 15	SW, SP, SM, SW-SM, SP-SM
F3	(a) Gravelly soils	Over 20	GM, GC
	(b) Sands, except very fine silty sands	Over 15	SM, SC
	(c) Clays, PI > 12	-	CL, CH
F4	(a) All silts	-	ML, MH
	(b) Very fine silty sands	Over 15	SM
	(c) Clays, PI < 12	-	CL, CL-ML
	(d) Varved clays and other fine-grained, banded sediments	-	CL and ML; CL, ML, and SM; CL, CH, and ML; CL, CH, ML, and SM

FIGURE 25-7 Summary of U.S. Army Corps of Engineers method of defining frost susceptibility of soils (ASTM 1993).

3.5 Seasonal Freezing and Thawing

In regions where there are prolonged winter periods with daily average air temperatures at or below freezing, the ground surface freezes. Conduction is the dominant ground heat-transfer process; however, the description of conductive heat flow must also be accompanied by an evaluation of the liberation or adsorption of heat resulting from phase changes in the water. Whenever a substance such as water changes from a solid to a liquid (by thawing) or from a liquid to a gas (by evaporation), considerable volumes of heat energy are required. Conversely, considerable heat energy is released when a gas converts to a liquid (by condensation) or from a liquid to a solid (by freezing). The term *latent heat* is used to define the energy components associated with these phase changes.

For estimation of seasonal freezing and thawing, ground surface temperature variations are

FIGURE 25-8 Hydraulic conductivity of frozen soils (Nixon 1992).

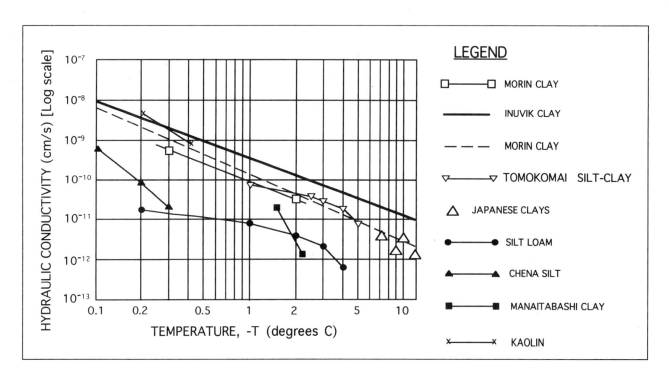

usually not considered in detail, but average values for the entire freezing or thawing period are used instead. In practice, the average seasonal temperatures are represented by the freezing or thawing index. The air freezing (I_{af}) and thawing (I_{at}) indexes are determined by computing the sum of the deviations of mean daily air temperatures from the freezing point during the thawing or freezing season. For time scales on the order of months, average ground surface temperatures can be correlated reasonably well with average air temperatures. For design purposes ground surface temperatures are estimated from air temperatures using an empirically defined n-factor, the ratio of the ground surface freezing or thawing index (I_{sf} or I_{st}) to the air freezing or thawing index (I_{af} or I_{at}):

$$n_f = I_{sf}/I_{af} \qquad n_t = I_{st}/I_{at} \qquad (25.1)$$

The magnitude of the freezing and thawing n-factors (n_f and n_t) depends on climate conditions as well as type of surface. In summer, vegetation-covered but treeless surfaces that remain wet will have average ground surface temperatures about equal to the corresponding average air temperatures, and the n-factors will thus be close to unity. Snow is an excellent insulator, and its presence causes the mean annual ground temperature to be several degrees warmer than it would be otherwise. Dry or well-drained surfaces, such as roadways, because of their sensitivity to variations in solar and long-wave radiation, have surface temperatures that vary over a wider range than do air temperatures. Care must be taken in estimating n-factor values (Lunardini 1978).

The depth of annual freezing is also dependent on the characteristics of the ground, especially its moisture content. The latent heat of a soil mass is dominated by the latent heat of water, which is the amount of heat that must be removed from water before it becomes ice and can cool to temperatures below 0°C. An identical amount of heat is required to melt ice. This amount of heat is much greater than the specific heat of water, the amount of heat it takes to warm the water 1 degree if there is no phase change. The latent heat of the mineral soil particles is zero since no phase change occurs. Thus, as the soil moisture content increases, soil masses freeze and thaw more slowly.

3.6 Ground Movements Resulting from Freezing

The process known as *frost creep* has been subjected to considerable study since it is readily measured in the field. The potential amount of downslope movement due to frost creep (C) results when the soil freezes and heaves a distance H perpendicular to the slope and a subsequent thaw causes the soil to settle vertically. The values of C and H and the slope angle θ are then related according to the following equation:

$$C = H \cdot \tan \theta \qquad (25.2)$$

Field measurements of frost creep suggest that it may be less important than first believed and that other mechanisms may combine with frost heave to create downslope movements. For example, Jahn (1975) measured 30 mm of downslope movement on a 4-degree slope in Spitsbergen, yet the measured frost heave of 0.15 m would only account for about 10 mm of the downslope movement. Most investigators suggest that viscous flowage of saturated soils is an important aspect of downslope movements of thawed soils, even on gentle slopes. This flow of saturated unfrozen earth material is called *solifluction*, which is discussed in more detail in Section 4.1.1.1.

3.7 Drainage Cutoff Caused by Freezing

Groundwater flow can be cut off or diverted by frozen zones with much lower permeabilities. This frequently occurs when the surface freezes and the groundwater is trapped beneath a surface cap and the permafrost table. Under these artesian conditions, quite high pressures can build up. These pressures may be further increased as the zones are progressively invaded by advancing freeze fronts. The buildup of pore-water pressure reduces the shear strength of the soil strata in which the pressures develop. Weaknesses in the frozen surface capping zone can result in seepage zones that may flow both summer and winter. In summer these zones form springs, but in the winter the discharging water freezes and builds up *icing deposits*—layers of solid ice that may block or damage transportation facilities or cover slopes, which may become unstable when the icing deposits melt in the following summer and provide a large supply of water (Figure 25-9).

FIGURE 25-9
Example of slope
icing.
R.G. TART, JR.

3.9 Shear-Strength Reduction During Thaw

Excess water is frequently drawn into freezing soil strata, causing these strata to become saturated and loose. When thawing occurs, the shear strengths of these loose saturated soils are very low until the soils have been allowed to drain and dry. Provision of adequate drainage is frequently a problem in the topographically low-relief regions common to many areas in the Arctic of North America. Low air temperatures inhibit rapid drying.

4. TYPES OF PERMAFROST SLOPE MOVEMENTS

In permafrost areas landslides are found in both frozen and unfrozen ground. Landslides involving unfrozen ground are mostly found in areas of recent sediments adjacent to bodies of water. Unfrozen Cretaceous shales are also affected by landsliding in Canada's Northwest Territories. The discussion in this chapter will be restricted to landslides in frozen or partially frozen ground. Most landsliding involving frozen ground begins when thawing occurs; shearing through frozen ground has been noted in some natural slopes but is relatively rare. There is also the potential that dynamic loadings related to earthquakes may cause liquefaction of thawed cohesionless soils situated between the frozen near-surface portions of the active layer and the underlying permafrost (Finn et al. 1978).

4.1 Classification of Permafrost Landslides

Typical landslide features and associated mass-wasting phenomena have been widely described in the literature (Bird 1967; MacKay 1966; MacKay and Mathews 1973; Washburn 1973). Because of the great degree of human activity along the Mackenzie River Valley of northwestern Canada, the characteristics and distribution of landslides in this region have received considerable attention (Isaacs and Code 1972; Chyurlia 1973; Code 1973; McRoberts and Morgenstern 1973). McRoberts and Morgenstern (1973, 1974a) proposed a classification of permafrost landslides based solely on descriptive features (Figure 25-10). The major divisions in this classification—flows, slides, and falls—closely correspond to the stan-

3.8 Ground Movements Resulting from Thawing

Frozen soils frequently contain moisture contents in excess of saturation, even for very loose soils. When these soils thaw, the soil matrix compresses as the excess moisture is forced out of the pores. This compaction is known as *thaw-consolidation*. The amount of compression is proportional to the ratio of the in-place frozen density to the after-thaw density. If the moisture content of the frozen soil can be determined, assuming that the soil has excess moisture, the frozen density can be calculated. The after-thaw density can be estimated by determining the soil type and the overburden pressure. Thaw strains of 25 percent are common in frozen fills and natural ground formations with high moisture contents. Thaw-consolidation should be anticipated in soils with frozen moisture contents of 20 percent or higher unless the soils contain a clay component. Taber (1943) described the role of thaw-consolidation in the processes of solifluction and skin flow, two major types of slope movement in originally frozen soils (see Section 4.1).

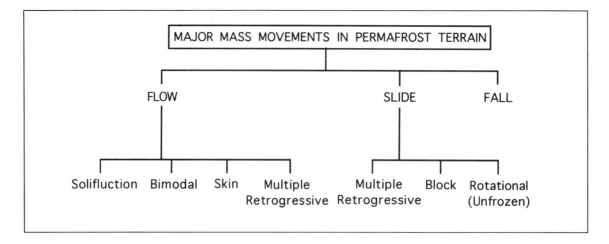

dard terminology for describing landslides, as defined in Chapter 3 of this report. However, the subdivisions of these categories contain some nonstandard terms that define some of the unique landsliding conditions found in permafrost areas.

4.1.1 Flows

Flow-dominated movements on natural and cut slopes are common in permafrost areas, especially when the slopes are composed of fine-grained frozen soils. Studies by Pufahl (1976) demonstrated the complexities of the heat balance on the surface of exposed and thawing permafrost. Denudation of vegetative cover on ice-rich permafrost slopes usually results in thawing. Because moisture contents are frequently well above 20 percent, flowing or creeping, or both, of the thawed zone is a natural consequence. This movement continues until the next winter, when freezing begins to restabilize the slope.

The McRoberts and Morgenstern classification (Figure 25-10) considered only naturally occurring landslides and included four types of flow movements: solifluction and skin, bimodal, and multiple retrogressive flows.

4.1.1.1 Solifluction

Solifluction is defined as "the slow downslope flow of saturated, unfrozen earth materials" (National Research Council of Canada 1988, 78). Solifluction has been observed on natural slopes and on slopes modified by human construction. In each case, freezing in winter seals the sloping ground surface, which then prevents surface drainage of pore water and causes water contents to reach levels of saturation and higher. As the saturated zone warms in spring and summer, the slope slowly creeps downward. As the upper zone thaws and drains, it becomes more stable but continues to move on top of the ice-rich materials, which remain frozen. A classic example of soli-

FIGURE 25-11
Solifluction lobes on alpine permafrost slope in eastern Kyrgystan.

R.G. TART, JR.

fluction is shown in Figure 25-11. Solifluction is similar to flow movements but is much slower and not as deep-centered. Figure 25-12 shows solifluction around a deeply embedded pipe pile. Inclinometers placed close to this pile showed the greatest movement at the surface with the magnitude of movement decreasing with depth (Figure 25-13). The movement is greatest during the period of spring thaw and slows as drainage occurs through the summer. It stops when the ground refreezes in the autumn.

4.1.1.2 Skin Flows

Skin flows involve the detachment of a thin cover of vegetation and mineral soil and subsequent movement of this mass over a planar undulating surface. These failures have much in common with the debris flows occurring in residual or colluvial soils (see Chapter 20). They form long ribbonlike shapes down the slope that may coalesce to involve large areas and are most common in ice-rich sediments in the discontinuous permafrost zone, particularly following a forest fire (Hardy and Morrison 1972; Hughes 1972; MacKay and Mathews 1973).

Skin flows develop when the pore pressures generated during thawing reduce the shearing resistance of the soil materials to low levels. Accordingly, skin flows may develop on slopes as flat as 6 degrees (McRoberts 1973), although they are frequently found on steep slopes.

4.1.1.3 Bimodal Flows

Bimodal flows were so named by McRoberts and Morgenstern to indicate that two modes of mass movement occur. These dual modes of movement are reflected by the characteristic low-angle tongue of flowed debris and the steep head scarp. Bimodal flows have many characteristics in common with the landslides occurring in sensitive clay deposits (see Chapter 24).

In most active bimodal flows, permafrost in the head scarp is exposed to the atmosphere and degrades by ablation. The surfaces of the tongues have very low slopes and the water contents of the materials forming the tongue may approach the liquid limit of these soils. Figure 25-14 shows plan and section views of a bimodal flow landslide along the Mackenzie River.

Bimodal flows are initiated by some process or combination of processes that removes the insulating cover of vegetation and thawed soil, expos-

ing the ice-rich permafrost in such a manner that ablation begins and is sustained. Following initiation, ablation can generate substantial rates of mass wasting. Observations by McRoberts (1973) suggested that head scarps may regress at rates ranging from 3 to 20 cm/day during the season of thaw. With such rates it is evident that bimodal flows can pose a threat to adjacent structures and transportation facilities.

4.1.1.4 Multiple Retrogressive Flows

Multiple retrogressive flows are dominated by flowage but may retain some portion of their prefailure relief. Their form suggests that a series of retrogressive failures have occurred in the head scarp. Rotational and nonrotational sliding may also be observed in portions of these flows, which may cover quite large areas and are usually relatively shallow.

4.1.2 Slides

McRoberts and Morgenstern subdivided permafrost slide movements into three classes: block, multiple retrogressive, and rotational slides (Figure 25-10). Block slides frequently involve a single large block that has moved downslope with little or no back tilting. Multiple retrogressive slides are characterized by a series of more or less arcuate blocks that are concave toward the toe and step upward from the toe to the head scarp. The blocks may exhibit back tilting.

FIGURE 25-12
Solifluction flow around 45-cm pipe pile.
R.G. TART, JR.

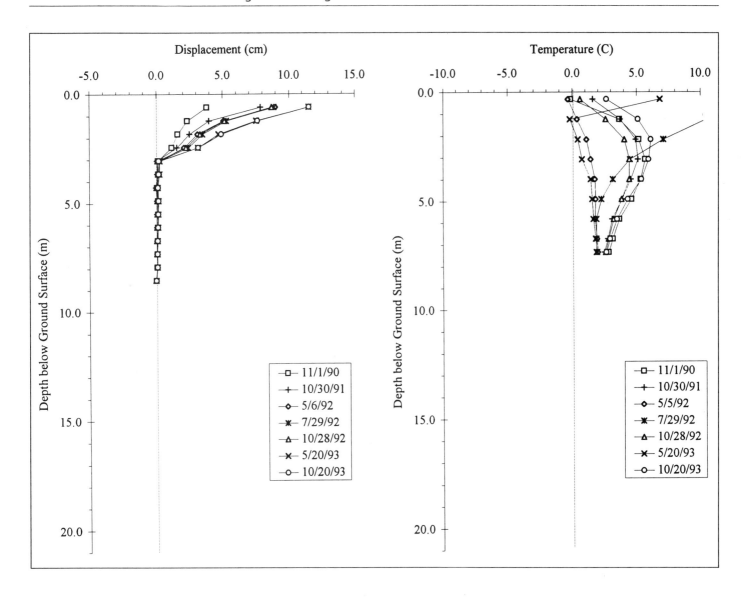

FIGURE 25-13
Inclinometer
displacements for
slope in
discontinuous
permafrost showing
solifluction flows.

Multiple retrogressive and block slides occur in materials whose water contents appear to be lower than those that result in flows. These slides are associated with shear failure in frozen and unfrozen soil and are frequently quite large features 45 to 60 m high. McRoberts and Morgenstern (1974b) discussed the mechanics of slides in frozen soil extensively and suggested that the base of these failures often lies in unfrozen clay, at least for the sites they studied in the Mackenzie River Valley.

Figure 25-15 shows one example of a multiple retrogressive slide. It appears to be the result of groundwater conditions in a talik beneath the active layer. Thawing presumably provided additional excess moisture to the sliding mass, and the result was an almost circular landslide. Figure 25-16

shows the presumed positions in the spring and in the fall of the groundwater source believed to have contributed to this landslide.

4.1.3 Falls

Permafrost-related falls involve the gravity-influenced rapid downward dropping of detached blocks of frozen materials. Falls are common along the banks of rivers and lakes and parts of the Arctic Ocean coastline where thawing and erosion undercut the permafrost-bonded bank materials. Large blocks of frozen material periodically break off, facilitating bank recession. Recession rates as high as 10 m/year have been reported (Walker and Arnborg 1963).

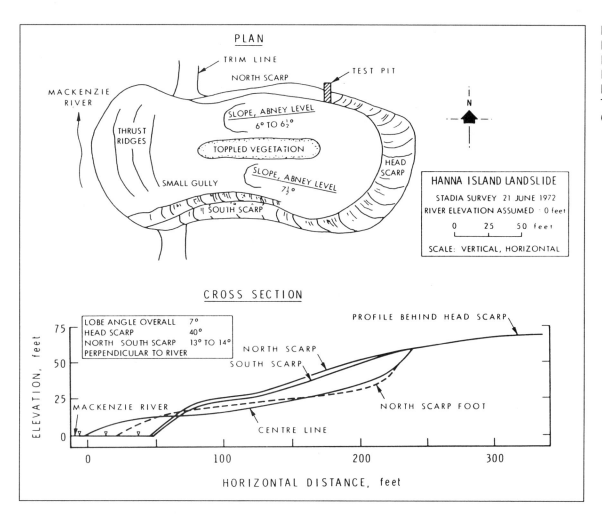

FIGURE 25-14
Bimodal flow near
Hanna Island,
Mackenzie River,
Northwest
Territories, Canada
(BROWN ET AL. 1981).

4.2 Selected Examples of Permafrost Landslides

In the following sections a number of slope failures in permafrost regions are discussed. Taken mostly from the author's own experience, they have been selected to illustrate typical slope stability problems encountered in permafrost regions.

4.2.1 Flow Movements in Mine Wastes

Flows frequently occur as a result of the placement of ice-rich overburden or waste materials from mining operations. In many cases, the waste is frozen when placed in a stockpile and can take more than one season to thaw completely. Rapid thaw can result in large flows, as shown in Figure 25-17. At this site ice-rich mine wastes stored on a sloping surface during winter began to flow as the waste thawed.

4.2.2 Flowage of Fill Material

Figures 25-18 and 25-19 show a flow failure at a location where a fill containing ice was placed on sloping natural ground, probably during the winter. As the fill thawed, the saturated mass began to flow. During subsequent winters the surface of the fill froze, confining incoming groundwater within the fill. During the spring and summer each year, as the thaw front moved into the loose ice-rich fill, the fill began to flow.

4.2.3 Solifluction and Creep Movements

Permafrost areas frequently include landslides that develop as a result of either shallow or deep-seated creep. Figure 25-20 presents the profile of a permafrost slope that is creeping and includes inclinometer deflections at various locations. The inclinometer deflections show the effects of deep

FIGURE 25-15
Multiple retrogres-
sive earth slide in
discontinuous
permafrost.
R.G. TART, JR.

creep and reduced movements in the the ground surface. In this case the ice-rich zone is sliding and creeping over a denser, drier zone. The creep failure surface developed at a depth at which the permafrost temperature had warmed sufficiently to creep but not enough to allow for adequate drainage and increased strength.

4.2.4 Landslides in Soil and Rock Cut Slopes

Figure 25-21 shows a permafrost rock cut slope that has become unstable as it has thawed, releasing both blocks of bedrock and masses of saturated, unstable soil. The darker materials in the center of Figure 25-21 are weathered rock and soils that were frozen at the time of construction but began sloughing and sliding as the slope thawed. This release is typical of a thawing composite soil-and-rock-slope failure. The slope appeared stable during construction, but the insulating overburden was removed, exposing the frozen and weathered bedrock to thawing. After the thaw front was able to penetrate the rock structure, blocks were released along the surfaces

of structural weakness, including both foliation surfaces and joints. Additional volumes of thawing soils added considerably to the total volume of unstable material.

Another example of this type of landslide can be seen in Figure 25-22. A mining operation developed a 110-m-high cut slope through old lake deposits to uncover a placer gold deposit in an ancient buried channel. The excavation approached maximum depth about midsummer. The lake deposits were mostly fine-grained soils with high water contents, although there were also some coarser-grained, and thus better-drained, strata. Seepage from the fine-grained strata could be observed during and after the landsliding, which is shown in more detail in Figure 25-23. The fine-grained soils with high water contents would become ice-rich when frozen. Although no confirming evidence is available at this time, it is possible that because of the high latent heat of the fine-grained ice-rich lake deposits, they were frozen at the time of cutting, whereas the well-drained deposits had thawed. For this reason, the fine-grained strata became unstable upon thawing, and the better-drained strata remained relatively stable.

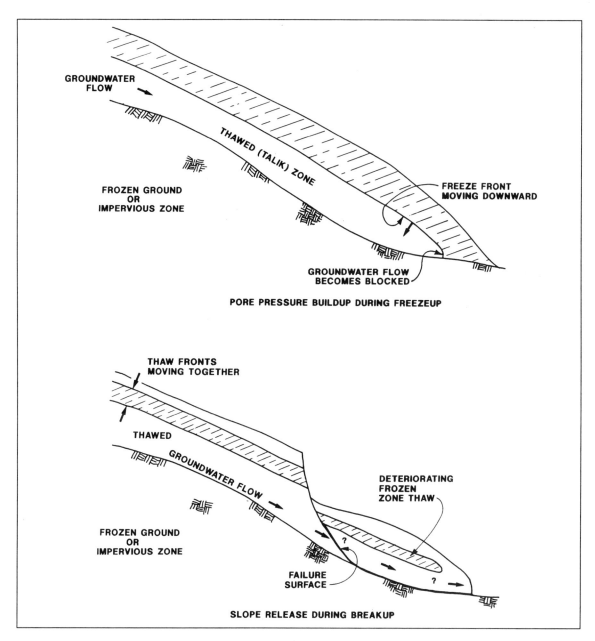

FIGURE 25-16
Schematic section
views showing prob-
able causes
of multiple
retrogressive slide.

Any ice-rich slope will begin to move as it thaws. The movements may include surficial sloughing and flow movements as well as more deep-seated landslides. Steeply inclined thawing slopes are frequently more stable than gently inclined slopes, and almost vertical slopes are often the most stable because vegetation mats fall more readily over the edges of steep slopes and begin to develop a new insulating cover that reduces thawing and increases stability. In the thawing permafrost slope pictured in Figure 25-24, minor surficial sloughing continues as a new vegetation mat slowly develops.

4.3 Associated Slope Stability Problems

The most common problem associated with slope stability in permafrost areas is the stabilization of earthworks, especially dams and dikes.

4.3.1 Stability of Embankments

If fill slopes are constructed in permafrost areas, the fill materials must be allowed to thaw before their placement if they cannot be protected from thawing after construction. It is virtually impossible to compact frozen chunks of soil unless the moisture

FIGURE 25-17
Flow of mine wastes.
R.G. TART, JR.

FIGURE 25-18
Flow failure of
pipeline workpad.
R.G. TART, JR.

content of these chunks is less than 5 percent. Even then compaction will be inefficient and possibly ineffective. Therefore, any earthwork project that will thaw during its lifetime must be thawed at the time of construction to prevent instabilities caused by subsequent thawing. An example of a flow failure of a fill was discussed in Section 4.2.2 and shown in Figures 25-18 and 25-19.

4.3.2 Thermal Piping or Erosion

Frozen fine-grained soils containing adequate water contents in the form of ice are quite impermeable. Naturally occurring deposits and constructed embankments containing such materials have often been used as dams in permafrost regions. These materials can thaw and then fail catastrophically by a process known as *thermal piping*. Erosion (or piping) of fine-grained soils refers to the progressive removal of smaller soil particles as flow passes through the erosive soil mass. If left unchecked, this well-known process may lead to complete destruction of the soil structure. Thermal piping refers to the addition of heat to a frozen mass by the infiltration of warmer ground-

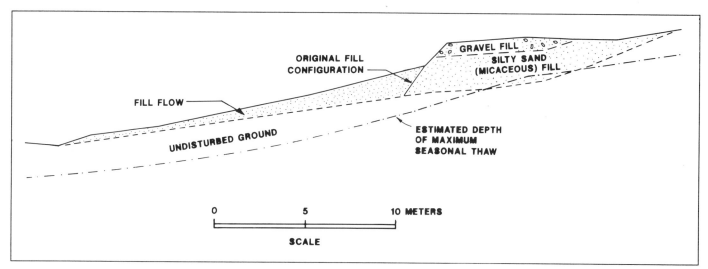

FIGURE 25-19
Profile of flow failure of pipeline workpad.

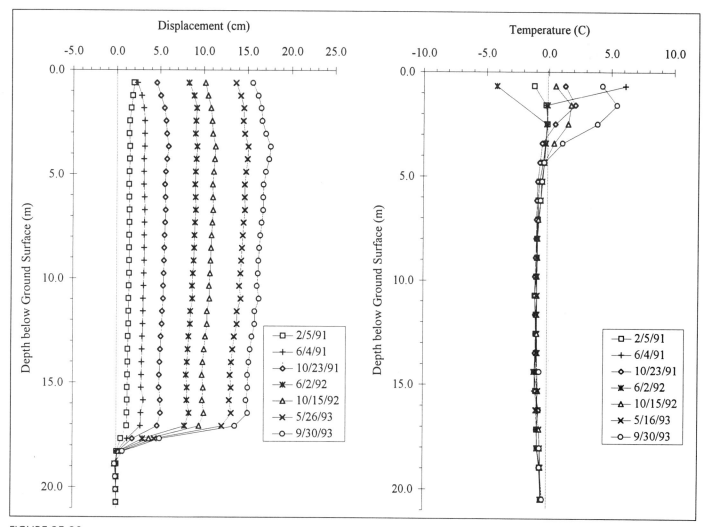

FIGURE 25-20
Inclinometer displacements on warm permafrost slope creeping near base of permafrost.

FIGURE 25-21
Cut slope failures
due to thawing
permafrost affecting
both soil and
bedrock (*upper left*),
which produced
rock falls that
damaged concrete
barriers (Jersey
barriers) placed
adjacent to highway
shown in foreground.
R.G. TART, JR.

water, which results in thawing of the ground ice. The ice itself may have formed a network of passages through which water may preferentially flow once the ice has thawed. If the soils are relatively fine-grained, these flows may also subsequently erode (or "pipe") the now highly erodible fine soil particles in the thawed soil mass.

In the area shown in Figure 25-25, a water supply lake was lost when thermal piping resulted in a breach in an embankment that retained the lake.

5. METHODS FOR CONTROLLING STABILITY OF PERMAFROST SLOPES

There are two primary types of methods for controlling the stability of permafrost slopes: physical restraint and thermal modification. In some cases, combinations of these two approaches can provide the most cost-effective means of control.

5.1 Physical Restraint Methods

Physical restraint methods are similar to those used in thawed soils and can include, but are not limited to, the following:

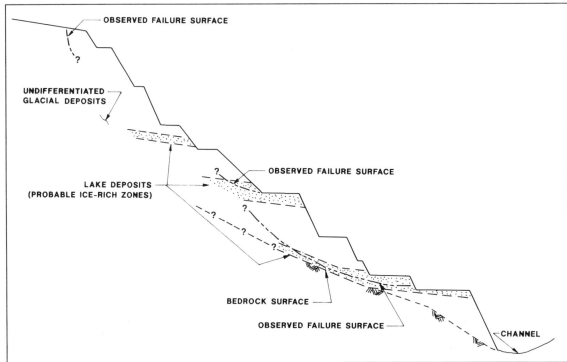

FIGURE 25-22
Open-pit mine cut-slope failure due to thawing discontinuous permafrost.

- Drainage improvements,
- Buttresses,
- Retaining walls,
- Slope geometry modifications, and
- A variety of other methods.

These methods are covered in Chapter 17 of this report.

5.2 Thermal Modification Methods

Thermally modified slopes are unique to cold regions. Thermal modifications involve

- Maintaining the current thermal state of the slope,
- Removing heat from the slope to increase freezing, or
- Adding heat to the slope to increase the rate of thawing.

In considering thermal modification, it is very important to know the thermal properties of the earth mass being modified and to perform the appropriate heat flow calculations to predict the rate and amount of modification that will take place.

FIGURE 25-23
Closer view of open-pit mine cut-slope failure shown in Figure 25-22.
R.G. TART, JR.

FIGURE 25-24
Partially stabilized cut slope in permafrost.
R.G. TART, JR.

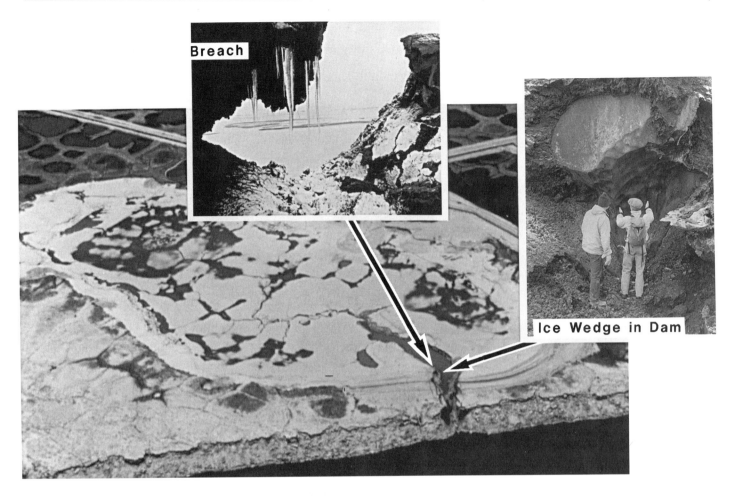

Breach

Ice Wedge in Dam

FIGURE 25-25
Permafrost dam on
North Slope of
Alaska breached by
thermal piping.
R.G. TART, JR.

5.2.1 Maintaining Current Thermal State

Thawing of permafrost is frequently caused by construction activities that disturb the ground cover of vegetation and organic materials. These materials are usually very good insulators and in some cases do not allow the thaw front to penetrate during the short summer seasons characteristic of the arctic regions. When these materials are crushed or removed and the underlying soils are allowed to thaw, slope instabilities and settlement usually result. This disturbance of the insulating cover can be mitigated in some instances by replacing the cover with new vegetation or some other type of insulating material.

An example of this type of modification is shown in Figure 25-26. Wood chips from a sawmill were placed on the surface of a slope in a permafrost area to provide additional insulation. Throughout the fall day on which this picture was taken, the wood chips remained frozen, but the surrounding ground surface was thawed, as indi-

cated by its darker color. Seepage was also observed on a section of the slope where the insulation was not being used. Additional protection at this location was provided by passive heat-extraction devices, discussed further in the following section.

5.2.2 Heat Removal from Ground

Heat may be removed from the ground by active or passive methods. Active refrigeration is one common heat-removal technique. This type of heat removal is used, for example, in indoor ice rinks; another example is the artificial freezing used to stabilize tunnels during construction in nonpermafrost areas. Because of its cost, however, it is usually used only for temporary stabilization. Several Trans-Alaska Pipeline pump stations that are located on permafrost have refrigerated foundations to prevent thaw settlement of the pump-station buildings. Passive methods include heat pipes and various types of passive air ventilation systems. Heat pipe systems have been used to stabilize slopes in Alaska, as discussed at the end of this section. Although other passive heat removal systems are in use, they have primarily been used to stabilize building foundations. The most common of the other systems is a passive air ventilation system in which a set of surface manifolds is oriented so as to passively maintain a sufficient pressure differential to cause a continuous flow of cold air through subsurface ducts.

Heat pipes are closed tubes filled with a fluid that boils at temperatures below freezing and then condenses at cooler temperatures. When the ground temperatures are near freezing, the fluid boils and rises to the top of the tube. During the winter, the cold air passing over the heat pipe fins causes the fluid to condense and flow to the bottom of the tube, where it receives heat from the ground to warm it, causing it to boil and rise once again. Thus these systems remove heat from the ground in a continuous cycle. In the summer, when the air temperatures are higher than the ground temperatures, the heat pipes are inactive. This is a passive heat extraction process because no external energy is required to make this system operate.

At the location shown in Figure 25-26, heat was also removed from the ground by passive thermal heat pipes. Two finned heat pipes can be observed on top of each pipe-pile supporting the pipeline and also in free-standing heat pipes between the piles. Each of these heat pipes removes heat from the ground in order to maintain the underlying permafrost and thereby increase the stability of the slope. Figure 25-27 shows finned heat pipes in a fill slope that overlies ice-rich permafrost near 0°C. Near the toe of the slope the underlying permafrost began to thaw and initiate slope movements. Finned heat pipes were installed along the toe of the slope to remove heat from the permafrost and retard the thawing, thus providing more stability to the slope.

5.2.3 Heat Input to Ground to Promote Thawing

In some cases, the addition of heat to the ground to promote thawing is the most cost-effective method of stabilization. In areas of discontinuous permafrost, it may be possible, by surface modification, to increase the heat flow into an area enough to cause it to thaw and thus allow thaw strain to occur before construction. This can be accomplished by

removing vegetation, covering the area with a dark material in summer or with a temporary heated facility, flooding, and other methods.

5.3 Design of Cut Slopes

Performance of highway cut slopes in permafrost soils has been studied in Alaska, the Yukon, and the Northwest Territories (Lotspeich 1971; Smith and Berg 1973; Pufahl et al. 1974; McPhail et al. 1976). These studies and experience in permafrost terrain led McRoberts (1978) to suggest that cut slope designs and configurations can be best evaluated in terms of four standard classes, which he termed Types I to IV.

5.3.1 Type I

Type I consists of low cuts, 1 to 3 m high, that are capable of self-stabilization provided there is a reasonably well-developed organic cover at least 0.2 m thick overlying the mineral soil. The cut should be made vertically. Any large trees should be removed from the upslope areas for a horizontal distance approximately 1.5 times the height of the cut. Removal of the trees promotes the movement of the organic mat downslope and over the face of

FIGURE 25-26 Permafrost slope insulated with wood chips (lighter-colored ground surface along pipeline), Trans-Alaska Pipeline. Heat pipes are also visible on each pipe pile supporting pipeline and as free-standing elements between piles.
R.G. TART, JR.

FIGURE 25-27
Finned heat pipes
being used to
stabilize slope
founded on warm,
ice-rich permafrost.
R.G. TART, JR.

the cut without tearing. The natural draping of the organic mat over the cut face acts as an insulator and promotes reestablishment of vegetation. In cuts higher than 1.5 m, a reinforcing mesh should be applied to the surface of the organic mat to provide it with an adequate tensile strength. Areas with thin organic cover may have to be designed so that they can slough without impeding the transportation facility, or they may be designed as Type IV cuts (see below) when ice-rich soils are involved.

5.3.2 Type II

Type II cuts are in frozen ice-poor soil or bedrock. Such cuts can be designed so that the slope angle is compatible with the thawed (unfrozen) properties of the materials. Higher rates of surface erosion can be expected because of severe freeze-thaw conditions and slope establishment of vegetation.

5.3.3 Type III

Type III cuts include any made in a terrain in which ground ice occurs as a matrix of ice wedges or poly-

gons in otherwise relatively ice-poor mineral soil. These cuts should be vertical if possible, since fewer ice wedges are likely to be encountered. If the ice wedges are relatively widely spaced and a thick organic mat is present, the cuts may self-heal as with Type I cuts. Considerable water may be produced during the melting of the ice wedges and the self-healing process. Revetments may be required along the toe of slope to stabilize it and allow the water to drain. Wider-than-normal ditches may be required to provide space for accumulations of slope debris, the revetments, and drainage. Since this extra width increases the amount of excavation required, a more economical alternative may be excavation of the ice wedges over some reasonable distance based on geocryologic considerations and backfilling with selected materials.

5.3.4 Type IV

Type IV cuts include those in ice-rich permafrost soils, which are the most difficult cuts to design and manage. These cuts may be subject to continuous exposure, melting, and retreat of the cut faces, essentially forming bimodal flows. Generally such slopes can only be stabilized by providing an adequate insulation cover to prevent, or at least reduce, the thawing process. The insulation must be free-draining and sufficiently flexible to accommodate thaw settlements; hence mixtures of sand, gravel, crushed rock, and artificial insulation materials are generally used.

Considerable experience is required to recognize terrain conditions and recommend appropriate cut-slope design responses. Conditions promoting the self-healing of cuts are often found in tills, alluvial or colluvial silts and sands, and bedrock. Conditions causing non-self-healing cuts are more common in continuous permafrost zones, ice-rich soils, and glaciolacustrine sediments. A clear understanding of the conditions within a particular cut may not emerge until the actual ground-ice conditions are exposed during construction. Inspection of adjacent natural slopes is often the best guide to the expected performance of cut slopes at any location.

REFERENCES

Andersland, O.B., and D.M. Anderson. 1978. *Geotechnical Engineering for Cold Regions.* McGraw-Hill, New York, 551 pp.

ASTM. 1993. *Annual Book of ASTM Standards*, Vol. 04.08: *Soil and Rock; Dimension Stone; Geosynthetics*, Philadelphia, Pa., pp. 617–621.

Bates, R.L., and J.A. Jackson, eds. 1980. *Glossary of Geology*, 2nd ed. American Geological Institute, Falls Church, Va., 751 pp.

Bird, J.B. 1967. *The Physiography of Arctic Canada.* Johns Hopkins Press, Baltimore, Md.

Brown, R.J.E. 1970. *Permafrost in Canada: Its Influence on Northern Development.* University of Toronto Press, Toronto, Ontario, Canada, 246 pp.

Brown, R.J.E., and T.L. Péwé. 1973. Distribution of Permafrost in North America and Its Relationship to the Environment: A Review, 1963–1973. In *Proc., Second International Conference on Permafrost*, Yakutsk, USSR, North American Contribution, National Academy of Sciences, Washington, D.C., pp. 71–100.

Brown, R.J.E., G.H. Johnston, J.R. MacKay, N.R. Morgenstern, and W.W. Shilts. 1981. Permafrost Distribution and Terrain Characteristics. In *Permafrost: Engineering Design and Construction* (G.H. Johnston, ed.), Associate Committee on Geotechnical Research, National Research Council of Canada, John Wiley & Sons, Toronto, pp. 31–72.

Casagrande, A. 1931. Discussion of Benkelman, A.C., and F.R. Olmstead, A New Theory of Frost Heaving, *HRB Proc.*, Vol. 11, Part 1, pp. 168–172.

Chyurlia, J. 1973. *Stability of River Banks and Slopes along the Liard River and Mackenzie River N.W.T.* Report 73-3. Environmental-Social Program, Northern Pipelines, Department of Indian and Northern Affairs Canada, Ottawa, pp. 111–152.

Code, J.A. 1973. *The Stability of Natural Slopes in the Mackenzie Valley.* Report 73-9. Environmental-Social Program, Northern Pipelines, Department of Indian and Northern Affairs Canada, Ottawa, 18 pp.

Eyles, N. 1983. *Glacial Geology: An Introduction for Engineers and Earth Scientists.* Pergamon Press, Elmsford, N.Y., 409 pp.

Finn, W.D., R.N. Yong, and K.W. Lee. 1978. Liquefaction of Thawed Layers in Frozen Soil. *Journal Geotechnical Engineering Division*, ASCE, Vol. 104, No. GT10, pp. 1243–1255.

Hardy, R.M., and H.A. Morrison. 1972. Slope Stability and Drainage Considerations for Arctic Pipelines. In *Proc., Canadian Northern Pipeline Research Conference*, Technical Memorandum 104, National Research Council of Canada, Ottawa, pp. 249–267.

Harris, C. 1987. Mechanics of Mass Movement in Periglacial Environments. In *Slope Stability* (M.G. Anderson and K.S. Richards, eds.), John Wiley and Sons, New York, pp. 531–559.

Hughes, O.L. 1972. Surficial Geology and Land Classification, Mackenzie Valley Transportation Corridor. In *Proc., Canadian Northern Pipeline Research Conference*, Technical Memorandum 104, National Research Council of Canada, Ottawa, pp. 17–24.

Hunter, J.A., A.S. Judge, H.A. Macauley, R.L. Good, R.M. Gagne, and R.A. Burns. 1976. *Permafrost and Frozen Sub-seabottom Materials in the Southern Beaufort Sea: Beaufort Sea Project.* Technical Report 22. Environment Canada, Ottawa, 177 pp.

Isaacs, R.M., and J.A. Code. 1972. Problems in Engineering Geology Related to Pipeline Construction. In *Proc., Canadian Northern Pipeline Research Conference*, Technical Memorandum 104, National Research Council of Canada, Ottawa, pp. 147–178.

Jahn, A. 1975. *Problems of the Periglacial Zone.* PWN-Polish Scientific Publishers, Warsaw, 223 pp. NTIS: PB 248 901.

Jessop, A.M. 1971. The Distribution of Glacial Perturbation of Heat Flow in Canada. *Canadian Journal of Earth Sciences*, Vol. 8, No. 2, pp. 162–166.

Jessberger, H.L. 1970. *Ground Frost: A Listing and Evaluation of More Recent Literature Dealing with the Effect of Frost on the Soil.* Report FSTC-HT-23-311-70 translated by U.S. Army Foreign Science and Technology Center, Charlottesville, Va., 494 pp.

Johnson, A.W. 1952. *Special Report 1: Frost Action in Roads and Airfields: A Review of the Literature.* HRB, National Research Council, Washington, D.C., 300 pp.

Johnston, G.H., ed. 1981. *Permafrost: Engineering Design and Construction.* Associate Committee on Geotechnical Research, National Research Council of Canada, John Wiley & Sons, Toronto, 540 pp.

Johnston, G.H., B. Ladanyi, N.R. Morgenstern, and E. Penner. 1981. Engineering Characteristics of Frozen and Thawing Soils. In *Permafrost: Engineering Design and Construction* (G.H. Johnston, ed.), Associate Committee on Geotechnical Research, National Research Council of Canada, John Wiley & Sons, Toronto, pp. 73–147.

Judge, A.S. 1973. Deep Temperature Observations in the Canadian North. In *Proc., Second International Conference on Permafrost*, Yakutsk, USSR, North American Contribution, National Academy of Sciences, Washington, D.C., pp. 35–40.

Krzewinski, T.G., and R.G. Tart. 1985. *Thermal Design Considerations in Frozen Ground Engineering.* American Society of Civil Engineers, New York, N.Y., pp. 42–43.

Ladanyi, B., and G.H. Johnston. 1978. Field Investigations of Frozen Ground. In *Geotechnical Engineering for Cold Regions* (O.B. Andersland and D.M. Anderson, eds.), McGraw-Hill, New York, N.Y., pp. 459–504.

Linnell, K.A., and C.W. Kaplar. 1959. The Factor of Soil and Material Type in Frost Action. *Bulletin 225*, HRB, National Research Council, Washington, D.C., pp. 81–126.

Linnell, K.A., and C.W. Kaplar. 1963. Description and Classification of Frozen Soils. In *Proc., First International Conference on Permafrost*, Lafayette, Ind., Publication 1287, National Academy of Sciences, Washington, D.C., pp. 481–487.

Lotspeich, F.B. 1971. *Environmental Guidelines for Road Construction in Alaska.* Environmental Protection Agency, College, Alaska.

Lunardini, V.J. 1978. Theory of n-Factors and Correlation of Data. In *Proc., Third International Conference on Permafrost*, Edmonton, Alberta, National Research Council of Canada, Ottawa, Vol. 1, pp. 41–46.

MacKay, J.R. 1966. Segregated Epigenetic Ice and Slumps in Permafrost, Mackenzie Delta Area, N.W.T. *Geographical Bulletin*, No. 8, pp. 59–80.

MacKay, J.R. 1972. Offshore Permafrost and Ground Ice, Southern Beaufort Sea, Canada. *Canadian Journal of Earth Sciences*, Vol. 9, No. 11, pp. 1550–1561.

MacKay, J.R., and W.H. Mathews. 1973. Geomorphology and Quaternary History of the Mackenzie River Valley near Fort Good Hope, N.W.T. *Canadian Journal of Earth Sciences*, Vol. 10, No. 1, pp. 26–41.

McPhail, J.F., W.B. McMullen, and A.W. Murfitt. 1976. Yukon River to Prudhoe Bay: Lessons in Arctic Design and Construction. *Civil Engineering* (New York City), Vol. 46, No. 2, pp. 78–82.

McRoberts, E.C. 1973. *Stability of Slopes in Permafrost.* Ph.D. Thesis. University of Alberta, Edmonton, Alberta, Canada.

McRoberts, E.C. 1978. Slope Stability in Cold Regions. In *Geotechnical Engineering for Cold Regions* (O.B. Andersland and D.M. Anderson, eds.), McGraw-Hill, New York, N.Y., pp. 363–404.

McRoberts, E.C., and N.R. Morgenstern. 1973. *Landslides in the Vicinity of the Mackenzie River, Mile 205 to 660.* Report 73-35. Environmental-Social Program, Northern Pipelines, Department of Indian and Northern Affairs Canada, Ottawa, 96 pp.

McRoberts, E.C., and N.R. Morgenstern. 1974a. The Stability of Thawing Slopes. *Canadian Geotechnical Journal*, Vol. 11, No. 4, pp. 447–469.

McRoberts, E.C., and N.R. Morgenstern. 1974b. Stability of Slopes in Frozen Soil, Mackenzie Valley, N.W.T. *Canadian Geotechnical Journal*, Vol. 11, No. 4, pp. 554–573.

Muller, S.W. 1947. *Permafrost or Permanently Frozen Ground and Related Engineering Problems.* J.W. Edwards, Inc., Ann Arbor, Mich., 231 pp.

National Academy of Sciences. 1983. *Fourth International Conference on Permafrost: Final Proceedings*, Fairbanks, Alaska, National Academy Press, Washington, D.C., 413 pp.

National Research Council of Canada. 1988. *Glossary of Permafrost and Related Ground-Ice Terms.* Technical Memorandum 142. Ottawa, Ontario, Canada.

Nixon, J.F. 1992. New Frost Heave Prediction Method for Design of Northern Pipelines. In *Proc., Second International Offshore and Polar Engineering Conference*, International Society of Offshore and Polar Engineers, Golden, Colo., pp. 32–39.

Nixon, J.F., and G. Lem. 1984. Creep and Strength Testing of Frozen Saline Fine Grained Soils. *Canadian Geotechnical Journal*, Vol. 21, pp. 518–529.

Osterkamp, T.E., and W.D. Harrison. 1976. Subsea Permafrost: Its Implications for Offshore Resource Development. *The Northern Engineer*, Vol. 8, No. 1, pp. 31–35.

Pihlainen, J.A., and G.H. Johnston. 1963. *Guide to a Field Description of Permafrost*. Technical Memorandum 79. Associate Committee on Soil and Snow Mechanics, National Research Council of Canada, Ottawa, 23 pp.

Pufahl, D.E. 1976. *The Behaviour of Thawing Slopes in Permafrost*. Ph.D. Thesis. University of Alberta, Edmonton, Alberta, Canada, 345 pp.

Pufahl, D.E., N.R. Morgenstern, and W.D. Roggensack. 1974. *Observations on Recent Highway Cuts in Permafrost*. Report 74-32. Environmental-Social Program, Northern Pipelines, Department of Indian and Northern Affairs Canada, Ottawa, 53 pp.

Samson, L., and F. Tordon. 1969. Experiences with Engineering Site Investigations in Northern Quebec and Northern Baffin Island. In *Proc., Third Canadian Conference on Permafrost*, Technical Memorandum 96, Associate Committee on Geotechnical Research, National Research Council of Canada, Ottawa, pp. 21–38.

Sellmann, P.V., and D.M. Hopkins. 1983. Subsea Permafrost Distribution on the Alaskan Shelf. In *Fourth International Conference on Permafrost: Final Proceedings*, Fairbanks, Alaska, National Academy Press, Washington, D.C., pp. 75–82.

Smith, N., and R. Berg. 1973. Encountering Massive Ground Ice During Road Construction in Central Alaska. In *Proc., Second International Conference on Permafrost*, Yakutsk, USSR, North American Contribution, National Academy of Sciences, Washington, D.C., pp. 730–736.

Taber, S. 1943. Perennially Frozen Ground in Alaska. *Geological Society of America Bulletin*, Vol. 54, pp. 1433–1548.

Walker, D.B.L., D.W. Haley, and A.C. Palmer. 1983. The Influence of Subsea Permafrost on Offshore Pipeline Design. In *Proc., Fourth International Conference on Permafrost*, Fairbanks, Alaska, National Academy Press, Washington, D.C., pp. 1338–1343.

Walker, H.J., and L. Arnborg. 1963. Permafrost and Ice Wedge Effects on River Bank Erosion. In *Proc., First International Conference on Permafrost*, Lafayette, Ind., Publication 1287, National Academy of Sciences, Washington, D.C., pp. 164–171.

Washburn, A.L. 1973. *Periglacial Processes and Environments*. Edward Arnold Publishers, London, England, 320 pp.

Appendix A

STEREOGRAPHIC PROJECTIONS FOR STRUCTURAL ANALYSIS

Chapter 15 introduces the application of stereographic projections to the analysis of structural fabric and the utility of such analysis in the identification and evaluation of potential rock-slope instability or failure mechanisms. Two of the most useful stereographic projections, or nets, are reproduced on the following pages:

- The equal-area Schmidt net (or Lambert net), and
- The equal-area polar (Billings) net.

The use of these nets is briefly discussed in Chapter 15. Additional information on the derivation and use of the nets may be found elsewhere (Dennison 1968; Ragan 1973). The principles of stereographic projection are discussed in most structural geology textbooks. Both these nets have been developed to maintain an equal-area property; this means that areas between lines of latitude and parallels of longitude remain equal over the entire region shown. This property is useful when comparing the frequency distribution of data by point density patterns or contoured results as in Figure 15-4.

The equal-area Schmidt net is always used as a graph underlying a transparent or semitransparent sheet attached to the net by a drawing pin placed through the center point. The various orientations of structural features may be shown by tracing great and small circles from the underlying net onto the overlay. Generally it is necessary to rotate the overlay about the net center in order to develop the desired orientations.

The equal-area polar (Billings) net is especially convenient for plotting linear data or poles of planar features in the field. A transparent sheet may be fixed on top of this polar net, and rotations of the sheet are not required. Diagrams prepared with these two nets may be interchanged.

REFERENCES

Dennison, J.M. 1968. *Analysis of Geologic Structures*. W.W. Norton & Company, New York, 209 pp.

Ragan, D.M. 1973. *Structural Geology: An Introduction to Geometrical Techniques*, 2nd ed. John Wiley and Sons, Inc., New York, 208 pp.

EQUAL-AREA SCHMIDT NET

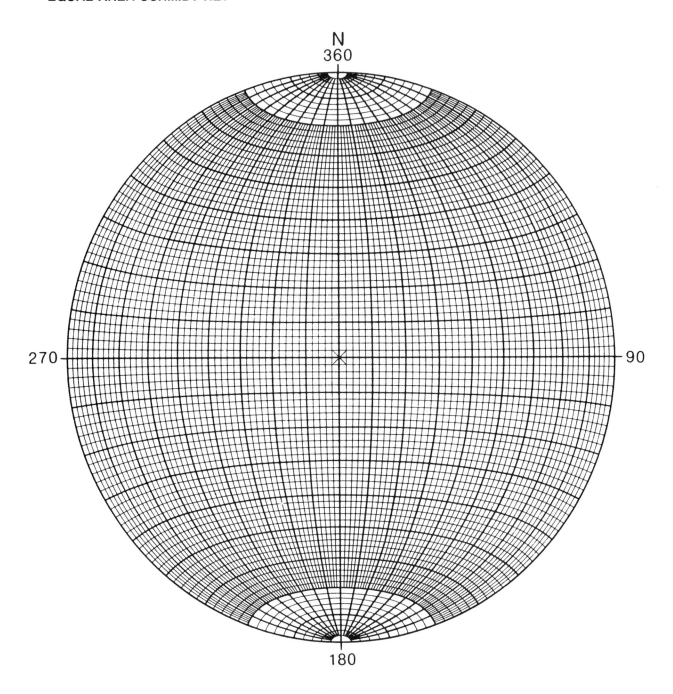

N
360

270

90

180

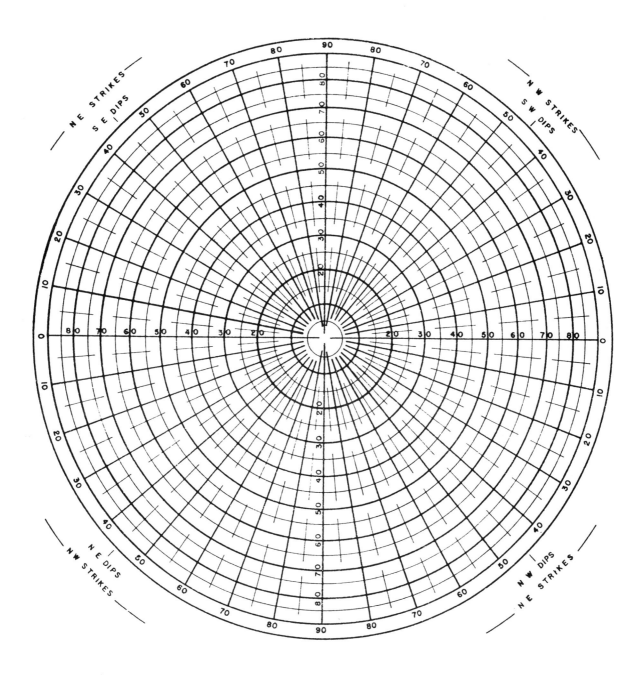

EQUAL- AREA POLAR (BILLINGS) NET

Graduated for poles of planar elements (lower hemisphere)

METRIC CONVERSION TABLE

SI units to inch/pound units

To convert from	To	Multiply by
	LENGTH	
millimeter (mm)	inch (in.)	0.03937
meter (m)	foot (ft)	3.281
	yard (yd)	1.094
kilometer (km)	mile (mi)	0.6214
	mile, nautical (nmi)	0.5400
	AREA	
meter^2 (m^2)	foot^2 (ft^2)	10.76
	yard^2 (yd^2)	1.196
	acre	0.0002471
hectometer^2 (km^2)	acre	2.471
kilometer^2 (km^2)	mile^2 (mi^2)	0.3861
	VOLUME	
centimeter^3 (cm^3)	inch^3 (in.^{3})	0.06102
decimeter^3 (dm^3)	inch^3 (in.^{3})	61.02
	pint (pt)	2.113
	quart (qt)	1.057
	gallon (gal)	0.2642
	foot^3 (ft^3)	0.03531
meter^3 (m^3)	foot^3 (ft^3)	35.31
	yard^3 (yd^3)	1.308
	gallon (gal)	264.2
	barrel (bbl), (petroleum, 1 bbl=42 gal)	6.290
	acre-foot (acre-ft)	0.0008107
hectometer^3 (hm^3)	acre-foot (acre-ft)	810.7
kilometer^3 (km^3)	mile^3 (mi^3)	0.2399

Inch/pound units to SI units

To convert from	To	Multiply by
	LENGTH	
inch (in.)	millimeter (mm)	25.4*
foot (ft)	meter (m)	0.3048
yard (yd)	meter (m)	0.9144*
mile (mi)	kilometer (km)	1.609
mile, nautical (nmi)	kilometer (km)	1.852*
	AREA	
foot^2 (ft^2)	meter^2 (m^2)	0.09290
yard^2 (yd^2)	meter^2 (m^2)	0.8361
acre	meter^2 (m^2)	4,047
	hectometer^2 (hm^2)	0.4047
mile^2 (mi^2)	kilometer^2 (km^2)	2.590
	VOLUME	
inch^3 (in.^{3})	centimeter^3 (cm^3)	16.39
	decimeter^3 (dm^3)	0.01639
foot^3 (ft^3)	decimeter^3 (dm^3)	28.32
	meter^3 (m^3)	0.02832
yard^3 (yd^3)	meter^3 (m^3)	0.7646
pint (pt)	decimeter^3 (dm^3)	0.4732
quart (qt)	decimeter^3 (dm^3)	0.9464
gallon (gal)	decimeter^3 (dm^3)	3.785
	meter^3 (m^3)	0.003785
barrel (bbl), (petroleum, 1bbl=42 gal)	meter^3 (m^3)	0.1590
acre-foot (acre-ft)	meter^3 (m^3)	1,233
	hectometer^3 (hm^3)	0.001233
mile^3 (mi^3)	kilometer^3 (km^3)	4.168

SI units to inch/pound units				Inch/pound units to SI units		
To convert from	To	Multiply by		To convert from	To	Multiply by
VOLUME PER UNIT TIME (included flow)				VOLUME PER UNIT TIME (included flow)		
decimeter^3 per second (dm^3/sec)	foot^3 per second (ft^3/sec)	0.03531		foot^3 per second (ft^3/sec)	decimeter^3 per second (dm^3/sec)	28.32
	gallon per minute (gal/min)	15.85			meter^3 per second (m^3/sec)	0.02832
	barrel per day (bbl/d), (petroleum)	543.4		gallon per minute (gal/min)	decimeter^3 per second (dm^3/sec)	0.06309
meter^3 per second (m^3/sec)	foot^3 per second (ft^3/sec)	35.31			meter^3 per second (m^3/sec)	0.00006309
	gallon per minute (gal/min)	15,850		barrel per day (bbl/d), (petroleum)	decimeter^3 per second (dm^3/sec)	0.001840
MASS				MASS		
gram (g)	ounce avoirdupois (oz avdp)	0.03527		ounce avoirdupois (oz avdp)	gram (g)	28.35
kilogram (kg)	pound avoirdupois (lb avdp)	2.205		pound avoirdupois (lb avdp)	kilogram (kg)	0.4536
megagram (Mg)	ton, short (2,000 lb)	1.102		ton, short (2,000 lb)	megagram (Mg)	0.9072
	ton, long (2,240 lb)	0.9842		ton, long (2,240 lb)	megagram (Mg)	1.016
PRESSURE				PRESSURE		
kilopascal (kPa)	pound-force per inch^2 (lbf/in.^{2})	0.1450		pound-force per inch^2 (lbf/in.^{2})	kilopascal (kPa)	6.895
	atmosphere, standard (atm)	0.009869		atmosphere, standard (atm)	kilopascal (kPa)	101.3
	bar	0.01*		bar	kilopascal (kPa)	100.*
	inch of mercury at 60°F (in Hg)	0.2961		inch of mercury at 60 F (in Hg)	kilopascal (kPa)	3.377
TEMPERATURE				TEMPERATURE		
kelvin (K)	degreee Fahrenheit (°F)	(1)		degree Fahrenheit (°F)	kelvin (K)	(3)
degree Celsius (°C)	degree Fahrenheit (°F)	(2)		degree Fahrenheit (°F)	degree Celsius (°C)	(4)

[1]Temp °F = 1.8 temp K − 459.67. [2]Temp °F = 1.8 temp °C +32.

[3]Temp K = (temp °F + 459.67)/1.8. [4]Temp °C = (temp °F − 32)/1.8.

NOTE: The International System of Units (SI) is a modernized metric system of measurement. In this table an asterisk after the last digit of the factor indicates that the conversion factor is exact and that all subsequent digits are zero; all other conversion factors have been rounded to four significant digits. Use of hectare (ha) as an alternative name for square hectometer (hm^2) is restricted to the measurement of small land or water areas. Use of liter (L) as a special name for cubic decimeter (dm^3) is restricted to the measurement of liquids and gases. No prefix other than "milli" should be used with liter.

SOURCE: W.R. Hansen, ed. 1991. *Suggestions to Authors of the Reports of the United States Geological Survey*, 7th ed. Government Printing Office, 289 pp.

◣ Author Biographical Information

A. Keith Turner, *Editor,* is Professor of Geology and Geological Engineering at the Colorado School of Mines in Golden. A member of the Colorado School of Mines faculty for more than 20 years, he was formerly Assistant Professor of Civil Engineering at the University of Toronto and Engineering Geologist with the Department of Public Works in Ottawa, Canada. He has had a long-standing interest in computer applications to geological and environmental studies, acting as consultant on computer applications in engineering geology for the United Nations and various national agencies, academic institutions, and private firms in several countries, including the United Kingdom, Poland, the Netherlands, Germany, and Mexico. Turner served as director of an advanced research workshop conducted by the North Atlantic Treaty Organization on three-dimensional geological modeling, which was held in Santa Barbara in 1989. Currently working with the U.S. geological Survey on developing three-dimensional geographic information system (GIS) approaches for data management, regional hydrogeologic modeling, and scientific visualization of geological data, Turner has conducted landslide investigations in Canada and Colorado. He has served as chairman of the TRB Committee on Engineering Geology and Committee on Exploration and Classification of Earth Materials as well as chairing the TRB Study Committee on Landslides: Analysis and Control.

Robert L. Schuster, *Editor,* is Research Geologist/ Civil Engineer in the Branch of Earthquake and Landslide Hazards, U.S. Geological Survey, Denver, Colorado. Formerly he was Chief of the Branch of Engineering Geology with the U.S. Geological Survey, Professor and Chairman of the Department of Civil Engineering at the University of Idaho, and Associate Professor and Professor of the Department of Civil Engineering at the University of Colorado. His major areas of professional specialization are engineering geology and geotechnical engineering.

George H. Beckwith is Chief Engineer and Vice President of AGRA Earth & Environmental, Inc., in Phoenix, Arizona. His fields of specialization include design of water storage dams, tailings dams, and other embankments; evaluation of static stability and seismic deformation of natural and constructed slopes; and geotechnical studies for highways, railroads, bridges, and tunnels that include the issues of earthquake-resistant design.

David M. Cruden is Professor of Civil Engineering and Geology at the University of Alberta in Edmonton, Alberta, Canada, where he has taught engineering geology, rock mechanics, and terrain analysis for the last 25 years. He was educated at the universities of Oxford and Alberta before receiving his Ph.D. from the Royal School of Mines in London. He has published some 150 technical

653

papers and coauthored two books. He currently chairs the Technical Committee on Landslides of the International Society of Soil Mechanics and Foundation Engineering and the Working Group on Landslides of the International Association of Engineering Geology (IAEG). He is Past Chairman of the IAEG Commission on Landslides and a fellow of the Geological Society of America and the Geological Society of London, a Professional Geologist, and a Chartered Geologist. He received the Thomas Roy Award of the Canadian Geotechnical Society in 1993 and has acted as a consultant to government agencies and transportation companies and as an expert witness.

Jerome V. DeGraff is Forest Geologist for the Sierra, Stanislaus, and Sequoia national forests, Clovis, California. He served as a geologist for the USDA Forest Service in Utah and California for 17 years. Coauthor of _Principles of Engineering Geology_ and _Engineering Geology—A Laboratory Manual,_ as well as having authored many articles published in technical and scientific publications, DeGraff was the recipient of the E. B. Burwell, Jr., Memorial Award from the Geological Society of America and the Claire P. Holdredge Award from the Association of Engineering Geologists (AEG). He was Manager of the AEG Committee on Landslides from 1983 to 1993. His fields of specialization include applied geomorphology, engineering geologic mapping, and assessment of risk from landslides and other mass movements.

J. Michael Duncan is University Distinguished Professor of Civil Engineering at the Virginia Polytechnic Institute and State University, Blacksburg, Virginia. He is a consultant on geotechnical engineering projects and was formerly Professor of Civil Engineering at the University of California, Berkeley. His professional specialty is geotechnical engineering, with emphasis on limit equilibrium analysis of slope stability, finite-element analysis of soil-structure interaction, strength and deformation properties of soils, and foundation engineering.

Herbert H. Einstein is Professor of Civil Engineering at the Massachusetts Institute of Technology in Cambridge. A graduate of Eidgenoessische Technische Hochschule (ETH) in Zurich, Switzerland, he has had extensive research and consulting experience in Europe and North America in rock mechanics, underground con-

struction in rock and soil, and engineering geology. Einstein has been associated with the MIT faculty and research programs for almost 30 years, with research periods at the Norwegian Geotechnical Institute and ETH-Lausanne. His major research areas involve the use of probabilistic approaches to the study of stability and fluid flow in fractured rock masses. Author or coauthor of more than 100 professional papers concerned with mechanical soil and rock properties, field measurements, grouting, analysis and design and project management of underground structures, landslides, and probabilistic methods in rock engineering and engineering geology, he currently chairs the International Society for Rock Mechanics Commission on Swelling Rock.

Jerry D. Higgins is Associate Professor of Geology and Geological Engineering at the Colorado School of Mines in Golden. Formerly Assistant Professor of Civil Engineering at Washington State University in Pullman, Higgins specializes in engineering geology and geotechnical engineering with emphasis on site investigation and site selection, geologic hazard analysis and mitigation, material properties of soil and rock, and engineering geologic mapping. He has conducted investigations of loess properties and stability in Washington State and Missouri.

Robert D. Holtz is Professor of Civil Engineering at the University of Washington in Seattle. Formerly with Purdue University, California Department of Water Resources, Swedish Geotechnical Institute, and the National Research Council of Canada, Holtz has had consulting engineering experience in the United States, France, and Italy. He began working with geotextiles in 1971 and since then has completed numerous research and consulting projects concerning almost all aspects of geosynthetics. He has coauthored or edited nine books and book chapters and has written more than 140 technical papers, discussions, and reports. A member of the American Society of Civil Engineers, TRB, International Society for Soil Mechanics and Foundation Engineering, ASTM, and the International Geosynthetics Society, he is the immediate Past President of the North American Geosynthetics Society and chairman of the TRB Technical Activities Division Section on Geology and Properties of Earth Materials.

G.P. Jayaprakash is Engineer of Soils, Geology, and Foundations with the Transportation Research Board. He has been a member of the TRB Technical Activities Division since 1987. Before that he was Research Geologist with the Kansas Department of Transportation and a research scientist at the University of Kansas. He has a master's degree in applied geology from Karnatak University, India, where he served as lecturer for two years, and a master's degree in geology from the University of California, Santa Barbara. His areas of interest are engineering geology and geotechnical engineering.

Jeffrey R. Keaton is Senior Engineering Geologist and Vice President of AGRA Earth & Environmental, Inc., in Salt Lake City, Utah, as well as Adjunct Professor of Civil and Environmental Engineering, Utah State University, Logan, and Adjunct Associate Professor of Geography, University of Utah, Salt Lake City. His fields of specialization include engineering geologic mapping, applied geomorphology, and quantitative assessment of risk from hazardous natural processes, particularly landslides, debris flows, liquefaction, fault rupture, and earthquake ground motion.

William J. Kockelman (deceased) was Earth Sciences Applications Planner with the U.S. Geological Survey in Menlo Park, California. Before his association with the USGS, he was Director of the New Mexico Council on Environmental Quality, Executive Director of the New Mexico State Planning Office, and Chief Planner of the Southeast Wisconsin Regional Planning Commission. His major areas of professional specialization were regional geologic hazard planning and transfer of scientific products into maps and reports to be used by planners.

Philip C. Lambe is Associate Professor of Civil Engineering at North Carolina State University in Raleigh, North Carolina. He was a consultant for five years with T.W. Lambe and Associates. His fields of professional specialization are slope stability in natural soil, earth dams, and embankments.

Guy Lefebvre received a degree in civil engineering and a doctorate in geotechnical engineering from Université Laval in Québec, Canada. Following a two-year period of research at the University of California, Berkeley, he joined the Civil Engineering Department at the Université de Sherbrooke, Canada. Since then he has been involved in teaching, research, and engineering projects dealing with slope stability, embankments, dams, earthquakes, and geoenvironmental engineering.

Verne C. McGuffey is a geotechnical engineer in Schenectady, New York. Formerly Assistant Director of the Soil Mechanics Bureau with the New York State Department of Transportation, he received a B.S. and an M.S. in civil engineering from Rensselaer Polytechnic Institute in Troy, New York. He has had more than 35 years of experience in geotechnical engineering investigations, including detailed evaluations of 20 to 50 landslides a year for over 20 years. McGuffey has authored numerous professional papers published by ASCE and TRB concerning landslides and geotechnical engineering. Long associated with TRB geotechnical committees, he currently chairs the Technical Activities Division Soil Mechanics Section.

P. Erik Mikkelsen is Vice President, Landslide Technology, in Bellevue, Washington. During his 28 years of geotechnical engineering experience, he has become an expert in instrumentation used in landslides, embankments, deep excavations, and deep foundations.

Victor A. Modeer, Jr., is Bureau Chief of Implementation with the Illinois Department of Transportation in Collinsville. Former positions within the Illinois Department of Transportation include materials engineer and geotechnical engineer. Before his association with the Illinois Department of Transportation, he was affiliated with private firms including Woodward-Clyde Consultants and Louis Capozzoli and Associates. His fields of professional specialization include geotechnical investigation and design for landslide mitigation, embankments and structures on soft soils, groundwater flow, seismic design of earth structures, and deep foundations.

Norman I. Norrish is Principal, Golder Associates, Inc., in Seattle, Washington. Born and educated in British Columbia, he is a graduate of the University of British Columbia with a bachelor's degree in geological engineering and a master's degree in mineral engineering. Norrish is a member of the Canadian Institute of Mining and Metallurgy, American Railway Engineering Association,

TRB, and the Association of Professional Engineers of British Columbia. His experience in the geotechnical aspects of surface mining, transportation, and civil construction projects in North America and internationally spans more than 20 years. His specialties include site investigations, slope design, evaluation of geologic hazards, remedial stabilization, slope reinforcement, and tunnel rehabilitation.

Paul M. Santi is Assistant Professor of Geological Engineering at the University of Missouri-Rolla. He has a bachelor's degree in geology and physics from Duke University, a master's degree in geology from Texas A&M University, and a doctorate in geological engineering from the Colorado School of Mines. He has conducted research into the kinematics of debris flow transport based on a 1984 debris flow event in Utah and is developing classification and testing procedures for weak and weathered rock materials. Santi has several years' experience in consulting engineering based in Denver and San Francisco, especially in the evaluation of landslides, rock falls, and slope stability; assessment of geologic hazards associated with development; and soil and groundwater investigations related to hazardous materials.

Robert Soeters is Geologist and Assistant Professor of the Department of Earth Resources Surveys at the International Institute for Aerospace Surveys and Earth Sciences (ITC), Enschede, the Netherlands. With more than 30 years of experience in remote-sensing applications to geological and engineering geological surveys, mainly in Europe and South America, Soeters' major professional specialization involves the mapping and analysis of landslides.

Wilson H. Tang is Professor of Civil Engineering at the University of Illinois at Champaign-Urbana. He received his bachelor's and master's degrees from Massachusetts Institute of Technology and his doctorate from Stanford University. He has authored or coauthored more than 100 technical publications, including two widely used texts on probability concepts in engineering planning and design. He is a registered professional engineer and structural engineer in Illinois. During the past 25 years his major research interest has been in the general area of risk, reliability, and decision analy-

sis. He has conducted numerous research projects to develop reliability methodologies for geotechnical, structural, offshore, and hydraulic engineering. A former member of the Geotechnical Board of the National Academy of Sciences, he currently chairs the Board's Committee on Reliability Methods for Risk Mitigation in Geotechnical Engineering. He is a Fellow of ASCE and vice-chair of the Technical Committee on Safety and Reliability for the ASCE Geotechnical Division. Tang is the recipient of several awards including the ASCE State-of-the-Art Award, the John Simon Guggenheim Fellowship, and the Campus Award for Excellence in Undergraduate Instruction.

Rupert G. Tart, Jr., is Principal, Golder Associates, Inc., Anchorage, Alaska. He has had more than 20 years of experience studying, designing, monitoring, and modifying slopes in permafrost. His first experience with permafrost slopes took place during the construction of the Trans-Alaska Pipeline, and he continues to monitor some of those same slopes today. He has been a member of the U.S. Committee of the International Permafrost Association for more than 10 years and has evaluated permafrost slopes for mining and road projects in Alaska, Russia, and Kyrgystan.

Cornelis J. van Westen is Geomorphologist and Senior Researcher of the Department of Earth Resources Surveys at the International Institute for Aerospace Surveys and Earth Sciences (ITC), Enschede, the Netherlands. His major professional activities have concentrated on the application of remote sensing and GIS in applied geomorphology and engineering geology, particularly in relation to slope processes.

David J. Varnes is Geologist in the Branch of Earthquake and Landslide Hazards, U.S. Geological Survey, Denver, Colorado. He has had more than 50 years of experience as a geologist with the U.S. Geological Survey, including serving as Chief of the Engineering Geology Branch. His fields of specialization include slope stability in soil and rock, engineering geologic mapping, and nonlinear dynamics of earthquakes.

Steven G. Vick is a consulting geotechnical engineer in Bailey, Colorado. He received a B.S. and an M.S. from Massachusetts Institute of Technology and has specialized in the design and assessment of tailings and mine waste structures. For more than 20 years, he has assessed the stability of these structures, including their liquefaction and landslide characteristics incorporating risk-based characterization. He is the author of *Planning, Design, and Analysis of Tailings Dams* and is currently chairman of the Tailings Dam Committee of the U.S. Committee on Large Dams, which recently published *Tailings Dam Incidents*, a case history compendium of tailings dam failures and stability problems. His practice centers on technical review of tailings and mine waste structures for government and industry throughout North America, South America, and Europe. Vick is currently a World Bank consultant to the government of Peru and is charged with establishing a program of tailings management in that country.

John L. Walkinshaw is a consulting geotechnical engineer in Walnut Creek, California. Born in Connecticut of British parents, he moved at an early age to Switzerland, where he later received a diploma from the Ecole Technique Supérieure in Geneva. Returning to the United States, Walkinshaw received a master's degree from Massachusetts Institute of Technology. He joined the Federal Highway Administration of the U.S. Department of Transportation, worked in several of their regional offices, and in the 1970s first became familiar with degradable shale problems in eastern Tennessee. Most recently he was Regional Geotechnical Engineer in the FHWA Region Nine office, San Francisco, where he was a technical specialist and consultant to FHWA engineers in California, Nevada, Arizona, and Hawaii. Walkinshaw's long-term career interests have been in ground reinforcement and structural foundations. He has been active in several TRB committees since 1973, having chaired the Committee on Soil and Rock Instrumentaion from 1988 to 1994.

Gerald F. Wieczorek is Research Civil Engineer, U.S. Geological Survey, in Reston, Virginia. His major areas of professional specialization are engineering geology, slope stability in rock and soil, and landslide hazard assessment.

Tien H. Wu is Professor Emeritus of Civil Engineering, The Ohio State University in Columbus. A graduate of St. John's University in China, he received an M.S. and Ph.D. from the University of Illinois. He served as professor at Michigan State University and as visiting professor at the National University of Mexico, the Norwegian Geotechnical Institute, the Royal Institute of Technology in Sweden, Punjab Agricultural University in India, and Tongji University and Southwest Jiaotong University in China. His research and consulting work has included problems in soil properties, soil erosion, slope stability and foundations, geotechnical risk and reliability, and glaciology. Wu has written two books and more than 60 papers for soil mechanics and geotechnical journals. He is a Fellow of ASCE and received their State-of-the-Art Award and has served as chairman of the Publications Committee of the Soil Mechanics Division. He has served as chairman of the TRB Committees on Soils Properties and Mechanics of Earth Masses.

Duncan C. Wyllie is Principal, Golder Associates, in Vancouver, Canada. He specializes in the field of rock mechanics and has worked on the design and construction of slopes, tunnels, and foundations for transportation, energy, urban development, and mining projects since 1967. Wyllie is the author of *Foundations on Rock*, published in 1992, and *Rock Slopes: Design, Excavation, Stabilization*, published in 1989.

◣ INDEX

AUTHOR

Dietrich, W.E. 78, 432, 535-536
Dikau, R. 154
Ditlevsen, O. 109
Dobecki, T.L. 253
Dobr, J. 69 fig.
Dobrin, M.B. 237, 240-241
Donaldson, P.R. 237
Doolittle, J.A. 242
Doornkamp, J.C. 154
Dowding, C.H. 121-123, 311, 311 fig.
Dracup, J.A. 25
Drucker, D.C. 319
Duffy, J.D. 496-497, 497 fig., 501
Duncan, J.M. 83, 122, 213, 215 fig.,
　　330-331, **337-371**, 350-351 fig.,
　　362, 365-366, 368, 417, 571
Dunlop, P. 331, 571
Dunne, T. 535
Dunnicliff, C.J. 203, 206, 278, 284, 289,
　　300, 301 tab., 303
Dunoyer, M. 138
Dupre, W.R. 86
Durr, D.L. 303
Dusek, C.J. 437
Dusseault, M.B. 399
Dutro, H.B. 293
Duvall, W.I. 242
Dyrness, C.T. 530, 540
Dywidag Systems International 484 fig.

Eastman, J.R. 165
Eaton, E.C. 529, 540
Eckel, E.B. 3-4
Eckersley, J.D. 60
Eden, D.N. 585
Eden, W.J. 607, 610
Edil, T.B. 444, 453
Egan, J.P. 460
Egger, P. 482
Einstein, H.H. **106-118**, 109, 111-112, 130
Eisbacher, G.H. 5-8, 19, 28, 30 tab., 37, 48,
　　49 fig.
Elderkin, B. 541 fig.
Elias, V. 456-458, 462-463
Eliot, T.S. 36
Ellen, S.D. 76, 78, 166, 528, 530-531,
　　533-534 fig., 535-537, 536 fig.
Elson, J.A. 609
Endo, T. 538-539
Environmental Systems Research Institute
　　165
Erban, P.-J. 383
Evans, S.G. 19, 29, 54, 493, 607, 608 fig.
Eyles, N. 620

Fabbri, A.G. 135
Fairbairn, W.A. 530, 540
Fairbridge, R.W. 608

Falcon, N.L. 48
Fang, H.Y. 461
Farmer, I.W. 485
Feckers, E. 378
Federal Emergency Management Agency
　　92-93
Federal Highway Administration 474,
　　491
Felio, G. 463
Fellenius, W. 339, 340 tab.
Feng, G. 368
Ferrer, M. 19
Filz, G.M. 364
Finn, W.D. 580, 629
Fleming, R.W. 15, 24, 100, 185, fig., 207,
　　219, 221 fig., 223 fig., 385 tab., 454,
　　456, 530-532, 534
Florsheim, J.L. 79
Fookes, P.G. 122, 138, 156, 474
Foott, R. 321, 329-330, 343-344
Forster, A. 184
Forster, J.W. 489
Forsyth, R.A. 303, 459-460
Foss, I. 332
Fragaszy, R.J. 587, 593, 602 fig., 604 fig.
France, Office of the Prime Minister 112
Franklin, J.A. 196, 556, 559-560, 562, 567,
　　568-572 fig., 571
Fredlund, D.G. 323, 326, 514, 515 fig.
Freeze, R.A. 265, 396
French, S.P. 101
Fritz, P. 59
Frost, D.D. 257
Frye, J.C. 585-586
Fukuoka, M. 28, 461, 536
Fulkert, R.J. 585

Gadd, N.R. 609
Gadsby, J. 582
Galloway, J.P. 86, 86 fig., 542
Galster, R.W. 186
Gamble, J. 436
Gano, D. 303
Gedney, D.S. 439, 442, 444-446 fig., 448,
　　452 fig., 453-455, 466-467
Geological Society Engineering Working
　　Party 191 tab.
Geological Society of America 529
Geological Society of London 262,
　　263-264 fig., 507, 509 fig.
Geological Survey of Canada 608 fig.
Geomation, Inc. 313
Gerasimov, I.P. 29
Gerber, B. 165
Gibbs, H.J. 326, 586-588, 587 fig., 591,
　　601
Gibson, R.E. 327
Gilbert, C. 281
Gill, K. 383

Gillon, M.D. 100, 278, 293, 305-307,
　　306 fig., 307 tab., 448, 450 fig.
Gilmore, J. 451, 548
Giraud, A. 48, 55
Girault, F. 159, 161
Glancy, P.A. 79
Gökce, A.O. 530
Golder, H.Q. 320, 330
Goldich, S.A. 557
Goodchild, M.F. 165
Goodlett, J.C. 535
Goodman, R.E. 48, 54-56, 379, 393, 412,
　　418, 418 fig., 419
Gostelow, T.P. 46
Goughnour, R.D. 17
Gould, J.P. 203, 206, 278, 284, 289, 455
Govi, M. 26
Graham, J. 447
Gray, D.H. 464-465, 465-466 fig., 538-539
Gray, I. 449
Green, G.E. 278, 282, 284-285 fig.,
　　287 fig., 289, 290-293 fig.
Greenfield, R.J. 242
Greenway, D.R. 464, 538, 539 fig.
Greer, D.J. 448
Gregersen, O. 19
Gress, J.C. 450
Griffiths, D.H. 233, 236, 240
Grondin, G. 615
Groupe d'Études des Falaises 474
Gryta, J.J. 76
Guerricchio, A. 140
Gunderson, D.E. 313
Guy, H.P. 529, 530 fig., 535

Hachich, W. 518
Hack, J.T. 535
Hackman, R.J. 542
Haji-Ahmad, K. 463
Halim, I.S. 107
Hallberg, G.R. 327
Haller, B. 497
Hamel, J.V. 385 tab.
Hamilton County Regional Planning
　　Commission 93
Hammond, C.J. 136, 165, 417
Hancox, G.T. 305, 306 fig.
Handy, R.L. 327, 586, 588-589 fig., 591
Haneberg, W.C. 530
Hanes, T.L. 540
Hanna, T.H. 486
Hansen, A. 92, 134, 142, 194
Hansen, J.A. 586, 591
Hansen, W.R. 20 fig., 57 fig., 97, 127,
　　650-651
Hanson, L.A. 463
Hardy, H.R., Jr. 242
Hardy, R.M. 631

SUBJECT

The **Transportation Research Board** is a unit of the National Research Council, which serves the National Academy of Sciences and the National Academy of Engineering. The Board's purpose is to stimulate research concerning the nature and performance of transportation systems, to disseminate the information produced by the research, and to encourage the application of appropriate research findings. The Board's program is carried out by more than 400 committees, task forces, and panels composed of nearly 4,000 administrators, engineers, social scientists, attorneys, educators, and others concerned with transportation; they serve without compensation. The program is supported by state transportation and highway departments, the modal administrations of the U.S. Department of Transportation, and other organizations and individuals interested in the development of transportation.

The National Academy of Sciences is a private, nonprofit, self-perpetuating society of distinguished scholars engaged in scientific and engineering research, dedicated to the furtherance of science and technology and to their use for the general welfare. Upon the authority of the charter granted to it by the Congress in 1863, the Academy has a mandate that requires it to advise the federal government on scientific and technical matters. Dr. Bruce M. Alberts is president of the National Academy of Sciences.

The National Academy of Engineering was established in 1964, under the charter of the National Academy of Sciences, as a parallel organization of outstanding engineers. It is autonomous in its administration and in the selection of its members, sharing with the National Academy of Sciences the responsibility for advising the federal government. The National Academy of Engineering also sponsors engineering programs aimed at meeting national needs, encourages education and research, and recognizes the superior achievements of engineers. Dr. Harold Liebowitz is president of the National Academy of Engineering.

The Institute of Medicine was established in 1970 by the National Academy of Sciences to secure the services of eminent members of appropriate professions in the examination of policy matters pertaining to the health of the public. The Institute acts under the responsibility given to the National Academy of Sciences by its congressional charter to be an adviser to the federal government and, upon its own initiative, to identify issues of medical care, research, and education. Dr. Kenneth I. Shine is president of the Institute of Medicine.

The National Research Council was organized by the National Academy of Sciences in 1916 to associate the broad community of science and technology with the Academy's purpose of furthering knowledge and advising the federal government. Functioning in accordance with general policies determined by the Academy, the Council has become the principal operating agency of both the National Academy of Sciences and the National Academy of Engineering in providing services to the government, the public, and the scientific and engineering communities. The Council is administered jointly by both the Academies and the Institute of Medicine. Dr. Bruce M. Alberts and Dr. Harold Liebowitz are chairman and vice chairman, respectively, of the National Research Council.